AF693038

Electromaterials for Environment & Energy: Volume I

Electromaterials for Environment & Energy: Volume I

Editors

Marc Cretin
Sophie Tingry
Zhenghua Tang

MDPI • Basel • Beijing • Wuhan • Barcelona • Belgrade • Manchester • Tokyo • Cluj • Tianjin

Editors

Marc Cretin
Institut Européen des
Membranes (IEMM, ENSCM
UM CNRS UMR5635)
France

Sophie Tingry
Institut Européen des
Membranes (IEMM, ENSCM
UM CNRS UMR5635)
France

Zhenghua Tang
South China University of
Technology
China

Editorial Office

MDPI
St. Alban-Anlage 66
4052 Basel, Switzerland

This is a reprint of articles from the Topic published online in the open access journals *Nanomaterials* (ISSN 2079-4991), *Materials* (ISSN 1996-1944), *Membranes* (ISSN 2077-0375), *Catalysts* (ISSN 2073-4344), and *Electrochem* (ISSN 2673-3293) (available at: https://www.mdpi.com/topics/materials_energy).

For citation purposes, cite each article independently as indicated on the article page online and as indicated below:

LastName, A.A.; LastName, B.B.; LastName, C.C. Article Title. *Journal Name* **Year**, *Volume Number*, Page Range.

Volume 1
ISBN 978-3-0365-6908-6 (Hbk)
ISBN 978-3-0365-6909-3 (PDF)

Volume 1-2
ISBN 978-3-0365-6499-9 (Hbk)
ISBN 978-3-0365-6500-2 (PDF)

Contents

About the Editors . . . . **ix**

Preface to "Electromaterials for Environment & Energy: Volume I" . . . . **xi**

Barbara Krystyna Wilk, Małgorzata Szopińska, Aneta Luczkiewicz, Michał Sobaszek, Ewa Siedlecka and Sylwia Fudala-Ksiazek
Kinetics of the Organic Compounds and Ammonium Nitrogen Electrochemical Oxidation in Landfill Leachates at Boron-Doped Diamond Anodes
Reprinted from: *Materials* **2021**, *14*, 4971, doi:10.3390/ma14174971 . . . . **1**

Artyom Plyushch, Nerijus Mačiulis, Aliaksei Sokal, Robertas Grigalaitis, Jan Macutkevič, Alexander Kudlash, Natalia Apanasevich, et al.
$0.7Pb(Mg_{1/3}Nb_{2/3})O_3$-$0.3PbTiO_3$ Phosphate Composites: Dielectric and Ferroelectric Properties
Reprinted from: *Materials* **2021**, *14*, 5065, doi:10.3390/ma14175065 . . . . **23**

Muhammad Syahmi Abd Rahman, Mohd Zainal Abidin Ab Kadir, Muhamad Safwan Abd Rahman, Miszaina Osman, Ungku Anisa Ungku Amirulddin, Shamsul Fahmi Mohd Nor and Noorlina Mohd Zainuddin
The Behavior of Polyurethane Foam-Filled Glass-Fiber-Reinforced Polymer Crossarm Subjected to Lightning Transient Voltage
Reprinted from: *Materials* **2021**, *14*, 5628, doi:10.3390/ma14195628 . . . . **33**

José Antonio Alonso, Paula Kayser, Bo-Kyung Hong, María Consuelo Álvarez-Galván, Francois Fauth and Carlos Alberto López
Mechano-Chemical Synthesis, Structural Features and Optical Gap of Hybrid $CH_3NH_3CdBr_3$ Perovskite
Reprinted from: *Materials* **2021**, *14*, 6039, doi:10.3390/ma14206039 . . . . **45**

Gbenro Folaranmi, Myriam Tauk, Mikhael Bechelany, Philippe Sistat, Marc Cretin and Francois Zaviska
Synthesis and Characterization of Activated Carbon Co-Mixed Electrospun Titanium Oxide Nanofibers as Flow Electrode in Capacitive Deionization
Reprinted from: *Materials* **2021**, *14*, 6891, doi:10.3390/ma14226891 . . . . **55**

Bipul Deka and Kyung-Hoon Cho
$BiFeO_3$-Based Relaxor Ferroelectrics for Energy Storage: Progress and Prospects
Reprinted from: *Materials* **2021**, *14*, 7188, doi:10.3390/ma14237188 . . . . **67**

Zhiyong Zhou, Haiwei Zhang, Weiyang Qin, Pei Zhu, Ping Wang and Wenfeng Du
Harvesting Energy from Bridge Vibration by Piezoelectric Structure with Magnets Tailoring Potential Energy
Reprinted from: *Materials* **2022**, *15*, 33, doi:10.3390/ma15010033 . . . . **93**

Joseph M. Gallet de St Aurin and Jonathan Phillips
Study of the Impact of Graphite Orientation and Ion Transport on EDLC Performance
Reprinted from: *Materials* **2022**, *15*, 155, doi:10.3390/ma15010155 . . . . **107**

Tongxin Li, Donglin Li, Qingbo Zhang, Jianhang Gao, Long Zhang and Xiaojiu Liu
Improving Fast Charging-Discharging Performances of Ni-Rich $LiNi_{0.8}Co_{0.1}Mn_{0.1}O_2$ Cathode Material by Electronic Conductor $LaNiO_3$ Crystallites
Reprinted from: *Materials* **2022**, *15*, 396, doi:10.3390/ma15010396 . . . . **123**

Eva Gil-González, Luis A. Pérez-Maqueda, Pedro E. Sánchez-Jiménez and Antonio Perejón
Flash Sintering Research Perspective: A Bibliometric Analysis
Reprinted from: *Materials* **2022**, *15*, 416, doi:10.3390/ma15020416 . . . . **137**

Junxue Chen, Sijia Li, Zizheng Qu, Zhonglin Li, Ding Wang, Jialong Shen and Yibing Li
Study on Oxygen Evolution Reaction Performance of Jarosite/C Composites
Reprinted from: *Materials* **2022**, *15*, 668, doi:10.3390/ma15020668 . . . . **157**

Alexandra Kosenko, Konstantin Pushnitsa, Artem Kim, Pavel Novikov and Anatoliy A. Popovich
Structural, Electrical, and Mechanical Properties Investigation of Open-Cell Aluminum Foams Obtained by Spark Plasma Sintering and Replication on Polyurethane Template
Reprinted from: *Materials* **2022**, *15*, 931, doi:10.3390/ma15030931 . . . . **167**

Viktor Chabanenko, Adam Nabiałek and Roman Puźniak
Multi-Steps Magnetic Flux Entrance/Exit at Thermomagnetic Avalanches in the Plates of Hard Superconductors
Reprinted from: *Materials* **2022**, *15*, 2037, doi:10.3390/ma15062037 . . . . **181**

Gideon Gwanzuwang Dankat and Laurentiu Marius Dumitran
Computation of the Electrical Resistance of a Low Current Multi-Spot Contact
Reprinted from: *Materials* **2022**, *15*, 2056, doi:10.3390/ma15062056 . . . . **201**

Peijie Liang, Zhiliang Pan, Liang Tang, Guoqi Zhang, Daoguo Yang, Siliang He and Haidong Yan
Molecular Dynamics Simulation of Sintering Densification of Multi-Scale Silver Layer
Reprinted from: *Materials* **2022**, *15*, 2232, doi:10.3390/ma15062232 . . . . **215**

Zhiyong Zhou, Haiwei Zhang, Weiyang Qin, Pei Zhu and Wenfeng Du
Improving Energy Harvesting from Bridge Vibration Excited by Moving Vehicles with a Bi-Stable Harvester
Reprinted from: *Materials* **2022**, *15*, 2237, doi:10.3390/ma15062237 . . . . **227**

Jaffer Saddique, Honglie Shen, Jiawei Ge, Xiaomin Huo, Nasir Rahman, Muhammad Mushtaq, Khaled Althubeiti, et al.
Synthesis and Characterization of $Sn/SnO_2/C$ Nano-Composite Structure: High-Performance Negative Electrode for Lithium-Ion Batteries
Reprinted from: *Materials* **2022**, *15*, 2475, doi:10.3390/ma15072475 . . . . **241**

Yi-Yueh Chen, Su-Jien Lin and Shou-Yi Chang
New n-p Junction Floating Gate to Enhance the Operation Performance of a Semiconductor Memory Device
Reprinted from: *Materials* **2022**, *15*, 3640, doi:10.3390/ma15103640 . . . . **251**

Yi-Yueh Chen, Su-Jien Lin and Shou-Yi Chang
Correction: Chen et al. New n-p Junction Floating Gate to Enhance the Operation Performance of a Semiconductor Memory Device. *Materials* 2022, *15*, 3640
Reprinted from: *Materials* **2022**, *15*, 6738, doi:10.3390/ma15196738 . . . . **261**

Changdong Chen, Kai Zhao, Ming La and Chenghao Yang
Insight into a Nitrogen-Doping Mechanism in a Hard-Carbon-Microsphere Anode Material for the Long-Term Cycling of Potassium-Ion Batteries
Reprinted from: *Materials* **2022**, *15*, 4249, doi:10.3390/ma15124249 . . . . **263**

Maximilien Coronas, Yaovi Holade and David Cornu
Review of the Electrospinning Process and the Electro-Conversion of 5-Hydroxymethylfurfural (HMF) into Added-Value Chemicals
Reprinted from: *Materials* **2022**, *15*, 4336, doi:10.3390/ma15124336 . . . . **275**

Bouchra Asbani, Yaovi Gagou, Said Ben Moumen, Jean-Luc Dellis, Abdelilah Lahmar, M'Barek Amjoud, Daoud Mezzane, et al.
Large Electrocaloric Responsivity and Energy Storage Response in the Lead-Free $Ba(Ge_xTi_{1-x})O_3$ Ceramics
Reprinted from: *Materials* **2022**, *15*, 5227, doi:10.3390/ma15155227 . . . . **297**

Alonso González, Mario Grágeda, Adrián Quispe, Svetlana Ushak, Philippe Sistat and Marc Cretin
Application and Analysis of Bipolar Membrane Electrodialysis for LiOH Production at High Electrolyte Concentrations: Current Scope and Challenges
Reprinted from: *Membranes* **2021**, *11*, 575, doi:10.3390/membranes11080575 . . . . **317**

Tzu-Sheng Huang, Tung-Li Hsieh, Chih-Ching Lai, Hsin-Yi Wen, Wen-Yao Huang and Mei-Ying Chang
Highly Proton-Conducting Membranes Based on Poly(arylene ether)s with Densely Sulfonated and Partially Fluorinated Multiphenyl for Fuel Cell Applications
Reprinted from: *Membranes* **2021**, *11*, 626, doi:10.3390/membranes11080626 . . . . **347**

Xin Guan, Lujun Pan and Zeng Fan
Flexible, Transparent and Highly Conductive Polymer Film Electrodes for All-Solid-State Transparent Supercapacitor Applications
Reprinted from: *Membranes* **2021**, *11*, 788, doi:10.3390/membranes11100788 . . . . **367**

Kaili Zhu, Yihu Xu, Xiao Yang, Wencai Fu, Wenhao Dang, Jinxia Yuan and Zhiwei Wang
Sludge Derived Carbon Modified Anode in Microbial Fuel Cell for Performance Improvement and Microbial Community Dynamics
Reprinted from: *Membranes* **2022**, *12*, 120, doi:10.3390/membranes12020120 . . . . **381**

Pengcheng Liu, Daijun Yang, Bing Li, Cunman Zhang and Pingwen Ming
Influence of Degassing Treatment on the Ink Properties and Performance of Proton Exchange Membrane Fuel Cells
Reprinted from: *Membranes* **2022**, *12*, 541, doi:10.3390/membranes12050541 . . . . **399**

Liangzhu Zhu and Anil V. Virkar
Conversion Kinetics and Ionic Conductivity in Na-β″-Alumina + YSZ (Naβ″AY) Sodium Solid Electrolyte via Vapor Phase Conversion Process
Reprinted from: *Membranes* **2022**, *12*, 567, doi:10.3390/membranes12060567 . . . . **417**

Etienne Dijoux, Nadia Yousfi Steiner, Michel Benne, Marie-Cécile Péra and Brigitte Grondin-Perez
Fault Structural Analysis Applied to Proton Exchange Membrane Fuel Cell Water Management Issues
Reprinted from: *Electrochem* **2021**, *2*, 38, doi:10.3390/electrochem2040038 . . . . **439**

About the Editors

Marc Cretin

Marc Cretin has been a Professor at the Faculty of Sciences, Montpellier University since 2012. His research interests focus on electrochemistry and material for applications in wastewater treatment and energy. In particular, he works on the development of electrochemical advanced oxidation processes (EAOP) and reactive electrochemical membranes (REM). He has published over 130 peer-reviewed papers in well-recognized scientific journals.

Sophie Tingry

Sophie Tingry is a CNRS Research Director at the European Membrane Institute in Montpellier. Her field of interest concerns the development of electrochemical systems for energy conversion and particularly the reactions at the interfaces of innovative electrocatalytic materials for energy production and storage.

Zhenghua Tang

Dr. Zhenghua Tang is currently is an Associate Professor at the School of Environment and Energy at South China University of Technology, China. He obtained his BS degree at Lanzhou University in 2005, and a PhD. degree at Georgia State University in 2012. After two years of postdoctoral training at the University of Miami, he started his faculty position at SCUT since 2014. His research focuses on the controllable synthesis and electrocatalytic applications of atomically precise metal nanoclusters and single-atom-based nanomaterials. He has published over 80 peer-reviewed papers in J. Am. Chem. Soc., Angew. Chem., Chem. Sci. and other well-recognized scientific journals.

Preface to "Electromaterials for Environment & Energy: Volume I"

In a developing world, the demands for energy and water and the damage to our environment are constantly increasing. Electrochemistry could be a great tool to solve these problems, with an impact that could minimize or at least control damage in our environment, since the main driver of the reaction is the electron that can be produced in a sustainable manner.

In electrochemical approach to energy conversion and production, drinking water production and wastewater treatment, the materials and interfaces are key elements that greatly affect a system's performance.

The objective of this topic is to propose a set of publications gathering the current research trends in the fields of energy, water and environmental pollution treatment, with a focus on the control of materials and interfaces, in view of process optimization.

Marc Cretin, Sophie Tingry, and Zhenghua Tang
Editors

materials

MDPI

Article

Kinetics of the Organic Compounds and Ammonium Nitrogen Electrochemical Oxidation in Landfill Leachates at Boron-Doped Diamond Anodes

Barbara Krystyna Wilk [1,*], Małgorzata Szopińska [1], Aneta Luczkiewicz [1], Michał Sobaszek [2], Ewa Siedlecka [3] and Sylwia Fudala-Ksiazek [4]

1 Department of Water and Wastewater Technology, Faculty of Civil and Environmental Engineering, Gdansk University of Technology, 11/12 Narutowicza St., 80-233 Gdansk, Poland; malszopi@pg.edu.pl (M.S.); aneta.luczkiewicz@pg.edu.pl (A.L.)
2 Department of Metrology and Optoelectronics, Faculty of Electronics, Telecommunication and Informatics, Gdansk University of Technology, 11/12 Narutowicza St., 80-233 Gdansk, Poland; michal.sobaszek@pg.edu.pl
3 Faculty of General and Inorganic Chemistry, University of Gdansk, Wita Stwosza 63 St., 80-306 Gdansk, Poland; ewa.siedlecka@ug.edu.pl
4 Department of Sanitary Engineering, Faculty of Civil and Environmental Engineering, Gdansk University of Technology, 11/12 Narutowicza St., 80-233 Gdansk, Poland; sksiazek@pg.edu.pl
* Correspondence: barbara.k.wilk@pg.edu.pl

Abstract: Electrochemical oxidation (EO) of organic compounds and ammonium in the complex matrix of landfill leachates (LLs) was investigated using three different boron-doped diamond electrodes produced on silicon substrate (BDD/Si)(levels of boron doping [B]/[C] = 500, 10,000, and 15,000 ppm—0.5 k; 10 k, and 15 k, respectively) during 8-h tests. The LLs were collected from an old landfill in the Pomerania region (Northern Poland) and were characterized by a high concentration of N-NH_4^+ (2069 ± 103 mg·L^{-1}), chemical oxygen demand (COD) (3608 ± 123 mg·L^{-1}), high salinity (2690 ± 70 mg Cl^-·L^{-1}, 1353 ± 70 mg SO_4^{2-}·L^{-1}), and poor biodegradability. The experiments revealed that electrochemical oxidation of LLs using BDD 0.5 k and current density (j) = 100 mA·cm^{-2} was the most effective amongst those tested (C_{8h}/C_0: COD = 0.09 ± 0.14 mg·L^{-1}, N-NH_4^+ = 0.39 ± 0.05 mg·L^{-1}). COD removal fits the model of pseudo-first-order reactions and N-NH_4^+ removal in most cases follows second-order kinetics. The double increase in biodegradability index—to 0.22 ± 0.05 (BDD 0.5 k, j = 50 mA·cm^{-2}) shows the potential application of EO prior biological treatment. Despite EO still being an energy consuming process, optimum conditions (COD removal > 70%) might be achieved after 4 h of treatment with an energy consumption of 200 kW·m^{-3} (BDD 0.5 k, j = 100 mA·cm^{-2}).

Keywords: boron-doped diamond electrode (BDD); advanced oxidation processes; electrooxidation (EO) kinetics; optimization of energy consumption

Citation: Wilk, B.K.; Szopińska, M.; Luczkiewicz, A.; Sobaszek, M.; Siedlecka, E.; Fudala-Ksiazek, S. Kinetics of the Organic Compounds and Ammonium Nitrogen Electrochemical Oxidation in Landfill Leachates at Boron-Doped Diamond Anodes. *Materials* **2021**, *14*, 4971. https://doi.org/10.3390/ma14174971

Academic Editor: Marc Cretin

Received: 26 July 2021
Accepted: 24 August 2021
Published: 31 August 2021

Publisher's Note: MDPI stays neutral with regard to jurisdictional claims in published maps and institutional affiliations.

1. Introduction

Despite a new solid waste strategy and enhanced recycling and utilization, landfilling still represents an important path for municipal solid waste management. Landfilling leads to the formation of landfill leachates (LLs), which are potentially toxic and complex liquids with high refractory contaminant levels that percolate through landfill [1–3]. LLs are most often classified by age. LLs from new cells (1–2 years old) usually have higher concentrations of biodegradable contaminants (ratio of five-day Biochemical Oxygen Demand to Chemical Oxygen Demand, BOD_5/COD) ratio is >0.5) due to the high concentration of volatile organics) [4,5]. LLs generated by older cells (>10 years old) are usually less prone to biological degradation with BOD_5/COD < 0.3 (due to the high concentration of refractory organic matter) [6]. However, the appropriate BOD_5/COD ratio itself does not ensure good biochemical degradation of contaminants [7]. Therefore, each LL treatment problem requires a comprehensive, individual approach. The problem of efficient and economically viable treatment of LLs is now one of the crucial, still-unresolved environmental issues globally.

According to the relevant legal requirements [8–10], LLs should be collected and subsequently treated to ensure that they are safe for the environment. However, there are no unique and flexible treatment methods available as of now [11]. A number of technologies and processes for LL treatment have been assessed in recent decades, including biological processes, adsorption, coagulation/flocculation, the membrane process, and advanced oxidation processes (AOPs) [12]. Biological processes (conventional activated sludge, sequencing batch reactors, membrane bioreactors, aerated lagoons, and upflow anaerobic sludge blanket) are most frequently used for pre-treatment of LLs generated by new landfill cells. The aim of the aforementioned processes is the simultaneous removal of nitrogen and easily biodegradable organic matter from the LLs. However, in LLs, biological treatment is usually hampered by certain toxic substances and inadequate C-to-N ratio. The probable presence of non-biodegradable emerging pollutants is also of concern. The outcome is thus unsatisfactory and requires the inclusion of an additional treatment step to meet the discharging standards [13–16]. Recently, AOPs are considered to be among the most efficient methods in disinfecting and purifying ground water and natural water, as well as being used in wastewater treatment [17,18]. The processes are becoming increasingly popular, particularly in industrial settings, as they are generally accepted to be environmentally friendly and very much in line with the Clean Production strategy (minimizing the volume of contaminants released into the environment) [19–21]. Among the most efficient and fastest developing AOPs are the electrochemical ones [22,23], which include: anodic oxidation, anodic oxidation with electrogenerated H_2O_2, electro-Fenton, photoelectro-Fenton, and solar photoelectro-Fenton [24].

Among the EO process, reagent-free anodic oxidation is increasingly popular. The main objective of these processes is to generate free hydroxyl radicals ($^{\bullet}OH$), which are non-selective, powerful oxidizing agents capable of oxidizing a wide variety of contaminants at nearly diffusion-limited rates. Anodic oxidation is conducted with the use of high-oxygen-overvoltage anodes (so-called "non-active" anodes), which include, among others, SnO_2, PbO_2, Ti_4O_7, and BDD. BDD has many unique vital characteristics: a broad range of electrochemical potentials in aqueous solutions (from −1.25 to 2.3 V, as compared to a standard hydrogen electrode [SHE]). The use of EO to treat wastewater by means of BDD anodes was reported in earlier studies [25–29]. In accordance with Comninellis' model, the degradation of organic compounds (R) in LLs by means of BDD anodes is mainly mediated by hydroxyl radicals through direct and/or indirect oxidation. In the case of direct oxidation, the contaminant first undergoes physical adsorption on the anode surface and is then oxidised via the intermediation of very reactive short-lived hydroxyl radicals (this occurs below the oxygen evolution reaction potential on the anodes) [30,31]. The indirect oxidation process is the consequence of the hydroxyl radicals (produced in the water discharge reaction) being physiosorbed in the vicinity of the anode surface [32], as shown in Equations (1) and (2).

$$H_2O + BDD \rightarrow BDD\,(^{\bullet}OH) + H^+ + 1e^- \quad (1)$$

$$BDD\,(^{\bullet}OH) + R \rightarrow BDD + CO_2 + H_2O + H^+ + 1e^- \quad (2)$$

Indirect anodic oxidation may also be mediated by electrogenerated oxidizing agents in the sample, such as hypochlorite, chlorine, ozone, or hydrogen peroxide [30]. As has been pointed out in the literature, the active chlorine removes $N\text{-}NH_4^+$ first [19,33,34]. Next, if there is surplus content of active chlorine in the LLs, intensive oxidation of organic compounds takes place. Most studies indicate [21,26,35] that the degradation of ammonium in LLs is a second-order reaction (first-order with respect to the concentration of ammonium, and first-order with respect to the concentration of active chlorine), while COD degradation fits a pseudo-first model, or pseudo-zeroth-order kinetics models. The kinetic model of

COD degradation is determined by the following factors: contaminant concentration level in the analyzed matrix, the applied current density (j), the electrode substrate, pH and temperature [26,36,37]. The degradation of contaminants in the LLs matrix is a very complex process that has not yet been fully understood. On the other hand, the possible high electrolytic conductivity of LLs (e.g., high concentrations of chloride and sulfate) should benefit the efficiency of the EO process.

EO has also a great potential for medical sector application, especially in the case of hospital wastewater treatment rich in non-biodegradable pharmaceuticals (e.g., antibiotics, antihistaminics, anti-inflammatories, antidepressants, antihypertensives, hormones, and antiulcer agents) [38]. Moreover, transformation of urine into nutrients using electrolytic oxidation was also evaluated [33]. Additional positive side effect of EO wastewater treatment is disinfection, which is described by Dbira and co-authors [34] as well as Kraft [35].

The aim of the study was (I) to determine the efficiency of BDD/Si electrodes for the removal of macropollutants present in LLs under different electrochemical conditions, and (II) to recognize the dependencies between the reaction parameters affecting the COD and $N\text{-}NH_4^+$ removals (e.g., by-product formation). To gain a better understanding of the contaminants' degradation kinetics in LLs samples by EO method, the different BDD/Si electrodes (levels of boron doping [B]/[C] = 500, 10,000, and 15,000 ppm—0.5 k; 10 k and 15 k, respectively) were studied. To date, the EO of old LLs has not been tested yet in laboratory-scale experiments with multiple iterations and a wide range of j values tested (in the range from 25 to 100 $mA\cdot cm^{-2}$). The EO reaction rates have been quantitatively discussed in order to optimize reaction condition. Furthermore, emphasis was placed on minimizing process costs and the preparatory protocol for carrying out the EO of LLs. Energy of removal of contaminants was calculated and optimized, additionally. Hence, this study can help to find the optimal conditions for the most cost-effective treatment of LLs with EO/BDD process.

2. Materials and Methods

2.1. LL Characteristics

The raw LLs used in this study were collected from the "Eko Dolina Lezyce" Municipal Solid Waste Plant (MSWP). The MSWP is located in the Pomerania region of north Poland, in Central Eastern Europe. The samples were collected from a closed, stabilized landfill cell that was operated from 2003 to 2011. While in operation, the landfill received mainly municipal waste, with unlimited disposal of organic wastes, including multi-material packaging or plastics [29].

The sanitary LLs used in this study were collected in 2019, before each experiment. Collected samples were transported in polyethylene containers (three bottles with a volume of 10 L each) to the laboratory (at 4 ± 1 °C). The samples were homogenized, then their physicochemical properties were evaluated. Each measurement was made three times.

The investigated raw LLs were rich in nitrogen (Total nitrogen, TN = 2148 ± 108 $mg\cdot L^{-1}$ and $N\text{-}NH_4^+$ = 2069 ± 103 $mg\cdot L^{-1}$) and organic substances (COD = 3608 ± 123 mg $O_2\cdot L^{-1}$ with relatively low values of BOD_{20} = 403 ± 54 mg $O_2\cdot L^{-1}$). The low BOD_{20}/COD ratio (0.12 ± 0.00) obtained is regarded as typical for LLs generated by old landfill cells. Furthermore, LLs in this study were characterized by low TSS (<90 $mg\cdot L^{-1}$) and high $N\text{-}NH_4^+$/TN ratio (from 0.95 to 0.98). All obtained data indicated that the tested LLs were refractory to biological treatment. On the other hand, inter alia, high concentrations of chloride (2690 ± 70 mg $Cl^-\cdot L^{-1}$), sulfate (1353 ± 70 mg $SO_4^{2-}\cdot L^{-1}$), and electrolytic conductivity ranging from 21.2 to 26.8 $mS\cdot cm^{-1}$ indicated the possible high efficiency of the EO process, if used for LL treatment.

2.2. Boron-Doped Diamond Electrodes (BDDs)

The BDD electrodes were synthesized in the Microwave Plasma Assisted Chemical Vapor Deposition (MWPACVD) process (SEKI Technotron AX5400S). Diamond films were deposited on two-inch Silicon wafers (ITME, Warsaw, Poland). The chamber pressure was kept at 50 Torr and the plasma was generated with microwave radiation (2.45 GHz 1300 W. The induction heating stage temperature was set to 700 °C, the total flow rate of gases reached 300 sccm and the molar ratio of methane was equal to 1%. Three different levels of boron doping in the gas phase, expressed as [B]/[C] ratio, was 500, 10,000, and 15,000 ppm using diborane (B_2H_6) as a dopant precursor. In total, three different electrodes were prepared. The deposition time was set at 12 h.

2.3. Experimental Apparatus and Procedure

In regard to a low amount of total suspended solids (70.8 ± 10.0 mg·L^{-1}) in the tested samples no pre-treatment nor pre-filtration was needed to conduce the electrolysis process. An LLs sample dilution of 1:1 (*V:V*) was applied. Electrooxidation runs were performed using 400 mL of sample in a 500-mL single-chambered reactor equipped with a magnetic stirrer (Electrochemical Stirrer, ES24, Wigo, Pruszkow, Poland) with a stirring speed of 300 rpm to keep the wastewater homogeny mixed. The temperature of the solution was maintained within 25 ± 1 °C (cooling bath). Assays were performed in the galvanostatic mode (power supply GW Instek GPD-23035, New Taipei City, Taiwan). Figure 1 shows the procedure used.

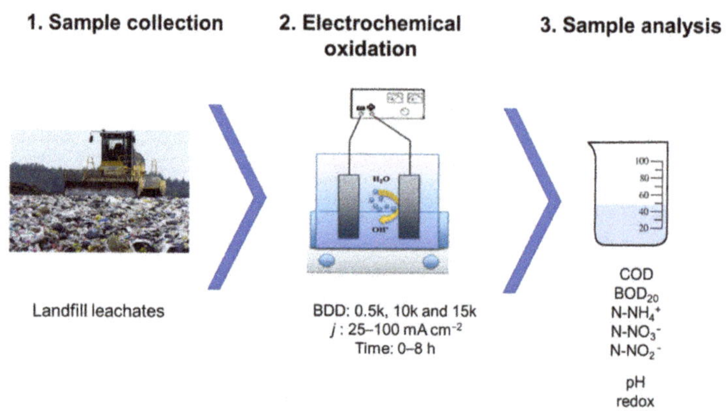

Figure 1. Procedure scheme.

Four different j of between 25 and 100 mA·cm^{-2} and three different BDD electrodes as anodes named 0.5 k, 10 k, and 15 k were tested (see point 2.2). Stainless-steel mesh served as the cathode. Both the anode and the cathode were used as flat with a geometric surface area of 10.5 cm^2 and the interelectrode distance ca. 2.0 cm. All assays were replicated three times, and samples (15 mL each) were collected every 2 h, until a final time of 8 h. Samples were degassed by mixing on a multipoint stirrer (Variomag, POLY 15 KOMED, Thermofisher Scientific, Port Orange, FL, USA) at 50 rpm for 15 min. This was necessary in order to perform a correct physico-chemical analysis of the samples. Table 1 presents a list of the physico-chemical analysis carried out in LLs, the methods of determinations used and the laboratory equipment used (all according to the APHA 2005 standard).

Table 1. List of the physico-chemical analysis applied.

Parameter	Method and Device Used
COD, N-NH_4^+, N-NO_3^-, N-NO_2^-, TP, P-PO_4^-, Cl^-, SO_4^{2-}, S^{2-}	XION 500 spectrophotometer (Dr Lange, GmbH, Düsseldorf, Germany)
BOD_{20}	manometric respirometric BOD OxiTop® method
pH, conductivity and ORP	portable multi-parameter meter (HL-HQ40d multi, HACH, Düsseldorf, Germany)
total (TSS), mineral (MSS), volatile suspended solids (VSSs)	the gravimetric method

2.4. Calculation

2.4.1. The Energy Consumption

One of the main challenges for AOPs is to reduce the energy consumption [36]. The energy consumption was expressed as W, kWh and was calculated in the following way: the applied current [A] was multiplied by electrolysis time [8 h] and the average cell voltage (E_{cell}) [V], then this value (W, kWh) was recalculated and expressed in kWh·m^{-3}. The specific energy consumptions in [kWh·kg^{-1} COD] and [kWh·kg^{-1} N-NH_4^+] was estimated by means of Equations (3) and (4):

$$EC_{COD} = (1000 \times E_{cell} \times I \times \Delta t)/(Vs\,[\Delta COD]) \quad (3)$$

$$EC_{N\text{-}NH_{4+}} = (1000 \times E_{cell} \times I \times \Delta t)/(Vs\,[\Delta N\text{-}NH4^+]) \quad (4)$$

where: 1000 is a conversion factor (in mg·g^{-1}), E_{cell} is the average cell voltage (in V), I is the applied current (in A), Δt is the electrolysis time (in h), Vs is the LL volume (in L), and (ΔCOD) is the experimental COD concentration decay (in mg·L^{-1}).

Additionally instantaneous current efficiency (%ICE) was calculated (see Table A1) according to the following Equation (5):

$$\%ICE = 100\% \times F \times Vs \times (COD_0 - COD_t)/\,(m \times I \times \Delta t) \quad (5)$$

where: F is the Faraday constant (96,485 C mol^{-1}), Vs is the total volume of the bulk electrolyte (L), COD_0 and COD_t (in gO_2L^{-1}) are the initial and final COD obtained before and after EO treatment, I is the current (A), m is the equivalent mass of oxygen (m = 8), and Δt is the electrolysis time (s).

2.4.2. Kinetic Study

The order of the reaction of removing selected pollutants from the LLs was determined graphically using Statistica 12 (StatSoft, Inc., Tulsa, OK, USA). To determine the order of reactions with regard to reactant, three plots were made (concentration of contaminant versus time—zeroth-order reaction, natural log of concentration versus time—first-order reaction and inverse concentration versus time—second-order reaction). The most linear graph (linear regression was used) with the highest coefficient of determination value showed the order of the reaction with regard to reactant. After determining the orders of reactions, the rate constants and half-life were calculated using the appropriate kinetic equations.

2.4.3. Biodegradability Index

Biodegradability indexes (BI) as BOD_{20} to COD ratios were determined for raw LL samples (before the electrochemical oxidation process) and for all samples after 8 h of treatment by means of EO (different current densities and different boron concentrations in anodes used).

2.5. Data Analysis

All plots and statistical analysis of data were performed using Statistica 12 (StatSoft, Inc., Tulsa, OK, USA), Microsoft Excel® 2016 (Microsoft, 2016, Redmond, Washington, USA), and OriginPro 9.0 (OriginLab Corporation, Northampton, MA, USA) software.

3. Results and Discussion

3.1. Electrodes Morphology and Composition

The deposited electrode with various boron content strongly influences surface morphology. The average grain size for the 0.5 k electrode was approximately 2 μm. In the case of highly boron-doped 15 k, the grain size decreased to approximately 0.5 μm. Boron introduces re-nucleation which results in the creation of smaller crystallites on primary higher diamond crystals. This significant change in surface morphology impacts film composition results with more non-diamond content [37], which also increases internal stress in the film [39]. Moreover, increasing boron level in the BDD films influences their electro-catalytic performance, resulting in time decrease of the electrolysis [40,41] (stronger generations of $^{\bullet}$OH radicals), but, in contrast, it reduces the width of the potential working window.

Experimental Raman spectra are shown in Figure 2. The sp^3 diamond lines are located at 1332, 1317, and 1310 cm^{-1} for the 0.5 k, 10 k and 15 k, respectively. The boron-doping affects the diamond line causing a shift towards lower wavenumbers. Additionally, for the 10 k and 15 k, the sp^3 peak is strongly asymmetric which is attributed to the Fano effect [42], a result of interference between the scattering by the zone-center phonon line and the scattering by an electronic continuum [43,44]. It is worth noting that high levels of doping reveal two broad peaks located at ca. 500 and 1200 cm^{-1}, which are attributed to the incorporation of boron in the diamond lattice [45]. According to the literature, these two broad bands are attributed to the maxima of the phonon densities of the states [46]. Only for the lowest doped sample (0.5 k), peaks assigned to the first order of silicon are visible, located at 520 cm^{-1} for one-phonon mode, and the second-order of silicon is visible at 970 cm^{-1}.

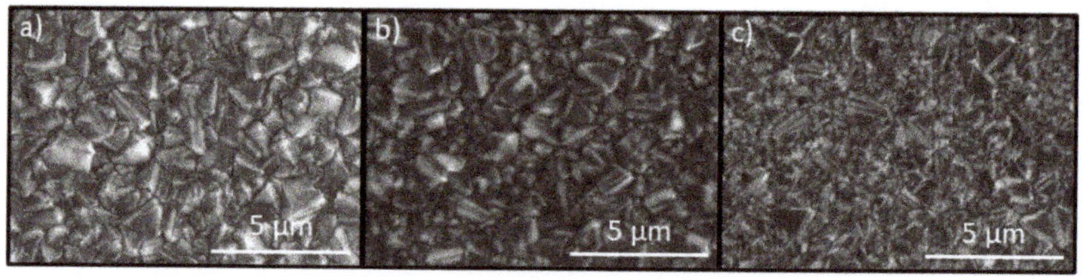

Figure 2. SEM images of BDD films deposited at different [B]/[C] ratios in plasma, namely, (**a**) 500 ppm, (**b**) 10,000 ppm, and (**c**) 15,000 ppm. Magnification of 10,000×.

Additionally, the Fano-shaped peaks located at ca. 500 cm^{-1} (BWF function #1), 1200 cm^{-1} (BWF function #2), and 1332, 1317, and 1310 cm^{-1} (BWF function #3) were modelled using the Breit–Wigner–Fano function (see Equation (6)) [47] which is shown in Figure 3a–c insets.

$$F_i(\omega) = \frac{A_i \times \left(q_i + \frac{\omega-\omega_i}{\Gamma_i}\right)^2}{1+\left(\frac{\omega-\omega_i}{\Gamma_i}\right)^2} \tag{6}$$

In the equation, A_i is the amplitude of the Fano-shaped peaks, q_i is the asymmetric parameter, ω_i is the width, and Γ_i is the position of the lines.

The fitting parameters are shown in Table 2. In the case of the 0.5 k, the sample results in low boron incorporation into the diamond lattice that in turn results in lack of BWF#1 and BWF#2 bands. The amplitude of the Fano-shaped sp^3 (BWF#3) peak is lower for the highly doped samples, but the peak width (ω) is wider, which is attributed to more boron incorporation into the diamond lattice [43]. According to an investigation by Mortet and co-authors [47], the asymmetric parameter of the diamond sp^3 line (q_3) can be used as a marker of the carrier concentration. Only the 10 k and 15 k samples indicated significant asymmetry (10 k q_3 = 1.77 and 15 k q_3 = 2.11), and it should be noted that the q_3 value of those electrodes proves high incorporation of boron atoms during MWPACVD growth.

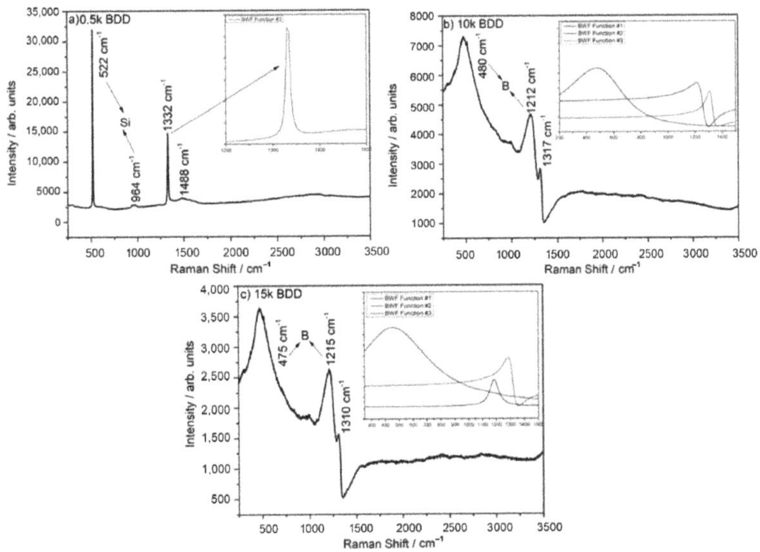

Figure 3. Raman spectra of (**a**) 0.5 k, (**b**) 10 k, and (**c**) 15 k BDD films. Insets shows Raman spectra modeled using Breit–Wigner–Fano function: Fano-shaped peaks (BWF #1 red line, BWF #2 blue line, and BWF #3 olive line).

Table 2. Fitting parameters of BDD electrodes. Indices 1, 2, and 3 correspond to the BWF#1, BWF#2, and BWF#3 bands, respectively.

Parameter	0.5 k BDD	10 k BDD	15 k BDD
A_1	-	3895	2232.13
q_1	-	5.3	4.3
ω_1	-	480.2	476.34
Γ_1	-	248.70	236.81
A_2	-	1309.63	1429
q_2	-	0.88	0.46
ω_2	-	1220	1214
Γ_2	-	42.99	42
A_3	11,860	1833	1039
q_3	0.127	1.77	2.11
ω_3	1332	1317	1309
Γ_3	4.60	26.51	33.8

Additionally, the band located at ca. 500 cm^{-1} can be used to calculate the amount of boron in the diamond film. In the case of the 10 k electrode, the boron content is ca. 8.32×10^{20} and in the case of the 15 k, it is 1.05×10^{21}. The slight difference in boron amount between 10 k and 15 k can be explained by a decrease in the effective doping of the diamond films during growth and an increase in the amount of boron in the gas phase.

The measurement of the electrochemical potential windows was carried on in three electrode setup in 1 M KCl at scan rate 100 mV/s, where the BDD electrode was as working, the Pt wire as counter and Ag/AgCl as reference. Figure 4 shows electrochemical potential windows of BDD electrodes with different boron doping. The increasing amount of boron content in diamond films narrows the width of electrochemical window. The 0.5 k doped sample have the wider window reaching up to 3.9 V, followed by 10 k sample with 3.12 V. The narrow window for the 15 k BDD, 2.23 V is connected with significant amount of defects in the diamond films.

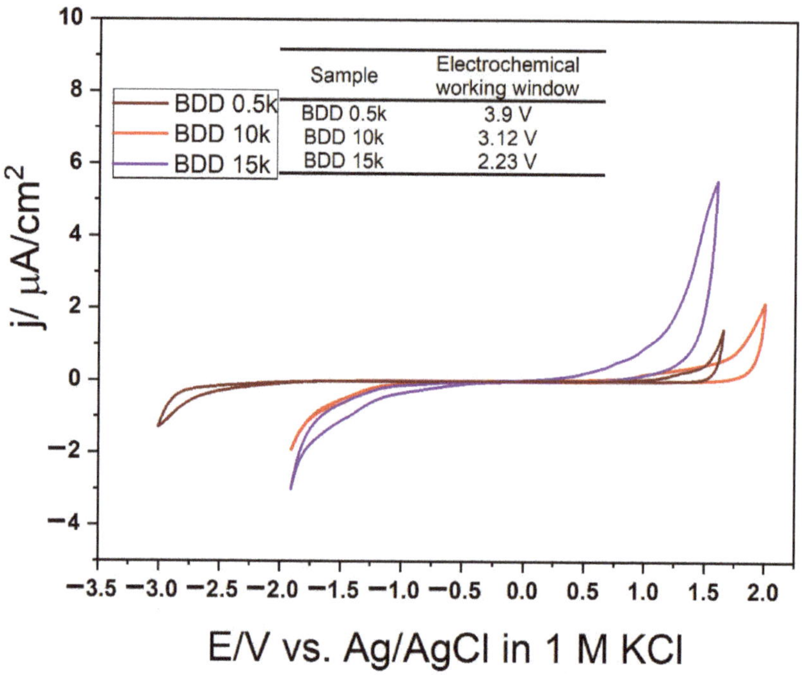

Figure 4. Cyclic voltammetry plots for BDD films with different boron doping recorded in 1 M KCl with 100 mV/s scan rate.

3.2. COD Electrooxidation

In this study, during EO of LLs, in general the COD and N-NH_4^+ elimination rate increased with increasing j and time, which is well in agreement with the literature [48–50]. However, the complete removal of studied contaminants was not reached. It was noticed that, using j = 50–100 mA·cm^{-2}, the increase in efficiency of COD removal was not as substantial as when using a lower j (see Figure 5a). It is generally considered that applying high j leads to the transfer of hydroxyl radicals ($^{\bullet}OH$) to H_2O_2 near the electrode surface, and consequently H_2O_2 is oxidized to O_2 [51]. Moreover, the energy is consumed not mainly for oxidation of organic compounds (expressed as COD), but to a large degree for oxidation of other ions in the LLs [52].

The 0.5 k anode was the most effective electrode with $C_{8h}/C_0 = 0.09 \pm 0.14$ for COD and 0.14 ± 0.01 for BOD_{20} after 8 h of process, at 100 mA·cm^{-2}. In the same conditions, using 15 k, C_{8h}/C_0 was 0.27 ± 0.31 for COD, and 0.23 ± 0.08 for BOD_{20}, while for 10 k, C_{8h}/C_0 was 0.3 ± 0.001 for COD and 0.26 ± 0.01 for BOD_{20}. COD was most effectively removed during the first two hours of the EO process, although removal efficiency decreased with time (for all tested electrodes and j).

EO treatment can also improve the biodegradability of LLs by oxidizing the molecular structure of refractory organics, and thus can be used as a pre-treatment process before, for example, further biological treatment. However, it should be noted that the best conditions ensuring good COD removal are BI > 0.5 [53]. Fernandez and co-authors [21,53], when testing the EO (BDD anodes, j = 700 mA·cm^{-2}, 36 h of process) for the treatment of sanitary LLs obtained an increase in biodegradability index (BI) from 0.18 to 0.84. Meanwhile, Nurhayati [54] got an increase in BI ranging from 0.06 in raw LLs up to 0.386 after electrooxidation process at j of 30 mA·cm^{-2} with a flow rate of 5 mL·s^{-1}. In this study, the calculated BI of raw and EO treated LLs is presented in Figure 6. BI in tested raw samples was equal to 0.11 ± 0.13.

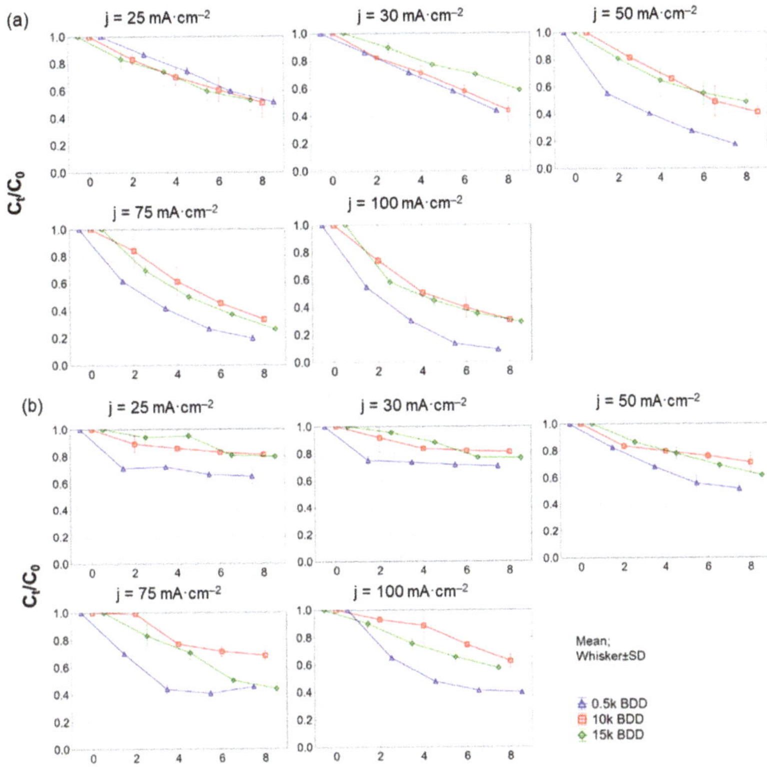

Figure 5. Removal of (**a**) COD and (**b**) $N-NH_4^+$ during 8 h test of LL treatment with different BDD anodes and current densities (25–100 mA·cm^{-2}) expressed as normalized concentration C_t/C_0.

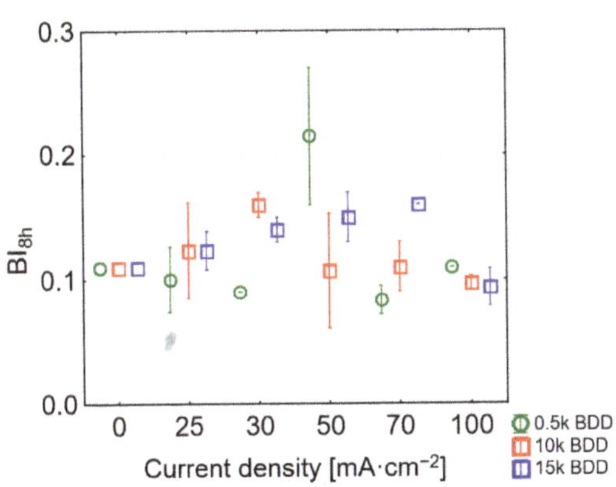

Figure 6. Influence of various current densities and electrodes used on BI (BI values after 8 h of LL treatment for different current densities applied).

Interestingly, the use of the highest j (100 mA·cm^{-2}) resulted in obtaining a BI in the treated LLs close to that calculated for the raw sample (BI for 0.5 k = 0.11 ± 0.00, 10 k = 0.097 ± 0.005, and 15 k = 0.090 ± 0.01). This can be explained by the large share of non-biodegradable intermediates in the sample. The highest BI = 0.22 ± 0.05 was obtained using a 0.5 k electrode and a j of 50 mA·cm^{-2}, successively using a 10 k electrode BI = 0.16 ± 0.01 and a current of 30 mA·cm^{-2}. Hence, following Deng and co-authors as well as McBeath and co-authors [20,55] in applying such a condition, EO might be used prior to biological treatment due to the doubling of BI. Generally, the results showed that BDD anodes are very effective in oxidation of organic matter, what can be attributed, inter alia, to their high oxygen evolution potential.

3.3. N-NH_4^+ Electrooxidation

For all anodes tested in this study, N-NH_4^+ removal was less than COD removal (e.g., 0.5 k, j = 100 mA·cm^{-2}, C_{8h}/C_0: COD = 0.09 ± 0.14 mg·L^{-1}, and N-NH_4^+ = 0.39 ± 0.05 mg·L^{-1}). Other researchers obtained similar results. After 6 h of process, Zhou and co-authors [56] achieved 87.5% COD and 74.06% N-NH_4^+ removal by the BDD/Nb electrodes using j = 50 mA·cm^{-2}. Luu and co-authors [57] explored EO of biologically treated LLs using Ti/BDD and Ti/RuO_2 anodes and observed the maximum COD and N-NH_4+ removal efficiency of 95.17% and 81.18%, respectively (after 8 h, at 83 mA·cm^{-2}). This phenomenon is characteristic of EO of LLs containing Cl^- in the range of concentrations from 150 to 4500 mg Cl^-·L^{-1} by means of BDD anodes (primarily COD is oxidized) [58]. Ammonia removal is mainly promoted by a reaction with active chlorine (chlorine-mediated pathway: Cl_2/$HOCl^-$ indirect oxidation) and the oxidization rate of N-NH_4^+ by hydroxyl radicals is lower than that of organics ("electrochemical combustion" process) [20]. Furthermore, BDD anodes are more suitable for $^{\bullet}OH$ radical generation than chlorine evolution.

Regarding N-NH_4 and TN removals from LLs, it can be assumed that j is a key experimental parameter affecting this process (Figures 5b and A3 (Appendix A)). The increase in j (according to Faraday's law) enhanced the electrochemical process and resulted in a satisfactory result in N-NH_4^+ and TN removal [59,60].

Removal of N-NH_4^+ is also strongly pH-dependent [20]. It was found that ammonia ions present in the initial sample were partially transformed to the nitrate or nitrite forms. However, based on Figure 7, it might be assumed that some ammonia nitrogen was transferred directly to gaseous nitrogen—approximately 417, 385, and 294 mg N-N_2·L^{-1} for BDD 0.5 k, 10 k, and 15 k, respectively.

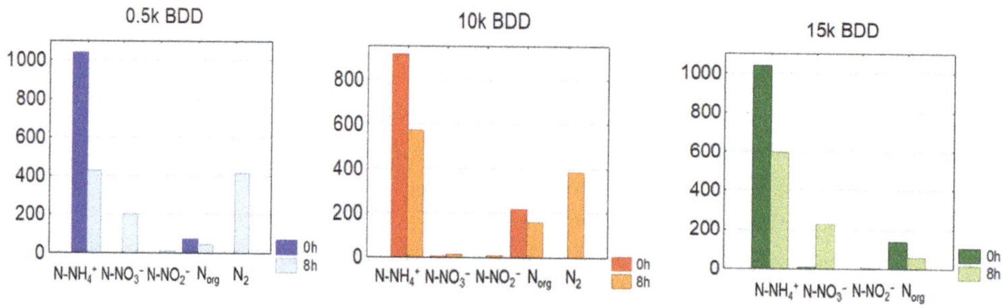

Figure 7. Average C_0 and C_{8h} concentrations of nitrogen compounds in treated LLs by means of different anodes used, j = 100 mA·cm^{-2} (N_{org} = TN − (N-NH_4^++ N-NO_3^-+ NO_2^-)).

Application of high current densities led to a fast increase in pH to above 8, which promoted the direct oxidation of free ammonia (gaseous) according to reactions (7) and (8) [61]:

$$2NH_3 + 3H_2O \rightarrow NO_3^- + 9H^+ + 8e^- \quad (7)$$

$$2NH_3 \rightarrow N_2 + 6H^+ + 6e^- \quad (8)$$

According to Zhou and co-authors [56], it might be assumed that higher ammonia removal efficiency might be achieved in acidic pH (where $HOCl^-$ are predominant species). Furthermore, removing ammonia from LLs via an active chlorine-mediated (indirect) path can be effective only if it occurs quickly, so good results can be obtained inter alia in high temperatures [62]. If these conditions are not met, the active chlorine is often converted to chlorate and perchlorate, and ammonium nitrogen is not removed effectively.

3.4. The Evolution of Nitrates and Nitrites during EO

Figures A1 and A2 (Appendix A) report the change of nitrate and nitrite concentration with time. In general, the concentration of N-NO_3^- increased intensively (or remained at the same level) during the whole electrochemical process due to partial oxidation of organic matter containing nitrogen and ammonia. For instance, at a j of 100 mA·cm^{-2} C_{8h}/C_0 for N-NO_3^- was 42.6 ± 3.0, 6.0 ± 1.8, and 39.2 ± 9 for 0.5 k, 10 k, and 15 k, respectively.

A similar dependency occurred in the case of N-NO_2^- concentrations (from 0.20 mg·L^{-1} up to 8.95 mg·L^{-1} for 0.5 k, j = 100 mA·cm^{-2}). N-NO_2^- ions were possibly formed through the EO of ammonia on the anode (Equation (9)) or electrochemical reduction of nitrate on the cathode (Equation (10)) [63]. However, oxidation of amine groups present in organic compounds needs to be taken into consideration during NO_2^- generation. The amine group in organic functional group redox series is very reductive and easily oxidized [64]:

$$NH_4^+ + 2H_2O \rightarrow NO_2^- + 8H^+ + 6e^- \quad (9)$$

$$NO_3^- + H_2O + 2e^- \rightarrow NO_2^- + 2OH^- \quad (10)$$

It should also be noted that, in our study, TN was mainly composed of N-NH_4^+. For this reason it was concluded that the oxidation of organic nitrogen had a negligible effect on the effects of TN removal in the tested samples.

3.5. Changes and Influence of pH on Electrooxidation

The initial pH of the tested raw LLs was 7.8 ± 0.1 and generally, with time of EO treatment, the pH of the LLs also increased (Figure A4, Appendix A). The high concentration of carbonate/bicarbonate ions in raw LLs, which are effective $^{\bullet}OH$ radical scavengers, may cause an increase in pH during EO treatment [65]. At the same time, the carbonate–bicarbonate equilibrium (Equation (11)) is shifting to the right.

$$CO_2 + H_2O \leftrightarrow H^+ + HCO_3^- \leftrightarrow 2H^+ + CO_3^{2-} \quad (11)$$

In this study, the exceptions were observed during EO processes using 0.5 k and 10 k electrodes with j = 25 mA·cm^{-2} and 10 k, 15 k using j = 30 mA·cm^{-2}. In these cases, a decrease in pH was observed with time. This phenomenon is supposed to be related to the electrochemical reactions intensively taking place on the surface of the anode [34,66] (please see Equations (12) and (13)). Another option could be that low-molecular-weight carboxylic acids present in LLs [67,68] undergo the following reactions presented in Equations (14) and (15) [69].

$$H_2O \rightarrow {}^{\bullet}OH + H^+ + 1e^- \quad (12)$$

$$2H_2O \rightarrow O_2 + 4H^+ + 4e^- \quad (13)$$

$$HCOOH \rightarrow 2CO_2 + 2H^+ + 2e^- \quad (14)$$

$$CH3COOH \text{ carboxylic acids} + 2H_2O \rightarrow 2CO_2 + 8H^+ + 8e^- \quad (15)$$

On the basis of the obtained pH results, it was also found that the $HOCl^-$ ion could play a large part in the oxidation of N-NH_4^+ (especially during the first 4 h of the oxidation process). Higher j shifted the pH to higher values, which led to the formation of the hypochlorite anion (ClO^-), the presence of which induced the formation of toxic chloramines and poor oxidation properties. It is supposed that one of the main reasons for the increase in pH was the reactions forming, inter alia, hydroxyl ions during the electrochemical reduction of nitrate [70]. Another reason could be the intense generation of sulfate radicals (reaction of the sulfate present in the LLs with the hydroxyl radicals). This reaction contributes to increasing the concentration of hydroxyl ions in LLs and increasing the pH value (Equations (16) and (17)) [71]:

$$SO_4^{2-} + {}^{\bullet}OH \rightarrow SO_4^{\bullet -} + OH^- \quad (16)$$

$$2H_2O + 4e^- \rightarrow H_2 + 2OH^- \quad (17)$$

However, it should be noted that at pH = 8 sulfate and chloride radicals are not permanent and the equilibrium is shifted in the ${}^{\bullet}OH$ direction (Equation (15)). Moreover the presence of chloride at the level of 2690 ± 70 mg·L^{-1} might have an 'inhibiting effect' on organics removal by sulfate radicals, because chlorine reactive species are expected to be the dominant oxidative species in the anodic oxidation of LLs [52].

3.6. COD and N-NH_4^+ Kinetic Evaluation

The results of the graphical method of the reaction kinetic order evaluation are presented in Figure 8 and Table 3. It should be noted that the rate constants in all cases (k) showed great dependence on the applied j: generally it increased with increasing applied j, and the value of the half-life ($T_{1/2}$) consistently decreased (Table 3). Ukundimana and co-authors [25] reported that a mechanism of COD removal from LLs (pre-treated in an ultrafiltration unit) by means of BDD electrodes fitted well with the pseudo-first-order kinetic model. A similar phenomenon has been observed in another study, in which Papastavrou and co-authors [72] treated stabilized LLs by means of BDD anodes and obtained the pseudo-first-order kinetic model for COD removal with a kinetic coefficient of 8.3 ± 1·10^{-3}·min^{-1} for current values 15 and 21 A. Conversely, studies by Cossu and co-authors [73] demonstrate that the rate constants (EO of LLs; PbO_2, and SnO_2 anodes) decrease over time due to the presence of compounds that are more easily oxidized than others in the initial LL sample. Another explanation could be that due to the very complex matrix of LLs (containing e.g., high-molecular-weight compounds such as humic and fulvic acids, and recalcitrant substances with a low molecular weight such as halogenated compounds), some compounds are more easily oxidized than others during the first stage of electrolysis [74].

Table 3. Electrochemical degradation kinetics for COD and N-NH_4^+ in LLs with different j and different electrodes used.

Parameter	Electrode	Current Density (mA·cm^{-2})	Order of Reaction	k *	$T_{1/2}$ (min)
COD	0.5 k BDD	25	pseudo-first-order	1.34 ± 0.24 ×10^{-3}	536.7 ± 124.7
		30	pseudo-zeroth-order	1.21 ± 0.01 ×10^{-3}	412.2 ± 3.9
		50	pseudo-zeroth-order	1.72 ± 0.01 ×10^{-3}	290.9 ± 2.4
		75	pseudo-first-order	3.46 ± 0.23 ×10^{-3}	203.6 ± 13.5
		100	pseudo-first-order	4.76 ± 0.71 ×10^{-3}	147.4 ± 2.2
	10 k BDD	25	pseudo-first-order	1.57 ± 0.36 ×10^{-3}	459.6 ± 101.6
		30	pseudo-zeroth-order	1.19 ± 0.17 ×10^{-3}	426.3 ± 61.9
		50	pseudo-second-order	2.38 ± 0.75 ×10^{-3}	449.4 ± 146.2
		75	pseudo-first-order	2.30 ± 0.07 ×10^{-3}	302.9 ± 9.5
		100	pseudo-first-order	2.47 ± 0.04 ×10^{-3}	280.1 ± 5.5
	15 k BDD	25	pseudo-first-order	1.38 ± 0.16 ×10−3	506.7 ± 57.3
		30	pseudo-zeroth-order	8.80 ± 0.10 ×10^{-4}	567.6 ± 4.5
		50	pseudo-first-order	1.86 ± 0.21 ×10^{-3}	374.5 ± 43.4
		75	pseudo-first-order	2.80 ± 0.14 ×10^{-3}	247.1 ± 12.1
		100	pseudo-second-order	5.22 ± 0.14 ×10^{-3}	190.7 ± 12.3
N-NH_4^+	0.5 k BDD	25	second-order	1.09 ± 0.22 ×10^{-3}	936.8 ± 191.3
		30	second-order	0.83 ± 0.07 ×10^{-3}	1205.4 ± 98.7
		50	second-order	2.12 ± 0.09 ×10^{-3}	472.1 ± 22.3
		75	zeroth-order	1.21 ± 0.01 ×10^{-3}	413.6 ± 3.2
		100	second-order	3.41 ± 0.001 ×10^{-3}	292.7 ± 8.5
	10 k BDD	25	second-order	0.49 ± 0.04 ×10^{-3}	1958.2 ± 62.4
		30	second-order	0.51 ± 0.01×10^{-3}	1959.0 ± 61.3
		50	second-order	0.84 ± 0.30 ×10^{-3}	1267.8 ± 151.7
		75	second-order	1.09 ± 0.19 × 10^{-3}	932.6 ± 161.4
		100	zeroth-order	0.80 ± 0.12 × 10^{-3}	628.68 ± 96.6
	15 k BDD	25	first-order	0.48 ± 0.14 × 10^{-3}	1450.9 ± 56.2
		30	first-order	0.55 ± 0.01× 10^{-3}	1247.5 ± 129.0
		50	second-order	1.33 ± 0.01 × 10^{-3}	749.6 ± 4.0
		75	zeroth-order	1.25 ± 0.01 × 10^{-3}	399.3 ± 3.5
		100	zeroth-order	0.93 ± 0.05× 10^{-3}	534.72 ± 26.7

k * units: 0th-order reactions = specific (C_t/C_0 units) × min^{-1}, 1st-order reactions = min^{-1}, 2nd-order reactions specific (C_t/C_0 units × min) $^{-1}$.

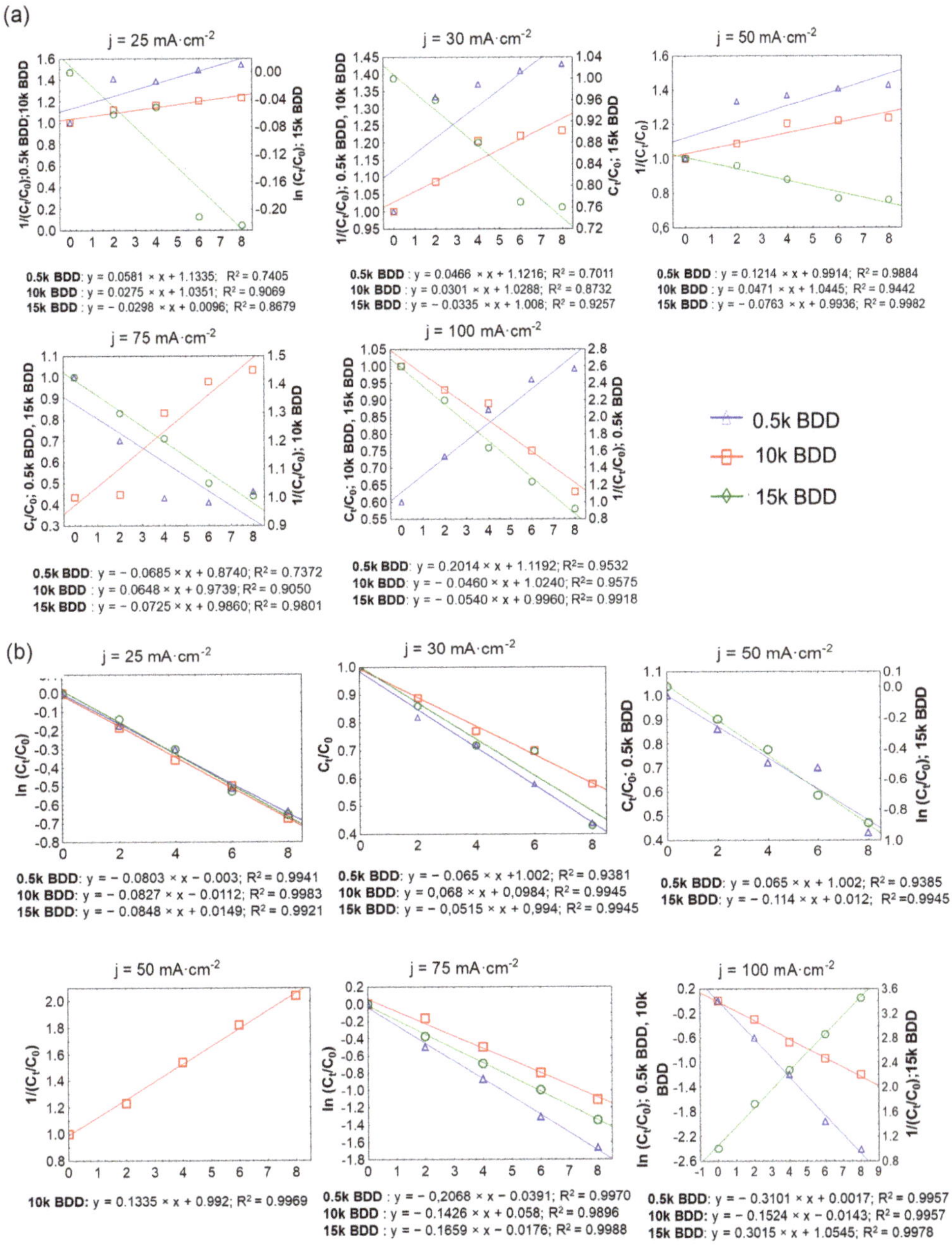

Figure 8. Determination of the order of removal of: (**a**) COD and (**b**) N-NH_4^+ by the graphical method.

According to the literature, kinetics of N-NH_4^+ removal from LLs usually shows sigmoidal ammonium concentration profiles and the order of reaction perfectly describes Equation (18) [75]:

$$\frac{d[N-NH_4]^+}{dt} = -k\,[N-NH_4]^+ \times [\mathrm{Cl}_2] \tag{18}$$

where: $[N\text{-}NH_4]^+$ is the ammonia concentration, k is the second-order rate constant, and $[Cl_2]$ is the concentration of dissolved active chlorine.

The results of this study showed that after 8 h of treatment with applied j of 100 mA·cm^{-2}, the average C_{8h}/C_0 for TN was 0.60, 0.66, and 0.75 using 0.5 k BDD, 10 k BDD, and 15 k BDD, respectively. In the same conditions, C_{8h}/C_0 for N-NH_4 was 0.39 (0.5 k BDD), 0.63 (10 k BDD), and 0.58 (15 k BDD). Figure 8 and Table 3 indicate that, in most cases, ammonia removal in different j followed second-order kinetics, but also first- and zeroth-order.

The differences in the trend of ammonium removal depended on the applied j and the electrode used, and might be attributed, inter alia, to the existence of different forms of active chlorine in different pH values. These results are in agreement with the results obtained by other authors [30,76–78]. Cabeza and co-authors [78,79] studied the EO of LLs by means of BDD electrodes using j ranging from 150 to 900 mA·cm^{-2}. The results showed that high values of the applied j caused elimination of N-NH_4^+ with zeroth-order kinetics, whereas, using low values of the applied j, exponential-like decaying ammonia concentration curves were obtained. On the other hand, according to Li and co-authors [50], removal of N-NH_4^+ from LLs by means of Ti/RuO_2–IrO_2 and Al electrodes indicated that the second-order equation fitted well. Basically, as in the case of COD removal, the rate constant of N-NH_4 $^+$ increased with increasing applied j, and the value of the half-life consistently decreased. The nitrogen was best removed by a 0.5 k electrode, which, due to having the lowest boron-doping level, was the most efficient in direct ammonium oxidation. The similar phenomenon was observed during PFOA and PFOS oxidation, where direct electrochemical reaction related with the perfluoroalkoxy radicals and hydroxyl radicals "production" on the electrode occurs more intensively on 0.5 k Nb/BDD than on 10 k Nb/BDD [29].

3.7. *Energy Consumtion Optimisation*

High energy consumption (EC) is one of the major drawbacks of EO, so energy consumption analysis should be an inherent part of all research scientific publications on the EO issue. Developing cost effective and stable anodes and application of renewable energy sources in EO treatment of LL is currently a major challenge [80]. Figure 9 shows the EC after 8 h of assays (with varying j application). In this study, EC_{COD} and $EC_{N\text{-}NH4+}$ increased and ICE decreased with increasing j and k (see also Figure 10 and Table A1), which is in good agreement with the literature [81–84].

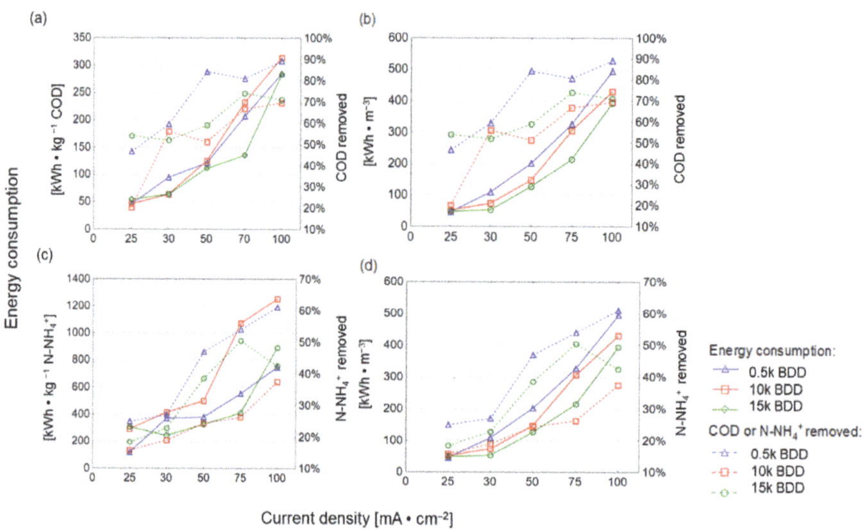

Figure 9. Energy consumption after 8 h of EO process using different BDD/Si electrodes and different current densities plotted against COD removed (**a**,**b**) and N-NH_4^+ removed (**c**,**d**).

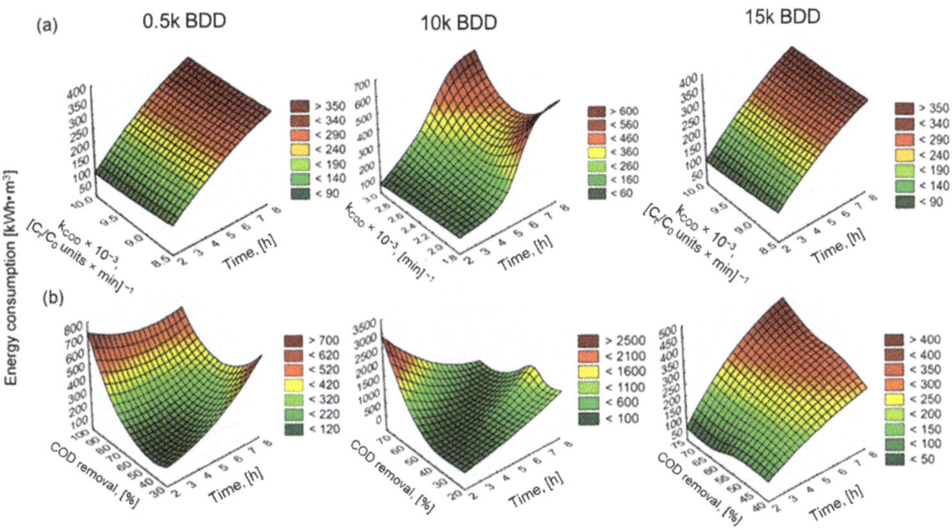

Figure 10. Surface plots (distance weighted least squares) of: (**a**) energy consumption against constant rate coefficient and time of EO and (**b**) energy consumption against COD removal efficacy and time of EO for j = 100 mA·cm^{-2} and different electrodes used.

For example, Panizza and co-workers [85] indicated an increase in energy consumption with increased j and achieved complete removal of COD after 7 h (flow rate 420 L·h^{-1}, pH = 8.2, j = 100 mA·cm^{-2}, PbO_2) by anode with EC = 220 kWh·m^{-3}. However, the initial concentration of the COD was low (780 mg·L^{-1}) and the chlorides were present in a very-high-concentration Cl^- (1800 mg·L^{-1}). In turn, the result of this study showed that the 15 k electrode consumed the lowest energy (expressed as kWh·kg^{-1} N-NH_4^+/COD), while the 0.5 k electrode consumed more, and the highest energy consumption was observed for the 10 k electrode (Figures 9 and 10). The electrode that was most effective in removing contamination (BDD 0.5 k, j = 100 mA·cm^{-2}, t = 8 h) had the following energy consumption: EC_{COD} = 285 kWh·kg^{-1} COD $EC_{N\text{-}NH4+}$ = 748 kWh·kg^{-1} N-NH_4^+ and 495 kWh·kg^{-1} (Figure 9). These results are analogous to results obtained by our research group testing BDD/Nb electrodes in terms of LL treatment [29]. Figure 10 clearly shows that the energy consumption increased with increasing k, indicating that it appears to be more cost-effective to remove COD than N-NH_4^+ (using 0.5 k). However, thanks to such a wide database, optimal conditions might be proposed: the lowest energy consumption (94.5 kWh·m^{-3}) for COD removal efficacy (>70%) may be achieved after 6 h of process using a 0.5 k electrode. Such a COD removal efficacy will be sufficient, e.g., in terms of EO application before biological treatment. Considering other types of electrodes (Ti/BDD and Ti/RuO_2) operated at 83 mA/cm^2 from 4 to 8 h, energy consumption ranged from 30 to 190 kWh·kg^{-1} COD. However, Tran Le Luu [57] pointed out that the energy consumption at Ti/BDD is less than Ti/RuO_2 anode under such a condition concluded Ti/BDD anode as a more favorable than Ti/RuO_2. Moreover, Zhou and co-authors [56] used ICE and EC to optimize the BDD/Nb EO process parameters. They showed that a current density of 50 mA/cm^2 lead to 87.5% COD and 74.06% N-NH_3 removal after 6 h of EO in flowing reactor with energy consumption of 223.2 kWh·m^{-3} and ICE equal to 35%. Under the same current density studied 0.5 k BDD/Si anode (after 8 h of process) in the batch reactors led to 83% COD and 48% N-NH_4^+ removal from LLs with energy consumption of 200 kWh·m^{-3}, 122 kWh kg^{-1} COD, and ICE = 45% (Appendix A, Table A1).

Despite parameter optimization, other solutions that will decrease energy consumption of advanced oxidation processes have been greatly explored. For instance, Cardoso and co-authors [86] indicated that a bubbling reactor with an ozone system resulted in reduced energy consumption compared to a combination of ozone with photocatalysis or photo-electrocatalysis. The bubbling reactor might also be considered in the optimization of EO reactor efficacy, possibly resulting in less energy consumption.

4. Conclusions

The application of new approaches in advanced treatment technologies such as EO can deliver environmental, economic, and social benefits. To be in accordance with the EU Green Deal and Water-Smart Society concept, EO application for LL treatment should be optimized in a 'fit-for-purpose' manner that is also affordable and practical for water-related sectors. This study presents that EO using BDD anodes is an effective technology to treat LLs in terms of macropollutant removal. At an optimized test, applied j of 100 mA·cm^{-2} with the 0.5 k electrode, the electro-oxidation process could nearly completely remove COD from LLs, with C_{8h}/C_0 = 0.09 ± 0.14, but the removal of N-NH_4^+ was less effective (C_{8h}/C_0 = 0.39). Thus, the proposed treatment process is instead dedicated to COD degradation. In most cases, COD removal fits the model of pseudo-first-order reactions with good linearity. Some conditions lead to a two-fold increase in biodegradability index (BI = 0.22 ± 0.05 was obtained using a 0.5 k electrode and a j of 50 mA·cm^{-2}). Such a phenomenon would suggest applying EO before biological treatment of LLs. EO of LLs may result in: (1) better subsequent effectiveness of biological processes due to the mineralization of poorly decomposable (non-biodegradable) COD and (2) an increase in N-NH_4^+ removal effectiveness.

The main drawback concerning applying the BDD electrodes for LLs treatment seems still to be the energy consumption. Considering using EO by means of BDD/Si in wider industrial and environmental applications appears complicated to implement for economic reasons. However, on the other hand, in this study, in a relatively short time, effective organic matter removal was obtained. Furthermore, the use of this method is also supported by the following factors: compact and modular reactor design, no addition of reagents needed, simple operation of devices, and the ability to adjust to variable organic loads in wastewater. To conclude, EO by means of BDD is an effective technology for the treatment of industrial wastewater such as LLs. Further studies are recommended to improve the ability of electrochemical conversion of ammonia to N_2 in LLs (inter alia by controlling the main operating parameters). Thus, it is stated that new developments in nanotechnology and material science, the widespread use of alternative energy sources, and reductions in material costs for the production of the electrodes now appear to be crucial in the wider implementation of EO.

Author Contributions: Conceptualization, B.K.W., A.L. and S.F.-K.; methodology, B.K.W., S.F.-K., A.L., M.S. (Małgorzata Szopińska) and M.S. (Michał Sobaszek); software, B.K.W.; validation, B.K.W. and M.S. (Małgorzata Szopińska); formal analysis, B.K.W. and M.S. (Małgorzata Szopińska); investigation, B.K.W.; resources, S.F.-K.; data curation, B.K.W.; writing—original draft preparation, B.K.W.; writing—review and editing, M.S. (Małgorzata Szopińska), E.S. and A.L.; visualization, B.K.W., M.S. (Michał Sobaszek) and M.S. (Małgorzata Szopińska); supervision, A.L. and S.F.-K.; project administration, M.S. (Małgorzata Szopińska); funding acquisition, S.F.-K. All authors have read and agreed to the published version of the manuscript.

Funding: This research was funded by the project "DIAOPS—effective removal of micropollutants from wastewater using electrochemical oxidation on nanocrystalline diamond anodes" funded by the Regional Fund for Environmental Protection and Water Management in Gdansk Poland (RX-15/13/2017).

Institutional Review Board Statement: Not applicable.

Informed Consent Statement: Not applicable.

Data Availability Statement: Data sharing is not applicable to this article.

Acknowledgments: The authors would like to thank the staff of the studied MSWP for the kind assistance with landfill leachate sampling.

Conflicts of Interest: The authors declare no conflict of interest. The funders had no role in the design of the study; in the collection, analyses, or interpretation of data; in the writing of the manuscript, or in the decision to publish the results.

Appendix A

Changes in N-NO_3^- in LLs with different BDD anodes and current densities (25–100 mA·cm^{-2}) (Figure A1), Changes in N-NO_2^- in LLs with different BDD anodes and current densities (25–100 mA·cm^{-2}) (Figure A2), Changes in TN in LLs with different BDD anodes and current densities (25–100 mA·cm^{-2}) (Figure A3), pH in LL during 8-h EO process (Figure A4) and The effect of using different B doped electrodes (0.5 k, 10 k, and 15 k BDD) on COD and N-NH_4^+removal, instantaneous current efficacy (ICE) and energy consumption EC [kWh kg^{-1} COD] and EC [kWh kg^{-1} N-NH_4^+] after 8 h of EO process (Table A1).

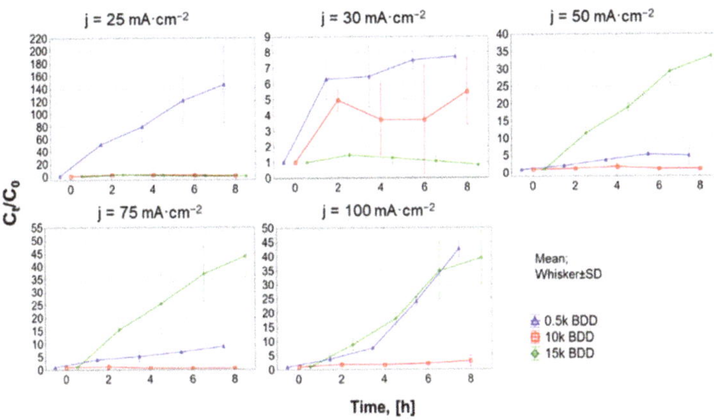

Figure A1. N-NO_3^- changes in LLs with different BDD anodes and current densities (25–100 mA·cm^{-2}).

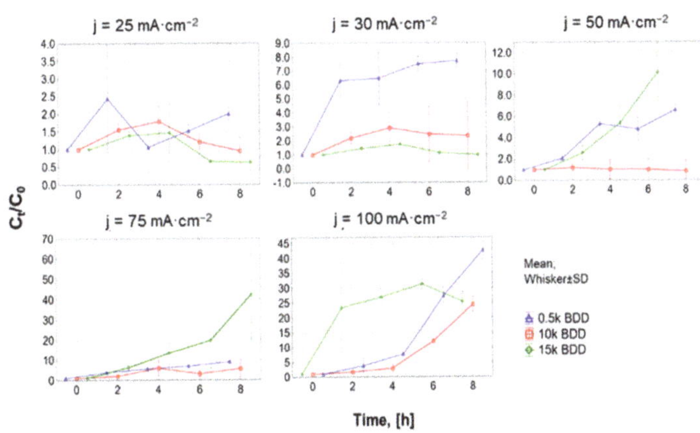

Figure A2. N-NO_2^- changes in LLs with different BDD anodes and current densities (25–100 mA·cm^{-2}).

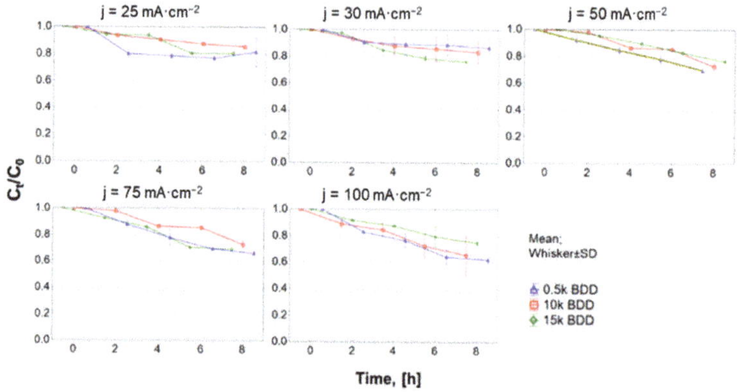

Figure A3. TN changes in LLs with different BDD anodes and current densities (25–100 mA·cm^{-2}).

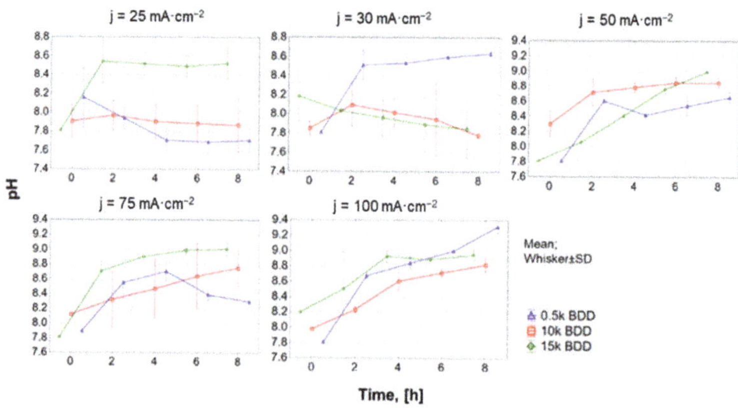

Figure A4. pH changes in LLs with different BDD anodes and current densities (25–100 mA·cm^{-2}).

Table A1. The effect of using different B doped electrodes (0.5 k, 10 k, and 15 k BDD) on COD and N-NH_4^+removal, instantaneous current efficacy (ICE) and energy consumption EC [kWh kg^{-1} COD] and EC [kWh kg^{-1} N-NH_4^+] after 8 h of EO process.

Anode Type	Current Density (mA/cm^2)	COD Removal (%)	ICE_{COD} (%)	EC (kWh kg^{-1} COD)	N-NH_4^+ Removal (%)	$ICE_{N-NH_4^+}$ (%)	EC (kWh kg^{-1} N-NH_4^+)
0.5 k BDD	25	46	63	46	35	24	121
	30	57	58	95	30	17	373
	50	83	51	122	48	17	381
	75	81	33	207	54	13	555
	100	91	28	285	60	10	748
10 k BDD	25	49	66	46	19	11	295
	30	56	61	64	19	9	416
	50	49	44	126	28	9	502
	75	67	28	233	31	6	1071
	100	69	22	314	37	5	1254
15 k BDD	25	48	57	55	18	4	310
	30	42	43	64	23	10	246
	50	59	36	112	38	8	328
	75	74	33	136	50	8	410
	100	71	22	284	42	5	891

References

1. Danley-Thomson, A.; Worley-Morse, T.; Contreras, S.U.J.; Herman, S.; Brawley, A.; Karcher, K. Determining the effects of Class I landfill leachate on biological nutrient removal in wastewater treatment. *J. Environ. Manag.* **2020**, *275*, 111198. [CrossRef]
2. Moradian, F.; Ramavandi, B.; Jaafarzadeh, N.; Kouhgardi, E. Effective treatment of high-salinity landfill leachate using ultraviolet/ultrasonication/ peroxymonosulfate system. *Waste Manag.* **2020**, *118*, 591–599. [CrossRef]
3. Wilk, B.K.; Fudala-Ksiazek, S.; Szopińska, M.; Luczkiewicz, A. Landfill leachates and wastewater of maritime origin as possible sources of endocrine disruptors in municipal wastewater. *Environ. Sci. Pollut. Res.* **2019**, *26*, 25690–25701. [CrossRef]
4. Mahtab, M.S.; Islam, D.T.; Farooqi, I.H. Optimization of the process variables for landfill leachate treatment using Fenton based advanced oxidation technique. *Eng. Sci. Technol. Int. J.* **2021**, *24*, 428–435. [CrossRef]
5. Surra, E.; Correia, M.; Figueiredo, S.; Silva, J.G.; Vieira, J.; Jorge, S.; Pazos, M.; Sanromán, M.Á.; Lapa, N.; Delerue-matos, C. Life cycle and economic analyses of the removal of pesticides and pharmaceuticals from municipal wastewater by anodic oxidation. *Sustainability* **2021**, *13*, 3669. [CrossRef]
6. Di Iaconi, C.; Ramadori, R.; Lopez, A. Combined biological and chemical degradation for treating a mature municipal landfill leachate. *Biochem. Eng. J.* **2006**, *31*, 118–124. [CrossRef]
7. Le, T.S.; Dang, N.M.; Tran, D.T. Performance of coupling electrocoagulation and biofiltration processes for the treatment of leachate from the largest landfill in Hanoi, Vietnam: Impact of operating conditions. *Sep. Purif. Technol.* **2021**, *255*, 117677. [CrossRef]
8. Council of the European Union. Council Directive 1999/31/EC of 26 April 1999 on the landfill of waste. *Off. J. Eur. Union* **1999**, *L182*, 1–19.
9. European Parliament, Council of the European Union. Directive 2008/98/EC of the European Parliament and of the Council of 19 November 2008 on waste and repealing certain Directives. *Off. J. Eur. Union* **2008**, *L312*, 3–30.
10. European Parliament, Council of the European Union. Directive 2000/60/EC of the European Parliament and of the Council of 23 October 2000 establishing a framework for Community action in the field of water policy. *Off. J. Eur. Communities* **2000**, *L327*, 1–73.
11. Torretta, V.; Ferronato, N.; Katsoyiannis, I.A.; Tolkou, A.K.; Airoldi, M. Novel and conventional technologies for landfill leachates treatment: A review. *Sustainability* **2017**, *9*, 9. [CrossRef]
12. Galvão, N.; de Souza, J.B.; de Sousa Vidal, C.M. Landfill leachate treatment by electrocoagulation: Effects of current density and electrolysis time. *J. Environ. Chem. Eng.* **2020**, *8*, 104368. [CrossRef]
13. Tałałaj, I.A.; Biedka, P.; Bartkowska, I. Treatment of landfill leachates with biological pretreatments and reverse osmosis. *Environ. Chem. Lett.* **2019**, *17*, 1177–1193. [CrossRef]
14. Žgajnar Gotvajn, A.; Pavko, A. Perspectives on Biological Treatment of Sanitary Landfill Leachate. In *Wastewater Treatment Engineering*; Samer, M., Ed.; Books on Demand: Norderstedt, Germany, 2015; pp. 115–151. [CrossRef]
15. Fudala-Ksiazek, S.; Sobaszek, M.; Luczkiewicz, A.; Pieczynska, A.; Ofiarska, A.; Fiszka-Borzyszkowska, A.; Sawczak, M.; Ficek, M.; Bogdanowicz, R.; Siedlecka, E.M. Influence of the boron doping level on the electrochemical oxidation of raw landfill leachates: Advanced pre-treatment prior to the biological nitrogen removal. *Chem. Eng. J.* **2018**, *334*, 1074–1084. [CrossRef]
16. Fudala-Ksiazek, S.; Luczkiewicz, A.; Fitobor, K.; Olanczuk-Neyman, K. Nitrogen removal via the nitrite pathway during wastewater co-treatment with ammonia-rich landfill leachates in a sequencing batch reactor. *Environ. Sci. Pollut. Res.* **2014**, *21*, 7307–7318. [CrossRef] [PubMed]
17. Arenhart Heberle, A.N.; García-Gabaldón, M.; Ortega, E.M.; Bernardes, A.M.; Pérez-Herranz, V. Study of the atenolol degradation using a Nb/BDD electrode in a filter-press reactor. *Chemosphere* **2019**, *236*, 124318. [CrossRef]
18. Polcaro, A.M.; Vacca, A.; Mascia, M.; Palmas, S.; Rodiguez Ruiz, J. Electrochemical treatment of waters with BDD anodes: Kinetics of the reactions involving chlorides. *J. Appl. Electrochem.* **2009**, *39*, 2083–2092. [CrossRef]
19. Cuerda-Correa, E.M.; Alexandre-Franco, M.F.; Fernández-González, C. Advanced oxidation processes for the removal of antibiotics from water. An overview. *Water (Switzerland)* **2020**, *12*, 102. [CrossRef]
20. Deng, Y.; Zhu, X.; Chen, N.; Feng, C.; Wang, H.; Kuang, P.; Hu, W. Review on electrochemical system for landfill leachate treatment: Performance, mechanism, application, shortcoming, and improvement scheme. *Sci. Total Environ.* **2020**, *745*, 140768. [CrossRef]
21. Díaz, V.; Ibáñez, R.; Gómez, P.; Urtiaga, A.M.; Ortiz, I. Kinetics of electro-oxidation of ammonia-N, nitrites and COD from a recirculating aquaculture saline water system using BDD anodes. *Water Res.* **2011**, *45*, 125–134. [CrossRef] [PubMed]
22. Deng, Y.; Zhao, R. Advanced Oxidation Processes (AOPs) in Wastewater Treatment. *Curr. Pollut. Reports* **2015**, *1*, 167–176. [CrossRef]
23. Ghime, D.; Ghosh, P. Removal of Organic Compounds Found in the Wastewater through Electrochemical Advanced Oxidation Processes: A Review 1. *Russ. J. Electrochem.* **2019**, *55*, 591–620. [CrossRef]
24. Moreira, F.C.; Soler, J.; Fonseca, A.; Saraiva, I.; Boaventura, R.A.R.; Brillas, E.; Vilar, V.J.P. Electrochemical advanced oxidation processes for sanitary landfill leachate remediation: Evaluation of operational variables. *Appl. Catal. B Environ.* **2016**, *182*, 161–171. [CrossRef]
25. Ukundimana, Z.; Omwene, P.I.; Gengec, E.; Can, O.T.; Kobya, M. Electrooxidation as post treatment of ultrafiltration effluent in a landfill leachate MBR treatment plant: Effects of BDD, Pt and DSA anode types. *Electrochim. Acta* **2018**, *286*, 252–263. [CrossRef]

26. Zhuo, Q.; Wang, J.; Niu, J.; Yang, B.; Yang, Y. Electrochemical oxidation of perfluorooctane sulfonate (PFOS) substitute by modified boron doped diamond (BDD) anodes. *Chem. Eng. J.* **2020**, *379*, 122280. [CrossRef]
27. Comninellis, C. Electrocatalysis in the electrochemical conversion/combustion of organic pollutants for waste water treatment. *Electrochim. Acta* **1994**, *39*, 1857–1862. [CrossRef]
28. Panizza, M.; Kapalka, A.; Comninellis, C. Oxidation of organic pollutants on BDD anodes using modulated current electrolysis. *Electrochim. Acta* **2008**, *53*, 2289–2295. [CrossRef]
29. Pierpaoli, M.; Szopińska, M.; Wilk, B.K.; Sobaszek, M.; Łuczkiewicz, A.; Bogdanowicz, R.; Fudala-Książek, S. Electrochemical oxidation of PFOA and PFOS in landfill leachates at low and highly boron-doped diamond electrodes. *J. Hazard. Mater.* **2021**, *403*, 123606. [CrossRef]
30. Abdelhay, A.; Jum'h, I.; Albsoul, A.; Abu Arideh, D.; Qatanani, B. Performance of electrochemical oxidation over BDD anode for the treatment of different industrial dye-containing wastewater effluents. *J. Water Reuse Desalin.* **2021**, *11*, 110–121. [CrossRef]
31. Linares-Hernández, I.; Barrera-Díaz, C.; Bilyeu, B.; Juárez-GarcíaRojas, P.; Campos-Medina, E. A combined electrocoagulation-electrooxidation treatment for industrial wastewater. *J. Hazard. Mater.* **2010**, *175*, 688–694. [CrossRef]
32. Alvarez-Pugliese, C.E.; Acuña-Bedoya, J.; Vivas-Galarza, S.; Prado-Arce, L.A.; Marriaga-Cabrales, N. Electrolytic regeneration of granular activated carbon saturated with diclofenac using BDD anodes. *Diam. Relat. Mater.* **2019**, *93*, 193–199. [CrossRef]
33. Bensalah, N.; Dbira, S.; Bedoui, A.; Ahmad, M.I. Electrolytic oxidation as a sustainable method to transform urine into nutrients. *Processes* **2020**, *8*, 460. [CrossRef]
34. Dbira, S.; Bensalah, N.; Ahmad, M.I.; Bedoui, A. Electrochemical oxidation/disinfection of urine wastewaters with different anode materials. *Materials (Basel)* **2019**, *12*, 1254. [CrossRef]
35. Kraft, A. Electrochemical water disinfection: A short review. *Platin. Met. Rev.* **2008**, *52*, 177–185. [CrossRef]
36. El Kateb, M.; Trellu, C.; Darwich, A.; Rivallin, M.; Bechelany, M.; Nagarajan, S.; Lacour, S.; Bellakhal, N.; Lesage, G.; Héran, M.; et al. Electrochemical advanced oxidation processes using novel electrode materials for mineralization and biodegradability enhancement of nanofiltration concentrate of landfill leachates. *Water Res.* **2019**, *162*, 446–455. [CrossRef]
37. Bogdanowicz, R.; Fabiańska, A.; Golunski, L.; Sobaszek, M.; Gnyba, M.; Ryl, J.; Darowicki, K.; Ossowski, T.; Janssens, S.D.; Haenen, K.; et al. Influence of the boron doping level on the electrochemical oxidation of the azo dyes at Si/BDD thin film electrodes. *Diam. Relat. Mater.* **2013**, *39*, 82–88. [CrossRef]
38. Calzadilla, W.; Espinoza, L.C.; Diaz-Cruz, M.S.; Sunyer, A.; Aranda, M.; Peña-Farfal, C.; Salazar, R. Simultaneous degradation of 30 pharmaceuticals by anodic oxidation: Main intermediaries and by-products. *Chemosphere* **2021**, *269*, 128753. [CrossRef]
39. Wang, W.L.; Polo, M.C.; Sánchez, G.; Cifre, J.; Esteve, J. Internal stress and strain in heavily boron-doped diamond films grown by microwave plasma and hot filament chemical vapor deposition. *J. Appl. Phys.* **1996**, *80*, 1846–1850. [CrossRef]
40. Feng, Y.; Lv, J.; Liu, J.; Gao, N.; Peng, H.; Chen, Y. Influence of boron concentration on growth characteristic and electro-catalytic performance of boron-doped diamond electrodes prepared by direct current plasma chemical vapor deposition. *Appl. Surf. Sci.* **2011**, *257*, 3433–3439. [CrossRef]
41. Salazar-Banda, G.R.; Andrade, L.S.; Nascente, P.A.P.; Pizani, P.S.; Rocha-Filho, R.C.; Avaca, L.A. On the changing electrochemical behaviour of boron-doped diamond surfaces with time after cathodic pre-treatments. *Electrochim. Acta* **2006**, *51*, 4612–4619. [CrossRef]
42. Ager, J.W.; Walukiewicz, W.; McCluskey, M.; Plano, M.A.; Landstrass, M.I. Fano interference of the Raman phonon in heavily boron-doped diamond films grown by chemical vapor deposition. *Appl. Phys. Lett.* **1995**, 616. [CrossRef]
43. Gonon, P.; Gheeraert, E.; Deneuville, A.; Fontaine, F.; Abello, L.; Lucazeau, G. Characterization of heavily B-doped polycrystalline diamond films using Raman spectroscopy and electron spin resonance. *J. Appl. Phys.* **1995**, *78*, 7059–7062. [CrossRef]
44. Mortet, V.; Živcová, Z.V.; Taylor, A.; Davydová, M.; Frank, O.; Hubík, P.; Lorincik, J.; Aleshin, M. Determination of atomic boron concentration in heavily boron-doped diamond by Raman spectroscopy. *Diam. Relat. Mater.* **2019**, *93*, 54–58. [CrossRef]
45. Zhang, R.J.; Lee, S.T.; Lam, Y.W. Characterization of heavily boron-doped diamond films. *Diam. Relat. Mater.* **1996**, *5*, 1288–1294. [CrossRef]
46. Bustarret, E.; Gheeraert, E.; Watanabe, K. Optical and electronic properties of heavily boron-doped homo-epitaxial diamond. *Phys. Status Solidi Appl. Res.* **2003**, *199*, 9–18. [CrossRef]
47. Mortet, V.; Gregora, I.; Taylor, A.; Lambert, N.; Ashcheulov, P.; Gedeonova, Z.; Hubik, P. New perspectives for heavily boron-doped diamond Raman spectrum analysis. *Carbon N. Y.* **2020**, *168*, 319–327. [CrossRef]
48. Silveira, J.E.; Zazo, J.A.; Pliego, G.; Bidóia, E.D.; Moraes, P.B. Electrochemical oxidation of landfill leachate in a flow reactor: Optimization using response surface methodology. *Environ. Sci. Pollut. Res.* **2015**, *22*, 5831–5841. [CrossRef] [PubMed]
49. Li, J.; Li, P.; Jin, C.; Zhao, Y.; Guo, L.; Ji, J. Treatment of Municipal Landfill Leachate by Electrochemical Oxidation: Assessing Operating Variables and Oxidation Products. *Environ. Eng. Sci.* **2019**, *36*, 1503–1513. [CrossRef]
50. Li, J.; Yang, Z.H.; Xu, H.Y.; Song, P.P.; Huang, J.; Xu, R.; Zhang, Y.J.; Zhou, Y. Electrochemical treatment of mature landfill leachate using Ti/RuO2-IrO2 and Al electrode: Optimization and mechanism. *RSC Adv.* **2016**, *6*, 47509–47519. [CrossRef]
51. Ammar, H.B.; Brahim, M.B.; Abdelhédi, R.; Samet, Y. Green electrochemical process for metronidazole degradation at BDD anode in aqueous solutions via direct and indirect oxidation. *Sep. Purif. Technol.* **2016**, *157*, 9–16. [CrossRef]
52. Agustina, F.; Bagastyo, A.Y.; Nurhayati, E. Electro-oxidation of landfill leachate using boron-doped diamond: Role of current density, pH and ions. *Water Sci. Technol.* **2019**, *79*, 921–928. [CrossRef]

53. Tchobanoglous, G.; Burton, F.L.; Stensel, H.D. *Wastewater Engineering: Treatment and Reuse*, 4th ed.; McGraw-Hill Science/Engineering/Math: New York, NY, USA, 2002; pp. 629–632, 983–1026.
54. Nurhayati, E.; Bagastyo, A.Y.; Hartatik, D.D. Effect of Flow Rate on Electrochemical Oxidation of Landfill Leachate Using DSA and BDD Anode. *IOP Conf. Ser. Earth Environ. Sci.* **2020**, *506*, 012038. [CrossRef]
55. McBeath, S.T.; Wilkinson, D.P.; Graham, N.J.D. Application of boron-doped diamond electrodes for the anodic oxidation of pesticide micropollutants in a water treatment process: A critical review. *Environ. Sci. Water Res. Technol.* **2019**, *5*, 2090–2107. [CrossRef]
56. Zhou, B.; Yu, Z.; Wei, Q.; Long, H.Y.; Xie, Y.; Wang, Y. Electrochemical oxidation of biological pretreated and membrane separated landfill leachate concentrates on boron doped diamond anode. *Appl. Surf. Sci.* **2016**, *377*, 406–415. [CrossRef]
57. Luu, T.L. Post treatment of ICEAS-biologically landfill leachate using electrochemical oxidation with Ti/BDD and Ti/RuO2 anodes. *Environ. Technol. Innov.* **2020**, *20*, 101099. [CrossRef]
58. Kjeldsen, P.; Barlaz, M.A.; Rooker, A.P.; Baun, A.; Ledin, A.; Christensen, T.H. Present and long-term composition of MSW landfill leachate: A review. *Crit. Rev. Environ. Sci. Technol.* **2002**, *32*, 297–336. [CrossRef]
59. Yao, J.; Mei, Y.; Xia, G.; Lu, Y.; Xu, D.; Sun, N.; Wang, J.; Chen, J. Process optimization of electrochemical oxidation of ammonia to nitrogen for actual dyeing wastewater treatment. *Int. J. Environ. Res. Public Health* **2019**, *16*, 2931. [CrossRef]
60. Ghimire, U.; Jang, M.; Jung, S.P.; Park, D.; Park, S.J.; Yu, H.; Oh, S.E. Electrochemical removal of ammonium nitrogen and cod of domestic wastewater using platinum coated titanium as an anode electrode. *Energies* **2019**, *12*, 883. [CrossRef]
61. Michels, N.L.; Kapałka, A.; Abd-El-Latif, A.A.; Baltruschat, H.; Comninellis, C. Enhanced ammonia oxidation on BDD induced by inhibition of oxygen evolution reaction. *Electrochem. commun.* **2010**, *12*, 1199–1202. [CrossRef]
62. Kapałka, A.; Joss, L.; Anglada, Á.; Comninellis, C.; Udert, K.M. Direct and mediated electrochemical oxidation of ammonia on boron-doped diamond electrode. *Electrochem. commun.* **2010**, *12*, 1714–1717. [CrossRef]
63. Huang, K.L.; Wei, K.C.; Chen, M.H.; Ma, C.Y. Removal of organic and ammonium nitrogen pollutants in swine wastewater using electrochemical advanced oxidation. *Int. J. Electrochem. Sci.* **2018**, *13*, 11418–11431. [CrossRef]
64. Ge, J. Amperometric Detection and Electrochemical Oxidation of Aliphatic Amines and Ammonia on Silver-Lead Oxide Thin-Film Electrodes. Master's Thesis, Iowa State University, Ames, IA, USA, 1996.
65. Buxton, G.V.; Elliot, A.J. Rate constant for reaction of hydroxyl radicals with bicarbonate ions. *Int. J. Radiat. Appl. Instrumentation. Part.* **1986**, *27*, 241–243. [CrossRef]
66. Zhang, T.; Lan, X.; Qiao, Z.; Wang, R.; Yu, X.; Xu, Q.; Wang, Z.; Jin, L.; Wang, Z. Role of the $(H_2O)n$ (n = 1-3) cluster in the HO2 + HO → 3O2 + H_2O reaction: Mechanistic and kinetic studies. *Phys. Chem. Chem. Phys.* **2018**, *20*, 8152–8165. [CrossRef] [PubMed]
67. Hamid, H.; Li, L.Y.; Grace, J.R. Review of the fate and transformation of per- and polyfluoroalkyl substances (PFASs) in landfills. *Environ. Pollut.* **2018**, *235*, 74–84. [CrossRef]
68. Solomon, D.; Kiflie, Z.; Hulle, S. Van Integration of sequencing batch reactor and homo - catalytic advanced oxidation processes for the treatment of textile wastewater. *Nanotechnol. Environ. Eng.* **2020**, *6*, 1–13. [CrossRef]
69. Arokiaraj, M.C.; Menesson, E. Novel anti-inflammatory and immunomodulation effects of Rose on the endothelium in normal and hypoxic invitro conditions. *Angiol. e Cir. Vasc.* **2019**, *15*, 238–248.
70. Ribeiro, M.C.E.; Couto, A.B.; Ferreira, N.G.; Baldan, M.R. Nitrate Removal By Electrolysis Using Cu/BDD Electrode Cathode. *ECS Meet. Abstr.* **2014**, *58*, 21–26. [CrossRef]
71. Vasilie, S.; Manea, F.; Baciu, A.; Pop, A. Dual use of boron-doped diamond electrode in antibiotics-containing water treatment and process control. *Process. Saf. Environ. Prot.* **2018**, *117*, 446–453. [CrossRef]
72. Papastavrou, C.; Mantzavinos, D.; Diamadopoulos, E. A comparative treatment of stabilized landfill leachate: Coagulation and activated carbon adsorption vs. electrochemical oxidation. *Environ. Technol.* **2009**, *30*, 1547–1553. [CrossRef]
73. Cossu, R.; Polcaro, A.M.; Lavagnolo, M.C.; Mascia, M.; Palmas, S.; Renoldi, F. Electrochemical treatment of landfill leachate: Oxidation at Ti/PbO2 and Ti/SnO2 anodes. *Environ. Sci. Technol.* **1998**, *32*, 3570–3573. [CrossRef]
74. Imai, A.; Onuma, K.; Inamori, Y.; Sudo, R. Biodegradation and adsorption in refractory leachate treatment by the biological activated carbon fluidized bed process. *Water Res.* **1995**, *29*, 687–694. [CrossRef]
75. Urtiaga, A.; Ortiz, I.; Anglada, A.; Mantzavinos, D.; Diamadopoulos, E. Kinetic modeling of the electrochemical removal of ammonium and COD from landfill leachates. *J. Appl. Electrochem.* **2012**, *42*, 779–786. [CrossRef]
76. Moraes, P.B.; Bertazzoli, R. Electrodegradation of landfill leachate in a flow electrochemical reactor. *Chemosphere* **2005**, *58*, 41–46. [CrossRef]
77. Szpyrkowicz, L.; Kelsall, G.H.; Kaul, S.N.; De Faveri, M. Performance of electrochemical reactor for treatment of tannery wastewaters. *Chem. Eng. Sci.* **2001**, *56*, 1579–1586. [CrossRef]
78. Cabeza, A.; Urtiaga, A.M.; Ortiz, I. Electrochemical Treatment of Landfill Leachates Using a Boron-Doped Diamond. *Ind. Eng. Chem. Res.* **2007**, 1439–1446. [CrossRef]
79. Cabeza, A.; Primo, Ó.; Urtiaga, A.M.; Ortiz, I. Definition of a clean process for the treatment of landfill leachates integration of electrooxidation and ion exchange technologies. *Sep. Sci. Technol.* **2007**, *42*, 1585–1596. [CrossRef]
80. Mandal, P.; Dubey, B.K.; Gupta, A.K. Review on landfill leachate treatment by electrochemical oxidation: Drawbacks, challenges and future scope. *Waste Manag.* **2017**, *69*, 250–273. [CrossRef]
81. Ambauen, N.; Weber, C.; Muff, J.; Hallé, C.; Meyn, T. Electrochemical removal of Bisphenol A from landfill leachate under Nordic climate conditions. *J. Appl. Electrochem.* **2020**, *50*, 1175–1188. [CrossRef]

82. Fernandes, A.; Catalão, E.; Ciríaco, L.; Pacheco, M.J.; Lopes, A.N.A. Electrochemical treatment of leachates from sanitary landfills. *J. Electrochem. Sci. Eng.* **2013**, *3*, 125–135. [CrossRef]
83. Fernandes, A.; Chamem, O.; Pacheco, M.J.; Ciríaco, L.; Zairi, M.; Lopes, A. Performance of electrochemical processes in the treatment of reverse osmosis concentrates of sanitary landfill leachate. *Molecules* **2019**, *24*, 2905. [CrossRef]
84. Saha, P.; Bruning, H.; Wagner, T.V.; Rijnaarts, H.H.M. Removal of organic compounds from cooling tower blowdown by electrochemical oxidation: Role of electrodes and operational parameters. *Chemosphere* **2020**, *259*, 1274912. [CrossRef]
85. Panizza, M.; Brillas, E.; Comninellis, C. Application of Boron-Doped Diamond Electrodes for Wastewater Treatment. *J. Environ. Eng. Manag.* **2008**, *18*, 139–153.
86. Cardoso, J.C.; Bessegato, G.G.; Boldrin Zanoni, M.V. Efficiency comparison of ozonation, photolysis, photocatalysis and photoelectrocatalysis methods in real textile wastewater decolorization. *Water Res.* **2016**, *98*, 39–46. [CrossRef]

materials

MDPI

Article

0.7Pb($Mg_{1/3}Nb_{2/3}$)O_3-0.3PbTiO_3 Phosphate Composites: Dielectric and Ferroelectric Properties

Artyom Plyushch [1,2,*], Nerijus Mačiulis [1], Aliaksei Sokal [3], Robertas Grigalaitis [1], Jan Macutkevič [1], Alexander Kudlash [3], Natalia Apanasevich [3], Konstantin Lapko [3], Algirdas Selskis [4], Sergey A. Maksimenko [2], Polina Kuzhir [2,5] and Juras Banys [1]

1 Faculty of Physics, Vilnius University, Sauletekio 9, LT-10222 Vilnius, Lithuania; nerijus.maciulis@ff.stud.vu.lt (N.M.); robertas.grigalaitis@ff.vu.lt (R.G.); jan.macutkevic@gmail.com (J.M.); juras.banys@ff.vu.lt (J.B.)
2 Institute for Nuclear Problems, Belarusian State University, 220006 Minsk, Belarus; sergey.maksimenko@gmail.com (S.A.M.); polina.kuzhir@uef.fi (P.K.)
3 Faculty of Chemistry, Belarusian State University, Nezalezhnastsi Ave. 4, 220030 Minsk, Belarus; sokolaa@bsu.by (A.S.); kudlash@bsu.by (A.K.); natalia_apanasevich@mail.ru (N.A.); LapkoKN@bsu.by (K.L.)
4 Department of Structural Analysis of Materials, Center for Physical Sciences and Technology, Sauletekio 3, LT-10257 Vilnius, Lithuania; algirdas.selskis@ftmc.lt
5 Department of Physics and Matematics, Institute of Photonics, University of Eastern Finland, Yliopistokatu 7, FI-80101 Joensuu, Finland
* Correspondence: artyom.plyushch@ff.vu.lt

Abstract: Composite materials with 83 wt.% of the 0.7Pb($Mg_{1/3}Nb_{2/3}$)O_3-0.3PbTiO_3 distributed in phosphate-bonded ceramics were prepared at three different pressures. A phosphate matrix comprises a mixture of an aluminum phosphate binder and melted periclase, MgO. All samples demonstrate a homogeneous distribution of the ferroelectric perovskite phase and are thermally stable up to 900 K. At higher temperatures, the pyrochlore cubic phase forms. It has been found that the density of the composites non-monotonously depends on the pressure. The dielectric permittivity and losses substantially increase with the density of the samples. The fabricated composites demonstrate diffused ferroelectric–paraelectric transition and prominent piezoelectric properties.

Keywords: phosphates; PMN-PT; composites; dielectric properties; ferroelectric properties; densification

Citation: Plyushch, A.; Mačiulis, N.; Sokal, A.; Grigalaitis, R.; Macutkevič, J.; Kudlash, A.; Apanasevich, N.; Lapko, K.; Selskis, A.; Maksimenko, S.; et al. 0.7Pb($Mg_{1/3}Nb_{2/3}$)O_3-0.3PbTiO_3 Phosphate Composites: Dielectric and Ferroelectric Properties. *Materials* **2021**, *14*, 5065. https://doi.org/10.3390/ma14175065

Academic Editor: Marc Cretin

Received: 14 July 2021
Accepted: 1 September 2021
Published: 4 September 2021

Publisher's Note: MDPI stays neutral with regard to jurisdictional claims in published maps and institutional affiliations.

1. Introduction

Composite materials filled with ferroelectric inclusions have received much attention in recent decades. Utilization of the different matrices and combinations with additional fillers allow to improve the properties of ferroelectrics and open wide perspectives for applications. In particular, composites with polymer matrices have potential applications in nanogeneration and energy harvesting [1–4], as well as ultrasonic transducers [5–7]. The addition of carbon nanotubes into such a system improves harvesting properties [8,9]. The composites comprising the mixture of ferroelectrics and ferromagnetics reveal the coupling of magnetic and electric polarizations [10,11]. The multiferroics have been widely used as sensors, transducers, memory elements, and spintronics [12–15]. Cement-based composites with ferroelectrics are used in structural health monitoring [16–18] and civil engineering fields [19,20].

Phosphate-bonded ceramics (PBCs) are intermediary materials between inorganic and polymer materials [21–23]. An important advantage of PBCs is the simplicity and low cost of the synthesis procedure. Phosphates are safe and eco-friendly. The preparation and curing steps occur at room temperature or with slight heating, without high-temperature sintering. Nowadays, PBCs can even be 3D printed [24]. After curing, the material demonstrates outstanding thermal stability and mechanical properties. Phosphates can act as a prospective host for a wide range of functional inclusions, which opens vast possibilities for biomedical [25,26], liquid wastes solidifying [27], drug delivery [24], and electromagnetic shielding applications [28,29]. Nevertheless, very few studies contribute to PBCs filled with ferroelectrics [30–32].

The only significant drawback of phosphate ceramics and cement for some utilizations is their relatively high porosity. However, this is a critical issue of ceramic materials in general. There are several approaches for densification, i.e., cold sintering process [33], microwave sintering [34], and increasing the pressure during material formation [35,36].

The present paper studied PBCs filled with a high concentration of ferroelectric inclusions. The composites were compressed under different pressures during the preparation procedure. The pressure-induced densification was studied for the case of chemically bonded material. Previously mentioned works study the densification of ceramics synthesized with sintering [35,36] that also impacts the density [37]. However, such densification is expected to apply to PBCs as well, despite the absence of sintering.

2. Materials and Methods

Ceramic composites consist of 3 components, i.e., a binder ($Al(H_2PO_4)_3$), a filler (MgO), and a functional filler ($0.7Pb(Mg_{1/3}Nb_{2/3})O_3$-$0.3PbTiO_3$). Commercially available by American Elements, lead magnesium niobate lead titanate (PMN-0.3PT, https://www.americanelements.com/printpdf/product/62022/datasheet, accessed on 3 September 2021) was used for the preparation of ceramic composites. The average grain size of the PMN-0.3PT powder is lower than 5 μm (see Figure 1). PMN-0.3PT is interesting as a filler for the composites due to its high dielectric permittivity, in combination with outstanding piezoelectric properties [38,39]. Commercially available by JSC Vostokogneupor (Russia, Yekaterinburg), melted periclase powder (MgO) was used as a filler. According to the supplier, the grain size of MgO is smaller than 63 μm.

Figure 1. Scanning electron microscopy of the $0.7Pb(Mg_{1/3}Nb_{2/3})O_3$-$0.3PbTiO_3$ powder.

The preparation procedure for the aluminum phosphate binder is as follows: Aluminum hydroxide powder $Al(OH)_3$ was dispersed in distilled water. Then, the concentrated (85 wt.%) H_3PO_4 was added to aluminum hydroxide suspension. The content of water was calculated to obtain the final concentration of orthophosphoric acid of 60 wt.%. The mixture reacted under constant stirring and heating up to 373 K. The synthesis time takes approximately 60–90 min. The reagents interacted according to the equation: $Al(OH)_3 + 3H_3PO_4 \rightarrow Al(H_2PO_4)_3 + 3H_2O$. After obtaining the transparent viscous dispersion, the mixture was cooled to room temperature and diluted with distilled water to a density of $\rho = 1.42$ g/cm^3.

The binder (0.3 g), filler (0.12 g), and PMN-0.3PT (2.01 g) were mixed in an agate mortar for 10 min and uniaxially pressed into the tablets with a diameter of 10 mm under 3, 6, and 8 US-tons. The prepared PBC/PMN-0.3PT composite samples are marked as 3 t,

6 t, and 8 t, respectively. Then, the composites were kept for 24 h at ambient temperature (293 K) and thermally treated up to 573 K with a heating rate of 1 K/min to speed up the curing process. As a result, the samples with a high content of PMN-0.3PT (83 wt.%) were prepared.

The scanning electron microscopy (SEM) with energy dispersive X-ray analysis (EDX) was performed on Helios NanoLab 650 microscope (Thermofisher Scientific, Hillsboro, OR, USA). Dielectric properties of the developed composites in the frequency range of 20 Hz–1 MHz were measured using an LCR HP4284A (Hewlett-Packard, Palo Alto, CA, USA), and at the frequencies of 1 MHz–300 MHz, a coaxial dielectric spectrometer with a vector network analyzer Agilent 8714ET Santa Clara, CA, USA) was employed. For temperature measurements, a homemade furnace and a cryostat with liquid nitrogen were used. AixaCCT TF2000 analyzer (Aachen, Germany) equipped with a 4 kV supply was applied for piezoelectric measurements. The silver paste was used for electrodes. The density of the samples was evaluated by measuring the volume and mass of precisely cut samples with a parallelepiped shape. The thermal gravimetric analysis with scanning differential calorimetry (TGA/DSC) was performed using NETSCH STA 449 (Selb, Germany). Samples were tested in an ambient air atmosphere with a heating rate of 10 K/min. The powder X-ray diffraction (XRD) analysis was carried out on DRON 3.0 diffractometer (BOUREVESTNIK, JSC, Saint-Petersburg, Russia) using CoKα radiation (λ = 1.78896 Å), IDDC database PDF4+ was applied for identification. For XRD measurements of thermally treated samples, the following protocol was used: heating up to 473 K, 573 K (heating rate 1 K/min); 973 K, 1073 K, and 1273 K (heating rate 5 K/min), followed by 30 min of isothermal treatment.

3. Results and Discussion

Scanning electron microscopy with energy dispersive X-ray analysis (Figure 2) of the 8 t sample reveals separate MgO grains surrounded with PMN-0.3PT grains (Table 1).

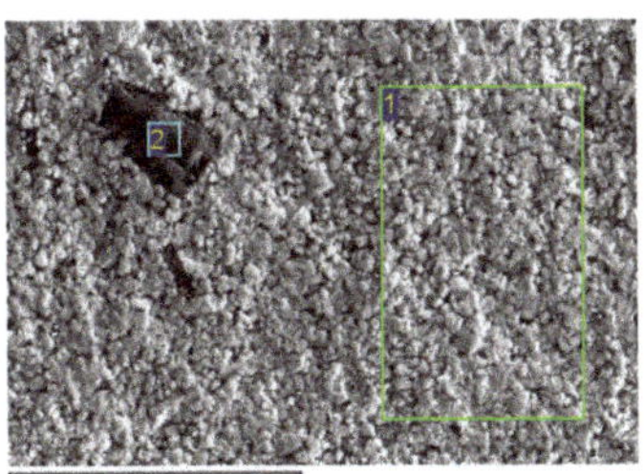

Figure 2. Scanning electron microscopy of the 8 t sample.

Table 1. Elemental content of the areas in Figure 2.

Area	O	Mg	Al	P	Ti	Nb	Pb
1	71.26	4.41	1	2.44	3.73	6.39	10.77
2	53.68	46.14	0.1	0.02	0.02	0.02	0.03

The result of TGA/DSC measurements of the 8 t composite material is presented in Figure 3. A total mass loss of approx. 3% occurs upon heating up to 750 K. In the temperature range of 300–450 K, the mass loss of 2%, accompanied by a sharp minimum in the DSC curve at 360 K, is associated with the absorbed water evaporation caused by the porous structure. The slight mass loss (approx. 1%) above 450 K occurs due to the evacuation of water obtained after the acid–base interaction processes in the PBC matrix. The series of peaks on the DSC curve at 900–1100 K without gravimetric effect is associated with high-temperature interactions between PMN-0.3PT and phosphate matrix. The developed composite remains stable up to 900 K.

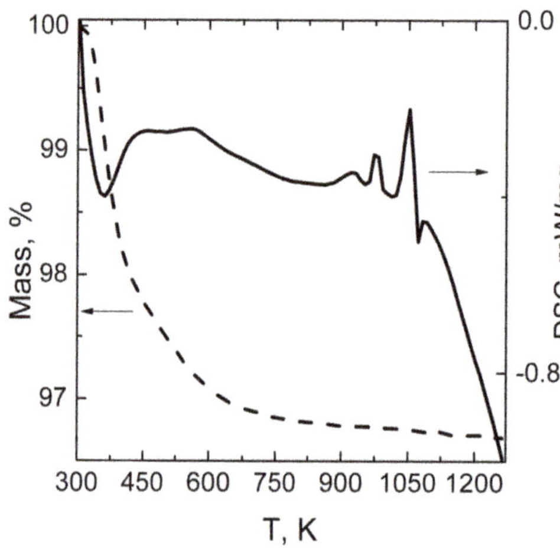

Figure 3. TG-DSC curves of the 8 t sample.

XRD analysis (Figure 4) of the as-synthesized material demonstrates the peaks from the perovskite PMN-0.3PT [01-088-1864] structure and small-intensity peaks from MgO [30-0794]. After high-temperature treatment, the intensity of the perovskite peaks gradually decreases. Additional reflections appear in the XRD spectra of samples treated at 1073 and 1273 K. These peaks are identified as pyrochlore $Pb_2Ti_2O_6$ [26-142] cubic phase. Since both PMN-0.3PT and PBC are stable in the studied temperature range [29,40], the formation of pyrochlore phase is associated with the interaction between PBC and PMN-0.3PT at 900–1100 K (see Figure 3).

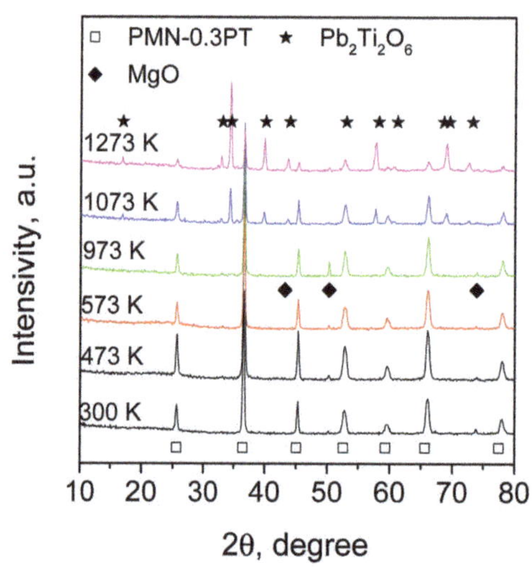

Figure 4. X-ray diffraction patterns of 8 t sample treated at different temperatures.

The temperature dependencies of the dielectric permittivity at different frequencies of 3 t and 6 t samples are presented in Figure 5. The results of the 8 t sample are close to 6 t, so only one frequency of 100 kHz is presented for comparison. The real part of ε' of the studied samples demonstrates a broad maximum near 460 K. The anomaly is related to the phase transition of $0.7Pb(Mg_{1/3}Nb_{2/3})O_3$-$0.3PbTiO_3$ from ferroelectric to paraelectric phase [41]. It is slightly different from the known peak position [41], which is probably due to possible defects of ferroelectric inclusions. Both samples demonstrate weak frequency dispersion of dielectric permittivity in the studied temperature range. The temperature corresponding to the maximum of ε' is frequency independent. Dielectric losses of the samples are not higher than 2 at temperatures up to 420 K. Upon further heating, the ε'' increases up to 2.5, and the strong frequency dispersion appears (Figure 5). This may indicate the Maxwell–Wagner relaxation at higher temperatures [42].

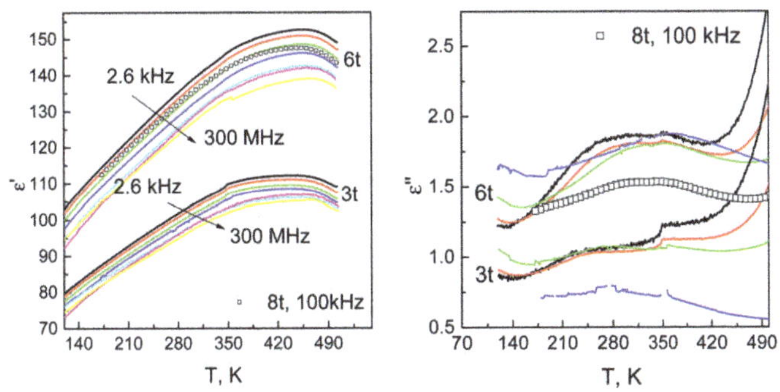

Figure 5. Real and imaginary parts of the dielectric permittivity as a function of temperature.

The densification effect is presented in Table 2. Several empirical models describe the densification behavior for conventional ceramics [35,36,43,44]. According to the mentioned literature, ceramics' density monotonously increases with pressure. In the studied case, an increase in density is observed up to 680 MPa, which is followed by saturation at 680–904 MPa.

The effective media approach links the porosity of ceramics and ε [45]. This makes dielectric permittivity a sensitive tool to verify density measurements [46]. Similar to the density, ε demonstrates a plateau for samples pressed at 680–904 MPa.

Table 2. Density of the samples as a function of the applied pressure.

Applied Pressure, MPa	340	680	904
Density, g/cm^3	5.48	6.10	5.94
ε, at 100 kHz, 450 K	109.5-1.05i	148.67-1.68i	147.56-1.41i

Several factors influence the density of ceramics. The increase in pressure better compacts the powder and improves the density. The densification process is limited and saturates at higher pressures when the theoretical density is achieved [36]. At the same time, high pressure introduces the elastic strain. The cracks may appear as a result of mechanical strain release [47]. The sintering step also impacts the density of the ceramics [37] due to the grain growth [48,49] and the healing of the cracks [50]. For PBCs, the saturation of density at the pressures of 680–904 MPa may indicate the achievement of theoretical density. Additionally, the absence of the sintering step somehow limits the relative densities of the ceramics.

The electromechanical properties and hysteresis loop of the samples were measured at the frequency of 10 Hz at room temperature (Figure 6). The shapes of the displacements at the maximal electric field are not sharp, compared to the pure PMN-0.3PT ceramics [38]. This difference might relate to the impact of the PBC on the overall elastic properties of the composite. The P-E loops for 8 t and 6 t samples are close to each other with similar values of remnant polarisations and coercive fields. The rounded shape of the P-E loops demonstrates the dilution of ferroelectric properties. Such shape of the P-E loop is not typical for the PMN-PT conventional ceramics [46]; however, it was previously reported for the PMN-PT-based composites [16]. Similarly, the difference might be attributed to the composite nature of the samples. The PBC matrix introduces additional losses to the system.

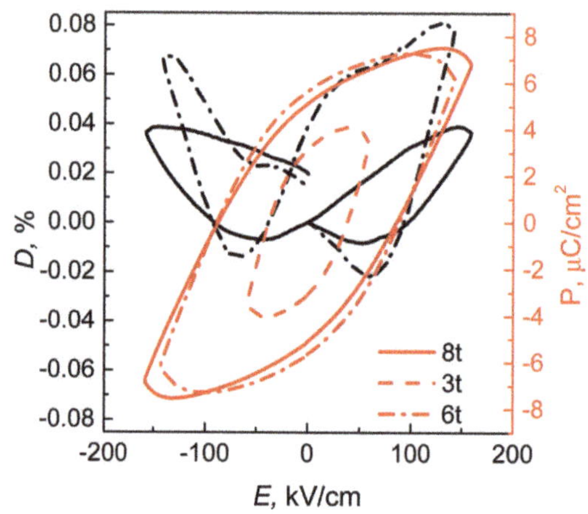

Figure 6. The strain and P-E hysteresis loops.

4. Conclusions

In contrast to polymer-based [1–3] or cement-based [17,18] composites, the usage of phosphate matrices allows researchers to successfully synthesize composite materials with substantially higher (83 wt.%) content of ferroelectrics. Obtained composites demonstrate a homogeneous distribution of PMN-0.3PT grains with inclusions of MgO grains. Thermal stability is another important feature of the presented materials. Polymer matrices are thermally degradable at comparatively low temperatures [51,52]. The cement also loses mechanical strength at high temperatures [53], and therefore, modified compounds are required [54]. In contrast, phosphate-bonded ceramics gain mechanical properties after thermal treatment [29]. The composites are stable up to 900 K, while the pyrochlore $Pb_2Ti_2O_6$ phase forms at a higher temperature.

The dependence of ceramic composites' density on the applied pressure was studied. Densification was observed upon the increase in the pressure from 340 to 680 MPa. The increase in density was verified with the measurements of dielectric permittivity. The densest composite has 1.5 times higher ε' than that of the least dense composite. The temperature dependence of ε was studied in a wide range of 20 Hz–300 MHz. The ε' shows the maximum at the temperature of 460 K, which is attributed to the ferroelectric—paraelectric phase transition. The increase in dielectric losses at higher temperatures indicates the onset of the Maxwell–Wagner relaxation. The 8 t sample demonstrates a high piezoelectric strain of 0.04%. The shape of the D-E dependence is not sharp in comparison with the conventional ferroelectrics. That is attributed to different elastic properties of the PMN-0.3PT and phosphates.

To summarize, we fabricated a prospective material for a wide range of applications that combines the advantages of both PBCs and ferroelectrics. The preparation is simple and environmentally friendly. The absence of the sintering step allows loading the samples with additional fillers such as carbon allotropes or ferromagnetics, but more importantly, avoiding interaction between them. Additionally, it allows dodging the evacuation of lead from lead-based ferroelectrics, which is extremely important due to the RoHS directive.

Author Contributions: Conceptualization, A.P., J.M. and R.G.; methodology, K.L., N.A. and A.S. (Aliaksei Sokal); validation, A.K. and N.M.; formal analysis, N.M.; investigation, N.M.; data curation, A.P.; writing—original draft preparation, A.P.; writing—review and editing, P.K.; visualization, A.S. (Algirdas Selskis); supervision, J.B. and S.A.M. All authors have read and agreed to the published version of the manuscript.

Funding: A.P. is supported by the European Social Fund under the No 09.3.3-LMT-K-712-19-0146 "Development of Competences of Scientists, other Researchers, and Students through Practical Research Activities". P.K. is supported by Horizon 2020 IF TURANDOT project 836816.

Institutional Review Board Statement: Not applicable.

Informed Consent Statement: Not applicable.

Data Availability Statement: Data is contained within the article.

Conflicts of Interest: The authors declare no conflict of interest.

References

1. Sriphan, S.; Charoonsuk, T.; Maluangnont, T.; Vittayakorn, N. High-performance hybridized composited-based piezoelectric and triboelectric nanogenerators based on $BaTiO_3$/PDMS composite film modified with $Ti_{0.8}O_2$ nanosheets and silver nanopowders cofillers. *ACS Appl. Energy Mater.* **2019**, *2*, 3840–3850. [CrossRef]
2. Suo, G.; Yu, Y.; Zhang, Z.; Wang, S.; Zhao, P.; Li, J.; Wang, X. Piezoelectric and triboelectric dual effects in mechanical-energy harvesting using $BaTiO_3$/polydimethylsiloxane composite film. *ACS Appl. Mater. Interfaces* **2016**, *8*, 34335–34341. [CrossRef] [PubMed]
3. Alluri, N.R.; Chandrasekhar, A.; Vivekananthan, V.; Purusothaman, Y.; Selvarajan, S.; Jeong, J.H.; Kim, S.J. Scavenging biomechanical energy using high-performance, flexible $BaTiO_3$ nanocube/PDMS composite films. *ACS Sustain. Chem. Eng.* **2017**, *5*, 4730–4738. [CrossRef]
4. Zeng, Z.; Gai, L.; Wang, X.; Lin, D.; Wang, S.; Luo, H.; Wang, D. A plastic-composite-plastic structure high performance flexible energy harvester based on PIN-PMN-PT single crystal/epoxy 2-2 composite. *Appl. Phys. Lett.* **2017**, *110*, 103501. [CrossRef]
5. Cheng, K.C.; Chan, H.L.; Choy, C.L.; Yin, Q.; Luo, H.; Yin, Z. Single crystal PMN-0.33PT/epoxy 1-3 composites for ultrasonic transducer applications. *IEEE Trans. Ultrason. Ferroelectr. Freq. Control* **2003**, *50*, 1177–1183. [CrossRef]
6. Li, Y.; Lu, G.; Chen, J.J.; Jing, J.C.; Huo, T.; Chen, R.; Jiang, L.; Zhou, Q.; Chen, Z. PMN-PT/Epoxy 1-3 composite based ultrasonic transducer for dual-modality photoacoustic and ultrasound endoscopy. *Photoacoustics* **2019**, *15*, 100138. [CrossRef] [PubMed]
7. Zhang, Y.; Wang, S.; Liu, D.; Zhang, Q.; Wang, W.; Ren, B.; Zhao, X.; Luo, H. Fabrication of angle beam two-element ultrasonic transducers with PMN–PT single crystal and PMN–PT/epoxy 1–3 composite for NDE applications. *Sens. Actuators A Phys.* **2011**, *168*, 223–228. [CrossRef]
8. Das, S.; Biswal, A.K.; Parida, K.; Choudhary, R.; Roy, A. Electrical and mechanical behavior of PMN-PT/CNT based polymer composite film for energy harvesting. *Appl. Surf. Sci.* **2018**, *428*, 356–363. [CrossRef]
9. Kang, S.W.; Cho, S.Y.; Bu, S.D.; Han, J.K.; Lee, G.J.; Lee, M.K. Effect of the Number of PZT Coatings on the Crystal Structure and Piezoelectric Properties in PZT-CNT Nanocomposites. *J. Korean Phys. Soc.* **2018**, *72*, 1209–1213. [CrossRef]
10. Singh, M.; Singh, J.; Kumar, M.; Kumar, S. Investigations on multiferroic properties of lead free (1-x)BCZT-xCZFMO based particulate ceramic composites. *Solid State Sci.* **2020**, *108*, 106380. [CrossRef]
11. Mane, S.M.; Tirmali, P.M.; Ranjit, B.; Khan, M.; Khan, N.; Tarale, A.N.; Kulkarni, S.B. Studies on magnetocapacitance, dielectric, ferroelectric, and magnetic properties of microwave sintered (1-x)($Ba_{0.8}Sr_{0.2}TiO_3$)-x($Co_{0.9}Ni_{0.1}Fe_2O_4$) multiferroic composite. *Solid State Sci.* **2018**, *81*, 43–50. [CrossRef]
12. Vopson, M.M. Fundamentals of multiferroic materials and their possible applications. *Crit. Rev. Solid State Mater. Sci.* **2015**, *40*, 223–250. [CrossRef]
13. Ma, J.; Hu, J.; Li, Z.; Nan, C.W. Recent progress in multiferroic magnetoelectric composites: From bulk to thin films. *Adv. Mater.* **2011**, *23*, 1062–1087. [CrossRef] [PubMed]
14. Sharma, S.; Paliwal, A.; Tomar, M.; Gupta, V. Multiferroic BFO/BTO multilayer structures based magnetic field sensor. *Phys. B Condens. Matter* **2019**, *571*, 1–4. [CrossRef]
15. Grigalaitis, R.; Petrović, M.V.; Bobić, J.; Dzunuzovic, A.; Sobiestianskas, R.; Brilingas, A.; Stojanović, B.; Banys, J. Dielectric and magnetic properties of $BaTiO_3$–$NiFe_2O_4$ multiferroic composites. *Ceram. Int.* **2014**, *40*, 6165–6170. [CrossRef]
16. Jaitanong, N.; Yimnirun, R.; Chaipanich, A. Effect of compressive stress on the ferroelectric hysteresis behavior in 0–3 PMN-PT/cement composites. *Ferroelectr. Lett.* **2011**, *38*, 11–17. [CrossRef]
17. Chaipanich, A.; Rianyoi, R.; Potong, R.; Suriya, W.; Jaitanong, N.; Chindaprasirt, P. Dielectric properties of 2-2 PMN-PT/cement composites. *Ferroelectr. Lett. Sect.* **2012**, *39*, 76–80. [CrossRef]
18. Chaipanich, A.; Jaitanong, N. Effect of PZT particle size on the electromechanical coupling coefficient of 0-3 PZT-cement composites. *Ferroelectr. Lett.* **2009**, *36*, 37–44. [CrossRef]
19. Newnham, R.E.; Amin, A. Smart systems: Microphones, fish farming, and beyond. *Chemtech* **1999**, *29*, 38–47.

20. Xin, C.; Shifeng, H.; Jun, C.; Zongjin, L. Piezoelectric, dielectric, and ferroelectric properties of 0-3 ceramic/cement composites. *J. Appl. Phys.* **2007**, *101*, 094110. [CrossRef]
21. Wagh, A.S.; Jeong, S.Y. Chemically bonded phosphate ceramics: I, a dissolution model of formation. *J. Am. Ceram. Soc.* **2003**, *86*, 1838–1844. [CrossRef]
22. Wagh, A.S.; Grover, S.; Jeong, S.Y. Chemically bonded phosphate ceramics: II, warm-temperature process for alumina ceramics. *J. Am. Ceram. Soc.* **2003**, *86*, 1845–1849. [CrossRef]
23. Wagh, A.S.; Jeong, S.Y. Chemically bonded phosphate ceramics: III, reduction mechanism and its application to iron phosphate ceramics. *J. Am. Ceram. Soc.* **2003**, *86*, 1850–1855. [CrossRef]
24. Trombetta, R.; Inzana, J.A.; Schwarz, E.M.; Kates, S.L.; Awad, H.A. 3D printing of calcium phosphate ceramics for bone tissue engineering and drug delivery. *Ann. Biomed. Eng.* **2017**, *45*, 23–44. [CrossRef] [PubMed]
25. Othman, Z.; Mohren, R.; Cillero-Pastor, B.; Shen, Z.; Lacroix, Y.; Guttenplan, A.; Birgani, Z.T.; Eijssen, L.; Luider, T.; van Rijt, S.; et al. Comparative proteomic analysis of human mesenchymal stromal cell behavior on calcium phosphate ceramics with different osteoinductive potential. *Mater. Today Bio* **2020**, *7*, 100066. [CrossRef]
26. Bouler, J.M.; Pilet, P.; Gauthier, O.; Verron, E. Biphasic calcium phosphate ceramics for bone reconstruction: A review of biological response. *Acta Biomater.* **2017**, *53*, 1–12. [CrossRef] [PubMed]
27. Tao, Y.; Zhenyu, L.; Chunrong, R.; Yuanyuan, W.; Zhichao, H.; Xin, H.; Jie, W.; Mengliang, L.; Qiubai, D.; Khan, K.; et al. Study on solidification properties of chemically bonded phosphate ceramics for cesium radionuclides. *Ceram. Int.* **2020**, *46*, 14964–14971. [CrossRef]
28. Plyushch, A.; Macutkevic, J.; Svirskas, S.; Banys, J.; Plausinaitiene, V.; Bychanok, D.; Maksimenko, S.; Selskis, A.; Sokal, A.; Lapko, K.; et al. Silicon carbide/phosphate ceramics composite for electromagnetic shielding applications: Whiskers vs. particles. *Appl. Phys. Lett.* **2019**, *114*, 183105. [CrossRef]
29. Apanasevich, N.; Sokal, A.; Lapko, K.; Kudlash, A.; Lomonosov, V.; Plyushch, A.; Kuzhir, P.; Macutkevic, J.; Banys, J.; Okotrub, A. Phosphate ceramics- carbon nanotubes composites: Liquid aluminum phosphate vs. solid magnesium phosphate binder. *Ceram. Int.* **2015**, *41*, 12147–12152. [CrossRef]
30. Bychanok, D.; Gorokhov, G.; Meisak, D.; Plyushch, A.; Kuzhir, P.; Sokal, A.; Lapko, K.; Sanchez-Sanchez, A.; Fierro, V.; Celzard, A.; et al. Exploring carbon nanotubes/$BaTiO_3$/Fe_3O_4 nanocomposites as microwave absorbers. *Prog. Electromagn. Res. C* **2016**, *66*, 77–85. [CrossRef]
31. Haily, E.; Bih, L.; Lahmar, A.; Elmarssi, M.; Manoun, B. Effect of $BaO–Bi_2O_3–P_2O_5$ glass additive on structural, dielectric and energy storage properties of $BaTiO_3$ ceramics. *Mater. Chem. Phys.* **2020**, *241*, 122434. [CrossRef]
32. Gittings, J.; Bowen, C.; Turner, I.; Baxter, F.; Chaudhuri, J. Characterisation of ferroelectric-calcium phosphate composites and ceramics. *J. Eur. Ceram. Soc.* **2007**, *27*, 4187–4190. [CrossRef]
33. Rubenis, K.; Zemjane, S.; Vecstaudza, J.; Bitenieks, J.; Locs, J. Densification of amorphous calcium phosphate using principles of the cold sintering process. *J. Eur. Ceram. Soc.* **2021**, *41*, 912–919. [CrossRef]
34. Wang, J.; Luo, P.; Wang, J.; Zhan, L.; Wei, Y.; Zhu, Y.; Yang, S.; Zhang, K. Microwave-sintering preparation and densification behavior of sodium zirconium phosphate ceramics with ZnO additive. *Ceram. Int.* **2020**, *46*, 3023–3027. [CrossRef]
35. Obradovic, N.; Dordevic, N.; Peleš, A.; Filipovic, S.; Mitrić, M.; Pavlović, V.B. The influence of compaction pressure on the density and electrical properties of cordierite-based ceramics. *Sci. Sinter.* **2015**, *47*, 15–22. [CrossRef]
36. Zhang, X.; Gao, H.; Zhang, Z.; Wen, R.; Wang, G.; Mu, J.; Che, H.; Zhang, X. Effects of pressure on densification behaviour, microstructures and mechanical properties of boron carbide ceramics fabricated by hot pressing. *Ceram. Int.* **2017**, *43*, 6345–6352. [CrossRef]
37. Luo, J.; Eitel, R. Aqueous tape casting of Al_2O_3 for multilayer co-fired ceramic based microfluidic chips with translucent windows. *Ceram. Int.* **2018**, *44*, 3488–3491. [CrossRef]
38. Wongmaneerung, R.; Rittidech, A.; Khamman, O.; Yimnirun, R.; Ananta, S. Processing and properties of $Pb(Mg_{1/3}Nb_2/3)O_3$-$PbTiO_3$-based ceramics. *Ceram. Int.* **2009**, *35*, 125–129. [CrossRef]
39. Bellaiche, L.; Vanderbilt, D. Intrinsic piezoelectric response in perovskite alloys: PMN-PT versus PZT. *Phys. Rev. Lett.* **1999**, *83*, 1347. [CrossRef]
40. Wang, L.; Xu, Z.; Li, Z.; Li, F. Investigation on the Thermal Stability of $Pb(Mg_{1/3}Nb_{2/3})O_3$-$PbTiO_3$ Single Crystals. *Ferroelectrics* **2010**, *402*, 187–192. [CrossRef]
41. Choi, S.; Shrout, T.R.; Jang, S.; Bhalla, A. Morphotropic phase boundary in $Pb(Mg_{1/3}Nb_{2/3})O_3$-$PbTiO_3$ system. *Mater. Lett.* **1989**, *8*, 253–255. [CrossRef]
42. Chen, Y.; Or, D. Effects of Maxwell-Wagner polarization on soil complex dielectric permittivity under variable temperature and electrical conductivity. *Water Resour. Res.* **2006**, *42*. [CrossRef]
43. Gallas, M.; Rosa, A.; Costa, T.; Da Jornada, J. High pressure compaction of nanosize ceramic powders. *J. Mater. Res.* **1997**, *12*, 764–768. [CrossRef]
44. Panelli, R.; Ambrozio Filho, F. A study of a new phenomenological compacting equation. *Powder Technol.* **2001**, *114*, 255–261. [CrossRef]
45. Nelson, S.O. Density-permittivity relationships for powdered and granular materials. *IEEE Trans. Instrum. Meas.* **2005**, *54*, 2033–2040. [CrossRef]

46. Kudrevičius, T.; Plyushch, A.; Ivanov, M.; Svirskas, Š.; Plaušinaitienė, V.; Selskis, A.; Kuzhir, P.; Banys, J. Aqueous tape casting of the $0.7Pb(Mg_{1/3}Nb_{2/3})O_3$-$0.3PbTiO_3$ ceramic films: Production optimization and properties. *J. Electroceramics* **2021**, 1–6. [CrossRef]
47. Coube, O.; Riedel, H. Numerical simulation of metal powder die compaction with special consideration of cracking. *Powder Metall.* **2000**, *43*, 123–131. [CrossRef]
48. Tsantilis, S.; Briesen, H.; Pratsinis, S.E. Sintering time for silica particle growth. *Aerosol Sci. Technol.* **2001**, *34*, 237–246. [CrossRef]
49. Ersoy, N.M.; Aydoğdu, H.M.; Değirmenci, B.Ü.; Çökük, N.; Sevimay, M. The effects of sintering temperature and duration on the flexural strength and grain size of zirconia. *Acta Biomater. Odontol. Scand.* **2015**, *1*, 43–50. [CrossRef]
50. Li, G.; Yang, G.; Li, C.; Li, C. A comprehensive sintering mechanism for thermal barrier coatings-Part III: Substrate constraint effect on healing of 2D pores. *J. Am. Ceram. Soc.* **2018**, *101*, 3636–3648. [CrossRef]
51. Meisak, D.; Macutkevic, J.; Selskis, A.; Kuzhir, P.; Banys, J. Dielectric relaxation spectroscopy and synergy effects in epoxy/MWCNT/Ni@C composites. *Nanomaterials* **2021**, *11*, 555. [CrossRef] [PubMed]
52. Santra, R.N.; Mukunda, P.; Nando, G.; Chaki, T. Thermogravimetric studies on miscible blends of ethylene-methyl acrylate copolymer (EMA) and polydimethylsiloxane rubber (PDMS). *Thermochim. Acta* **1993**, *219*, 283–292. [CrossRef]
53. Zhu, J. Prevent Cement Strength Retrogression Under Ultra High Temperature. In Proceedings of the Abu Dhabi International Petroleum Exhibition and Conference, Abu Dhabi, United Arab Emirates, 14 November 2019. Available online: https://onepetro.org/SPEADIP/proceedings-pdf/19ADIP/4-19ADIP/D041S119R002/1117384/spe-197739-ms.pdf (accessed on 3 September 2021). [CrossRef]
54. Pernites, R.B.; Santra, A.K. Portland cement solutions for ultra-high temperature wellbore applications. *Cem. Concr. Compos.* **2016**, *72*, 89–103. [CrossRef]

materials

MDPI

Article

The Behavior of Polyurethane Foam-Filled Glass-Fiber-Reinforced Polymer Crossarm Subjected to Lightning Transient Voltage

Muhammad Syahmi Abd Rahman [1,*], Mohd Zainal Abidin Ab Kadir [2], Muhamad Safwan Abd Rahman [1], Miszaina Osman [1], Ungku Anisa Ungku Amirulddin [1], Shamsul Fahmi Mohd Nor [1] and Noorlina Mohd Zainuddin [1]

1 Institute of Power Engineering, Universiti Tenaga Nasional, Kajang 43000, Selangor, Malaysia; asafwan@uniten.edu.my (M.S.A.R.); miszaina@uniten.edu.my (M.O.); anisa@uniten.edu.my (U.A.U.A.); shamsulfahmi.mohdnor@gmail.com (S.F.M.N.); noorlina.zainuddin@uniten.edu.my (N.M.Z.)
2 Advanced Lightning, Power and Energy Research Centre (ALPER), Universiti Putra Malaysia, Serdang 43400, Selangor, Malaysia; mzk@upm.edu.my
* Correspondence: syahmiarahman@yahoo.com

Citation: Abd Rahman, M.S.; Ab Kadir, M.Z.A.; Abd Rahman, M.S.; Osman, M.; Ungku Amirulddin, U.A.; Mohd Nor, S.F.; Zainuddin, N.M. The Behavior of Polyurethane Foam-Filled Glass-Fiber-Reinforced Polymer Crossarm Subjected to Lightning Transient Voltage. *Materials* **2021**, *14*, 5628. https://doi.org/10.3390/ma14195628

Academic Editors: Marc Cretin and Ricardo J. C. Carbas

Received: 30 June 2021
Accepted: 15 September 2021
Published: 28 September 2021

Publisher's Note: MDPI stays neutral with regard to jurisdictional claims in published maps and institutional affiliations.

Abstract: The demand for composite materials in high-voltage electrical insulation is escalating over the last decades. In the power system, the composite glass-fiber-reinforced polymer has been used as an alternative to wood and steel crossarm structures due to its superior properties. As a composite, the material is susceptible to multi-aging factors, one of which is the electrical stress caused by continuous and temporary overvoltage. In order to achieve a better insulation performance and higher life expectancy, the distribution of the stresses should firstly be studied and understood. This paper focuses on the simulation work to better understand the stress distribution of the polyurethane foam-filled glass-fiber-reinforced polymer crossarm due to the lightning transient injection. A finite-element-based simulation was carried out to investigate the behavior of the electric field and voltage distribution across the sample using an Ansys Maxwell 3D. Electrical stresses at both outer and inner surfaces of the crossarm during the peak of lightning were analyzed. Analyses on the electric field and potential distribution were performed at different parts of the crossarm and correlated to the physical characteristics and common discharge location observed during the experiment. The results of the electric field on the crossarm indicate that both the outer and internal parts of the crossarm were prone to high field stress.

Keywords: GFRP composite crossarm; fiberglass; polyurethane foam; electric field; critical flashover voltage lightning impulse voltage; finite element method

1. Introduction

Nowadays, composite materials have been used in broad applications. Glass-fiber-reinforced polymer (GFRP) composites have been used in recent decades for high-voltage (HV) applications, especially for transmission tower design to replace wooden crossarm [1]. GFRP has been recognized to have an excellent mechanical strength-to-weight ratio, corrosion resistance, longer lifespan, and cost-saving capabilities in installation and replacement, which have driven many designers to use it for electrical insulation. However, it is still vulnerable to aging, responsible for the composite's long-term performance [2,3]. Some of the factors contributing to aging include electrical, salt fog, rain, ultraviolet (UV), and hygrothermal caused degradation as summarized in Figure 1 [4].

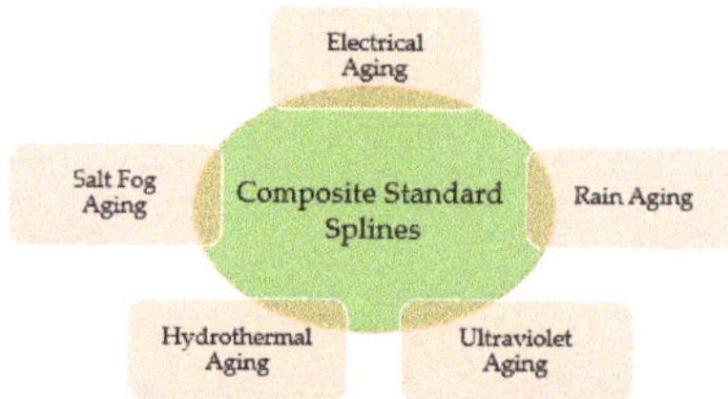

Figure 1. The multifactor aging of composites material [4].

The extensive use of composites in HV equipment has motivated much further research on their mechanical and electrical performance as well [5–8]. The GFRP crossarm consists of a hollow tubes structure, which is a product of the pultrusion process. The process combines layers of fiberglass fabric by thermoset special polyester resin and is covered by polyurethane (PU). The hollow tube structure is typically filled with a closed-cell PU foam to eliminate the void spaces, which can cause moisture accumulation, thereby increasing the risk of electrical bridging or fires [9,10].

Previous studies evaluated the insulation performance of composite GFRP crossarm through laboratory testing to obtain AC withstand and critical flashover (CFO) voltages [11,12]. In the studies, the performance of differently aged crossarm and new crossarm were compared. At the same time, the GFRP crossarm under different conditions and polarity considering the crossarm length have been thoroughly discussed. The authors have found that the CFO per unit length of the GFRP crossarm tends to decrease as the length increased. In a separate study, the high voltage and direct current (HVDC) performance of GFRP samples were also investigated [13].

In different studies, the issue of bubble formation inside the PU foam has been discussed. The formation of the bubbles may lead to partial discharges and probably cause permanent damage to the insulation when they reach a certain threshold [14,15]. Moreover, the pore diameters of the PU foam have been inclusively studied in [14]. The authors have highlighted the challenge posed by different pore sizes that caused variation in permittivity and volume resistivity of the material, thus influencing the electrical stress distribution. The finite element method (FEM) has been widely used to evaluate the stress distribution on composite crossarms and insulators in the event of lightning [16–19]. These studies have satisfactorily predicted and located the electrical stress by its potential and E-field distributions. Simulation conducted in [20] indicated that some parts of the crossarm experienced irregular distribution of E-field along the surfaces and triple junctions. Based on the previous studies, this paper aims to understand the stress distribution and its magnitude on the external and internal parts of the GFRP crossarm, considering the lightning overvoltages.

2. Materials and Methods

Two crossarm samples with a dimension of 100 mm × 127 mm × 127 mm were obtained from the main member of a newly pultruded 275 kV GFRP crossarm. The pultruded material comprised ten layers of fiberglass consisting of four layers of fiberglass fabric, three layers of fiberglass roving, and three layers of fiberglass chopped strand mat (CSM), with a total thickness of 7.00 mm to 7.40 mm (refer to Figure 2). The hollow samples were filled with PU foam and sealed at both ends with a special polyester resin using a cold-curing method for moisture resistance, as practiced in the field (refer to Figure 3). Later, these samples will be used as a reference for current and future studies. The methodology of the current study is divided into two sections, i.e., experimental and simulation works.

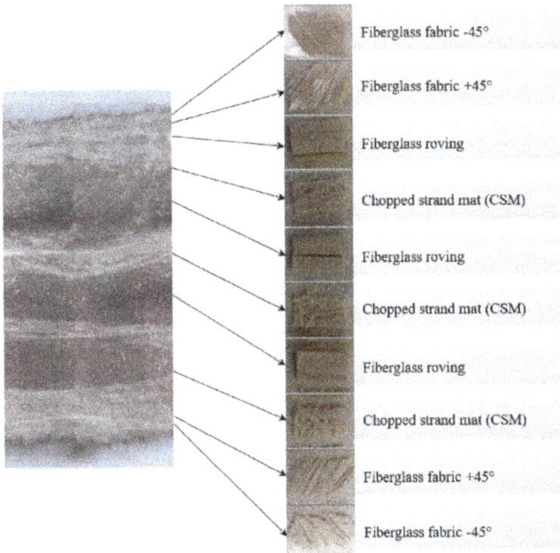

Figure 2. The construction layers of the GFRP crossarm sample [21].

Figure 3. Hollow crossarm sample filled with closed-cell foam, sealed with resin sealer.

2.1. Experimental Work

A lightning impulse test was preliminarily carried out to determine the CFO voltage of the sample by adopting the up-and-down method, as suggested in IEC 60060-1 [22]. The sample was sandwiched between two parallel plate electrodes during the test considering a uniform field, as shown in Figure 4. In total, 20 shots of standard 1.2/50 μs impulse voltage were energized at one of the electrodes, while another electrode was earthed with a 50 mΩ current shunt connected. Both positive and negative impulse polarity was considered in this study. A sample was tested three times in each polarity to investigate the loss of performance due to the repeated flashovers. The average CFO voltages obtained in the study were corrected to the standard atmospheric condition.

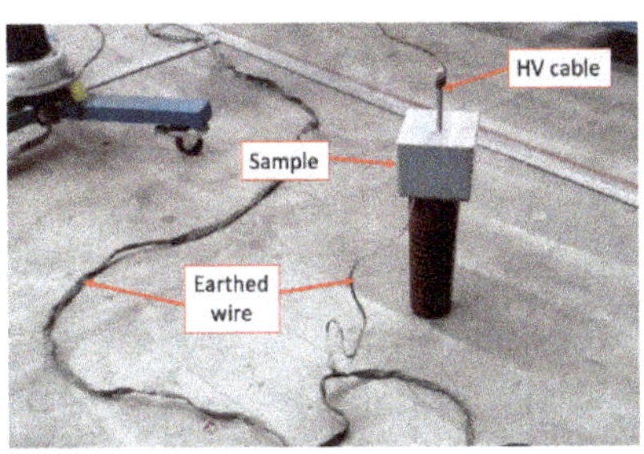

Figure 4. The experimental setup.

2.2. Finite Elements Simulation

A FEM-based simulation was carried out using an Ansys Maxwell (Electronic Desktop 2020R2, Ansys Inc., Canonsburg, PA, USA) numerical package to simulate the sample's 3-dimensional (3D) model. The 3D model, as illustrated in Figure 5, is the direct representation of the actual sample, consisting of five different materials: hollow GFRP tube, PU foam, steel electrodes, air, and polyester resin sealer. Each material was differentiated by the electrical characteristics, i.e., permittivity and bulk conductivity (S/m) as indicated in Table 1. In addition, Table 2 describes the characteristics of PU foam, which reflects the material's permittivity and bulk conductivity.

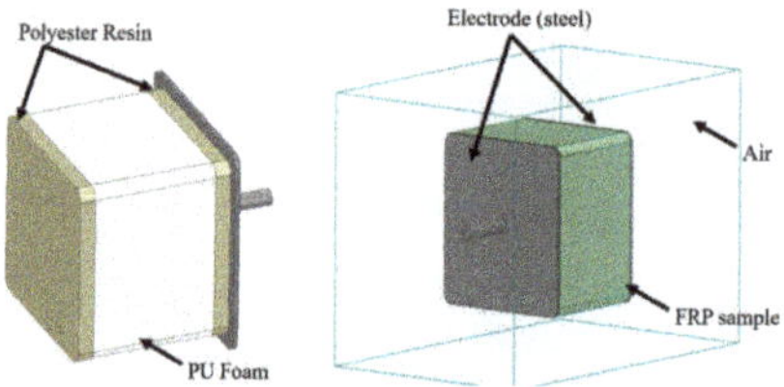

Figure 5. The experimental setup.

Table 1. Material-based parameters.

Parts	Material	Relative Permittivity (ε_r)	Volume Conductivity (σ) S/m	Ref.
Air	Air	1.0006	0	Ansys Lib.
Electrodes	Steel	1.0	2.00×10^{6}	Ansys Lib.
Crossarm	GFRP	4.6	1.11×10^{-14}	[23,24]
Foam	Polyurethane	1.8	5.56×10^{-15}	[25]
Sealer	Polyester resin	3.2	0	Ansys Lib.

Table 2. Characteristics of PU foam used in the simulation [25].

Characteristics	Description
Pore Geometry	Spherical
Density, g/cm^3	0.2510
Average Diameter, µm	200
Pore Porosity, %	73

An optimum number of elements at 696,471 were obtained by the adaptive meshing technique, which was repeatedly refined until convergence was achieved, as shown in Figure 6. The convergence was set based on a pre-defined percentage error at 1%. Figure 7 shows the fine mesh plot of the model that was based on the finer mesh theory; a finer mesh would produce more precise and accurate results [26].

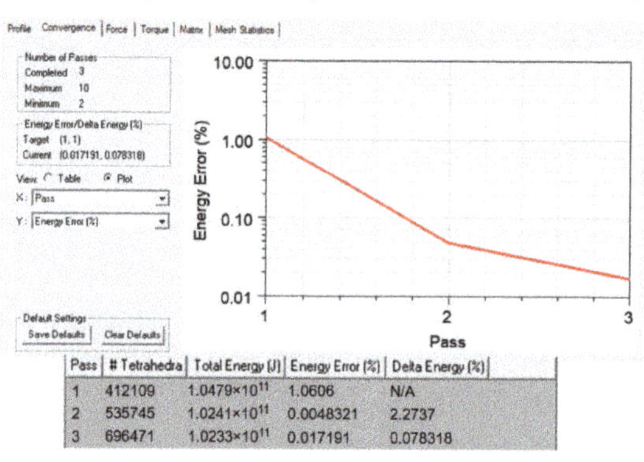

Figure 6. Convergence graph and iterated mesh elements.

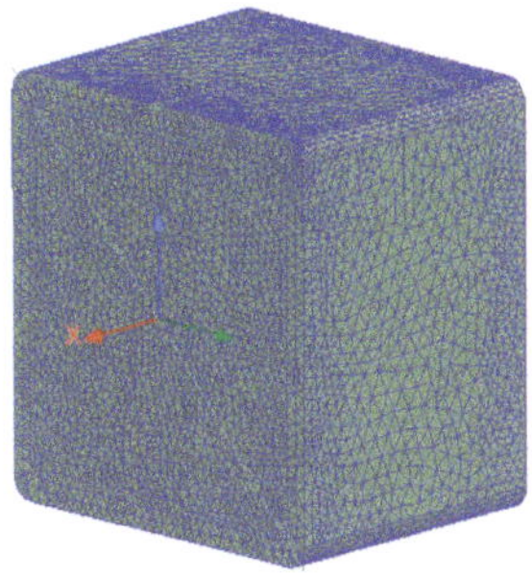

Figure 7. Mesh plot of the simulation model.

The electric transient analysis was adopted, considering the impulse voltage in the time domain. One of the electrodes was energized with a standard lightning impulse voltage with a peak of −94.5 kV for 100 μs, which is equivalent to the CFO voltage obtained in the experiment, while another electrode was subjected to 0 kV for the same duration. The energization with CFO voltages shall reflect the maximum stress immediately before the breakdown occurs. The 1.2/50 μs lightning impulse voltage can be expressed by the double exponential function as follows [27]:

$$u(t) = u_o k(e^{-\alpha t} - e^{-\beta t}), \quad (1)$$

where u_o is the peak value of lightning impulse voltage, and α is the attenuation coefficient of the wavefront, which is set to 1.473×10^4. Meanwhile, β is the attenuation coefficient of the wave tail set to 2.08×10^6, and k is the correction coefficient set to 1.043. The whole flow of the simulation study is summarized in Figure 8.

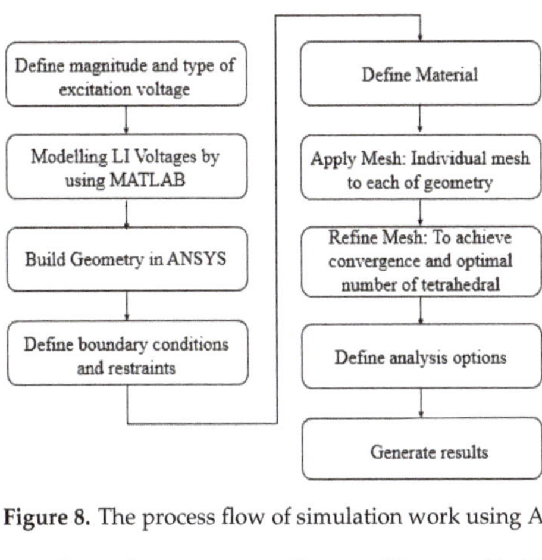

Figure 8. The process flow of simulation work using ANSYS.

Several measurement lines, as illustrated in Figure 9, were introduced at different parts of the model. X1, X2, and X3 are the lines across the length (x-axis) laid on the outer and inner surface of GFRP and inside the foam, respectively, while Y1, Y2, and Y3 laid across the width (y-axis). In this study, E-field and potential distributions were generated at these lines with respect to the distance. Meanwhile, position A was introduced to investigate the variation of stress to time.

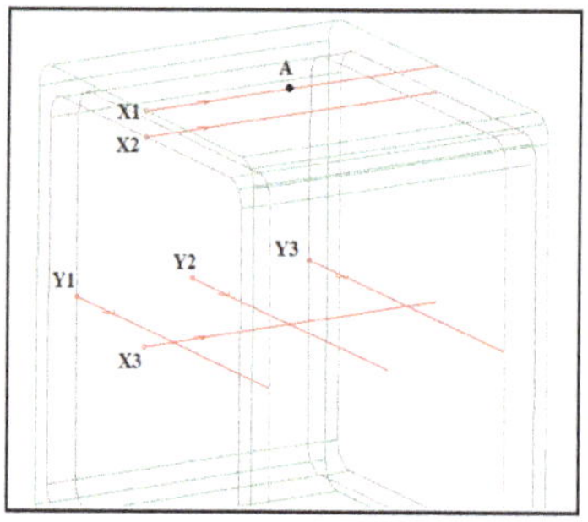

Figure 9. Measurement lines across the simulation model.

3. Results and Discussion

3.1. Critical Flashover Voltage

In the experimental works, the critical flashover voltage (CFO) of the 10 cm crossarm sample was determined at 85.2 kV and −94.5 kV for positive and negative impulse polarity, respectively. It can be found that the CFO under negative polarity was much higher than positive polarity, indicating a 10.9% difference. Investigation on the repeated test is summarized in Table 3.

Table 3. CFO voltages of crossarm sample.

Polarity	Sample	CFO Voltage, U_{50} (kV)	Standard Deviation (± %)	No. of Withstand	No. of Flashover
Positive	Sample 1	85.2	3.43	10	10
		84.6	2.43	9	11
		83.9	2.84	8	12
Negative	Sample 2	−94.5	1.47	10	10
		−93.7	1.81	10	10
		−92.6	1.59	9	10

It was recorded that the CFO value was slightly reduced after the subsequent test. This perhaps indicates a minor degradation of the crossarm insulation after the repeated flashovers. The observation showed that most of the flashovers occurred on the insulation surface in the air, whereas some were not visible, indicating internal flashover occurred. Most of the discharges on the surfaces happened nearer to the edges of the samples, as shown in Figure 10. For the internal flashover, it is believed that the flashover happens across the internal surface (interfaces between GFRP and foam) or within the lamination interfaces, where air voids may occasionally exist due to poor manufacturing. In this case, the non-self-healing nature of the composite is a significant concern because the post-flashover may leave conductive charred trails, yet the test conducted is not sufficient to reflect the permanent damages caused by the internal flashover.

Figure 10. Flashover path on the surface.

3.2. FEM Analyses

Measurement of E-field strength at point A shows that the electrical stress was directly proportional to the voltage applied. The E-field profile generated at point A is illustrated in Figure 11, where the maximum E-field at 9.00×10^5 V/m occurred during the peak of the lightning voltage. As the voltage reduced over time, the E-field continued to reduce relatively.

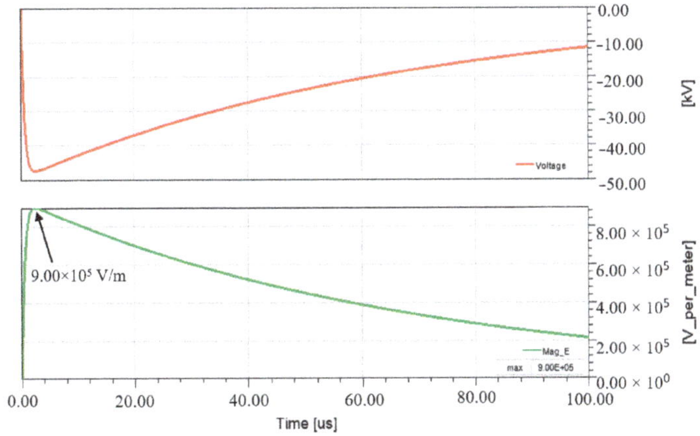

Figure 11. Voltage and E-field strength on GFRP surface (at position A).

In the flashover event, the E-field on the crossarm surface was predicted to be at the maximum sufficient to sustain an arc across the insulation surface. Based on the theory of the uniform field, flashover could occur in the air at 30 kV/cm, which the threshold might vary according to the surrounding conditions [28]. Moreover, the E-fields distribution within the material is similarly important. Flashover might occur internally when the E-field exceeds the breakdown strength of the material.

In the current study, the conducted simulation had satisfactorily predicted E-field at the cross section of the crossarm, as illustrated in Figure 12. It showed that the field stress was abnormally distributed around the structure, which was anticipated to be evenly distributed based on the theory of parallel plate distribution. The highest E-field was primarily concentrated inside the foam up to 9.99×10^5 V/m and gradually decreased away from the center.

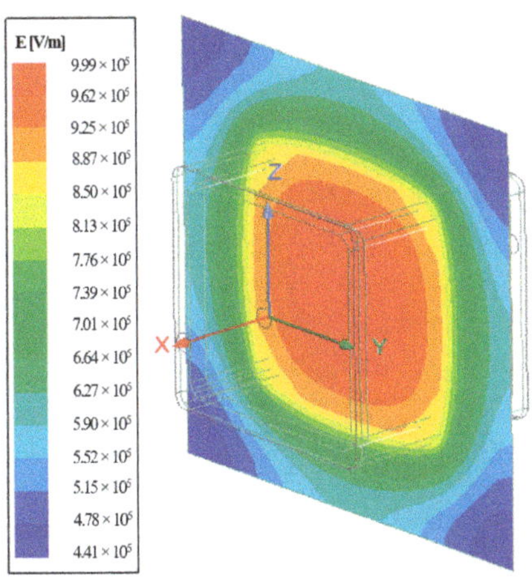

Figure 12. E-field distribution at the cross section of the sample.

Investigation of the E-field distribution at different time moments of the lightning transient revealed that the E-fields on the crossarm developed from the center of the foam outwards. As illustrated in Figure 13, the maximum E-field could be observed as early as 1.2 μs (front time) and continued to expand to the entire foam region until time equal to 2.3 μs

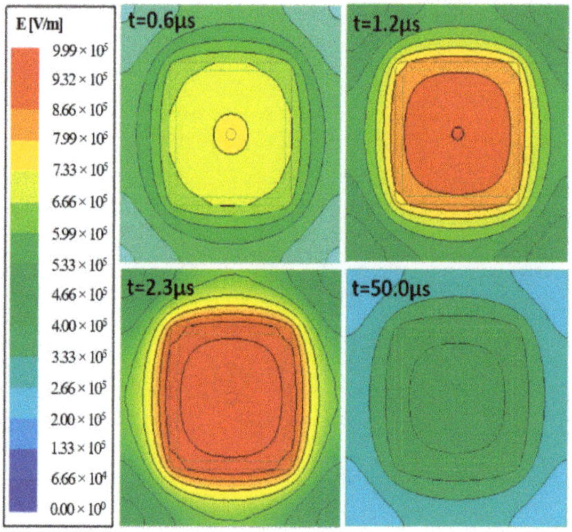

Figure 13. E-field at different time moments.

Translated into curves, the sample's E-field measurements across the *y*-axis (or *y* cross section) were labeled Y1, Y2, and Y3 and represented by a "dome shape" (refer to Figure 14). The profiles showed that the maximum E-field on the outer surface of GFRP was slightly lower than the internal surface, marking a 1.32% difference. In contrast, 10.02% of the difference was measured between the outer surface and foam. In general, it could be observed that the magnitude of the E-field presented in Figure 14 exceeded the typical streamer threshold that is usually initiated at a magnitude of 0.50 MV/m to 0.60 MV/m [29–31].

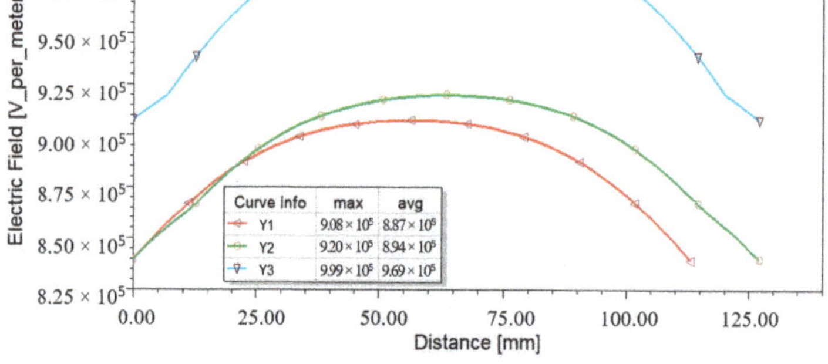

Figure 14. E-field profiles across Y1, Y2, and Y3 during the peak of impulse voltage.

In general, the E-field intensity is closely associated with the voltage gradient across the insulating material. For example, the voltage distribution across X3 is presented in Figure 15. The voltage was non-linearly distributed across the distance where the gradient changes could be seen at 10 mm and 90 mm at which the interface of sealer and foam material existed.

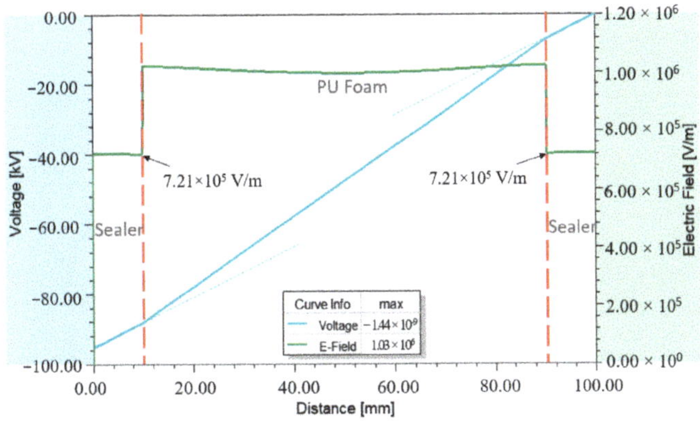

Figure 15. Voltage distribution versus E-field distribution across X3.

Based on Figure 16, three distinct E-field profiles could be found across X1, X2, and X3. It showed that the non-homogenously distributed fields caused higher electrical stress on both ends of the samples. A maximum field up to 2.60 × 10^6 V/m was recorded on the surface (X1) at which a triple junction between air, GFRP, and steel electrodes existed. It should be noted that this value was proximate to the breakdown threshold of air. In addition, the triple junction between polyester sealer, GFRP, and PU foam showed a minor impact on the field distribution at X2.

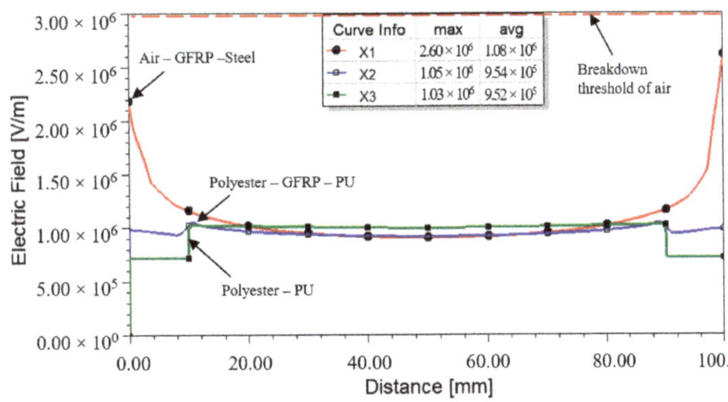

Figure 16. E-field profiles across X1, X2, and X3 during the peak of voltage.

A maximum field up to 1.03 × 10^6 V/m was projected across the PU foam (X3). As a comparison, this value was one-sixth of the average lightning breakdown strength of a typical closed-cell foam, which occurs typically at 6.09 × 10^6 V/m [14]. It is unlikely for a flashover to happen across the foam material at this level. However, it should not be disregarded as the indicated value might promote partial discharges that lead to the progressive aging of the material, as the value surpassed the threshold of streamer development.

Field plot on the outer surface of the crossarm sample revealed that the higher E-field stress was confined at every corner, as shown in Figure 17, whereby the highest E-field on the surface was 2.60 × 10^6 V/m. The field distribution explained why most of the flashover occurred at the corners, as revealed in Figure 10.

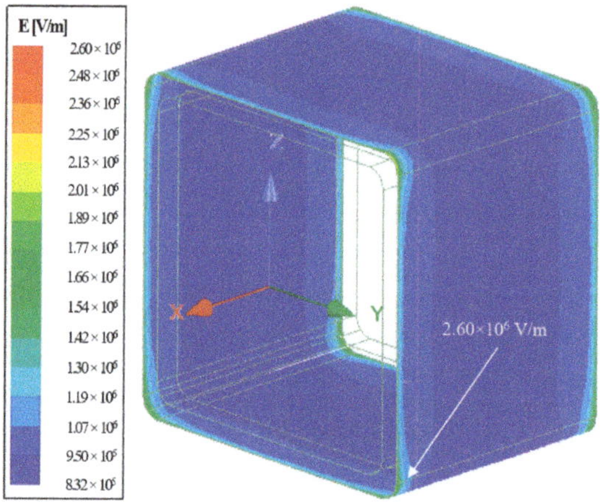

Figure 17. E-field stress on the external surfaces of GFRP crossarm sample.

Similarly, the internal GFRP surface indicated the exact behavior, whereby localized E-field was more significant at all corners, having a maximum E-field up to 1.14×10^6 V/m (see Figure 18). Circled in red is the material interfaces with localized E-field.

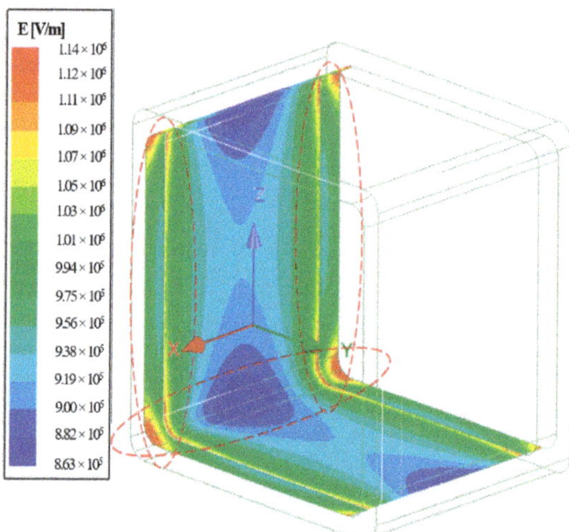

Figure 18. E-field stress on the internal surfaces of GFRP crossarm sample.

4. Conclusions

The experimental works successfully determined the CFO voltage of the 10 cm crossarm samples at 85.2 kV and −94.5 kV, where the polarity effect caused 10.9% of the difference. In addition, a reduction trend of CFO voltages was observed when the crossarm was subjected to a repeated lightning impulse test. It is believed that minor degradation occurred due to the electrical arc.

Analysis of the E-field distribution was carried out at different parts of the sample and correlated to the physical characteristic, and common discharge location was observed during the experiment. The FEM analysis revealed that the external GFRP surface suffered maximum electrical stress at both ends, which was consistent with the location of triple junctions. However, it significantly reduced toward the center of the sample, forming a

U-shape profile distribution. Internally, the stress was found slightly lower at both sample ends due to the presence of sealer.

A dome-shape distribution could be observed across the width (y-axis) of the sample, where the maximum field could be found at the center. Therefore, it revealed that the middle span of the sample suffered the highest field in the PU foam, followed by the internal and external surface of GFRP. Meanwhile, the overall analysis of the sample showed that the edges suffered the highest E-field, which is understood to be caused by the pointed shape and the triple junctions, correlated to the flashover arc's location.

In the current study, the way in which the composite material reacts to the E-field was therefore considered to understand their long-term performance. It was also suggested that fair attention should be given to the insulation properties of the foam during designing since the E-field that developed during the lightning transient exceeded the typical threshold of streamers. Even though the localized E-field is not responsible for an instant loss of insulation, it may lead to partial discharges that might cause a preliminary fault in the crossarm and degradation over some time. By knowing the magnitude of stresses on the crossarm, an appropriate selection of insulation material, properties, shapes, etc. can be planned to keep stress at a minimum level.

Author Contributions: Conceptualization, M.S.A.R. (Muhammad Syahmi Abd Rahman); re-sources, S.F.M.N. and U.A.U.A.; investigation, M.S.A.R. (Muhammad Syahmi Abd Rahman); writing—original draft preparation, M.S.A.R. (Muhammad Syahmi Abd Rahman); writing—review and editing, M.S.A.R. (Muhamad Safwan Abd Rahman), M.O. and N.M.Z.; super-vision, M.Z.A.A.K. All authors have read and agreed to the published version of the manuscript.

Funding: This research was funded by Universiti Tenaga Nasional through UNITEN Bold Grant and URND for RA Scheme.

Institutional Review Board Statement: Not applicable.

Informed Consent Statement: Not applicable.

Data Availability Statement: Not applicable.

Acknowledgments: The authors would like to thank Universiti Tenaga Nasional for the BOLD Scholarship and URND for RA Scheme. Special thanks to High Voltage Test Lab Sdn. Bhd. for providing testing facilities support.

Conflicts of Interest: The authors declare no conflict of interest. The funders had no role in the design of the study; in the collection, analyses, or interpretation of data; in the writing of the manuscript, or in the decision to publish the results.

References

1. Rawi, I.M.; Abd Rahman, M.S.; Ab Kadir, M.Z.A.; Izadi, M. Wood and fiberglass crossarm performance against lightning strikes on transmission towers. In Proceedings of the International Conference on Power System, Seoul, Korea, 26–29 June 2017; pp. 1–6.
2. Yang, B.; Zhang, J.; Zhou, L.; Lu, M.; Liang, W.; Wang, Z. Effect of fiber surface modification on water absorption and hydrothermal aging behaviors of GF/pCBT composites. *Compos. Part B* **2015**, *82*, 84–91. [CrossRef]
3. Mlyniec, A.; Korta, J.; Kudelski, R.; Uhl, T. The influence of the laminate thickness, stacking sequence and thermal aging on the static and dynamic behavior of carbon/epoxy composites. *Compo. Struct.* **2014**, *118*, 208–216. [CrossRef]
4. Shao, J.; Wang, J.; Long, M.; Li, J.; Ma, Y. 5000 h multi-factor accelerated aging test of frp made transmission tower: Characterization, thermal decomposition and reaction kinetics study. *Polymers* **2017**, *9*, 170. [CrossRef]
5. Asyraf, M.R.M.; Ishak, M.R.; Sapuan, S.M.; Yidris, N. Utilization of Bracing Arms as Additional Reinforcement in Pultruded Glass Fiber-Reinforced Polymer Composite Cross-Arms: Creep Experimental and Numerical Analyses. *Polymers* **2021**, *13*, 620. [CrossRef]
6. Mohamad, D.; Syamsir, A.; Beddu, S.; Abas, A.; Ng, F.; Razali, M.; Seman, S.A.H.A. Numerical study of composite fiberglass cross arms under statics loading and improvement with sleeve installation. *IOP Conf. Ser. Mater. Sci. Eng.* **2019**, *530*, 012027. [CrossRef]
7. Sá, M.F.; Gomes, A.M.; Correia, J.R.; Silvestre, N. Creep behavior of pultruded GFRP elements–Part 1: Literature review and experimental study. *Compos. Struct.* **2011**, *93*, 2450–2459. [CrossRef]
8. Xin, H.; Liu, Y.; Mosallam, A.S.; He, J.; Du, A. Evaluation on material behaviors of pultruded glass fiber reinforced polymer (GFRP) laminates. *Compos. Struct.* **2017**, *182*, 283–300. [CrossRef]
9. Britt, W.F., Jr. Composite Structural Support Arm. U.S. Patent 9790704 B2, 17 January 2017.

10. Pieper, R.J. The use of mechanical testing, photomicrography an electron microscopy to characterize an insulating fiberglass composite post-electrical arc failure. *Microsc. Microanal.* **2016**, *22*, 1830–1831. [CrossRef]
11. Grzybowski, S. Electrical performance of 115 kV transmission lines fiberglass crossarms aged in field. In Proceedings of the 5th International Conference on Properties and Applications of Dielectric Materials, Seoul, Korea, 25–30 May 1997; pp. 774–777.
12. Grzybowski, S.; Disyadej, T. Electrical performance of fiberglass crossarm in distribution and transmission lines. In Proceedings of the 2008 IEEE/PES Transmission and Distribution Conference and Exposition, Bogota, Columbia, 21–24 April 2008; pp. 1–5.
13. Laninga, J.; Amer, M.; Swatek, D.; McDermid, W.; Kordi, B. HVDC flashover performance of fibreglass reinforced (FRP) hot sticks considering space charges. In Proceedings of the 2017 IEEE Conference on Electrical Insulation and Dielectric Phenomenon (CEIDP), Fort Worth, TX, USA, 22–25 October 2017; pp. 609–612.
14. Karady, G.G.; Argin, M.; Shi, B.; Rahmatian, F.; Rose, A.H. Electrical properties of rigid pour polyurethane foam applied for high voltage insulation. In Proceedings of the 2003 IEEE PES Transmission and Distribution Conference and Exposition (IEEE Cat. No 03CH37495), Dallas, TX, USA, 7–12 September 2003; pp. 870–874.
15. Strauchs, A.; Mashkin, A.; Schnettler, A.; Podlazly, J.; Freiheit-Jensen, B. Investigations on the partial discharge behavior of syntactic foam under uniform field stress. In Proceedings of the IEEE International Symposium on Electrical Insulation, San Diego, CA, USA, 6–9 June 2010; pp. 1–5.
16. Abd Rahman, M.S.; Ab Kadir, M.Z.A.; Ab-Rahman, M.; Osman, M.; Nor, S.F.M. Lightning impulse strength of 275 kV and 132 kV Tower with composite crossarm. In Proceedings of the 11th Asia-Pacific International Conference on Lightning (APL), Hong Kong, China, 12–14 June 2019; pp. 1–6.
17. Abd Rahman, M.S.; Ab Kadir, M.Z.A.; Ab-Rahman, M.; Osman, M.; Nor, S.F.M.; Mohd Zainuddin, N. Effects of a Crossarm Brace Application on a 275 kV Fiberglass-Reinforced Polymer Crossarm Subjected to a Lightning Impulse. *Energies* **2020**, *13*, 6248. [CrossRef]
18. Jahangiri, T.; Wang, Q.; Bak, C.L.; da Silva, F.F.; Skouboe, H. Electric stress computations for designing a novel unibody composite cross-arm using finite element method. *IEEE Trans. Dielectr. Electr. Insul.* **2017**, *24*, 3567–3577. [CrossRef]
19. Shanmugam, G.; Karakkad, S. Influence of the insulator geometry on the streamer propagation characteristics in polymeric insulators under positive polarity lightning impulse voltages. *IET Sci. Meas. Technol.* **2018**, *12*, 1082–1088. [CrossRef]
20. Zachariades, C.; Rowland, S.M.; Cotton, I.; Peesapati, V.; Chambers, D. Development of electric-field stress control devices for a 132 kV insulating cross-arm using finite-element analysis. *IEEE Trans. Power Deliv.* **2015**, *31*, 2105–2113. [CrossRef]
21. Rosli, A.N. Numerical Modelling of Glass Fiber Reinforced Polymer (GFRP) Crossarm in Transmission Tower Subjected to Static Loading. Master's Dissertation, Universiti Tenaga Nasional, Kajang, Malaysia, 2020.
22. *IEC60060-1: High Voltage Test Techniques—Part 1: General Definition and Test Requirements*; International Electrotechnical Commission: Geneva, Switzerland, 2010.
23. Tuncer, E.; Sauers, I.; James, D.R.; Ellis, A.R. Electrical insulation characteristics of glass fiber reinforced resins. *IEEE Trans. Appl. Supercond.* **2009**, *19*, 2359–2362. [CrossRef]
24. Yamano, Y.; Takahashi, M.; Kobayashi, S.; Hanada, M.; Ikeda, Y. Surface discharge related properties of fiberglass reinforced plastic insulator for use in neutral beam injector of JT-60U. *Rev. Sci. Instrum.* **2008**, *79*, 02A524. [CrossRef]
25. Liang, X.; Shen, Y.; Liu, Y.; Wang, J.; Gao, Y.; Li, S.; Wang, M.; Gao, S. Investigations on the basic electrical properties of Polyurethane foam material. In Proceedings of the 2015 IEEE 11th International Conference on the Properties and Applications of Dielectric Materials (ICPADM), Sydney, Australia, 19–22 July 2015; pp. 863–866.
26. Huang, D.; Ruan, J.; Chen, Y.; Huo, F.; Yu, S.; Liu, S. Calculation and measurement of potential and electric field distribution along 1000 kV AC transmission line composite insulator. In Proceedings of the ICEMS 2008: International Conference on Electrical Machines and Systems, Wuhan, China, 17–20 October 2008; pp. 428–433.
27. Yang, X.; Wang, Q.; Wang, H.; Zhang, S.; Peng, Z. Transient electric field computation for composite cross-arm in 750 kV AC transmission line under lightning impulse voltage. *IEEE Trans. Dielectr. Electr. Insul.* **2016**, *23*, 1942–1950. [CrossRef]
28. Ryan, H.M. Fundamental aspects of air breakdown. In *High Voltage Engineering and Testing*, 3rd ed.; The Institution of Engineering and Technology: London, UK, 2013; pp. 631–674.
29. Shanmugam, G.; Samajdar, G.; Karakkad, S. Surface Charging and its Influence on Lightning Impulse Flashover Characteristics of Polymeric Insulator. In Proceedings of the 2019 IEEE International Conference on Electrical, Computer and Communication Technologies (ICECCT), Tamil Nadu, India, 20–22 February 2019; pp. 1–5.
30. Thione, L.; Pigini, A.; Allen, N.; Aro, M.; Baker, A. *Guidelines for the Evaluation of the Dielectric Strength of External Insulation*; CIGRE Brochure: Paris, France, 1992.
31. Liu, L.; Becerra, M. An efficient model to simulate stable glow corona discharges and their transition into streamers. *J. Phys. D Appl. Phys.* **2017**, *50*, 105204. [CrossRef]

materials

MDPI

Article

Mechano-Chemical Synthesis, Structural Features and Optical Gap of Hybrid $CH_3NH_3CdBr_3$ Perovskite

José Antonio Alonso [1,*], Paula Kayser [1,2], Bo-Kyung Hong [1,3], María Consuelo Álvarez-Galván [3], Francois Fauth [4] and Carlos Alberto López [5,*]

1 Instituto de Ciencia de Materiales de Madrid, Consejo Superior de Investigaciones Científicas, Cantoblanco, 28049 Madrid, Spain; Pakayser@ucm.es (P.K.); sally.bkhong@ies.upm.es (B.-K.H.)
2 Departamenton de Química Inorgánica, Facultad de Ciencias Químicas, Universidad Compluense de Madrid, 28040 Madrid, Spain
3 Instituto de Catálisis y Petroleoquímica, Consejo Superior de Investigaciones Científicas, Cantoblanco, 28049 Madrid, Spain; c.alvarez@icp.csic.es
4 CELLS—ALBA Synchrotron Light Facility, Cerdanyola del Valles, E-08290 Barcelona, Spain; ffauth@cells.es
5 Instituto de Investigación en Tecnología Química (UNSL-CONICET), Facultad de Química, Bioquímica y Farmacia, Universidad Nacional de San Luis, Almirante Brown 1455, San Luis 5700, Argentina
* Correspondence: ja.alonso@icmm.csic.es (J.A.A.); calopez@unsl.edu.ar (C.A.L.)

Abstract: Hybrid methyl-ammonium (MA:$CH_3NH_3^+$) lead halide $MAPbX_3$ (X = halogen) perovskites exhibit an attractive optoelectronic performance that can be applied to the next generation of solar cells. To extend the field of interest of these hybrid materials, we describe the synthesis by a solvent-free ball-milling procedure, yielding a well crystallized, pure and moisture stable specimen of the Cd tribromide counterpart, $MACdBr_3$, which contains chains of face-sharing $CdBr_6$ octahedra in a framework defined in the $Cmc2_1$ (No 36) space group. The details of the structural arrangement at 295 K have been investigated by high angular resolution synchrotron x-ray diffraction (SXRD), including the orientation of the organic MA units, which are roughly aligned along the c direction, given the acentric nature of the space group. UV-vis spectra unveil a gap of 4.6 eV, which could be useful for ultraviolet detectors.

Keywords: methylammonium cadmium tribromide; acentric crystal structure; polar $CH_3NH_3^+$ orientation; ultraviolet pigment and solar cells

Citation: Alonso, J.A.; Kayser, P.; Hong, B.-K.; Álvarez-Galván, M.C.; Fauth, F.; Alberto López, C. Mechano-Chemical Synthesis, Structural Features and Optical Gap of Hybrid $CH_3NH_3CdBr_3$ Perovskite. *Materials* **2021**, *14*, 6039. https://doi.org/10.3390/ma14206039

Academic Editors: Marc Cretin, Sophie Tingry and Zhenghua Tang

Received: 25 August 2021
Accepted: 8 October 2021
Published: 13 October 2021

Publisher's Note: MDPI stays neutral with regard to jurisdictional claims in published maps and institutional affiliations.

1. Introduction

Hybrid organic-inorganic halide perovskites are suitable as light absorbers in solar cells [1–4], with an efficiency of power conversion (PCE) of about 23%, similar to silicon-based devices. In particular, methyl-ammonium lead iodide $CH_3NH_3PbI_3$ ($CH_3NH_3^+$: MA) is the prototype of light harvester in hetero-junction solar cells [5–7]. Hybrid perovskites $MAPbX_3$ (X = I, Br and Cl) also exhibit properties such as ambipolar charge mobility, low exciton binding energy and tolerance to structural defects [8–14]. Other divalent elements instead of Pb^{2+} have been evaluated; experimental [15] and theoretical [16,17] studies have been carried out for other divalent cations, such as those of Group 12 (Zn, Cd and Hg) that also are closed-shell divalent cations. However, structural and electronic studies on these materials are scarce. In particular, the Cd-containing phase $MACdBr_3$ is little studied with only structural and theoretical analysis [17–19]. The addition of Cd^{2+} is paradigmatic since it is responsible for interesting phenomena; the electroluminescence is not inhibited, while a composite with polystyrene is effective in preventing the degradation of cadmium halide due to humidity [20].

The crystal structure of $MACdBr_3$ differs from those of $MAPbX_3$ (X = I, Br and I), most of which can be described in the cubic arystotype perovskite structure, defined in the space group $Pm\overline{3}m$. Given the smaller size of Cd^{2+} (0.95 Å) vs. Pb^{2+} (1.19 Å), the tolerance factor of the hypothetical perovskite would be higher than unity, and therefore, instead of the classical framework of corner-sharing octahedra, this Cd compound presents chains of face-sharing $CdBr_6$ octahedra in a quasi 1D arrangement. It was described by Hassen et al. [19] from a single crystal prepared by solution chemistry. Additionally, Kallel et al. [18] described a phase transition at 170 K from the above described structure to a complex orthorhombic unit cell. Regarding the absorption properties, theoretical calculations in an unrealistic cubic structure yields a gap of 1.3 eV [17]. The experimental optical gap has not been reported in the literature.

In this paper, we describe the preparation of $MACdBr_3$ by an alternative mechano-synthesis procedure, with green credentials since the utilization of organic solvents is not required. A sample with an excellent crystallinity was obtained by ball-milling under N_2 atmosphere, characterized by laboratory x-ray diffraction (XRD), and the crystal structure was refined from synchrotron x-ray diffraction (SXRD) data. We propose the H positions, unveiling the H-bond interactions between the $CH_3NH_3^+$ units and $CdBr_6$ octahedral chains, where the particular distribution of ADP parameters is a result of N–H···Br interactions. The diffuse reflectance spectra of the sample were measured for the first time, observing a band gap of ~4.6 eV, shifted to the UV region with respect to the lead-containing $MAPbBr_3$ counterpart.

2. Materials and Methods

$MACdBr_3$ was prepared in polycrystalline form by ball milling (mechano-chemical synthesis) starting from stoichiometric proportions of $CdBr_2$ and MABr. The mixture of bromides (totalling 1.5 g), was set into a zirconia-lined jar together with 30 zirconia balls (5 mm diameter) and sealed in a N_2-filled glove box. The mechanically activated reaction was carried out in a Retsch (Haan, Germany) PM100 mill for 4 h at 400 rpm. Laboratory XRD was used for assessing phase purity; the XRD patterns were recorded in a Bruker (Germany) D5 diffractometer with K_αCu (λ = 1.5418 Å) radiation; the 2θ range was 4° up to 90° with increments of 0.05°.

The crystal structure of $MACdBr_3$ was investigated at RT (295 K) by synchrotron X-ray powder diffraction (SXRD) using the MSPD station at the ALBA facility, Barcelona (Spain). Radiation with 38 keV energy, λ = 0.32511 Å, was selected in the high angular resolution mode (MAD set-up) [21]. The sample was measured in a glass capillary of 0.7 mm diameter, which rotated during data acquisition. The structural refinement by the Rietveld method [22] was carried out with the *Fullprof* software [23]. The full refinement of the profiles included the zero-point error; scale factor; background coefficients; unit-cell parameters; pseudo-Voigt shape parameters; atomic coordinates; anisotropic displacements for the metal and halogen atoms and isotropic for C and N from the methyl-ammonium groups.

Field-effect scanning electron microscopy (FE-SEM) pictures were collected on an FEI-Nova microscope, with an acceleration potential of 5 kV. The optical diffuse-reflectance spectrum was recorded at room temperature in a UV-vis spectrophotometer Varian Cary 5000. A photodetector device was fabricated by drop-casting the phase solution in dimetilformamide onto Au/Cr pre-patterned electrodes with a gap of 10 μm, and drying in a hot plate at 100 °C. The illumination power was 20 μW.

3. Results and Discussions

3.1. Initial Characterization: FE-SEM

The hybrid $CH_3NH_3CdBr_3$ compound was obtained as a white polycrystalline material. High-resolution FE-SEM images were obtained to get an insight into the microstructure of this product obtained by ball milling (Figure 1). Figure 1a illustrates an overall view with low magnification (800×), showing irregular-shaped clusters of particles of different sizes. Figure 1b,c unveil the presence of microcrystals of uneven form, with flat facets in the micrometer-size range (e.g., the crystal in Figure 1c has a width of 13 μm), which are grown during the ball milling process. Figure 1d illustrates the homogeneity of the crystals, with no particular microstructural features, in a large magnification view (24,000×). EDX analysis coupled to the FE-SEM images yields an atomic composition close to 1:3 for the Cd/Br ratio. A typical EDX spectrum is included in Figure S1 in the Supplementary Information; other SEM images are included in Figure S2.

Figure 1. FE-SEM images of the $MACdBr_3$ collected with magnifications of (**a**) 800×, (**b**) 6000×, (**c**) 12,000× and (**d**) 24,000×.

3.2. Structural Characterization

High angular resolution SXRD data allowed us a precise structural characterization. This is essential to accurately define the symmetry and crystallographic features. The pattern can be indexed in an orthorhombic unit cell with a = 7.91722(5) Å, b = 13.7108(1) Å, c = 6.89374(2) Å, and the crystal structure can be defined in the acentric $Cmc2_1$ (No 36) space group, confirming the work by Ben Hassen et al. from single-crystal x-ray diffraction [19].

Cd^{2+} cations are allocated at 4a (0,y,z) Wyckoff sites, while Br1 and Br2 atoms are placed at 8b (x,y,z) and 4a sites, respectively. C and N atoms are also located at 4a positions. Figure 2 illustrates the quality of the fit from SXRD data, and Figure 3 displays two views of the crystal structure, including displacement ellipsoids for Cd and Br atoms. Figure 3b highlights the face-sharing of $CdBr_6$ octahedra along the c axis. Table 1 contains the main crystallographic parameters after the Rietveld refinement.

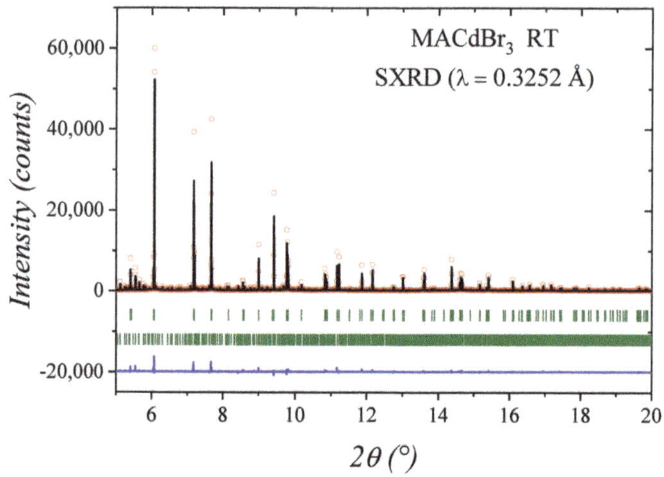

Figure 2. SXRD profiles for $CH_3NH_3CdBr_3$ at RT. The experimental points are represented by red circles, while calculated profile is a black full line and the blue lower line is the difference. The first series of green markers correspond to the allowed Bragg positions for the main phase ($Cmc2_1$ space group). The second series of markers corresponds to a minor impurity MA_2CdBr_4, ($P2_1/c$ space group).

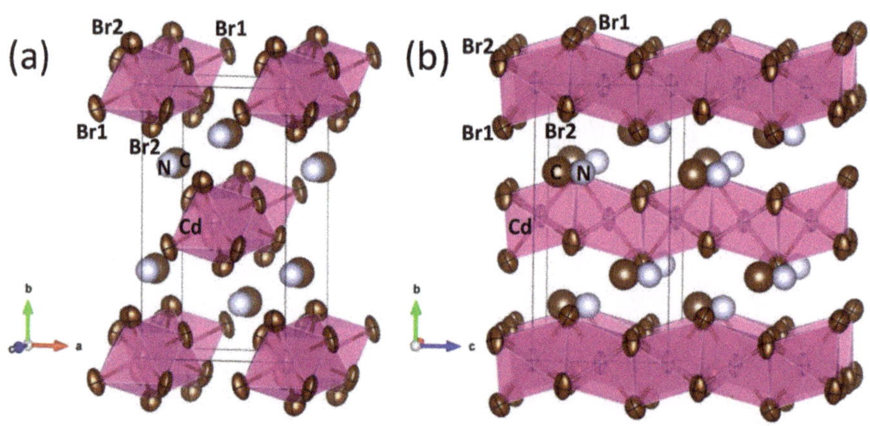

Figure 3. Two views of the crystal structure, showing the anisotropic atomic displacement parameters (ADP) of Br site and the face-sharing octahedral chains along c axis. (**a**) along c-axis direction, (**b**) along a-axis direction.

In comparison with previous results [19], a subtle increase in the unit-cell parameters is observed in the present sample. This volume expansion can be associated with the presence of structural defects (vacancies), as it was observed previously in both hybrid and all-inorganic halide systems [24,25]. Additional refinements considering atomic vacancies did not lead to an improvement of the discrepancy factors. Considering that the volume cell increase is small ($\approx$0.4%) it is possible to infer that the vacancy level is also low. These structural differences also are observed in the Cd-Br distances, as illustrated in Table 2. The distance increment mainly affects the Cd–Br2 bonds, which allows supposing that this bromine site is more plausible to present vacancies.

Table 1. Structural parameters for $CH_3NH_3CdBr_3$ from the Rietveld refinement in the orthorhombic *Cmc*2_1 (No 36) space group, with a = 7.91722(5) Å, b = 13.7108(1) Å, c = 6.89374(2) Å, V = 748.33(1) Å^3, Z = 4, from SXRD data at 295 K.

Atoms	*Wyckoff Sites*	x	y	z	U_{iso}*/U_{eq}	*Occ*
Cd	4*a*	0	0.0036(3)	1	0.035(3)	1
Br1	8*b*	0.2627(7)	0.4199(6)	0.248(3)	0.051(5)	1
Br2	4*a*	0	0.1612(7)	0.262(2)	0.056(9)	1
C	4*a*	0	0.32(1)	0.67(2)	0.12(1) *	1
N	4*a*	0	0.317(8)	0.85(2)	0.078(8) *	1
	U_{11}	U_{22}	U_{33}	U_{12}	U_{13}	U_{23}
Cd	0.029(3)	0.058(5)	0.017(1)	0	0	−0.001(5)
Br1	0.034(4)	0.082(6)	0.038(6)	−0.018(3)	−0.012(3)	0.002(5)
Br2	0.057(6)	0.063(7)	0.05(1)	0	0	0.011(7)
Discrepancy factors: R_p: 8.46%; R_{wp}: 11.6%; χ^2: 4.27; R_{Bragg}: 6.09%; MA_2CdBr_4: 16.7(2)%w/w						

The symbol * is indicated in the heading Uiso*.

Table 2. Atomic distances for $CH_3NH_3CdBr_3$ from the Rietveld refinements in comparison with those reported by Hassen et al.

	Present Structure	**Hassen et al. [19]**
Cd–Br1	2.765 Å	2.774 Å
Cd–Br1	2.787 Å	2.783 Å
Cd–Br2	2.816 Å	2.769 Å
Cd–Br2	2.792 Å	2.786 Å

With the collected data (both angular resolution and high Q range), it is possible to reasonably refine anisotropic displacement parameters (ADP). The displacement factors of Br atoms are quite anisotropic, as shown in Figure 3, and they display a flattened shape (oblate type) with the disks perpendicular to the Cd-Br chemical bonds. In this configuration, quite standard in perovskites, the thermal vibrations are allowed in the perpendicular plane to the covalent Cd-Br bonding. The anisotropy of Br1 is much superior, with r.m.s. (root mean square) displacements of 0.11 Å parallel to the chemical bond and 0.30 Å and 0.22 Å perpendicular to it (for Br2, r.m.s.'s are 0.23 Å, 0.25 Å and 0.26 Å, respectively). As also shown in Figure 3, $CH_3NH_3^+$ groups are in the space between the chains, roughly aligned along **c** axis. H atoms could not be localized from SXRD data. However, the MA position and the r.m.s. displacements suggest that the H-bond interaction of MA is stronger with Br2 atoms. To stand out this, the H atoms were added considering: (a) the expected distances and angles for the MA molecule and (b) the closeness with bromine atoms, with the aim to visualize the probable H-bond interactions. Figure 4 shows the environments of a MA unit, and the dashed lines highlight the H···Br distances less than 3 Å. This figure reveals that the disk-like ADP bromine is perpendicular to the H···Br directions, hence, it is possible to speculate that the particular distribution of ADP parameters is a result of N–H···Br interactions. Finally, a highly polar character is expected for this compound, since all the $CH_3NH_3^+$ units lie along the same direction, giving the acentric nature of the space group.

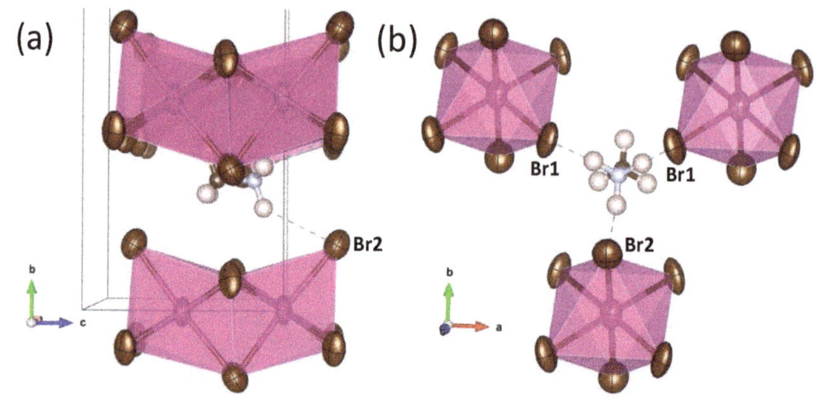

Figure 4. Views of MA with the expected hydrogen positions and H-bond interactions. (**a**) along a-axis direction, (**b**) along c-axis direction.

3.3. Optical Gap by UV-Vis Spectra and Photocurrent Properties

The absorption capacity of $MACdBr_3$ powder prepared by mechano-chemistry was investigated by diffuse reflectance UV/vis spectroscopy. Figure 5a shows the UV-vis absorption spectrum, where there are two relative maximum absorption regions placed in the ranges 300–330 and 700–740 nm. The slope disruption at 300–330 nm corresponds to an exciton absorption peak that reveals the UV light activity of the sample; this exciton peak is due to the formation of an electron-hole pair that is favoured in nanostructured samples, containing nanoplates, as observed here (Figure 1) [26,27]. The onset at ~750 nm shows the absorption by minor MABr impurities [28].

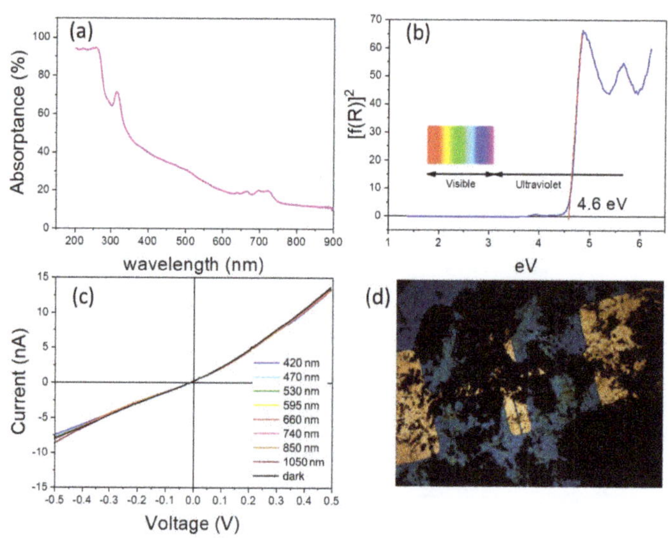

Figure 5. (**a**) Absorptance vs. wavelength of the incident radiation and (**b**) Kubelka–Munk transformed diffuse reflectance spectrum for $CH_3NH_3CdBr_3$. (**c**) Photocurrent measurements in 420–1050 nm wavelength range and (**d**) optical image of optoelectronic device.

Figure 5b illustrates the optical absorption coefficient related to the Kubelka–Munk function ($F(R) = (1 - R)^2/2R$), being R the reflectance of each sample, vs. wavelength in eV.

The band gap has been calculated by extrapolating the linear region to the abscissa. The value obtained for $MACdBr_3$ (~4.6 eV; ~270 nm) is shifted to the UV region in comparison to the corresponding lead composition ($MAPbBr_3$), which presents a band gap around 2.2 eV [29]. This shift will hamper its use in photovoltaic devices but could make this material useful for optical applications with short-wavelength UV radiation.

The optoelectronic properties of this phase were evaluated from photocurrent measurements. Figure 5c,d show the I vs. V curves at different wavelengths and an image of the used device, respectively. Figure 5c plots a typical ohmic behavior and display a negligible effect of illumination at different energies within the visible light range. This result confirms that this system is not an appropriate material for solar cells technology. However, as was mentioned above, it may find applications in the ultraviolet spectra range.

4. Conclusions

In contrast with the well-known $MAPbBr_3$ hybrid perovskite, structurally consisting of a 3D framework of corner-sharing $PbBr_6$ octahedra, the Cd counterpart contains infinite chains of face-sharing $CdBr_6$ octahedra, with the $CH_3NH_3^+$ units lying in between. The acentric nature of the $Cmc2_1$ space group implied that the organic units are all aligned along the same direction, conferring a polar character to this compound. Starting from the obtained atomic positions and anisotropic displacement factors for Cd, Br, C and N, it was possible propose the H positions, unveiling the H-bond interactions between the $CH_3NH_3^+$ units and $CdBr_6$ octahedral chains. The different structural arrangement is a consequence of the much smaller size of Cd^{2+} vs. Pb^{2+}, implying a tolerance factor greater than unity for the hypothetical perovskite. The large band gap and photocurrent results disable this material for solar cell applications, but it may find interest as a UV detector.

Supplementary Materials: The following are available online at https://www.mdpi.com/article/10.3390/ma14206039/s1, Figure S1: EDX spectrum and quantitative results for $MACdBr_3$ obtained with 18 kv of acceleration potential. H, N and C could not be observed in the presence of heavy Cd and Br atoms. The theoretical Cd/Br ratio of 1:3 is close to that found of 26.32%/73.68%, Figure S2: FE-SEM images of $MACdBr_3$ with different magnifications, illustrating the overall aspect of this material obained by ball milling.

Author Contributions: Conceptualization, J.A.A. and M.C.Á.-G.; Synthesis of the specimen, P.K. and B.-K.H.; SXRD data collection, P.K., J.A.A. and F.F.; Structure refinement and analysis; C.A.L.; Uv-vis spectrum collection, M.C.Á.-G.; writing—original draft preparation, J.A.A. and C.A.L.; writing—review and editing, J.A.A., P.K., B.-K.H., M.C.Á.-G., F.F. and C.A.L. All authors have read and agreed to the published version of the manuscript.

Funding: This research was funded by the Spanish Ministry of Economy and Competitiveness (grant No. MAT2017-84496-R) co-founded by FEDER. C.A.L. acknowledges UNSL and ANPCyT for financial support (projects PROICO 2-2320 and PICT2017-1842), Argentine.

Institutional Review Board Statement: Not applicable.

Informed Consent Statement: Not applicable.

Data Availability Statement: Not applicable.

Acknowledgments: Authors are grateful to ALBA for making the synchrotron light beamtime available through proposal number 2017072260.

Conflicts of Interest: The authors declare no conflict of interest.

References

1. Grätzel, M. The light and shade of perovskite solar cells. *Nat. Mater.* **2014**, *13*, 838–842. [CrossRef]
2. Green, M.A.; Ho-Baillie, A.; Snaith, H. The emergence of perovskite solar cells. *Nat. Photon* **2014**, *8*, 506–514. [CrossRef]
3. Snaith, H.J. Perovskites: The Emergence of a New Era for Low-Cost, High-Efficiency Solar Cells. *J. Phys. Chem. Lett.* **2013**, *4*, 3623–3630. [CrossRef]
4. Park, N.-G. Organometal Perovskite Light Absorbers toward a 20% Efficiency Low-Cost Solid-State Mesoscopic Solar Cell. *J. Phys. Chem. Lett.* **2013**, *4*, 2423–2429. [CrossRef]
5. Etgar, L.; Gao, P.; Xue, Z.; Peng, Q.; Chandiran, A.K.; Liu, B.; Nazeeruddin, K.; Grätzel, M. Mesoscopic $CH_3NH_3PbI_3/TiO_2$ Heterojunction Solar Cells. *J. Am. Chem. Soc.* **2012**, *134*, 17396–17399. [CrossRef] [PubMed]
6. Kim, H.-S.; Lee, C.-R.; Im, J.-H.; Lee, K.-B.; Moehl, T.; Marchioro, A.; Moon, S.-J.; Humphry-Baker, R.; Yum, J.-H.; Moser, J.-E.; et al. Lead Iodide Perovskite Sensitized All-Solid-State Submicron Thin Film Mesoscopic Solar Cell with Efficiency Exceeding 9%. *Sci. Rep.* **2012**, *2*, 591. [CrossRef] [PubMed]
7. Hsu, H.-Y.; Ji, L.; Ahn, H.S.; Zhao, J.; Yu, E.T.; Bard, A.J. A Liquid Junction Photoelectrochemical Solar Cell Based on p-Type $MeNH_3PbI_3$ Perovskite with 1.05 V Open-Circuit Photovoltage. *J. Am. Chem. Soc.* **2015**, *137*, 14758–14764. [CrossRef]
8. Kojima, A.; Ikegami, M.; Teshima, K.; Miyasaka, T. Highly Luminescent Lead Bromide Perovskite Nanoparticles Synthesized with Porous Alumina Media. *Chem. Lett.* **2012**, *41*, 397–399. [CrossRef]
9. Gao, P.; Grätzel, M.; Nazeeruddin, M.K. Organohalide lead perovskites for photovoltaic applications. *Energy Environ. Sci.* **2014**, *7*, 2448–2463. [CrossRef]
10. Saparov, B.; Mitzi, D.B. Organic–Inorganic Perovskites: Structural Versatility for Functional Materials Design. *Chem. Rev.* **2016**, *116*, 4558–4596. [CrossRef] [PubMed]
11. Zhang, M.; Yu, H.; Lyu, M.; Wang, Q.; Yun, J.-H.; Wang, L. Composition-dependent photoluminescence intensity and prolonged recombination lifetime of perovskite $CH_3NH_3PbBr_3-xClx$films. *Chem. Commun.* **2014**, *50*, 11727–11730. [CrossRef]
12. Alvarez-Galván, M.C.; Alonso, J.A.; López, C.A.; Lopez-Linares, E.; Contreras, C.; Lazaro, M.J.; Fauth, F.; Huerta, M.V.M. Crystal Growth, Structural Phase Transitions, and Optical Gap Evolution of $CH_3NH_3Pb(Br1–xClx)_3$ Perovskites. *Cryst. Growth Des.* **2018**, *19*, 918–924. [CrossRef]
13. Lopez, C.A.; Alvarez-Galvan, M.C.; Huerta, M.V.M.; Fernandez-Diaz, M.T.; Alonso, J.A. Dynamic Disorder Restriction of Methylammonium (MA) Groups in Chloride-Doped MAPbBr 3 Hybrid Perovskites: A Neutron Powder Diffraction Study. *Chem. A Eur. J.* **2019**, *25*, 4496–4500. [CrossRef]
14. López, C.A.; Alvarez-Galván, M.C.; Martínez-Huerta, M.V.; Fauth, F.; Alonso, J.A. Crystal structure features of $CH_3NH_3PbI_3-xBrx$ hybrid perovskites prepared by ball milling: A route to more stable materials. *CrystEngComm* **2020**, *22*, 767–775. [CrossRef]
15. Dirin, D.; Dreyfuss, S.; Bodnarchuk, M.; Nedelcu, G.; Papagiorgis, P.; Itskos, G.; Kovalenko, M.V. Lead Halide Perovskites and Other Metal Halide Complexes As Inorganic Capping Ligands for Colloidal Nanocrystals. *J. Am. Chem. Soc.* **2014**, *136*, 6550–6553. [CrossRef] [PubMed]
16. Li, Y.; Yang, K. High-throughput computational design of organic–inorganic hybrid halide semiconductors beyond perovskites for optoelectronics. *Energy Environ. Sci.* **2019**, *12*, 2233–2243. [CrossRef]
17. Koliogiorgos, A.; Baskoutas, S.; Galanakis, I. Electronic and gap properties of lead-free perfect and mixed hybrid halide perovskites: An ab-initio study. *Comput. Mater. Sci.* **2017**, *138*, 92–98. [CrossRef]
18. Kallel, A.; Mlik, Y.; Courseille, C.; Couzi, M. Structural phase transition in single-crystal $CH_3NH_3CdBr_3$: I. Experimental studies. *J. Phys. Condens. Matter* **1992**, *4*, 7179–7190. [CrossRef]
19. Ben Hassen, R.; Ben Salah, A.; Kallel, A.; Daoud, A.; Jaud, J. Crystal structure of monomethylammonium tribromocadmate(II). *J. Chem. Crystallogr.* **2002**, *32*, 427–430. [CrossRef]
20. Vassilakopoulou, A.; Papadatos, D.; Koutselas, I. Polystyrene based perovskite light emitting diode. *Appl. Mater. Today* **2018**, *12*, 15–20. [CrossRef]
21. Fauth, F.; Boer, R.; Gil-Ortiz, F.; Popescu, C.; Vallcorba, O.; Peral, I.; Fullà, D.; Benach, J.; Juanhuix, J. The crystallography stations at the Alba synchrotron. *Eur. Phys. J. Plus* **2015**, *130*, 160. [CrossRef]
22. Rietveld, H.M. A profile refinement method for nuclear and magnetic structures. *J. Appl. Crystallogr.* **1969**, *2*, 65–71. [CrossRef]
23. Rodríguez-Carvajal, J. Recent advances in magnetic structure determination by neutron powder diffraction. *Phys. B Condens. Matter* **1993**, *192*, 55–69. [CrossRef]
24. López, C.A.; Abia, C.; Rodrigues, J.E.F.S.; Serrano-Sánchez, F.; Nemes, N.M.; Martinez, J.L.; Fernandez-Díaz, M.T.; Biškup, N.; Alvarez-Galván, C.; Carrascoso, F.; et al. Enhanced stability in $CH_3NH_3PbI_3$ hybrid perovskite from mechano-chemical synthesis: Structural, microstructural and optoelectronic characterization. *Sci. Rep.* **2020**, *10*, 11228. [CrossRef] [PubMed]
25. López, C.A.; Abia, C.; Alvarez-Galván, M.C.; Hong, B.-K.; Martínez-Huerta, M.V.; Serrano-Sánchez, F.; Carrascoso, F.; Castellanos-Gómez, A.; Fernández-Dıaz, M.T.; Alonso, J.A. Crystal Structure Features of $CsPbBr_3$ Perovskite Prepared by Mechanochemical Synthesis. *ACS Omega* **2020**, *5*, 5931–5938. [CrossRef]
26. Watthage, S. Solution-Processed Fabrication of Hybrid Organic-Inorganic Perovskites & Back Interface Engineering of Cadmium Telluride Solar Cells, University of Toledo. 2017. Available online: http://rave.ohiolink.edu/etdc/view?acc_num=toledo1512390043951256 (accessed on 10 June 2021).
27. Khokhra, R.; Bharti, B.; Lee, H.-N.; Kumar, R. Visible and UV photo-detection in ZnO nanostructured thin films via simple tuning of solution method. *Sci. Rep.* **2017**, *7*, 1–14. [CrossRef]

28. Yang, M.; Zhang, T.; Schulz, P.; Li, Z.; Li, G.; Kim, D.H.; Guo, N.; Berry, J.J.; Zhu, K.; Zhao, Y. Facile fabrication of large-grain $CH_3NH_3PbI_{3-x}Br_x$ films for high-efficiency solar cells via CH_3NH_3Br-selective Ostwald ripening. *Nat. Commun.* **2016**, *7*, 12305. [CrossRef] [PubMed]
29. López, C.A.; Martínez-Huerta, M.V.; Alvarez-Galván, M.C.; Kayser, P.; Gant, P.; Castellanos-Gomez, A.; Fernández-Díaz, M.T.; Fauth, F.; Alonso, J.A. Elucidating the Methylammonium (MA) Conformation in $MAPbBr_3$ Perovskite with Application in Solar Cells. *Inorg. Chem.* **2017**, *56*, 14214–14219. [CrossRef]

materials

MDPI

Article

Synthesis and Characterization of Activated Carbon Co-Mixed Electrospun Titanium Oxide Nanofibers as Flow Electrode in Capacitive Deionization

Gbenro Folaranmi, Myriam Tauk, Mikhael Bechelany, Philippe Sistat, Marc Cretin * and Francois Zaviska *

Institut Européen des Membranes, IEM, UMR-5635, Université de Montpellier, ENSCM, CNRS, Place Eugène Bataillon, CEDEX 5, 34095 Montpellier, France; gbenro.folaranmi@umontpellier.fr (G.F.); myriam.tauk@umontpellier.fr (M.T.); mikhael.bechelany@umontpellier.fr (M.B.); philippe.sistat@umontpellier.fr (P.S.)

* Correspondence: marc.cretin@umontpellier.fr (M.C.); francois.zaviska@umontpellier.fr (F.Z.)

Abstract: Flow capacitive deionization is a water desalination technique that uses liquid carbon-based electrodes to recover fresh water from brackish or seawater. This is a potential second-generation water desalination process, however it is limited by parameters such as feed electrode conductivity, interfacial resistance, viscosity, and so on. In this study, titanium oxide nanofibers (TiO_2NF) were manufactured using an electrospinning process and then blended with commercial activated carbon (AC) to create a well distributed flow electrode in this study. Field emission scanning electron microscope (FESEM), X-ray diffraction (XRD), Raman spectroscopy, X-ray photoelectron spectroscopy (XPS), and energy dispersive X-ray (EDX) were used to characterize the morphology, crystal structure, and chemical moieties of the as-synthesized composites. Notably, the flow electrode containing 1 wt.% TiO_2NF (ACTiO_2NF 1 wt.%) had the highest capacitance and the best salt removal rate (0.033 mg/min·cm^2) of all the composites. The improvement in cell performance at this ratio indicates that the nanofibers are uniformly distributed over the electrode's surface, preventing electrode passivation, and nanofiber agglomeration, which could impede ion flow to the electrode's pores. This research suggests that the physical mixture could be used as a flow electrode in capacitive deionization.

Keywords: flow electrode capacitive deionization; electrospinning; activated carbon; desalination

Citation: Folaranmi, G.; Tauk, M.; Bechelany, M.; Sistat, P.; Cretin, M.; Zaviska, F. Synthesis and Characterization of Activated Carbon Co-Mixed Electrospun Titanium Oxide Nanofibers as Flow Electrode in Capacitive Deionization. *Materials* **2021**, *14*, 6891. https://doi.org/10.3390/ma14226891

Academic Editor: Zhenghua Tang

Received: 6 October 2021
Accepted: 9 November 2021
Published: 15 November 2021

1. Introduction

One of the growing challenges of the 21st century is the availability of fresh water. Water contamination because of anthropogenic activities such as industrialization, demographic change, and global warming has resulted in a significant increase in demand for safe drinking water. Water desalination technology could help to alleviate this problem by delivering high quality, pure water. Most desalination technologies, such as multiple effect desalination (MED), reverse osmosis (RO), and others, have high capital costs (when considering plant setup) and energy consumption (when considering pre- and post-treatment of water, as in RO), necessitating the development of a new desalination technique [1].

Capacitive deionization (CDI) is a growing desalination technology attracting attention as an energy-efficient, cost-effective, and ecofriendly water treatment technology. The first and most widely used CDI form involves a pair of porous carbon electrodes separated by a space in which salt water flows as an influent perpendicular to the applied electric field direction [2]. A fundamental variation of this basic CDI form emerged with the unfolding of membrane CDI (MCDI). In this architecture, ion exchange membranes were added to the CDI cell configuration to block co-ions from carrying parasitic current, which improves charge efficiency and can increase the charge storage in the electrodes porous structure [3]. In the past decade, a new class for CDI based on MCDI was developed which introduced carbon flow electrodes that can be pumped through the electrode compartments. Flow electrode CDI (FCDI) is a promising second-generation water desalination method based on the principle of ions adsorption. When a specific voltage is applied, it entails using a polarized flow electrode (liquid electrode) travelling via a flow channel to adsorb ions from brackish or seawater via electrical double layer (EDL) formation (interface between electrolyte and electrode) [4].

One of the challenges in FCDI is to improve the conductivity and capacitance of carbon materials used as electrodes, as well as reducing the resistance between the electrode and the electrolyte. This can be accomplished by improving the surface charge of activated carbon for better EDL formation [5].

Because of its beneficial qualities such as high surface area, porosity, availability, and low cost, graphitic carbon, materials such as activated carbon are frequently used as flow electrodes [6,7]. However, for more effective EDL formation and better ion storage, the resistance at the electrode-electrolyte interface should be lowered. As a result, surface modification is required to raise the surface charge of activated carbon.

Metal oxides such as zinc oxide (ZnO), tin oxide (SnO), zirconium oxide (ZrO), and titanium oxide (TiO_2) have been used to modify carbon in this concept [8,9]. Among all these oxides, TiO_2 has a high surface charge that can be used in conjunction with AC to create new types of electrodes. The aggregation of TiO_2 particles, however, is one of the additive's downsides, limiting its performance [10–12]. Small quantities of TiO_2 nanofibers are used in activated carbon flow electrodes to alleviate the agglomeration problem since they have a favorable large axial ratio morphology [13,14].

Due to the high resistance of the liquid medium utilized for carbon dispersion for slurry formation, charge transfer resistance is a strong factor hindering the ease of ions diffusion into the pores of carbon electrodes in flow electrodes [15]. As a result, methods for reducing interfacial resistance by adding conducting additives or introducing oxygenated functional groups to activated carbon have been reported in the literature [15–18], but no report on the properties of TiO_2 nanofibers in reducing interfacial resistance and their effect on the rheological properties of carbon-based flow electrodes has been found to date.

Nanofiberous materials of high surface area and porosity are considered recently to be used for generating sustainable and green energy resources. Electrospinning method is a low-cost method used to create nanofiberous materials suited for many energy-related applications such as supercapacitors, and Li-ion batteries [19,20].

Motivated by this, we have synthesized TiO_2 nanofibers by electrospinning process and mixed it with commercial AC. The proposed electrodes which consisted of different weight percentage of TiO_2NF were labelled TiO_2NF-x where x represents the weight percentage of TiO_2NF in the composite (x = 0.5, 1.0, 1.5, 2.0, 2.5, and 5.0 wt.% TiO_2NF). The electrodes were characterized by FESEM, XRD, Raman spectroscopy, XPS, EDX, and N_2 adsorption/desorption. Dynamic viscosity of the mixtures in suspension to build the flow electrodes was also investigated. Hence, the electrochemical properties of the as-prepared flow electrodes (AC and TiO_2NF-x) were analyzed by cyclic voltammetry (CV) to determine their capacitance after which the flow electrodes were tested in a laboratory-made FCDI cell for desalination.

2. Materials and Methods

2.1. Materials

Activated carbon, Darco (CAS no: 7440-44-0, MW 12.01 g mol^{-1}), titanium (IV) isopropoxide ($C_{12}H_{28}O_4Ti$, CAS no: 546-68-9, MW 284.22 g mol^{-1}, 97%), poly(vinyl pyrrolidone) (PVP, CAS no: 9003-39-8, MW 1,300,000 g mol^{-1}), and sodium chloride (NaCl, CAS no: 7647-14-5, MW 58.44 g mol^{-1}, 99%) were obtained from Sigma Aldrich, Steinheim, Germany. Acetic acid (CH3COOH MW 60.052 g mol^{-1}, 99.5% assay), absolute ethanol (C2H5OH MW 46.07 g mol^{-1}, 99.99%,) were obtained from VWR chemicals, Fontenay-sous Bois, France. Polyvinylidenefluoride (PVDF) (CAS no: 24937-79-9) was obtained from Alfa Aesar, Erlenbachweg 2, Kendel, Germany. All reagents were used without any further purification. Cationic and anionic exchange membranes were purchased from Membranes International Inc. (Ringwood, NJ, USA) and Deionized water (18 MΩ·cm) was used to prepare standard solutions and suspensions.

2.1.1. Synthesis of TiO_2 Nanofibers by Electrospinning

The TiO_2 nanofibers were synthesized by electrospinning technique. Briefly, solution A containing 3 mL of ethanol was added to 0.3 g of polyvinyl pyrrolidone (PVP) and then stirred until dissolution. Then, solution B consisting of 2 mL ethanol, 2 mL of acetic acid, and 3 mL of titanium isopropoxide was stirred for 45 min. This is then followed by the addition of solution B into clear solution of A and then further stirred for 50 min to obtain a sol gel solution. Electrospinning of the sol–gel solution was carried out at 22.30 kV. The distance between the collector and the syringe tip was maintained at 10 cm, and the injection speed was 0.5 mL min^{-1}. The obtained electrospun nanofiber materials were sintered in air atmosphere at 400 °C for 4 h with a ramp rate of 1 °C min^{-1}.

2.1.2. Preparation of AC and $ACTiO_2NF$–x Flow Electrodes

For flow electrode preparation, certain amount of powdered commercial AC and $ACTiO_2NF$-x (in which x corresponds to 1.0, 1.5, 2.0, 2.5, and 5.0 wt.% of TiO_2NF) were weighed and dispersed in deionized water as presented in Table 1. The mixtures were sonicated for 3 h and stirred for 1 h before being fed into the cell. The tank containing the slurry electrode was continuously stirred on a magnetic stirrer during the course of the experiment.

Table 1. Composition of flow electrodes.

Electrode Material	FE (10 wt.%) (g)	DH_2O (mL)	TiO_2NF (g)
AC	7.80	70	0
$ACTiO_2NF$-0.5	7.76	70	0.04
$ACTiO_2NF$-1.0	7.72	70	0.08
$ACTiO_2NF$-1.5	7.68	70	0.12
$ACTiO_2NF$-2.0	7.64	70	0.16
$ACTiO_2NF$-2.5	7.60	70	0.20
$ACTiO_2NF$-5.0	7.40	70	0.40

Note: FE: Feed electrode; DH_2O: deionized water.

2.2. Physical Characterization

FESEM was used to analyze the morphology of the samples (Hitachi S4800, Tokyo, Japan). The structural properties were studied by using Raman spectroscopy (HORIBA Xplora, Tokyo, Japan) and XRD (Pan Analytical X'pert Phillips, Almelo, The Netherlands). XPS (ESCALAB 250 Thermo Electron, Montigny Le Bretonneux, France) and EDX (X-Max, Oxford, UK) were used to investigate the atomic composition and chemical moieties of the materials. For the XPS analysis, the excitation source was a monochromatic source Al Kα anode with photoenergy that was observed at 1486.6 eV. The analyzed surface has a diameter of 500 μm. The photoelectron spectra were calibrated in terms of bond energy with respect to the energy of the C=C component of carbon C1s at 284.4 eV. Surface area was obtained by using N_2 adsorption/desorption at 77 K (Micromeritics ASAP, Verneuil en Halatte, France). Dynamic viscosity was measured using Anton Paar Rheometer Physica MCR 301 (Anton Paar, Graz, Austria).

2.3. Electrochemical Characterizations

Solid electrodes for electrochemical characterization of pristine AC and $ACTiO_2NF$ were made by combining activated carbon powder (0.32 g), carbon black (0.04 g), and poly (vinylidene fluoride fluoride PVdF, 0.04 g) in 3 mL N-Methyl-2-pyrrolidone (NMP). To establish homogeneity, the mixture was agitated for 2 h and then sonicated for 40 min.

After that, the slurry was applied on a graphite sheet. The coated electrode was dried in an oven for 1–2 h at 80 °C. The as-synthesized titanium oxide nanofibers were added (0.5, 1.0, 1.5, 2.0, 2.5, and 5.0 wt.%) to AC containing carbon black and PVDF, and the mixture was agitated for several hours in 3 mL NMP for $ACTiO_2NF$ synthesis. After that, the mixture was sonicated for 40 min. The slurry was subsequently immobilized by depositing it on a graphite sheet. It was then dried in an oven at 80 °C for 1–2 h to produce solid electrodes.

The electrochemical properties of the prepared electrodes were examined by using CV. CV tests were performed using a three-electrode system. The mixtures were deposited on a graphite sheet as support with an exposed surface area of 1 cm^2, while a platinum mesh and a saturated 3 M KCl, Ag/AgCl electrode served as counter and reference electrodes respectively. A molar NaCl solution was used as electrolyte. Voltammetry measurements were performed with Origalys potentiostat (OGF01A, Origalys Electrochem SAS, Rillieux-la-Pape, France) at an operating window from −0.4 to 0.6 V vs ref (to ensure electrochemical stability of the electrolyte and prevent water splitting i.e., oxygen and hydrogen evolution) in 1 M NaCl electrolyte.

The double-layer capacitance was determined using cyclic voltammetry at different scan rates by considering the charging and the discharging currents at 0.1 V vs. ref. The determined double-layer capacitance of the system was the average of the absolute value of the slope of the linear plot of charging and discharging currents fitted to the data. CDL as the double layer capacitance was determined using Equation (1):

$$i = \upsilon\, C_{DL}. \tag{1}$$

The double-layer charging current density i ($A \cdot cm^{-2}$) is equal to the product of the scan rate υ ($V \cdot s^{-1}$), and the electrochemical double-layer capacitance C_{DL} ($F \cdot cm^{-2}$).

2.4. FCDI Measurement

The schematic diagram of the close loop experimental set-up is shown in Figure 1 in which the cell was powered by a potentiostat and the feed solution (FS, 5 $g \cdot L^{-1}$) was made to pass through a spacer sandwiched in between cation and anion exchange membranes. The feed electrodes (FE) stored in a reservoir (were made to pass (by pumping) through flow channels and as they exit the channels, they are fed back to the reservoir and then re-circulated; this allows co-mixing of opposite charged ions outside the cell. Figure 2 shows the breakdown of the cell components with the current collector (6 mm width and 0.9 mm depth channel), the ions exchanges membranes and spacer (0.9 mm thick). The flow rate of the electrodes was operated at 40 mL min^{-1}. Each desalination experiment was conducted for 30 min. The initial conductivity of the salt solution and that of the effluent was monitored at room temperature by ion conductivity meter (Hanna Instruments SRL) all along the 30 min of the experiment. A constant voltage difference of 1.2 V was applied to the FCDI unit cell using an Origalys potentiostat (OGF01A, Origalys Electrochem SAS) for desalination experiments. The current intensity passing through the FCDI unit cell was consequently measured by the potentiostat during the experiment.

In the present work, the following indicators defined FCDI performances:

Salt removal rate in $mg \cdot min^{-1} \cdot cm^{-2}$ (SRR) relates to the mass of salt adsorbed (mg) per FE-FS contact area (cm^2) per unit of time (min). It is calculated by Equation (2):

$$SRR = \frac{(Co - Cf) * V}{A * t} \tag{2}$$

Co and Cf are the initial and final (at t = 30 min) concentration ($mol \cdot L^{-1}$), respectively, V is the volume of the solution (L), A is the contact area between FE: FS, and t is the charging time.

The salt removal efficiency in % (SRE) was calculated using Equation (3):

$$SRE = \frac{(Co - Cf)}{Co} * 100 \tag{3}$$

Charge efficiency (CE) in % which relates to the ratio of salt adsorbed to the quantity of charge passed into the system was calculated by Equation (4):

$$CE = \frac{z\,(Co - Cf)\,V\,F}{M \int Idt} * 100 \quad (4)$$

where z is the equivalent charge of the ions, F is the Faradaic constant, M is the molar weight, and $\int Idt$ is the integrated quantity of charge passed to the system as a function of time. CE is a good indicator for the energy efficiency of the system and will directly affect the operating cost of the system (OPEX).

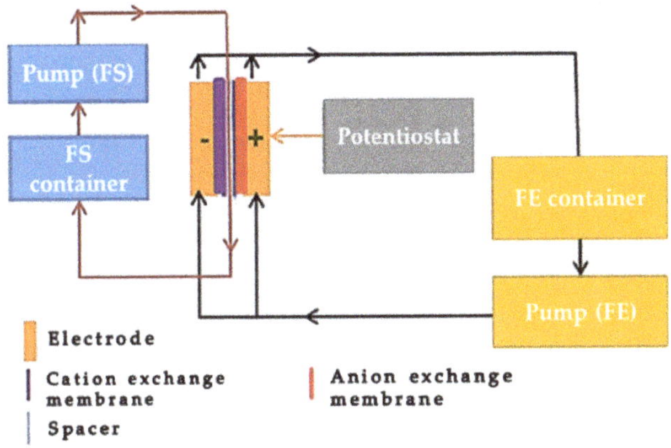

Figure 1. Schematic diagram of FCDI setup is a figure.

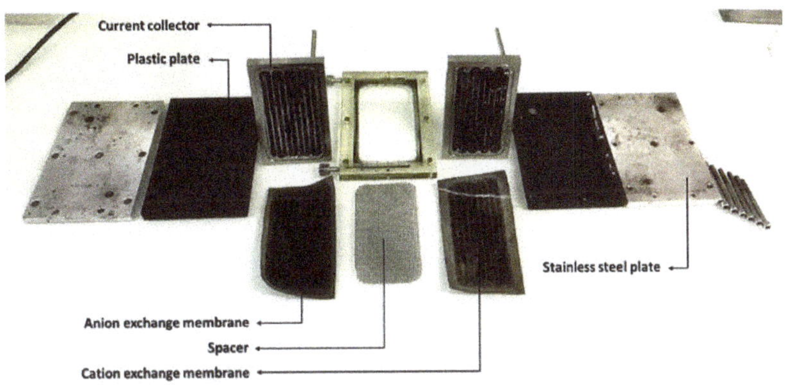

Figure 2. Individual components of FCDI cell.

3. Results and Discussion

3.1. Morphology

The morphology of as-synthesized titanium oxide nanofibers (TiO_2NF), pristine AC, and the ACTiO_2NF-x electrodes are shown in Figure 3a–h respectively. From Figure 3a, it is apparent that the electrospun TiO_2 showed fiber-like morphology with no beads formation. This shows the successful nanofibrous morphology of TiO_2 formation by electrospinning process. Figure 3b shows the morphology of pristine AC. It is clear that it has no defined shape with rough or uneven surface characteristics while Figure 3c–h reveals the presence of TiO_2NF on the surface of AC, indicating that the additive was successfully introduced by co-mixing.

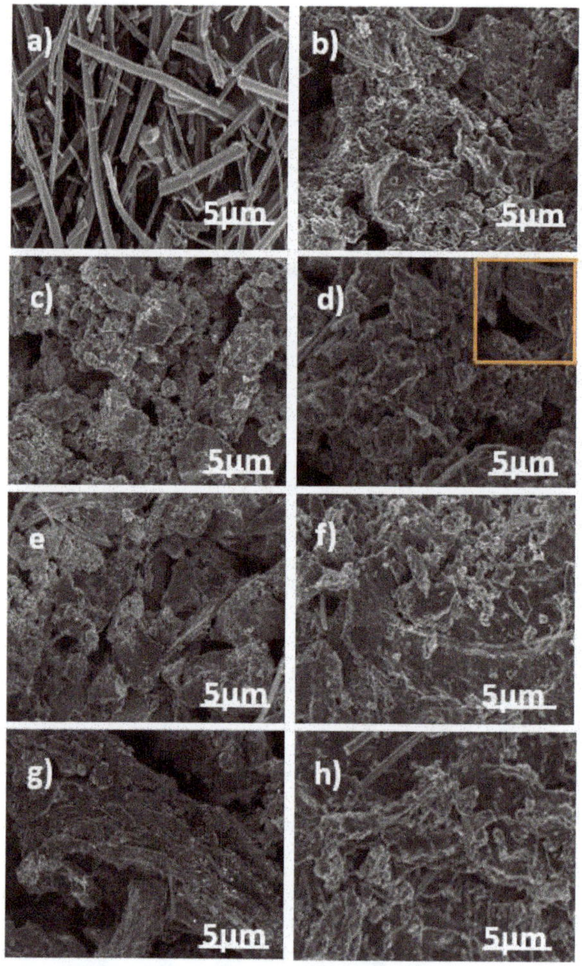

Figure 3. FESEM top view images of (**a**) TiO_2NF (**b**) AC (**c–h**) ACTiO_2NF–x, where (x = 0.5, 1.0, 1.5, 2.0, 2.5 and 5.0 wt.% TiO_2NF) respectively.

3.2. Structural Properties

Figure 4a,b shows the Raman spectra and diffractogram of the TiO_2NF, pristine AC and ACTiO_2NF s respectively. The Raman spectrum of the pristine AC and ACTiO_2NF in Figure 4a conforms to a typical graphitic carbon with distinguishable peaks at 1350 cm^{-1} and 1590–1610 cm^{-1} corresponding to D and G bands respectively [21]. D band arises from a defect that is based on out of plane vibration while G band relates to the ordered structure of graphite crystals [13]. For the as-synthesized TiO_2NF, Figure 4a, major peaks are observed at 142, 388, 516, and 638 cm^{-1} indicating anatase phase characteristics [8]. Distinguishable peaks of TiO_2NF were observed in some of the composites especially for those with high percentage of TiO_2NF as shown in Figure 4a. This indicates the successful mixing of the nanofibers with the AC.

Structural investigation was carried out to understand the crystalline nature of the materials. Figure 4b shows the diffractogram of pure TiO_2NF, pristine AC, and its composites. The TiO_2NF crystals are predominantly dominated with definite and sharp diffraction peaks at $2\theta = 25°$, 39° and 43.5° relating to 101, 004, and 200 planes of anatase phase respectively with the presence of rutile phase at $2\theta = 27.5°$, 36°, and 41° relating to 110, 101, and

111 planes respectively [8]. Typical diffraction peaks of all graphite material is observed for pristine AC and $ACTiO_2NF$ at $2\theta = 26°$ and 43.5 ° corresponding to 002 and 100 or 101 planes of graphite respectively. The sharp diffraction peaks observed at 002 planes indicates the presence of graphite microcrystalline structure in the AC [13]. In comparison with pristine AC, peaks of TiO_2NF were detected at 25°, 27.5°, and 48° diffraction peaks of TiO_2NF in all the composites thus showing successful doping of the nanofibers in the AC. Furthermore, using Debye-Scherrer equation, ($D = K\lambda/\beta, \cos\theta$) where K is the constant value of 0.9, λ is the radiation of the XRD machine (0.1541 nm), β is the full width at half maximum of the diffraction peak in radian, and θ is the diffraction angle in radian, no changes were observed in the crystallite size (10 ± 0.830 nm) of the as-synthesized TiO_2 nanofibers and $ACTiO_2NF$.

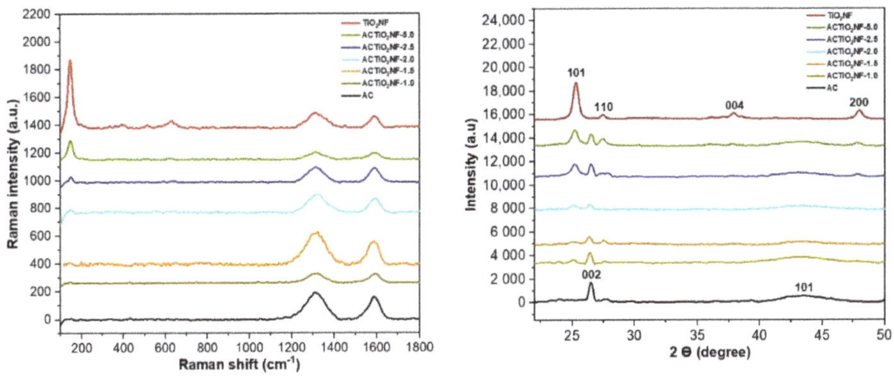

(**a**) (**b**)

Figure 4. (**a**) Raman spectra of Titanium oxide nanofibers TiO_2NF, pristine AC, and its composites ($ACTiO_2NF$–x) and (**b**) diffractograms of titanium oxide nanofibers (TiO_2NF), pristine AC and its composites ($ACTiO_2NF$–x where x = 0.5, 1.0, 1.5, 2.0, 2.5, and 5.0 wt.% TiO_2NF).

3.3. EDX and XPS Studies

Figure 5a,b shows the EDX spectra that were obtained in order to identify the composition of the pristine AC and the composite electrode. Evidently from EDX, the elements detected at highest percentage in our materials are C and O. Titanium was detected among other elements in little quantity in the composite as shown in Figure 5b. The presence of fluorine was also detected due to the addition of PVDF (binder) added during electrode fabrication. Moreover, EDX mapping was further used to investigate the distribution of the additive in the mixture. From Figure S1a,b it can be seen that the additive distribution at lower concentration (1 wt.%) is homogenous (even dispersion) while at higher concentration (5 wt.%), its distribution is concentrated within a particular region. The results confirm the formation of well-dispersed nanofibrous within the carbon structure at low percentage concentration.

Further investigation was carried out using XPS in order to verify any change in the chemical composition of the pristine AC and its composites. Figure 5c shows two prominent peaks at 458.69 and 464.44 eV belonging to Ti $2P_{3/2}$ and Ti $2P_{1/2}$ respectively [22,23]. From Figure 5c, it is shown that Ti (IV) is present in normal state in the composites due to the observed spin orbital splitting corresponding to 5.76 eV that is obtained between Ti $2P_{1/2}$ and Ti $2P_{3/2}$. [23]. As shown in Figure 5d, no titanium was detected in the XPS spectra survey scan of the pristine AC on comparison with $ACTiO_2NF$; a complementary result with that of EDX. The O1s peak in the composite increased a little bit when compared to pristine AC due to the influence of oxygen content of the additive. According to XPS, the atomic composition of the $ACTiO_2NF$-1.0 consisted of C1s 90.6 ± 0.11%, O1s 7.9 ± 0.12%, and Ti 2p 1.0 ± 33.33% while that of pristine AC is C1s 94 ± 0.84% and O1s 5 ± 3.84%.

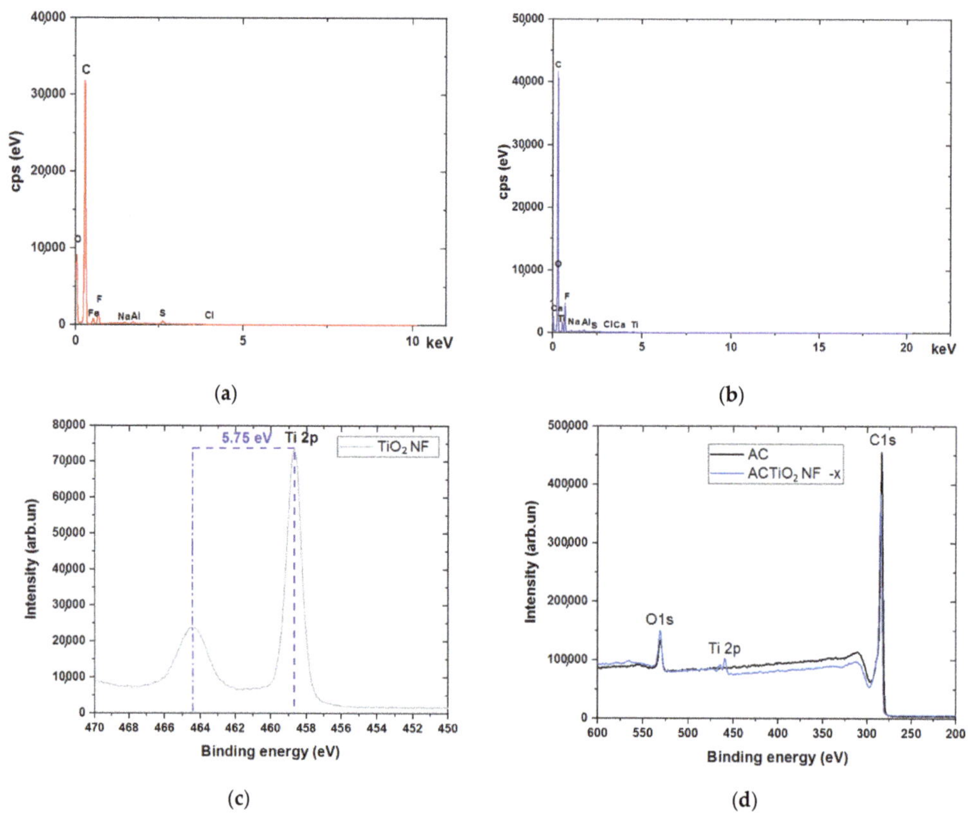

Figure 5. EDX spectra of (**a**) pristine AC (**b**) $ACTiO_2NF$–x and XPS spectra of (**c**) Ti 2p for TiO_2NF (**d**) the composite pristine AC and $ACTiO_2NF$ where x = 1.0 wt.% TiO_2NF.

3.4. Rheology Study

Rheology property reveals the flow nature of the slurry used under applied force. The viscosity nature of the slurry electrode was measured as a function of shear rate. Here, rheology property is in terms of dynamic viscosity, which is used to describe resistance to flow of liquid while shear rate describes the speed of deformation of the slurry under applied force. The dynamic viscosity was determined at a constant concentration of 10 wt.% carbon content in the slurry. From Figure 6a, it is obvious that the slurry follows a non-Newtonian fluid (shear thinning effect) in which the viscosity of the slurries decreases with increasing shear rate. To understand the effect of the additive, the viscosities of both pristine AC and the $ACTiO_2NF$ mixture were measured and compared as shown in Figure 6a. It is obvious from the rheogram curves that viscosity increases as the additive content increases in the composite. However, at low percentage of additive, there seems to be no significant difference in the viscosities of both the pristine AC and the $ACTiO_2NF$ mixture but at i.e., $TiO_2NF \geq 2.0$ wt.%, a sharp increase in viscosity was observed. The increase in viscosity can be attributed to the high specific surface area (increase frictions) of titanium nanofibers as well as its surface charge. The surface charge of the nanofibers (whether positive or negative) leads to creation of repulsive forces existing within the nanoparticles and as a result, they tend to move further apart due to strengthened force of repulsion; consequently, viscosity increases with increasing repulsive forces [24–28]. Therefore, as the concentration of the additive in the electrodes increases, a distinguishable jump in their viscosity was observed due to increase in force of repulsion as shown in Figure 6b. This implies that the presence of additive at high content level could lead to potential clogging of the feed electrodes in the cell.

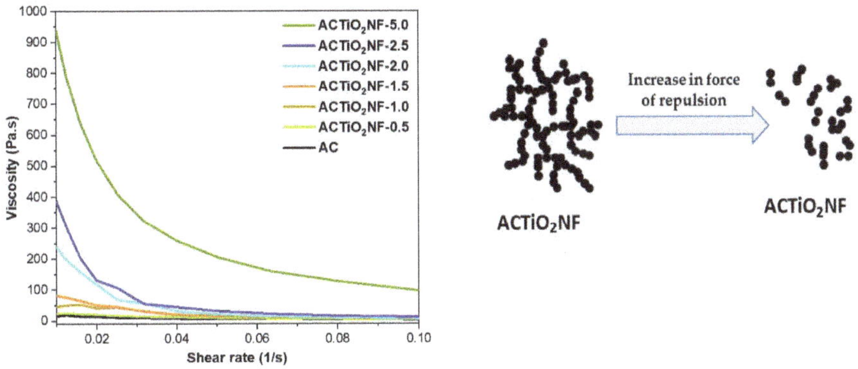

(a) (b)

Figure 6. (**a**) Rheological properties of the flow-electrodes. (**b**) Schematic diagram of the $ACTiO_2NF$ mixture behavior under repulsive forces.

3.5. Electrochemical Properties

Electrochemical behavior of the pristine AC electrode and $ACTiO_2NF$ electrodes were carried out using CV at different scan rates in a potential window from −0.4 to 0.6 V vs. Ref (to ensure electrochemical stability of the electrolyte and prevent water splitting i.e., oxygen and hydrogen evolution). The experiment was conducted in 1 M NaCl aqueous solution to investigate the influence of the additive at different ratios. CV is an important technique to probe the capacitive nature of EDL [29,30].

Based on the cyclic voltammetry cycles, the EDL capacitance of the electrodes was calculated [31] and is shown in Figure 7.

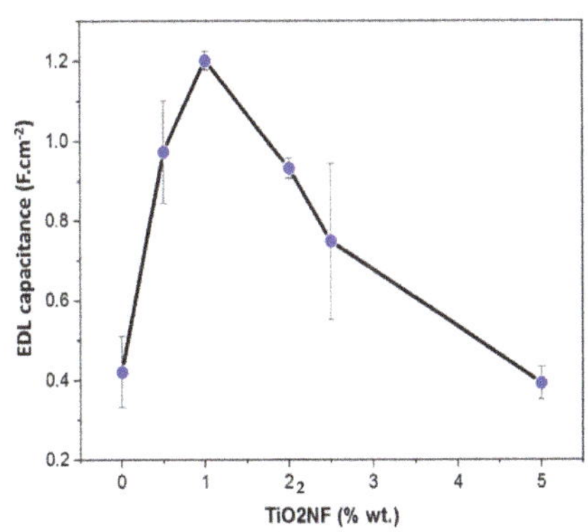

Figure 7. Electrical double-layer capacitance of the system for AC and $ACTiO_2NF$-x composite electrodes ($ACTiO_2NF$–x where x = 0.5, 1.0, 2.0, 2.5, and 5.0 wt.% TiO_2NF).

Thus, incorporating TiO_2NF at low rate into our carbon material enhances its double layer capacitance due to the formation of a uniform network distribution of TiO_2 nanofibers between AC particles. However, as the presence of TiO_2NF increases in the composite, there seem to be exhibition of poor electrochemical behavior possibly due to the increase in resistance to easy flow of ions into the pores of the electrode because of TiO_2NF agglomeration [14]. Our results correlate to the findings reported in literature [8,14].

3.6. Desalination Performance

As reported in literature, the electrosorption performance of carbon-based materials is linked to their capacitive properties amid other factors [9,29]. As explained earlier, TiO_2NF was added at different wt.% to influence the capacitance of commercial AC and this factor was verified through the electrochemical characteristics of the electrodes. Furthermore, in order to confirm the link between electrosorption performances and capacitive properties, desalination experiments were carried out under a semi-continuous system in a flow channel. Desalination was conducted at an operational cell potential $\Delta E = 1.2$ V for 30 min using 5000 mg·L^{-1} NaCl as the feed solution. Difference in conductivity was monitored and recorded during the experiment. FCDI performance indicators such as desalination efficiency (DE) in Figure 8a, salt removal rate (SRR), and charge efficiency (CE) in Figure 8b were used to evaluate the performance of the flow electrodes.

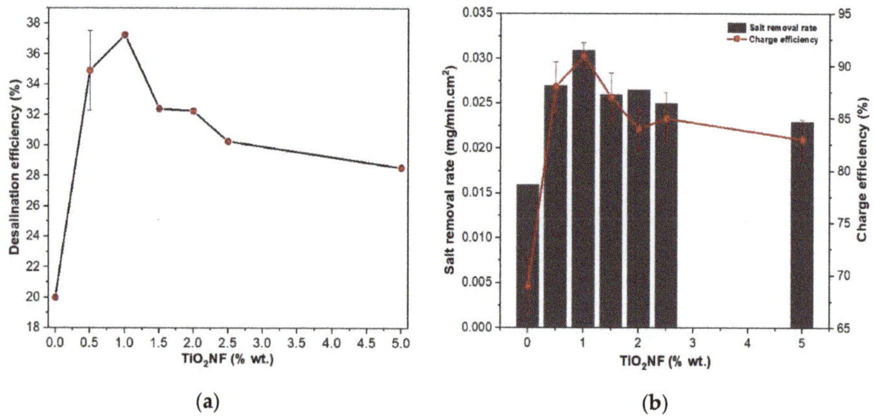

Figure 8. (**a**) Desalination efficiency; (**b**) salt removal rate and charge efficiency of AC and ACTiO_2NF−x, where (x = 0.5, 1.0, 1.5, 2.0, 2.5, and 5.0 wt.% TiO_2NF) respectively.

Important increase in DE is noticed in Figure 8a. Notably, the ACTiO_2NF electrodes show higher DE than pristine AC and among all the electrodes, ACTiO_2NF-1.0 exhibited the best DE, which implies fast ion mobility to the pores of the flow electrode thus leading to quick salt removal; a consequent effect on CE and SRR.

Low desalination behavior of ACTiO_2NF-x ($x \geq 2.0$ wt.% TiO_2NF) at high percentage could be due to the fact that at this ratio, the nanofibers tend to agglomerate among themselves (not well dispersed or less uniform) and as such, making the pores of the electrode (surface hindrance) not easily accessible for ions adsorption. In addition, electrode passivation is likely to occur due to the high presence of the nanofibers [32]. This will make the electrode surface not easily permeable for ions (impermeable layer formation on electrode surface) thus significantly affecting the electrosorption performance of the electrodes as observed in our materials, (whitish layers on the surface of flow electrodes). It has to be mentioned that only a small fraction of titanium nanofiber ($\leq$1% weight) is sufficient to significantly increase FCDI performances without increasing much viscosity.

4. Conclusions

To summarize, we used an electrospinning procedure to create titanium nanofibers. Without any further post-treatment, the as-spun nanofibers were rationally mixed with AC using a simple agitating procedure to generate hybrid composites of ACTiO_2NF-x (x = 0.5,

1.0, 1.5, 2.0, 2.5, and 5.0 wt.% ACTiO_2NF). The composites were subsequently described and tested for the first time as electrodes in FCDI. Introduction of TiO_2 nanofibers into AC improved the electrochemical characteristics of the material. The addition of TiO_2 nanofibers to the composite electrodes increased their performance, with ACTiO_2NF-1.0 demonstrating distinguishing and remarkable properties with the best desalination behavior. The composites' nanofibrous shape allows for greater anchoring within the AC network, allowing for improved ion transport and migration to the pores. Finally, the technology described has the potential to produce carbon-based flow electrodes with better shape and desalination performance in the FCDI methodology.

Supplementary Materials: The following are available online at https://www.mdpi.com/article/10.3390/ma14226891/s1. Figure S1: EDX mapping of (a) ACTiO_2NF-5.0 and (b) ACTiO_2NF-1.0.

Author Contributions: Conceptualization, G.F.; methodology, G.F. and M.B.; software, G.F. and M.T.; validation, M.B., F.Z., P.S. and M.C.; investigation, G.F.; resources, F.Z., M.B., P.S. and M.C.; data curation, G.F. and M.T.; writing—original draft preparation, G.F. and M.T.; writing—review and editing, G.F., M.T., F.Z., M.B. and M.C.; visualization, G.F., M.T., F.Z. and M.B.; supervision, M.B., F.Z., P.S. and M.C.; project administration, M.B., F.Z. and M.C. All authors have read and agreed to the published version of the manuscript.

Funding: This research was funded by "Axe H_2O" and "Axe Energie" in European Institute of Membranes (IEM). And the Federal Government of Nigeria through Tertiary Education Trust fund (TETFUND) and Campus France (CF) for the Ph.D. funding of Gbenro Folaranmi with the funding number CAMPUS FRANCE 914886H.

Data Availability Statement: The data presented in this study are available on request from the corresponding authors.

Acknowledgments: Special thanks to Fida Tanos (IEM) for guidance in titanium nanofibers synthesis and Mahmoud Abid (IEM) for furnace operation.

Conflicts of Interest: The authors declare no conflict of interest.

References

1. Folaranmi, G.; Bechelany, M.; Sistat, P.; Cretin, M.; Zaviska, F. Towards Electrochemical Water Desalination Techniques: A Review on Capacitive Deionization, Membrane Capacitive Deionization and Flow Capacitive Deionization. *Membranes* **2020**, *10*, 96. [CrossRef]
2. Ahmed, M.A.; Tewari, S. Capacitive deionization: Processes, materials and state of the technology. *J. Electroanal. Chem.* **2018**, *813*, 178–192. [CrossRef]
3. Biesheuvel, P.M.; van der Wal, A. Membrane capacitive deionization. *J. Membr. Sci.* **2010**, *346*, 256–262. [CrossRef]
4. Zhang, C.; Ma, J.; Wu, L.; Sun, J.; Wang, L.; Li, T.; Waite, T.D. Flow Electrode Capacitive Deionization (FCDI): Recent Developments, Environmental Applications, and Future Perspectives. *Environ. Sci. Technol.* **2021**, *55*, 4243–4267. [CrossRef]
5. Qu, D. Studies of the activated carbons used in double-layer supercapacitors. *J. Power Sources* **2002**, *109*, 403–411. [CrossRef]
6. Porada, S.; Zhao, R.; van der Wal, A.; Presser, V.; Biesheuvel, P.M. Review on the science and technology of water desalination by capacitive deionization. *Prog. Mater. Sci.* **2013**, *58*, 1388–1442. [CrossRef]
7. Zou, L.; Li, L.; Song, H.; Morris, G. Using mesoporous carbon electrodes for brackish water desalination. *Water Res.* **2008**, *42*, 2340–2348. [CrossRef] [PubMed]
8. Yasin, A.S.; Mohamed, I.M.A.; Mousa, H.M.; Park, C.H.; Kim, C.S. Facile synthesis of TiO_2/ZrO_2 nanofibers/nitrogen co-doped activated carbon to enhance the desalination and bacterial inactivation via capacitive deionization. *Sci. Rep.* **2018**, *8*, 541. [CrossRef]
9. El-Deen, A.G.; Boom, R.M.; Kim, H.Y.; Duan, H.; Chan-Park, M.B.; Choi, J.-H. Flexible 3D Nanoporous Graphene for Desalination and Bio-decontamination of Brackish Water via Asymmetric Capacitive Deionization. *ACS Appl. Mater. Interfaces* **2016**, *8*, 25313–25325. [CrossRef]
10. Ao, Y.; Xu, J.; Fu, D.; Shen, X.; Yuan, C. Low temperature preparation of anatase TiO_2-activated carbon composite film. *Appl. Surf. Sci.* **2008**, *254*, 4001–4006. [CrossRef]
11. Ao, Y.; Xu, J.; Fu, D.; Yuan, C. A simple route for the preparation of anatase titania-coated magnetic porous carbons with enhanced photocatalytic activity. *Carbon* **2008**, *46*, 596–603. [CrossRef]
12. Xing, B.; Shi, C.; Zhang, C.; Yi, G.; Chen, L.; Guo, H.; Huang, G.; Cao, J. Preparation of TiO_2/Activated Carbon Composites for Photocatalytic Degradation of RhB under UV Light Irradiation. *J. Nanomater.* **2016**, *2016*, e8393648. [CrossRef]

13. Mohamed, I.M.A.; Dao, V.-D.; Barakat, N.A.M.; Yasin, A.S.; Yousef, A.; Choi, H.-S. Efficiency enhancement of dye-sensitized solar cells by use of ZrO_2-doped TiO_2 nanofibers photoanode. *J. Colloid. Interface Sci.* **2016**, *476*, 9–19. [CrossRef] [PubMed]
14. El-Deen, A.G.; Choi, J.-H.; Khalil, K.A.; Almajid, A.A.; Barakat, N.A.M. A TiO_2 nanofiber/activated carbon composite as a novel effective electrode material for capacitive deionization of brackish water. *RSC Adv.* **2014**, *4*, 64634–64642. [CrossRef]
15. Park, H.; Choi, J.; Yang, S.; Kwak, S.J.; Jeon, S.; Han, M.H.; Kim, D.K. Surface-modified spherical activated carbon for high carbon loading and its desalting performance in flow-electrode capacitive deionization. *RSC Adv.* **2016**, *6*, 69720–69727. [CrossRef]
16. Yang, S.; Choi, J.; Yeo, J.; Jeon, S.; Park, H.; Kim, D.K. Flow-Electrode Capacitive Deionization Using an Aqueous Electrolyte with a High Salt Concentration. *Environ. Sci. Technol.* **2016**, *50*, 5892–5899. [CrossRef]
17. Cho, Y.; Yoo, C.-Y.; Lee, S.W.; Yoon, H.; Lee, K.S.; Yang, S.; Kim, D.K. Flow-electrode capacitive deionization with highly enhanced salt removal performance utilizing high-aspect ratio functionalized carbon nanotubes. *Water Res.* **2019**, *151*, 252–259. [CrossRef]
18. Liang, P.; Sun, X.; Bian, Y.; Zhang, H.; Yang, X.; Jiang, Y.; Liu, P.-P. Optimized desalination performance of high voltage flow-electrode capacitive deionization by adding carbon black in flow-electrode. *Desalination* **2017**, *420*, 63–69. [CrossRef]
19. Martin, C.R. Nanomaterials: A Membrane-Based Synthetic Approach. *Science* **1994**, *266*, 1961–1966. [CrossRef]
20. Kiehl, J.T.; Trenberth, K.E. Earth's Annual Global Mean Energy Budget. *Bull. Am. Meteor. Soc.* **1997**, *78*, 197–208. [CrossRef]
21. Pant, B.; Park, M.; Park, S.-J. TiO_2 NPs Assembled into a Carbon Nanofiber Composite Electrode by a One-Step Electrospinning Process for Supercapacitor Applications. *Polymers* **2019**, *11*, 899. [CrossRef] [PubMed]
22. Xiang, Q.; Yu, J.; Jaroniec, M. Nitrogen and sulfur co-doped TiO_2 nanosheets with exposed {001} facets: Synthesis, characterization and visible-light photocatalytic activity. *Phys. Chem. Chem. Phys.* **2011**, *13*, 4853–4861. [CrossRef] [PubMed]
23. Wang, C.; Shao, C.; Zhang, X.; Liu, Y. SnO_2 Nanostructures-TiO_2 Nanofibers Heterostructures: Controlled Fabrication and High Photocatalytic Properties. *Inorg. Chem.* **2009**, *48*, 7261–7268. [CrossRef]
24. Sizochenko, N.; Syzochenko, M.; Gajewicz, A.; Leszczynski, J.; Puzyn, T. Predicting Physical Properties of Nanofluids by Computational Modeling. *J. Phys. Chem. C* **2017**, *121*, 1910–1917. [CrossRef]
25. Yu, F.; Chen, Y.; Liang, X.; Xu, J.; Lee, C.; Liang, Q.; Tao, P.; Deng, T. Dispersion stability of thermal nanofluids. *Prog. Nat. Sci. Mater. Int.* **2017**, *27*, 531–542. [CrossRef]
26. Kaggwa, A.; Carson, J.K.; Atkins, M.; Walmsley, M. The effect of surfactants on viscosity and stability of activated carbon, alumina and copper oxide nanofluids. *Mater. Today Proc.* **2019**, *18*, 510–519. [CrossRef]
27. Kriegseis, S.; Vogl, A.Y.; Aretz, L.; Tonnesen, T.; Telle, R. Zeta potential and long-term stability correlation of carbon-based suspensions for material jetting. *Open Ceram.* **2020**, *4*, 100037. [CrossRef]
28. Hill, A.; Carrington, S. Understanding the links between rheology and particle parameters. *Am. Lab. News* **2010**, *93*, 199–203.
29. Folaranmi, G.; Bechelany, M.; Sistat, P.; Cretin, M.; Zaviska, F. Comparative Investigation of Activated Carbon Electrode and a Novel Activated Carbon/Graphene Oxide Composite Electrode for an Enhanced Capacitive Deionization. *Materials* **2020**, *13*, 5185. [CrossRef]
30. Hatzell, K.B.; Hatzell, M.C.; Cook, K.M.; Boota, M.; Housel, G.M.; McBride, A.; Kumbur, E.C.; Gogotsi, Y. Effect of Oxidation of Carbon Material on Suspension Electrodes for Flow Electrode Capacitive Deionization. *Environ. Sci. Technol.* **2015**, *49*, 3040–3047. [CrossRef]
31. Suryanto, B.H.R.; Chen, S.; Duan, J.; Zhao, C. Hydrothermally Driven Transformation of Oxygen Functional Groups at Multiwall Carbon Nanotubes for Improved Electrocatalytic Applications. *ACS Appl. Mater. Interfaces* **2016**, *8*, 35513–35522. [CrossRef] [PubMed]
32. Hanssen, B.L.; Siraj, S.; Wong, D.K.Y. Recent strategies to minimise fouling in electrochemical detection systems. *Rev. Anal. Chem.* **2016**, *35*, 1–28. [CrossRef]

materials

MDPI

Review

$BiFeO_3$-Based Relaxor Ferroelectrics for Energy Storage: Progress and Prospects

Bipul Deka [1,2,3,*] and Kyung-Hoon Cho [1,2,*]

1 Research Institute of Advanced Materials, Kumoh National Institute of Technology, Gumi 39177, Korea
2 School of Materials Science and Engineering, Kumoh National Institute of Technology, Gumi 39177, Korea
3 Department of Physics, Pub Kamrup College, Kamrup, Assam 781381, India
* Correspondence: bipul.deka@kumoh.ac.kr (B.D.); khcho@kumoh.ac.kr (K.-H.C.)

Abstract: Dielectric capacitors have been widely studied because their electrostatic storage capacity is enormous, and they can deliver the stored energy in a very short time. Relaxor ferroelectrics-based dielectric capacitors have gained tremendous importance for the efficient storage of electrical energy. Relaxor ferroelectrics possess low dielectric loss, low remanent polarization, high saturation polarization, and high breakdown strength, which are the main parameters for energy storage. This article focuses on a timely review of the energy storage performance of $BiFeO_3$-based relaxor ferroelectrics in bulk ceramics, multilayers, and thin film forms. The article begins with a general introduction to various energy storage systems and the need for dielectric capacitors as energy storage devices. This is followed by a brief discussion on the mechanism of energy storage in capacitors, ferroelectrics, anti-ferroelectrics, and relaxor ferroelectrics as potential candidates for energy storage. The remainder of this article is devoted to reviewing the energy storage performance of bulk ceramics, multilayers, and thin films of $BiFeO_3$-based relaxor ferroelectrics, along with a discussion of strategies to address some of the issues associated with their application as energy storage systems.

Keywords: energy storage; $BiFeO_3$; relaxor ferroelectrics; domain engineering; polymorphic nanodomain

Citation: Deka, B.; Cho, K.-H. $BiFeO_3$-Based Relaxor Ferroelectrics for Energy Storage: Progress and Prospects. *Materials* **2021**, *14*, 7188. https://doi.org/10.3390/ma14237188

Academic Editors: Marc Cretin, Sophie Tingry and Zhenghua Tang

Received: 14 October 2021
Accepted: 23 November 2021
Published: 25 November 2021

1. Introduction

Global warming poses potential threats to the planet Earth's future. The continuous burning of fossil fuels has increased the concentration of CO_2 and other greenhouse gases in the Earth's atmosphere, leading to a warmer atmosphere and climate change. In addition, the depletion of fossil fuel resources, the dominant candidate in the energy market, poses the risk of an energy crisis in a world where the number of consumers is increasing day by day. Considering the serious threats to the lives on Earth and the risk of an energy crisis posed by the use of fossil fuel resources, the transition to clean energy is a serious consideration. However, renewable energy sources such as solar, wind, tides, and geothermal energy are intermittent by nature. Therefore, harnessing and storing renewable energy for future access is a challenging task.

Electrical energy harvested from renewable sources offers enormous opportunities for meeting future energy demands and the feasibility of the transition to clean energy. However, the usefulness of the electrical energy generated depends on its efficient storage, which is necessary for around-the-clock use. An efficient electrical energy storage (EES) system is the heart of the commercial and residential grid-based utilization of electrical energy. Therefore, the development of advanced EES systems is critically important for meeting the growing energy demands and effectively leveling the cyclic nature of such energy sources [1]. For over 200 years, batteries have been widely used in EES systems and are still being widely used. Solid oxide fuel cells (SOFCs), electrochemical capacitors (ECs), superconducting magnetic energy storage (SMES) systems, flywheels, and electrostatic capacitors (dielectric capacitors) are common current energy storage technologies [2].

A perfect energy storage device is characterized by high energy and power densities. A comparison of the storage efficiency of the technologically relevant candidates for EES systems can be realized from the Ragone plot shown in Figure 1, which displays the status of EES systems according to their energy and power densities. As can be seen, SOFCs and batteries exhibit a high energy density with low power density, while dielectric capacitors exhibit the opposite behavior, that is, high power density and low energy density. The ECs, SMES, and flywheel have medium power and energy densities. In addition to the high power and energy density, the charge/discharge rate is a deciding factor for EES systems. The energy storage and delivery in SOFCs/batteries are based on the chemical reaction process and may take 1–100 h of time. Dielectric capacitors typically exhibit fast charge/discharge rates, between µs and ms, whereas those for ECs, flywheels, and SMES are between 1 s and 1 h. The fast charge/discharge rate of dielectric capacitors is associated with the separation of comparatively fewer heavy bound charges under the influence of the electric field to be stored. Thus, to summarize, no individual EES candidate possesses both high power and energy densities simultaneously. Therefore, the technological relevance of each candidate as an EES may be determined by the final requirements. However, among the various potential candidates for EES, dielectric capacitors have the advantage of withstanding high-voltage and large-scale applications because of their lower cost [3,4].

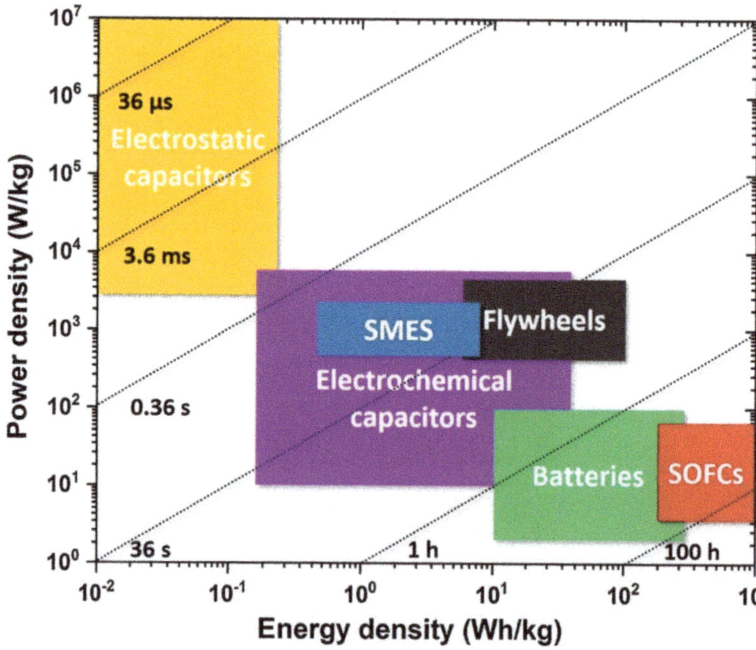

Figure 1. Ragone plot of various energy storage devices: electrostatic capacitors, electrochemical capacitors, SMES, flywheels, batteries, and SOFCs. The straight dashed lines and the associated times correspond to the characteristic times. Reused with permission from [2]. © 2021 Elsevier Ltd.

Dielectric capacitors are one of the key components in modern communication technology, with applications in electronic circuits, warfare, distributed power systems, hybrid electric vehicles, clean energy storage, high-power applications, etc. [5–11]. Among the various EES systems, dielectric capacitors exhibit the fastest discharge speed; therefore, they can generate intense pulse power [11–14]. The fast discharge speed and high fatigue resistance of dielectric capacitors enable their potential applications in various electronic systems. This includes medical equipment such as defibrillators, pacemakers, surgical lasers, and X-ray units; scientific research equipment such as high-power accelerators and high-intensity magnetic fields; commercial system devices such as camera flash, underground oil and gas exploration, avionics, transportation (hybrid cars, space shuttle power systems), transversely excited atmospheric lasers, and advanced electromagnetic systems [15].

The current review discusses the recent progress on the development of high-energy storage dielectric capacitors based on the relaxor ferroelectric (RFE) of $BiFeO_3$. The two important figures of a capacitor that determine its energy storage performance are the recoverable energy density (U_{rec}) and energy efficiency (η), which depend on the saturation polarization (P_{max}), remnant polarization (P_r), and breakdown strength (BDS) of the materials. Linear dielectric (LD), ferroelectric (FE), and anti-ferroelectric (AFE) materials are widely used for the fabrication of ceramic capacitors. Although the LDs possess excellent values of BDS, their U_{rec} values are quite low due to weak polarization. FE materials, on the other hand, possess quite a large P_{max} and P_r. As can be seen in Section 2, the large values of both P_{max} and P_r result in a low U_{rec} value. Moreover, the ferroelectric materials suffer large hysteresis loss, which has a significant detrimental effect on the energy efficiency of the capacitors. AFE materials behave like LDs in the low field regime, undergoing a field-induced FE state, yielding a high P_{max} at a high electric field with large hysteresis loss. Therefore, the issues associated with the AFE for their applications in the high-energy storage application are essentially similar to that of LDs at low electric fields and to that of FEs at high electric fields. However, the problem of hysteresis loss inherent to FE and AFE materials is found to be minimized in RFEs while maintaining a significantly large value of P_{max}. This motivates the scientific community to turn their heads towards the RFEs in search of high-energy storage capacitors.

Ceramic dielectric capacitors based on $BiFeO_3$ have recently gained interest in the field of energy storage applications because of the high polarization (~90 $\mu C\ cm^{-2}$) predicted in $BiFeO_3$, along with its high ferroelectric Curie temperature (T_C) (~830 °C) [16]. The advantage of having a high T_C is that the materials do not lose their ferroelectric (FE) nature at such high temperatures, which is essential for applications in the high-temperature regime. Temperature stability is an important issue that needs to be addressed while designing a capacitor for operation in the high-temperature regime. For example, ceramic-polymer composites have excellent storage performance; however, their performance degrades very rapidly as the temperature approaches 100 °C [15]. Pb-based ferroelectrics often have the disadvantage of adverse harmful effects on the environment and human beings. Pb-oxide, which is a main component of Pb-based ferroelectrics, is highly toxic and volatile at high temperatures, causing environmental pollution during the fabrication process. Disposal and recycling of Pb-based materials and devices at an industrial scale also creates atmospheric problems due to the difficulty of Pb removal. Exposure to the heavy metal Pb causes detrimental effects such as kidney and brain damages, and chronic exposure may lead to damages to the central nervous system and affect blood pressure, vitamin D metabolism, etc. Young children are more vulnerable to Pb exposure, as the absorption of Pb in children's bodies is 4–5 times higher than in an adult body. Therefore, to reduce the use of hazardous materials such as Pb, various countries have adopted different restrictions on hazardous materials [17,18]. Although there are a few reviews on dielectrics for energy storage in general, to the best of our knowledge, there has been no such review for $BiFeO_3$-based relaxor ferroelectrics. Here, we present a review of the recent progress on $BiFeO_3$-based relaxor ferroelectric for energy storage, discussing various issues to meet practical applications. We first discuss the fundamentals of energy storage in dielectrics and the pros and cons of various nonlinear dielectrics with respect to their applications in energy storage. We then discuss the characteristics of relaxor ferroelectrics and their importance in energy storage, followed by a brief discussion of the basic properties of $BiFeO_3$. Following this, we present the recent progress in energy storage studies on $BiFeO_3$ and strategies for further enhancement.

2. Fundamentals of the Energy Storage Mechanism in Dielectrics

The energy storage mechanism of dielectrics is based on their polarization under the application of an electric field. A dielectric under an applied electric field is polarized such that equal amounts of positive and negative charges accumulate at the surfaces of the dielectrics. In other words, an electric field opposite to the applied field is induced inside the dielectric. The strength of the induced field grows exponentially with time until its magnitude is equal to that of the external field. This process is known as charging the capacitor. Thus, the induced electrostatic energy is stored in the dielectric and can be used for application upon discharge through a load. The amount of stored energy (U) can be obtained from the potential difference (V) across the dielectrics and the charge (q) induced at the electrode on the surface of the dielectrics using the following equation:

$$U = \int_0^{q_{max}} V dq \tag{1}$$

where q_{max} is the maximum amount of charge accumulated at the electrode when the capacitor is fully charged, and dq is the increment of charge during charging. A figure-of-merit (FOM), which signifies the energy storage performance of a capacitor, is represented in terms of energy storage density (U_{st}), defined as the energy stored per unit volume. Mathematically,

$$U_{st} = \frac{\int_0^{q_{max}} V dq}{Ad} = \int_0^{D_{max}} E dD \tag{2}$$

where A is the electrode area of the capacitor, d is the distance between the electrodes (thickness of the dielectric layer), and D is the electric displacement of the capacitor. For a weak electric field, D is related to the external electric field (E) and polarization (P) as follows:

$$D = P + \varepsilon_0 E \tag{3}$$

where ε_0 represents the permittivity of a vacuum. Materials obeying Equation (3) are classified as linear dielectrics. For a linear dielectric, P is assumed to be a linear function of E:

$$P = \varepsilon_0 \chi E \tag{4}$$

where the quantity χ is termed the linear dielectric susceptibility. At a high electric field, it is necessary to consider the nonlinear contribution of susceptibility, and Equation (4) takes the most general form as:

$$P = \varepsilon_0 \left(\chi E + \chi^{(2)} E^2 + \chi^{(3)} E^3 + \ldots \right) \tag{5}$$

where $\chi^{(2)}$ and $\chi^{(3)}$ are higher-order susceptibilities, giving rise to nonlinear effects. Using linear approximation, the stored energy density of a dielectric material with a high dielectric constant ($D \approx P$) can be calculated as follows:

$$U_{st} = \int_0^{P_{max}} E dP \tag{6}$$

Equation (6) indicates that the electric polarization as a function of the electric field should be measured to calculate U_{st}. In other words, it is necessary to measure the polarization-electric field ($P-E$) hysteresis loop to obtain the stored energy density, as shown in Figure 2. Therefore, the shape and size of the $P-E$ loop and the nature of the dipole/domain structures determine the energy storage performance of dielectric materials. However, the dynamics of the polarization vector, growth of domains, and domain wall movements for $E = 0 \rightarrow E_{max}$ and $E = E_{max} \rightarrow 0$ directions in the $P-E$ measurement protocol are different from each other. This leads to a non-zero value of polarization, even at $E = 0$, known as remanent polarization (P_r). As a result, a part of the stored energy is lost, which appears as the hysteresis of the $P-E$ loop. In other words, it is impossible to recover the stored energy density to its fullest amount when it is discharged. The loss part of the stored energy or energy loss density (U_{loss}) is given by the area of the loop. The recoverable energy density is calculated as follows:

$$U_{rec} = \int_{P_r}^{P_{max}} EdP \tag{7}$$

Another FOM signifying the energy storage performance is the efficiency (η), which represents the amount of stored energy density available for use as recoverable energy density. It is defined as the ratio of the recoverable energy density to the total stored energy density:

$$\eta = \frac{U_{rec}}{U_{st}} \times 100\% = \frac{U_{rec}}{U_{rec} + U_{loss}} \times 100\% \tag{8}$$

Equation (7) suggests that a combination of high P_{max}, low P_r, and high breakdown strength are necessary to obtain a high U_{rec} value. In addition, Equation (8) requires a dielectric with low hysteresis loss to obtain a large efficiency value. Therefore, Equations (7) and (8) are often considered the governing equations for designing dielectric materials for high-performance energy storage. However, a dielectric with high ε usually features high dielectric loss, leading to heat generation during electric field cycling and the possibility of thermal breakdown during operation.

Typical P–E loops of LDs, FEs, AFEs, and RFEs are shown in Figure 2. LDs are characterized by very low values of polarization and a high BDS. Some of the widely studied LDs are $CaTiO_3$ [19,20], $SrTiO_3$ [21], and $CaTiO_3$-$CaHfO_3$ [22]. Because of the low value of polarization, the recoverable energy density of LD is quite low. Therefore, LDs are not suitable for application in the field of high-energy storage application. Over the years, FE and anti-ferroelectric (AFE) materials have been extensively studied for application in energy storage systems, and efforts to enhance their performance have surged. FE materials exhibit spontaneous polarization, a large value of P_{max}, and a coercive field (E_c). A typical P–E loop of FE materials is shown in Figure 2b. On the microscopic scale, FEs are composed of a large number of domains separated by domain walls. The dipoles in a domain are oriented in the same direction, and the directions of the domain polarizations can be switched by applying an electric field. However, the energy loss density is quite high in the FEs because of their high coercivity. Moreover, the P_r and P_{max} values have the same order of magnitude, resulting in a very small value of $P_{max}-P_r$. Therefore, the U_{rec} and η of the FEs are not promising. Because of this, single-phase FEs have not gained much interest in energy storage devices. Among the various FEs, representative compositions studied for energy storage are based on $(Bi,Na)TiO_3$ [23–26], $Ba(Zr,Ti)O_3$ [27–29], $BaTiO_3$ [30,31], and $(K,Na)NbO_3$ [32]. Unlike FEs, AFE materials lack a net polarization because of the anti-parallel alignment of the spontaneous polarization vectors in their domain. The typical $P-E$ loops of AFE materials are shown in Figure 2c. The electric dipoles align anti-parallel to each other in the AFE domain, as shown in the inset of Figure 2c. At a low electric field, the polarization of the AFE materials varies linearly with the applied field. At a sufficiently high electric field, the electric dipoles in a domain rotate to align in the parallel direction, and the AFE behaves similarly to an FE with a further increase in the field strength. This is known as the field-induced AFE-FE transition. Once the electric field is removed, the induced FE phase reverts to the AFE state, thereby producing double hysteresis in the $P-E$ loop. The high electric field for the AFE-FE phase transition (E_{AFE-FE}) coupled with the high P_{max} and low P_r indicates the possibility of achieving high storage capacity in AFE materials. The most intensively studied AFE systems are based on $(Pb, Zr)O_3$ [33–36], $(Bi, Na)TiO_3$ [37–39], and $AgNbO_3$ [40–42]. However, the E_{AFE-FE} for some AFEs is higher than their BDS at room temperature, signifying a breakdown before the transition to the highly polarized FE phase. Moreover, AFE materials cannot withstand large charge-discharge cycles, which is an important aspect in practical operations because such cycling leads the materials to undergo several alternate AFE-FE transitions, leading to physical cracks [43]. Moreover, the high-field FE phase often suffers from severe energy loss, which is particularly observed in $AgNbO_3$-based AFEs [44].

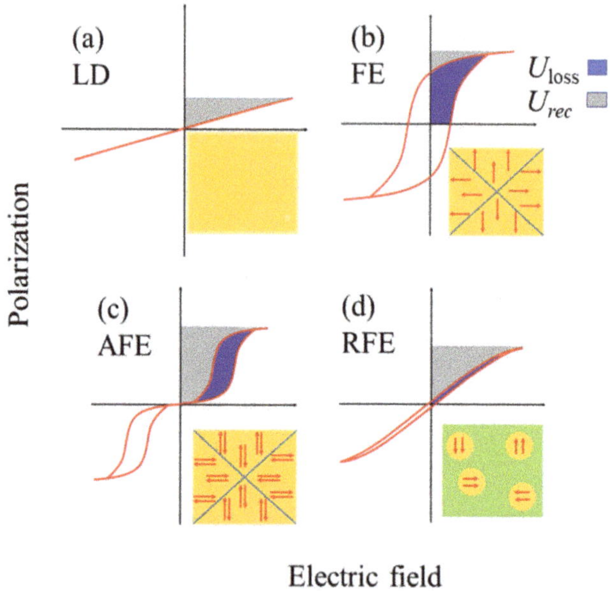

Figure 2. Typical *P–E* loops for (**a**) LD (**b**) FE, (c)AFE, and (**d**) RFE. Insets are schematics of domains with the alignment of polarization vectors (arrowheads). In RFEs, polar nanoregions (shown in circular patches) are sparingly distributed in a non-ferroelectric matrix (green area in the inset).

3. Relaxor Ferroelectrics

Relaxor ferroelectrics (RFEs) are an important class of materials that have attracted significant interest in energy storage applications. RFEs exhibit nanosized polar regions embedded in a nonpolar matrix. The polar nanoregions (PNRs) exhibit spontaneous polarization; however, the inter-PNR interaction is very weak [45]. The typical size of a PNR is 2–10 nm. PNRs are highly dynamic and sensitive to external stimuli. Because of the lack of inter-PNR interactions, PNRs under an electric field evolve independently of nearby PNRs. Therefore, the polarization state can return to its initial state after the electric field is removed. The $P-E$ loop for a typical RFE is shown in Figure 2d, which features $P_r \sim 0$, a considerably high P_{max}, and a small hysteresis loop.

RFEs feature (i) a broad maximum in ε around T_m (maximum temperature in the ε-T curve); (ii) strong frequency dispersion of ε and loss tangent (tan δ) peaks, i.e., shifting of the peaks toward higher temperatures while measured at lower to higher frequencies; and (iii) low P_r [46,47]. Therefore, RFEs are FE materials that simultaneously exhibit dielectric relaxation and ferroelectricity. However, unlike normal FEs, where the paraelectric-FE phase transition can be explained by Curie's law, the temperature dependence of χ in RFEs in the paraelectric phase obeys the following:

$$\chi = \frac{C}{(T - T_c)^{\gamma}} \tag{9}$$

where the parameter γ ($1 < \gamma < 2$) represents the broadness of the dielectric peak. For a normal FE, $\gamma = 1$. Several models have been proposed to explain the peculiar characteristics of RFEs, such as the diffuse phase transition model [48], super paraelectric model [49], dipolar glass model [50], random-field model [51], random-site model [52], bi-relaxation model [53], and spherical random-bond-random-field model [54]. However, the underlying mechanism of RFEs is yet to be clearly understood.

4. Energy Storage Performance of $BiFeO_3$-Based Relaxor Ferroelectrics

$BiFeO_3$ exhibits a distorted perovskite structure, as shown in Figure 3a. It possesses a rhombohedral structure (point group: R3c) at room temperature with an $a^-a^-a^-$ tilt system, in which the neighboring oxygen octahedra rotate anti-clockwise about the [111] direction [55,56]. The rhombohedral unit cell is described with lattice constants $a = b = c = 3.965$ Å and $\alpha = \beta = \gamma = 89.3$–$89.4°$ [57]. There are two formula units of $BiFeO_3$ in the rhombohedral cell, with three atoms in its asymmetric unit occupying Wyckoff positions: $6a$ (Bi^{3+} and Fe^{3+}) and $18b$ (O^{2-}) [58]. in a hexagonal frame of reference with the hexagonal c-axis parallel to the diagonals of the cubic perovskite with the lattice constants $a_{hex} = 5.58$ and $c = 13.90$ Å [57].

The $R3c$ symmetry permits long-range FE order in $BiFeO_3$ along the threefold axis [111]. Various experiments have confirmed the ferroelectricity in $BiFeO_3$ below $T_C = 1143$ K [55,59]. The constituent atoms Bi, Fe, and O are displaced from their centrosymmetric positions along the threefold axis, and Bi ions have the largest displacement with respect to O ions [55]. The lone-pair-active Bi ions in $BiFeO_3$ are displaced to a large extent in comparison with other FE compounds with non-lone-pair-active cations. Therefore, a large value of spontaneous polarization, on the order of 90 μC cm^{-2}, has been predicted in $BiFeO_3$ from ab initio calculations [16,60]. However, a polarization value close to the calculated values could not be obtained until recently [61], after a series of initial experimental failures to achieve spontaneous polarization of $BiFeO_3$, as predicted by theory [62–67]. Lebeugle et al. [61] measured a very large saturated polarization (approximately 60 μC cm^{-2}) in a high-quality single-crystal $BiFeO_3$ at room temperature, as shown in Figure 3b.

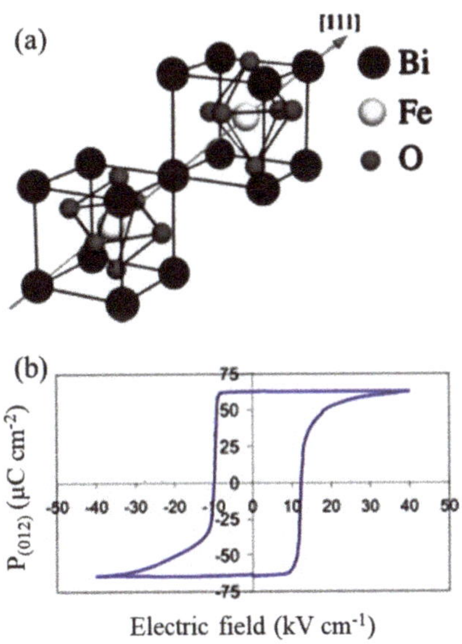

Figure 3. (**a**) Structure of R3c $BiFeO_3$. (**b**) P-E loop of $BiFeO_3$ bulk single crystal. Figures reproduced with permission from [55] and [61], respectively. © 2021 American Physical Society (**a**) and 2007 American Institute of Physics (**b**).

$BiFeO_3$ possesses the highest values of spontaneous polarization and T_C among Pb-free FEs. As discussed in Section 2, high P_{max} and large $P_{max} - P_r$ are among the most important factors for obtaining a high storage capacity. Therefore, $BiFeO_3$ with a large P_r in its naturally occurring FE phase is not suitable for energy storage applications, which is a drawback of all FE materials in general. One way to obtain a small P_r is to break the long-range FE order such that it becomes an RFE. Many researchers have reported this method to create PNRs embedded in a non-FE matrix and obtain a significant reduction in the P_r value. In addition, there are always issues pertaining to the leakage current in pure $BiFeO_3$, which eventually limits its high breakdown strength. Various successful methods to enhance the resistivity in phase-pure $BiFeO_3$, such as doping at the A-site and the addition of Mn, have been discussed in the literature. In the following sections, considering these issues, the energy storage performance of $BiFeO_3$-based materials, with special emphasis on the RFEs, are reviewed.

4.1. $BiFeO_3$-Based Binary System

$BiFeO_3-BaTiO_3$ solid solutions have been widely investigated as promising candidates in the field of ceramic dielectrics-based energy storage materials. A remarkable feature of $BiFeO_3$-based solid solutions is the morphotropic phase boundary (MPB), where the solid solution displays a composition-driven structural transition in its phase diagram. The crystal structure changes abruptly across the MPB, and various physical properties, such as piezoelectric coefficients and polarization, are maximal at the MPB. The MPB of the $BiFeO_3-BaTiO_3$ system is shown in the phase diagram in Figure 4, where the rhombohedral and tetragonal phases coexist in the MPB region [68–70]. Careful optimization of the $BaTiO_3$ content produces excellent ferroelectric and piezoelectric properties in MPB compositions.

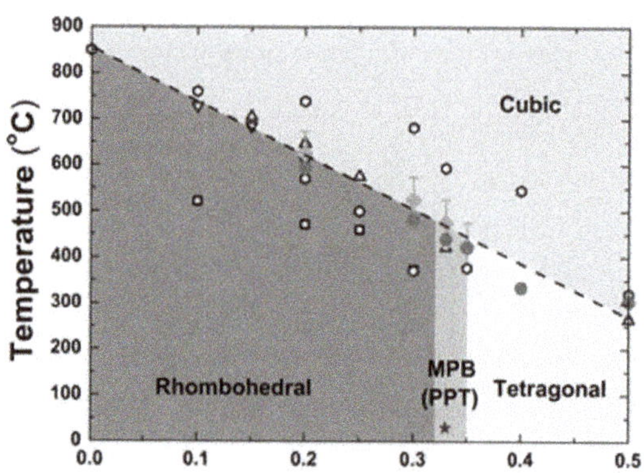

Figure 4. Phase diagram of (1-x)$BiFeO_3$-x$BaTiO_3$ solid solution. Reused with permission from [68]. © 2021 Willey-VCH Verlag GmbH & Co.

However, $BiFeO_3$-$BaTiO_3$ also possesses a high P_r [71] and high dielectric loss, which are detrimental for energy storage. Previous studies have shown that doping small amounts of La_2O_3, MnO_2, and Nb_2O_5 in $BiFeO_3$ can significantly enhance the electrical resistivity and energy loss density [72,73], which is beneficial for energy storage. Wang et al. [74] reported a large enhancement of resistivity in Nb_2O_5-modified $BiFeO_3$-$BaTiO_3$, i.e., (1-x)(0.65$BiFeO_3$-0.35$BaTiO_3$)–xNb_2O_5 (x = 0, 1, 3, 5 mol%) by several orders (~10^{10}–10^{14} Ω cm) compared with the undoped $BiFeO_3$-$BaTiO_3$ (~10^8 Ω cm). The compounds with x = 0.01 and 0.03 exhibited slimmer $P-E$ loops, similar to RFEs, with x = 0.03 exhibiting the highest P_{max} (25.21 μC cm^{-2}) and lowest P_r (5.53 μC cm^{-2}). They obtained a maximum U_{rec} of 0.71 J cm^{-3} at E = 90 kV cm^{-1}. Zhu et al. [75] found a significant improvement in the BDS of 0.52$BiFeO_3$-0.48$BaTiO_3$ ceramic up to 130–140 kV cm^{-1} by adding La_2O_3 and MnO_2. Under this condition, they obtained U_{rec} = 1.22 J cm^{-3} with

η = 58% for $0.52Bi_{0.98}La_{0.02}FeO_3$-$0.48BaTiO_3$ + 0.3 wt.% MnO_2 compound, whereas undoped $0.52BiFeO_3$-$0.48BaTiO_3$ exhibited U_{rec} = 1.08 J cm^{-3} with η = 49%. They found that the addition of La_2O_3 and MnO_2 increased the amount of the FE phase, reduced the grain size, and facilitated densification, which helped to induce large ΔP ($P_{max} - P_r$) as well as BDS compared with undoped compounds. This phenomenon was found to be more pronounced when Nd was substituted for Bi sites in MnO_2-added $BiFeO_3$-$BaTiO_3$ solid solution, as shown in Figure 5. Wang et al. [76] synthesized highly dense (relative density ρ_r = 95% to 97.6%) $0.75BiFeO_3$-$0.25BaTiO_3$ ceramics by Nd substitution with 0.1 wt.% MnO_2 addition that could endure a high electric field up to 180 kV cm^{-1}, as shown in Figure 5. The solid solution with 15 mol% Nd content featured the highest value of U_{st} = 4.1 J/cm^3 and U_{rec} = 1.82 J cm^{-3}; however, it had a low value of η = 41.3%. Recently, Chen et al. [77] successfully synthesized highly dense Sm-doped $BiFeO_3$-$BaTiO_3$ binary ceramics that can endure a very high electric field of up to 200 kV cm^{-1}. Sm substitution significantly reduced the grain size and enhanced the density, which is believed to be the reason for the high BDS. The binary solid solutions exhibit excellent U_{st} and U_{rec}; however, their low efficiency limits practical device applications.

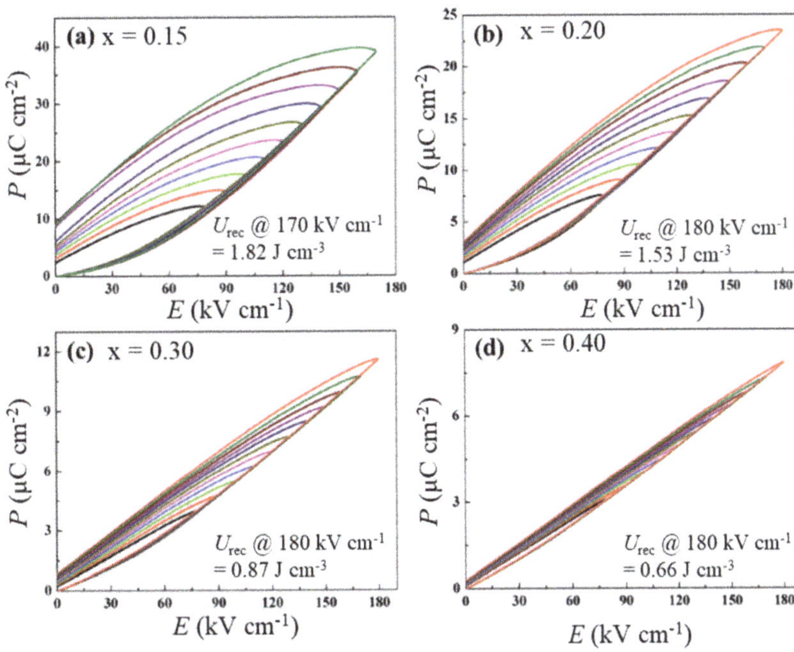

Figure 5. Unipolar P-E loops for $0.75Bi_{1-x}Nd_xFeO_3$-$0.25BaTiO_3$ + 0.1 wt% MnO_2 system: (**a**) x = 0.15, (**b**) x = 0.20, (**c**) x = 0.30, and (**d**) x = 0.40. Reproduced with permission from [76]. © 2021 The Royal Society of Chemistry.

4.2. $BiFeO_3$-Based Ternary System

4.2.1. Bulk Ceramics

The addition of a third perovskite oxide to binary $BiFeO_3$-$MTiO_3$ (M = Ba and/or Sr) as the end member of the ternary system has been found to be very promising for inducing the relaxor phase and enhancing the energy storage performance [78–84]. Zheng et al. [78] reported the successful induction of the relaxor phase in $BiFeO_3$-$BaTiO_3$ as a result of the substitution of $BaMg_{1/3}Nb_{2/3}O_3$. The relaxor $0.61BiFeO_3$-$0.33BaTiO_3$-$0.06BaMg_{1/3}Nb_{2/3}O_3$ exhibited a U_{rec} of 1.56 J cm^{-3} at E = 125 kV cm^{-1}, with η ~ 75%. This compound also exhibited good temperature stability for energy storage and efficiency in the temperature range of 25 °C to 190 °C. In a similar effort, Zheng et al. [79] reported an improved storage performance with a U_{rec} of 1.66 J cm^{-3} at 130 kV cm^{-1} and η ~ 82% in highly dense (ρ_r > 97%) 0.61BiFeO$_3$-0.33BaTiO$_3$-0.06LaMg$_{1/2}$Ti$_{1/2}$O$_3$ ceramics. Meanwhile, Liu et al. [80] reported enhancement of relaxor characteristics in terms of broader peaks of dielectric permittivity (Figure 6) and significant enhancement of energy storage performance in (0.66 − x)BiFeO$_3$-0.34BaTiO$_3$-xBaZn$_{1/3}$Ta$_{2/3}$O$_3$ for x > 0. They reported slim *P–E* loops for x > 0 with the highest BDS of E = 160 kV cm^{-1} and a high U_{rec} of 2.56 J cm^{-3} for the x = 0.06 composition (Figure 7). Tang et al. [84] reported a BDS of 180 kV cm^{-1} and U_{rec} = 1.62 J cm^{-3} for 0.85(0.65BiFeO$_3$-0.35BaTiO$_3$)-0.15Ba(Zn$_{1/3}$Nb$_{2/3}$)O$_3$ bulk ceramics. Sun et al. [85] reported similar values of U_{rec} = 2.11 J cm^{-3} at E = 195 kV cm^{-1} with η = 84% in a highly dense 0.56BiFeO$_3$-0.30BaTiO$_3$-0.14AgNbO$_3$+5 mol% CuO system prepared by a modified thermal quenching technique.

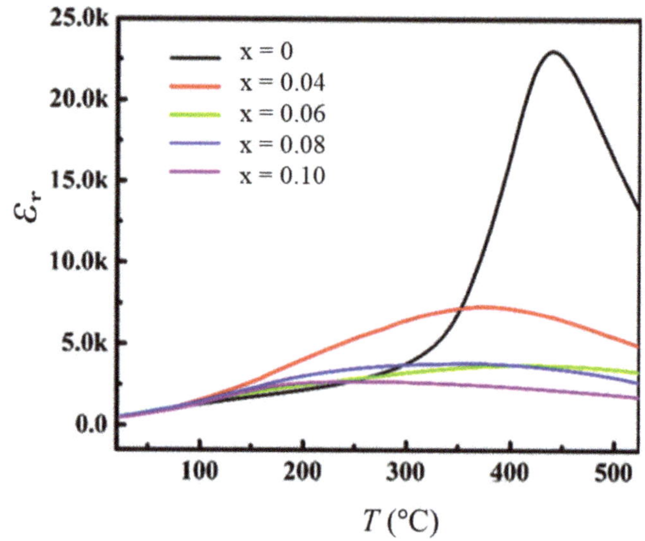

Figure 6. Induction of relaxor phase characterized by diffused phase transition in (0.66-x)BiFeO$_3$-0.34BaTiO$_3$-xBaZn$_{1/3}$Ta$_{2/3}$O$_3$ systems. Reproduced with permission from [80]. © 2021 The Royal Society of Chemistry.

Yu et al. [86] studied the effect of the microstructure on the energy storage performance of a BiFeO$_3$-BaTiO$_3$-Bi(Mg$_{2/3}$Nb$_{1/3}$)O$_3$ system. They prepared coarse-grained (grain size ~2 to 4 μm) and fine-grained (~0.55 to 0.9 μm) microstructures using planetary ball milling and high-energy ball milling processes, respectively. BiFeO$_3$-BaTiO$_3$-Bi(Mg$_{2/3}$Nb$_{1/3}$)O$_3$ solid solutions with a fine-grained microstructure exhibited higher ΔP (~ 30 μC cm^{-2}) and BDS (~ 110 kV cm^{-1}) than the coarse-grained samples (ΔP ~ 10 μC cm^{-2}, BDS ~ 50 kV cm^{-1}). Under such drastic microstructural evolution, they reported a higher U_{rec} of 1.26 J cm^{-3} in fine-grained samples, compared with U_{rec} = 0.16 J cm^{-3} for coarse-grained samples. Yang et al. [87] showed that utilizing a liquid-phase sintering mechanism can significantly enhance the BDS while maintaining the relaxor characteristics and high dielectric permittivity. Using 2 wt% BaCu(B$_2$O$_5$) (BCB) as the low melting point additive in 0.1 wt% MnO$_2$-added (0.67−x)BiFeO$_3$–0.33(Ba$_{0.8}$Sr$_{0.2}$)TiO$_3$-xLa(Mg$_{2/3}$Nb$_{1/3}$)O$_3$ solid solution, they achieved a BDS of 230 kV cm^{-1} for x = 0.06. With the use of such a high field, the compound exhibited U_{rec} = 3.38 J cm^{-3} with η = 59%. BCB formed large amounts of liquid phase at the grain boundaries during sintering and significantly reduced the average grain size down to the submicron range, as shown in Figure 8, by impeding grain growth at lower sintering temperatures. The high density of grain boundaries in the microstructure of the submicron grain size offered high electrical resistivity, resulting in enhanced BDS (Figure 8). Moreover, low-temperature sintering with the addition of MnO_2 helped to decrease $Fe^{3+} \leftrightarrow Fe^{2+}$ valence fluctuations by minimizing Bi_2O_3 loss during synthesis, which was also critical for enhancing the BDS. This compound exhibited good temperature stability, with U_{rec} = 1.15–1.27 J cm^{-3} in the temperature range of 30 °C to 170 °C. Using the PVA-assisted viscous polymer process (VPP) route, Liu et al. [88] obtained high ρ_r ~ 99% 15 mol% $Sr_{0.7}Bi_{0.3}FeO_3$–modified 0.85(0.65$BiFeO_3$-0.35$BaTiO_3$) system with a fine grain microstructure. The ultra-high ρ_r and uniform submicron grains significantly enhanced the BDS, with a value of 330 kV cm^{-1}, compared with that of the sample prepared without VPP (180 kV cm^{-1}). Under this condition, they obtain an ultra-high U_{rec} of 4.95 J cm^{-3} with η ~ 73%. The calculated U_{rec} from the charge-discharge cycling test was 2.36 J cm^{-3} at 300 kV cm^{-1}. Likewise, Sm doping has been found to be very effective in reducing the grain size and increasing the density of sintered ceramics. Chen et al. [78] reported a significant decrease in the average grain size in the 0.67$Bi_{1-x}Sm_xFeO_3$-0.33$BaTiO_3$ system with an increase in x, resulting in a BDS of 200 kV cm^{-1} at x = 0.1. This resulted in an U_{rec} = 2.8 J cm^{-3} for x = 0.1; however, it had low efficiency, η = 55.8%. In another report, Li et al. [89] reported an ultra-high U_{rec} = 3.2 J cm^{-3} at E = 206 kV cm^{-1} with high efficiency, η = 92%, in a highly dense (ρ_r ~ 98%) solid solution of (1−x)$Bi_{0.83}Sm_{0.17}Fe_{0.95}Sc_{0.05}O_3$-x(0.85$BaTiO_3$-0.15Bi($Mg_{0.5}Zr_{0.5}$)$O_3$) with x = 0.75. A similar $U_{rec} \approx$ 3.06 J cm^{-3} at E = 167 kV cm^{-1} with $\eta \approx$ 92% associated with the improvement of BDS has been reported in (1−x)$BiFeO_3$-x(0.85$BaTiO_3$-0.15Bi ($Sn_{0.5}Zn_{0.5}$)O_3) with x = 0.65 [90].

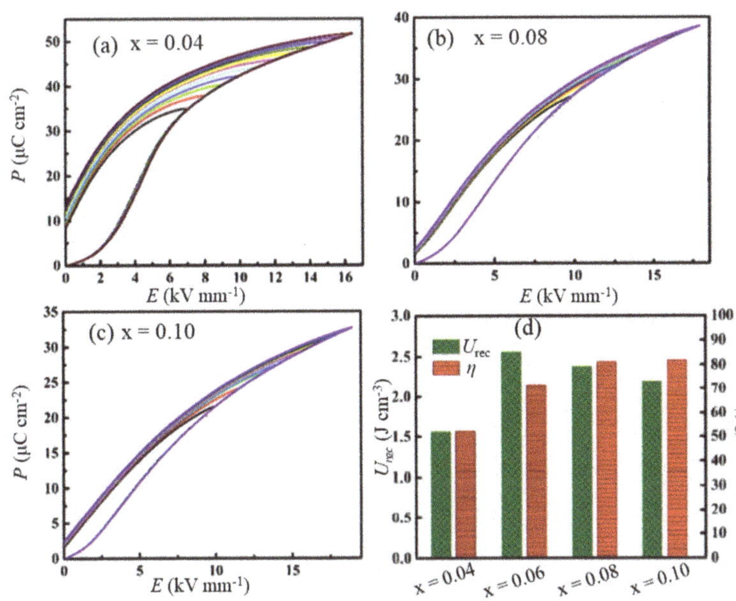

Figure 7. $BaZn_{1/3}Ta_{2/3}O_3$ modification to $BiFeO_3$-$BaTiO_3$ leads to slimmer *P–E* loops. Unipolar *P–E* loops for (0.66-x)$BiFeO_3$-0.34$BaTiO_3$-x$BaZn_{1/3}Ta_{2/3}O_3$ with (**a**) x = 0.04, (**b**) x = 0.08, and (**c**) x = 0.10. (**d**) The optimum recoverable energy density and efficiencies. Reproduced with permission from [80]. © 2021 The Royal Society of Chemistry.

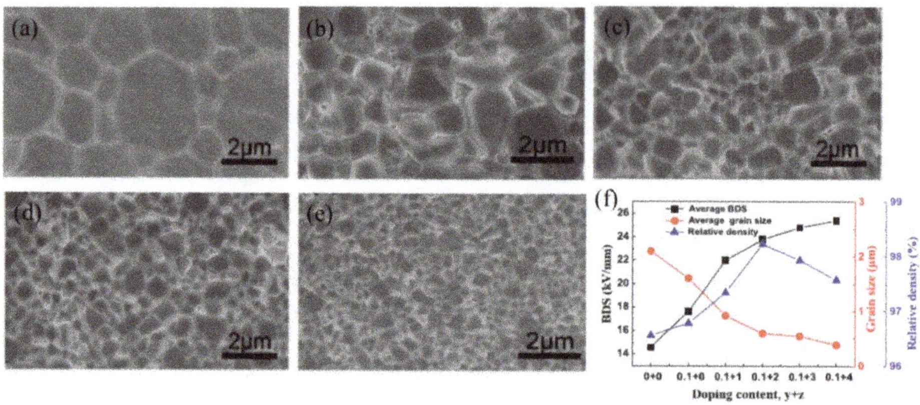

Figure 8. SEM images of $0.61BiFeO_3$-$0.33(Ba_{0.8}Sr_{0.2})TiO_3$-$0.06La(Mg_{2/3}Nb_{1/3})O_3$ + y wt.% MnO_2 + z wt.% $BaCu(B_2O_5)$ solid solution sintered at their optimum temperatures: (**a**) y = z = 0, (**b**) y = 0.1 and z = 1, (**c**) y = 0.1 and z = 2, (**d**) y = 0.1 and z = 3, and (**e**) y = 0.1 and z = 4. (**f**) Variations of BDS, grain size, and relative density with y + z. Reused with permission from [87]. © 2021 Elsevier Ltd.

Designing a specific microstructure has been found to be highly effective for enhancing energy storage performance. Microstructure strongly influences the BDS of dielectric materials and their relaxor characteristics. Wang et al. [81] designed core-shell microstructure (Figure 9a) with $BaTiO_3$-rich shells and $BiFeO_3$-rich cores in (0.7-x)$BiFeO_3$-$0.30BaTiO_3$-x$Bi(Zn_{2/3}Nb_{1/3})O_3$+0.1wt%Mn_2O_3 ceramics that could withstand an electric field as high as 190 kV cm^{-1}. They found that the shells and cores in the solid solutions were different in structure and electrical characteristics. The shell exhibited a pseudo-cubic structure, which is paraelectric in nature, whereas the core parts had a ferroelectric R3c structure. The core-shell structure and the cationic charge disorder at the B-sites helped to establish the relaxor phase in $0.70BiFeO_3$-$0.30BaTiO_3$ (Figure 9a). Substitution of $Bi(Zn_{2/3}Nb_{1/3})O_3$ further exacerbated the long-range order, thereby inducing a highly disordered RFE phase while maintaining the polarizability as high as P_{max} = 36.7 μC cm^{-2} and ΔP = 32.8 μC cm^{-2}. They obtained U_{st} = 3.7 J cm^{-3}, U_{rec} = 2.06 J cm^{-3} at 180 kV cm^{-1} and U_{st} = 2.9 J cm^{-3}, U_{rec} = 1.98 J cm^{-3} at 190 kV cm^{-1} for 5 mol% and 8 mol% $Bi(Zn_{2/3}Nb_{1/3})O_3$-doped compounds, respectively. The solid solutions could also successfully deliver discharge energy within 0.5 μs. Wang et al. [82] reported relaxor behavior in chemically inhomogeneous, but electrically homogeneous (0.7-x)$BiFeO_3$-$0.3BaTiO_3$-x$Nd(Zr_{0.5}Zn_{0.5})O_3$ for x = 0.05, 0.08, and 0.10, with a core-shell structure, and studied the energy storage performance at room temperature both in ceramics and multilayers prepared by a solid-state reaction route. The substitution of $Nd(Zr_{0.5}Zn_{0.5})O_3$ induced chemical inhomogeneity, revealed as a (Bi, Fe)-rich core for x = 0.05 and (Ba,Ti)-rich cores for x = 0.1, and complex multiphase microstructures with both (Bi, Fe)-rich and (Ba,Ti)-rich cores for x = 0.08. A high degree of chemical inhomogeneity and better electrical homogeneity of grains existed at x = 0.08, which led to a high ΔP and provided a difficult current path for electrical breakdown. As a result, x = 0.08 exhibited U_{rec} ~2.45 J cm^{-3} at E = 240 kV cm^{-1} with η = 72%. Lu et al. [91] demonstrated excellent energy storage properties in a series of solid solutions composed of $BiFeO_3$, $SrTiO_3$, Nb_2O_5, and $BiMg_{2/3}Nb_{1/3}O_3$ exhibiting a core-shell structure. Their study successfully demonstrated that Nb_2O_5 doping into $BiFeO_3$-$SrTiO_3$ and employing the third perovskite end member $BiMg_{2/3}Nb_{1/3}O_3$ (BMN) could induce an insulating relaxor phase at room temperature without reducing the average ionic polarizability of the solid solution. It was found that the substitution of 1–3% Nb^{5+} for Ti^{4+} (B sites) in $0.6BiFeO_3$-$0.4SrTiO_3$ suppressed the formation of oxygen vacancies and significantly reduced p-type conductivity compared with that of the undoped compound. A similar reduction in the p-type conductivity was observed for the $0.56BiFeO_3$-$0.4SrTiO_3$-$0.04BiMg_{2/3}Nb_{1/3}O_3$-x$Nb_2O_5$

(x = 0–0.05) solid solution, and an enhanced BDS of 360 kV cm^{-1} was obtained at an optimized Nb_2O_5 content (x = 0.03). Such a high BDS was attributed to two factors: (i) the improved insulating character caused by Nb doping at B-sites related to the suppression of the formation of oxygen vacancies, and (ii) a core-shell microstructure with electrical homogeneity throughout the grains. The $x = 0.03$ composition exhibited a high performance of U_{rec} = 6 J cm^{-3} with η = 74.6%. Then, a much-improved U_{rec} value with similar η could be achieved by optimizing the $BiMg_{2/3}Nb_{1/3}O_3$ content in (0.6-y)$BiFeO_3$-0.4$SrTiO_3$-0.03Nb_2O_5-y$BiMg_{2/3}Nb_{1/3}O_3$ (y = 0.02–0.12) solid solutions. For y = 0.1, the BDS was enhanced further up to 460 kV cm^{-1}, producing U_{rec} = 8.2 J cm^{-3} and η = 74.6%. A core-shell microstructure design provides a large BDS and large U_{rec}, but at the same time, it has low efficiency because of the not-very-fast response of the PNRs to the electric field.

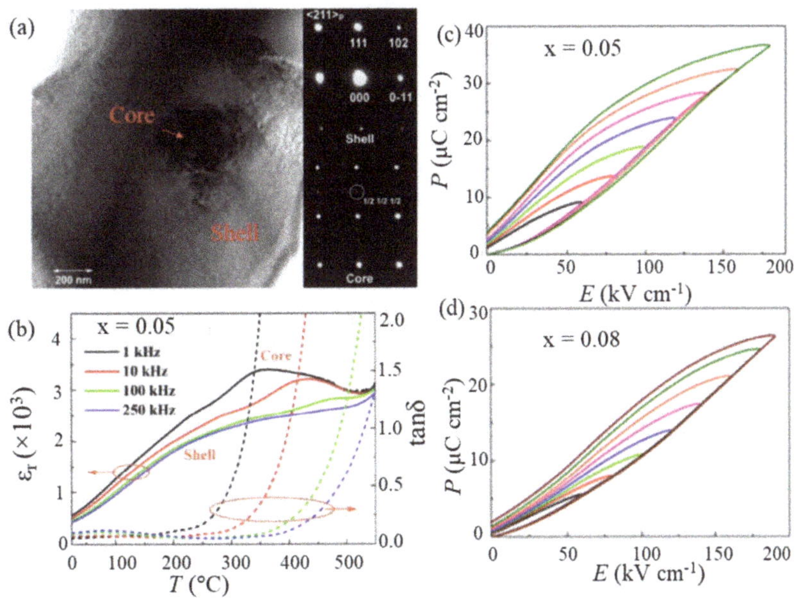

Figure 9. (**a**) Bright-field TEM image of a grain in (0.70-x)$BiFeO_3$-0.30$BaTiO_3$-x$Bi(Zn_{2/3}Nb_{1/3})O_3$ with x = 0.05, illustrating a $BiFeO_3$-rich core and $BaTiO_3$-rich shell. (**b**) Broad dielectric anomalies associated with the core-shell structure. P-E loops for (**c**) x = 0.05 and (**d**) x = 0.08, respectively. Reproduced with permission from [81]. © 2021 American Chemical Society.

Qi et al. [92] employed a domain engineering technique to optimize the microstructure at the domain level and showed that this technique was very effective for enhancing U_{rec} and η. Using it, they obtained large P_{max}, large ΔP, and superior energy storage performance in 0.57$BiFeO_3$-0.33$BaTiO_3$-0.1$NaNbO_3$ with the addition of 0.1 wt% MnO_2 and 2 wt% $BaCu(B_2O_5)$. Nanodomain engineering produced stripe-like PNRs embedded in featureless nanodomains. The stripe-like PNRs were rich in $BiFeO_3$, whereas the matrix domains were rich in $BaTiO_3$ and $NaNbO_3$. An HR-TEM image of the solid solution is shown in Figure 10a. Stripe-like PNRs rich in $BiFeO_3$ were dispersed in $BaTiO_3$- and $NaNbO_3$-rich featureless nanodomains. This structural heterogeneity at the domain level led to a rapid polarization response to the external E-field, unlike in the core-shell microstructure, and produced a hysteresis-free $P-E$ loop, as well as a large P_{max}. Moreover, $NaNbO_3$ substitution increased the bandgap and helped to obtain a uniform and fine-grained microstructure, which was beneficial to enhance BDS up to 360 kV cm^{-1}, producing $U_{rec} \approx 8.12$ J cm^{-3} with $\eta \approx 90\%$. This solid solution also exhibited excellent thermal stability, with U_{rec} = 8.12 J cm^{-3} ± 10% in the temperature range −25 °C to 250 °C, and an ultrafast discharge rate (< 100 ns).

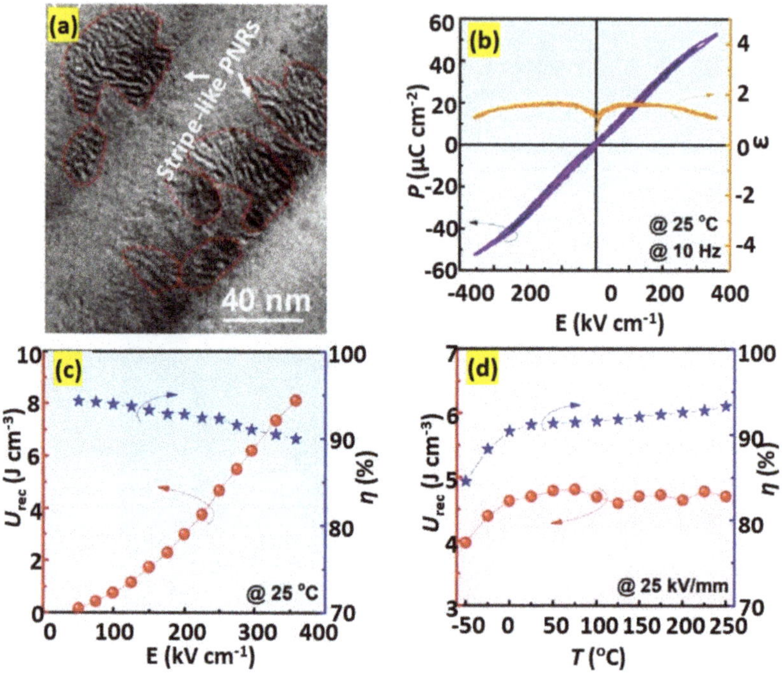

Figure 10. (**a**) Domain morphology exhibiting stripe-like PNRs in $0.57BiFeO_3$-$0.33BaTiO_3$-$0.1NaNbO_3$ with the addition of 0.1 wt.% MnO_2 and 2 wt.% $BaCu(B_2O_5)$. (**b**) Room-temperature *P–E* loops and ε measured under various electric fields for the same compound. U_{rec} and η measured at (**c**) various electric fields and (**d**) temperatures. Figures reused with permission from [92]. © 2021 WILEY-VCH Verlag GmbH & Co. KGaA, Weinheim.

4.2.2. Thin Films

In the past few decades, thin films have gained enormous scientific importance in a plethora of applications. The obvious advantages thin films possess over their bulk counterparts are that they save material and have reduced weight. In addition, it is well known that material properties drastically change when they are deposited in thin film forms, which paves the way for industrial applications. Dielectric thin films with thicknesses on the nano- or submicron scale have shown promising potential in the field of energy storage for low-power small devices. This is because of their extraordinarily high BDS (1 MV/cm) and energy density compared with bulk dielectrics.

Correia et al. [93] deposited a thin film of $0.4BiFeO_3$-$0.6SrTiO_3$ onto a $SrRuO_3$-electroded (100)-$SrTiO_3$ substrate using the pulsed laser deposition (PLD) method, which can endure an electric field as large as 0.972 MV cm^{-1}. Under such a high electric field, the film exhibited U_{rec} = 18.6 J cm^{-3} with $\eta > 85\%$. It also featured a small temperature coefficient of capacitance (TCC < 11%) over a wide range of temperatures up to 200 °C. Pan et al. [94] demonstrated that the BDS of the same material could be enhanced more than by three times, up to 3.6 MV cm^{-1}, when doped with Mn and deposited onto a Nb-doped $SrTiO_3$ (001) substrate by the PLD technique. The high BDS of the Mn-doped $0.4BiFeO_3$-$0.6SrTiO_3$ thin film was attributed to (i) a dense, uniform, and crack-free microstructure, (ii) high-quality epitaxial growth, and (iii) low leakage current by reducing the Fe^{3+}/Fe^{2+} valence fluctuation with Mn substitution. They obtained a colossal U_{rec} = 51 J cm^{-3} and η = 64%, which are comparable to those of lead-based thin films (U_{rec} = 61 J cm^{-3} and η = 33%) [95]. The thin film also featured high fatigue endurance quality over 2×10^7 cycles and good thermal stability over a wide range of temperatures, from −40 °C to 140 °C. In another report, Pan et al. [96] utilized a domain engineering method to obtain strong relaxor behavior in $(1-x)BiFeO_3$-$xSrTiO_3$ thin films (x = 0.3–0.75) deposited by the PLD technique and found that the BDS of the films could be enhanced further, up to 4.46 MV cm^{-1}, by increasing the $SrTiO_3$ content. Atomic-scale microstructure analysis based on the STEM of $BiFeO_3$-$SrTiO_3$ films revealed that $SrTiO_3$ disrupted the long-range FE order, and this disruption cascaded with an increase in $SrTiO_3$ content. The incorporation of $SrTiO_3$ could transform the micrometer-scale FE domains into nanoscale PNRs. Paraelectric $SrTiO_3$ acts as a matrix for embedded PNRs and separates the PNRs such that the inter-PNR interaction almost vanishes. Because there are no inter-PNR interactions, the PNRs are very dynamic under an external electric field and produce slim *P–E* loops with very small P_r values. The size of the PNRs continued to decrease with increasing $SrTiO_3$ content, forming an almost domain-less feature for x = 0.75. This domain evolution reduced the domain switching energy, producing a slim *P–E* loop, while maintaining a large P_{max} value and a colossal BDS. In addition to the domain perspective, $SrTiO_3$ substitution enhanced the insulating character of the films by stabilizing the Fe^{3+}/Fe^{2+} valence fluctuation and reducing the formation of oxygen vacancies, leading to further enhancement of BDS. For example, the BDS value of $(1\text{-}x)BiFeO_3$-$xSrTiO_3$ films increased from 2.77 MV cm^{-1} to 4.46 MV cm^{-1} as the $SrTiO_3$ content increased from for $x = 0.3$ to $x = 0.75$. At such a high electric field, the films with $x = 0.6$ and 0.75 exhibited a giant $U_{rec} \sim 70$ J cm^{-3}.

Domain engineering techniques have been found to be more fruitful in thin films that exhibit polymorphic nanodomains, e.g., rhombohedral (*R*) and tetragonal (*T*) domains in a cubic paraelectric matrix. If these polymorphs have competitive free energy, Landau phenomenological theory predicts the weakening of polarization anisotropy and lowering of the energy barrier between the *R* and *T* polarization states [97]. It facilitates a flatter energy profile for polymorphic nanodomain RFEs compared with classic FEs and nanodomain RFEs, which minimizes the hysteresis while maintaining a high polarization (Figure 11). Pan et al. [97] demonstrated that the polymorphic domain engineering technique can produce ultra-high energy density with high efficiency in thin films of a $(0.55-x)BiFeO_3$-$xBaTiO_3$-$0.45SrTiO_3$ solid solution (x = 0 to 0.4). Here, $BiFeO_3$ and $BaTiO_3$ were the hosts for the *R* and *T* domains, whereas $SrTiO_3$ provided a cubic paraelectric matrix. The relaxor nature and energy storage performance of the $(0.55-x)BiFeO_3$-$xBaTiO_3$-$0.45SrTiO_3$ solid solutions are shown in Figure 12. The incorporation of $BaTiO_3$ gradually enhanced the relaxor nature, as can be seen from the wider peaks in the ε–T plots (Figure 12a), as well as the BDS for higher $BaTiO_3$ contents. The BDS increased up to 4.9 MV cm^{-1} for x = 0.3 and 5.3 MV cm^{-1} for x = 0.4, compared with 3.2 MV cm^{-1} for the x = 0 compound. This resulted in a maximum U_{rec} of 112 J cm^{-3} and 110 J cm^{-3} for x = 0.3 and 0.4, respectively, with $\eta > 80\%$ (Figure 12). Moreover, the enhancement of the relaxor nature resulted in good temperature stability for x = 0.3 and 0.4 regarding their energy storage performance over a wide range of temperatures. Kurusumovic et al. [98] employed a combined defect engineering method to explore the energy storage performance of relaxor thin films of $BiFeO_3$-$BaTiO_3$ solid solutions doped with Mn. The combined approach of defect engineering consisted of an interval mono-layer by mono-layer deposition (LLD) and Mn addition. The LLD produced highly stoichiometric and perfectly crystalline films compared with standard deposited films, while the addition of Mn reduced the leakage current by creating vacancy trap centers. Using this method, they obtained an ultra-high value of $U_{rec} = 80$ J cm^{-3} at a BDS of 3.1 MV cm^{-1} with $\eta = 78\%$ in 2.5 mol.% Mn-doped $0.25BiFeO_3$-$0.75BaTiO_3$ thin films.

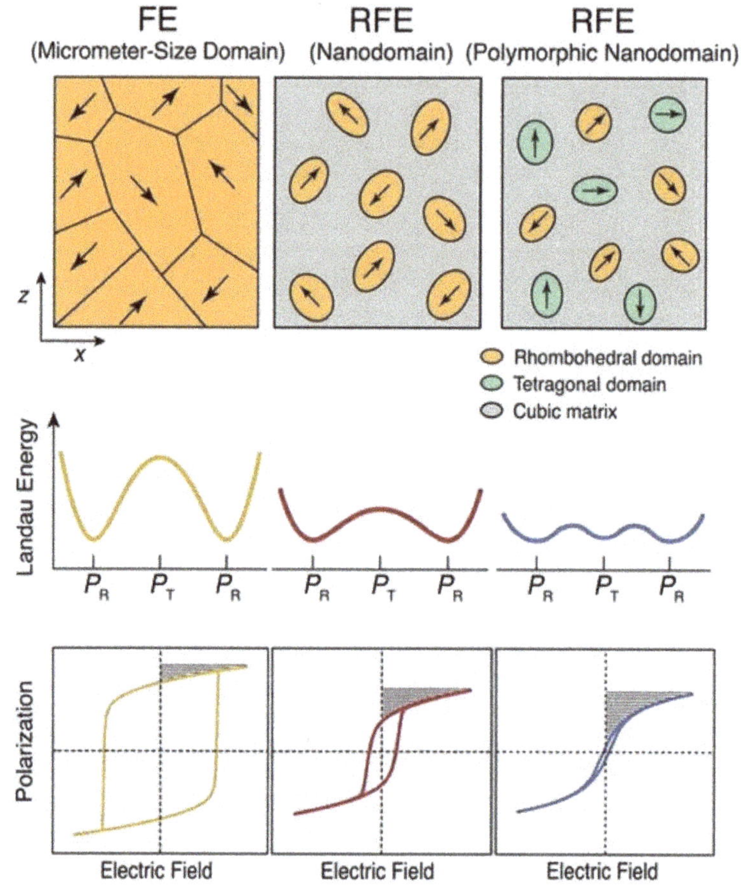

Figure 11. Design of RFEs via polymorphic nanodomain design. Comparative display of Landau energy profiles and $P-E$ loops of an FE with micrometer-sized domains, an RFE with nanodomains, and an RFE with polymorphic nanodomains. The P_R represents the polarization states along the rhombohedral (*R*) directions, and P_T is along the tetragonal (*T*) direction. The shadowed area in the *P-E* loops represents the recoverable energy density. Figures reproduced with permission from [97].

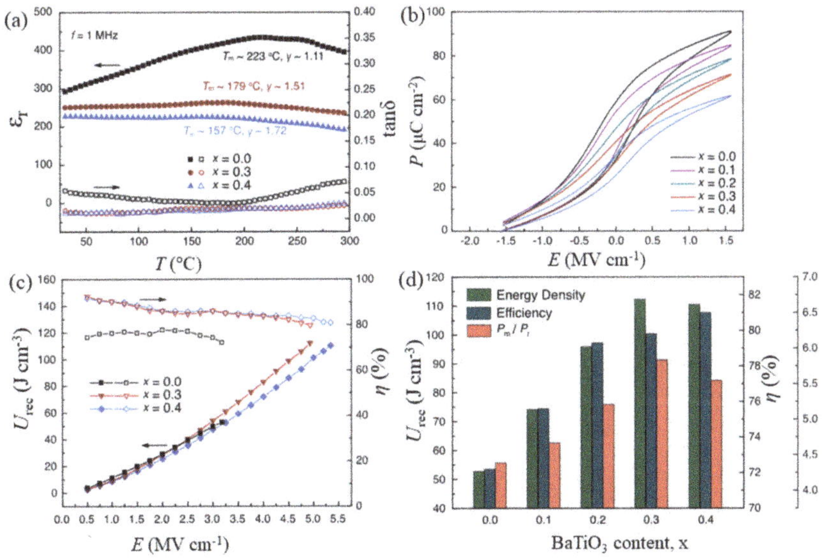

Figure 12. Characteristics of $(0.55-x)BiFeO_3$-$xBaTiO_3$-$0.45SrTiO_3$ (x = 0.0–0.4) films. (**a**) Temperature-dependent dielectric permittivity and loss tangent at a frequency of 1 MHz, (**b**) first-order reversal curve (FORC) $P-E$ loops, (**c**) energy density and efficiency values with respect to applied electric fields up to breakdown fields, and (**d**) comparison of the energy storage performance at breakdown fields. Figures reproduced with permission from [97]. © 2021 The Authors, some rights reserved; exclusive licensee American Association for the Advancement of Science.

4.3. Multilayered Structure

Although the energy density of thin films is superior to that of bulk ceramics, the usability of thin films is limited for low-power applications because of their small volume. In this context, multilayered structures have received great attention because the technology behind them is well known and inexpensive. A multilayered ceramic consists of a number of thin ceramic layers with thicknesses on the micrometer scale and internal electrode layers stacked in parallel and connected through terminal electrodes. It features both high BDS (on the MV cm^{-1} level) and large volume; therefore, it is very promising for practical high-power energy storage applications.

Wang et al. [76] reported a large improvement in the energy storage performance of a multilayered ceramic composed of $0.75(Bi_{0.85}Nd_{0.15}FeO_3)$-$0.25BaTiO_3$ + 0.1wt.% MnO_2 and Pt internal electrodes, compared with its bulk ceramic counterpart, as shown in Figure 13. With nine active layers having a total thickness of ~ 0.78 mm, they obtained a significant enhancement of the BDS to 540 kV cm^{-1} with U_{st} ~ 8.75 J cm^{-3}, U_{rec}~ 6.74 J cm^{-3}, and η~77%. Yan et al. [99] reported similar energy storage properties in (1-x)$(0.67BiFeO_3$-$0.33BaTiO_3)$-$xNa_{0.73}Bi_{0.09}NbO_3$ multilayered ceramics with A-site cation vacancies. For an optimized content of $Na_{0.73}Bi_{0.09}NbO_3$, i.e., x = 0.12, the multilayer exhibited U_{rec} = 5.57 J cm^{-3} at E = 410 kV cm^{-1} with η = 83.8%. A multilayer ceramic with 7-μm-thick $0.61BiFeO_3$-$0.33Ba_{0.8}Sr_{0.02}TiO_3$-$0.06La(Mg_{2/3}Nb_{1/3})O_3$ dielectric layers was reported to exhibit a much higher BDS, > 740 kV cm^{-1} [100]. This device featured a high U_{rec} of 10 J cm^{-3} with η ~ 72% at E = 730 kV cm^{-1}. Wang et al. reported a large ΔP of ~ 34 μC cm^{-2} and a high BDS of 700 kV cm^{-1} in a multilayered ceramic composed of 16-μm-thick electrically homogeneous $0.62BiFeO_3$-$0.3BaTiO_3$-$0.08NdZr_{1/2}Zn_{1/2}O_3$ ceramic layers with a core-shell microstructure [82]. With seven ceramic layers having an active electrode area of 33 mm^2, U_{rec} as high as 10.5 J cm^{-3} with η = 87% was obtained. Lu et al. [91] obtained BDS > 1 MV cm^{-1} in a multilayered ceramic with 10 mol% $Bi(Mg_{2/3}Nb_{1/3})O_3$ and 3 mol% Nb_2O_5-doped $0.6BiFeO_3$-$0.4SrTiO_3$ with a core-shell microstructure. The device, with 8-μm-thick dielectric layers, exhibited U_{rec} = 15.8 J cm^{-3} under E > 1 MV cm^{-1} with η = 75.2%. Wang et al. [101] reported a high BDS of 953 kV cm^{-1} in a multilayered ceramic with 8-μm-thick 0.57BiFeO$_3$-0.3BaTiO$_3$-0.13Bi(Li$_{0.5}$Nb$_{0.5}$)O$_3$ layers (Figure 14) having a core-shell structure and a 5-mm^2 active electrode area, which was greatly improved compared with that of the bulk ceramic counterpart (260 kV cm^{-1}). While the bulk ceramic featured a U_{rec} ~ 3.64 J cm^{-3} at E ~ 260 kV cm^{-1} with η ~ 75%, the multilayered ceramic exhibited a U_{rec} = 13.8 J cm^{-3} with η = 81% because of the large enhancement of the BDS. The multilayered ceramic showed good temperature stability (< 10%) and frequency independence (< 5%) of U_{rec} from 0.01 to 1 Hz, as well as fatigue resistance (< 5%) during 10^4 cycles of unipolar P−E loop tests in the temperature range from RT to 100 °C at E = 400 kV-cm^{-1}.

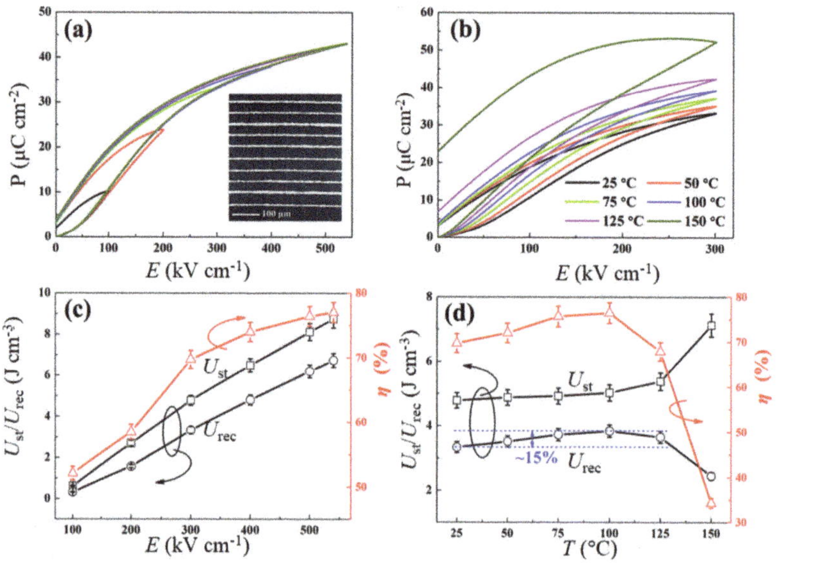

Figure 13. Energy storage performance of 0.75(Bi$_{0.85}$Nd$_{0.15}$)FeO$_3$-0.25BaTiO$_3$ multilayers. (**a**) P–E loops at various electric fields at room temperature, (**b**) P–E loops at various temperatures. U_{st}, U_{rec}, and η with (**c**) electric field and (**d**) temperature. An SEM image of the multilayers is shown in the inset of (**a**). Figures reproduced with permission from [76]. © 2021 The Royal Society of Chemistry.

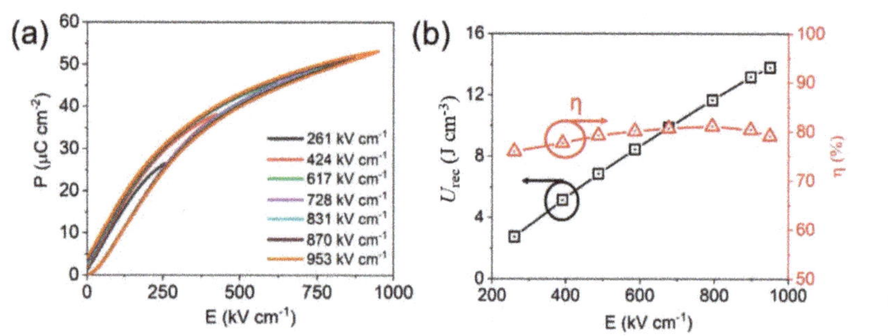

Figure 14. (**a**) Room temperature unipolar P–E loops at various electric fields and (**b**) calculated energy storage properties of 0.57BiFeO$_3$-0.3BaTiO$_3$-0.13Bi(Li$_{0.5}$Nb$_{0.5}$)O$_3$ multilayers. Figures reused under Creative Commons Attribution 3.0 Unported Licence [101]. © 2021 The Royal Society of Chemistry.

5. Critical Issues and Strategy

5.1. Leakage Current Control

High discharge energy density and energy efficiency are the primary requirements of an EES system. However, most $BiFeO_3$-based RFEs suffer from a low efficiency (<85%) despite having ultra-high discharge energy densities. Leakage-related high conductivity in $BiFeO_3$-based RFEs produces apparent large values of P_r and wide hysteresis loops, which lead to a higher loss density. The conductivity usually increases at high temperatures, limiting their potential for high-temperature applications.

The high leakage current in $BiFeO_3$-based compounds is associated with the loss of Bi_2O_3 during its synthesis at elevated temperatures (950–1050 °C), which creates ionic vacancies and Fe^{3+}/Fe^{2+} valence fluctuation [102–105]. Aliovalent doping in $BiFeO_3$-based compounds has been effective for reducing the leakage current density by several orders of magnitude [104,105]. However, these compounds are fired at high temperatures for synthesis in both bulk ceramics and multilayered structures, exposing them to Bi_2O_3 loss. Adopting a low-temperature sintering method is expected to reduce the Bi_2O_3 loss and the associated leakage current densities by a few more orders. The addition of appropriate glass additives and/or low melting point compounds is expected to significantly reduce the sintering temperature [106], thereby reducing the possibility of Bi_2O_3 volatilization. Low-temperature sintering is also beneficial for the enhancement of BDS, as discussed in Section 5.2.

5.2. Microstructure Engineering

In Figure 15, we plot U_{rec} as a function of BDS for $BiFeO_3$-based binary and ternary solid solutions fabricated in bulk ceramics, ceramic multilayers, and thin films. A high BDS usually provides a higher recoverable energy density. Therefore, enhancing the BDS is one of the most effective ways to increase the energy storage performance of RFEs. Microstructure engineering, such as the design of core-shell structures and domain engineering, has been found to be very effective for enhancing the BDS and realizing a dynamic relaxor ferroelectric phase. The BDS depends on the grain size (BDS $\propto 1/\sqrt{G}$ (G = average grain size)) and size distribution. Therefore, it is imperative to design a fine-grained microstructure with a uniform size distribution to obtain a high BDS. Another way to improve the BDS is to prepare a highly dense microstructure. Pores or cavities present in microstructures are usually filled with gaseous or liquid phases with lower permittivity than those of solid dielectrics. As a result, the voltage across the cavities or pores (V_c) is enhanced as per the following relation [107]:

$$V_c = \frac{V_{app}}{\left[1 + \frac{\varepsilon_c}{\varepsilon_r}\left(\frac{d}{t} - 1\right)\right]} \quad (10)$$

where V_{app} is the applied voltage, ε_c and ε_r are the permittivity of the cavity and dielectric, respectively, and t and d are the size of the cavity and the thickness of the dielectric, respectively. A small pore can create a large electric field across it and cause local breakdown and internal discharge even at a low external voltage. Therefore, a highly dense microstructure is essential. Because $BiFeO_3$-based materials are prone to the loss of Bi and Fe valence fluctuations during sintering at high temperatures, resulting in high electrical conductivity, sintering at low temperatures could be very beneficial. However, a detailed investigation of the microstructure control for the purpose of enhancing the energy storage performance employing the low-temperature sintering technique in $BiFeO_3$-based dielectrics is yet to be conducted. The two-step sintering (TSS) method modified by Chen and Wan [108] is one of the most cost-effective and simple methods to produce ultra-high-density materials and fine grains. In this technique, high-temperature heating is performed for a few minutes, and then the material is allowed to cool to the sintering temperature, where the sample is sintered for a prolonged time. However, the sintering temperature should be wisely chosen so that densification without further grain growth occurs.

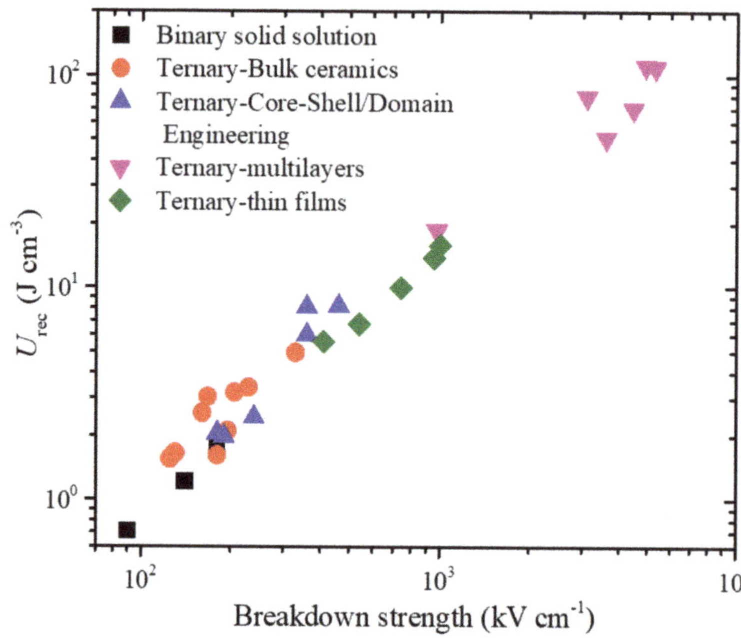

Figure 15. Recoverable energy density as a function of breakdown strength for $BiFeO_3$-based bulk ceramics of binary solid solutions and bulk ceramics, microstructure-designed bulk ceramics, and multilayers and thin films of ternary solid (TS) solutions.

5.3. Band Gap Engineering

Electronic breakdown is a crucial intrinsic mechanism for the breakdown of a dielectric in a large electric-field regime. Above a certain electric field, electrons in the valence band gain sufficient energy to jump to the conduction band. This results in an increase in the electron density in the conduction band and leads to a large current discharge. Therefore, an insulator exhibiting a wide forbidden energy gap can withstand a high electric field. The empirical relation between the BDS and energy band gap is given by the following relation [109]:

$$E_B = 24.442\, exp\left(0.315\sqrt{E_g\omega_{max}}\right) \tag{11}$$

where E_g is the energy bandgap, and ω_{max} is the maximum phonon frequency. Pure $BiFeO_3$ has a direct bandgap of 3 eV [110]. Only a few studies have attempted to enhance the BDS of $BiFeO_3$-based RFEs via modulation of the energy bandgap. Qi et al. [92] reported the enhancement of BDS by doping of high-band-gap materials such as $NaNbO_3$ (E_g ~ 3.28 eV) in $0.67BiFeO_3$-$0.33BaTiO_3$ solid solutions. They demonstrated that the bandgap of the solid solutions (0.67-x)$BiFeO_3$-$0.33BaTiO_3$-x$NaNbO_3$ ($0 \leq x \leq 0.15$) increases monotonically with increases in x from ~ 2.6 eV for x = 0 to ~ 2.95 eV for x = 0.15. The enhancement of BDS followed a similar trend with an increase in $NaNbO_3$ content, where the average BDS increased from 230 kV cm^{-1} for x = 0 to 420 kV cm^{-1} for x = 0.15. Further studies on bandgap engineering can be explored to enhance the BDS and energy storage performance of $BiFeO_3$-based relaxor ferroelectrics.

5.4. Electromechanical Breakdown

Another issue that has not gained much attention but is very critical in $BiFeO_3$-based relaxor ferroelectrics is the electromechanical breakdown. Because a high electric field is required to obtain a large recoverable energy density, the dielectrics are under extreme electrostrictive strain, increasing the mechanical breakdown. This issue should be taken very seriously from the device viewpoint, as the devices undergo a large number of charge-discharge cycles. Although high recoverable energy density with high efficiency is obtained at high electric fields under laboratory conditions, the same samples might not be suitable for application purposes where they totally collapse because of mechanical failure under a large number of charge-discharge cycles. However, systematic studies on the issues of electromechanical breakdown in $BiFeO_3$-based dielectrics are lacking. Rare-earth-ion doped $BiFeO_3$ compounds with their composition lying across the morphotropic phase boundary could be interesting in this regard as these compounds feature high electromechanical strain. For example, Walker et al. [111] reported cycle-dependent large electromechanical strain in Sm-doped $BiFeO_3$ polycrystalline samples, associated with the electric field-induced phase transition and ferroelectric/ferroelastic domain switching.

6. Summary

$BiFeO_3$-based relaxor ferroelectrics are projected to be potential Pb-free candidates for application in the field of high-energy-density storage and high-power-delivery systems. They are mostly fabricated in the form of ceramics, multilayers, and thin films from binary or ternary solid solutions with other perovskite oxides. $BiFeO_3$-$BaTiO_3$ and $BiFeO_3$-$SrTiO_3$, with their compositional ratios lying in the morphotropic phase boundary, are the two most widely studied $BiFeO_3$-based binary solid solutions for high-energy-density storage. However, the typical value of recovering energy density of bulk ceramics of $BiFeO_3$-based binary solid solutions is well below 3 J cm^{-3}, with low efficiency (typically below 50%). Ternary solid solutions exhibiting specific microstructures, such as core-shell structures and stripe-like nanodomains, have shown significant enhancement of the energy density properties. For example, $0.5BiFeO_3$-$0.4SrTiO_3$-$0.03Nb_2O_5$-$0.1BiMg_{2/3}Nb_{1/3}O_3$ bulk ceramic with a core-shell microstructure exhibited excellent energy storage properties, with U_{rec} ~ 8.2 J cm^{-3} and η ~ 74%. The nanodomain-engineered bulk ceramic $0.57BiFeO_3$-$0.33BaTiO_3$-$0.1NaNbO_3$ with the addition of 0.1 wt% MnO_2 and 2 wt% $BaCu(B_2O_5)$ also featured a similar value of U_{rec} (~ 8.12 J cm^{-3}); however, it had a higher value of η ~ 90%. The microstructure design approach has been found to be attractive for enhancing the energy storage properties of thin films and multilayers. The BDS values of thin films and multilayers can be as high as MV cm^{-1}, which is very helpful for further increasing the recoverable energy density. Thin films of a $(0.55-x)BiFeO_3$-$xBaTiO_3$-$0.45SrTiO_3$ solid solution (x = 0–0.4) containing polymorphic nanodomains showed a recoverable energy density of ~ 110 J cm^{-3} at a BDS ~ 5 MV/cm with an efficiency of ~ 80%, which is the highest among $BiFeO_3$-based systems. The achievement of such excellent energy storage properties is very encouraging for applications in low-power, small-size devices. However, for large-scale applications, it is necessary to focus on the fabrication of multilayers, as they deliver high power and are easy to fabricate. Although relatively few studies have been conducted on $BiFeO_3$-based multilayers, their energy storage performance is very encouraging (U_{rec} ~ 15 J cm^{-3}), with a high BDS of ~ 1 MV/cm. More studies should be conducted in the field of $BiFeO_3$-based multilayers to optimize structural parameters such as the thickness of dielectric layers and number of layers so that a better energy storage device that can deliver high output power can be fabricated at a low cost.

Author Contributions: B.D. and K.-H.C. equally contributed to this work. K.-H.C. conceptualized the idea, B.D. reviewed literatures, drafted and wrote the manuscript. B.D. and K.-H.C. read and edited the manuscript for final submission. K.-H.C. acquired the funding. All authors have read and agreed to the published version of the manuscript.

Funding: This work was supported by the National Research Foundation (NRF) of Korea funded by the Ministry of Education [NRF-2019R1I1A3A01058105, NRF-2018R1A6A1A03025761] and Ministry of Science and ICT (Grand Information Technology Research Center support program) [IITP-2021-2020-0-01612] supervised by the Institute for Information & Communications Technology Planning & Evaluation.

Conflicts of Interest: The authors declare no conflict of interest.

References

1. Goodenough, J.B.; Abruna, H.D.; Buchanan, M.V. *Basic Research Needs for Electrical Energy Storage. Report of the Basic Energy Sciences Workshop on Electrical Energy Storage, 2–4 April 2007*; U.S. Department of Energy Office of Scientific and Technical Information: Washington, DC, USA, 2007. [CrossRef]
2. Yang, L.; Kong, X.; Li, F.; Hao, H.; Cheng, Z.; Liu, H.; Li, J.-F.; Zhang, S. Perovskite lead-free dielectrics for energy storage applications. *Prog. Mater. Sc.* **2019**, *102*, 72. [CrossRef]
3. Kusko, A.; Dedad, J. Stored energy—Short-term and long-term energy storage methods. *IEEE Ind. Appl. Mag.* **2007**, *13*, 66. [CrossRef]
4. Yao, K.; Chen, S.; Rahimabady, M.; Mirshekarloo, M.S.; Yu, S.; Tay, F.E.H.; Sritharan, T.; Lu, L. Nonlinear dielectric thin films for high-power electric storage with energy density comparable with electrochemical supercapacitors. *IEEE Trans. Ultrason. Ferroelectr. Freq. Control.* **2011**, *58*, 1968. [PubMed]
5. Jow, T.R.; MacDougall, F.W.; Ennis, J.B.; Yang, X.H.; Schneider, M.A.; Scozzie, C.J.; White, J.D.; MacDonald, J.R.; Schalnat, M.C.; Cooper, R.A.; et al. Pulsed power capacitor development and outlook. *IEEE* **2015**, unpublished.
6. Sarjeant, W.J.; Zirnheld, J.; MacDougall, F.W. Capacitors. *IEEE Trans. Plasma Sci.* **1998**, *26*, 1368. [CrossRef]
7. Irvine, J.T.S.; Sinclair, D.C.; West, A.R. Electroceramics: Characterization by impedance spectroscopy. *Adv. Mater.* **1990**, *2*, 132. [CrossRef]
8. Sarjeant, W.J.; Clelland, I.W.; Price, R.A. Capacitive components for power electronics. *Proc. IEEE.* **2001**, *89*, 846. [CrossRef]
9. Bell., A.J. Ferroelectrics: The role of ceramic science and engineering. *J. Eur. Ceram. Soc.* **2008**, *28*, 1307. [CrossRef]
10. Tan, Q.; Irwin, P.; Cao, Y. Advanced dielectrics for capacitors. *IEEJ Trans. Fund. Mater.* **2006**, *126*, 1152. [CrossRef]
11. Chu, B.J.; Zhou, X.; Ren, K.; Neese, B.; Lin, M.; Wang, Q.; Bauer, F.; Zhang, Q.M. A dielectric polymer with high electric energy density and fast discharge speed. *Science* **2006**, *313*, 334. [CrossRef]
12. Khanchaitit., P.; Han., K.; Gadinski, M.R.; Li, Q.; Wang, Q. Ferroelectric polymer networks with high energy density and improved discharged efficiency for dielectric energy storage. *Nat. Commun.* **2013**, *4*, 2845. [CrossRef] [PubMed]
13. Li, Q.; Han, K.; Gadinski, M.R.; Zhang, G.; Wang, Q. High Energy and Power Density Capacitors from Solution-Processed Ternary Ferroelectric Polymer Nanocomposites. *Adv. Mater.* **2014**, *26*, 6244. [CrossRef]
14. Li, Q.; Liu, F.; Yang, T.; Gadinski, M.R.; Zhang, G.; Chen, L.-Q.; Wang, Q. Sandwich-structured polymer nanocomposites with high energy density and great charge–discharge efficiency at elevated temperatures. *Proc. Natl. Acad. Sci. USA* **2016**, *113*, 9995. [CrossRef] [PubMed]
15. Peddigari, M.; Palneedi, H.; Hwang, G.-T.; Ryu, J. Linear and nonlinear dielectric capacitors for high-power energy storage capacitor applications. *J. Korean Ceram. Soc.* **2019**, *56*, 1. [CrossRef]
16. Ravindran, P.; Vidya, R.; Kjekshus, A.; Fjellvåg, H.; Eriksson, O. Theoretical investigation of magnetoelectric behavior in $BiFeO_3$. *Phys. Rev. B* **2006**, *74*, 224412. [CrossRef]
17. Yang, J.; Li, X.; Xiong, Z.; Wang, M.; Liu, Q. Environmental Pollution Effect Analysis of Lead Compounds in China Based on Life Cycle. *Int. J. Environ. Res. Public Health* **2020**, *17*, 2184. [CrossRef]
18. Tchounwou, P.B.; Yedjou, C.G.; Patlolla, A.K.; Sutton, D.J. Heavy metal toxicity and the environment. In *Molecular, Clinical and Environmental Toxicology*; Springer: Berlin/Heidelberg, Germany, 2012; pp. 133–164.
19. Zhou, H.Y.; Liu, X.Q.; Zhu, X.L.; Chen, X.M. $CaTiO_3$ Linear Dielectric Ceramics with Greatly Enhanced Dielectric Strength and Energy Storage Density. *J. Am. Ceram. Soc.* **2017**, *101*, 1999. [CrossRef]
20. Zhou, H.Y.; Zhu, X.N.; Ren, G.R.; Chen, X.M. Enhanced Energy Storage Density and Its Variation Tendency in $CaZr_xTi_{1-x}O_3$ Ceramics. *J. Alloys Compd.* **2016**, *688*, 687. [CrossRef]
21. Yang, H.; Yan, F.; Lin, Y.; Wang, T. Enhanced Recoverable Energy Storage Density and High Efficiency of $SrTiO_3$-Based Lead-Free Ceramics. *Appl. Phys. Lett.* **2017**, *111*, 253903. [CrossRef]
22. Shay, D.P.; Podraza, N.J.; Randall, C.A. High Energy Density, High Temperature Capacitors Utilizing Mn-Doped 0.8CaTiO3–0.2CaHfO3 Ceramics. *J. Am. Ceram. Soc.* **2012**, *95*, 1348. [CrossRef]
23. Lv, J.; Li, Q.; Li, Y.; Tang, M.; Jin, D.; Yan, Y.; Fan, B.; Jin, L.; Liu, G. Significantly improved energy storage performance of NBT-BT based ceramics through domain control and preparation optimization. *Chem. Eng. J.* **2021**, *420*, 129900. [CrossRef]
24. Gao, F.; Dong, X.; Mao, C.; Cao, F.; Wang, G. c/a Ratio-Dependent Energy-Storage Density in (0.9−x)Bi0.5Na0.5TiO3–xBaTiO3–0.1K0.5Na0.5NbO3 Ceramics. *J. Am. Ceram. Soc.* **2011**, *94*, 4162. [CrossRef]
25. Viola, G.; Ning, H.; Reece, M.J.; Wilson, R.; Correia, T.M.; Weaver, P.; Cain, M.G.; Yan, H. Reversibility in electric field-induced transitions and energy storage properties of bismuth-based perovskite ceramics. *J. Phys. D Appl. Phys.* **2012**, *45*, 355302. [CrossRef]
26. Xu, Q.; Xie, J.; He, Z.; Zhang, L.; Cao, M.; Huang, X.; Lanagan, M.T.; Hao, H.; Yao, Z.; Liu, H. Energy-storage properties of Bi0.5Na0.5TiO3-BaTiO3-KNbO3 ceramics fabricated by wet-chemical method. *J. Eur. Ceram. Soc.* **2017**, *37*, 99. [CrossRef]

27. Puli, V.S.; Pradhan, D.K.; Chrisey, D.B.; Tomozawa, M.; Sharma, G.L.; Scott, J.F.; Katiyar, R.S. Structure, dielectric, ferroelectric, and energy density properties of (1 − x)BZT–xBCT ceramic capacitors for energy storage applications. *J. Mater. Sci.* **2013**, *48*, 2151. [CrossRef]
28. Puli, V.S.; Pradhan, D.K.; Riggs, B.C.; Chrisey, D.B.; Katiyar, R.S. Structure, Ferroelectric, Dielectric and Energy Storage Studies of Ba0.70Ca0.30TiO3, Ba(Zr0.20Ti0.80)O3 Ceramic Capacitors. *Integr. Ferroelectr.* **2014**, *157*, 139. [CrossRef]
29. Zhang, Y.; Li, Y.; Zhu, H.; Fu, Z.; Zhang, Q. Sintering temperature dependence of dielectric properties and energy-storage properties in (Ba,Zr)TiO3 ceramics. *J. Mater. Sci. Mater. Electron.* **2017**, *28*, 514. [CrossRef]
30. Zhu, C.; Wang, X.; Zhao, Q.; Cai, Z.; Chen, Z.; Li, L. Effects of gtain size and temperature on the energy storage and dielectric tunablity of non-reducible $BaTiO_3$-based bulk ceramics. *J. Euro. Ceram. Soc.* **2019**, *39*, 1142. [CrossRef]
31. Liu, B.; Wang, X.; Zhao, Q.; Li, L. Improved energy storage properties of fine crystalline $BaTiO_3$ ceramics by coating powders with Al_2O_3 and SiO_2. *J. Am. Ceram. Soc.* **2015**, *98*, 2641. [CrossRef]
32. Qu, B.; Du, H.; Yang, Z.; Liu, Q. Large recoverable energy storage density and low sintering temperature in potassium-sodium niobate-based ceramics for multilayer pulsed power capacitors. *J. Am. Ceram. Soc.* **2017**, *100*, 1517. [CrossRef]
33. Ciuchi, I.V.; Mitoseriu, L.; Galassi, C. Antiferroelectric to Ferroelectric Crossover and Energy Storage Properties of (Pb1−xLax)(Zr0.90Ti0.10)1−x/4O3 ($0.02 \leq x \leq 0.04$) Ceramics. *J. Am. Ceram. Soc.* **2016**, *99*, 2382. [CrossRef]
34. Dan, Y.; Xu, H.; Zou, K.; Zhang, Q.; Lu, Y.; Chang, G.; Huang, H.; He, Y. Energy storage characteristics of $(Pb,La)(Zr,Sn,Ti)O_3$ antiferroelectric ceramics with high Sn content. *Appl. Phys. Lett.* **2018**, *113*, 063902. [CrossRef]
35. Jo, H.R.; Lynch, C.S. A high energy density relaxor antiferroelectric pulsed capacitor dielectric. *J. Appl. Phys.* **2016**, *119*, 024104. [CrossRef]
36. Xu, C.; Liu, Z.; Chen, X.; Yan, S.; Cao, F.; Dong, X.; Wang, G. High charge-discharge performance of Pb0.98La0.02(Zr0.35Sn0.55Ti0.10)0.995O3 antiferroelectric ceramic. *J. Appl. Phys.* **2016**, *120*, 074107. [CrossRef]
37. Li, Q.; Zhou, C.; Xu, J.; Yang, L.; Zhang, X.; Zeng, W.; Yuan, C.; Chen, G.; Rao, G. Tailoring antiferroelectricity with high energy-storage properties in Bi0.5Na0.5TiO3–BaTiO3 ceramics by modulating Bi/Na ratio. *J. Mater. Sci. Mater. Electron.* **2016**, *27*, 10810.
38. Yang, H.; Yan, F.; Lin, Y.; Wang, T. Improvement of dielectric and energy storage properties in $SrTiO_3$-based lead-free ceramics. *J. Alloys. Compd.* **2017**, *728*, 780. [CrossRef]
39. Cao, W.; Li, W.; Feng, Y.; Bai, T.; Qiao, Y.; Hou, Y.; Zhang, T.; Yu, Y.; Fei, W. Defect dipole induced large recoverable strain and high energy-storage density in lead-free Na0.5Bi0.5TiO3-based systems. *Appl. Phys. Lett.* **2016**, *108*, 202902. [CrossRef]
40. Zhao, L.; Liu, Q.; Gao, J.; Zhang, S.; Li, J.-F. Lead-Free Antiferroelectric Silver Niobate Tantalate with High Energy Storage Performance. *Adv. Mater.* **2017**, *29*, 1701824. [CrossRef]
41. Zhao, L.; Gao, J.; Liu, Q.; Zhang, S.; Li, J.-F. Silver Niobate Lead-Free Antiferroelectric Ceramics: Enhancing Energy Storage Density by B-Site Doping. *ACS Appl. Mater. Interfaces* **2018**, *10*, 819. [CrossRef]
42. Li, J.; Jin, L.; Tian, Y.; Chen, C.; Lan, Y.; Hu, Q.; Li, C.; Wei, X.; Yan, H. Enhanced energy storage performance under low electric field in Sm^{3+} doped $AgNbO_3$ ceramics. *J. Materiomics* **2021**, in press. [CrossRef]
43. Zhao, Y.; Xu, J.; Zhou, C.; Yuan, C.; Li, Q.; Chen, G.; Wang, H.; Yang, L. High energy storage properties and dielectric behavior of (Bi0.5Na0.5)0.94Ba0.06Ti1−x(Al0.5Nb0.5)xO3 lead-free ferroelectric ceramics. *Ceram. Int.* **2016**, *42*, 2221. [CrossRef]
44. Liu, Z.; Lu, T.; Ye, J.; Wang, G.; Dong, X.; Withers, R.; Liu, Y. Antiferroelectrics for energy storage applications: A review. *Adv. Mat. Technol.* **2018**, *3*, 1800111. [CrossRef]
45. Palneedi, H.; Peddigari, M.; Hwang, G.T.; Jeong, D.W.; Ryu, J. High-Performance Dielectric Ceramic Films for Energy Storage Capacitors: Progress and Outlook. *Adv. Funct. Mater.* **2018**, *28*, 1803665. [CrossRef]
46. Marqués, M.I.; Aragó, C. Microscopic model for the formation of nanodomains in relaxor materials. *Phys. Rev. B* **2010**, *81*, 064114. [CrossRef]
47. Li, F.; Zhang, S.; Damjanovic, D.; Chen, L.-Q.; Shrout, T.R. Local Structural Heterogeneity and Electromechanical Responses of Ferroelectrics: Learning from Relaxor Ferroelectrics. *Adv. Func. Mater.* **2018**, *28*, 1801504. [CrossRef]
48. Kirillov, V.V.; Isupov, V.A. Relaxation polarization of $PbMg_{1/3}Nb_{2/3}O_3$ (PMN)-A ferroelectric with a diffused phase transition. *Ferroelectrics* **1973**, *5*, 3. [CrossRef]
49. Cross, L.E. Relaxor ferroelectrics. *Ferroelectrics* **1987**, *76*, 241. [CrossRef]
50. Viehland, D.; Jang, S.J.; Cross, L.E.; Wuttig, M. Freezing of the polarization fluctuations in lead magnesium niobate relaxors. *J. Appl. Phys.* **1990**, *68*, 2916. [CrossRef]
51. Westphal, V.; Kleemann, W.; Glinchuk, M.D. Diffuse phase transitions and random-field-induced domain states of the relaxor ferroelectric $PbMg_{1/3}Nb_{2/3}O_3$. *Phys. Rev. Lett.* **1992**, *68*, 847. [CrossRef]
52. Akbas, M.A.; Davies, P.K. Domain Growth in $Pb(Mg_{1/3}Ta_{2/3})O_3$ Perovskite Relaxor Ferroelectric Oxides. *J. Am. Ceram. Soc.* **1997**, *80*, 2933. [CrossRef]
53. Cheng, Z.Y.; Katiyar, R.S.; Yao, X.; Guo, A. Dielectric behavior of lead magnesium niobate relaxors. *Phys. Rev. B* **1997**, *55*, 8165. [CrossRef]
54. Pirc, R.; Blinc, R. Spherical random-bond–random-field model of relaxor ferroelectrics. *Phys. Rev. B* **1999**, *60*, 13470. [CrossRef]
55. Neaton, J.B.; Ederer, C.; Waghmare, U.V.; Spaldin, N.A.; Rabe, K.M. First-principles study of spontaneous polarization in multiferroic $BiFeO_3$. *Phys. Rev. B* **2005**, *71*, 014113. [CrossRef]

56. Moreau, J.M.; Michel, C.; Gerson, R.; James, W.J. Ferroelectric $BiFeO_3$ X-ray and neutron diffraction study. *J. Phys. Chem. Solids* **1971**, *32*, 1315. [CrossRef]
57. Kubel, F.; Schmid, H. Structure of a ferroelectric and ferroelastic monodomain crystal of the perovskite $BiFeO_3$. *Acta Cryst. B* **1990**, *46*, 698. [CrossRef]
58. Fischer, P.; Polomska, M.; Sosnowska, I.; Szymanski, M. Temperature dependence of the crystal and magnetic structures of $BiFeO_3$. *J. Phys. C Solid State Phys.* **1980**, *13*, 1931. [CrossRef]
59. Smith, R.T.; Achenbach, G.D.; Gerson, R.; James, W.J. Dielectric Properties of Solid Solutions of $BiFeO_3$ with $Pb(Ti, Zr)O_3$ at High Temperature and High Frequency. *J. Appl. Phys.* **1968**, *39*, 70. [CrossRef]
60. Ederer, C.; Spaldin, N.A. Effect of Epitaxial Strain on the Spontaneous Polarization of Thin Film Ferroelectrics. *Phys. Rev. Lett.* **2005**, *95*, 257601. [CrossRef] [PubMed]
61. Lebeugle, D.; Colson, D.; Forget, A.; Viret, M. Very large spontaneous electric polarization in $BiFeO_3$ single crystals at room temperature and its evolution under cycling fields. *Appl. Phys. Lett.* **2007**, *91*, 022907. [CrossRef]
62. Teague, J.R.; Gerson, R.; James, W.J. Dielectric hysteresis in single crystal $BiFeO_3$. *Solid State Commun.* **1970**, *8*, 1073. [CrossRef]
63. Wang, Y.P.; Zhou, L.; Zhang, M.F.; Chen, X.Y.; Liu, J.-M.; Liu, Z.G. Room-temperature saturated ferroelectric polarization in $BiFeO_3$ ceramics synthesized by rapid liquid phase sintering. *Appl. Phys. Lett.* **2004**, *84*, 1731. [CrossRef]
64. Ueda, K.; Tabata, H.; Kawai, T. Coexistence of ferroelectricity and ferromagnetism in $BiFeO_3$–$BaTiO_3$ thin films at room temperature. *Appl. Phys. Lett.* **1999**, *75*, 555. [CrossRef]
65. Palkar, V.R.; John, J.; Pinto, R. Observation of saturated polarization and dielectric anomaly in magnetoelectric $BiFeO_3$ thin films. *Appl. Phys. Lett.* **2002**, *80*, 1628. [CrossRef]
66. Wang, J.; Neaton, J.B.; Zheng, H.; Nagarajan, V.; Ogale, S.B.; Liu, B.; Viehland, D.; Vaithyanathan, V.; Schlom, D.G.; Waghmare, U.V.; et al. Epitaxial $BiFeO_3$ Multiferroic Thin Film Heterostructures. *Science* **2003**, *299*, 1719. [CrossRef]
67. Yun, K.Y.; Noda, M.; Okuyama, M. Prominent ferroelectricity of $BiFeO_3$ thin films prepared by pulsed-laser deposition. *Appl. Phys. Lett.* **2003**, *83*, 3981. [CrossRef]
68. Lee, M.H.; Kim, D.J.; Park, J.S.; Kim, S.W.; Song, T.K.; Kim, M.-H.; Kim, W.-J.; Do, D.; Jeong, I.-K. High-Performance Lead-Free Piezoceramics with High Curie Temperatures. *Adv. Mater.* **2015**, *27*, 6976. [CrossRef]
69. Kumar, M.M.; Srinivas, A.; Suryanarayana, S.V. Structure property relations in $BiFeO_3/BaTiO_3$ solid solutions. *J. Appl. Phys.* **2000**, *87*, 855. [CrossRef]
70. Zhaludkevich, D.V.; Latushka, S.I.; Latushka, T.V.; Sysa, A.V.; Shaman, Y.P.; Dronova, D.A.; Chobot, A.N.; Chobot, G.M.; Nekludov, K.N.; Silibin, M.V.; et al. Crystal structure and magnetic properties of (1-x) $BiFeO_3$-x$BaTiO_3$ ceramics across the phase boundary. *Nanomaterials Sci. Eng.* **2020**, *2*, 93.
71. Wei, Y.; Wang, X.; Jia, J.; Wang, X. Multiferroic and piezoelectric properties of 0.65BiFeO3–0.35BaTiO3 ceramic with pseudo-cubic symmetry. *Ceram. Int.* **2012**, *38*, 3499. [CrossRef]
72. Zhou, C.; Yang, H.; Zhou, Q.; Cen, Z.; Li, W.; Yuan, C.; Wang, H. Dielectric, ferroelectric and piezoelectric properties of La-substituted $BiFeO_3$–$BaTiO_3$ ceramics. *Ceram. Int.* **2013**, *39*, 4307. [CrossRef]
73. Leontsev, S.O.; Eitel, R.E. Dielectric and Piezoelectric Properties in Mn-Modified $(1-x)BiFeO_3$–$xBaTiO_3$ Ceramics. *J. Am. Ceram. Soc.* **2009**, *92*, 2957. [CrossRef]
74. Wang, T.; Jin, L.; Tian, Y.; Shu, L.; Hu, Q.; Wei, X. Microstructure and ferroelectric properties of Nb_2O_5-modified $BiFeO_3$-$BaTiO_3$ lead-free ceramics for energy storage. *Mater. Lett.* **2014**, *137*, 79. [CrossRef]
75. Zhu, L.-F.; Lei, X.-W.; Zhao, L.; Hussain, M.I.; Zhao, G.-Z.; Zhang, B.-P. Phase structure and energy storage performance for $BiFeO_3$–$BaTiO_3$ based lead-free ferroelectric ceramics. *Ceram. Int.* **2019**, *45*, 20266. [CrossRef]
76. Wang, D.; Fan, Z.; Zhou, D.; Khesro, A.; Murakami, S.; Feteira, A.; Zhao, Q.; Tan, X.; Reaney, I.M. Bismuth ferrite-based lead-free ceramics and multilayers with high recoverable energy density. *J. Mater. Chem. A* **2018**, *6*, 4133. [CrossRef]
77. Chen, Z.; Bai, X.; Wang, H.; Du, J.; Bai, W.; Li, L.; Wen, F.; Zheng, P.; Wu, W.; Zheng, L.; et al. Achieving high-energy storage performance in 0.67Bi1-xSmxFeO3-0.33BaTiO3 lead-free relaxor ferroelectric ceramics. *Ceram. Int.* **2020**, *46*, 11549. [CrossRef]
78. Zheng, D.; Zuo, R.; Zhang, D.; Li, Y. Novel $BiFeO_3$–$BaTiO_3$–$Ba(Mg_{1/3}Nb_{2/3})O_3$ Lead-Free Relaxor Ferroelectric Ceramics for Energy-Storage Capacitors. *J. Am. Ceram. Soc.* **2015**, *98*, 2692. [CrossRef]
79. Zheng, D.; Zuo, R. Enhanced energy storage properties in $La(Mg_{1/2}Ti_{1/2})O_3$-modified $BiFeO_3$-$BaTiO_3$ lead-free relaxor ferroelectric ceramics within a wide temperature range. *J. Eur. Ceram. Soc.* **2017**, *37*, 413. [CrossRef]
80. Liu, N.; Liang, R.; Zhou, Z.; Dong, X. Designing lead-free bismuth ferrite-based ceramics learning from relaxor ferroelectric behavior for simultaneous high energy density and efficiency under low electric field. *J. Mater. Chem. C* **2018**, *6*, 10211. [CrossRef]
81. Wang, D.; Fan, Z.; Li, W.; Zhou, D.; Feteira, A.; Wang, G.; Murakami, S.; Sun, S.; Zhao, Q.; Tan, X.; et al. High Energy Storage Density and Large Strain in $Bi(Zn_{2/3}Nb_{1/3})O_3$-Doped $BiFeO_3$–$BaTiO_3$ Ceramics. *ACS Appl. Energy Mater.* **2018**, *1*, 4403. [CrossRef]
82. Wang, G.; Li, J.; Zhang, X.; Fan, Z.; Yang, F.; Feteira, A.; Zhou, D.; Sinclair, D.C.; Ma, T.; Tan, X.; et al. Ultrahigh energy storage density lead-free multilayers by controlled electrical homogeneity. *Energy Environ. Sci.* **2019**, *12*, 582. [CrossRef]
83. Liu, N.; Liang, R.; Zhao, X.; Xu, C.; Zhou, Z.; Dong, X. Novel bismuth ferrite-based lead-free ceramics with high energy and power density. *J. Am. Ceram. Soc.* **2018**, *101*, 3259. [CrossRef]
84. Tang, M.; Yu, L.; Wang, Y.; Lv, J.; Dong, J.; Guo, B.; Chen, F.; Ai, Q.; Luo, Y.; Li, Q.; et al. Dielectric, ferroelectric, and energy storage properties of $Ba(Zn_{1/3}Nb_{2/3})O_3$-modfied $BiFeO_3$–$BaTiO_3$ Pb-Free relaxor ferroelectric ceramics. *Ceram. Int.* **2021**, *47*, 3780. [CrossRef]

85. Sun, H.; Wang, X.; Sun, Q.; Zhang, X.; Ma, Z.; Guo, M.; Sun, B.; Zhu, X.; Liu, Q.; Lou, X. Large energy storage density in $BiFeO_3$-$BaTiO_3$-$AgNbO_3$ lead-free relaxor ceramics. *J. Eur. Ceram. Soc.* **2020**, *40*, 2929. [CrossRef]
86. Yu, Z.; Zeng, J.; Zheng, L.; Rousseau, A.; Li, G.; Kassiba, A. Microstructure effects on the energy storage density in $BiFeO_3$-based ferroelectric ceramics. *Ceram. Int.* **2021**, *47*, 12735. [CrossRef]
87. Yang, H.; Qi, H.; Zuo, R. Enhanced breakdown strength and energy storage density in a new $BiFeO_3$-based ternary lead-free relaxor ferroelectric ceramic. *J. Eur. Ceram. Soc.* **2019**, *39*, 2673. [CrossRef]
88. Liu, G.; Tang, M.; Hou, X.; Guo, B.; Lv, J.; Dong, J.; Wang, Y.; Li, Q.; Yu, K.; Yan, Y.; et al. Energy storage properties of bismuth ferrite based ternary relaxor ferroelectric ceramics through a viscous polymer process. *Chem. Eng. J.* **2020**, *412*, 127555. [CrossRef]
89. Li, Q.; Ji, S.; Wang, D.; Zhu, J.; Li, L.; Wang, W.; Zeng, M.; Hou, Z.; Gao, X.; Lu, X.; et al. Simultaneously enhanced energy storage density and efficiency in novel $BiFeO_3$-based lead-free ceramic capacitors. *J. Eur. Ceram. Soc.* **2021**, *41*, 387. [CrossRef]
90. Ji, S.; Li, Q.; Wang, D.; Zhu, J.; Zeng, M.; Hou, Z.; Fan, Z.; Gao, X.; Lu, X.; Li, Q.; et al. Enhanced energy storage performance and thermal stability in relaxor ferroelectric (1-x)BiFeO3-x(0.85BaTiO3-0.15Bi(Sn0.5Zn0.5)O3) ceramics. *J. Am. Ceram. Soc.* **2021**, *104*, 2426. [CrossRef]
91. Lu, Z.; Wang, G.; Bao, W.; Li, J.; Li, L.; Mostaed, A.; Yang, H.; Ji, H.; Li, D.; Feteira, A.; et al. Superior energy density through tailored dopant strategies in multilayer ceramic capacitors. *Energy Environ. Sci.* **2020**, *13*, 2938. [CrossRef]
92. Qi, H.; Xie, A.; Tian, A.; Zuo, R. Superior Energy-Storage Capacitors with Simultaneously Giant Energy Density and Efficiency Using Nanodomain Engineered $BiFeO_3$-$BaTiO_3$-$NaNbO_3$ Lead-Free Bulk Ferroelectrics. *Adv. Energy Mater.* **2020**, *10*, 1903338. [CrossRef]
93. Correia, T.M.; McMillen, M.; Rokosz, M.K.; Weaver, P.M.; Gregg, J.M.; Viola, G.; Cain, M.G. A Lead-Free and High-Energy Density Ceramic for Energy Storage Applications. *J. Am. Ceram. Soc.* **2013**, *96*, 2699. [CrossRef]
94. Pan, H.; Zeng, Y.; Shen, Y.; Lin, Y.-H.; Ma, J.; Li, L.; Nan, C.-W. $BiFeO_3$–$SrTiO_3$ thin film as a new lead-free relaxor-ferroelectric capacitor with ultrahigh energy storage performance. *J. Mater. Chem. A* **2017**, *5*, 5920. [CrossRef]
95. Hu, Z.; Ma, B.; Koritala, R.E.; Balachandran, U. Temperature dependent energy storage properties of antiferroelectric Pb_0. *96La0.04Zr0.98Ti0.02O3 thin films. Appl. Phys. Lett.* **2014**, *104*, 263902.
96. Pan, H.; Ma, J.; Ma, J.; Zhang, Q.; Liu, X.; Guan, B.; Gu, L.; Zhang, X.; Zhang, Y.-J.; Li, L.; et al. Giant energy density and high efficiency achieved in bismuth ferrite-based film capacitors via domain engineering. *Nat. Commun.* **2018**, *9*, 1813. [CrossRef] [PubMed]
97. Pan, H.; Li, F.; Liu, Y.; Zhang, Q.; Wang, M.; Lan, S.; Zheng, Y.; Ma, J.; Gu, L.; Shen, Y.; et al. Ultrahigh–energy density lead-free dielectric films via polymorphic nanodomain design. *Science* **2019**, *365*, 578. [CrossRef] [PubMed]
98. Kursumovic, A.; Li, W.-W.; Cho, S.; Curran, P.J.; Tjhe, D.H.L.; MacManus-Driscoll, J.L. Lead-free relaxor thin films with huge energy density and low loss for high temperature applications. *Nano Energy* **2020**, *71*, 104536. [CrossRef]
99. Yan, F.; Shi, Y.; Zhou, X.; Zhu, K.; Shen, B.; Zhai, J. Optimization of polarization and electric field of bismuth ferrite-based ceramics for capacitor applications. *Chem. Eng. J.* **2020**, *417*, 127945. [CrossRef]
100. Wang, G.; Lu, Z.; Li, J.; Ji, H.; Yang, H.; Li, L.; Sun, S.; Feteira, A.; Yang, H.; Zuo, R.; et al. Lead-free (Ba,Sr)TiO_3—$BiFeO_3$ based multilayer ceramic capacitors with high energy density. *J. Eur. Ceram. Soc.* **2020**, *40*, 1779. [CrossRef]
101. Wang, G.; Lu, Z.; Yang, H.; Ji, H.; Mostaed, A.; Li, L.; Wei, Y.; Feteira, A.; Sun, S.; Sinclair, D.C.; et al. Fatigue resistant lead-free multilayer ceramic capacitors with ultrahigh energy density. *J. Mater. Chem. A* **2020**, *8*, 11414. [CrossRef]
102. Qi, X.; Dho, J.; Tomov, R.; Blamire, M.G.; MacManus-Driscoll, J.L. Greatly reduced leakage current and conduction current mechanism in aliovalent-ion-doped $BiFeO_3$. *Appl. Phys. Lett.* **2005**, *86*, 062903. [CrossRef]
103. Hu, G.D.; Fan, S.H.; Yang, C.H.; Wu, W.B. Low leakage and current and enhanced ferroelectric properties of Ti and Zn codoped $BiFeO_3$ thin films. *Appl. Phys. Lett.* **2008**, *92*, 192905. [CrossRef]
104. Singh, S.K.; Maruyama, K.; Ishiwara, H. Reduced leakage current in La and Na codoped $BiFeO_3$ thin films. *Appl. Phys. Lett.* **2007**, *91*, 112913. [CrossRef]
105. Hu, G.D.; Cheng, X.; Wu, W.B.; Yang, C.H. Effect of Gd substitution on structure and ferroelectric properties of BiFeO3 thin films prepared using metal organic decomposition. *Appl. Phys. Lett.* **2007**, *91*, 232909. [CrossRef]
106. Veerapandian, V.; Benes, F.; Gindel, T.; Deluca, M. Strategies to develop energy storage properties of perovskite lead free relaxor ferroelectrics: A review. *Materials* **2020**, *13*, 5742. [CrossRef]
107. Ray, S. *An Introduction to High Voltage Engineering*; PHI Learning Pvt. Ltd.: Delhi, India, 2013.
108. Chen, I.W.; Wang, X.H. Sintering dense nanocrystalline ceramics without final-stage grain-growth. *Nature* **2000**, *404*, 168. [CrossRef] [PubMed]
109. Kim, C.; Pilania, G.; Ramprasad, R. From organized high-throughput data to phenomenological theory using machine learning: The example of dielectric breakdown. *Chem. Mater.* **2016**, *28*, 1304. [CrossRef]
110. Moubah, R.; Schmerber, G.; Rousseau, O.; Colson, D.; Viret, M. Photoluminescence Investigation of Defects and Optical Band Gap in Multiferroic $BiFeO_3$ Single Crystals. *Appl. Phys. Express* **2012**, *5*, 035802. [CrossRef]
111. Walker, J.; Simons, H.; Alikin, D.O.; Turygin, A.P.; Shur, V.Y.; Kholkin, A.L.; Ursic, H.; Bencan, A.; Malic, B.; Nagarajan, V.; et al. Dual strain mechanisms in a leadfree morphotropic phase boundary ferroelectric. *Sci. Rep.* **2016**, *6*, 19630. [CrossRef] [PubMed]

materials

MDPI

Article

Harvesting Energy from Bridge Vibration by Piezoelectric Structure with Magnets Tailoring Potential Energy

Zhiyong Zhou [1,2], Haiwei Zhang [1], Weiyang Qin [3], Pei Zhu [1,*], Ping Wang [4,*] and Wenfeng Du [1]

1 School of Civil Engineering and Architecture, Henan University, Kaifeng 475004, China; 10160091@vip.henu.edu.cn (Z.Z.); hwzhang@henu.edu.cn (H.Z.); 10160016@vip.henu.edu.cn (W.D.)
2 School of Power and Energy, Northwestern Polytechnical University, Xi'an 710072, China
3 Department of Engineering Mechanics, Northwestern Polytechnical University, Xi'an 710072, China; qinweiyang@nwpu.edu.cn
4 AECC Hunan Aviation Powerplant Research Institute, Zhuzhou 412002, China
* Correspondence: 10160096@vip.henu.edu.cn (P.Z.); wp608@sina.com (P.W.)

Abstract: Bridges play an increasingly more important role in modern transportation, which is why many sensors are mounted on it in order to provide safety. However, supplying reliable power to these sensors has always been a great challenge. Scavenging energy from bridge vibration to power the wireless sensors has attracted more attention in recent years. Moreover, it has been proved that the linear energy harvester cannot always work efficiently since the vibration energy of the bridge distributes over a broad frequency band. In this paper, a nonlinear energy harvester is proposed to enhance the performance of harvesting bridge vibration energy. Analyses on potential energy, restoring force, and stiffness were carried out. By adjusting the separation distance between magnets, the harvester could own a low and flat potential energy, which could help the harvester oscillate on a high-energy orbit and generate high output. For validation, corresponding experiments were carried out. The results show that the output of the optimal configuration outperforms that of the linear one. Moreover, with the increase in vehicle speed, a component of extremely low frequency is gradually enhanced, which corresponds to the motion on the high-energy orbit. This study may give an effective method of harvesting energy from bridge vibration excited by moving vehicles with different moving speeds.

Keywords: energy harvesting; bridge vibration; vehicle; mono-stable energy harvester; linear energy harvester; moving speed

Citation: Zhou, Z.; Zhang, H.; Qin, W.; Zhu, P.; Wang, P.; Du, W. Harvesting Energy from Bridge Vibration by Piezoelectric Structure with Magnets Tailoring Potential Energy. *Materials* **2022**, *15*, 33. https://doi.org/10.3390/ma15010033

Academic Editors: Marc Cretin, Sophie Tingry and Zhenghua Tang

Received: 2 December 2021
Accepted: 17 December 2021
Published: 21 December 2021

1. Introduction

Bridge safety has drawn great concern in recent decades owing to the developments in the cross-sea bridge, cable-stayed bridge, and suspension bridge [1]. To ensure the safety of the bridge, a sensor network should be incorporated into it so that many node sensors are distributed to monitor the state of the bridge to prevent hazards caused by degradation [2]. For the node sensors, the replacement of the battery is a difficult challenge [3]. Frequent battery replacement is excessively expensive and nearly impossible in some dangerous and special hard-to-reach areas [4]. To tackle the difficulty, harvesting energy from bridge vibration to power sensors is emerging as a promising solution. In principle, the vibration energy of structure can be converted to electric energy through three methods, i.e., the triboelectric, electro-magnetic, and piezoelectric mechanisms [5].

The piezoelectric scenario generally scavenges energy from bridge vibration excited by travelling vehicles. Developing energy harvesting devices coupled with the bridge structure has been drawing more and more attention from scholars [6]. Assadi et al. [7] attached a piezoelectric patch to a simply supported beam with a mass moving on it. When the moving mass travelled on the beam, the piezoelectric patch could produce voltage due to deflection. The analytical and experimental results indicated that the speed of the moving mass had a significant influence on voltage output. Ali et al. [8] used a linear single-degree-of-freedom model as an energy harvester to harvest the vibration energy of a simply supported beam excited by the motion of a point load. Galchev et al. [9] designed an inertial energy harvester to convert traffic-induced bridge vibrations to electric energy. The average power of this fabricated device was measured at different positions of a suspension bridge. Peigney and Siegert [10] designed and tested a cantilever piezoelectric harvester to harvest vibration energy from traffic-induced vibrations in bridges. Erturk [11] derived analytical expressions of transient response in the time domain of piezoelectric output power. Zhang et al. [12] conducted a numerical parametric study and derived the exact analytical solution of a piezoelectric energy harvester excited by beam vibration under moving harmonic loads. Amini [13] numerically investigated the harvested power from the vibrations of a beam excited by multi-moving loads. The results showed that the efficiency of the harvester increased with an increase in moving velocity. Xiang et al. [14] investigated a pavement system under moving vehicles to scavenge traffic-induced energy using piezoelectric transducers. When the velocity of a moving vehicle was close to a critical value, the power output of the piezoelectric harvester was optimal. Romero et al. [15] analyzed the energy harvesting performance of railway bridges in different operational conditions. The results showed that the passage of trains had a major effect on energy harvesting performance. Therefore, the typical cantilevered (linear) harvester always had an enormous challenge in scavenging energy from bridge vibration since the abundant vibration energy of a bridge excited by different vehicles exists within a broad frequency band.

At present, harvesting energy from base vibrations is being widely investigated, in which the magnetic or axial loading forces are often used to enhance harvesting performance. Accordingly, the nonlinearity caused by the introduced magnetic forces was studied in depth so as to improve the enhancement in harvesting ability [16]. With the development of the study, the configurations with different kinds of stability, e.g., mono-stability [17], bi-stability, tri-stability [18], quad-stability [19], and even penta-stability, were proposed and proved to be able to broaden the operating bandwidth effectively. For example, Fan et al. [20] proposed a mono-stable piezoelectric energy harvester to improve the efficiency of energy extraction from low-level excitations by two stoppers and four magnets. Naseer et al. [21] theoretically investigated harvesting energy from vortex-induced vibrations by using nonlinear attractive magnetic forces. Jang and Chen [22] derived the Fokker–Plank–Kolmogorov equation of a mono-stable Duffing oscillator with piezoelectric coupling. The effects of the bandwidth, initial conditions, and cubic nonlinearity on voltage were numerically studied. Erturk and Inman [23] conducted the theoretical and experimental studies on the high-energy orbits of a bi-stable energy harvester. Moreover, the relative advantages and trade-offs of bi-stable and mono-stable harvesters were presented by Zhao and Erturk [24]. Although the mono-stable energy harvester is more robust and reliable than the bi-stable one under stochastic excitations, the bi-stable energy harvester should be designed deliberately to work more efficiently for a given excitation. Kim and Seok [25] developed the mathematical model of an energy harvester with multi-stable (mono-, bi-, and tri-stable) characteristics. The results of the simulation showed that the tri-stable energy harvester had a wider bandwidth than both the bi-stable and linear ones. Wang et al. [26] utilized the tri-stable characteristics to enhance the energy harvesting performance of a galloping piezoelectric energy harvester.

In previous studies, the linear vibration energy harvesters had been adopted to scavenge energy from bridge vibration, but it was found that they could not work efficiently under broadband excitations. Hence, nonlinearity was introduced into the design of the energy harvester under different excitation types, e.g., the harmonic, stochastic, or flow-induced vibrations. The multi-stability created by magnetic interaction has been proved to be particularly effective. Striking improvement in energy harvesting performance has been demonstrated. However, in the multi-stable structures, the energy required to cross the potential barrier and produce a big deflection is fairly large. In this study, we tried using nonlinear magnetic forces to obtain an optimal shape for potential energy. To the best of the authors' knowledge, using magnetic nonlinearity to tailor potential energy has not been studied in the harvesting of bridge vibration energy. These techniques can produce large amplitudes by creating a desired potential energy shape rather than a multi-stable shape. In this paper, a scenario of harvesting the energy of a bridge by mono-stability is proposed, in which the magnetic forces are introduced to form a desirable potential energy and a high-energy orbit. The mono-stable piezoelectric energy harvester (MPEH) can produce a large output if it oscillates in the high-energy orbit. It was proved that the MPEH could reach a high efficiency in harvesting the vibration energy of a bridge. Firstly, for the particular mono-stable characteristic, corresponding analyses on potential energy, restoring force, and stiffness were carried out. The results could help determine the optimal shape for potential energy. Then, experimental studies were conducted, and the dynamic responses and electrical outputs were investigated under different conditions. Finally, a summary is presented and some conclusions are drawn at the end of this paper.

2. Energy Harvester for Bridge Vibration Excited by Moving Vehicles

The schematic of MPEH is shown in Figure 1, which is composed of a cantilever beam with two piezoelectric patches attached to the root, a tip magnet fixed at the beam's free end, and two fixed magnets. The three magnets are placed such that the tip and fixed magnets are magnetically attractive. By adjusting the separation distance (d) and gap distance (d_g), the potential energy can be tailored to own a flat-shaped bottom, which is beneficial for the MPEH to execute a large-amplitude vibration. In particular, the directions of the magnetic and elastic forces are opposite to each other, which could move the operating frequency band toward the low frequency, thus reducing the energy required to produce a large deflection of the harvester and improving the voltage and power outputs. The classical linear piezoelectric energy harvester (LPEH) often shows inefficiency since the abundant vibration energy excited by moving vehicles at different speeds exist in the form of a broad frequency band with small magnitude. In order to collect more electric energy, a large LPEH (especially a very long substrate layer with a length L_s) is required; however, it is difficult to meet the energy supply requirements of the densely distributed sensors. Therefore, scavenging energy from bridge vibration by linear structures has proved to be an enormous challenge. Compared with LPEH, the MPEH can attain the high-energy orbit under the excitation of passing vehicles and generate a high output.

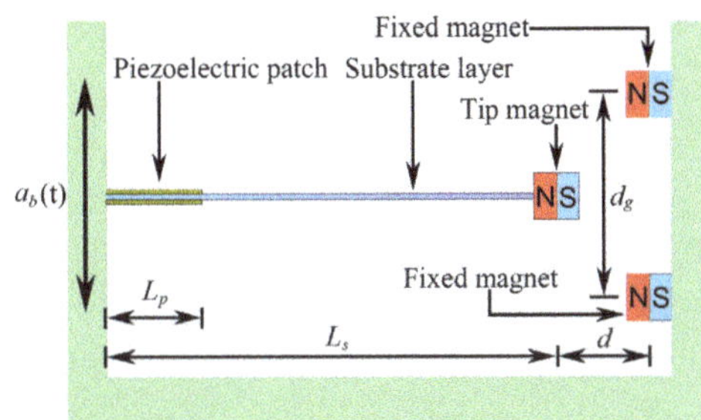

Figure 1. Schematic diagram of MPEH.

As Figure 2 shows, the MPEH is fixed to a bridge with the length of L_b and the thickness of T_b. Then, as a moving vehicle travels on the bridge at speed v, its load excites the elastic bridge to oscillate with the acceleration $a_b(t)$. Thus, the MPEH oscillates and generates electric power through piezoelectric materials. According to previous related studies, the bridge can be modeled as an Euler–Bernoulli beam, which is a simply supported elastic beam with infinite degrees of freedom. The moving vehicle travelling on this beam with hinged-hinged boundary conditions can be simplified as a moving mass.

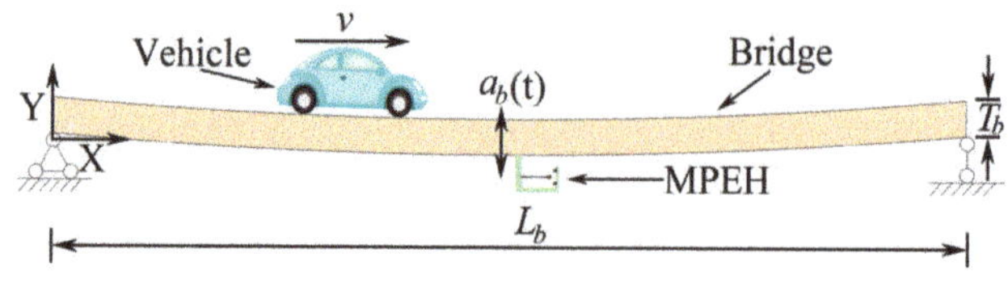

Figure 2. Schematic diagram of energy harvesting from bridge vibration excited by a moving vehicle.

3. Analysis of Potential Energy, Restoring Force, and Stiffness

The total potential energy of MPEH consists of two parts: the elastic potential energy of the substrate and the magnetic potential energy that originated from the magnets.

According to the Euler–Bernoulli beam theory, the strain of the piezoelectric cantilever is proportional to the second spatial derivative of deflection. Hence, the potential energy calculation in this paper has been based on the following two assumptions: (a) the flat section perpendicular to the center line of the beam before deformation still stays flat after deformation; (b) the plane of the deformed cross section is perpendicular to the deformed axis. Then, the elastic potential energy of the piezoelectric cantilever beam can be given by the following equation [27]:

$$U_e = \frac{1}{2}E_s I_s \int_0^{L_s} \left(\frac{\partial^2 w(x,t)}{\partial x^2}\right)^2 dx + E_p I_p \int_0^{L_p} \left(\frac{\partial^2 w(x,t)}{\partial x^2}\right)^2 dx \tag{1}$$

where E_s is the Young modulus of the substrate; I_s is the inertia moment of the substrate given by $I_s = \frac{W_s T_s^3}{12}$ (W_s and T_s are the width and thickness of the substrate, respectively); L_s is the length of the substrate; E_p is the bending stiffness of the piezoelectric patch; I_p is the total inertia moment of two piezoelectric patches on both sides of the substrate, which can be given by $I_p = \frac{W_p T_p (4T_p^2 + 6T_p T_s + 3T_s^2)}{6}$ (W_p and T_p are the width and thickness of the piezoelectric patch, respectively); L_p is the length of the piezoelectric patch; $w(x,t)$ is the displacement of the cantilever beam.

The magnetic potential energy can be obtained based on the assumption of a magnetic dipole. Thus, the magnetic potential energy generated by two fixed magnets upon the tip of the magnet can be expressed as follows [28]:

$$U_m = -\frac{\mu_0 a_1 a_2}{2\pi}\left\{\left(w(L_s,t) - \frac{d_g}{2}\right)^2 + d^2\right\}^{-\frac{3}{2}} - \frac{\mu_0 a_1 a_3}{2\pi}\left\{\left(w(L_s,t) + \frac{d_g}{2}\right)^2 + d^2\right\}^{-\frac{3}{2}} \tag{2}$$

where μ_0 is the permeability constant; a_1 and a_2 (a_3) are the effective magnetic moment of the tip magnet and the fixed magnets, respectively; d is the horizontal distance from the center of the tip magnet to the center of the fixed magnets; d_g is the gap distance between the two fixed magnets.

Since the potential energy is determined by the separation distance d, the total potential energy is computed with the distance d varying so as to illustrate the influence of d. The results are illustrated in Figure 3. The system parameters are given in Table 1. It can be seen from Figure 3 that as d decreases from 90 mm to 14 mm, the potential energy experiences the linear stable state, the mono-stable state, and then the bi-stable state. Specifically, when d is quite large, the system acts nearly as a linear one resulting from a very weak magnetic force. With a decrease in the separate distance d, the magnetic force will become large. The potential energy could own a steep bottom, implying that the MPEH may have a relatively large amplitude of vibration under external excitations. The ideal mono-stable configuration is expected to make a strong magnetic coupling between the magnetic force and the elastic force. This desirable case can provide a large beam deflection and then a high-voltage output under low-level bridge excitations. Especially if d is smaller than a critical value, i.e., 14 mm, the magnetic force will be larger than the elastic restoring force.

With two stable equilibrium positions emerging, the MPEH becomes a bi-stable system. However, in the bi-stable state, two potential wells and a potential barrier will appear, which would need more excitation energy. Thus, for weak excitations, the bi-stable MPEH may not be the best option. In contrast, the mono-stable state near the bi-stable one is the most appropriate choice.

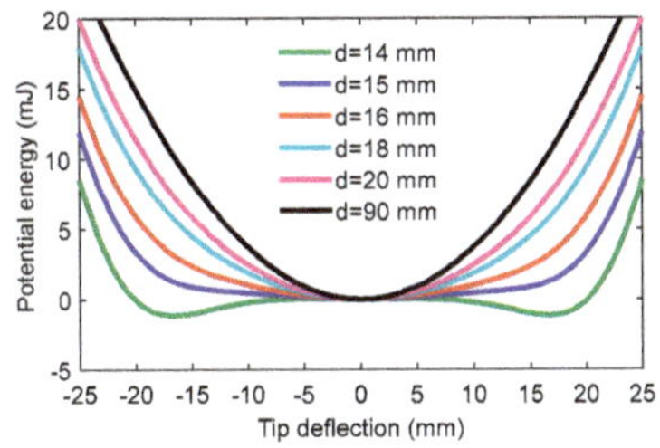

Figure 3. Potential energy for different values of separation distance.

Table 1. Properties for the analysis of potential energy function.

Symbol	Parameter	Value
Substrate		
L_s	Length	190 mm
T_s	Thickness	1 mm
W_s	Width	10 mm
E_s	Young modulus	205 GPa
Piezoelectric patch		
L_p	Length	5 mm
T_p	Thickness	5 mm
W_p	Width	0.25 mm
E_p	Young modulus	56 GPa
Magnet		
$a_1(a_2, a_3)$	Effective magnetic moment	0.218 Am^2
d_g	Gap distance between the two fixed magnets	45 mm
μ_0	Permeability constant of the magnet	4×10^{-7} NA^{-2}

To illustrate the variation of the MPEH's characteristics with respect to *d*, the restoring force and stiffness at different separation distances are depicted in Figure 4. It is obvious that the restoring force and stiffness are greatly reduced with a decrease in *d*. The reason is that the attractive magnetic force will keep increasing when *d* decreases and will reach nearly an equal value to the elastic force at a critical distance. The MPEH will have an extremely low restoring force and bending stiffness. Therefore, even under the weak vibration of the bridge, the large deflection of the MPEH is likely to happen, thereby producing a high-voltage output. In contrast, the linear harvester only has the satisfactory harvesting efficiency while the excitation frequency matches the resonance frequency. However, in the practical environment, the speeds and weights of vehicles passing the bridge usually exhibit stochastic characteristics. Therefore, the vibration energy generated by passing vehicles distributes over a fairly broad frequency band. Thus, in terms of practical excitation, the MPEH may be superior to the LPEH.

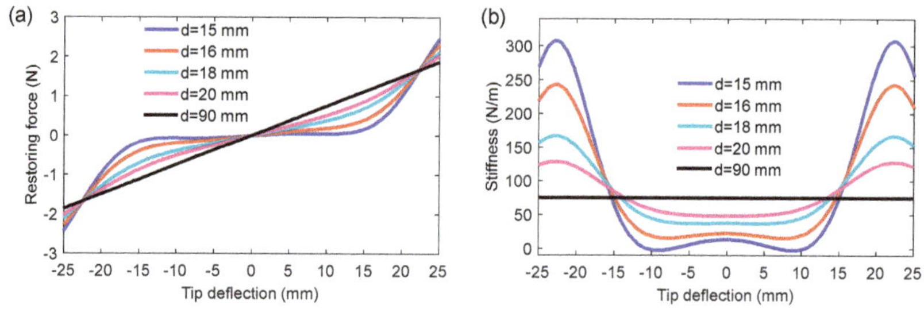

Figure 4. (**a**) Restoring force; (**b**) stiffness at various separation distances.

4. Experiment Setup

In order to show the dynamic response and performance of the electric output of the MPEH, a prototype of the MPEH is designed and fabricated. The experimental setup is shown in Figure 5. The model bridge is made from an acrylic sheet with the dimensions of 1300 mm × 130 mm × 8 mm (seen in Figure 5a). The moving vehicle is simulated by a moving steel ball with the mass of 1.003 kg, which can be accelerated from different heights of the acceleration track, as shown in Figure 5b. The guardrails are installed on both sides of the bridge to prevent the steel ball from falling off the bridge. The schematic diagram of the experiment equipment is presented in Figure 5c. The MPEH is fixed in the middle of the bridge, as shown Figure 6. The related parameter values are listed in Table 1. The Piezoelectric patch pasted onto the substrate of the MPEH is connected to a resistance box. The strain sensor (120-5AA) is attached to the substrate of the MPEH near the piezoelectric patch. When it passes the bridge, the moving load will excite the bridge to oscillate. As for the three magnets, one is attached to the free end of the piezoelectric beam, and the other two magnets are fixed at the fixture. The separation distance (d) and gap distance (d_g) are set to 17 mm and 45 mm, respectively. The LPEH is fabricated as well, as shown in Figure 7, which is similar to the MPEH in configuration but does not have the two fixed magnets. In the experiments, the LPEH and MPEH were put in the same testing environment for comparison. The strain response and voltage output of the harvester under bridge vibration were displayed and recorded by a digital signal acquisition device (DH5922N, Dong Hua). The value of the sampling frequency was set to 1000 Hz. In each test, to simulate different speeds of vehicles, the steel ball was released from different heights of the acrylic track.

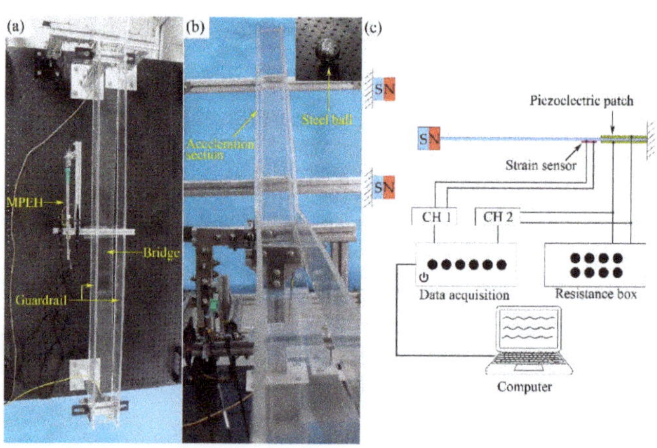

Figure 5. The experimental setup: (**a**) test section, (**b**) acceleration section, and (**c**) schematic diagram of equipment.

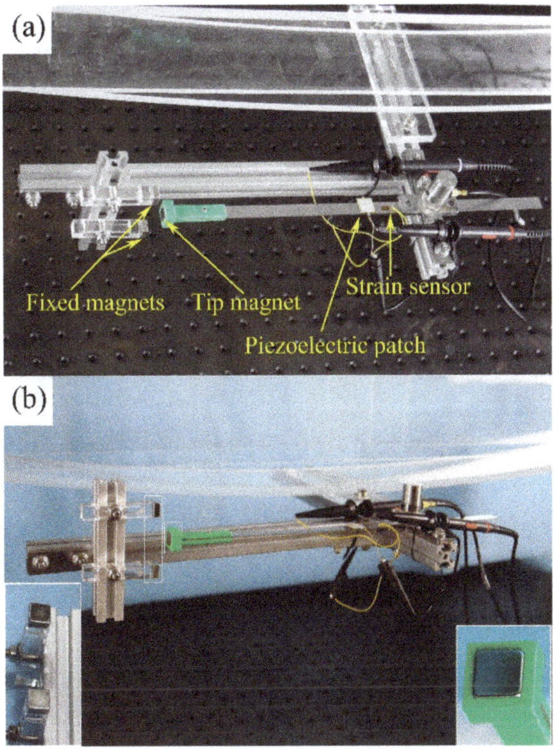

Figure 6. Prototype of the MPEH: (**a**) top view and (**b**) side view.

Figure 7. Prototype of the LPEH.

5. Results and Discussions

In the experiment, both the LPEH and MPEH were installed in the middle span of the bridge. A strain sensor and a piezoelectric patch were bonded to the root of the substrate so we could obtain the strain response of the system and the dynamic output voltage. Figure 8 gives the variance of strain and power density (W/m^3) at the load resistance of 0.9 MΩ for different moving speeds. To control the moving speed, the steel ball was put on an inclined track and released at different heights. The running interval was controlled to be between 0.50 s and 1.07 s, corresponding to the practical interval of a vehicle passing through bridges. From Figure 8, it is clear that for both the LPEH and MPEH, their variances of strain and power density increase with the moving speed. Especially for the speeds higher than 1.82 m/s, the MPEH's power density increases rapidly and outperforms that of LPEH, and so does the MPEH's strain.

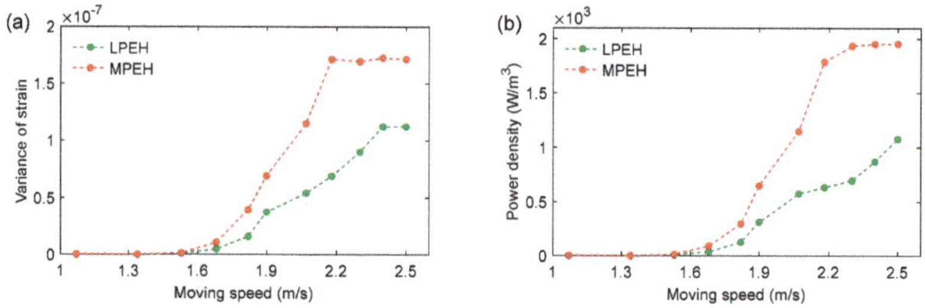

Figure 8. (**a**) Variance of strain; (**b**) power density of LPEH and MPEH for different moving speeds.

To show the dynamics of the MPEH and LPEH, the time histories of strain are illustrated in Figure 9, in which six moving speeds (v = 1.07, 1.53, 1.82, 2.07, 2.30, and 2.50 m/s) are considered. In Figure 9, at t = 0 s, the vehicle arrives at the bridge, while the dashed line denotes the moment the vehicle leaves the bridge. The label "On bridge" represents the time interval of the vehicle moving on the bridge, while the label "Leaving bridge" represents the time interval after the vehicle leaves the bridge. In all cases, it is apparent that the strain of the MPEH is much larger than that of the LPEH when the vehicle travels on the bridge. This can be explained by the fact that an impact load acts on the bridge at the instant when the vehicle just enters the bridge. The MPEH is designed to have two magnets, with which the potential energy can be tailored to have a wide and flat bottom. This potential energy is beneficial for the occurrence of a large deflection. Thus, the small vibration of the bridge can trigger the large deflection of the MPEH, which is desirable in energy harvesting. Specifically, at a slow speed, e.g., v = 1.07 m/s, the maximum strain value of the MPEH can reach 3.6×10^{-5}, which is 164% higher than that of the LPEH (1.36×10^{-5}). In both the MPEH and LPEH, the maximum strain increases with the moving speed. In any case, the vibration period of the MPEH is much longer than that of the LPEH, which is beneficial to harvesting weak vibration energy. After the vehicle leaves the bridge, the responses of the LPEH and MPEH exhibit a characteristic of decay vibration. Although the amount of time that the vehicle moves on the bridge is not very long due to its relatively fast speed (such as v = 2.07 m/s, 2.30 m/s, or 2.5 m/s), the frequency and amplitude of a large strain is satisfactory. From the comparison of strains between the MPEH and LPEH, it can be concluded that the deflection of the MPEH is much larger than that of the LPEH, particularly during the period when the vehicle is moving on the bridge.

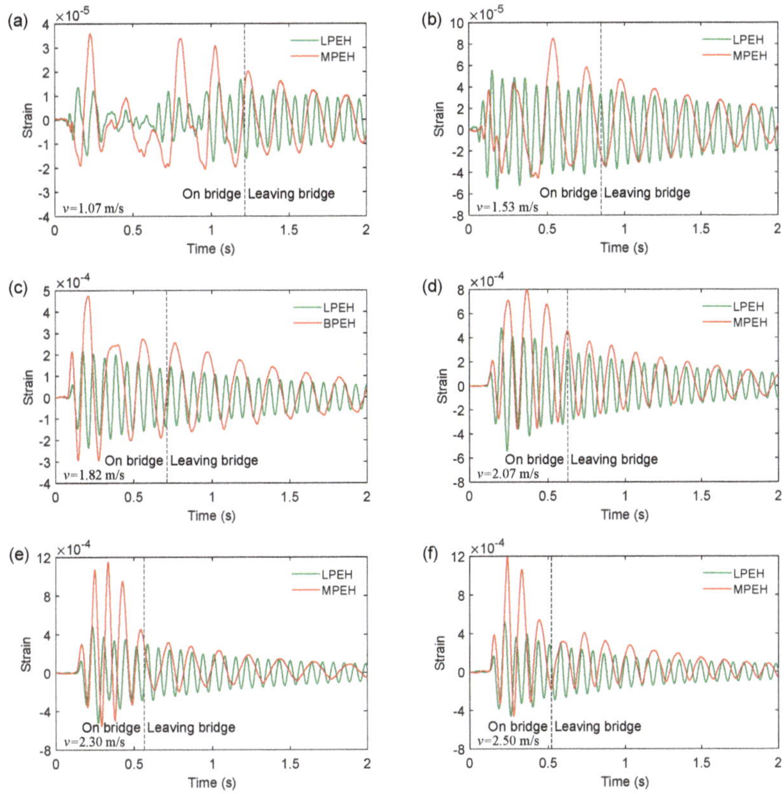

Figure 9. Strain responses of the LPEH and the MPEH for the vehicle travelling on and leaving the bridge: (**a**) v = 1.07 m/s, (**b**) v = 1.53 m/s, (**c**) v = 1.82 m/s, (**d**) v = 2.07 m/s, (**e**) v = 2.30 m/s and (**f**) v = 2.50 m/s.

Figure 10 shows the open-circuit voltages of the LPEH and MPEH when the vehicle travels on and leaves the bridge. By comparing Figure 9 with Figure 10, it can be concluded that the deflection of the harvester is closely related to the output voltage. Moreover, from Figure 10, it can be seen that the voltage of the MPEH is significantly larger than that of the LPEH. In all cases, the large output voltage is generated during the period when the vehicle is travelling on the bridge, in which the vibration amplitude makes a major contribution to the voltage output. For a slower speed of 1.53 m/s, the voltage amplitudes of the LPEH and MPEH are 0.41 V and 0.57 V, respectively. The vibrations caused by the impact load and the moving vehicle are very weak, which make the LPEH execute a small amplitude vibration. For the MPEH, owing to the nonlinear magnetic forces, the vibration amplitude is relative large under weak excitations. Thus, the MPEH could improve the voltage output at slower speeds. By increasing the moving speed, the bridge vibration caused by the moving vehicle becomes violent. Thus, the amplitude of the voltage output increases accordingly, resulting in a high harvesting efficiency. It should be noted that when the speed of the vehicle reaches a certain level (such as 2.30 m/s or 2.50 m/s), there is no significant increase in the voltage amplitude. For example, the maximum voltage of the MPEH reaches 6.94 V for the speed of 2.30 m/s. However, as the moving speed increases to 2.50 m/s, the maximum voltage only has a small increase and reaches 7.49 V. This is because the MPEH's two magnets are prompted to protect the piezoelectric material from damage when the beam deflection is too large.

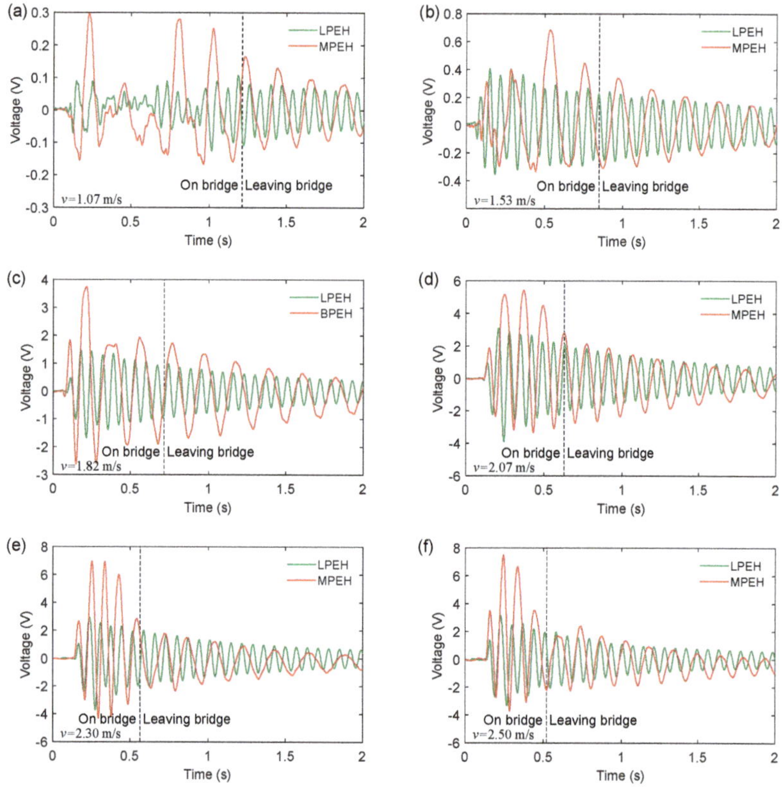

Figure 10. Voltage output of the LPEH and MPEH for the vehicle travelling on and leaving the bridge: (**a**) v = 1.07 m/s, (**b**) v = 1.53 m/s, (**c**) v = 1.82 m/s, (**d**) v = 2.07 m/s, (**e**) v = 2.30 m/s and (**f**) v = 2.50 m/s.

In order to better understand the dynamic characteristics of the LPEH and MPEH, their frequency spectra of output voltage are shown Figure 11. The peak of the LPEH is located near 14.5 Hz. In contrast, the MPEH's peak changes with the vehicle speed. For different moving speeds, unlike the LPEH, the MPEH has a different frequency spectrum distribution. More specifically, at v = 1.07 m/s, the frequency spectrum of the MPEH output is located at 4.90 Hz, and the voltage amplitude of the MPEH is significantly greater than that of the LPEH. As the moving speed increases to 2.07 m/s, the main peak of voltage shifts to 7.32 Hz, and the response energy mainly distributes over a wide frequency range of 0–10.0 Hz (Figure 9d). It should be noted that a peak appears at the extremely low frequency of 0.25 Hz, corresponding to the motion between the two fixed magnets or to the high-energy orbit. Then, by further increasing the moving speed, the frequency distribution of the voltage is changed, but the peak at the extremely low frequency increases steadily, implying that the response between the two fixed magnets enhances gradually. In particular, at v = 2.30 m/s and v = 2.50 m/s, from the frequency spectra of the voltage, it is apparent that the vibration energy is concentrated in the range of 0–15 Hz, with a peak of the extremely low frequency, as shown in Figures 9 and 10 (the vibration period of the MPEH is significantly larger than that of the LPEH). Therefore, for the MPEH, since the potential energy is controlled to have an optimal shape, the energy required to produce a large deflection is greatly reduced due to a comparatively strong coupling between the elastic force and the magnetic force. Especially in practice, the moving vehicles consecutively travel on a bridge, so the MPEH will be excited continuously. Then, the

MPEH can keep oscillating in a high-energy orbit and produce a consecutive large output. Thus, the MPEH is suitable for harvesting vibration energy from a practical traffic bridge.

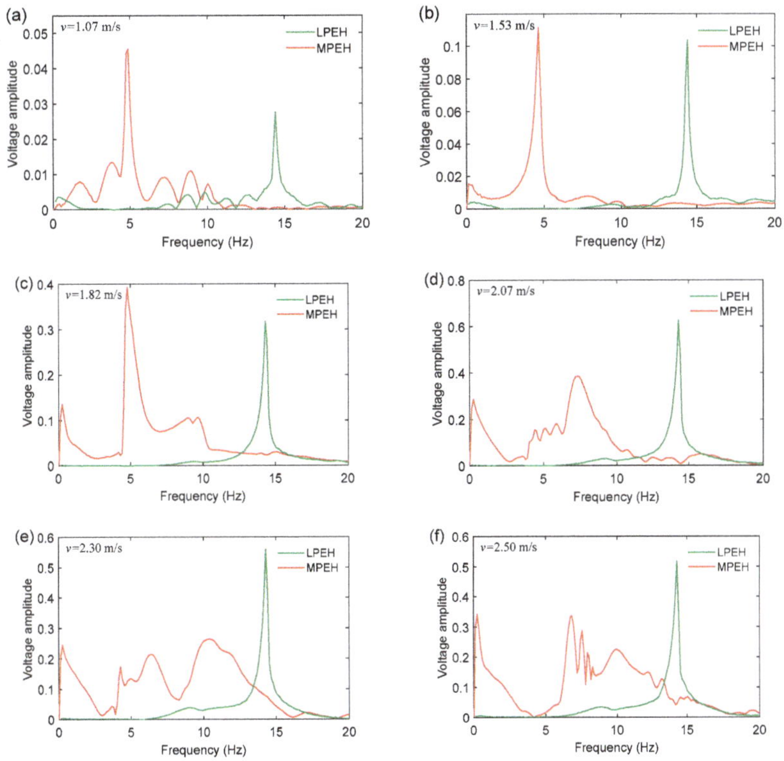

Figure 11. Frequency spectra of the output voltage for a vehicle travelling on and leaving the bridge: (**a**) v = 1.07 m/s, (**b**) v = 1.53 m/s, (**c**) v = 1.82 m/s, (**d**) v = 2.07 m/s, (**e**) v = 2.30 m/s and (**f**) v = 2.50 m/s.

6. Conclusions

In summary, this paper reports a novel concept on how to improve energy harvesting from traffic bridges. The magnets are introduced to tailor the potential energy rather than to create bi-stability. By adjusting the magnets' positions, the ideal mono-stable configuration with an extremely low restoring force and bending stiffness can be obtained. This desirable configuration can make the beam oscillate in a high-energy orbit and gives a large output under weak bridge excitations. The experimental results show that the energy harvesting performance of MPEH is significantly higher than the LPEH when the vehicle speed exceeds 1.82 m/s. For different moving speeds, the MPEH has an extreme peak in frequency spectrum, corresponding to a high-energy orbit. This study may open a new way for energy harvesters designed for traffic bridges. However, some further investigations are still needed to maximize the energy harvesting output. The separation distance and gap distance are two key factors for potential energy, the restoring force, and stiffness. The performance of the MPEHs can be promoted further by optimizing the tip magnet and the fixed magnet.

Author Contributions: Conceptualization, Z.Z. and H.Z.; methodology, P.Z.; software, Z.Z.; validation, Z.Z., W.Q. and W.D.; formal analysis, P.W.; investigation, H.Z.; resources, P.Z.; data curation, W.D. and P.W.; writing—original draft preparation, Z.Z., H.Z.; writing—review and editing, W.Q. and P.W.; visualization, P.W.; supervision, P.Z. and P.W.; project administration, Z.Z.; funding acquisition, Z.Z. and P.Z. All authors have read and agreed to the published version of the manuscript.

Funding: This work was supported by the National Natural Science Foundation of China (Grant No. 52005155), China Postdoctoral Science Foundation (Grant No. 2020M673470), Key Scientific Research Project of Colleges and Universities in Henan Province (Grant No. 20A130001), and Key Research Development and Promotion Project in Henan Province (Grant Nos. 202102310249, 212102310952).

Institutional Review Board Statement: Not applicable.

Informed Consent Statement: Not applicable.

Data Availability Statement: The data presented in this study are available on request from the corresponding author.

Conflicts of Interest: The authors declare no conflict of interest.

References

1. Reitsema, A.D.; Luković, M.; Grünewald, S.; Hordijk, D.A. Future Infrastructural Replacement Through the Smart Bridge Concept. *Materials* **2020**, *13*, 405. [CrossRef] [PubMed]
2. Iqbal, M.; Khan, F.U. Hybrid vibration and wind energy harvesting using combined piezoelectric and electromagnetic conversion for bridge health monitoring applications. *Energy Convers. Manag.* **2018**, *172*, 611–618. [CrossRef]
3. Fang, S.; Zhou, S.; Yurchenko, D.; Yang, T.; Liao, W.-H. Multistability phenomenon in signal processing, energy harvesting, composite structures, and metamaterials: A review. *Mech. Syst. Signal Process.* **2022**, *166*, 108419. [CrossRef]
4. Yang, K.; Wang, J.L.; Yurchenko, D. A double-beam piezo-magneto-elastic wind energy harvester for improving the galloping-based energy harvesting. *Appl. Phys. Lett.* **2019**, *115*, 193901. [CrossRef]
5. Wang, J.; Gu, S.; Zhang, C.; Hu, G.; Chen, G.; Yang, K.; Li, H.; Lai, Y.; Litak, G.; Yurchenko, D. Hybrid wind energy scavenging by coupling vortex-induced vibrations and galloping. *Energy Convers. Manag.* **2020**, *213*, 112835. [CrossRef]
6. Hou, W.; Li, Y.; Zheng, Y.; Guo, W. Multi-frequency energy harvesting method for vehicle induced vibration of rail transit continuous rigid bridges. *J. Clean. Prod.* **2020**, *254*, 119981. [CrossRef]
7. Assadi, H.; Hamani, I.D.; Tikani, R.; Ziaei-Rad, S. An experimental and analytical piezoelectric energy harvesting from a simply supported beam with moving mass. *J. Intell. Mater. Syst. Struct.* **2016**, *27*, 2408–2415. [CrossRef]
8. Ali, S.F.; Friswell, M.I.; Adhikari, S. Analysis of energy harvesters for highway bridges. *J. Intell. Mater. Syst. Struct.* **2011**, *22*, 1929–1938. [CrossRef]
9. Galchev, T.V.; Mccullagh, J.; Peterson, R.L.; Najafi, K. Harvesting traffic-induced vibrations for structural health monitoring of bridges. *J. Micromech. Microeng.* **2011**, *21*, 104005. [CrossRef]
10. Peigney, M.; Siegert, D. Piezoelectric energy harvesting from traffic-induced bridge vibrations. *Smart Mater. Struct.* **2013**, *22*, 095019. [CrossRef]
11. Erturk, A. Piezoelectric energy harvesting for civil infrastructure system applications: Moving loads and surface strain fluctuations. *J. Intell. Mater. Syst. Struct.* **2011**, *22*, 1959–1973. [CrossRef]
12. Zhang, Z.; Xiang, H.; Shi, Z. Mechanism exploration of piezoelectric energy harvesting from vibration in beams subjected to moving harmonic loads. *Compos. Struct.* **2017**, *179*, 368–376. [CrossRef]
13. Amini, Y.; Heshmati, M.; Fatehi, P.; Habibi, S. Piezoelectric energy harvesting from vibrations of a beam subjected to multi-moving loads. *Appl. Math. Model.* **2017**, *49*, 1–16. [CrossRef]
14. Xiang, H.J.; Wang, J.J.; Shi, Z.F.; Zhang, Z.W. Theoretical analysis of piezoelectric energy harvesting from traffic induced deformation of pavements. *Smart Mater. Struct.* **2013**, *22*, 095024. [CrossRef]
15. Romero, A.; Cámara-Molina, J.; Moliner, E.; Galvín, P.; Martínez-Rodrigo, M. Energy harvesting analysis in railway bridges: An approach based on modal decomposition. *Mech. Syst. Signal Process.* **2021**, *160*, 107848. [CrossRef]
16. Miao, G.; Fang, S.; Wang, S.; Zhou, S. A low-frequency rotational electromagnetic energy harvester using a magnetic plucking mechanism. *Appl. Energy* **2022**, *305*, 117838. [CrossRef]
17. Fan, K.; Tan, Q.; Liu, H.; Zhang, Y.; Cai, M. Improved energy harvesting from low-frequency small vibrations through a monostable piezoelectric energy harvester. *Mech. Syst. Signal Process.* **2019**, *117*, 594–608. [CrossRef]
18. Zhou, S.; Zuo, L. Nonlinear dynamic analysis of asymmetric tristable energy harvesters for enhanced energy harvesting. *Commun. Nonlinear Sci. Numer. Simul.* **2018**, *61*, 271–284. [CrossRef]
19. Zhou, Z.; Qin, W.; Zhu, P. Harvesting performance of quad-stable piezoelectric energy harvester: Modeling and experiment. *Mech. Syst. Signal Process.* **2018**, *110*, 260–272. [CrossRef]
20. Fan, K.; Tan, Q.; Zhang, Y.; Liu, S.; Cai, M.; Zhu, Y. A monostable piezoelectric energy harvester for broadband low-level excitations. *Appl. Phys. Lett.* **2018**, *112*, 123901. [CrossRef]

21. Naseer, R.; Dai, H.; Abdelkefi, A.; Wang, L. Piezomagnetoelastic energy harvesting from vortex-induced vibrations using monostable characteristics. *Appl. Energy* **2017**, *203*, 142–153. [CrossRef]
22. Jiang, W.-A.; Chen, L.-Q. Energy harvesting of monostable Duffing oscillator under Gaussian white noise excitation. *Mech. Res. Commun.* **2013**, *53*, 85–91. [CrossRef]
23. Erturk, A.; Inman, D. Broadband piezoelectric power generation on high-energy orbits of the bistable Duffing oscillator with electromechanical coupling. *J. Sound Vib.* **2011**, *330*, 2339–2353. [CrossRef]
24. Zhao, S.; Erturk, A. On the stochastic excitation of monostable and bistable electroelastic power generators: Relative advantages and tradeoffs in a physical system. *Appl. Phys. Lett.* **2013**, *102*, 103902. [CrossRef]
25. Kim, P.; Seok, J. Dynamic and energetic characteristics of a tri-stable magnetopiezoelastic energy harvester. *Mech. Mach. Theory* **2015**, *94*, 41–63. [CrossRef]
26. Wang, J.; Geng, L.; Zhou, S.; Zhang, Z.; Lai, Z.; Yurchenko, D. Design, modeling and experiments of broadband tristable galloping piezoelectric energy harvester. *Acta Mech. Sin.* **2020**, *36*, 592–605. [CrossRef]
27. Zhu, Y.; Zu, J.; Su, W.-J. Broadband energy harvesting through a piezoelectric beam subjected to dynamic compressive loading. *Smart Mater. Struct.* **2013**, *22*, 045007. [CrossRef]
28. Tang, L.; Yang, Y.; Soh, C.-K. Improving functionality of vibration energy harvesters using magnets. *J. Intell. Mater. Syst. Struct.* **2012**, *23*, 1433–1449. [CrossRef]

materials

MDPI

Article

Study of the Impact of Graphite Orientation and Ion Transport on EDLC Performance

Joseph M. Gallet de St Aurin [1] and Jonathan Phillips [2,*]

[1] US Navy, Naval Postgraduate School, GSEAS, Monterey, CA 93943, USA; joegallet@gmail.com
[2] Capacitor Foundry, Sand City, CA 93955, USA
* Correspondence: jonathan@capacitorfoundry.com; Tel.: +1-(505)-577-6584

Abstract: A model study of electric double layer capacitor (EDLC)-style capacitors in which the electrodes were composed of low surface area-oriented flakes of graphite that compressed to form a paper-like morphology has suggested that ion transport rates significantly impact EDLC energy and power density. Twelve capacitors were constructed, each using the same model electrode material and the same aqueous NaCl electrolyte, but differing in relative electrode orientation, degree of electrode compression, and presence/absence of an ionic transport salt bridge. All were tested with a galvanostat over a range of discharge currents. Significant differences in energy and power density and estimated series resistance were found as a function of all the factors listed, indicating that capacitor performance is not simply a function of the electrode surface area. This simple postulation was advanced and tested against data: net ion (Na^+, Cl^-) 'velocity' during both charge and discharge significantly impacts capacitive performance.

Keywords: dielectric; oriented graphite; electrolyte; EDLC; ion flow; capacitance

Citation: de St Aurin, J.M.G.; Phillips, J. Study of the Impact of Graphite Orientation and Ion Transport on EDLC Performance. *Materials* **2022**, *15*, 155. https://doi.org/10.3390/ma15010155

Academic Editors: Marc Cretin, Sophie Tingry and Carlos Manuel Silva

Received: 12 November 2021
Accepted: 22 December 2021
Published: 26 December 2021

1. Introduction

The development of capacitors with high energy and/or power density will help enable the proposed switch from combustion-based energy systems to 'green' electric systems that are recharged with renewables. Ideally, and theoretically, capacitor energy density could surpass that of batteries, allowing capacitors to replace batteries. Realistically, it is unlikely that capacitors will ever match the energy density of batteries, but the superior power output and durability of capacitors ensures that capacitors will have a number of niche rolls, including load leveling in systems for which batteries are the primary source (e.g., electric vehicles), battery life extension applications such as for satellites, collecting energy from high power surge sources such as regenerative braking, and providing pulsed power for inherently high power applications, as in [1–3].

Given the increased urgency of efforts to create a fully electric power system using only renewable energy sources, resources for research to improve capacitors, particularly electric double layer capacitors (EDLC), also known as Supercapacitors, have increased dramatically in the last decade. In 'carbon only' systems, as per this report, virtually all the increased research has been directed at exploring the effect of the microstructure (e.g., graphene, CNT) and the impact of various additives on carbon energy density, power density, and electric conductivity [1,2,4–12]. There are some general caveats to much of this research. As pointed out elsewhere, the fact that there are no universal protocols for capacitor testing makes comparisons and claims of superior performance difficult to verify [1,13–16]. In addition, other factors, including ionic conductivity, fabrication method, absolute electrode dimensions, even measurement set-up, are likely to impact performance [17]. There has been relatively little work designed to quantify the impact of these factors on carbon-based capacitor performance [13,17,18].

The novelty of the present work is the test of the impact of two overlooked factors on EDLC performance: the orientation of graphite planes in the electrodes, and enhanced pathways for ionic transport. Study of these factors show, for the first time, that they dramatically impact energy storage and power production. Regarding orientation, in the absence of any other modification, the orientation of the graphite flake-based electrodes relative to each other led to a factor of >3× change in energy density. Regarding enhancing ion transport, the use of a 'salt bridge' to enhance ion transport between electrodes increased both energy density and power production significantly. These results, from the simple 'model' system tested, provide experimentally quantitative support for the general postulation that many factors impact EDLC performance.

2. Materials and Methods

EDLC Microstructure: The electrodes were created from a commercial material, Grafoil, composed of flakes of graphite compressed to create a 'paper-like' material with a very modest surface area [19,20]. In prior work, it was demonstrated that the graphite flake basal planes are strongly oriented parallel to the macroscopic surface of the paper [21].

Two types of Grafoil were employed in capacitor electrodes: (1) Grafoil provided by the manufacture directly; (2) Grafoil mechanically compressed between stainless steel platens to 2000 psi using an Instron SATEC machine (Instron Corp, Norwood, MA, USA).

One simple measure of the impact of compression was the reduction in measured thickness. The measured thickness of the 'raw' material was 0.4 mm, which was reduced to 0.36 mm by compression. To determine the effect of compression on microstructure, both surface area and relative basal plane orientation were determined. A NOVA 4200e Surface Area and Pore Size analyzer (Quantachrome Instruments, Boynton Beach, FL, USA) was used to determine the BET surface area of uncompressed (15.5 m^2/g) and compressed (12.7 m^2/g), material. Notably, the measured value for the uncompressed material was similar to that reported by the manufacture, 18.5 m^2/g, indicating that the measurement technique yields reasonably good quantitative data. In sum, compression at 2000 psi reduces surface area moderately, ~20%.

To test for changes in orientation, anticipated on the basis of earlier studies, relative X-ray diffraction peak heights were determined with a Rigaku MiniFlex 600 benchtop X-ray diffractometer (Rigaku Analytical Devices, Wilmington, MA, USA) (Figure 1). Based on a comparison of intensity ratios between specific crystallographic planes for graphite, it was determined that compression did not modify the relative basal plane alignment of the Grafoil by a statistically significant amount. Both the compressed and uncompressed samples were found to have more than 90% of the graphitic crystals aligned with the macroscopic surface. This is not consistent with earlier work, though might reflect changes in the Grafoil manufacturing process over the decades between studies. That is, the commercial material presently available shows greater alignment of the basal planes and lower surface area than that recorded for the 'as received' material in prior decades.

EDLC Electrode Structure: To test the impact of orientation, electrodes were made of Grafoil that was folded to form an accordion structure, in which each fold was 1 cm across and 3 cm long (Figure 2).

Once the surface incisions were made, the Grafoil was folded along these incisions in an alternating pattern such that the incision formed the outer spine of the fold. The opposite side of each initial incision was folded into each "valley". Each electrode, dry, weighed approximately 1.5 g.

After the accordion-fold was complete, the inside surface of each valley fold was lightly scoured to promote liquid adhesion when the electrolyte was added. Prior to testing of each configuration, two drops of 3% NaCl electrolyte (approximately 0.05 mL/drop) were added to each valley fold on both faces of the electrode. The accordion-fold electrodes were then lightly compressed to close each of the folds.

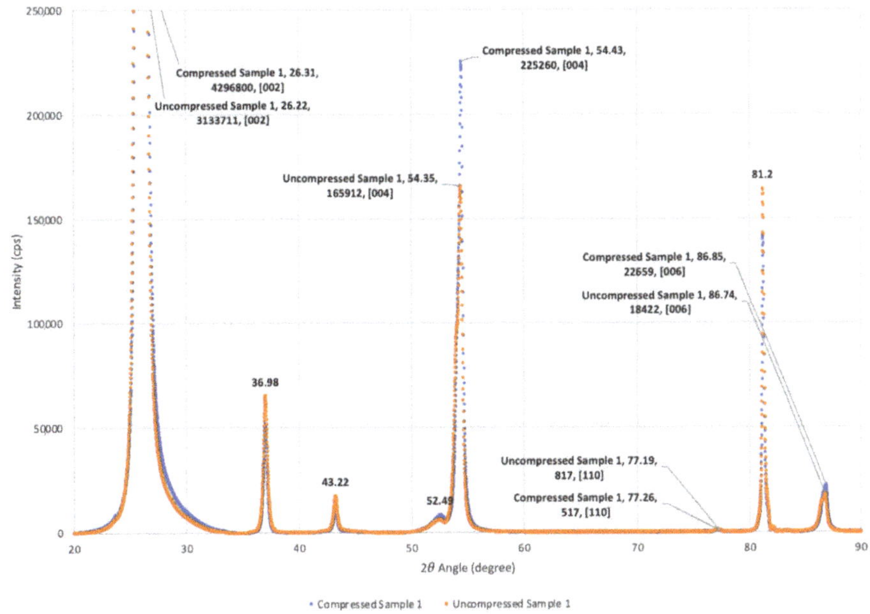

Figure 1. *Compressed and Uncompressed Grafoil XRD.* Both compressed and uncompressed XRD show high orientation of [002] graphite basal planes parallel to surface. For the true powder pattern diffraction pattern arising from ground Grafoil, the [110] (77.19° 2θ) and [112] (81.2° 2θ) diffraction lines are nearly equal in intensity. In contrast to the powder pattern, for both the compressed and uncompressed materials employed in this study, the [110] peak is nearly invisible, but the [112] peak is quite strong [21]. (Note: Only ~6% of the [002] reflection is shown in order to allow the relative intensity of the other peaks to be clearly shown).

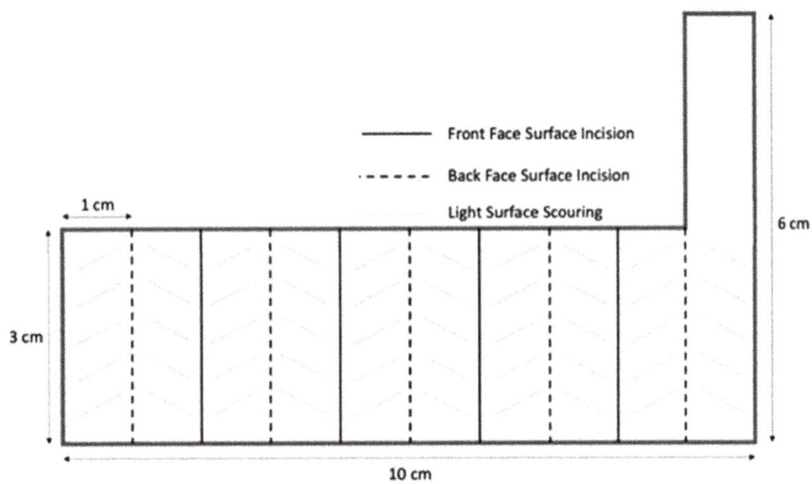

(A)

Figure 2. *Cont.*

(B)

(C)

Figure 2. *Accordion Electrode Structure*: (**A**) Dimensions shown. For each electrode, a section of Grafoil, 10 cm × 3 cm + a tab at one end, was cut, scored, and folded as shown. (**B**) After initial folding, ~0.01 cm^3 of 3% NaCl/DI electrolyte was added to each 'valley'. (**C**) The completed electrode.

Capacitor Construction: Following the construction of two Grafoil accordion electrodes, the next step in capacitor construction was to employ an electrically insulating, commercially woven nylon blend separator (84% nylon and 16% spandex) with a measured thickness of 0.4 cm and 50% open space, employed previously in parallel plate capacitors [22], as an electrically insulating, ion transport 'friendly' media between two, well oriented, electrodes. To enable ion transport, the woven nylon was wetted with 5 drops of the 3 wt% NaCl solution. In prior work with parallel plate capacitors, this electrolyte was found to qualify as an SDM. Indeed, in earlier work [18,22–24], aqueous NaCl solution generally had dielectric constants $>10^8$ at low frequency (ca. <1 Hz).

The three 'specific orientations' of electrodes that were employed are illustrated in Figure 3. In each configuration, Grafoil surface planes are oriented in a different manner relative to the separator. In the Basal Plane Orientation (BPO), all the macroscopic surfaces of the folded Grafoil are parallel to the separator. Ions diffusing through the separator need to either climb 'up and over' a Grafoil layer, or through one, to reach the next valley. In the Edge Plane Orientation (EPO), the Grafoil surface are perpendicular to the separator.

Ions diffusing through the separator could directly transport into all 'valleys' between the Grafoil sheets. In the Offset Orientation (OO), one electrode is arranged per the BPO, and one per the EPO.

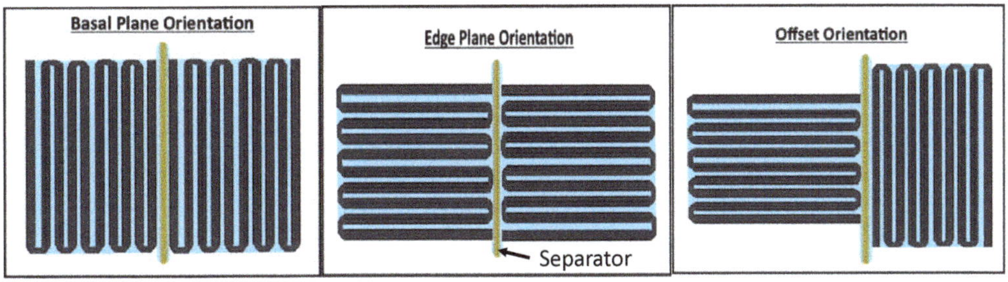

Figure 3. *Edge View of the Grafoil Sheet Orientation.* Capacitors were constructed using accordion folded Grafoil with three 'specific orientations' relative to the separator, as illustrated.

The next step was to create a 'compressed', mechanically stable, capacitor by using rubber bands to 'attach' the electrodes to a standard sized light microscope slide. In Figure 4, a capacitor in the offset configuration is shown with a 'salt bridge'. Indeed, each capacitor was tested in two configurations: with and without a 'salt bridge'. The net was twelve capacitors built and tested, each with a unique arrangement of the materials, orientations, and 'salt bridge' level: three different orientations × two different compression levels × two salt bridge configurations (Table 1).

Figure 4. *Salt Bridge.* Capacitor in offset configuration with molded 'salt bridge' at one end.

The salt bridge was prepared with a gel prepared by mixing 1.8 g of fumed silica (Sigma Aldrich, 0.007 µm avg. particle size, St. Louis, MO, USA) to which a NaCl (saltwater) solution, 0.6 g NaCl and 20 g DI, was slowly added. Similar gels were previously shown to be effective for ion transport in parallel plate-type capacitors [23,24]. Its use as a salt bridge between electrodes requires the same properties. Additionally, the gel held forms such that it was capable of being shaped. As shown, it was 'molded' as a relatively thick layer on one end of the capacitors, creating an avenue for ion transport between the two electrodes.

The final step, to prevent drying, was to place the capacitor, along with a small amount of wetted paper towel, inside a gallon-sized, low density polypropylene bag. Small slits allowed the tabs on the electrodes (Figure 2) to be connected to the galvanostat with alligator clips.

Table 1. Summary of Construction and Testing Protocol for 12 Capacitors.

NPS Configuration			Experiment Protocol						
Electrode	**Salt Bridge?**	**Orientation**	**Sequences per Technique**	**CCC**	**CVH**	**CCD Technique 1**	**2**	**3**	**4**
Uncompressed Grafoil	No Salt Bridge	BPO	6 Positive 6 Negative	25 mA	2.5 v, 1000 s	6 mA	3 mA	1 mA	0.5 mA
		EPO							
		OO							
	With Salt Bridge	BPO							
		EPO							
		OO							
Compressed Grafoil	No Salt Bridge	BPO							
		EPO							
		OO							
	With Salt Bridge	BPO							
		EPO							
		OO							

CCC—Constant current charge. CVH—Constant voltage hold. CCD—Constant current discharge. (Illustrated, Figure 4, bottom.).

Measurement Protocol: The test apparatus used in each experiment was the BioLogic VSP-300, a multichannel potentiostat/galvanostat. Each capacitor was studied using the same three step /bidirectional constant current method, a protocol similar to that used to characterize commercial capacitors [25]. Step 1: Charge to a selected positive voltage at constant current. Step 2: Hold capacitor at that voltage for a select period. Step 3: Discharge at a given constant current. A cycle is completed by repeating the Steps 1–3, except Step 1 is done to a negative voltage. Each cycle is repeated six times. Next, the discharge current in Step 3 is changed, and a second set of data collected. In total, six cycles, each at a different discharge current, were measured for each capacitor (Table 1). A typical cycle for one capacitor is shown in its entirety in Figure 5.

As noted earlier, there is no generally accepted protocol for permitting claims of superior performance, or even an absolute comparison of any capacitor properties [1,13–16]; however, for this study, that is not a significant concern. This study is not about creating a superior capacitor or comparison with prior published reports, but rather designing a methodology which allows for quantitative comparisons for energy storage and power delivery between capacitors with different orientations and different ion transport paths/mechanisms. The method employed undoubtedly serves this purpose well; moreover, it is consistent with the body of research conducted at the Naval Postgraduate School [1,18,22–24]. Prediction: any and all alternative approached to capacitive property measurement would lead to the same basic conclusions regarding the impact of graphite plane orientation and ion transport significance.

All techniques for measuring capacitance are imperfect [1]. An illustrative example are the inherent limits of impedance spectroscopy. This technique is generally limited at 300 K to study capacitive behavior below 25 mV, as higher temperatures lead to non-linear behavior. In fact, the fundamental non-linearity of capacitors is reflected in the need to model capacitor behavior with 'equivalent circuits', which are often very complex. The selection of constant current discharge for the present study was made because the results, using this approach, for low frequency discharge are the least convoluted, most 'transparent', available over wide voltage and current ranges, and comparable to a large body of literature [1,13–16,18,22–24].

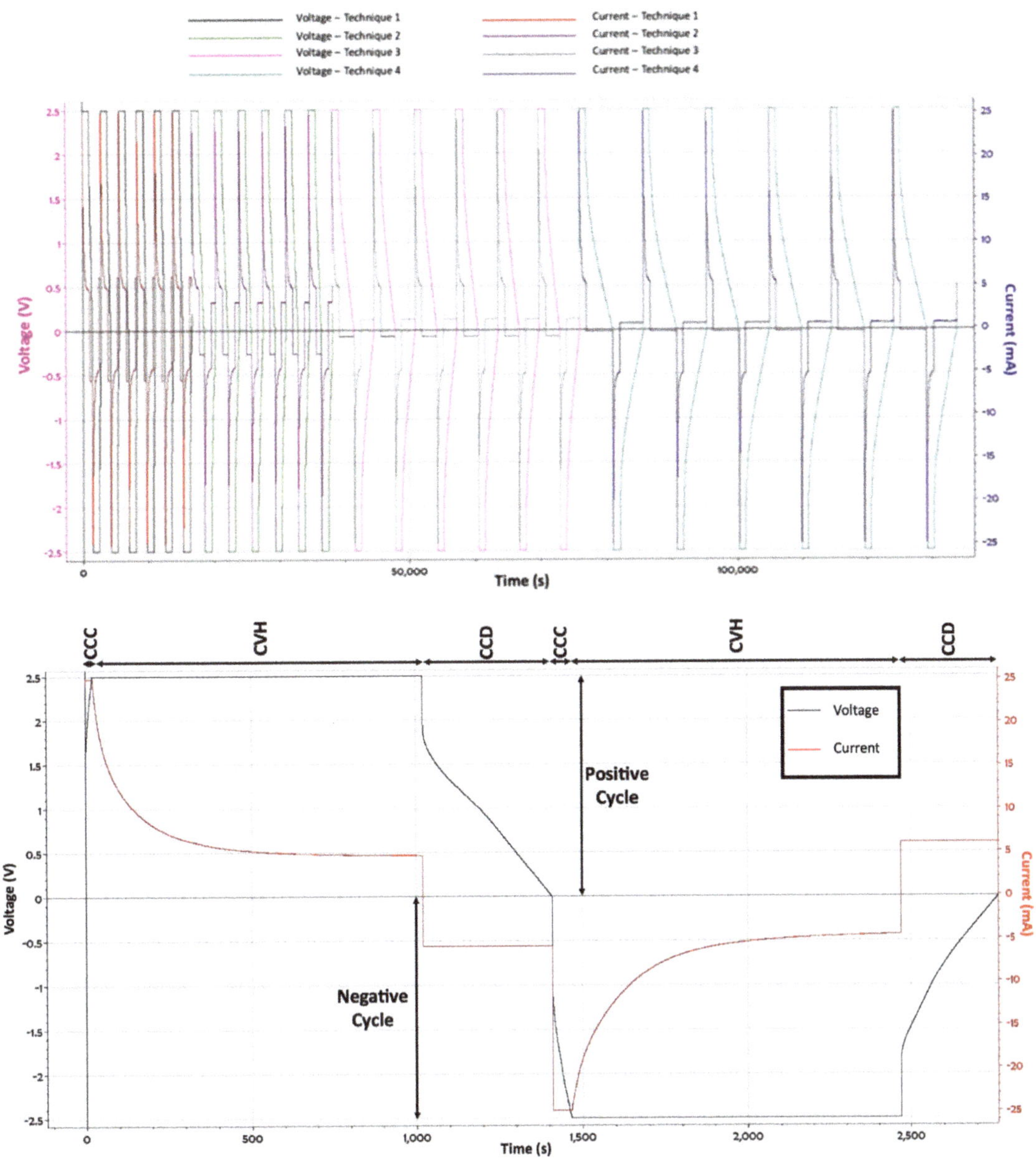

Figure 5. *Example Full Data Set for One Capacitor:* Top: All the data from one capacitor: four discharge currents and six cycles of each (Table 1). Bottom: An expansion of data showing a single cycle. Note: Theoretically, for a capacitor of constant capacitance, the discharge is linear with time. For these capacitors, that was approximately true below ~1.9 Volts. Above that value, voltage dropped sharply, and little energy was delivered to the load.

Three primary values obtained from the data, specific energy (J/kg), specific power density (W/kg), and capacitance below 1 volt (F/g), are reported here. The reported/plotted specific energy density (J/kg) value for each discharge current was the average of energy density values obtained from the twelve discharge curves (six positive voltage, six negative voltage) collected at that current, divided by the total weight of the dry electrodes. The energy density for each individual discharge curve was computed to be the integrated area under the entire discharge curve (V × sec) multiplied by the discharge current. 'Power density' (J/s* kg), a value appropriate for the quantitative comparison of 'power' between different capacitors, as established in earlier works [18,23,24], was obtained from each discharge curve by dividing the energy of the discharge by the total discharge time and total electrode mass. The capacitance below 1 volt was computed in the standard fashion [1] for constant current discharge, by dividing the current by the near linear value of the discharge curve (dV/dt). The ESR for each CHD sequence was determined using the industry standard method [25]: dividing the voltage drop ΔV that occurs within the first 10 ms by the constant current discharge rate I. The reported/graphed value at any given discharge current is the average obtained from twelve discharges.

3. Results

Overall, the data for the model system showed many of the qualitative results expected for capacitors, including: (1) The higher the discharge current, the higher the power density; (2) The higher the electrode surface area/kg, the higher the energy and power density. Three other results are not anticipated by standard theory, and have not been previously reported: (1) The higher the discharge current, the *higher* the energy density; (2) The geometry of the electrodes significantly impacts both energy and power density (3); A 'salt bridge' increases energy and power density. Indeed, the difference in energy density as a function of configuration is virtually eliminated by the salt bridge.

Specific Energy: The specific energies of each of the three configurations with no salt bridge, using either uncompressed or compressed Grafoil, are shown in Figure 6. The electrode configuration clearly impacts performance. BPO configurations resulted in considerably poorer performances than EPO and OO configurations. For example, at a discharge time of 100 s, the uncompressed BPO (~220 J/kg) had a measured specific energy approximately 28% that of the uncompressed EPO (~800 J/kg).

Another clear finding is that, for any particular configuration using compressed Grafoil, electrodes resulted in lower specific energy than its uncompressed counterparts. In the case of the BPO configuration, the difference is very significant. For example, at a discharge time of 1000 s, the compressed BPO electrodes have only about 25 percent the energy density of the uncompressed electrodes. The difference in performance between compressed and uncompressed Grafoil electrodes is significant, but not as dramatic, in the other two configurations. In both the OO and EPO configurations, the uncompressed energy density at a discharge time of 1000 s is <2× larger than that of the compressed electrode configuration.

It is notable that the energy vs. discharge time curves are rather 'flat', and, more significantly, have a 'negative' slope. That is, for EDLC and other capacitors, the energy delivered to the load generally increases with longer discharge times. In this work, the energy densities decreased as the discharge current was decreased, resulting in an increase in discharge time.

In Figure 7, the impact of the salt bridge on specific energy for the uncompressed Grafoil electrodes in the three configurations is shown. It is evident that the salt bridge increases the energy density in all cases. However, the most important finding is that the salt bridge nearly removes the difference in energy density as a function of configuration. With a salt bridge in place, the three configurations show very similar behavior. Indeed, with the salt bridge in place, the BPO and EPO behaviors are virtually identical and the OO performance is only ~25% better. Finally, it should be noted that the 'negative slope' as a function of discharge time is not impacted by a salt bridge.

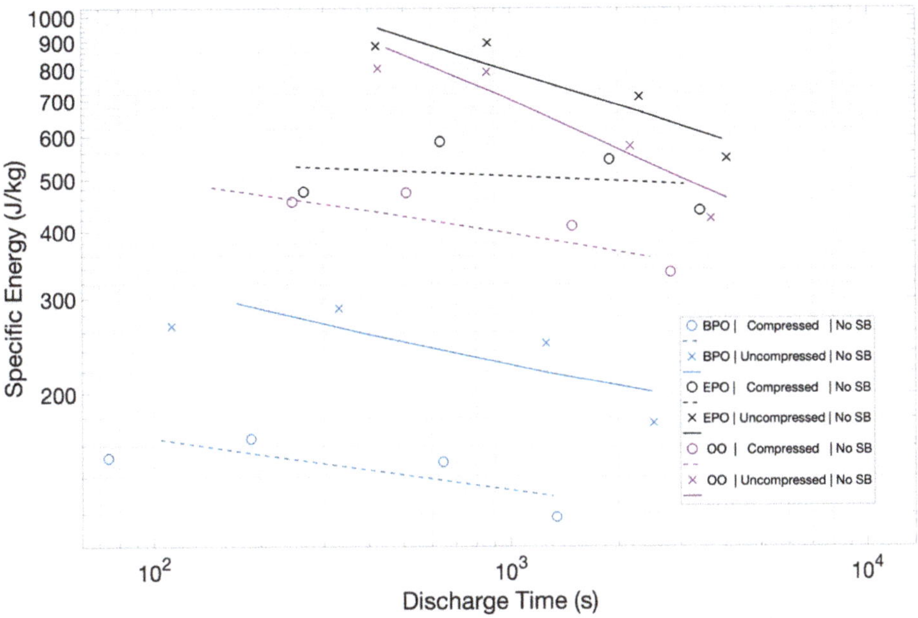

Figure 6. *Summary of Specific Energy for Salt Bridge Free Capacitors.* The energy density data for the six capacitors is shown. One previously unreported trend is apparent: The energy density is a strong function of electrode orientation. It is also clear that compressing/reducing the surface area of the graphitic electrode material reduces specific energy, and energy density decreases slowly with decreasing discharge time.

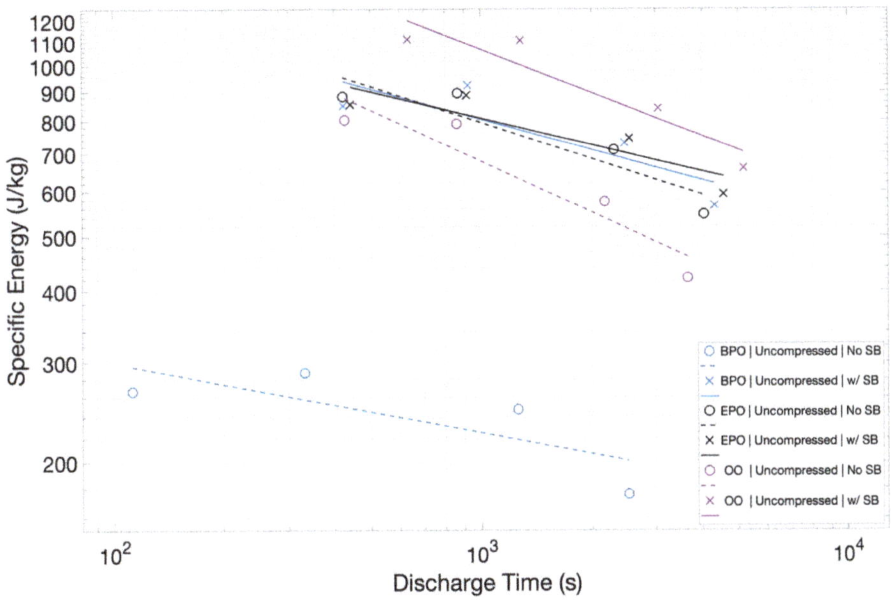

Figure 7. *Impact of Salt Bridge on Specific Energy.* The use of a salt bridge (solid lines) improves performance for all three configurations. More significantly, it 'collapses' the differences in energy density as a function of configuration.

It is reasonable to compare the capacitance (below 1 volt), and, by inference, energy density, of these electrodes with other pure carbon materials. In earlier studies, a graph of capacitance, as a function of the carbon surface area, demonstrated a linear relationship [2,26]. The slope of the line can be employed to predict capacitance on the basis of the measured surface area. For uncompressed Grafoil with a measured surface area of 15.5 m^2/g, this suggests a capacitance of 0.93 F/g. As shown in Figure 8, this value is about a factor of two better than that observed in the uncompressed material in the OO configuration. The figure also shows that comparisons of 'capacitance' are fraught with confusion, as so many parameters, such as voltage range, over which capacitance is measured, measurement frequency, ion identity, and details of construction, which are not simply carbon surface area, impact this value. In sum, the results of this study are qualitatively consistent with the anticipated capacitance based on earlier correlations. That is, the energy density per unit area of the electrodes is in close quantitative agreement with earlier findings.

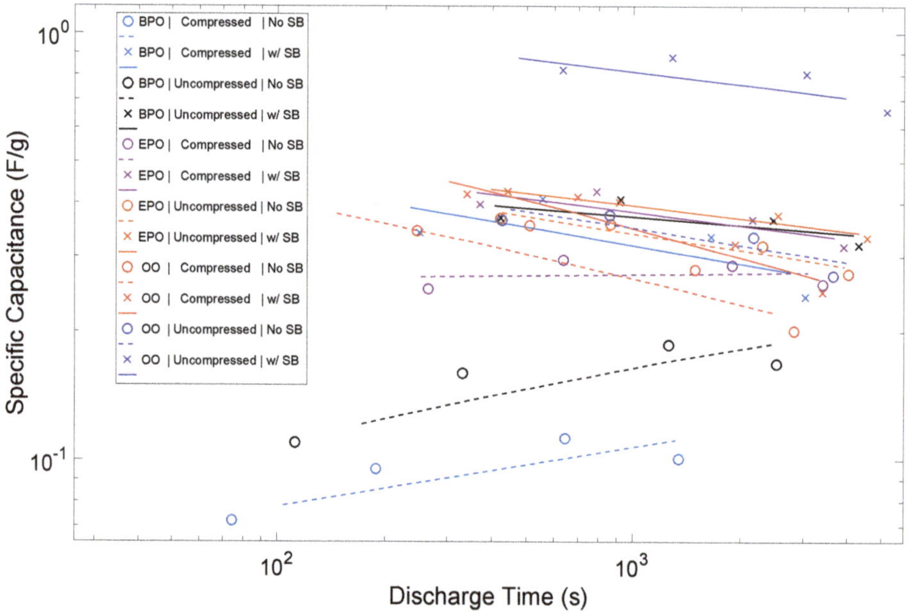

Figure 8. *Capacitance as a Function of Discharge Time.* Consistently, the presence of a salt bridge (solid lines) improves capacitance, all other factors were unchanged.

Specific Power: The specific power (W/kg) for the six capacitors, made with the uncompressed Grafoil, all three configurations with and without salt bridge, are shown (Figure 9). For the three capacitors without a salt bridge, as with the energy density results, the power density is clearly a function of the configuration. The BPO configuration without a salt bridge delivers the least power. For example, for a 1000 s discharge, it would deliver ~2.5×10^{-1} W/kg, which is about 30% of the power delivery (~8.0×10^{-1} W/kg) for the uncompressed EPO configuration.

As with the case for energy density, the power density differences collapse for the capacitors built with salt bridges. All three configurations with a salt bridge have very similar power density curves as a function of discharge time. Finally, it is notable that the power density curves are of standard slope. That is, as the discharge times decrease, and the delivered power increases.

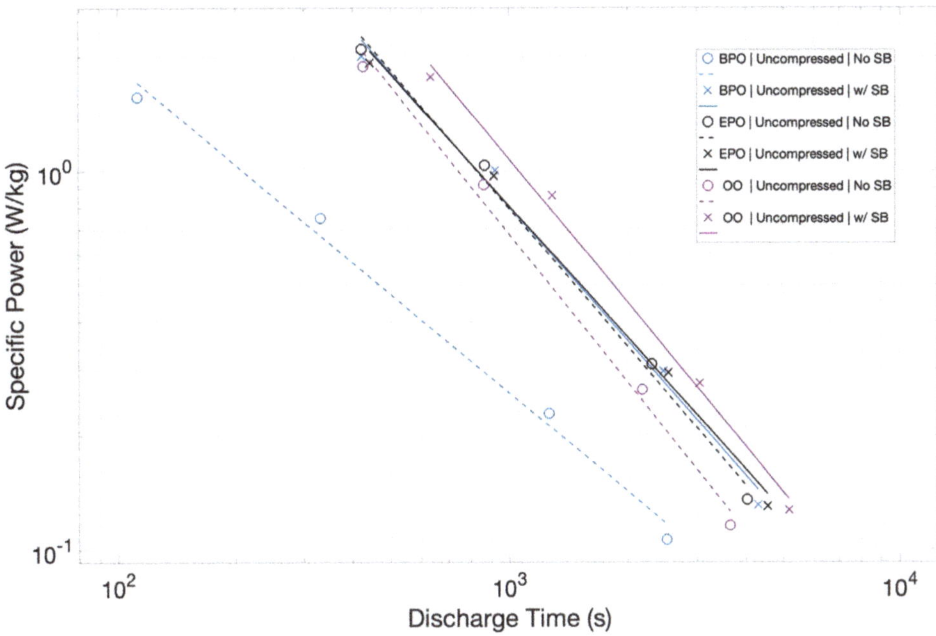

Figure 9. *Specific Power Increases with Higher Current/Lower Discharge Time.* Configuration impacts power significantly in the absence of a salt bridge, but the specific power value differences are almost independent of configuration when a salt bridge (solid lines) is added.

Estimated Series Resistance: The finding that the energy and power density values of all the capacitors studied herein increases with shorter discharges may reflect the unusual ESR behavior. As shown in Figure 10, the ESR values are not constant, but rather a function of the discharge time/current. For all the capacitors, compressed and uncompressed, with and without a salt bridge, ESR values are higher for smaller currents/longer discharges. Given that the load and the output resistance form a voltage divider, increased output resistance will concomitantly reduce the energy drop across the load. Hence, the ESR trend with the discharge time is both consistent with, and explains, the observed drop in the delivered energy at longer discharge times.

The ESR data show clear trend lines for any particular capacitor, but certainly do not suggest a clear quantifiable formula linking ESR to orientation, the state of compression, or presence of a salt bridge. Table 2 does suggest some weak qualitative correlations. First, for any given orientation, compression increases ESR, and second a salt bridge reduces ESR. It is also clear that there is not a strong orientation dependence. For example, all three configurations are the same with compressed/no salt bridge electrodes within experimental error. Uncompressed/salt bridge capacitors are all essentially identical.

Ragone Plots: Presenting the data in Ragone format helps illustrate the unique behavior of all these capacitors (Figure 11). In particular, plotting the data in the standard Ragone format shows that the 'slopes' are inverted. In general, Ragone charts show that, for both batteries and capacitors, delivering more power comes at the expense of reducing the energy delivered. Thus, the lines of conventional capacitors and batteries, plotted as per Figure 11, show a negative slope. For all capacitors studied herein, there is no trade-off. Both energy and power increase as the discharge time is reduced, yielding positive slopes.

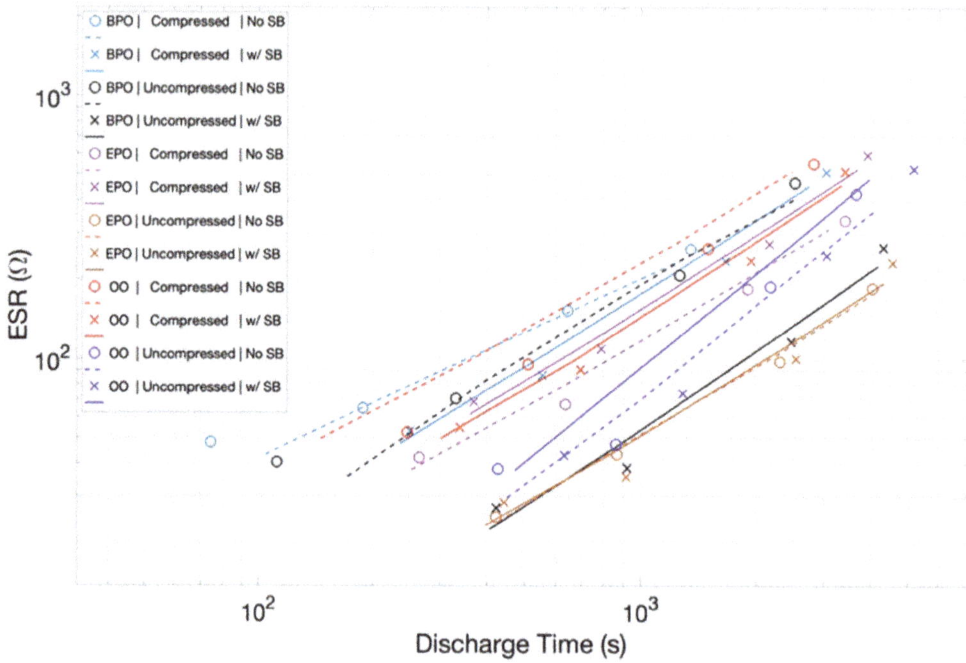

Figure 10. *ESR of all capacitors as a Function of Discharge Rate.* It is notable that ESR is a strong function of discharge time for all capacitors studied.

Table 2. ESR as a function of orientation, compression, and salt bridge.

ORIENTATION	STRUCTURE	RESISTANCE, Ohms (@1000 s)
BPO	Uncompressed/No salt bridge	150
BPO	Compressed/No salt bridge	200
BPO	Uncompressed/Salt bridge	30
BPO	Compressed/Salt Bridge	190
EPO	Uncompressed/No salt bridge	60
EPO	Compressed/No salt bridge	200
EPO	Uncompressed/Salt bridge	30
EPO	Compressed/Salt Bridge	100
OO	Uncompressed/No salt bridge	70
OO	Compressed/No salt bridge	210
OO	Uncompressed/Salt bridge	30
OO	Compressed/Salt Bridge	25

The power delivered curves also show a remarkably high slope. This reflects the fact that, unlike standard capacitive behavior, the energy delivered increases as the discharge time is reduced. The power delivered is the ratio of the energy delivered to the time of delivery. Conventionally, the numerator term decreases with shorter discharge time. Still, typically, this value does not decrease as rapidly as the denominator, hence the power increases with the decreasing discharge time/higher current. For the capacitors studied herein, the numerator increased, and the denominator decreased, leading to a sharp slope.

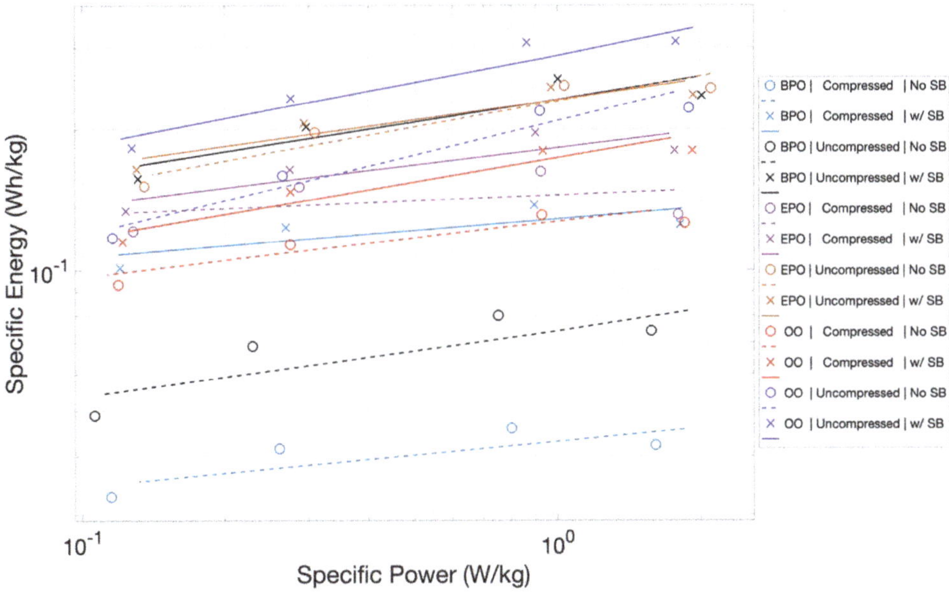

Figure 11. *Ragone Plot.* These plots show the same configuration and salt bridge impacts observed in the specific energy and power plots. Also notable is that the plots are all positive slopes, whereas, in general, for capacitors, the slopes on a Ragone chart are negative.

4. Discussion

This study was conducted to determine, on a fundamental level, whether either or both orientation and ion transport 'velocity' can impact the performance of EDLC capacitors. In order to minimize the number of complicating factors, the electrodes were created with low surface area, pure graphite flakes, strongly oriented with basal planes parallel to the surface.

The study, as intended, produced clear empirical findings regarding the target questions of the impact of electrode orientation and the potential impact of ion transport. Regarding electrode orientation: As shown in Figure 6, the orientation can change the energy density by a factor of more than 3×, and, as shown in Figure 8, the orientation can impact the power density by more than a factor of 2. Regarding ion transport: In all cases, the use of a salt bridge increased performance, although the magnitude varied as a function of electrode orientation. Capacitors with electrodes in orientations that yielded higher energy and power density even without salt bridges were marginally improved. The orientation with the lowest energy and power density without a salt bridge, BPO, improved significantly. In net, using a salt bridge effectively made all electrode orientations roughly equivalent performers. Postulate: Ion transport in EDLC can be performance determining. Note: It is understood that improved ion transport only improves performance to a finite limit.

Other data supports the suggestion that ion transport, and not always electron transport, can be performance limiting. For example, it was found that compression of the Grafoil led to a 25% reduction in surface area, but far larger fractional reductions in energy and power density in all cases. Indeed, the reduction was as great as an order of magnitude decrease in these values. Yet, compression, logically, should increase electron transport by enhancing contact between graphitic plates and concomitantly reduce ion transport by 'shrinking' or eliminating channels for ion transport. Thus, the large reduction in performance to Grafoil compression can logically be attributed to reductions in ion transport, but not to a reduction in the electron conductivity of the electrodes.

Can the velocity of ion transport somehow explain the inverted slope for the energy vs. discharge time curves (Figures 6, 8 and 10)? Postulate: The energy and voltage of the charged species, electrons, on the electrodes are higher if more 'ionic' dipoles (Na^+ or Cl^-) are present in the electric double layer. This concept is a variation on the recently postulated Theory of Superdielectric Materials (T-SDM), as discussed elsewhere [27–29]. Thus, given a constant charging period, a system with faster ion transport will allow more ions to travel and 'add' to the electric double layer than a system with lower ion transport rates. Conversely, this same ion transport 'advantage' can become disadvantageous the greater the discharge time. For lower discharge currents/longer discharge times, more ions 'retreat' and neutralize (e.g., $Na^+ + Cl^- \Rightarrow NaCl$). This reduces the number of electric dipoles in the boundary layer, and hence reduces the energy and voltage of the charges remaining on the electrodes. This implies that, the faster the electrons are removed, the fewer ions, which clearly move much more slowly than electrons, will neutralize. Consequently, rapid discharge should lead to more net energy falling on the load, as observed.

In order to gain additional insight into the impact of various factors on 'transport', the ESR was measured for all capacitors. The impact of a salt bridge on the ESR, an addition to the capacitor which should only impact ion transport, is revealing. In all cases the salt bridge reduced the ESR value. This trend raises this question: why should ion transport change net resistance? Is not net resistance a function of electron transport? A postulated answer: Ion and electron transport are 'coupled'. If ion transport is enhanced, so too is electron transport. The physics behind this proposed coupling is not obvious, or is at least 'complicated', and was not be considered here.

5. Conclusions

In conclusion, the key empirical finding is that, in carbon-based EDLC, enhanced ion transport in the electrolytic material improves performance. Increasing ion transport rates via the proper orientation of electrodes, the inclusion of a salt bridge, or an increased porosity also increases energy and power density. Moreover, it appears that ion and electron transport are linked. If ion transport is enhanced, electron transport is as well. The empirical findings are certain, and a simplistic model of improved performance linking to ion transport enhancement appears to be at least consistent with all observations. Yet, the underlying physics is not clear. Why are ion transport and electron transport coupled? As the ions retreat/neutralize, does this reduce the energy of the charges remaining on the electrodes?

Author Contributions: Concepetulalization: J.P.; methodology, J.P. and J.M.G.d.S.A.; software J.M.G.d.S.A.; validation, J.P. and J.M.G.d.S.A.; formal analysis, J.P. and J.M.G.d.S.A.; investigation, J.M.G.d.S.A.; resources, J.P.; data curation, J.M.G.d.S.A.; writing—original draft preparation, J.M.G.d.S.A.; writing—review and editing, J.P.; visualization, J.M.G.d.S.A.; supervision, J.P. project administration, J.P.; funding acquisition, J.P. and J.M.G.d.S.A. All authors have read and agreed to the published version of the manuscript.

Funding: No external funding supported this work.

Institutional Review Board Statement: Not applicable.

Informed Consent Statement: Not applicable.

Data Availability Statement: A great deal of the source data can be found here: Joseph Gallet de St. Aurin, M.S. Thesis: FOLDED CARBON ELECTRODES: A NOVEL APPROACH TO ENHANCING SUPERCAPACITOR PERFORMANCE Naval Postgraduate School. 2021.

Acknowledgments: Claudia C. Luhrs, MAE Dept., Naval Postgraduate School, assisted with XRD and BET measurements.

Conflicts of Interest: The authors declare no conflict of interest.

References

1. Cortes, F.J.Q.; Suarez, A.; Phillips, J. *Innovations in Engineered Porous Materials for Energy Generation and Storage Application*; Ranjusha, R., Avinash, B., Eds.; CRC Press: New York, NY, USA, 2018; Chapter 13; p. 343.
2. Beguin, F.; Raymundo-Pinero, E.; Frackowiak, E. *Electrical Double-Layer Capacitors and Pseudocapacitors in Carbons or Electrochemical Energy Storage Conversion Systems*; Beguin, F., Frackowiak, E., Eds.; CRC Press: New York, NY, USA, 2010.
3. Naval Sea Systems Command. *Naval Power and Energy Systems Technology Development Roadmap*; Naval Sea Systems Command: Washington, DC, USA, 2019. Available online: https://www.navsea.navy.mil/Portals/103/Documents/2019_NPES_TDR_Distribution_A_Approved_Final.pdf?ver=2019-06-26-132556-223 (accessed on 18 October 2021).
4. Abioye, A.M.; Nasir, F. Recent development in the production of activated carbon electrodes from agricultural waste biomass for supercapacitors: A review. *Renew. Sustain. Energy Rev.* **2015**, *52*, 1282–1293. [CrossRef]
5. Raymundo-Pinero, A.; Khomenko, V.; Frackowiak, E.; Beguin, F.J. Performance of manganese oxide/CNTs composites as electrode materials for electrochemical capacitors. *J. Electrochem. Soc.* **2005**, *152*, A229. [CrossRef]
6. Mombeshora, E.T.; Nyamori, V.O. A Review of the use of carbon nanostructured materials in electrochemical capacitors. *Int. J. Energy Res.* **2015**, *39*, 1955–1980. [CrossRef]
7. Lokhande, V.C.; Lokhande, A.C.; Lokhande, C.D.; Kim, J.H.; Ji, T. Supercapacitor Composite metal oxide electrodes formed with carbon metal oxides and conducting polymers. *Alloys Compd.* **2016**, *682*, 381–403. [CrossRef]
8. Deng, Y.; Ji, X. Review on recent advances in nitrogen-doped carbons: Preparations and applications in supercapacitors. *J. Mater. Chem. A* **2016**, *4*, 1144–1173. [CrossRef]
9. Chee, W.K.; Lim, H.N.; Zainal, Z.; Huang, N.M.; Harrison, I.; Andou, Y. Flexible Graphene-based Supercapacitors: A Review. *J. Phys. Chem. C* **2016**, *120*, 4153–4172. [CrossRef]
10. Borenstein, A.; Hanna, O.; Attias, R.; Luski, S.; Thierry, S.; Brousse, T.; Aurbach, D. Carbon-based composite materials for supercapacitor electrodes: A review. *J. Mater. Chem. A* **2017**, *5*, 12653–12672. [CrossRef]
11. Saikia, B.K.; Benoy, S.M.; Bora, M.; Tamuly, J.; Pandey, M.; Bhattacharya, D. A brief review on supercapacitor energy storage devices and utilization of natural carbon resources as their electrode materials. *Fuel* **2020**, *282*, 118796. [CrossRef]
12. Vinodh, R.; Gopi, C.V.M.; Kummara, V.G.R.; Atchudan, R.; Ahamad, T.; Sambasivam, S.; Yi, M.; Obaidat, I.M.; Kim, H.J. A review on porous carbon electrode material derived from hypercrosslinked polymers for supercapacitor applications. *J. Energy Storage* **2020**, *32*, 101831. [CrossRef]
13. Gogotsi, Y.; Simon, P. True performance metrics in electrochemical energy storage. *Sci. Mag.* **2011**, *334*, 917–918. [CrossRef]
14. Shao, Y.; El-Kady, M.F.; Sun, J.; Li, Y.; Zhang, Q.; Zhu, M.; Wang, H.; Dunn, B.; Kaner, R.B. Design and mechanisms of asymmetric supercapacitors. *Chem. Rev.* **2018**, *118*, 9233. [CrossRef]
15. Stoller, M.D.; Ruoff, R.S. Best practice method for determining an electrode material performance for ultracapacitors. *Energy Environ. Sci.* **2010**, *3*, 1294. [CrossRef]
16. Laheaar, A.; Przygocki, P.; Abbas, Q.; Beguin, F. Appropriate methods for evaluation the efficiency and capacitive behavior of different types of supercapacitors. *Electrochem. Commun.* **2015**, *60*, 21. [CrossRef]
17. Zhang, S.; Pan, N. Supercapacitor performance evaluation. *Adv. Energy Mater.* **2015**, *5*, 1401401. [CrossRef]
18. Lombardo, S.M.; Phillips, J. *Supercapacitors: Theoretical and Practical Solutions*; Liudvinavicius, L., Ed.; InTech: London, UK, 2017; Chapter 4.
19. Phillips, J.; Clausen, B.; Dumesic, J.S. Iron pentacarbonyl decomposition over grafoil. production of small metallic iron particles. *J. Phys. Chem.* **1980**, *84*, 1814. [CrossRef]
20. Phillips, J.; Dumesic, J.A. Iron pentacarbonyl decomposition over grafoil ii: Effect of sample outgassing on decomposition kinetics. *Appl. Surf. Sci.* **1981**, *7*, 215. [CrossRef]
21. Wu, N.L.; Phillips, J. XRD evidence of preferential orientation of platinum crystallites on graphite. *Surf. Sci.* **1987**, *184*, 463. [CrossRef]
22. Phillips, J. Novel superdielectric materials: Aqueous salt solution saturated fabric. *Materials* **2016**, *9*, 918. [CrossRef] [PubMed]
23. Jenkins, N.; Petty, C.; Phillips, J. Investigation of fumed silica/aqueous nacl superdielectric material. *Materials* **2016**, *9*, 118. [CrossRef]
24. Petty, C.; Phillips, J. Super dielectric material based capacitors: Punched membrane/gel. *Electron. Mater.* **2018**, *47*, 4902. [CrossRef]
25. Note: Typical Measurement Waveform Characterization in Industry Found Here. Available online: http://52ebad10ee97eea25d5e-d7d40819259e7d3022d9ad53e3694148.r84.cf3.rackcdn.com/UK_MAX_16V-Small-Cell-Module_DS_CAT-20.pdf (accessed on 15 September 2021).
26. Barbieri, O.; Han, M.; Herzog Kotz, R. Capacitance of Activated Carbide Derived Carbons. *Carbon* **2005**, *43*, 1303. [CrossRef]
27. Phillip, J.; Roman, A. Understanding dielectrics: Impact of External Salt wWater Bath. *Materials* **2019**, *12*, 2033. [CrossRef] [PubMed]
28. Phillips, J. Toward an improved understanding of the role of sielectrics in capacitors. *Materials* **2018**, *11*, 1519. [CrossRef] [PubMed]
29. Phillips, J. Theoretical and experimental basis for the super dielectric model of dielectric materials. *Phys. Essays* **2020**, *33*, 306. [CrossRef]

materials

MDPI

Article

Improving Fast Charging-Discharging Performances of Ni-Rich $LiNi_{0.8}Co_{0.1}Mn_{0.1}O_2$ Cathode Material by Electronic Conductor $LaNiO_3$ Crystallites

Tongxin Li, Donglin Li *, Qingbo Zhang, Jianhang Gao, Long Zhang and Xiaojiu Liu

New Energy Materials and Devices Laboratory, School of Materials Science and Engineering, Chang'an University, Xi'an 710064, China; 2018031001@chd.edu.cn (T.L.); 2019131013@chd.edu.cn (Q.Z.); 2019231009@chd.edu.cn (J.G.); 2020131022@chd.edu.cn (L.Z.); 2021231096@chd.edu.cn (X.L.)

* Correspondence: dlli@chd.edu.cn

Abstract: Fast charging-discharging is one of the important requirements for next-generation high-energy Li-ion batteries, nevertheless, electrons transport in the active oxide materials is limited. Thus, carbon coating of active materials is a common method to supply the routes for electron transport, but it is difficult to synthesize the oxide-carbon composite for $LiNiO_2$-based materials which need to be calcined in an oxygen-rich atmosphere. In this work, $LiNi_{0.8}Co_{0.1}Mn_{0.1}O_2$ (NCM811) coated with electronic conductor $LaNiO_3$ (LNO) crystallites is demonstrated for the first time as fast charging-discharging and high energy cathodes for Li-ion batteries. The $LaNiO_3$ succeeds in providing an exceptional fast charging-discharging behavior and initial coulombic efficiency in comparison with pristine NCM811. Consequently, the NCM811@3LNO electrode presents a higher capacity at 0.1 C (approximately 246 mAh g^{-1}) and a significantly improved high rate performance (a discharge specific capacity of 130.62 mAh g^{-1} at 10 C), twice that of pristine NCM811. Additionally, cycling stability is also improved for the composite material. This work provides a new possibility of active oxide cathodes for high energy/power Li-ion batteries by electronic conductor $LaNiO_3$ coating.

Keywords: Li-ion batteries; fast charging; $LiNi_{0.8}Co_{0.1}Mn_{0.1}O_2$; $LaNiO_3$; electron transport; rate capability

Citation: Li, T.; Li, D.; Zhang, Q.; Gao, J.; Zhang, L.; Liu, X. Improving Fast Charging-Discharging Performances of Ni-Rich $LiNi_{0.8}Co_{0.1}Mn_{0.1}O_2$ Cathode Material by Electronic Conductor $LaNiO_3$ Crystallites. *Materials* **2022**, *15*, 396. https://doi.org/10.3390/ma15010396

Academic Editor: Sophie Tingry

Received: 3 December 2021
Accepted: 3 January 2022
Published: 5 January 2022

1. Introduction

Li-ion batteries (LIBs) have received increasing attention for electric vehicles (EVs) and portable electronics due to their high energy density and long lifespan [1–3]. As a result, the development of LIBs with high capacity, high power/energy density, as well as long cycle life is necessary. The power density and energy density are significant parameters for LIBs [4]. Except for improving energy density, novel technologies and materials are needed to resolve the requirement for high power densities by enabling rapid charging-discharging rates without sacrificing cycling stability and energy densities. In particular, power density is critical for most applications, such as power grid stabilization and fast-charging EVs. Recently, despite the charging technology of LIBs having been intensively investigated, the current charging capability is still far from offering consumers the same refueling experience as conventional vehicles [5]. This is a significant reason, causing "range anxiety" for EVs owners and potential customers. Consequently, a higher charging rate with a shorter charging time is essential to achieve fast charging in the future [6–8].

Based on previous reports, multiple properties of the applied cathode, anode, and electrolyte materials affect the fast-charging capability of LIBs. Fast charging technology mainly depends on the transport rate of electrons and Li^+ between the LIBs components. To improve the fast-charging capability of LIBs, numerous research have been devoted toward reducing the diffusion length of Li^+ and electrons by nanotechnology (such as nanofibers, nanotubes, or nanoparticles), and increasing the electronic and ionic conductivity by hybrid-composite design, Li-ion diffusion control, surface modification, and dopant manipulation [9,10]. Previous reports have demonstrated that electronic or ionic conductors can improve electronic or ionic conductivity, providing a positive impact on the fast charging of LIBs [11].

Compared with conductive carbon anode materials, the study on the influence of fast charging on the cathodes is still in its infancy [12]. At present, developing cathodes with high capacity, high operating voltage, and long lifespan is of great importance for practical application in LIBs. Traditional cathodes, such as $LiCoO_2$, $LiFePO_4$, and $LiMn_2O_4$, have limited reversible capacity (<200 mAh g^{-1}), which cannot satisfy the requirement for high power and energy density LIBs [13,14]. Among numerous cathodes, $LiNi_{0.8}Co_{0.1}Mn_{0.1}O_2$ (NCM811) is considered as one of the most promising cathodes due to its high specific capacity (>200 mAh g^{-1}), low cost, and high average redox potential [15,16]. Nevertheless, the inferior rate performance of NCM811 could be ascribed to the Li^+/Ni^{2+} cation mixing, caused by the similar ionic radius between Ni^{2+} (0.69 Å) and Li^+ (0.76 Å), resulting in reduced electrode kinetics and specific capacity [17–19]. Additionally, the poor cycling performance is closely related to moisture sensitivity and detrimental side reactions [20,21]. More importantly, the intrinsic poor electronic conductivity (10^{-5} S cm^{-1}) of NCM811 restricts its transport kinetics (e.g., rate performance) and cycling stability [22]. Moreover, these problems will worsen at high current density. To resolve the above problems, surface modification is considered as an efficient strategy to enhance the electrochemical performances of NCM811 cathodes. Among various coating materials, the majority of the lithium ionic conductors (Li_3PO_4 [23], Li_2ZrO_3 [24], Li_3VO_4 [25]) are focused on increasing the ionic conductivity of Ni-rich cathodes.

Furthermore, previous reports have demonstrated that electronic conductivity plays a crucial role in initiating the electrochemical process, and electron transport is critical to improving the electrode kinetics that dominate the power density of LIBs [26]. As a consequence, it is necessary to enhance the electronic conductivity of NCM811 cathode. It is well known that the high electronic conductivity carbon is beneficial to supply the electron transport pathway in the manufacturing process of LIBs. Thus, it is essential to add conductive agent (carbon black) into the conventional electrode materials. Moreover, many studies have reported that effective improvement of the electronic conductivity for NCM811 cathodes could be achieved by coating the carbon or conductive polymer. More recently, Sim et al. [27] prepared carbon-modified $LiNi_{0.8}Co_{0.1}Mn_{0.1}O_2$ cathodes using carbon black (Super P) as the carbon source, resulting in superior electrochemical performances. Furthermore, Zha et al. reported an efficient method to decorate the surface of $LiNi_{0.8}Co_{0.1}Mn_{0.1}O_2$ (NCM811) by combining with polyimide and carbon nanotubes. Compared with the modified-NCM811 (199.6 mAh g^{-1}), NCM811 exhibits a higher initial discharge capacity (201.1 mAh g^{-1}). However, it can be found that surface modification with carbon and conductive polymers is usually realized under an inert atmosphere and high temperature, leading to a poor rate capability or a loss in capacity. To improve the rate performance and specific capacity, it is necessary to achieve the coating process under an oxygen or air atmosphere, and the oxides could satisfy this demand.

The perovskite oxide $LaNiO_3$ (LNO) has been intensively investigated in various applications in ferroelectric devices due to its highly electronic properties [28]. The rhombohedral structure of $LaNiO_3$ oxide is metallic at all temperatures, and its high electronic conductivity (over 100 S cm^{-1}) enough to act as an electrode [29–33]. Furthermore, $LaNiO_3$ has been reported as a novel anode for LIBs, which exhibits superior electrochemical performances. More importantly, $LaNiO_3$ could be annealed under an oxygen atmosphere. In this paper, we report the electronic conductor $LaNiO_3$ as a coating layer to decorate the NCM811 surface for fast charging-discharging LIBs. The conductive $LaNiO_3$ coating provides an effective electron transport pathway and serves as a protective layer that restrains the interfacial side reactions between the electrolyte and the NCM811 surface. Additionally, the impact of the $LaNiO_3$ coating on the NCM811 cathodes is studied in detail.

2. Materials and Methods

2.1. Synthesis of NCM811 Cathode Materials

NCM811 was prepared via a typical sol-gel method as follows. First of all, stoichiometric amounts of $Co(COOCH_3)_2{\cdot}4H_2O$, $Ni(COOCH_3)_2{\cdot}4H_2O$, $Mn(COOCH_3)_2{\cdot}4H_2O$, and $LiNO_3$ (with 5% excess) were dissolved together in ethyl alcohol to obtain a uniform solution. An excessive amount of $LiNO_3$ was employed to compensate for possible lithium loss at high temperature. Afterward, acetylacetone (the molar ratio of transition metal ions to acetylacetone was 1:1) was added into the metal solution. The mixed solution was then evaporated with stirring in an 80 °C water bath. The obtained gel was dried at 100 °C and annealed at 450 °C for 2 h. After that, the acquired powder was ground, and calcined at 800 °C for 12 h in oxygen (denoted as pristine NCM811).

2.2. Synthesis of $LaNiO_3$ Surface-Modified NCM811

A simple wet chemical method was employed to prepare the $LaNiO_3$-modified NCM811 cathodes. Firstly, deionized water was used to dissolve $Ni(NO_3)_2{\cdot}6H_2O$ and $La(NO_3)_3{\cdot}6H_2O$ with the stoichiometric ratio of 1:1. Citric acid was then added under vigorous stirring to obtain the $LaNiO_3$ transparent solution. Subsequently, the as-synthesized NCM811 sample was added into a required amount of LNO solution (from 0, 1, to 3 wt%), and the obtained suspension was stirred continuously at 80 °C to evaporate the water. Finally, the powder was dried at 100 °C and subsequently calcined at 700 °C for 3 h under flowing O_2 to obtain the $LaNiO_3$ surface-modified NCM811 materials. Based on the ratio of LNO to NCM811, the LNO-modified NCM811 materials were labeled as pristine NCM811, NCM811@LNO, and NCM811@3LNO, respectively.

2.3. Material Characterization

Powder X-ray diffraction (XRD, Bruker D8 ADVANCE) using a Cu target under 40 mA and 40 kV was used to characterize the crystal structure of samples. The XRD patterns were collected over the 2θ range of 15–90° with a step size of 0.02°, and the scanning rate was 2.4° min^{-1}. Furthermore, Rietveld refinement program—General Structure Analysis System (GSAS) software was used to further analyze the XRD data. Scanning electron microscopy (SEM, Hitachi S-4800) was employed to characterize the morphology. Elemental distribution on the surface of samples was analyzed by energy dispersive X-ray spectroscopy (EDS).

2.4. Electrochemical Measurements

Furthermore, 20 wt% carbon black, 10 wt% polyvinylidene fluoride (PVDF), and 70 wt% active material were dissolved together in N-methyl pyrrolidone (NMP), forming a slurry to prepare the electrodes. The slurry was cast on Al foil and dried at 100 °C. The mass loading of the active material was approximately 1–2 mg cm^{-2}. The as-prepared electrode as the working electrode, Li metal as the reference electrode, and microporous polypropylene membrane (Celgard 2500) as the separator, assembling the CR2025-type coin cells in an argon gas-filled glove box. The electrolyte was 1 M $LiPF_6$ dissolved in dimethyl carbonate (DMC) and ethylene carbonate (EC) (1:1 vol/vol). After aging, a multichannel battery testing system (Neware Technology Co., Ltd., Shenzhen, China) was used to measure the electrochemical performances between 2.8 and 4.3 V. Cyclic voltammetry (CV) and electrochemical impedance spectroscopy (EIS) tests were carried out on a Princeton Applied Research VersaSTAT 3 electrochemical workstation. The EIS tests were conducted between 100 kHz and 10 mHz, and the voltage amplitude was 5 mV. The cyclic voltammetry was tested in the potential range between 2.8 and 4.5 V, and the scan rate was 0.1 mV s^{-1}. The differential capacity versus voltage curves (dQ dV^{-1}) were obtained according to the charge/discharge testing data of individual cycles.

3. Results

3.1. Physical Characterizations

The crystal structure of pristine NCM811, NCM811@LNO, and NCM811@3LNO samples were investigated by powder X-ray diffraction (Figure 1a) combined with Rietveld refinement (Figure 1b–d). The main diffraction peaks of NCM811@LNO and NCM811@3LNO samples are similar to those of pristine NCM811, which belong to hexagonal layered α-$NaFeO_2$ structure (R−3m space group), suggesting that the LNO coating does not affect the crystal structure of NCM811 [34]. Additionally, all three samples exhibit clear splitting of (108)/(110) and (006)/(102), which implies a well-ordered layered structure [35]. Compared with pristine NCM811, NCM811@LNO and NCM811@3LNO samples exhibit a negligible change in lattice parameters (Table 1), suggesting that La^{3+} is not doped into the bulk structure. The (003)/(104) peak intensity ratio is closely related to the degree of Li^+/Ni^{2+} cation mixing in the Li layer according to the literature [36,37]. Interestingly, NCM811@LNO and NCM811@3LNO samples exhibit higher $I_{(003)}/I_{(104)}$ values compared to the pristine NCM811, suggesting that the $LaNiO_3$-modified $LiNi_{0.8}Co_{0.1}Mn_{0.1}O_2$ samples have lower Li^+/Ni^{2+} disordering. Previous research results demonstrated that Li/Ni disordering is harmful to the kinetic diffusion of Li ion during electrochemical cycling, thus deteriorating the rate capability and discharge capacity [38]. Obviously, in the NCM811@LNO sample, no diffraction peak corresponding to the $LaNiO_3$ can be seen, which may be caused by the low content of LNO material. In addition, the relatively weaker diffraction peaks between 30 and 35° could be identified as the crystalline $LaNiO_3$ (JCPDS #33−0711, labeled with *) in the NCM811@3LNO sample, suggesting the successful introduction of $LaNiO_3$ nanocrystals to the NCM811 [39]. Based on these results, we can speculate that the NCM811@LNO sample contains crystalline LNO owing to the identical preparation method.

Table 1. The lattice parameters of all three samples.

Parameter	Pristine NCM811	NCM811@LNO	NCM811@3LNO
a (Å)	2.8731	2.8709	2.8720
c (Å)	14.1963	14.1951	14.1959
c/a	4.9411	4.9444	4.9428
$I_{(003)}/I_{(104)}$	1.5893	1.8820	1.6045

All three samples are composed of well crystalline particles, and the average particle size is approximately 300–700 nm. Clearly, the pristine NCM811 exhibits smooth and clear particle surfaces (Figure 2a,b), whereas some small nanoparticles can be seen on the rougher surface of NCM811@LNO (Figure 2c,d) and NCM811@3LNO (Figure 2e,f) samples. Clearly, with increasing LNO coating content, the amount of small nanoparticles increases gradually on the NCM811 particles surface. Additionally, Figure 3 presents the EDS elemental mapping of the NCM811@3LNO sample. Obviously, Mn, Ni, Co, O, and La elements are homogeneous distribution on the particle's surface. In addition, the atomic percentages (at.%) of La, Ni, Mn, and Co are 1.51, 38.68, 4.62, and 4.71 at.%, respectively. According to these results, it is concluded that the $LaNiO_3$ can be evenly coated on the NCM811 surface by a simple wet chemical process, forming a conductive coating layer. As a consequence, the electronic conductor LNO crystallites coating can facilitate the electron transport on the surface, restrain the direct contact between the electrolyte and the cathode surface, and thus reduce the transition metal ions dissolution and the interfacial side reactions.

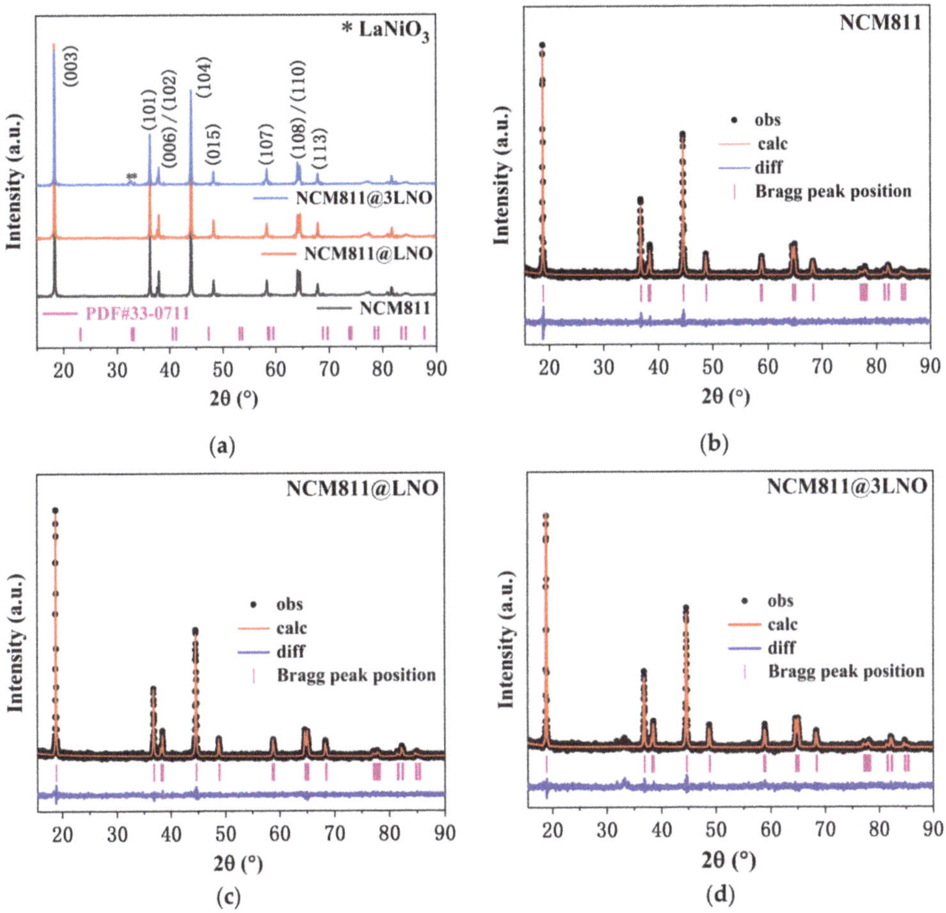

Figure 1. (**a**) XRD patterns of all three samples; the Rietveld refinement of (**b**) pristine NCM811, (**c**) NCM811@LNO, and (**d**) NCM811@3LNO samples.

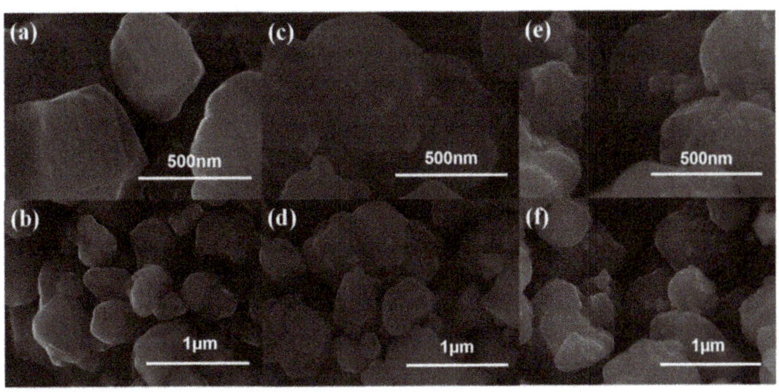

Figure 2. SEM images of (**a**,**b**) pristine NCM811, (**c**,**d**) NCM811@LNO, and (**e**,**f**) NCM811@3LNO samples.

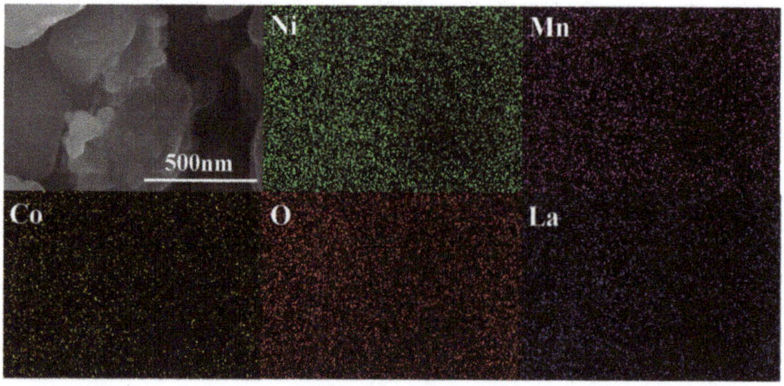

Figure 3. EDS elemental mapping of NCM811@3LNO sample.

3.2. Electrochemical Performance

The first charge/discharge voltage profiles of pristine NCM811, NCM811@LNO, and NCM811@3LNO electrodes at 0.1 C (1 C = 180 mA g^{-1}) are presented in Figure 4a. Compared with the charge and discharge profiles of pristine NCM811, NCM811@LNO and NCM811@3LNO electrodes do not show any additional voltage plateau. Table 2 shows the first coulombic efficiencies and charge/discharge specific capacities of all three electrodes. It is conspicuous that the NCM811@3LNO electrode exhibits an excellent coulombic efficiency of 85.10% and an ultrahigh first discharge capacity of 246.39 mAh g^{-1} at 0.1 C, far surpassing the pristine NCM811 (82.12% and 194.67 mAh g^{-1}). The significantly enhanced coulombic efficiency and specific capacity of the NCM811@3LNO electrode can be attributed to the electronic conductor $LaNiO_3$ coating layer that provides the electronic conduction pathway between particles, leading to fast electron transport.

Table 2. The electrochemical performances of all three electrodes in the initial cycle at 0.1 C.

Electrode	First Charge Capacity (mAh g^{-1})	First Discharge Capacity (mAh g^{-1})	First Coulombic Efficiency (%)
pristine NCM811	237.06	194.67	82.12
NCM811@LNO	218.24	192.28	88.11
NCM811@3LNO	289.53	246.40	85.10

To further investigate the phase transition behavior during the first charge and discharge process, Figure 4b–d presents the corresponding dQ dV^{-1} curves of pristine NCM811, NCM811@LNO, and NCM811@3LNO electrodes, respectively. All the dQ dV^{-1} curves exhibit the phase transitions from hexagonal to monoclinic (H1 to M) and then to other hexagonal (H2 and H3). Clearly, NCM811@LNO (0.0068 V) and NCM811@3LNO (0.0032 V) electrodes demonstrate the lower potential difference of anodic-cathodic peaks compared to that of the pristine NCM811 (0.0096 V). The decreased potential difference of NCM811@3LNO electrode implies improved electrode reversibility and reduced electrochemical polarization according to the literature [40,41]. The above results indicate that surface modification with electronic conductor $LaNiO_3$ crystallites is beneficial to improve the electrode kinetics, leading to increasing the first coulombic efficiency and charge/discharge capacity and decreasing the electrochemical polarization degree of NCM811 cathodes.

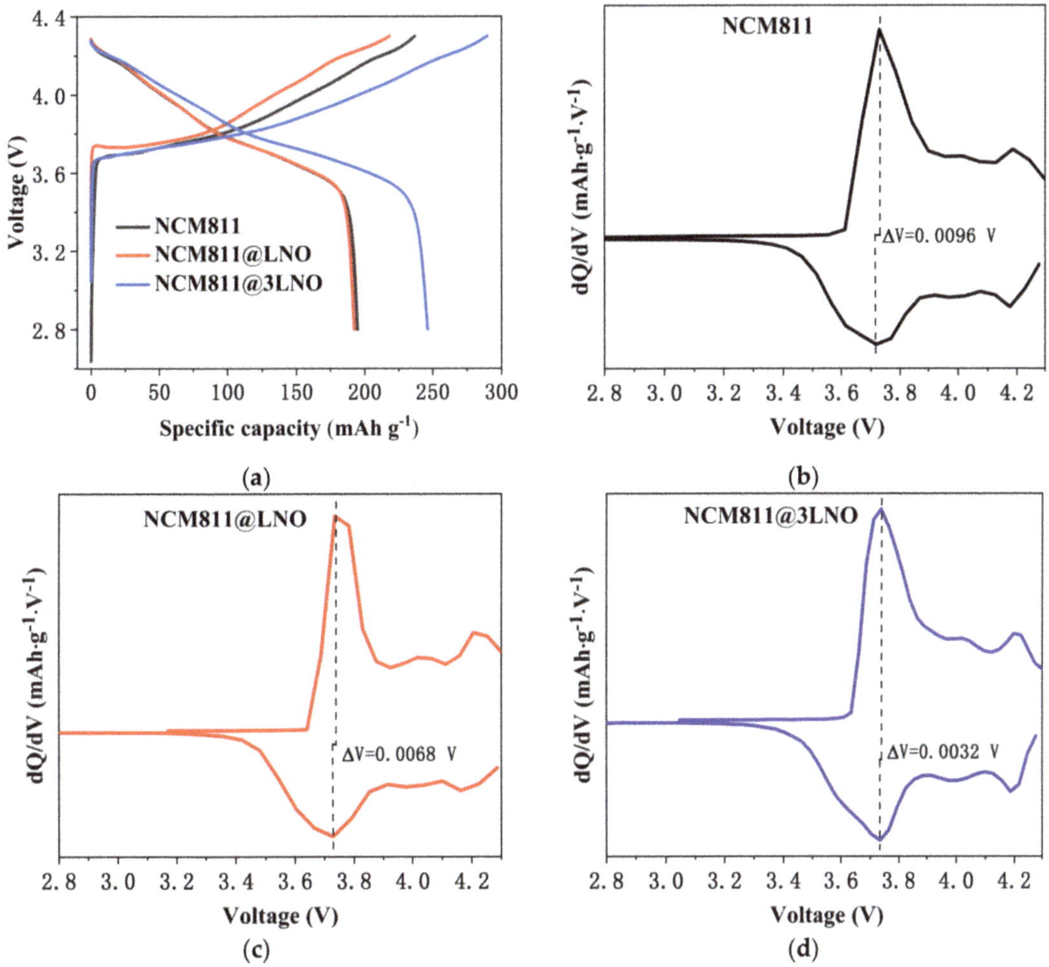

Figure 4. (**a**) First charge and discharge voltage profiles, and corresponding dQ dV^{-1} curves of (**b**) pristine NCM811, (**c**) NCM811@LNO, and (**d**) NCM811@3LNO electrodes at 0.1 C rate.

Figure 5a exhibits the rate capability of pristine NCM811, NCM811@LNO, and NCM811@3LNO electrodes. Additionally, the charge/discharge voltage profiles of all three electrodes at different rates are displayed in Figure 5b–d. Compared with the pristine NCM811, LNO surface-modified NCM811 electrodes exhibit obviously improved high-rate performance. It is conspicuous that a higher discharge capacity of 130.62 mAh g^{-1} is retained after 35 cycles at 10 C for the NCM811@3LNO electrode, corresponding to a capacity loss of 1.12% per cycle from an initial discharge capacity at 0.1 C. In contrast, pristine NCM811 electrode decreases dramatically to 69.50 mAh g^{-1}, corresponding to a capacity loss of 1.81% at the same condition. More importantly, NCM811@3LNO exhibits a superior discharge capacity of 213.48 mAh g^{-1} when the current rate recovers back to 0.1 C, far surpassing the pristine NCM811 (147.63 mAh g^{-1}). The above results demonstrate that the conductive $LaNiO_3$ surface-modified NCM811 cathodes exhibit significantly improved high-rate charge-discharge performance and electrochemical reversibility, suggesting easier electron transport in the $LaNiO_3$-modified NCM811.

The cycling performances of pristine NCM811, NCM811@LNO, and NCM811@3LNO electrodes at 0.5 C are displayed in Figure 6a. At 0.5 C, the NCM811@3LNO electrode exhibits an ultrahigh first discharge capacity of 213.23 mAh g^{-1}, significantly surpassing the pristine NCM811 (177.86 mAh g^{-1}). Pristine NCM811 exhibits obvious capacity fading with increasing cycle number, maintaining only 62.92% after 50 cycles, and the synchronous decay in the dQ dV^{-1} peaks (Figure 6c) can be observed. In contrast, the NCM811@3LNO electrode shows excellent capacity retention of 87.87% at the same condition, and the corresponding dQ dV^{-1} profiles (Figure 6d) overlap well among various cycles, suggesting outstanding electrochemical reversibility. Besides, Figure 6b shows the cycle performances of all three electrodes at a higher rate of 2 C. Remarkably, the NCM811@LNO electrode maintains 84.95% of its original capacity after 100 cycles, far surpassing 58.31% of the pristine NCM811. These results suggest that the electronic conductor $LaNiO_3$ crystallites can provide the electronic conduction pathway between particles, restrain the direct contact between the electrolyte and the cathode surface, thus decreasing the transition metal ions dissolution and the interfacial side reactions, leading to superior cycling stability of the NCM811 cathodes.

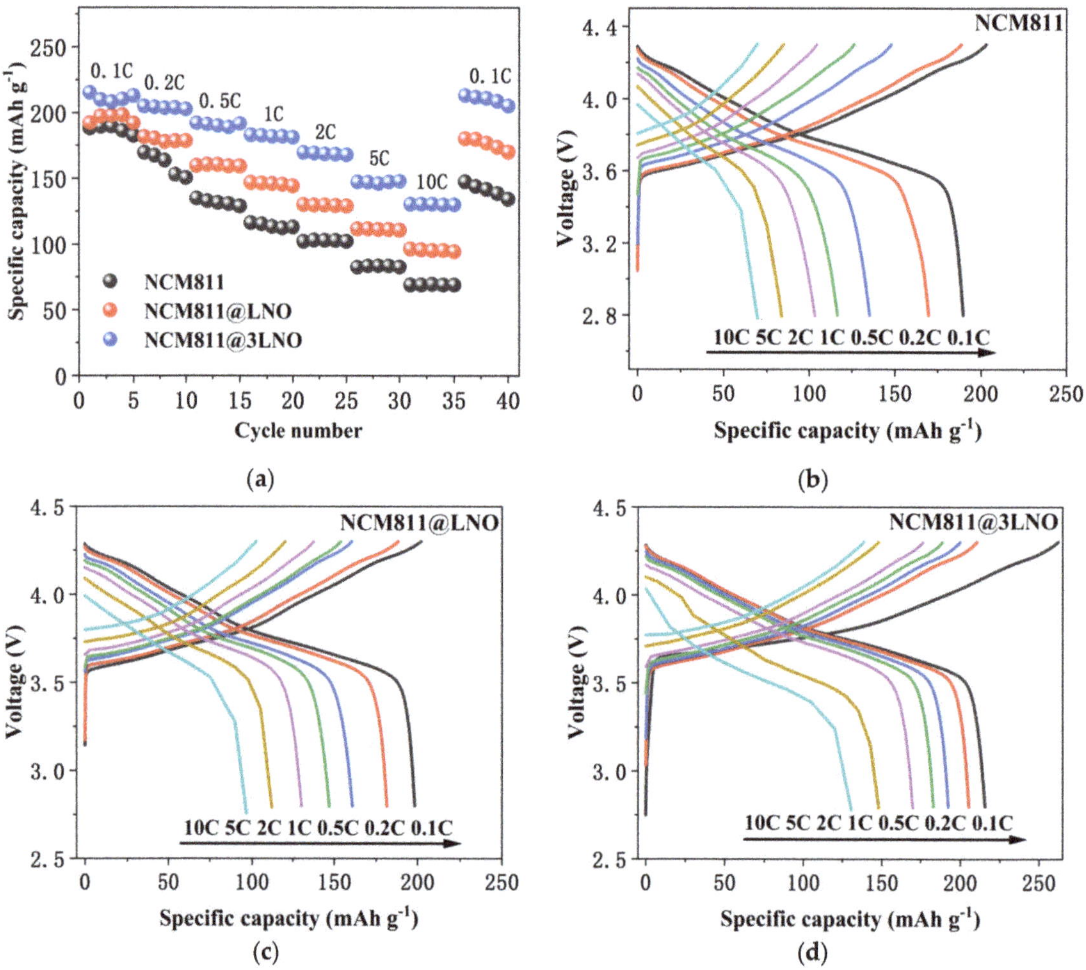

Figure 5. (**a**) Rate performances comparison; charge and discharge profiles of (**b**) pristine NCM811, (**c**) NCM811@LNO, and (**d**) NCM811@3LNO electrodes at different rates.

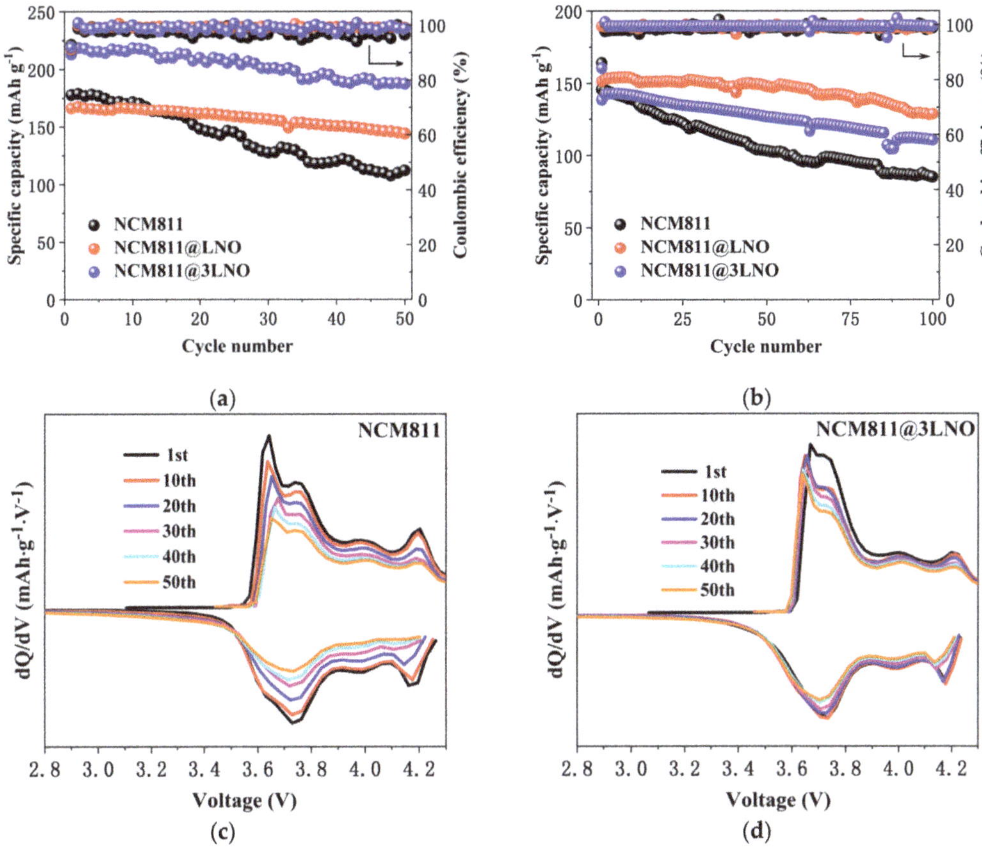

Figure 6. Cycling performances of pristine NCM811, NCM811@LNO, and NCM811@3LNO electrodes at (**a**) 0.5 C and (**b**) 2 C; the corresponding dQ dV^{-1} curves of (**c**) pristine NCM811 and (**d**) NCM811@3LNO electrodes at various cycles at 0.5 C.

To further investigate the redox process and the electrochemical reversibility of NCM811 cathodes during charge-discharge cycling, the first three cyclic voltammograms (Figure 7) were recorded between 2.8 and 4.5 V, and the scan rate was 0.1 mV s^{-1}. All electrodes exhibit three redox couples. Compared with pristine NCM811, NCM811@LNO and NCM811@3LNO electrodes exhibit similar CV features, suggesting that the redox process of the NCM811 is not affected by the presence of $LaNiO_3$ coating. The potential difference (ΔV) of oxidation-reduction peaks is closely related to the polarization degree of the electrode materials and the reversibility of the electrochemical redox reaction according to the literature [42]. Significantly, the potential difference of the pristine NCM811 electrode is 0.052 V, which surpassed the NCM811@LNO (0.044 V) and NCM811@3LNO (0.023 V) electrodes. Furthermore, compared with pristine NCM811, the CV curves of NCM811@LNO and NCM811@3LNO electrodes overlap well in the 2nd and 3rd cycles, suggesting that the $LaNiO_3$-modified $LiNi_{0.8}Co_{0.1}Mn_{0.1}O_2$ electrodes have quasi-reversible electrochemical kinetics during the Li^+ insertion and extraction processes [43]. These results suggest that the electronic conductor $LaNiO_3$ coating are conducive to improving the electrochemical reversibility and reducing the polarization degree of the NCM811 cathodes, which are consistent with the results from superior cycling performance and high-rate performance of the NCM811@LNO and NCM811@3LNO electrodes.

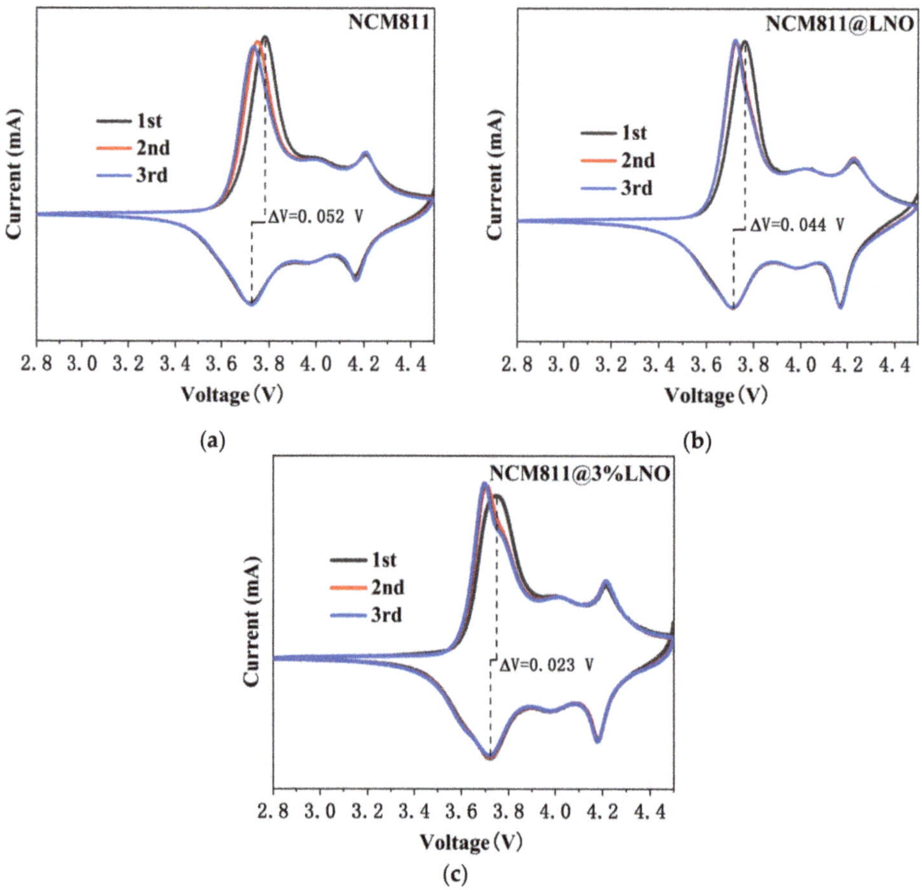

Figure 7. CV curves of (**a**) pristine NCM811, (**b**) NCM811@LNO, and (**c**) NCM811@3LNO electrodes.

To study the electrochemical kinetic behaviors for the high-rate performance, EIS of the pristine NCM811, NCM811@LNO, and NCM811@3LNO electrodes were measured after 100 cycles at 2 C. Two semicircles and an inclined line can be seen in all Nyquist plots (Figure 8a). Generally, the solution resistance (R_s) is represented by the high-frequency intercept at the real axis [44,45]. The surface-film resistance (R_{sf}) is represented by the first semicircle at high-frequency, while the charge-transfer resistance (R_{ct}) between the electrolyte and cathode materials is represented by the second semicircle at medium-frequency [46]. The inclined line at low-frequency stands for the Warburg impedance (Z_w), which is related to the lithium ions diffusion in the bulk of cathode materials [47]. Furthermore, the following equation was used to calculate the Li^+ diffusion coefficient (D_{Li^+}) from the Nyquist plots in the low-frequency region:

$$D_{Li^+} = \frac{R^2T^2}{2A^2n^4F^4C^2\sigma^2} \quad (1)$$

where C represents the concentration of Li^+ in the NCM811 cathode, F represents the Faraday constant, A represents the surface area of the electrode, T represents the absolute temperature, n represents the amount of the electrons per molecule participating in the electronic transfer reaction, R represents the gas constant, and σ represents the Warburg coefficient, which can be obtained by the following equation from the slope of the linear fitting of resistance (Z') vs. the reciprocal square roots of the frequency ($\omega^{-1/2}$) in the low-frequency region [48]:

$$Z' = R_s + R_{ct} + \sigma\omega^{-1/2} \quad (2)$$

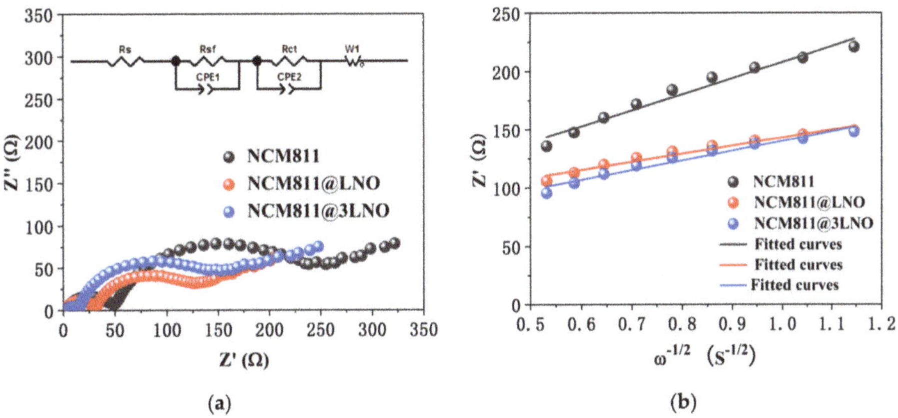

Figure 8. (**a**) Nyquist plots with an equivalent circuit (inset) of pristine NCM811, NCM811@LNO, and NCM811@3LNO electrodes; (**b**) the relationship plots between Z' and $\omega^{-1/2}$ in the low-frequency region.

Figure 8b presents the relationship between Z' and $\omega^{-1/2}$. Moreover, Table 3 exhibits the fitting resistance and calculated D_{Li^+} values. Compared with pristine NCM811, NCM811@LNO and NCM811@3LNO electrodes exhibit lower R_{sf} and R_{ct} values. Additionally, the calculated D_{Li^+} value for the pristine NCM811 electrode is 1.17×10^{-14} cm^2 s^{-1} after 100 cycles, whereas the D_{Li^+} value for NCM811@LNO and NCM811@3LNO is 4.61×10^{-14} cm^2 s^{-1} and 3.21×10^{-14} cm^2 s^{-1}, respectively. These results are consistent with the results from the cycling performances at 2 C (Figure 6b), indicating that the high conductive LNO coating layer is conducive to facilitating the electron transport, suppressing the side reactions between the electrolyte and NCM811 surface, subsequently improving the electrode kinetics and reducing the interfacial resistance.

Table 3. Impedance parameters of pristine NCM811, NCM811@LNO, and NCM811@3LNO electrodes after fitting.

Electrode	R_s (Ω)	R_{sf} (Ω)	R_{ct} (Ω)	D_{Li^+} (cm^2 s^{-1})
pristine NCM811	4.35	39.94	173.30	1.17×10^{-14}
NCM811@LNO	2.88	24.94	74.88	4.61×10^{-14}
NCM811@3LNO	4.06	8.96	108.50	3.21×10^{-14}

4. Conclusions

In this paper, electronic conductor $LaNiO_3$ crystallite surface-modified $LiNi_{0.8}Co_{0.1}Mn_{0.1}O_2$ cathodes were prepared, and the effects of conductive $LaNiO_3$ coating on the $LiNi_{0.8}Co_{0.1}Mn_{0.1}O_2$ cathodes were studied. The present work indicates that $LaNiO_3$ nanoparticles are uniformly distributed on the NCM811 particle surface, which is beneficial towards improving the electrode kinetics caused by the fast electron transport and restraining the direct contact between the electrolyte and cathode materials, leading to excellent cycling performance and high-rate capability of $LaNiO_3$-modified $LiNi_{0.8}Co_{0.1}Mn_{0.1}O_2$ cathodes. As a result, surface modification with high electronic conductivity oxide is an effective method to improve the electrochemical performances of Ni-rich $LiNi_{0.8}Co_{0.1}Mn_{0.1}O_2$ cathodes, which can also be extended to other electrodes for fast charging-discharging LIBs.

Author Contributions: Conceptualization, D.L.; methodology, T.L. and Q.Z.; software, J.G. and L.Z.; formal analysis, Q.Z. and J.G.; investigation, T.L. and X.L.; data curation, L.Z. and X.L.; writing—original draft preparation, T.L.; writing—review and editing, D.L.; supervision, D.L.; project administration, T.L.; funding acquisition, D.L. All authors have read and agreed to the published version of the manuscript.

Funding: This research was funded by the National Natural Science Foundation of China (Grant No. 22179011 and 21473014).

Institutional Review Board Statement: Not applicable.

Informed Consent Statement: Not applicable.

Data Availability Statement: The data presented in this study are available from the corresponding authors upon reasonable request.

Conflicts of Interest: The authors declare no conflict of interest.

References

1. An, J.; Shi, L.Y.; Chen, G.; Li, M.; Liu, H.J.; Yuan, S.; Chen, S.; Zhang, D. Insights into the stable layered structure of a Li-rich cathode material for lithium-ion batteries. *J. Mater. Chem. A* **2017**, *5*, 19738–19744. [CrossRef]
2. Sun, Z.; Xu, L.; Dong, C.; Zhang, H.; Zhang, M.; Liu, Y.; Zhou, Y.; Han, Y.; Chen, Y. Enhanced cycling stability of boron-doped lithium-rich layered oxide cathode materials by suppressing transition metal migration. *J. Mater. Chem. A* **2019**, *7*, 3375–3383. [CrossRef]
3. Park, K.J.; Lim, B.B.; Choi, M.H.; Jung, H.G.; Sun, Y.K.; Haro, M.; Vicente, N.; Bisquert, J.; Garcia-Belmonte, G. A high-capacity $Li[Ni_{0.8}Co_{0.06}Mn_{0.14}]O_2$ positive electrode with a dual concentration gradient for next-generation lithium-ion batteries. *J. Mater. Chem. A* **2015**, *3*, 22183–22190. [CrossRef]
4. Braun, P.V.; Cho, J.; Pikul, J.H.; King, W.P.; Zhang, H. High power rechargeable batteries. *Curr. Opin. Solid State Mater. Sci.* **2012**, *16*, 186–198. [CrossRef]
5. Liu, Y.Y.; Zhu, Y.Y.; Cui, Y. Challenges and opportunities towards fast-charging battery materials. *Nat. Energy* **2019**, *4*, 540–550. [CrossRef]
6. Wang, X.; Ding, Y.L.; Deng, Y.P.; Chen, Z. Ni-Rich/Co-Poor Layered Cathode for Automotive Li-Ion Batteries: Promises and Challenges. *Adv. Energy Mater.* **2020**, *10*, 1903864. [CrossRef]
7. Tomaszewska, A.; Chu, Z.Y.; Feng, X.N.; O'Kane, S.; Liu, X.H.; Chen, J.Y.; Ji, C.Z.; Endler, E.; Li, R.H.; Liu, L.S.; et al. Lithium-ion battery fast charging: A review. *eTransportation* **2019**, *1*, 100011. [CrossRef]
8. Rodrigues, M.T.F.; Son, S.B.; Colclasure, A.M.; Shkrob, I.A.; Trask, S.E.; Bloom, I.D.; Abraham, D.P. How Fast Can a Li-Ion Battery Be Charged? Determination of Limiting Fast Charging Conditions. *ACS Appl. Energy Mater.* **2021**, *4*, 1063–1068. [CrossRef]
9. Kang, B.; Ceder, G. Battery materials for ultrafast charging and discharging. *Nature* **2009**, *458*, 190–193. [CrossRef] [PubMed]
10. Tang, Y.X.; Zhang, Y.Y.; Li, W.L.; Ma, B.; Chen, X.D. Rational material design for ultrafast rechargeable lithium-ion batteries. *Chem. Soc. Rev.* **2015**, *46*, 5926–5940. [CrossRef] [PubMed]
11. Yoo, G.W.; Son, J.T. Improvement of Electrochemical Properties and Thermal Stability of a Ni-rich Cathode Material by Polypropylene Coating. *J. Electrochhem. Sci. Technol.* **2016**, *7*, 179–184. [CrossRef]
12. Hu, J.W.; Fan, F.S.; Zhang, Q.; Zhong, S.W.; Ma, Q.X. Effects of long-term fast charging on a layered cathode for lithium-ion batteries. *J. Energy Chem.* **2021**, *67*, 604–612. [CrossRef]
13. Ding, X.; Li, Y.X.; Deng, M.M.; Wang, S.; Aqsa, Y.; Hu, Q.; Chen, C.H. Cesium doping to improve the electrochemical performance of layered $Li_{1.2}Ni_{0.13}Co_{0.13}Mn_{0.54}O_2$ cathode material. *J. Alloys Compd.* **2019**, *791*, 100–108. [CrossRef]
14. Zou, P.J.; Lin, Z.H.; Fan, M.N.; Wang, F.; Liu, Y.; Xiong, X.H. Facile and efficient fabrication of Li_3PO_4-coated Ni-rich cathode for high-performance lithium-ion battery. *Appl. Surf. Sci.* **2020**, *504*, 144506. [CrossRef]
15. Xin, F.X.; Zhou, H.; Chen, X.B.; Zuba, M.; Chernova, N.; Zhou, G.W.; Whittingham, M.S. Li-Nb-O coating/substitution enhances the electrochemical performance of $LiNi_{0.8}Mn_{0.1}Co_{0.1}O_2$ (NMC811) cathode. *ACS Appl. Mater. Interfaces* **2019**, *11*, 34889–34894. [CrossRef]
16. Becker, D.; Börner, M.; Nölle, R.; Diehl, M.; Klein, S.; Rodehorst, U.; Schmuch, R.; Winter, M.; Placke, T. Surface Modification of Ni-rich $LiNi_{0.8}Co_{0.1}Mn_{0.1}O_2$ Cathode Material by Tungsten Oxide Coating for Improved Electrochemical Performance in Lithium Ion Batteries. *ACS Appl. Mater. Interfaces* **2019**, *11*, 18404–18414. [CrossRef]
17. Yang, H.; Wu, H.; Ge, M.; Li, L.; Yuan, Y.; Yao, Q.; Chen, J.; Xia, L.; Zheng, J.; Chen, Z.; et al. Simultaneously Dual Modification of Ni-Rich Layered Oxide Cathode for High-Energy Lithium-Ion Batteries. *Adv. Funct. Mater.* **2019**, *29*, 1808825. [CrossRef]
18. Hou, P.; Yin, J.; Ding, M.; Huang, J.; Xu, X. Surface/Interfacial Structure and Chemistry of High-Energy Nickel-Rich Layered Oxide Cathodes: Advances and Perspectives. *Small* **2017**, *13*, 1701802. [CrossRef] [PubMed]

19. Su, Y.F.; Chen, G.; Chen, L.; Li, W.K.; Zhang, Q.Y.; Yang, Z.R.; Lu, Y.; Bao, L.Y.; Tan, J.; Chen, R.J.; et al. Exposing the {010} Planes by Oriented Self-Assembly with Nanosheets To Improve the Electrochemical Performances of Ni-Rich $Li[Ni_{0.8}Co_{0.1}Mn_{0.1}]O_2$ Microspheres. *ACS Appl. Mater. Interfaces* **2018**, *10*, 6407–6414. [CrossRef] [PubMed]
20. Chen, Z.; Nguyen, H.; Zarrabeitia, M.; Liang, H.; Geiger, D.; Kim, J.; Kaiser, U.; Passerini, S.; Iojoiu, C.; Bresser, D. Lithium Phosphonate Functionalized Polymer Coating for High-energy $Li[Ni_{0.8}Co_{0.1}Mn_{0.1}]O_2$ with Superior Performance at Ambient and Elevated Temperatures. *Adv. Funct. Mater.* **2021**, *31*, 2105343. [CrossRef]
21. Jamil, S.; Wang, G.; Yang, L.; Xie, X.; Cao, S.; Liu, H.; Chang, B.; Wang, X. Suppressing H2–H3 phase transition in high Ni–low Co layered oxide cathode material by dual modification. *J. Mater. Chem. A* **2020**, *8*, 21306–21316. [CrossRef]
22. Liu, Y.; Tang, L.B.; Wei, H.X.; Zhang, X.H.; He, Z.J.; Li, Y.J.; Zheng, J.C. Enhancement on structural stability of Ni-rich cathode materials by in-situ fabricating dual-modified layer for lithium-ion batteries. *Nano Energy* **2019**, *65*, 104043. [CrossRef]
23. Lee, S.; Kim, M.; Jeong, J.; Kim, D.; Chung, K.; Roh, K.; Kim, K. Li_3PO_4 surface coating on Ni-rich $LiNi_{0.6}Co_{0.2}Mn_{0.2}O_2$ by a citric acid assisted sol-gel method: Improved thermal stability and high-voltage performance. *J. Power Sources* **2017**, *360*, 206–214. [CrossRef]
24. Song, B.H.; Li, W.D.; Oh, S.M.; Manthiram, A. Long-Life Nickel-Rich Layered Oxide Cathodes with a Uniform Li_2ZrO_3 Surface Coating for Lithium-Ion Batteries. *ACS Appl. Mater. Interfaces* **2017**, *9*, 9718–9725. [CrossRef]
25. Ran, Q.W.; Zhao, H.Y.; Hu, Y.Z.; Shen, Q.Q.; Liu, W.; Liu, J.T.; Shu, X.H.; Zhang, M.L.; Liu, S.S.; Tan, M.; et al. Enhanced electrochemical performance of dual-conductive layers coated Ni-rich $LiNi_{0.6}Co_{0.2}Mn_{0.2}O_2$ cathode for Li-ion batteries at high cut-off voltage. *Electrochim. Acta* **2018**, *289*, 82–93. [CrossRef]
26. Li, D.; Xie, R.; Tian, M.; Ma, S.; Gou, L.; Fan, X.; Shi, Y.; Yong, H.; Hao, L. Improving high-rate performance of mesoporous $Li_2FeSiO_4/Fe_7SiO_{10}/C$ nanocomposite cathode with a mixed valence Fe_7SiO_{10} nanocrystal. *J. Mater. Chem. A* **2014**, *2*, 4375–4383. [CrossRef]
27. Sim, S.J.; Lee, S.H.; Jin, B.S.; Kim, H.S. Use of carbon coating on $LiNi_{0.8}Co_{0.1}Mn_{0.1}O_2$ cathode material for enhanced performances of lithium-ion batteries. *Sci. Rep.* **2020**, *10*, 11114. [CrossRef]
28. Hu, Z.G.; Li, W.W.; Li, Y.W.; Zhu, M.; Zhu, Z.Q.; Chu, J.H. Electronic properties of nanocrystalline $LaNiO_3$ and $La_{0.5}Sr_{0.5}CoO_3$ conductive films grown on silicon substrates determined by infrared to ultraviolet reflectance spectra. *Appl. Phys. Lett.* **2009**, *94*, 488. [CrossRef]
29. Zhang, X.D.; Hao, J.J.; Wu, L.C.; Guo, Z.M.; Ji, Z.H.; Luo, J.; Chen, C.G.; Shu, J.F.; Long, H.M.; Yang, F.; et al. Enhanced electrochemical performance of perovskite $LaNiO_3$ coating on $Li_{1.2}Mn_{0.54}Ni_{0.13}Co_{0.13}O_2$ as cathode materials for Li-ion batteries. *Electrochim. Acta* **2018**, *283*, 1203–1212. [CrossRef]
30. Hofer, H.E.; Schmidberger, R. Electronic Conductivity in the $La(Cr, Ni)O_3$ Perovskite System. *J. Electrochem. Soc.* **1994**, *141*, 782–786. [CrossRef]
31. Chen, M.S.; Wu, T.B.; Wu, J.M. Effect of textured $LaNiO_3$ electrode on the fatigue improvement of $Pb(Zr_{0.53}Ti_{0.47})O_3$ thin films. *Appl. Phys. Lett.* **1996**, *68*, 1430–1432. [CrossRef]
32. Rajeev, K.P.; Shivashankar, G.V.; Raychaudhuri, A.K. Low-Temperature Electronic Properties of a Normal Conducting Perovskite Oxide ($LaNiO_3$). *Solid State Commun.* **1991**, *79*, 591–595. [CrossRef]
33. Fowlie, J.; Gibert, M.; Tieri, G.; Gloter, A.; Iniguez, J.; Filippetti, A.; Catalano, S.; Gariglio, S.; Schober, A.; Guennou, M.; et al. Conductivity and Local Structure of $LaNiO_3$ Thin Films. *Adv. Mater.* **2017**, *29*, 1605197. [CrossRef] [PubMed]
34. Babulal, L.M.; Yang, C.C.; Wu, S.H.; Chien, W.C.; Lue, S.J. Enhanced performance of a Ni-rich $LiNi_{0.8}Co_{0.1}Mn_{0.1}O_2$ cathode material formed through Taylor flow synthesis and surface modification with Li_2MoO_4. *Chem. Eng. J.* **2020**, *413*, 127150. [CrossRef]
35. Li, S.M.; Wu, J.; Li, J.Y.; Liu, G.B.; Liu, H. Facilitated Coating of Li_3PO_4 on the Rough Surface of $LiNi_{0.85}Co_{0.1}Mn_{0.05}O_2$ Cathodes by Synchronous Lithiation. *ACS Appl. Energy Mater.* **2021**, *4*, 2257–2265. [CrossRef]
36. Ryu, H.H.; Park, G.T.; Chong, S.Y.; Sun, Y.K. Suppressing detrimental phase transitions via tungsten doping of $LiNiO_2$ cathode for next-generation lithium-ion batteries. *J. Mater. Chem. A* **2019**, *7*, 18580–18588. [CrossRef]
37. Si, Z.; Shi, B.Z.; Huang, J.; Yu, Y.; Han, Y.; Zhang, J.L.; Li, W. Titanium and fluorine synergetic modification improves the electrochemical performance of $Li(Ni_{0.8}Co_{0.1}Mn_{0.1})O_2$. *J. Mater. Chem. A* **2021**, *9*, 9354–9363. [CrossRef]
38. Zheng, J.X.; Ye, Y.K.; Liu, T.C.; Xiao, Y.G.; Wang, C.M.; Wang, F.; Pan, F. Ni/Li Disordering in Layered Transition Metal Oxide: Electrochemical Impact, Origin, and Control. *Acc. Chem. Res.* **2019**, *52*, 2201–2209. [CrossRef]
39. Retuerto, M.; Pereira, A.G.; Pérez-Alonso, F.J.; PenA, M.A.; Fierro, J.L.; Alonso, J.A.; Fernández-Díaz, M.T.; Pascual, L.; Rojas, S. Structural effects of $LaNiO_3$ as electrocatalyst for the oxygen reduction reaction. *Appli. Catal. B Environ.* **2017**, *203*, 363–371. [CrossRef]
40. Ding, G.Y.; Li, Y.H.; Gao, Y.; Wang, Q.L.; Zhu, Z.; Jing, X.G.; Yan, F.Q.; Yue, Z.H.; Li, X.M.; Sun, F.G. Uniform Coating of Se on Selenophilic Surfaces of Nickel-Rich Layered Oxide Cathode Materials for High Performance Li-Ion Batteries. *ACS Sustain. Chem. Eng.* **2020**, *8*, 9632–9640. [CrossRef]
41. Huang, Y.P.; Yao, X.; Hu, X.C.; Han, Q.Y.; Wang, S.Q.; Ding, L.X.; Wang, H.H. Surface coating with Li-Ti-O to improve the electrochemical performance of Ni-rich cathode material. *Appl. Surf. Sci.* **2019**, *489*, 913–921. [CrossRef]
42. Zhang, D.K.; Liu, Y.; Wu, L.; Feng, L.W.; Jin, M.L.; Zhang, R.; Jin, M.L. Effect of Ti ion doping on electrochemical performance of Ni-rich $LiNi_{0.8}Co_{0.1}Mn_{0.1}O_2$ cathode material. *Electrochim. Acta* **2019**, *328*, 135086. [CrossRef]

43. Mofid, W.E.; Ivanov, S.; Konkin, A.; Bund, A. A high performance layered transition metal oxide cathode material obtained by simultaneous aluminum and iron cationic substitution. *J. Power Sources* **2014**, *268*, 414–422. [CrossRef]
44. Song, X.; Liu, G.X.; Yue, H.F.; Luo, L.; Yang, S.Y.; Huang, Y.Y.; Wang, C.R. A Novel Low-Cobalt Long-Life $LiNi_{0.88}Co_{0.06}Mn_{0.03}Al_{0.03}O_2$ Cathode Material for Lithium Ion Batteries. *Chem. Eng. J.* **2020**, *407*, 126301. [CrossRef]
45. Yuan, J.; Wen, J.W.; Zhang, J.B.; Chen, D.M.; Zhang, D.W. Influence of calcination atmosphere on structure and electrochemical behavior of $LiNi_{0.6}Co_{0.2}Mn_{0.2}O_2$ cathode material for lithium-ion batteries. *Electrochim. Acta* **2017**, *230*, 116–122. [CrossRef]
46. Yang, Z.G.; Xiang, W.; Wu, Z.G.; He, F.R.; Zhang, J.; Xiao, Y.; Zhong, B.H.; Guo, X.D. Effect of niobium doping on the structure and electrochemical performance of $LiNi_{0.5}Co_{0.2}Mn_{0.3}O_2$ cathode materials for lithium ion batteries. *Ceram. Int.* **2017**, *43*, 3866–3872. [CrossRef]
47. Zhang, L.J.; Jiang, J.C.; Zhang, C.P.; Wu, B.R.; Wu, F. High-rate layered lithium-rich cathode nanomaterials for lithium-ion batteries synthesized with the assist of carbon spheres templates. *J. Power Sources* **2016**, *331*, 247–257. [CrossRef]
48. Meng, J.X.; Xu, H.Z.; Ma, Q.X.; Li, Z.F.; Xu, L.S.; Chen, Z.J.; Cheng, B.M.; Zhong, S.W. Precursor pre-oxidation enables highly exposed plane {010} for high-rate Li-rich layered oxide cathode materials. *Electrochim. Acta* **2019**, *309*, 326–338. [CrossRef]

materials

MDPI

Perspective

Flash Sintering Research Perspective: A Bibliometric Analysis

Eva Gil-González [1,2,*], Luis A. Pérez-Maqueda [1,*], Pedro E. Sánchez-Jiménez [1,3,*] and Antonio Perejón [1,3]

1 Instituto de Ciencia de Materiales de Sevilla, Consejo Superior de Investigaciones Científicas—Universidad de Sevilla, Calle Américo Vespucio 49, 41092 Sevilla, Spain; antonio.perejon@icmse.csic.es
2 Departamento de Ingeniería Química, Universidad de Sevilla, Escuela Politécnica Superior, Calle Virgen de África, 7, 41011 Sevilla, Spain
3 Departamento de Química Inorgánica, Facultad de Química, Universidad de Sevilla, 41071 Sevilla, Spain
* Correspondence: eva.gil@icmse.csic.es (E.G.-G.); maqueda@icmse.csic.es (L.A.P.-M.); pedro.enrique@icmse.csic.es (P.E.S.-J.)

Abstract: Flash Sintering (FS), a relatively new Field-Assisted Sintering Technique (FAST) for ceramic processing, was proposed for the first time in 2010 by Prof. Rishi Raj's group from the University of Colorado at Boulder. It quickly grabbed the attention of the scientific community and since then, the field has rapidly evolved, constituting a true milestone in materials processing with the number of publications growing year by year. Moreover, nowadays, there is already a scientific community devoted to FS. In this work, a general picture of the scientific landscape of FS is drawn by bibliometric analysis. The target sources, the most relevant documents, hot and trending topics as well as the social networking of FS are unveiled. A separate bibliometric analysis is also provided for Reaction or Reactive Flash Sintering (RFS), where not only the sintering, but also the synthesis is merged into a single step. To the best of our knowledge, this is the first study of this nature carried out in this field of research and it can constitute a useful tool for researchers to be quickly updated with FS as well as to strategize future research and publishing approaches.

Keywords: flash sintering; bibliometric analysis; field assisted sintering; knowledge structure; ceramic materials

Citation: Gil-González, E.; Pérez-Maqueda, L.A.; Sánchez-Jiménez, P.E.; Perejón, A. Flash Sintering Research Perspective: A Bibliometric Analysis. *Materials* **2022**, *15*, 416. https://doi.org/10.3390/ma15020416

Academic Editors: Mattia Biesuz and Peter Tatarko

Received: 9 November 2021
Accepted: 2 January 2022
Published: 6 January 2022

1. Introduction

Flash Sintering (FS), an electric Field-Assisted Sintering Technique (FAST) [1] for the densification of ceramic materials at a greatly reduced temperature and time, has gained widespread attention since it was established in 2010 by Prof. Rishi Raj's group from the University of Colorado at Boulder. It basically consists of simultaneously applying heat and a modest electric field to a green body [2] placed on a furnace, allowing the current to totally flow through the sample. The main material requirement is that it should possess a negative temperature coefficient of electrical resistance so that the electrical conductivity increases while heating. At a given applied electric field, there is a critical temperature at which there is a sudden non-lineal rise of the conductivity of the material, which is normally accompanied by instantaneous densification as well as photoluminescence [3]. This signals the flash event and it is now accepted that it is initiated by a thermal runaway induced by Joule Heating [4,5]. Much effort is also being devoted to understanding the underlying mechanisms of FS. A few driven mechanisms have been proposed, but none of them can solely explain the flash phenomenon, which still remains elusive [6–11].

Nevertheless, FS has many practical advantages. For instance, it has been proven to be an ecofriendly and versatile methodology, as a wide range of materials from insulators to conductors can be sintered within seconds at furnace temperature much lower than those employed in conventional processing [12–17], thereby minimizing the energy footprint. Additionally, in comparison to other FAST techniques such as Spark Plasma Sintering (SPS) [1], it does not require any sophisticated experimental setup; basically just a furnace and a power supply are the two essential components to carry out a FS experiment [18].

An inner working atmosphere is not a prerequisite either, but indeed it is possible to tune it to study its effect over the final properties of the material [14,19,20]. Moreover, some flash-sintered materials have been granted special properties [21,22] and it has been shown that it is possible to sinter unstable oxides with volatile components and complex composition while preserving their stoichiometry and properties [23–26]. Very recently, in 2018, it was reported that sintering and synthesis can be merged into a single step, giving rise to what has been named Reaction or Reactive Flash Sintering (RFS) and constitutes [27], by itself, another important branch of research.

Given the advantages of FS, the widespread attention that it has received not only from the scientific community but also from the industry is not surprising [28]. Since it was proposed for the first time a decade ago, this research field has deeply evolved, constituting a true milestone in ceramic processing. The number of peer-reviewed articles has continued dramatically growing year by year with an annual growth rate of about 46% (see Figure 1). To the best of our knowledge, bibliometric analysis [29], which is a powerful tool to quantitatively identify essential variables in a particular research topic (such as top authors, institutions, trends of publication, collaboration, networking, etc.) has never been carried out on FS or RFS. Hence, the aim of this work is to gain insight into the knowledge structure, research perspective, and trends of this fascinating research field. It is noteworthy that the readings of these review articles are highly encouraged [10,18,28,30,31], as a comprehensive literature review is out of the scope of this work. The Biblioshiny web interface of the Bibliometrix R package [32] has been employed to conduct the bibliometric analysis. Peer-reviewed articles and reviews retrieved from the Web of Science (WoS) database have been analyzed based on several bibliometric indicators [33], i.e., quantity (productivity), quality (impact), and structural (networking) indicators. The results derived from this bibliometric analysis allow one to identify the target sources and trend topics in order to strategize future research and publishing approaches. Those publications that most influence this research field have been pointed out. Authors, institutions, and countries in terms of impact and the number of publications have also been identified. Additionally, the social structure of FS is shown by the clustering and collaboration networking from countries to institutions to authors. Moreover, a brief bibliometric analysis is also presented independently for RFS, given its impact and the scientific growth that it has experienced in a very short lifetime (just 3 years).

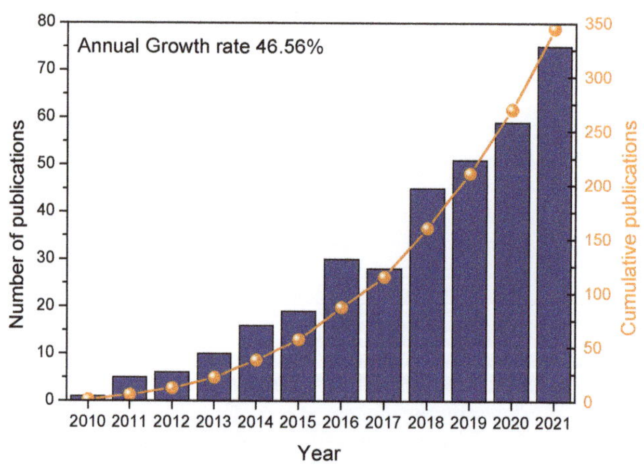

Figure 1. FS annual and cumulative publications from 2010 to December 2021.

This article is structured in several sections. The section herein provides a basic and brief introduction about flash sintering, referring to the most relevant literature. The second section explains in detail the methodology used to carry out the bibliometric analysis. The third section provides the most relevant results of this work (target sources, authors, most influential publications, and hot and trending topics) along with data visualization and social networking. The last part of this section is dedicated to RFS. Finally, the conclusions and final remarks are presented.

2. Methods

This section presents the design of this bibliometric study, which is schematically depicted in Scheme 1. As explained as follows, it implicitly contains the standard stages of bibliometric analysis, that is to say, the study design, data collection, data analysis, visualization, and interpretation [34,35].

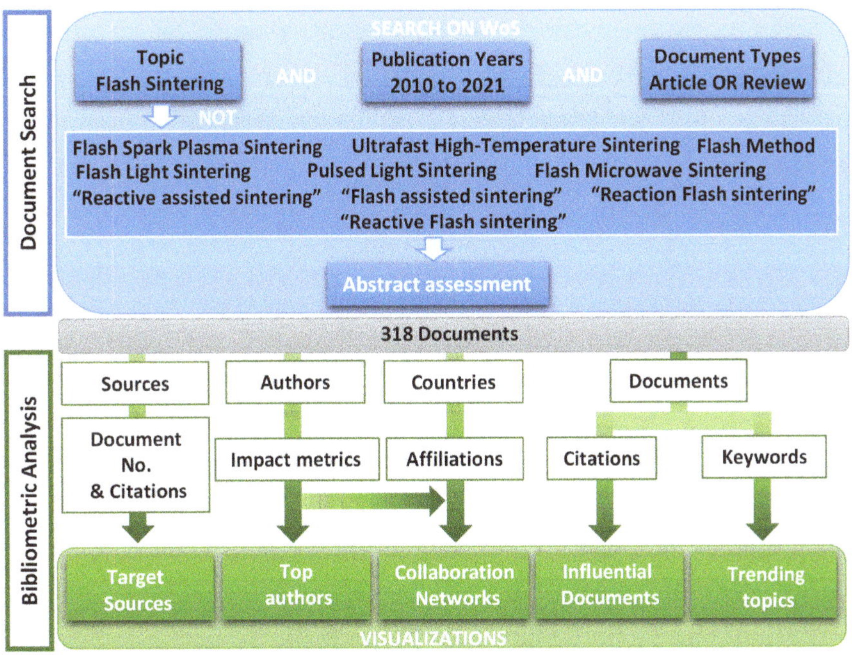

Scheme 1. Study design and workflow diagram.

2.1. Document Search

Clarivate Analytics Web of Science (WoS), one of the largest databases of peer-reviewed articles in different disciplines, was used for the data collection. The documents included in WoS related to the topic FS were retrieved according to the search strategy shown in Scheme 1, which also includes the specific search query words and Boolean operators. As a search topic, the words Flash Sintering were used, which implies that WoS documents that contain these words in their titles, abstracts, or keywords were scanned. The search was restricted to Flash Sintering, excluding other derived hybrid methodologies such as Flash Spark Plasma Sintering (F-SPS) [36,37], Flash Microwave Sintering [38,39], or Ultrafast High-Temperature Sintering [40], as has been properly indicated in Scheme 1. The reaction or reactive flash sintering documents were also excluded, as a separate bibliometric analysis is provided about this topic (see Table S5). The document types were limited to Articles or Reviews and the time span from 2010 to 15 July 2021, whereas the language was set to English. Finally, manual abstract assessment and screening were performed for each document, resulting in a list of 318 documents, of which the meta-data fields were compiled to conduct the bibliometric analysis.

2.2. Bibliometric Analysis

The data analysis was performed with the web interface Biblioshiny of the Bibliometrix R package (Version 1.4). The meta-data files extracted from the 318 documents were converted into R-readable files. Biblioshiny allows an analysis of three metric levels: Sources, authors, and documents. Hence, the most relevant sources, top authors, and documents in terms of productivity and impact can be identified. Additionally, the knowledge structure can be conceptual, intellectual, and socially analyzed, so that thematic mapping and evolution, co-citation, and collaboration networks can be built, among other things.

The metrics analyzed and the derived results from this analysis are schematically depicted in the workflow diagram in Scheme 1. The source analysis points out the target sources, which are the journals where most of the documents are published. The author analysis as well as their countries and affiliations allow one to identify not only the most prolific authors but also the social structure of the FS community, showing co-authorship, institutional, and international networking. The document-level study identifies the most important publications in the field, while the keyword analysis provides a general overview of the most-studied topics and evolution over the years, which makes it possible to identify mainstream and trend topics.

3. Results and Discussion

3.1. Flash Sintering (FS)

3.1.1. General Descriptive Information

Table 1 summarizes the essential information extracted from the analysis of the set of 318 documents retrieved from WoS from 2010 to 2021, using the word search query indicated in Scheme 1. Detailed information about each metric level is commented as follows.

Table 1. Main information.

Description	Results
Period	2010–2021
Documents	318
Sources (Journals, Books, etc.)	63
Average citations per documents	25.07
References	4852
Authors	
Authors	670
Author appearances	1498
Authors of single-authored documents	7
Authors of multi-authored documents	663
Authors Collaboration	
Single-authored documents	17
Documents per Author	0.475
Authors per Document	2.11
Co-Authors per Documents	4.71
Collaboration Index	2.2

3.1.2. Scientific Production

FS has constituted a breakthrough in material processing, and since its inception in 2010, it has become a hot topic and a new paradigm in ceramic processing. As it is shown in Figure 1, the number of publications has kept on growing, with an average of approximately 32 documents per year and an annual growth rate of 46.56%. Note that hybrid methodologies such as F-SPS and Reaction or Reactive FS (RFS) have been excluded and only peer-reviewed articles and reviews have been considered. Indeed, the cumulative publications since 2010 follow an exponential trend, as Figure 1 shows. If the topic development continues along the same trend, by a simple fitting to an exponential growth equation, the predicted number of publications in 10 years will be 4400, one order of magnitude higher than nowadays.

There are probably several reasons behind such dramatic growth in such a short period of its lifetime. The main reason is, of course, the scientific relevance of FS for the scientific community, mainly for those interested in ceramic research. Moreover, the simplicity of the FS experimental setup is significant, and as mentioned, just a furnace and a power supply are strictly needed, unlike other FAST methodologies that require complex and expensive equipment [1]. Despite its simplicity, FS is an extremely powerful sintering method that can be successfully used for most ceramic materials, from dielectrics ($BaTiO_3$ [41–47] or $(Bi_{0.2}Na_{0.2}K_{0.2}Ba_{0.2}Ca_{0.2})TiO_3$ [48]) to ionic (Zirconia, YSZ [2,49–52], CeO_2 or doped-CeO_2 [53–58]) or electronic (TiO_2 [19,22,59–62], $BiFeO_3$ or substituted-$BiFeO_3$ [24,27]) conductors. Interestingly, it can be also applied for processing ceramic composites of complex stoichiometry, metastable phases, or materials constituted by volatile species at the temperatures required for their sintering such as YSZ-Al_2O_3 composites [63–65], different types of solid state electrolytes [25,66,67], $BiFeO_3$ [68,69], or $K_{0.5}Na_{0.5}NbO_3$ [26,70–73]. Moreover, ceramics prepared by FS present very interesting properties rarely reported for materials obtained by convectional procedures. For example, it has been observed that FS specimens deform plastically before fracture when compressed at high strain, due to their extraordinarily high density of defects, such as stacking faults, dislocations, and twins [22,74] or that chemically inert ceramics are converted into active catalytic compounds by enhancing the concentration and reactivity of the ionic species [21,75]. Furthermore, the understanding of the flash phenomena still remains elusive and requires contributions from different scientific fields. This challenge has raised the interest of many researchers with different expertise, including theoreticians and experimentalists [76–79]. Another significant feature of FS is its ecofriendly nature and potential to possibly scale-up, as it requires less energy than conventional processing processes [80] and, therefore, it contributes to reducing CO_2 emissions. The possibility of working under continuous FS with rolling electrodes has been proposed in the literature [81]. This idea is already being explored at a larger scale by the company Lucideon [82] in a pilot plant. Very recently, the possibility of homogenously sintering 3D-complex shaped ceramics by the application of a three-phase power supply has been reported, giving rise to what has been named Multiphase Flash Sintering (MPFS) [83]. MPFS is presented as a feasible option to overcome the shape restrictions of conventional FS specimens. Moreover, the capabilities of FS combined with other FAST techniques are evolving in other interesting "flash-based methodologies" that have experienced great development as well. That is the case of Flash Spark Plasma Sintering (F-SPS), where pressure and pulsed currents are simultaneously applied, enabling the homogeneous and energetically efficient sintering (by different electrodes architectures) of both electric conductive and insulating materials [36,37,84–88]. Contactless Flash Sintering (Contactless-FS) is another flash-based methodology. Plasma electrodes are used instead of the traditional metallic wires. The plasma not only heats the material but also carries the current to trigger the flash, minimizing some of the thermal management issues encountered in conventional flash sintering due to the sample–electrodes contact [76]. Similarly, Flame-assisted Flash Sintering (FAFS) uses a flame as an electrode and heating source [89]. It has proved to be an effective technique for the sintering of ceramic coatings on metallic substrates. The combination of FS with Cold Sintering has resulted in Cold Flash Sintering (CFS), where the presence of relatively small amounts of liquids, such as water of acetic acid, on the pellets are used as electrolytes and enables the flash event even at room temperature [90,91]. Last but not least, the efforts made by the pioneers in the field, Prof. Raj and others, have been relevant by spreading the topic through the scientific community by inviting visitors to their labs, collaborating with other groups, giving lectures, and organizing successful International Conferences on the topic "Electromagnetic/Electric Fields in materials processing" such as those held on 2016 and 2019 in Tomar (Portugal) or the symposiums arranged by the Materials Research Society (MRS).

3.1.3. Source-Level Analysis

As mentioned, just peer-reviewed articles have been considered in this literature set. The analysis reveals that the documents have been published in 63 different journals (Table 1). Nevertheless, a close examination of Figure 2, which includes the sources where most of the articles have been published, shows that FS documents are concentrated in a few journals, i.e., *Journal of European Ceramic Society* and *Journal of the American Ceramic Society*. Indeed, those two are the core journals where more than one-third of the entire collection has been published. Additionally, these two journals together with *Scripta Materialia (Letters journal of Acta Materialia)*, *Ceramics International*, and *Acta Materialia* have published more than 65% of the analyzed set of documents. These journals are mainly targeted towards ceramic materials as well as their relationship between processing, microstructure, and properties, which makes sense as the vast majority of flash-sintered materials are ceramics provided that a negative temperature coefficient of electrical resistance is possessed [28]. It is noteworthy that these five journals are top-ranked journals of the first quartile according to the Journal Citation Report (2020) in the category of Material Science–Ceramics (*Journal of European Ceramic Society*, *Ceramics International*, and *Journal of the American Ceramic Society*) or Metallurgy and Metallurgical Engineering-Science (*Scripta* and *Acta Materialia*). Reciprocally, these journals are also the most locally cited sources in the set of documents analyzed (see Table S1). A locally cited source is a journal included in at least one of the reference lists of the analyzed document collection. Additionally, there are also articles devoted to FS in general high-impact journals such as *Nature Communications* [74] or *Science Advances* [22].

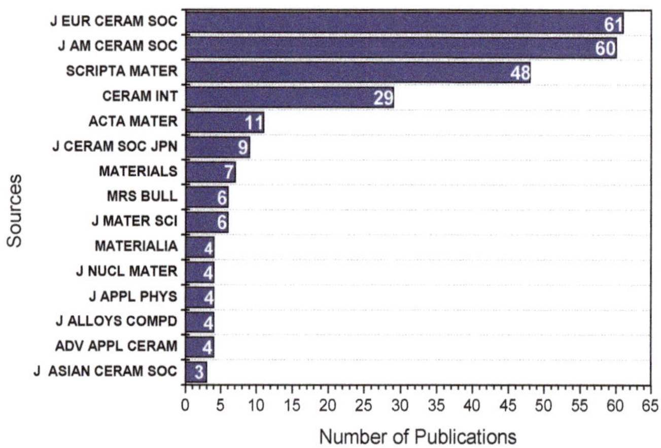

Figure 2. Top 15 sources with 3 or more published papers.

This simple source-level analysis shows that FS research is mainly published in high-ranked, peer-reviewed journals and provides an idea about the high quality of the research carried out in this field and the interest that it generates and attracts within the scientific community.

3.1.4. Author-Level Analysis and Networking

As shown in Table 1, the author-level analysis reveals that this set of documents involved 670 authors from 29 different countries. Seven of those authors published single-authored documents, whereas the rest participated in co-authored documents. Generally speaking, each document is written by 2.11 authors on average and the Collaboration Index, defined as the total authors of multi-authored documents divided by the total number of multi-authored articles, is 2.2. Additionally, Figure S1 (Supplementary Materials) depicts the country's scientific production map, where it is qualitatively shown that the USA, China, and Italy are the most involved countries.

Table 2 includes a list of authors that have published, so far, more than 10 articles directly related to FS along with their local *h*-index. The local *h*-index has been calculated using the conventional procedure (number *h* of publications that have been cited *h* times or more [92]) but only considering the set of 318 documents analyzed here. More details about authors with the highest numbers of local citations (citations included in this set of 318 documents) and global citations can be found in Table S2. It must be noted that those lists have been exclusively elaborated with the information retrieved from articles directly related to FS, as explained in the methods section. Those dealing with other similar techniques, such as F-SPS and others, have not been contemplated in the lists. In any case, Table 2 and Table S2 should not be considered as any type of author's ranking, as most authors in those lists have a significantly larger number of publications and citations in different related scientific topics that have not been included here. Moreover, some authors have been involved in FS almost since the early days, while others have become involved in recent times. As a way of example, the production of the authors included in Table 2 over the 11-year period of lifetime of FS is depicted in Figure S2, where the bubble size is proportional to the number of documents and the color darkness to the citations. Regardless of this, Prof. Raj not only introduced the FS topic for the first time in 2010, but he is the most prolific author in terms of the quantity of papers, citations, and the local *h*-index.

Table 2. Authors with the highest number of publications in the analyzed documents set along with their local *h*-index.

Authors	Articles	*h*-Index
Raj R	53	27
Sglavo VM	36	17
Biesuz M	30	14
Jha SK	22	15
Yamamoto T	18	8
Tsakalakos T	16	9
Wang YG	16	9
Wang HY	16	8
Wang H	15	9
Yoshida H	15	9
Charalambous H	14	9
Phuah XL	14	7
Tokunaga T	14	6
Liu JL	13	8
Lebrun JM	12	11
Liu DG	12	8
Chaim R	12	7
Muccillo R	12	7
Grasso S	12	6
Luo J	11	9

The high level of interaction among authors working in FS is very relevant, as shown in the collaboration maps of Figure 3. This type of figure displays the knowledge structure of the research field by providing a general overview about its social structure or, in other words, how the FS scientific community interacts at different levels (authors, institutions, and countries) [35]. Qualitatively, the bubble size in these collaboration maps represents the number of documents, while the strength of the relationship is represented by the links; the thicker, the stronger. Moreover, the position represents the influence, placing the most influential items, i.e., author, institution, or country, at the center of the maps. The different colors or clusters denote common collaboration networks or sub-networks. Figure 3a includes the collaboration map just among authors included in Table 2. Interestingly, all authors in Table 2, apart from Prof. Chaim, have very strong collaborations with others from the same list. Actually, in some cases, they even belonged to the same research group. That is the case of Biesuz who was a former PhD student in Sglavo's lab (Trento University, Trento, Italy), Charalambous in Tsakalakos' lab (Rutgers University, New Brunswick, NJ, USA), H Wang and Phuah in Haiyan Wang's lab (Purdue University, West Lafayette, IN, USA), while Lebrun was a postdoc in Raj's lab (University of Colorado at Boulder, Boulder, CO, USA). Moreover, there is a significant movement of authors from one institution into another that is helping to spread the topic. For instance, Jha who is currently an Assistant Professor at the Indian Institute of Technology Kanpur was a former PhD student of Raj and, later on, a postdoc in Tsakalakos's lab. Dianguang Liuwas a former PhD student at Northwestern Polytechnical University (working with Wang, Yiguang) and then moved to Southwest Jiaotong University where he works with Jinling Liu. Grasso is currently a professor at the Southwest Jiaotong University after having worked at the Queen Mary University of London. Moreover, as seen in Figure 3a by the solid connectors, there are many fruitful collaborations among different groups, resulting in co-authored publications. As mentioned, the behavior observed in Figure 3a is limited to authors from Table 2. Nevertheless, it is quite general and can be extrapolated to the whole FS community, as explained as follows by Figure 3b,c, which represents the countries and institutional collaboration maps, respectively, and aims to provide a broader overview of the scientific landscape of FS. As depicted in Figure 3b, there are authors from 26 different countries with at least one mutual publication. Note that the whole FS community involved authors from 29 countries, as is shown in Figure S1. This highlights, once again, the strong interconnection of this scientific community. Similar to Figure 3a, the USA is the country with the highest number of publications and the strongest collaboration network. Authors from China and Italy also have important scientific production and collaboration networks. Interestingly, even authors from countries such as Germany, Spain, India, and France, among others, have become involved in FS more recently and, therefore, while their number of publications is not that high, they have very strong collaboration networks. For example, in the whole document set, only five papers have been issued by corresponding authors from Spain, all of them being internationally collaborative publications. Figure 3c shows the collaboration network at an institutional level. The FS community involves authors from 217 institutions. Thus, for the sake of clarity and visualization, the institutional collaboration map has been limited to those 18 institutions with the highest number of published papers. Analogously to Figure 3a, the bubble size of the University of Colorado at Boulder (USA) and Trento University (Italy) is not surprising; as mentioned, the most-productive authors in terms of publications belong to these institutions, i.e., Raj, Sglavo, and Biesuz. Moreover, both institutions have strong collaboration networks worldwide. For instance, Trento University, Southwest Jiaotong University, one of the most prolific Chinese institutions, and Queen Mary University of London (UK) have established a strong collaboration sub-network. As commented for Figure 3a, Southwest Jiaotong University and the Queen Mary University of London are the current and previous affiliations of Prof. Grasso, respectively. At the same time, other researchers from Southwest Jiaotong University (Liu Dg and Liu Jl among others) work with researchers from other academic centers, such as the Northwestern Polytechnical University (China) and the University of Central Florida (USA). Additionally, Figure 3c also unveils a strong collaboration sub-network within American institutions composed of Rutgers University, Purdue University, Argonne National Laboratory, the University of California San Diego, and the University of California Davis. In albeit extreme simplification, this subnetwork is partially shown in Figure 3a, proving again its similarities with the whole FS scientific community. Most of the works of this American co-authorship sub-network involve some kind of in situ measurements during FS or to flash-sintered samples, such as energy-dispersive X-ray diffraction [15,53,69,93].

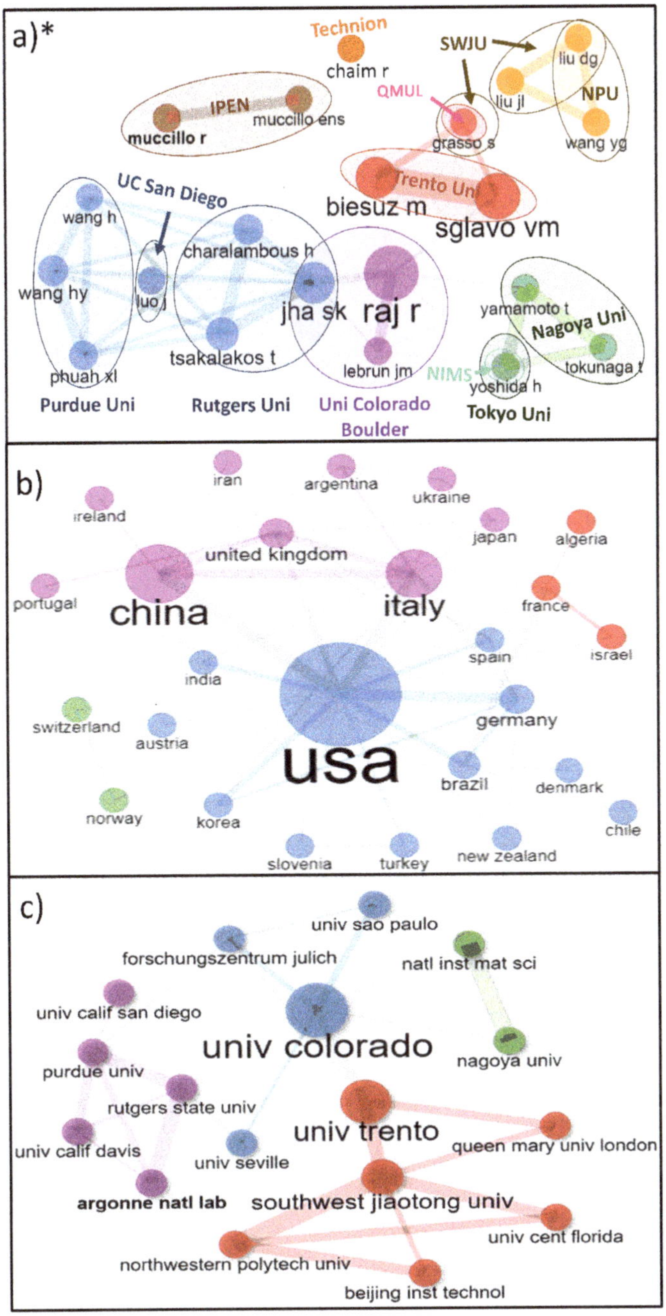

Figure 3. (**a**) Selected authors from Table 2, (**b**) countries, and (**c**) institution's collaboration networks. * UC San Diego: University of California San Diego, IPEN: Instituto de Pesquisas Energéticas e Nucleares, QMUL: Queen Mary University of London, SWJU: Southwest Jiaotong University, NPU: Northwestern Polytechnical University, NIMS: National Institute for Materials Science, Technion: Technion–Israel Institute of Technology.

All in all, Figure 3 highlights the significant level of interaction among research scientists and groups in FS. It is evidenced that authors involved in this research field are quite eager to collaborate with others from different institutions and even countries. Thus, internationalization is at the core of FS. This practice may be related to the complexity of the FS process that involves physical and chemical phenomena. Thus, its understanding and implementation demand an interdisciplinary approach from Physics to Chemistry and Engineering, and it requires of the participation of both experimentalists and theoreticians. Moreover, collaborations with experts from large scientific facilities (National Laboratories) in the monitoring of the FS process under in situ conditions are also quite noticeable. This strong collaborative networking as well as the International Conferences in Tomar (Portugal) and MRS symposiums, mentioned above, are helping to create a sense of community among researchers worldwide, and it is probably contributing to the fast development of this relatively new research field.

3.1.5. Document-Level Analysis. Influential Documents and Trending Topics

This section studies the most influential documents of the collection with respect to their number of citations as well as the most commonly tackled topics in FS by keyword analysis.

Table S3 includes the top 20 most-cited documents, where the number of local and global citations as well as their ratio are shown. A local citation refers to the number of citations that one document has received from the analyzed document set (in this case, 318 documents and all of them related to FS), whereas global citations are the total number of citations from WoS that may include documents that are not necessarily associated with the analyzed research topic. Therefore, in principle, the higher local citations, the more important and influential the document is in a particular research field. This list is ranked by the paper of Cologna, Rashkova, and Raj reporting the first demonstration of FS, which triggered the development of this research field [2]. This paper constitutes a true milestone in ceramic processing and to date has received more than 474 citations, 270 of which correspond to local citations. This is followed by a single-authored paper of Raj, addressing the role of Joule Heating during FS [8]. It entails one of the first works devoted to explaining the underlying mechanisms of FS, concluding that the fast sintering rates achieved in FS cannot be solely explained by Joule Heating. This work closely links with another top-ranked paper (fourth place) authored by Todd et al. [4]. The authors modelled the electrical and thermal response of 3YSZ under FS conditions and established that the flash event is triggered by a thermal runaway caused by Joule Heating. Indeed, it is now well-accepted within the scientific community that the thermal runaway induced by Joule Heating is the actual phenomenon that initiates the flash. The third most-cited paper both globally and locally corresponds to the work of Cologna et al. published in 2011 [13], which deals with the demonstration of FS in alumina, a highly insulating material. Finally, special attention is deserved for the paper ranked fifth, as this is the second literature review dedicated integrally to FS [28] (to the best of our knowledge, the first review was published by Dancer in 2016 [30]). It was published in *Advances in Applied Ceramics* in 2017 and, to date, has received more than 207 citations, where half of them correspond to local citations. It is also worth mentioning that 75% of the top 20 most-cited documents in FS are published in either the *Journal of the European Ceramic Society* or the *Journal of the American Ceramic Society*. As mentioned in the source-level analysis section, both are high-ranked journals dedicated to the study of ceramic materials, which once again highlights the quality of the research carried out by the scientific community of this field.

In order to provide some perspective about influential works recently published, we carried out a similar analysis while refining the time span to 2020 and 2021. Thus, Table S4 presents the most global and locally cited documents in 2020 and 2021. Most of these publications deal with the understanding of the FS mechanisms in ceria and titania [62,94] or the correlation between the effect of the experimental parameters and the induced defects on the final properties and microstructure of the flash-sintered materials [51,95–98]. Many of these documents are co-authored by early-stage researchers, such as Lavagnini [51] and Storion [96], both PhD students at the University of São Paulo (Brazil), or Phuah [95,98] and Mishra [94,97] who recently obtained their PhD degrees from Purdue University (USA) and Forschungszentrum Jülich (Germany), respectively. This is an indication that many early-stage researchers are developing their careers in FS and are making important contributions to the field.

Keyword analysis is commonly used in bibliometric analysis to systematically identify the document content, trend topics, and research hotspots of a particular research field. Table 3 contains the top 10 Keywords Plus and Authors' Keywords. Authors' keywords are provided by the authors themselves, whereas Keywords Plus are generated by an algorithm, extracting words that frequently appear in the title's references and not necessarily in the title of the articles or as Author Keywords [99]. Very recently, a study carried out by Zhang J. et al. reveals that both types of keywords identify very similar research trends and knowledge structures [100]. Indeed, the list of words included in Table 3 for both categories as well as the prevalence order is quite similar. Note that trivial keywords, such as Flash Sintering or Flash Sintered, have been cleaned. An examination of these keywords reveals that zirconia is the most-studied material. As mentioned, the first demonstration of FS was carried out with this material [2]. Since then, zirconia powders of different compositions have been widely used as a model material to study the underlying mechanisms of FS as well as the driving sources triggering the flash event, which is also linked to other top keywords such as Joule Heating, thermal runaway or defects. As a way of example, a few works are cited herein [4,50,101,102]. Another important part of the research carried out in FS deals with the study of the properties of the flash-sintered materials, such as their microstructure, abnormal grain growth, defect structures, etc. [22,103,104], some of them granted with special properties. This also explains some of the top keywords obtained, e.g., "microstructure", "grain growth", "defect structures".

Table 3. Top 10 Keywords Plus and Authors' Keywords.

Keywords Plus	Frequency	Author's Keywords	Frequency
Zirconia/YSZ	160	Zirconia	32
Electrical Conductivity	105	Microstructure	29
Densification	73	Joule Heating	19
Grain-growth	73	Grain Growth	16
Thermal Runaway	71	ZnO	14
Alumina	65	Impedance Spectroscopy	12
Ceramics	49	Ceramics	11
Microstructure	41	Defects	11
ZnO	30	Electrical Conductivity	9
Defect Structure	15	Alumina	8

Figure 4a shows the Keywords Plus dynamic, representing the frequency of each keyword as a function of time (from 2010 to 2020), which allows for identifying trend topics and research hotspots. For instance, it can be observed that zirconia has always been a hot topic in FS. As mentioned above, it is now well accepted that the flash event is triggered by a thermal runaway. Since it was proposed in 2015 by Todd et al. in the already-mentioned paper entitled "Electrical characteristics of flash sintering: thermal runaway of Joule heating" [4], the keyword thermal runaway has continued growing. Therefore, the contribution of Todd R.I. et al. constitutes an important milestone in FS. From Figure 4a, is also depicted that ZnO is currently a trending topic, and in the last few years its annual occurrence has been increasing. In a similar way to zirconia, ZnO has been widely employed as a model material in FS [103,105,106]. This Keyword Plus dynamic analysis agrees well with Figure 4b, which includes the top 10 most-studied materials during 2020 and 2021. As expected, Zirconia leads the ranking, with more than 38 publications, followed by ZnO. Figure 4b also reveals that another important body of work in FS is dedicated to the sintering of materials with technological interest, which development is being hampered by the high temperatures required or other kinds of difficulties in their processing such as the volatilization of some of their components. That is the case of different sodium and lithium ion conductive ceramics for solid-state batteries [67,107], the lead-free piezoelectric ceramic potassium sodium niobate [70,73], or ZnO-Bi_2O_3-based varistor ceramics [108–110]. We expect that this trend will probably be maintained during the next few years. That is to say that Zirconia and ZnO will continue being hot topics. As mentioned, the FS mechanisms are still clouded, and further studies about their understanding will certainly be carried out. Therefore, it is quite likely that these two materials will continue to be used as models in future works dealing with the underlying nature of FS. On the other hand, besides the reduced temperatures and times offered by the FS technique, it has also proved to be an engineering tool to grant special and unexpected properties to materials [22,75]. Thus, researchers will continue exploring these FS capabilities and, therefore, the study of the properties of the materials prepared by FS as well as the preparation of new ceramic materials that are hard or impossible to prepare by conventional procedures will be another mainstream area of research.

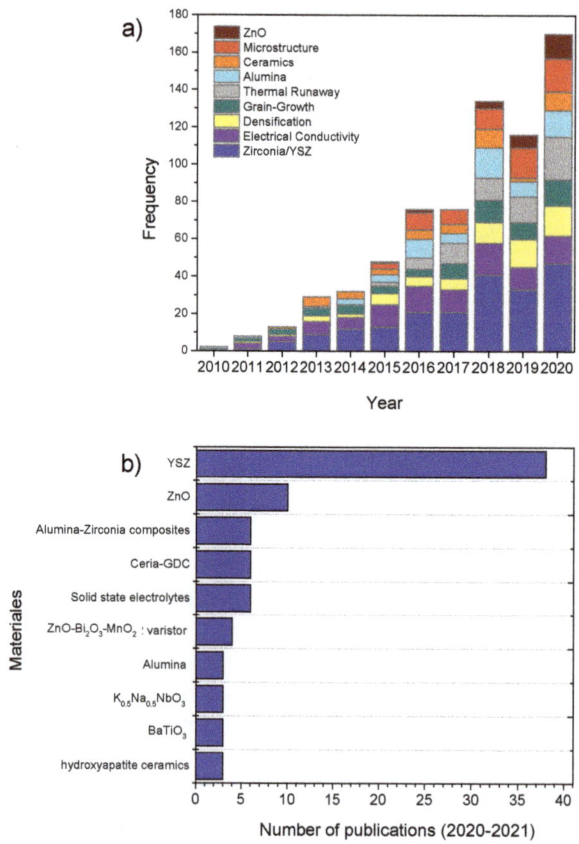

Figure 4. (**a**) Keywords Plus dynamic and (**b**) top 10 studied materials during the two-year period 2020–2011.

All in all, we would like to emphasize that this keyword analysis provides a general and brief overview of the document content as well as of the major tackled topics and future research prospects in FS. It is probably quite trivial for researchers who have been working in the field for a while but may be useful to strategize future research as well as a starting point for those researchers who are not acquainted but keen on FS.

3.2. Reactive Flash-Sintering (RFS)

As mentioned in the introduction, another important branch of research directly related to FS is Reaction or Reactive Flash sintering (RFS). The foundations and the working mode are quite similar but instead synthesis and sintering are merged in a single step. RFS was reported for the first time in 2018, showing that a highly dense and pure metastable oxide can be prepared in a matter of seconds from a mixture of its basic constituents [27]. RFS soon garnered the attention of the scientific community and, since then, many documents have been published exploiting the capabilities of RFS. Indeed, Figure 5a shows the scientific production for RFS with a remarkable number of 32 publications in just three years, which implies a dramatic annual scientific production growth rate of 100%. Thus, due to the relevance of RFS, we decided to carry out a separate bibliometric analysis to sketch the general scientific map for RFS. The specific word search query as well as the general descriptive information from the analyzed document set for RFS can be found in Tables S5 and S6, respectively.

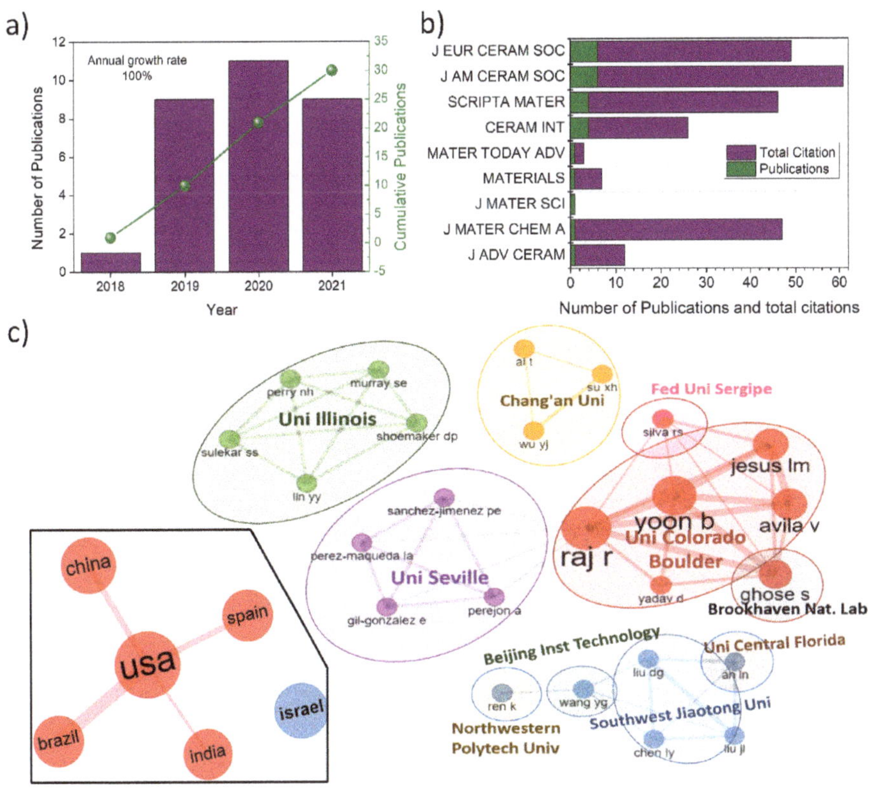

Figure 5. (**a**) Annual and cumulative publications, (**b**) number of documents and total citations per source, and (**c**) country, author, and institution collaboration networks in RFS. The author and institution collaboration networks are limited to those with more than two publications and at least one co-authored document.

As can be observed from Figure 5b, RFS follows the same trend as FS in terms of targeted sources. *The Journal of the American Ceramic Society* and *the Journal of the European*

Ceramic Society are where almost 50% of the entire collection have been published. Those two journals together with *Ceramics International* and *Scripta Materialia* published 80% of the entire collection. All of them are highly ranked journals according to Journal Citation Reports (2020) and accumulate most of the citations, which highlights, once again, the quality and relevance of the topic.

Nevertheless, the most-relevant publication of this set of documents is probably from *the Journal of Materials Chemistry A*. This journal published just one paper about RFS, but, by itself, it accumulates 20% of the global citations of the entire set. This document was authored by Gil-Gonzalez et al. in 2018 [27] and constitutes, to the best of our knowledge, the first demonstration of RFS. It has received 46 global citations with an average of 11.50 citations per year. Additionally, it is also the most-cited document of the set. Seventeen documents out of twenty-five have cited it, which makes this paper the most relevant one for the research topic of RFS. Details about the historical direct citation network in RFS can be found in Figure S3. Further details about authors, document impact, in terms of quality and quantity, as well as the most studied materials in RFS can be found in SI (Tables S7–S10). It is worth mentioning that according to Table S10, RFS has been mostly dedicated to the preparation of materials with technological interest. This is probably due to the advantages of reduced temperatures and times offered by the technique. A significant number of works have been dedicated to the preparation of high-entropy oxides [111–115], followed by ceramics such as the multiferroic $BiFeO_3$ and related materials [27,80,116], solid electrolytes [117–119], or the lead-free piezoelectric potassium sodium niobate [120,121]. Analogously to Zirconia or ZnO in FS, the underlying mechanisms in RFS have been studied by high-resolution in situ measurements in the reaction of MgO and Al_2O_3 to form the spinel $MgAl_2O_4$ [122,123].

Figure 5c represents the collaboration networks in RFS from countries to institutions and authors. The RFS community is formed by 82 authors and just one of them published single-authored documents (Prof. Chaim from Technion Israel Institute of Technology). On average, each document is co-authored by 5.04 authors with a collaboration index of 3.38 (See Table S5). As shown in the inset of Figure 5c, the documents come from six different countries: The USA, China, Brazil, Spain, India, and Israel. All authors internationally collaborate, besides authors from Israel with two single-country publications [124,125]. More details about author clusters and institutions are shown in Figure 5c. Note that for the sake of visualization, the collaboration network is limited to authors with more than two publications and at least one co-authored document, and unfortunately, Figure 5c does not show the whole RFS networking. From a simple visual inspection, it can be identified that the most prolific authors in number of publications are Prof. Raj and Yoon from the University of Colorado at Boulder, with seven and six publications, respectively (see also Tables S7 and S8). Note that the bubble size is related to the number of publications. Additionally, Figure 5c also unveils the collaboration sub-networks within the RFS community. Analogously to FS, internationalization and collaboration are at the core of this scientific community. For instance, authors from the University of Colorado at Boulder strongly collaborate with Ghose from Brookhaven National Laboratory [117,118,122,123] and authors from Seville University (Spain) [27] and the Federal University of Sergipe (Brazil) [117]. Indeed, the most relevant works of RFS have been a result of these collaboration sub-networks such as the already-mentioned document of *the Journal of Materials Chemistry A* by Gil-Gonzalez et al. [27] or that by Yoon et al. [123], where RFS of MgO and α-Al_2O_3 were studied by in situ synchrotron measurements in Brookhaven National Laboratory. This document received 19 global citations, 12 of which are local citations, being one of the most-relevant documents for RFS. Another strong collaboration network is formed by authors from Chinese institutions such as Southwest Jiaotong University, Beijing Institute of Technology, and Northwestern Polytechnical University, who work with An from the University of Central Florida. Their research is primarily dedicated to the preparation of high-entropy oxides by RFS [112,113]. Researchers from Chang'an University working on the preparation of piezoelectric materials also form another cluster [121,126]. It is worth mentioning that China is the second most-prolific country in the number of RFS publications, just behind the USA, at 9 vs. 11 documents. Finally, another important cluster of authors is formed by researchers affiliated to the University of Illinois at Urbana-Champaign, whose works have been devoted to study the transformation of manganese oxides during RFS [127,128].

4. Conclusions

FS has constituted a truly breakthrough in materials processing, and since it was proposed for the first time by Prof Raj in 2010, the number of publications has grown exponentially. In this work, we report the scientific landscape of FS by bibliometric analysis, identifying key aspects and peculiarities of the FS community. The target journals where most of the FS papers are published have been pointed out. All of them are dedicated almost exclusively to ceramic materials and are highly ranked journals of the first quartile according to JCR. This highlights the quality of the research carried out in the field. Socially speaking, the knowledge structure has been depicted at different levels, from countries to institutions and authors. A detailed analysis of the interaction of the authors with the highest number of publications is provided. This unveils important collaboration subnetworks that describe the general social structure of the FS scientific community. One of the most striking features is the large number of fruitful national and international collaborations among authors involved in FS with co-authored publications. It seems to be part of the core of this research field. It may be related to the complexity of the flash event that requires the contribution of experts from different disciplines for its understanding and development. The most influential documents in terms of local and global citations have been also identified. A brief description about the topics addressed in those documents is presented as well. Finally, the most recurrent keywords that best describe the document content have been analyzed along with the most-studied materials in 2020 and 2021, identifying the most-tackled and mainstream topics in FS. They reveal that zirconia has always been a hot topic in FS and the documents frequently deal with the understanding of the underlying mechanisms of FS or the non-typical properties granted to flash-sintered materials, such as the abnormal grain growth or defects. As a future research prospect, it is predicted that these topics will be maintained in the next few years. Finally, due to the dramatic scientific growth experienced in RFS in a very short period of time, a bibliometric analysis is provided separately for RFS. A detailed collaboration network is laid out, interestingly showing that the most influential works in the field of RFS are the results of international or national collaboration between authors from different institutions.

All in all, the aim of this work is to draw a general picture of the scientific landscape of FS and RFS by a bibliometric analysis, where the target sources, the most relevant documents, hot and trending topics, and social networking have been identified. To the best of our knowledge, this is the first study of this nature carried out in the field of FS. We believe that this work can be of interest not only for researchers working in the field but also for those who are keen on but not acquainted with FS and RFS. It can be a useful tool to strategize future research and publishing approaches as well as to be quickly updated with this research field, in spite of the high number of scientific publications and the dramatic growth that the field is experiencing.

Supplementary Materials: The following supporting information can be downloaded at: https://www.mdpi.com/article/10.3390/ma15020416/s1, Figure S1: Country scientific production map; Figure S2: Production over time of authors included in Table 2; Figure S3: Historical direct citation network in RFS; Table S1: Most local cited sources; Table S2: Most local cited authors along with their global citations; Table S3: Top-20 most cited documents; Table S4: Most cited documents published in the last two years (2020–2021); Table S5: Word search query for RFS; Table S6: Main information about RFS document sets; Table S7: Authors with three or more publications and local h-index in RFS; Table S8: Most local and global cited authors in RFS; Table S9: Top-5 most cited documents in RFS; Table S10: Top Materials in RFS.

Author Contributions: Conceptualization, E.G.-G. and L.A.P.-M.; methodology, E.G.-G.; software, E.G.-G.; validation, L.A.P.-M., P.E.S.-J. and A.P.; formal analysis, E.G.-G. and L.A.P.-M.; investigation, E.G.-G.; resources, L.A.P.-M.; data curation, E.G.-G.; writing—original draft preparation, E.G.-G and L.A.P.-M.; writing—review and editing, P.E.S.-J.; visualization, E.G.-G.; supervision, L.A.P.-M., P.E.S.-J.; project administration, L.A.P.-M.; funding acquisition, L.A.P.-M.; All authors have read and agreed to the published version of the manuscript.

Funding: This research was funded by the Spanish Government Agency Ministerio de Ciencia, Innovación y Universidades and FEDER grant numbers CTQ2017–83602-C2–1-R and CTQ2017–83602-C2–2-R), Junta de Andalucía-Consejería de Economía, Conocimiento, Empresas y Universidad grant number P18-FR-1087, FEDER grant number US-1262507 and INTRAMURAL-CSIC grant numbers 201960E092 and 202060I004.

Institutional Review Board Statement: Not applicable.

Informed Consent Statement: Not applicable.

Data Availability Statement: Data are contained within the article.

Conflicts of Interest: The authors declare no conflict of interest.

References

1. Guillon, O.; Gonzalez-Julian, J.; Dargatz, B.; Kessel, T.; Schierning, G.; Räthel, J.; Herrmann, M. Field-Assisted Sintering Technology/Spark Plasma Sintering: Mechanisms, Materials, and Technology Developments. *Adv. Eng. Mater.* **2014**, *16*, 830–849. [CrossRef]
2. Cologna, M.; Rashkova, B.; Raj, R. Flash Sintering of Nanograin Zirconia in <5 s at 850 °C. *J. Am. Ceram. Soc.* **2010**, *93*, 3556–3559. [CrossRef]
3. Naik, K.; Jha, S.K.; Raj, R. Correlations between conductivity, electroluminescence and flash sintering. *Scr. Mater.* **2016**, *118*, 1–4. [CrossRef]
4. Todd, R.I.; Zapata-Solvas, E.; Bonilla, R.S.; Sneddon, T.; Wilshaw, P.R. Electrical characteristics of flash sintering: Thermal runaway of Joule heating. *J. Eur. Ceram. Soc.* **2015**, *35*, 1865–1877. [CrossRef]
5. Zhang, Y.; Jung, J.-I.; Luo, J. Thermal runaway, flash sintering and asymmetrical microstructural development of ZnO and ZnO–Bi_2O_3 under direct currents. *Acta Mater.* **2015**, *94*, 87–100. [CrossRef]
6. Narayan, J. A new mechanism for field-assisted processing and flash sintering of materials. *Scr. Mater.* **2013**, *69*, 107–111. [CrossRef]
7. Chaim, R. Liquid Film Capillary Mechanism for Densification of Ceramic Powders during Flash Sintering. *Materials* **2016**, *9*, 280. [CrossRef]
8. Raj, R. Joule heating during flash-sintering. *J. Eur. Ceram. Soc.* **2012**, *32*, 2293–2301. [CrossRef]
9. Raj, R.; Kulkarni, A.; Lebrun, J.-M.; Jha, S. Flash sintering: A new frontier in defect physics and materials science. *MRS Bull.* **2021**, *46*, 36–43. [CrossRef]
10. Becker, M.Z.; Shomrat, N.; Tsur, Y. Recent Advances in Mechanism Research and Methods for Electric-Field-Assisted Sintering of Ceramics. *Adv. Mater.* **2018**, *30*, 1706369. [CrossRef]
11. Chaim, R. Particle surface softening as universal behaviour during flash sintering of oxide nano-powders. *Materials* **2017**, *10*, 179. [CrossRef]
12. Yoshida, H.; Sakka, Y.; Yamamoto, T.; Lebrun, J.-M.; Raj, R. Densification behaviour and microstructural development in undoped yttria prepared by flash-sintering. *J. Eur. Ceram. Soc.* **2014**, *34*, 991–1000. [CrossRef]
13. Cologna, M.; Francis, J.S.C.; Raj, R. Field assisted and flash sintering of alumina and its relationship to conductivity and MgO-doping. *J. Eur. Ceram. Soc.* **2011**, *31*, 2827–2837. [CrossRef]
14. Zhang, Y.; Luo, J. Promoting the flash sintering of ZnO in reduced atmospheres to achieve nearly full densities at furnace temperatures of <120 °C. *Scr. Mater.* **2015**, *106*, 26–29. [CrossRef]
15. Bicer, H.; Beyoglu, B.; Ozdemir, T.E.; Okasinski, J.; Tsakalakos, T. Direct in situ observation of electric field assisted densification of ZnO by energy dispersive X-ray diffraction. *Ceram. Int.* **2019**, *45*, 7614–7618. [CrossRef]
16. Prette, A.L.G.; Cologna, M.; Sglavo, V.; Raj, R. Flash-sintering of Co_2MnO_4 spinel for solid oxide fuel cell applications. *J. Power Sources* **2011**, *196*, 2061–2065. [CrossRef]
17. Zapata-Solvas, E.; Bonilla, S.; Wilshaw, P.R.; Todd, R.I. Preliminary investigation of flash sintering of SiC. *J. Eur. Ceram. Soc.* **2013**, *33*, 2811–2816. [CrossRef]
18. Biesuz, M.; Sglavo, V.M. Flash sintering of ceramics. *J. Eur. Ceram. Soc.* **2019**, *39*, 115–143. [CrossRef]
19. Wang, Q.; Watts, C.S.; Athanasiou, C.E.; Dai, Z.; Hu, M.; Sheldon, B.W.; Padture, N.P. The effect of atmosphere on the flash-sintering of nanoscale titania ceramics. *Scr. Mater.* **2021**, *199*, 113894. [CrossRef]
20. Hwang, C.; Yun, J. Effect of processing conditions on the flash onset temperature in hydroxyapatite. *Materials* **2021**, *14*, 5229. [CrossRef] [PubMed]

21. Klauke, K.; Kayaalp, B.; Biesuz, M.; Iannaci, A.; Sglavo, V.M.; D'Arienzo, M.; Lee, S.; Seo, J.; Jung, W.; Mascotto, S. Enhancement of the $SrTiO_3$ Surface Reactivity by Exposure to Electric Fields. *ChemNanoMat* **2019**, *5*, 948–956. [CrossRef]
22. Li, J.; Cho, J.; Ding, J.; Charalambous, H.; Xue, S.; Wang, H.; Phuah, X.L.; Jian, J.; Wang, X.; Ophus, C.; et al. Nanoscale stacking fault–assisted room temperature plasticity in flash-sintered TiO_2. *Sci. Adv.* **2019**, *5*, eaaw5519. [CrossRef]
23. Perez-Maqueda, L.A.; Gil-Gonzalez, E.; Perejon, A.; Lebrun, J.-M.; Sanchez-Jimenez, P.E.; Raj, R. Flash sintering of highly insulating nanostructured phase-pure $BiFeO_3$. *J. Am. Ceram. Soc.* **2017**, *100*, 3365–3369. [CrossRef]
24. Gil-González, E.; Perejón, A.; Sánchez-Jiménez, P.E.; Raj, R.; Pérez-Maqueda, L.A. Processing and properties of Bi0.98R0.02FeO3 (R = La, Sm, Y) ceramics flash sintered at ~650 °C in <5 s. *J. Am. Ceram. Soc.* **2020**, *103*, 136–144. [CrossRef]
25. Clemenceau, T.; Andriamady, N.; Kumar M.K., P.; Badran, A.; Avila, V.; Dahl, K.; Hopkins, M.; Vendrell, X.; Marshall, D.; Raj, R. Flash sintering of Li-ion conducting ceramic in a few seconds at 850 °C. *Scr. Mater.* **2019**, *172*, 1–5. [CrossRef]
26. Serrazina, R.; Ribeiro, C.; Costa, M.E.; Pereira, L.; Vilarinho, P.M.; Senos, A.M.O.R. Particle characteristics' influence on FLASH sintering of potassium sodium niobate: A relationship with conduction mechanisms. *Materials* **2021**, *14*, 1321. [CrossRef] [PubMed]
27. Gil-González, E.; Perejón, A.; Sánchez-Jiménez, P.E.; Sayagués, M.J.; Raj, R.; Pérez-Maqueda, L.A. Phase-pure $BiFeO_3$ produced by reaction flash-sintering of Bi_2O_3 and Fe_2O_3. *J. Mater. Chem. A* **2018**, *6*, 5356–5366. [CrossRef]
28. Yu, M.; Grasso, S.; Mckinnon, R.; Saunders, T.; Reece, M.J. Review of flash sintering: Materials, mechanisms and modelling. *Adv. Appl. Ceram.* **2017**, *116*, 24–60. [CrossRef]
29. Pritchard, A. Statistical Bibliography or Bibliometrics? *J. Doc.* **1969**, *25*, 348–349.
30. Dancer, C.E.J. Flash sintering of ceramic materials. *Mater. Res. Express* **2016**, *3*, 1–25. [CrossRef]
31. Zhou, H.; Li, X.; Zhu, Y.; Liu, J.; Wu, A.; Ma, G.; Wang, X.; Jia, Z.; Wang, L. Review of flash sintering with strong electric field. *High Volt.* **2021**, 1–11. [CrossRef]
32. Aria, M.; Cuccurullo, C. bibliometrix: An R-tool for comprehensive science mapping analysis. *J. Informetr.* **2017**, *11*, 959–975. [CrossRef]
33. Durieux, V.; Gevenois, P.A. Bibliometric indicators: Quality measurements of scientific publication. *Radiology* **2010**, *255*, 342–351. [CrossRef]
34. Zupic, I.; Čater, T. Bibliometric Methods in Management and Organization. *Organ. Res. Methods* **2014**, *18*, 429–472. [CrossRef]
35. Morris, S.A.; Van der Veer Martens, B. Mapping research specialties. *Annu. Rev. Inf. Sci. Technol.* **2008**, *42*, 213–295. [CrossRef]
36. Grasso, S.; Saunders, T.; Porwal, H.; Milsom, B.; Tudball, A.; Reece, M.; Chen, I.-W. Flash Spark Plasma Sintering (FSPS) of α and β SiC. *J. Am. Ceram. Soc.* **2016**, *99*, 1534–1543. [CrossRef]
37. Grasso, S.; Saunders, T.; Porwal, H.; Cedillos-Barraza, O.; Jayaseelan, D.D.; Lee, W.E.; Reece, M.J. Flash spark plasma sintering (FSPS) of pure ZrB_2. *J. Am. Ceram. Soc.* **2014**, *97*, 2405–2408. [CrossRef]
38. Manière, C.; Lee, G.; Zahrah, T.; Olevsky, E.A. Microwave flash sintering of metal powders: From experimental evidence to multiphysics simulation. *Acta Mater.* **2018**, *147*, 24–34. [CrossRef]
39. Bykov, Y.V.; Egorov, S.V.; Eremeev, A.G.; Kholoptsev, V.V.; Rybakov, K.I.; Sorokin, A.A. Flash Microwave Sintering of Transparent Yb:$(LaY)_2O_3$ Ceramics. *J. Am. Ceram. Soc.* **2015**, *98*, 3518–3524. [CrossRef]
40. Wang, C.; Ping, W.; Bai, Q.; Cui, H.; Hensleigh, R.; Wang, R.; Brozena, A.H.; Xu, Z.; Dai, J.; Pei, Y.; et al. A general method to synthesize and sinter bulk ceramics in seconds. *Science* **2020**, *368*, 521–526. [CrossRef]
41. Shi, R.; Pu, Y.; Wang, W.; Shi, Y.; Li, J.; Guo, X.; Yang, M. Flash sintering of barium titanate. *Ceram. Int.* **2019**, *45*, 7085–7089. [CrossRef]
42. M'Peko, J.-C.; Francis, J.S.C.; Raj, R. Field-assisted sintering of undoped $BaTiO_3$: Microstructure evolution and dielectric permittivity. *J. Eur. Ceram. Soc.* **2014**, *34*, 3655–3660. [CrossRef]
43. Yoshida, H.; Uehashi, A.; Tokunaga, T.; Sasaki, K.; Yamamoto, T. Formation of grain boundary second phase in $BaTiO_3$ polycrystal under a high DC electric field at elevated temperatures. *J. Ceram. Soc. Jpn.* **2016**, *124*, 388–392. [CrossRef]
44. Uehashi, A.; Yoshida, H.; Tokunaga, T.; Sasaki, K.; Yamamoto, T. Enhancement of sintering rates in $BaTiO_3$ by controlling of DC electric current. *J. Ceram. Soc. Jpn.* **2015**, *123*, 465–468. [CrossRef]
45. Nakagawa, Y.; Yoshida, H.; Uehashi, A.; Tokunaga, T.; Sasaki, K.; Yamamoto, T. Electric current-controlled synthesis of $BaTiO_3$. *J. Am. Ceram. Soc.* **2017**, *100*, 3843–3850. [CrossRef]
46. Ren, K.; Huang, S.; Cao, Y.; Shao, G.; Wang, Y. The densification behavior of flash sintered $BaTiO_3$. *Scr. Mater.* **2020**, *186*, 362–365. [CrossRef]
47. Umemura, R.; Tokunaga, T.; Yamamoto, T. Flash sintering for $BaTiO_3$ with square alternating current field including zero-field duration. *J. Ceram. Soc. Jpn.* **2020**, *128*, 1018–1023. [CrossRef]
48. Liu, J.; Ren, K.; Ma, C.; Du, H.; Wang, Y. Dielectric and energy storage properties of flash-sintered high-entropy $(Bi_{0.2}Na_{0.2}K_{0.2}Ba_{0.2}Ca_{0.2})TiO_3$ ceramic. *Ceram. Int.* **2020**, *46*, 20576–20581. [CrossRef]
49. Jha, S.K.; Terauds, K.; Lebrun, J.-M.; Raj, R. Beyond flash sintering in 3 mol % yttria stabilized zirconia. *J. Ceram. Soc. Jpn.* **2016**, *124*, 283–288. [CrossRef]
50. Downs, J.A.; Sglavo, V.M.; Industriale, I. Electric Field Assisted Sintering of Cubic Zirconia at 390 °C. *J. Am. Ceram. Soc.* **2013**, *1344*, 1342–1344. [CrossRef]
51. Lavagnini, I.R.; Campos, J.V.; Ferreira, J.A.; Pallone, E.M.J.A. Microstructural evolution of 3YSZ flash-sintered with current ramp control. *J. Am. Ceram. Soc.* **2020**, *103*, 3493–3499. [CrossRef]

52. Biesuz, M.; Pinter, L.; Saunders, T.; Reece, M.; Binner, J.; Sglavo, V.M.; Grasso, S. Investigation of electrochemical, optical and thermal effects during flash sintering of 8YSZ. *Materials* **2018**, *11*, 1214. [CrossRef]
53. Jha, S.K.; Charalambous, H.; Wang, H.; Phuah, X.L.; Mead, C.; Okasinski, J.; Wang, H.; Tsakalakos, T. In-situ observation of oxygen mobility and abnormal lattice expansion in ceria during flash sintering. *Ceram. Int.* **2018**, *44*, 15362–15369. [CrossRef]
54. Phuah, X.L.; Wang, H.; Qi, Z.; Misra, S.; Kalaswad, M.; Wang, H. Field-assisted heating of Gd-doped ceria thin film. *J. Am. Ceram. Soc.* **2020**, *103*, 2309–2314. [CrossRef]
55. Mishra, T.P.; Lenser, C.; Raj, R.; Guillon, O.; Bram, M. Development of a processing map for safe flash sintering of gadolinium-doped ceria. *J. Am. Ceram. Soc.* **2021**, 1–13. [CrossRef]
56. Guan, L.; Li, J.; Song, X.; Bao, J.; Jiang, T. Graphite assisted flash sintering of Sm_2O_3 doped CeO_2 ceramics at the onset temperature of 25 °C. *Scr. Mater.* **2019**, *159*, 72–75. [CrossRef]
57. Gaur, A.; Sglavo, V.M. Tuning the flash sintering characteristics of ceria with $MnCo_2O_4$. *Mater. Sci. Eng. B Solid-State Mater. Adv. Technol.* **2018**, *228*, 160–166. [CrossRef]
58. Spiridigliozzi, L.; Pinter, L.; Biesuz, M.; Dell'Agli, G.; Accardo, G.; Sglavo, V.M. Gd/Sm-Pr co-doped ceria: A first report of the precipitation method effect on flash sintering. *Materials* **2019**, *12*, 1218. [CrossRef]
59. Yoon, B.; Yadav, D.; Raj, R.; Sortino, E.P.; Ghose, S.; Sarin, P.; Shoemaker, D. Measurement of O and Ti atom displacements in TiO_2 during flash sintering experiments. *J. Am. Ceram. Soc.* **2018**, *101*, 1811–1817. [CrossRef]
60. Zhang, Y.; Nie, J.; Luo, J. Effects of phase and doping on flash sintering of TiO_2. *J. Ceram. Soc. Jpn.* **2016**, *124*, 296–300. [CrossRef]
61. Charalambous, H.; Jha, S.K.; Phuah, X.L.; Wang, H.; Wang, H.; Okasinski, J.S.; Tsakalakos, T. In situ measurement of temperature and reduction of rutile titania using energy dispersive x-ray diffraction. *J. Eur. Ceram. Soc.* **2018**, *38*, 5503–5511. [CrossRef]
62. Jongmanns, M.; Wolf, D.E. Element-specific displacements in defect-enriched TiO_2: Indication of a flash sintering mechanism. *J. Am. Ceram. Soc.* **2020**, *103*, 589–596. [CrossRef]
63. Biesuz, M.; Ometto, A.; Sglavo, V.M. Flash sintering of YSZ/Al_2O_3 composites: Effect of processing and testing conditions. *Materials* **2021**, *14*, 1031. [CrossRef]
64. Syed, K.; Xu, M.; Ohtaki, K.K.; Kok, D.; Karandikar, K.K.; Graeve, O.A.; Bowman, W.J.; Mecartney, M.L. Correlations of grain boundary segregation to sintering techniques in a three-phase ceramic. *Materialia* **2020**, *14*. [CrossRef]
65. Biesuz, M.; Ometto, A.; Tyrpekl, V.; Vilémová, M.; Sglavo, V.M. On the power density at the onset for flash sintering in ceramic composites. *Scr. Mater.* **2021**, *201*, 113984. [CrossRef]
66. Sun, K.; Zhang, J.; Jiang, T.; Qiao, J.; Sun, W.; Rooney, D.; Wang, Z. Flash-Sintering and Characterization of $La_{0.8}Sr_{0.2}Ga_{0.8}Mg_{0.2}O_{3-\delta}$ Electrolytes for Solid Oxide Fuel Cells. *Electrochim. Acta* **2016**, *196*, 487–495. [CrossRef]
67. Ren, K.; Cao, Y.; Chen, Y.; Shao, G.; Dai, J.; Wang, Y. Flash sintering of $Na_3Zr_2(SiO_4)_2(PO_4)$ solid-state electrolyte at furnace temperature of 700 °C. *Scr. Mater.* **2020**, *187*, 384–389. [CrossRef]
68. Perez-Maqueda, L.A.; Gil-Gonzalez, E.; Wassel, M.A.; Jha, S.K.; Perejon, A.; Charalambous, H.; Okasinski, J.; Sanchez-Jimenez, P.E.; Tsakalakos, T. Insight into the $BiFeO_3$ flash sintering process by in-situ energy dispersive X-ray diffraction (ED-XRD). *Ceram. Int.* **2019**, *45*, 2828–2834. [CrossRef]
69. Wassel, M.A.B.; Pérez-Maqueda, L.A.; Gil-Gonzalez, E.; Charalambous, H.; Perejon, A.; Jha, S.K.; Okasinski, J.; Tsakalakos, T. Anisotropic lattice expansion determined during flash sintering of $BiFeO_3$ by in-situ energy-dispersive X-ray diffraction. *Scr. Mater.* **2019**, *162*, 268–291. [CrossRef]
70. Şavklıyıldız, İ.; Okur, Ç.; Akdoğan, E.K. Flash sintering and dielectric properties of $K_{0.5}Na_{0.5}NbO_3$. *J. Am. Ceram. Soc.* **2021**, *105*, 469–480. [CrossRef]
71. Corapcioglu, G.; Ali Gulgun, M.; Kisslinger, K.; Sturm, S.; Jha, S.K.; Raj, R. Microstructure and microchemistry of flash sintered $K_{0.5}Na_{0.5}NbO_3$. *J. Ceram. Soc. Jpn.* **2016**, *124*, 321–328. [CrossRef]
72. Serrazina, R.; Dean, J.S.; Reaney, I.M.; Pereira, L.; Vilarinho, P.M.; Senos, A.M.O.R. Mechanism of densification in low-temperature FLASH sintered lead free potassium sodium niobate (KNN) piezoelectrics. *J. Mater. Chem. C* **2019**, *7*, 14334–14341. [CrossRef]
73. Serrazina, R.; Senos, A.M.O.R.; Pereira, L.; Dean, J.S.; Reaney, I.M.; Vilarinho, P.M. The Role of Particle Contact in Densification of FLASH Sintered Potassium Sodium Niobate. *Eur. J. Inorg. Chem.* **2020**, *2020*, 3720–3728. [CrossRef]
74. Cho, J.; Li, Q.; Wang, H.; Fan, Z.; Li, J.; Xue, S.; Vikrant, K.S.N.; Wang, H.; Holland, T.B.; Mukherjee, A.K.; et al. High temperature deformability of ductile flash-sintered ceramics via in-situ compression. *Nat. Commun.* **2018**, *9*, 2063. [CrossRef] [PubMed]
75. Kayaalp, B.; Klauke, K.; Biesuz, M.; Iannaci, A.; Sglavo, V.M.; D'Arienzo, M.; Noei, H.; Lee, S.; Jung, W.; Mascotto, S. Surface Reconstruction under the Exposure of Electric Fields Enhances the Reactivity of Donor-Doped $SrTiO_3$. *J. Phys. Chem. C* **2019**, *123*, 16883–16892. [CrossRef]
76. Saunders, T.; Grasso, S.; Reece, M.J. Ultrafast-Contactless Flash Sintering using Plasma Electrodes. *Sci. Rep.* **2016**, *6*, 27222. [CrossRef] [PubMed]
77. Lebrun, J.-M.; Hellberg, C.S.; Jha, S.K.; Kriven, W.M.; Steveson, A.; Seymour, K.C.; Bernstein, N.; Erwin, S.C.; Raj, R. In-situ measurements of lattice expansion related to defect generation during flash sintering. *J. Am. Ceram. Soc.* **2017**, *100*, 4965–4970. [CrossRef]
78. Kathiria, R.; Wolf, D.E.; Raj, R.; Jongmanns, M. Frenkel pairs cause elastic softening in zirconia: Theory and experiments. *New J. Phys.* **2021**, *23*, 53013. [CrossRef]
79. Jongmanns, M.; Raj, R.; Wolf, D.E. Generation of Frenkel defects above the Debye temperature by proliferation of phonons near the Brillouin zone edge. *New J. Phys.* **2018**, *20*, 93013. [CrossRef]

80. Taibi, A.; Chaguetmi, S.; Sánchez-Jiménez, P.E.; Perejón, A.; García, J.E.; Satha, H.; Pérez-Maqueda, L.A. Pure perovskite $BiFeO_3$-$BaTiO_3$ ceramics prepared by reaction flash sintering of Bi_2O_3-Fe_2O_3-$BaTiO_3$ mixed powders. *Ceram. Int.* **2021**. [CrossRef]
81. Sortino, E.; Lebrun, J.-M.; Sansone, A.; Raj, R. Continuous flash sintering. *J. Am. Ceram. Soc.* **2018**, *101*, 1432–1440. [CrossRef]
82. Lucideon Materials Development and Commercialization [Homepage on the Internet]. Available online: https://www.lucideon.com/ceramics/technologies/flash-sintering (accessed on 22 October 2021).
83. Molina-Molina, S.; Gil-González, E.; Durán-Olivencia, F.J.; Valverde, J.M.; Perejón, A.; Sánchez-Jiménez, P.E.; Pérez-Maqueda, L.A. A novel Multi-Phase Flash Sintering (MPFS) technique for 3D complex-shaped ceramics. *Appl. Mater. Today* **2022**, *26*, 101274. [CrossRef]
84. Manière, C.; Lee, G.; Olevsky, E.A. All-Materials-Inclusive Flash Spark Plasma Sintering. *Sci. Rep.* **2017**, *7*, 15071. [CrossRef]
85. Niu, B.; Zhang, F.; Zhang, J.; Ji, W.; Wang, W.; Fu, Z. Ultra-fast densification of boron carbide by flash spark plasma sintering. *Scr. Mater.* **2016**, *116*, 127–130. [CrossRef]
86. Vasylkiv, O.; Borodianska, H.; Sakka, Y.; Demirskyi, D. Flash spark plasma sintering of ultrafine yttria-stabilized zirconia ceramics. *Scr. Mater.* **2016**, *121*, 32–36. [CrossRef]
87. Manière, C.; Harnois, C.; Riquet, G.; Lecourt, J.; Bilot, C.; Marinel, S. Flash spark plasma sintering of zirconia nanoparticles: Electro-thermal-mechanical-microstructural simulation and scalability solutions. *J. Eur. Ceram. Soc.* **2022**, *42*, 216–226. [CrossRef]
88. Hérisson de Beauvoir, T.; Ghomari, Z.; Chevallier, G.; Flaureau, A.; Weibel, A.; Elissalde, C.; Mauvy, F.; Chaim, R.; Estournès, C. Flash Spark Plasma Sintering of 3YSZ: Modified sintering pathway and impact on grain boundary formation. *J. Eur. Ceram. Soc.* **2021**, *41*, 7762–7770. [CrossRef]
89. Johnson, S.L.; Venugopal, G.; Hunt, A.T. Flame-assisted flash sintering: A noncontact method to flash sinter coatings on conductive substrates. *J. Am. Ceram. Soc.* **2018**, *101*, 536–541. [CrossRef]
90. Kermani, M.; Biesuz, M.; Dong, J.; Deng, H.; Bortolotti, M.; Chiappini, A.; Reece, M.J.; Sglavo, V.M.; Hu, C.; Grasso, S. Flash cold sintering: Combining water and electricity. *J. Eur. Ceram. Soc.* **2020**, *40*, 6266–6271. [CrossRef]
91. Dong, J.; Kermani, M.; Hu, C.F.; Grasso, S. Flash cold sintering of Nb_2O_5: Polarity and electrolyte effects. *J. Asian Ceram. Soc.* **2021**, *9*, 934–939. [CrossRef]
92. Hirsch, J.E. An index to quantify an individual's scientific research output. *Proc. Natl. Acad. Sci. USA* **2005**, *102*, 16569–16572. [CrossRef] [PubMed]
93. Cho, J.; Phuah, X.L.; Li, J.; Shang, Z.; Wang, H.; Charalambous, H.; Tsakalakos, T.; Mukherjee, A.K.; Wang, H.; Zhang, X. Temperature effect on mechanical response of flash-sintered ZnO by in-situ compression tests. *Acta Mater.* **2020**, *200*, 699–709. [CrossRef]
94. Mishra, T.P.; Neto, R.R.I.; Speranza, G.; Quaranta, A.; Sglavo, V.M.; Raj, R.; Guillon, O.; Bram, M.; Biesuz, M. Electronic conductivity in gadolinium doped ceria under direct current as a trigger for flash sintering. *Scr. Mater.* **2020**, *179*, 55–60. [CrossRef]
95. Phuah, X.L.; Cho, J.; Akriti; Dou, L.; Rheinheimer, W.; García, R.E.; Zhang, X.; Wang, H. Field-assisted growth of one-dimensional ZnO nanostructures with high defect density. *Nanotechnology* **2021**, *32*, 95603. [CrossRef]
96. Storion, A.G.; Ferreira, J.A.; Maestrelli, S.C.; Maria de Jesus Agnolon Pallone, E. Influence of the forming method on flash sintering of ZnO ceramics. *Ceram. Int.* **2021**, *47*, 965–972. [CrossRef]
97. Mishra, T.P.; Neto, R.R.I.; Raj, R.; Guillon, O.; Bram, M. Current-rate flash sintering of gadolinium doped ceria: Microstructure and Defect generation. *Acta Mater.* **2020**, *189*, 145–153. [CrossRef]
98. Phuah, X.L.; Cho, J.; Tsakalakos, T.; Mukherjee, A.K.; Wang, H.; Zhang, X. Defects in flash-sintered ceramics and their effects on mechanical properties. *MRS Bull.* **2021**, *46*, 44–51. [CrossRef]
99. Xie, H.; Zhang, Y.; Wu, Z.; Lv, T. A Bibliometric Analysis on Land Degradation. *Land* **2020**, *9*, 28. [CrossRef]
100. Zhang, J.; Yu, Q.; Zheng, F.; Long, C.; Lu, Z.; Duan, Z. Comparing keywords plus of WOS and author keywords: A case study of patient adherence research. *J. Assoc. Inf. Sci. Technol.* **2016**, *67*, 967–972. [CrossRef]
101. Yang, B. Flash sintering of additively manufactured 3YSZ gears. *J. Am. Ceram. Soc.* **2021**, 3828–3832. [CrossRef]
102. Steil, M.C.; Marinha, D.; Aman, Y.; Gomes, J.R.C.; Kleitz, M. From conventional ac flash-sintering of YSZ to hyper-flash and double flash. *J. Eur. Ceram. Soc.* **2013**, *33*, 2093–2101. [CrossRef]
103. Phuah, X.L.; Wang, H.; Charalambous, H.; Jha, S.K.; Tsakalakos, T.; Zhang, X.; Wang, H. Comparison of the grain growth behavior and defect structures of flash sintered ZnO with and without controlled current ramp. *Scr. Mater.* **2019**, *162*, 251–255. [CrossRef]
104. Chaim, R.; Chevallier, G.; Weibel, A.; Estournès, C. Grain growth during spark plasma and flash sintering of ceramic nanoparticles: A review. *J. Mater. Sci.* **2018**, *53*, 3087–3105. [CrossRef]
105. Nie, J.; Zhang, Y.; Chan, J.M.; Huang, R.; Luo, J. Water-assisted flash sintering: Flashing ZnO at room temperature to achieve ~98% density in seconds. *Scr. Mater.* **2018**, *142*, 79–82. [CrossRef]
106. Charalambous, H.; Jha, S.K.; Christian, K.H.; Lay, R.T.; Tsakalakos, T. Flash Sintering using Controlled Current Ramp. *J. Eur. Ceram. Soc.* **2018**, *38*, 3689–3693. [CrossRef]
107. Lachal, M.; El Khal, H.; Bouvard, D.; Chaix, J.-M.; Bouchet, R.; Steil, M.C. Flash sintering of cationic conductive ceramics: A way to build multilayer systems. *J. Am. Ceram. Soc.* **2021**, *104*, 3845–3854. [CrossRef]
108. Mei, Y.; Pandey, S.; Long, W.; Liu, J.; Zhong, S.; Zhang, L.; Du, S.; Xu, D. Processing and characterizations of flash sintered ZnO-Bi_2O_3-MnO_2 varistor ceramics under different electric fields. *J. Eur. Ceram. Soc.* **2020**, *40*, 1330–1337. [CrossRef]

109. Cui, B.; Niu, J.; Peng, P.; Shi, L.; Du, S.; Liu, J.; Xu, D. Flash sintering preparation and electrical properties of ZnO–Bi_2O_3-M (M = Cr_2O_3, MnO_2 or Co_2O_3) varistor ceramics. *Ceram. Int.* **2020**, *46*, 14913–14918. [CrossRef]
110. Niu, J.; She, H.; Liu, Z.; Cheng, M.; Xu, J.; Liu, J.; Chen, G.; Tang, B.; Xu, D. A current-controlled flash sintering processing leading to dense and fine-grained typical multi-element ZnO varistor ceramics. *J. Alloys Compd.* **2021**, *876*, 160124. [CrossRef]
111. Mao, H.-R.; Guo, R.-F.; Cao, Y.; Jin, S.-B.; Qiu, X.-M.; Shen, P. Ultrafast densification of high-entropy oxide ($La_{0.2}Nd_{0.2}Sm_{0.2}Eu_{0.2}Gd_{0.2})_2Zr_2O_7$ by reactive flash sintering. *J. Eur. Ceram. Soc.* **2021**, *41*, 2855–2860. [CrossRef]
112. Wang, K.; Ma, B.; Li, T.; Xie, C.; Sun, Z.; Liu, D.; Liu, J.; An, L. Fabrication of high-entropy perovskite oxide by reactive flash sintering. *Ceram. Int.* **2020**, *46*, 18358–18361. [CrossRef]
113. Liu, D.; Peng, X.; Liu, J.; Chen, L.; Yang, Y.; An, L. Ultrafast synthesis of entropy-stabilized oxide at room temperature. *J. Eur. Ceram. Soc.* **2020**, *40*, 2504–2508. [CrossRef]
114. Li, W.; Chen, L.; Liu, D.; Liu, J.; An, L. Ultra-low temperature reactive flash sintering synthesis of high-enthalpy and high-entropy $Ca_{0.2}Co_{0.2}Ni_{0.2}Cu_{0.2}Zn_{0.2}O$ oxide ceramics. *Mater. Lett.* **2021**, *304*, 130679. [CrossRef]
115. Yoon, B.; Avila, V.; Raj, R.; Jesus, L.M. Reactive flash sintering of the entropy-stabilized oxide $Mg_{0.2}Ni_{0.2}Co_{0.2}Cu_{0.2}Zn_{0.2}O$. *Scr. Mater.* **2020**, *181*, 48–52. [CrossRef]
116. Gil-González, E.; Perejón, A.; Sánchez-Jiménez, P.E.; Román-González, D.; Pérez-Maqueda, L.A. Control of experimental conditions in reaction flash-sintering of complex stoichiometry ceramics. *Ceram. Int.* **2020**, *46*, 29413–29420. [CrossRef]
117. Avila, V.; Yoon, B.; Ingraci Neto, R.R.; Silva, R.S.; Ghose, S.; Raj, R.; Jesus, L.M. Reactive flash sintering of the complex oxide $Li_{0.5}La_{0.5}TiO_3$ starting from an amorphous precursor powder. *Scr. Mater.* **2020**, *176*, 78–82. [CrossRef]
118. Avila, V.; Yoon, B.; Ghose, S.; Raj, R.; Jesus, L.M. Phase evolution during reactive flash sintering of $Li_{6.25}Al_{0.25}La_3Zr_2O_{12}$ starting from a chemically prepared powder. *J. Eur. Ceram. Soc.* **2021**, *41*, 4552–4557. [CrossRef]
119. Avila, V.; Raj, R. Reactive flash sintering of powders of four constituents into a single phase of a complex oxide in a few seconds below 700 °C. *J. Am. Ceram. Soc.* **2019**, *102*, 6443–6448. [CrossRef]
120. Su, X.; Jiao, Z.; Fu, M.; An, G.; Wu, Y.; Tian, Q.; Xu, P.; Wu, W.; Chang, X.; Liu, J. Ultrafast synthesis and densification of ZrO_2 doped KNN ceramics by reactive flash sintering. *Int. J. Appl. Ceram. Technol.* **2021**, *18*, 1999–2009. [CrossRef]
121. Wu, Y.; Su, X.; An, G.; Hong, W. Dense $Na_{0.5}K_{0.5}NbO_3$ ceramics produced by reactive flash sintering of $NaNbO_3$-$KNbO_3$ mixed powders. *Scr. Mater.* **2020**, *174*, 49–52. [CrossRef]
122. Yoon, B.; Yadav, D.; Ghose, S.; Sarin, P.; Raj, R. On the synchronicity of flash sintering and phase transformation. *J. Am. Ceram. Soc.* **2019**, *102*, 3110–3116. [CrossRef]
123. Yoon, B.; Yadav, D.; Ghose, S.; Raj, R. Reactive flash sintering: MgO and α-Al_2O_3 transform and sinter into single-phase polycrystals of $MgAl_2O_4$. *J. Am. Ceram. Soc.* **2019**, *102*, 2294–2303. [CrossRef]
124. Chaim, R. Reactive flash sintering (RFS) in oxide systems: Kinetics and thermodynamics. *J. Mater. Sci.* **2021**, *56*, 278–289. [CrossRef]
125. Chaim, R.; Amouyal, Y. Liquid-Film Assisted Mechanism of Reactive Flash Sintering in Oxide Systems. *Materials* **2019**, *12*, 1494. [CrossRef]
126. Jia, Y.; Su, X.; Wu, Y.; Bai, G.; Wang, Z.; Yan, X.; Ai, T.; Zhao, P. Fabrication of lead zirconate titanate ceramics by reaction flash sintering of PbO-ZrO_2-TiO_2 mixed oxides. *J. Eur. Ceram. Soc.* **2019**, *39*, 3915–3919. [CrossRef]
127. Murray, S.E.; Jensen, T.J.; Sulekar, S.S.; Lin, Y.-Y.; Perry, N.H.; Shoemaker, D.P. Propagation of the contact-driven reduction of Mn_2O_3 during reactive flash sintering. *J. Am. Ceram. Soc.* **2019**, *102*, 7210–7216. [CrossRef]
128. Murray, S.E.; Lin, Y.-Y.; Sulekar, S.S.; Gebre, M.S.; Perry, N.H.; Shoemaker, D.P. Predicting transformations during reactive flash sintering in CuO and Mn_2O_3. *J. Am. Ceram. Soc.* **2021**, *104*, 76–85. [CrossRef]

materials

MDPI

Article

Study on Oxygen Evolution Reaction Performance of Jarosite/C Composites

Junxue Chen †, Sijia Li †, Zizheng Qu, Zhonglin Li, Ding Wang, Jialong Shen * and Yibing Li *

Department of Materials Science and Engineering, Guilin University of Technology, Guilin 541000, China; cchenjunxue@gmail.com (J.C.); lisj701023@gmail.com (S.L.); dadpro01@gmail.com (Z.Q.); dahe121133@gmail.com (Z.L.); dingnvhuang@gmail.com (D.W.)

* Correspondence: Jialong.Shen@glut.edu.cn (J.S.); lybgems@glut.edu.cn (Y.L.)

† These authors contributed equally to this work.

Abstract: In the electrolysis of water process, hydrogen is produced and the anodic oxygen evolution reaction (OER) dominates the reaction rate of the entire process. Currently, OER catalysts mostly consist of noble metal (NM) catalysts, which cannot be applied in industries due to the high price. It is of great importance to developing low-cost catalysts materials as NM materials substitution. In this work, jarosite ($AFe_3(SO_4)_2(OH)_6$, A = K^+, Na^+, NH^{4+}, H_3O^+) was synthesized by a one-step method, and its OER catalytic performance was studied using catalytic slurry (the weight ratios of jarosite and conductive carbon black are 2:1, 1:1 and 1:2). Microstructures and functional groups of synthesized material were analyzed using XRD, SEM, FI-IR, etc. The OER catalytic performance of $(NH_4)Fe_3(SO_4)_2(OH)_6$/conductive carbon black were examined by LSV, Tafel, EIS, ECSA, etc. The study found that the OER has the best catalytic performance when the weight ratio of $(NH_4)Fe_3(SO_4)_2(OH)_6$ to conductive carbon black is 2:1. It requires only 376 mV overpotential to generate current densities of 10 mA cm^{-2} with a small Tafel slope (82.42 mV dec^{-1}) and large C_{dl} value (26.17 mF cm^{-2}).

Keywords: jarosite; ammoniojarosite; electrocatalyst; oxygen evolution reaction; stability

Citation: Chen, J.; Li, S.; Qu, Z.; Li, Z.; Wang, D.; Shen, J.; Li, Y. Study on Oxygen Evolution Reaction Performance of Jarosite/C Composites. *Materials* **2022**, *15*, 668. https://doi.org/10.3390/ma15020668

Academic Editor: Marc Cretin

Received: 30 November 2021
Accepted: 11 January 2022
Published: 17 January 2022

1. Introduction

With the increasingly negative impact of fossil fuels on the environment, it is with a huge demand that modern science and technology need to pursue clean and sustainable energy [1,2]. Hydrogen production by water splitting, as a technology for producing clean energy, has caused extensive researches [3,4]. The electrochemical water splitting process includes the oxygen evolution reaction (OER) at the anode and the hydrogen evolution reaction (HER) at the cathode. The HER reaction is a two-electron transfer process, while OER is a four-electron transfer process, whose higher energy barrier dominates the rate of the cathodic hydrogen production [5]. Therefore, the development and research of cost-effective OER catalysts with high activity and long-periodic cycle stability is a primary task.

Noble-metal-based materials, including IrO_2 and RuO_2, are state-of-the-art OER electrocatalysts because of their high electrocatalytic OER activity both in alkaline and acidic solutions [6–8]. However, the large-scale application of IrO_2 and RuO_2 in OER is severely limited not only by the high cost but also by the scarcity of Ir and Ru [9]. Thus far, considerable research efforts have been devoted to the exploration of low-cost and highly active noble-metal-free catalysts to replace expensive and scarce precious catalysts [10–12]. Especially for transition metal Fe-based materials, including oxides/hydroxides [13–15], chalcogenides [16], phosphides [16,17], and nitrides [18], which have been investigated extensively as promising candidates for the OER.

Compared with these compounds, Fe-based polyanionic compounds [1–4], such as jarosite, are an earth-abundant natural mineral that belongs to the alunite supergroup with the formula $AFe_3(SO_4)_2(OH)_6$, where A represents different monovalent cations, such as

K^+, Na^+, NH_4^+, and H_3O^+. At present, jarosite has been extensively studied by the acid leach mining industry due to the precipitation of jarosite in acidic media—a crucial step that allows for the physical separation of Fe^{3+} and other cations from the leach solution [19–23]. As a result, these refining plants produce large amounts of environmentally hazardous jarosite wastes that currently provide no commercial value. Fortunately, several researchers have explored the use of jarosite as a cathode in the lithium and sodium-ion battery [24–26]. According to our best efforts, there is no research on jarosite in OER so far. Therefore, exploring and tapping the potential of these environmentally harmful jarosite wastes in the field of OER is an attractive strategy to provide economic advantages for jarosite.

In this work, the design of applying jarosite to OER catalytic material is proposed. A simple one-step method was used to synthesize four various types of jarosite materials (($NaFe_3(SO_4)_2(OH)_6$), $KFe_3(SO_4)_2(OH)_6$), $(NH_4)Fe_3(SO_4)_2(OH)_6$, and (($(H_3O)Fe_3(SO_4)_2(OH)_6$)). We explored the OER performance of these four catalysts under acidic, neutral and alkaline conditions. The electrochemical test results suggest that $(NH_4)Fe_3(SO_4)_2(OH)_6$ shows the best OER activity among the four catalysts. When the weight ratio of $(NH_4)Fe_3(SO_4)_2(OH)_6$ to conductive carbon black is 2:1, the overpotential of $(NH_4)Fe_3(SO_4)_2(OH)_6$ is 376 mV at a current density of 10 mA cm^{-2} in alkaline conditions. Although this performance is not comparable to that of precious metals (IrO_2), there is a lot of room to improve the performance, which is our future research task. In short, exploring the application of jarosite waste in OER can not only protect the environment but also improve its economic value.

2. Experimental Section

2.1. Materials

$Fe(SO_4)\cdot 7H_2O$ (99.0~101.0%), K_2CO_3 (99%), Na_2CO_3 (99.8%), $(NH_4)_2SO_4$ (98.0%), $NH_3\cdot 7H_2O$ (25.0~28.0%) and C_2H_5OH (99.7%) were purchased from Guangxi Dragon Technology Company (Guangxi, China). Conductive carbon black (Ketjenblack) and Nafion solution (5 wt%) were purchased from Suzhou Yilongsheng Energy Technology Co., Ltd. (Suzhou, China) All chemicals used were of analytical grade and there was no need to further purification.

2.2. Preparation of $(A)Fe_3(SO_4)_2(OH)_6$, ($A = K^+$, Na^+ and NH_4^+)

$FeSO_4$ and $(NH_4)_2SO_4$ were dissolved in 250 mL and 100 mL of deionized water to form solution A (0.02 M $FeSO_4$) and solution B (0.5 M $(NH_4)_2SO_4$), respectively. Dilute H_2O_2 solution was added dropwise to solution A to oxidize Fe^{2+} to Fe^{3+} and then solution B was injected into the above solution with a water bath at a constant temperature of 95 °C and stirred magnetically for 3 h, meanwhile, the pH value of the mixed solution was maintained throughout at 1.5–2.0 with 1 M ammonia solution. Finally, the yellow precipitate $(NH_4)Fe_3(SO_4)_2(OH)_6$ catalyst was collected, washed with deionized water and then dried in a vacuum at 80 °C. The main synthesis procedure for $Na/KFe_3(SO_4)_2(OH)_6$ is similar to that for $(NH_4)Fe_3(SO_4)_2(OH)_6$, corresponding to the use of 0.5 M Na_2CO_3 and K_2CO_3 instead of 0.5 M $(NH_4)_2SO_4$ solution and the replacement of the pH adjuster with 1 M Na_2CO_3 and K_2CO_3 solution, respectively.

2.3. Preparation of $(H_2O)Fe_3(SO_4)_2(OH)_6$

A certain amount of $FeSO_4$ was dispersed in 70 mL of deionized water, with stirring at 95 °C. H_2O_2 was used to oxidize Fe^{2+} to Fe^{3+}. After it was completely oxidized, the reactants were transferred to a 100 mL reactor at 120 °C and kept for 12 h. The final product was collected after filtration and washed with deionized water several times.

2.4. Preparation of Working Electrode

Add 10 mg NH_4-Fe_3@KB-1($(NH_4)Fe_3(SO_4)_2(OH)_6$ and conductive carbon black with a mass ratio of 2:1). These were dispersed into a mixed solvent of Nafion (30 µL), anhydrous ethanol (400 µL) and deionized water (600 µL). Form a uniform dispersion after ultrasonic for 30 min, then use a pipette gun to take the dispersion (4 µL), add to the glassy carbon electrode and dry. The working electrodes of other catalytic materials are prepared by using the same method.

2.5. Characterization

Scanning electron microscopy (SEM, Hitachi Works, Ltd., Tokyo, Japan) was conducted on S-4800 at 5 kV to perform microstructure analysis. The phase structure of the sample was analyzed by a X-ray powder diffractometer (XRD, X'Pert PRO, PANalytical B.V, Cu Kα, 40 kV, 40 mA, λ = 1.54056 Å, PANalytical B.V., Almelo, The Netherlands) at a scanning rate of $5^{\circ}\cdot min^{-1}$. The functional groups of the samples were analyzed by Fourier transform infrared (FTIR, Thermo Nexus 407 spectrometer, White Bear Lake, MN, USA) and the laser Raman confocal microscope Raman spectrometer (Raman, Thermo Fisher Scientific DXR, thermoelectric company, 532 nm, White Bear Lake, MN, USA). Transmission electron microscopy (TEM, JEOL, Beijing, China) images were collected on Titan G260-300 at an acceleration voltage of 200 kV. Before BET and BJH measured via Surface area and pore porosimetry analyzer NoVA 1200e (Quantachrome Instruments, Shanghai, China), all samples were degassed for 5 h at 100 °C.

2.6. Electrochemical Measurements

All electrochemical measurements were conducted on a computer-controlled CHI 760E electrochemical workstation with a conventional three-electrode system. The glassy carbon electrode with a diameter of 3 mm was used as a working electrode, Ag/AgCl electrode and Hg/HgO electrode were respectively used as the reference electrode for the acidic (neutral) system and alkaline system, and graphite rods were used as a counter electrode. The measurements were performed in three different electrolytes, 0.05 M H_2SO_4, 1 M KOH, and 1 M PBS. All the potentials were converted to reversible hydrogen electrode (RHE) based on the formula $E_{RHE} = E_{Hg/HgO} + 0.0591 \times pH + 0.098$ and $E_{RHE} = E_{Ag/AgCl} + 0.0591 \times pH + 0.1976$. The polarization curves were measured at 5 mV s^{-1} and iR-corrected. Tafel plots were calculated using the Tafel formula $\eta = b \log j + a$, where j is the current density, b is the Tafel slope, and a is the intercept relative to the exchange current density. EIS measurements were conducted under a particular applied potential in the frequency range 0.1 Hz to 100 kHz. The electrochemically active surface area (ECSA) was estimated by the double-layer capacitance (C_{dl}). The time-current curve was measured at a fixed voltage corresponding to 10 mA cm^{-2} of current density. All tests were performed at room temperature.

3. Results and Discussion

3.1. Characterization of Samples

Four various types of jarosite were synthesized, named $NaFe_3(SO_4)_2(OH)$, $KFe_3(SO4)_2(OH)_6$, $(NH_4)Fe_3(SO_4)_2(OH)_6$ and $(H_3O)Fe_3(SO_4)_2(OH)_6$, respectively. The crystal structure of various types of jarosite was firstly investigated by X-ray diffraction (XRD). As shown in Figure 1a, the diffraction pattern of the as-prepared jarosite can be indexed to the hexagonal system with a space group of R3m, suggesting the successful preparation of the jarosite samples.

Figure 1b shows the infrared spectrum test chart of the jarosite. The infrared absorption peaks of different jarosite appear at similar positions. The peaks appearing at 469 cm^{-1} and 502 cm^{-1} are Fe-O peaks. The corresponding peaks at 624 cm^{-1}, 1082 cm^{-1}, and 1204 cm^{-1} are SO_4^{2-}. The broad and strong absorption peaks at 1004 cm^{-1} and 3416~3700 cm^{-1} are the stretching vibrations of –OH and the weaker absorption peak at 1638 cm^{-1} is caused by the bending vibration of H_2O [27,28]. A sharp peak appears at 1425 cm^{-1} in the infrared spectrum of $(NH_4)Fe_3(SO_4)_2(OH)_6$, which is regarded as the absorption of the $-NH_4$ peak. According to the findings in the report [15], transition metal hydroxides have good OER catalytic performance and the presence of hydroxyl groups in jarosite makes it possible to have OER catalytic performance. This view is confirmed in the following electrochemical performance test.

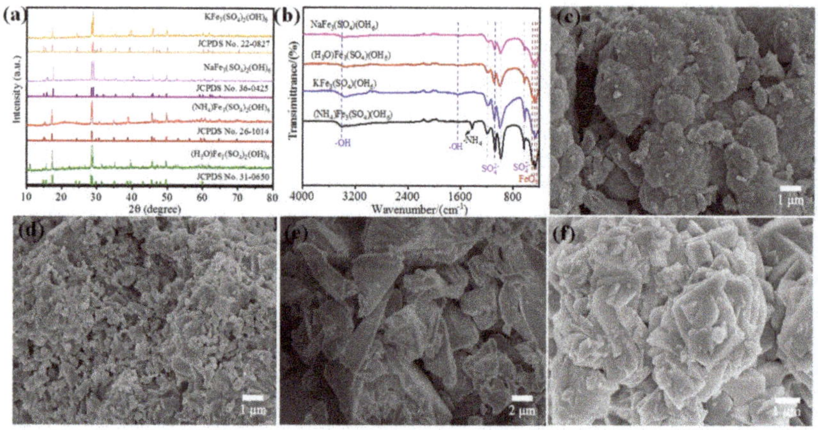

Figure 1. (**a**) XRD patterns and (**b**) FT-IR spectra of various types of jarosite. SEM images of (**c**) $KFe_3(SO_4)_2(OH)_6$, (**d**) $NaFe_3(SO_4)_2(OH)_6$, (**e**) $(H_3O)Fe_3(SO_4)_2(OH)_6$ and (**f**) $(NH_4)Fe_3(SO_4)_2(OH)_6$.

It can be seen from the figure that $KFe_3(SO_4)_2(OH)_6$ (Figure 1c) and $NaFe_3(SO_4)_2(OH)_6$ (Figure 1d) have similar morphologies. Both of them are densely packed. The precipitation rate of $KFe_3(SO_4)_2(OH)_6$ is fast and the sample morphology has not yet been completely formed before it settles together. In the morphology of $NaFe_3(SO_4)_2(OH)_6$, it can be observed that they are stacked together in a rhombic structure, which has not yet been completely formed. It can be seen from Figure 1e that the particle diameter is larger and the shape is irregular. $(NH_4)Fe_3(SO_4)_2(OH)_6$ (Figure 1f) is uniformly distributed in lumps of different sizes while particles do not appear to pile up, showing a larger specific surface area.

The nitrogen adsorption-desorption isotherm curves of the as-synthesized catalyst under various pressures were characterized with a Surface Area and Pore Porosimetry Analyzer NoVA 1200e., and the specific surface area and pore size distribution were calculated via Brumaire-Emmett-Teller(BET) and Barret-Joyner-Hallender (BJH) methods. As shown in Figure 2, $(NH_4)Fe_3(SO_4)_2(OH)_6$ displays the highest BET surface areas of 6.5845 $m^2\ g^{-1}$, which is higher than that of $KFe_3(SO_4)_2(OH)_6$ (4.6879 $m^2\ g^{-1}$), $NaFe_3(SO_4)_2(OH)_6$ (4.1587 $m^2\ g^{-1}$), and $(H_3O)Fe_3(SO_4)_2(OH)_6$ (2.5179 $m^2\ g^{-1}$). The pore size distribution curves of the four materials in Figure 2b suggest the existence of a mesoporous structure (~8 nm). The large specific surface area of the catalyst is very beneficial for the exposure of catalytic active sites for OER.

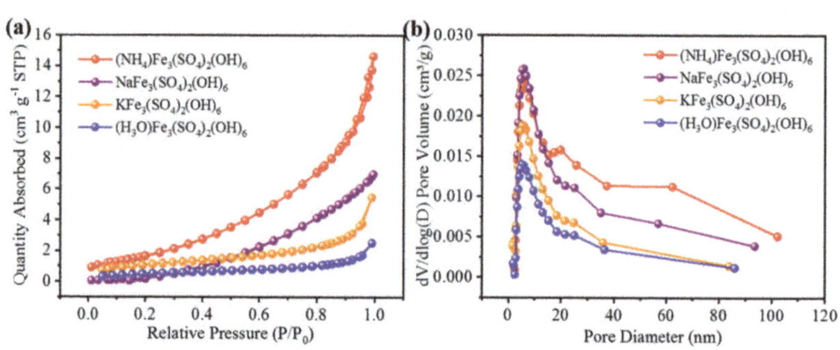

Figure 2. (**a**) N_2 adsorption-desorption isotherm curves, (**b**) the corresponding pore size distribution curves.

Transmission electron microscopy (TEM) was carried out to further identify the details of samples. Figure 3a,b shows the lamellar structure of the sample. It shows some branch-like structures, which can provide more active sites. The selected-area electron-diffraction (SAED) pattern (inset of Figure 3b) of $(NH_4)Fe_3(SO_4)_2(OH)_6$ was also recorded. It displays the weak diffraction rings, which further explained how the prepared $(NH_4)Fe_3(SO_4)_2(OH)_6$ possesses poor crystallization form. Figure 3c shows the recorded high-resolution transmission electron microscopy (HRTEM) image of the $(NH_4)Fe_3(SO_4)_2(OH)_6$. The interplanar spacing of 0.287 nm was indexed matching the (006) crystal plane of $(NH_4)Fe_3(SO_4)_2(OH)$, which is in good agreement with the XRD spectra. Furthermore, the high-angle annular dark-field scanning-TEM (HAADF–STEM) and its corresponding mapping were employed to analyze the distribution of the elements in the $(NH_4)Fe_3(SO_4)_2(OH)_6$ catalyst. It shows that Fe, N, O, and S are evenly distributed across the entire nanoparticles without any noticeable segregation.

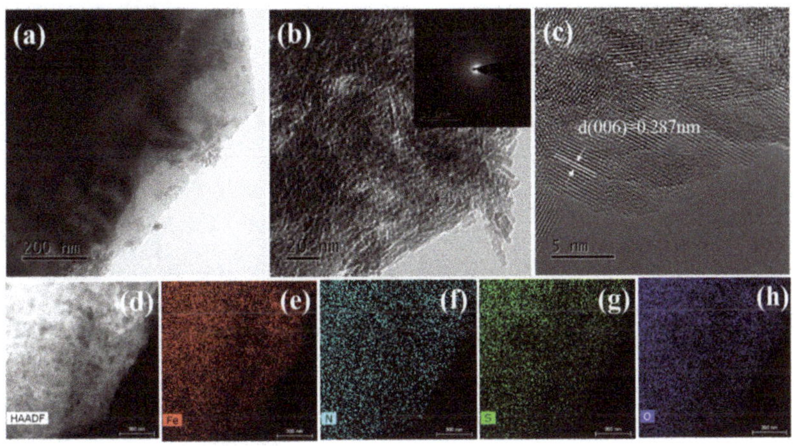

Figure 3. (**a**,**b**) TEM images (the inset shows SAED), (**c**) HRTEM image, and (**d**–**h**) HAADF-TEM diagrams of $(NH_4)Fe_3(SO_4)_2(OH)_6$ and the corresponding EDS elemental mapping images.

3.2. *Electrochemical Analysis*

To increase the electronic conductivity of the jarosite, the catalyst slurry with a weight ratio of 1:1 (jarosite to conductive carbon black) was prepared and an OER polarization curve performance test was conducted. As shown in Figure 4a, when the current density is 10 mA cm^{-2}, the overpotentials of $KFe_3(SO_4)_2(OH)_6$, $NaFe_3(SO_4)_2(OH)_6$, $(H_3O)Fe_3(SO_4)_2(OH)_6$, and$(NH_4)Fe_3(SO_4)_2(OH)_6$ are 412 mV, 400 mV, 424 mV, and 394 mV, respectively. Meanwhile, the Tafel slope of the $(NH_4)Fe_3(SO_4)_2(OH)_6$ is 127.31 mV dec^{-1}, which is smaller than those of the $NaFe_3(SO_4)_2(OH)_6$ (135.86 mV dec^{-1}), $KFe_3(SO_4)_2(OH)_6$ (144.81 mV dec^{-1}) and $(H_3O)Fe_3(SO_4)_2(OH)_6$ (148.85 mV dec^{-1}), which indicates that $(NH_4)Fe_3(SO_4)_2(OH)_6$ shows an excellent OER activity among the four jarosite catalysts.

In addition, OER tests were carried out on four catalyst materials in the acidic (pH = 1 H_2SO_4) and neutral (pH = 7 PBS) solution. As shown in Figure 5c,d, the catalytic performance of $(NH_4)Fe_3(SO_4)_2(OH)_6$ in the acidic and neutral solution is better than the other three materials. However, the OER performance of $(NH_4)Fe_3(SO_4)_2(OH)_6$ in the acidic and neutral condition is far inferior to that in the alkaline condition. Therefore, we will take alkaline conditions as an example to focus on the $(NH_4)Fe_3(SO_4)_2(OH)_6$ catalyst.

The electrochemical double-layer capacitance (C_{dl}) approach was applied to estimate the electrocatalytic active surface area (ECSA) from cyclic voltammetry curves at various scan rates over a small potential range. The $(NH_4)Fe_3(SO_4)_2(OH)_6$ electrode possesses the largest C_{dl} of 15.49 mF cm^{-2} compared to those of $KFe_3(SO_4)_2(OH)_6$ (6.69 mF cm^{-2}), $NaFe_3(SO_4)_2(OH)_6$ (13.56 mF cm^{-2}), and $(H_3O)Fe_3(SO_4)_2(OH)_6$ (4.28 mF cm^{-2})

(Figures 5 and 6), showing indeed that a larger ECSA of $(NH_4)Fe_3(SO_4)_2(OH)_6$ allows for more exposed active sites to promote OER performance.

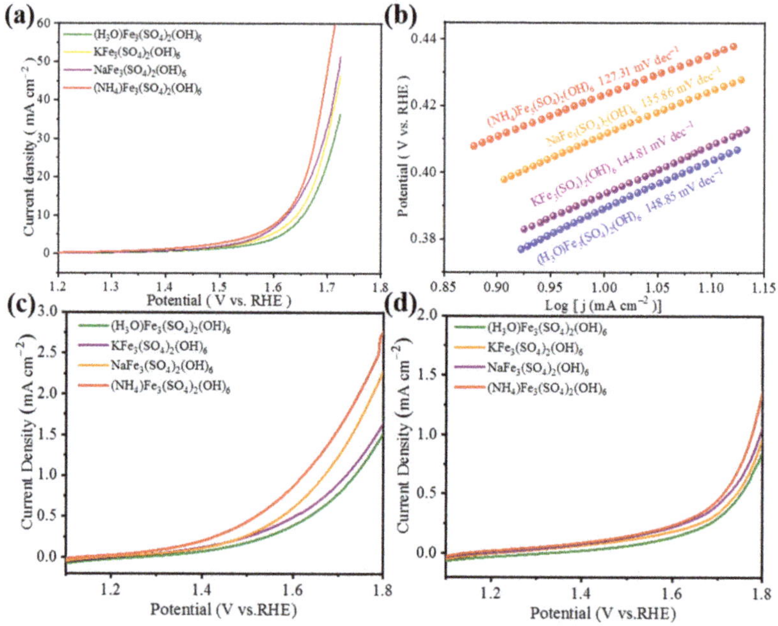

Figure 4. Polarization curves for jarosite to conductive carbon black ratio of 1:1, (**a**) 1 M KOH (pH = 14) polarization curves, (**b**) Tafel plots derived from the Ph = 14 polarization curves, (**c**) 0.05 M H_2SO_4 (pH = 1) polarization curves, (**d**) 1 M PBS (pH = 7) polarization curves.

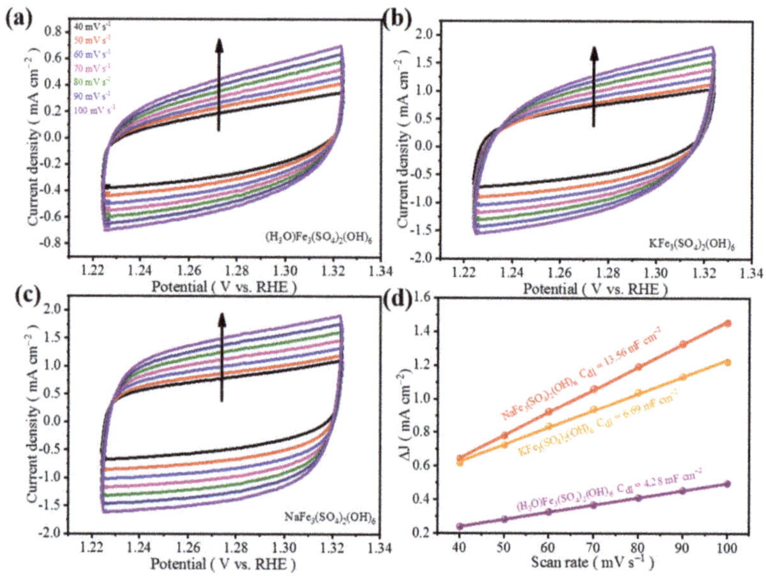

Figure 5. CV curves of (**a–c**) $(H_3O)Fe_3(SO_4)_2(OH)_6$, $KFe_3(SO_4)_2(OH)_6$, and $NaFe_3(SO_4)_2(OH)_6$ at different scan rates, (**d**) C_{dl} diagram of $(H_3O)Fe_3(SO_4)_2(OH)_6$, $KFe_3(SO_4)_2(OH)_6$, and $NaFe_3(SO_4)_2(OH)_6$.

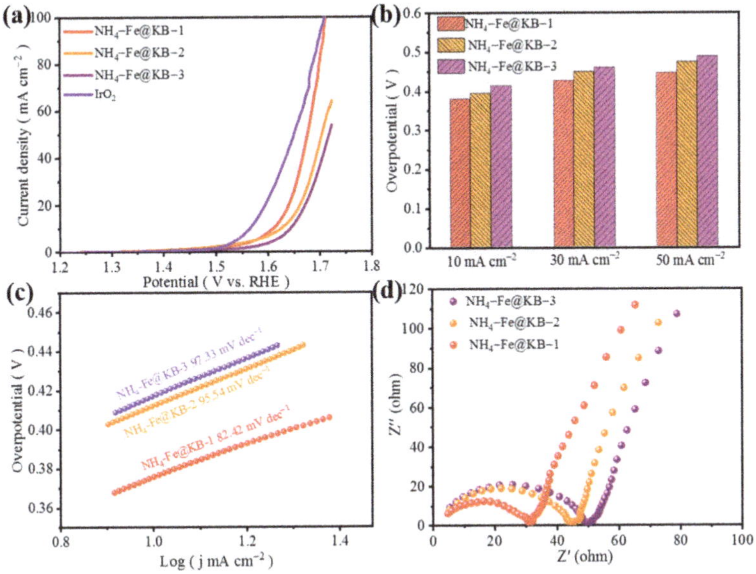

Figure 6. (**a**) NH_4-Fe_3@KB-1, NH_4-Fe_3@KB-2, NH_4-Fe_3@KB-3 and IrO_2 polarization curves; (**b**) NH_4-Fe_3@KB-1, NH_4-Fe_3@KB-2, NH_4-Fe_3@KB-3 at overpotentials reaching current densities of 10, 50, and 100 mA cm^{-2}; (**c**,**d**) the Tafel slope and EIS diagram of NH_4-Fe_3@KB-1, NH_4-Fe_3@KB-2, NH_4-Fe_3@KB-3.

Different ratios of catalyst powder and conductive carbon black may affect the results. The different weight ratios of $(NH_4)Fe_3(SO_4)_2(OH)_6$ and conductive carbon black (2:1, 1:1, 1:2) are prepared and the total mass of 10 mg is guaranteed. The samples are referred to as NH_4-Fe_3@KB-1, NH_4-Fe@KB-2, NH_4-Fe@KB-3 and IrO_2. The OER polarization curve test was performed on them in 1 M KOH electrolyte saturated with oxygen, and the test results are shown in Figure 6a. Additionally, NH_4-Fe_3@KB-1 has better OER catalytic performance. When the current density is 10 mA cm^{-2}, the overpotential of NH_4-Fe_3@KB-1 is 379 mV, and it is 15 mV and 34 mV lower than NH_4-Fe_3@KB-2 and NH_4-Fe_3@KB-3, respectively. Furthermore, $(NH_4)Fe_3(SO_4)_2(OH)_6$ and IrO_2 have the same overpotential when the current density is 100 mA cm^{-2}. When the current density is 30 mA cm^{-2} and 50 mA cm^{-2}, the overpotential of NH_4-Fe_3@KB-1 is still the lowest (Figure 6b). NH_4-Fe_3@KB-1 has a higher current density with the same measurement conditions.

To get insight into the OER kinetics, the Tafel slope values were calculated from the steady-state OER polarization curves. As shown in Figure 6c, NH_4-Fe_3@KB-1 (82.42 mV dec^{-1}) has the smallest Tafel slope. Figure 6d is the AC impedance (EIS) test results of NH_4-Fe_3@KB-1, NH_4-Fe_3@KB-2 and NH_4-Fe_3@KB-3. The charge transfer resistance of NH_4-Fe_3@KB-1 is significantly smaller than that of NH_4-Fe_3@KB-2 and NH_4-Fe_3@KB-3, which suggests the catalytic interface and the electrolyte have a faster charge transfer rate.

The slope was calculated to get the C_{dl} value and the test result is shown in Figure 7. NH_4-Fe_3@KB-1 has the largest C_{dl} value of 26.17 mF cm^{-2}, indicating that the ECSA of NH_4-Fe_3@KB-1 is large. This is allowing more active sites to be exposed and promotes the catalytic process of OER. This also explains the good OER catalytic performance of NH_4-Fe_3@KB-1.

Additionally, durability was another significant parameter of the catalyst for OER. Through the i-*t* test, the stability of NH_4-Fe_3@KB-1 was evaluated, and the test was carried out for 48 h at a constant voltage of 0.68 V (vs. Hg/HgO) with a current density equal to 10 mA cm^{-2}. The test results are shown in Figure 8a. With the increase in test time, the current density of NH_4-Fe_3@KB-1 increases slightly around 10 mA cm^{-2}, which may be caused by the burst of oxygen bubbles generated during the test. Overall, NH_4-Fe_3@KB-1 still shows good stability. The structure and composition of $(NH_4)Fe_3(SO_4)_2(OH)_6$ after the stability test was also studied in detail, with the SEM image of $(NH_4)Fe_3(SO_4)_2(OH)_6$ (Figure 8b) after stability test displaying a newly formed rice-like structure, indicating that the $(NH_4)Fe_3(SO_4)_2(OH)_6$ catalyst may have undergone surface reconstruction during electrolysis. In addition, the XRD pattern (Figure 8c) of the $(NH_4)Fe_3(SO_4)_2(OH)_6$ catalyst showed an amorphous feature after OER. As with many reported works [29,30], the catalyst undergoes a surface reconstruction accompanied by the appearance of an amorphous structure (e.g., oxyhydroxide species) during the OER process, and the observed amorphous feature was further analyzed by Raman spectroscopy. As shown in Figure 8d, the four Raman bands at 215, 275, 390 and 599 cm^{-1} represent the phase of FeOOH, which is well-matched with the literature reports [31–33]. Therefore, it might be reasonable to conclude that $(NH_4)Fe_3(SO_4)_2(OH)_6$ was transformed into amorphous FeOOH during the OER process.

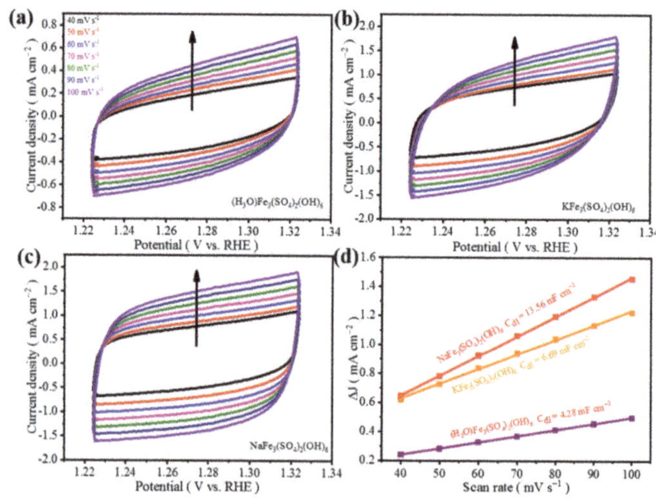

Figure 7. CV curves of (**a**) NH_4-Fe_3@KB-1, (**b**) NH_4-Fe_3@KB-2, and (**c**) NH_4-Fe_3@KB-3 at different scan rates; (**d**) C_{dl} diagram of NH_4-Fe_3@KB-1, NH_4-Fe_3@KB-2, and NH_4-Fe_3@KB-3.

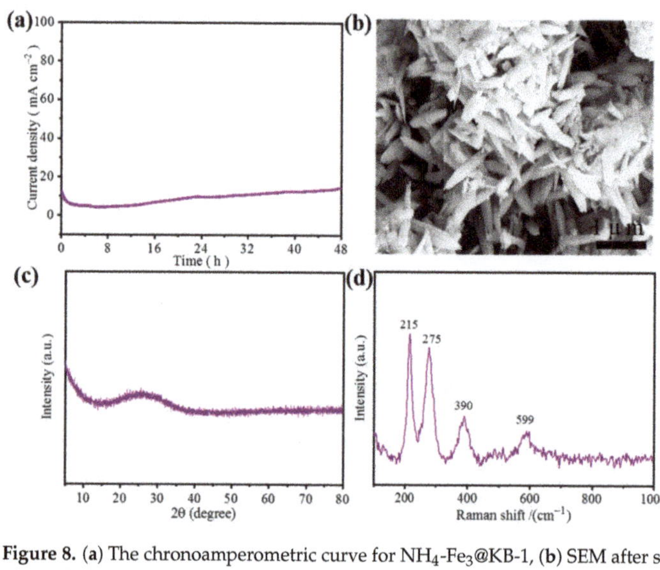

Figure 8. (**a**) The chronoamperometric curve for NH_4-Fe_3@KB-1, (**b**) SEM after stability test, (**c**) XRD after stability test, (**d**) Raman after stability test.

4. Conclusions

In this work, a simple hydrothermal method was used to successfully prepare the jarosite. Furthermore, $(NH_4)Fe_3(SO_4)_2(OH)_6$ shows the best catalytic performance. The OER catalytic performance of $(NH_4)Fe_3(SO_4)_2(OH)_6$ and conductive carbon black with different weight ratios were further explored. The OER catalytic performance is best when the weight ratio of $(NH_4)Fe_3(SO_4)_2(OH)_6$ to conductive carbon black is 2:1. Additionally, NH_4-Fe_3@KB-1 has a lower starting potential of 1.42 V (vs. RHE) and Tafel slope (82.42 mV dec^{-1}). It also showed a small charge transfer resistance and a large C_{dl} (26.17 mF cm^{-2}). The raw materials for preparing the synthetic are easily obtained and are low in price. Experimental results show that jarosite has a broad development space and further research is needed to improve its OER performance.

Author Contributions: Conceptualization, J.C., S.L. and Y.L.; methodology, J.C. and S.L.; software, J.C. and S.L.; validation, Z.Q., Z.L. and D.W.; formal analysis, J.C. and S.L.; investigation, J.C. and S.L.; resources, Y.L.; data curation, S.L.; writing—original draft preparation, J.C. and S.L.; writing—review and editing, J.C.; visualization, J.S. and Y.L.; supervision, J.S. and Y.L.; project administration, Y.L. All authors have read and agreed to the published version of the manuscript.

Funding: This research received no external funding.

Institutional Review Board Statement: Not applicable.

Informed Consent Statement: Not applicable.

Data Availability Statement: The data used to support the findings of this study are available from the corresponding author upon request.

Acknowledgments: The authors are grateful for the Key Laboratory of New Processing Technology for Nonferrous Metals and Materials, Ministry of Education, Guilin University of Technology.

Conflicts of Interest: The authors declare no conflict of interest.

References

1. Zhao, D.D.; Zhang, N.; Bu, L.Z.; Shao, Q.; Huang, X.Q. The latest development of non-noble metal electrocatalytic oxygen evolution catalysts. *Electrochemistry* **2018**, *24*, 455–465.
2. Guo, Y.X.; Shang, C.S.; Li, J.; Wang, E.K. Research progress in electrocatalytic hydrogen evolution, oxygen evolution, and oxygen reduction. *Sci. China Chem.* **2018**, *48*, 926–940.
3. Shao, Q.; Yang, J.; Huang, X. The Design of Water Oxidation Electrocatalysts from Nanoscale Metal-Organic Frameworks. *Chem. Eur. J.* **2018**, *24*, 15143–15155. [CrossRef] [PubMed]
4. Goodenough, J.B.; Park, K.S. The Li-ion rechargeable battery: A perspective. *J. Am. Chem. Soc.* **2013**, *135*, 1167–1176. [CrossRef] [PubMed]
5. Wang, H.; Wang, W.; Xu, Y.Y.; Dong, S.; Xiao, J.; Wang, F.; Liu, H.; Xia, B.Y. Hollow nitrogen-doped carbon spheres with Fe_3O_4 nanoparticles encapsulated as a highly active oxygen-reduction catalyst. *ACS Appl. Mater. Interfaces* **2017**, *9*, 10610–10617. [CrossRef] [PubMed]
6. Long, X.; Xiao, S.; Wang, Z.; Zheng, X.; Yang, S. Co intake mediated formation of ultrathin nanosheets of transition metal LDH-an advanced electrocatalyst for oxygen evolution reaction. *Chem. Commun.* **2015**, *51*, 1120–1123. [CrossRef]
7. Luo, Y.; Wu, Y.; Wu, D.; Huang, C.; Xiao, D.; Chen, H.; Zheng, S.; Chu, P.K. NiFe-Layered Double Hydroxide synchronously activated by heterojunctions and Vacancies for the Oxygen Evolution Reaction. *Appl. Mater. Interfaces* **2020**, *12*, 42850–42858. [CrossRef] [PubMed]
8. Zhang, X.; Zhang, X.; Xu, H.; Wu, Z.; Wang, H.; Liang, Y. Iron-Doped cobalt monophosphide Nanosheet/Carbon nanotube hybrids as active and stable electrocatalysts for water splitting. *Adv. Funct. Mater.* **2017**, *27*, 1606635. [CrossRef]
9. Xu, K.; Chen, P.; Li, X.; Tong, Y.; Ding, H.; Wu, X.; Chu, W.; Peng, Z.; Wu, C.; Xie, Y. Metallic Nickel Nitride nanosheets realizing enhanced electrochemical water oxidation. *J. Am. Chem. Soc.* **2015**, *137*, 4119–4125. [CrossRef] [PubMed]
10. Huo, J.; Wang, Y.; Yan, L.; Xue, Y.; Li, S.; Hu, M.; Jiang, Y.; Zhai, Q.G. In situ semi-transformation from heterometallic MOFs to Fe-Ni LDH/MOF hierarchical architectures for boosted oxygen evolution reaction. *Nanoscale* **2020**, *12*, 14514. [CrossRef] [PubMed]
11. Lu, X.; Yim, W.L.; Suryanto, B.H.; Zhao, C. Electrocatalytic oxygen evolution at Surface-Oxidized multiwall carbon nanotubes. *J. Am. Chem. Soc.* **2015**, *137*, 2901–2907. [CrossRef] [PubMed]
12. Lu, J.; Zhou, W.; Wang, L.; Jia, J.; Ke, Y.; Yang, L.; Zhou, K.; Liu, X.; Tang, Z.; Li, L.; et al. Core-shell nanocomposites based on gold nanoparticle @Zinc–iron- embedded porous carbons derived from metal-organic frameworks as efficient dual catalysts for oxygen reduction and hydrogen evolution reactions. *ACS Catal.* **2016**, *6*, 1045–1053. [CrossRef]
13. Li, X.; Hao, X.; Wang, Z.; Abudula, A.; Guan, G. In-situ intercalation of NiFe LDH materials: An effificient approach to improve electrocatalytic activity and stability for water splitting. *J. Power Sources* **2017**, *347*, 193–200. [CrossRef]

14. Smith, R.D.; Sporinova, B.; Fagan, R.D.; Trudel, S.; Berlinguette, C.P. Facile photochemical preparation of amorphous iridium oxide films for water oxidation catalysis. *Chem. Mater.* **2014**, *26*, 1654–1659. [CrossRef]
15. Su, L.W.; Zhou, Z.; Shen, P.W. Core-shell Fe@Fe_3C/C Nanocomposites as Anode Materials for Li Ion Batteries. *Electrochim. Acta* **2013**, *87*, 180–185. [CrossRef]
16. Boyanov, S.; Bernardi, J.; Gillot, F.; Dupont, L.; Womes, M.; Tarascon, J.M.; Monconduit, L.; Doublet, M.L. FeP: AnotherAttractive Anode for the Li-Ion Battery Enlisting a Reversible Two-Step Insertion/Conversion Process. *Chem. Mater.* **2006**, *18*, 3531–3538. [CrossRef]
17. Boyanov, S.; Womes, M.; Monconduit, L.; Zitoun, D. Mossbauer Spectroscopy and Magnetic Measurements As Comple-mentary Techniques for the Phase Analysis of FeP Electrodes Cycling in Li-Ion Batteries. *Chem. Mater.* **2009**, *21*, 3684–3692. [CrossRef]
18. Ding, X.; Yin, S.; An, K.; Luo, L.; Shi, N.; Qiang, Y.; Pasupathi, S.; Pollet, B.G.; Shen, P.K. FeN stabilized FeN@Pt core-shell nanostructures for oxygen reduction reaction. *J. Mater. Chem. A* **2015**, *3*, 4462–4469. [CrossRef]
19. Cai, Z.; Zhou, D.; Wang, M.; Bak, S.M.; Wu, Y.; Wu, Z.; Tian, Y.; Xiong, X.; Li, Y.; Liu, W.; et al. Introducing Fe^{2+} into nickel-iron layered double hydroxide: Local structure modulated water oxidation activity. *Angew. Chem.-Int. Ed.* **2018**, *130*, 9536–9540. [CrossRef]
20. Basciano, L.C.; Peterson, R.C. The crystal structure of ammoniojarosite, $(NH_4)Fe_3(SO_4)W(OH)_6$ and the crystal chemistry of the ammoniojarosite-hydronium jarosite solid-solution series. *Mineral. Mag.* **2007**, *71*, 427–441. [CrossRef]
21. Frost, R.; Wills, R.A.; Kloprogge, J.; Martens, W. Thermal decomposition of ammonium jarosite $(NH_4)Fe_3(SO_4)_2(OH)_6$. *J. Therm. Anal. Calorim.* **2006**, *84*, 489–496. [CrossRef]
22. Li, H.J.; Yang, H.Y.; Chen, G.B. Catalytic performance of biological method seeds on jarosite process. *Trans. Nonferrous Met. Soc. China* **2016**, *26*, 557–564. [CrossRef]
23. Liu, F.; Shi, J.; Duan, J.; Zhou, L.; Xu, J.; Hao, X.; Fan, W. Significance of jarosite dissolution from the biooxidized pyrite surface on further biooxidation of pyrite. *Hydrometallurgy* **2018**, *176*, 33–41. [CrossRef]
24. Xu, W.; Xie, Z.; Cui, X.; Zhao, K.; Zhang, L.; Mai, L.; Wang, Y. Direct growth of an economic green energy storage material: A monocrystalline jarosite-$KFe_3(SO_4)_2(OH)_6$-nanoplates@rGO hybrid as a superior lithium-ion battery cathode. *J. Mater. Chem. A* **2016**, *4*, 3735–3742. [CrossRef]
25. Wu, N.; Tian, W.; Shen, J.; Qiao, X.; Sun, T.; Wu, H.; Zhao, J.; Liu, X.; Zhang, Y. Facile fabrication of a jarosite ultrathin $KFe_3(SO_4)_2(OH)_6$@rGO nanosheet hybrid composite with pseudocapacitive contribution as a robust anode for lithiumion batteries. *Inorg. Chem. Front.* **2019**, *6*, 192–198. [CrossRef]
26. Wang, X.L.; Dong, L.Z.; Qiao, M.; Tang, Y.J.; Liu, J.; Li, Y.; Li, S.L.; Su, J.X.; Lan, Y.Q. Exploring the performance improvement of the oxygen evolution reaction in a stable bimetal-organic framework system. *Angew. Chem.-Int. Ed.* **2018**, *57*, 1–6.
27. Gao, K.; Jiang, M.; Guo, C.; Zeng, Y.; Fan, C.; Zhang, J.; Reinfelder, J.R.; Huang, W.; Lu, G.; Dang, Z. Reductive dissolution of jarosite by a sulfate reducing bacterial community: Secondary mineralization and microflflora development. *Sci. Total Environ.* **2019**, *690*, 1100–1109. [CrossRef] [PubMed]
28. Liu, C.; Ling, Z.; Zhang, J.; Bi, X.; Xin, Y. Laboratory Raman and VNIR spectroscopic studies of jarosite and other secondary mineral mixtures relevant to Mars. *J. Raman Spectrosc.* **2020**, *51*, 1575–1588. [CrossRef]
29. Duan, Y.; Lee, J.Y.; Xi, S.; Sun, Y.; Ge, J.; Ong, S.J.H.; Chen, Y.; Dou, S.; Meng, F.; Diao, C.; et al. Anodic Oxidation Enabled Cation Leaching for Promoting Surface Reconstruction in Water Oxidation. *Angew. Chem.-Int. Ed.* **2021**, *133*, 7418–7425. [CrossRef]
30. Liu, X.; Meng, J.; Ni, K.; Guo, R.; Xia, F.; Xie, J.; Li, X.; Wen, B.; Wu, P.; Li, M.; et al. Complete Reconstruction of Hydrate Pre-Catalysts for Ultrastable Water Electrolysis in Industrial-Concentration Alkali Media. *Cell Rep. Phys. Sci.* **2020**, *1*, 100241. [CrossRef]
31. Hu, J.; Li, S.; Chu, J.; Niu, S.; Wang, J.; Du, Y.; Li, Z.; Han, X.; Xu, P. Understanding the Phase-Induced Electrocatalytic Oxygen Evolution Reaction Activity on FeOOH Nanostructures. *ACS Catal.* **2019**, *9*, 10705–10711. [CrossRef]
32. Xiao, T.; Yang, C.; Lu, Y.; Zeng, F. One-pot hydrothermal synthesis of rod-like FeOOH/reduced graphene oxide composites for supercapacitor. Journal of Materials Science. *Mater. Electron.* **2014**, *25*, 3364–3374. [CrossRef]
33. Ohtsuka, T.; Tanaka, S. Monitoring the development of rust layers on weathering steel using in situ Raman spectroscopy under wet-and-dry cyclic conditions. *J. Solid State Electrochem.* **2015**, *19*, 3559–3566. [CrossRef]

materials MDPI

Article

Structural, Electrical, and Mechanical Properties Investigation of Open-Cell Aluminum Foams Obtained by Spark Plasma Sintering and Replication on Polyurethane Template

Alexandra Kosenko [1,]*, Konstantin Pushnitsa [1], Artem Kim [1], Pavel Novikov [1] and Anatoliy A. Popovich [1]

Institute of Machinery, Materials, and Transport, Peter the Great Saint Petersburg Polytechnic University, Politechnicheskaya ul. 29, 195251 Saint Petersburg, Russia; pushnitsa.k@gmail.com (K.P.); artem_7.kim@mail.ru (A.K.); novikov.p.a@gmail.com (P.N.); popovicha@mail.ru (A.A.P.)

* Correspondence: alxndra.kosenko@gmail.com; Tel.: +7-929-105-73-38

Abstract: The present paper illustrates a comparison of open-cell aluminum foams. The foams were fabricated by two different methods: spark plasma sintering and replication on a polyurethane template. The influence of pressure, temperature, and diameter of space holding material on foam obtained by the spark plasma sintering method was investigated. Additionally, the aluminum powder content in slurry and atmosphere during thermal processing of foam prepared by the replication technique were studied. The morphology and structure of obtained samples were analyzed by scanning electron microscopy and X-ray diffraction analysis. Supplementarily, mechanical properties and electrical conductivity were studied. The porosity of obtained samples was 83% for the SPS sample and 85% for the replication sample. The results of the studies carried out gave us an understanding that the SPS method is more promising for using the obtained foams as cathode current collectors in lithium-ion batteries due to excessive aluminum oxidation during sintering in the furnace.

Keywords: open-cell foams; porous materials; spark plasma sintering; replication technique; PU foam template

Citation: Kosenko, A.; Pushnitsa, K.; Kim, A.; Novikov, P.; Popovich, A.A. Structural, Electrical, and Mechanical Properties Investigation of Open-Cell Aluminum Foams Obtained by Spark Plasma Sintering and Replication on Polyurethane Template. *Materials* **2022**, *15*, 931. https://doi.org/10.3390/ma15030931

Academic Editors: Marc Cretin and Zhenghua Tang

Received: 13 December 2021
Accepted: 21 January 2022
Published: 26 January 2022

1. Introduction

Metal foams are currently a special class of lightweight materials that can be used in various fields of the industry due to their physical, chemical, and mechanical properties [1–3]. Using metal foam instead of solid metal is more efficient and reduces the weight and cost of the final product [2]. Metal foam might be able to substitute a variety of conventional materials. Currently it is used in a wide range of engineering applications. Therefore, this fact allows applying these materials in various fields such as automobiles, aerospace, and naval industries [3]. It is possible to manufacture metal foam from pure metals, for example, titanium, nickel, and aluminum [4,5], and in the same way from different alloys [6,7]. Due to the lightweight nature of this type of material, metal foam is an extremely promising material for the automotive industry; in particular, open-cell foams made of aluminum and Al alloys are of special interest. Remarkable properties of open-cell aluminum foam such as low density, high specific electrical and thermal conductivity, and high surface area allow to be applied as a component of the lithium-ion battery, specifically to use aluminum foam instead of aluminum foil as cathode current collectors [8–11].

A variety of fabrication processes have been developed to produce metal foams, in particular aluminum foams [12–20]. The common processes are either via the addition of a foaming agent into molten aluminum alloy or through bubbling gas into a molten aluminum alloy. The main drawbacks are that aluminum foams produced by these methods are limited in terms of control of the pore content and size and microstructure of cell walls. The methods of production for open-cell aluminum foam can be divided into three classes:

solid-state methods, liquid metallurgy methods, and vapor state methods [21]. Solid-state methods include the sintering of metal powder. It is well known that spark plasma sintering (SPS) decreases the sintering time and temperature, compared with sintering in an electric furnace, resulting in suppression of grain growth during sintering [22–25]. Therefore, it is worthwhile to investigate the microstructure and mechanical properties of aluminum foam processed by SPS. In addition, one of the promising ways to manufacture open-cell metal foams is the sponge replication technique, and this method has been successfully applied for the production of open-cell titanium, copper, and aluminum foams [26–33]. The main disadvantage of the replication method is the high oxidation rates of powders during the sintering step [34,35].

In the present paper, open-cell aluminum foams, fabricated by spark plasma sintering and by sponge replication technique, are investigated. The necessity of the research performed was in changing the standard current collector in lithium-ion batteries, aluminum foil, to aluminum foam, due to aluminum foam being a more lightweight and effective type of material. The main difference between this research and previously published ones is that no additives, such as foaming and thickener agents, were used. This fact can lead to a low impurity content in composition of the resulting foams, which is important in the further use of the foam. The investigation focused on foam production with a highly open structure, with low impurity content, a lower amount of oxide film on the surface, high electrical conductivity, and sufficient mechanical strength. The last two points are high priority in this research. Mechanical strength is necessary since the aluminum foam must not collapse when the active cathode material is applied. High electrical conductivity is important for low internal resistance and high specific energy capacity of the battery.

2. Materials and Methods

2.1. Materials

Figure 1 represents an SEM image of the aluminum alloy powder used for SPS method (a), sodium chloride (b), and aluminum powder used for replication technique (c) obtained in the backscattered electron detection mode. Aluminum alloy powder particles are predominantly rounded, the particle size distribution is as follows: d_{10} = 20.9 μm, d_{50} = 51.6 μm, d_{90} = 95.9 μm. It means that 10% of particles have a size less than 20.9 μm, 50% of particles have a size less than 51.6 μm, and 90% of particles have a size less than 95.9 μm. Additionally, there are some submicron size particles. The powder has a homogeneous structure. In Figure 1b sodium chloride particles are presented. Particles have mostly spheroidal and round form, the powder was sieved and fraction 500–1000 μm was chosen. Figure 1c shows the aluminum powder particles, all of the particles have a spherical shape, a homogeneous structure, and the particle size distribution is d_{10} = 2.4 μm, d_{50} = 5.5 μm, d_{90} = 10.8 μm.

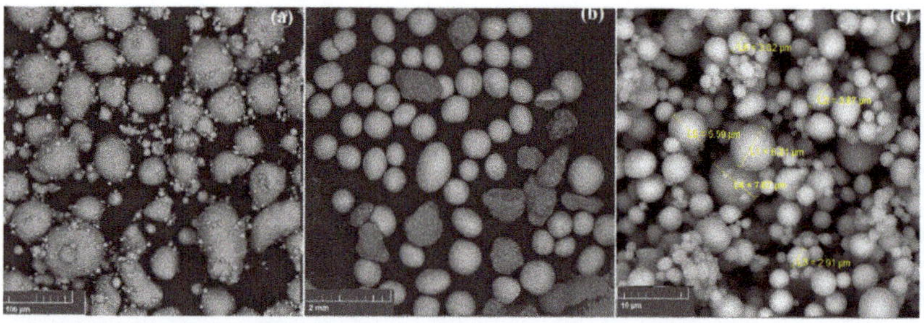

Figure 1. (**a**) Aluminum alloy powder, (**b**) sodium chloride, (**c**) aluminum powder.

2.2. Spark Plasma Sintering Method

The commercially available aluminum alloy powder (AMD5, supplier Normin LTD., Oxford, UK) with the content of Al 94.8 wt.%, Mg 4.8 wt.%, Ti 0.4 wt.%, with an average particle size of 90 μm, was used as a starting material. It is well known that Al–Mg alloys are characterized by a combination of satisfactory strength, good ductility, very good weldability and corrosion resistance [36]. Titanium in Al alloys also enhances microstructural and mechanical properties, including specific strength, oxidation and corrosion resistance [37]. Sodium chloride (supplier LenReactiv JSC, Saint Petersburg, Russia) with average particle size 500–1000 μm, was used as a space holder for SPS. The weight ratio of the aluminum powder to NaCl was determined to be 10:90 and the volume ratio 8:92. To select the optimal weight ratio of aluminum powder content and porosity of the final Al foam, mixtures with a content of 5% and 10% were prepared and foams with the same sintering conditions were made. In the case of 5%, the sample retains its shape, but begins to crumple with a slight pressure with tweezers, in the case of samples of 10% this is not observed. The spark plasma sintering method is schematically presented in Figure 2. In general, the process consists of 4 steps: mixing the aluminum powder and sodium chloride, pressing, sintering, and leaching in water. Firstly, the aluminum powder and space-holding material were mixed in the vibrating mixer (C50.0 Vibrotechnik LLC, Saint Petersburg, Russia) for 90 min. After the ingredients were homogeneously mixed, the resulting powder was pressed in graphite mold at a pressure of 5 MPa. Then, sintering was carried out on the equipment for spark sintering FCT system HPD 25 with the following parameters: sintering temperature 550 °C, pressure 38 MPa, sintering time 5 min, pulse duration 2–4 ms, pause duration 2 ms. The sintered sample after the sintering was placed into hot water for 12 h to leach out the embedded sodium chloride particles, to obtain the aluminum foam with porous structure.

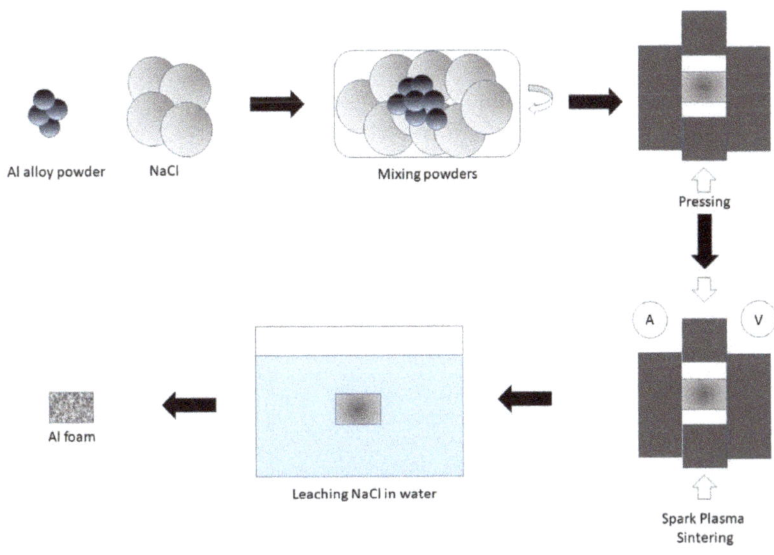

Figure 2. Schematic illustration of spark plasma sintering (SPS) method.

2.3. Sponge Replication Technique

The commercially available aluminum powder with purity 99% and with particle size <10 μm was used (ASD6, supplier Normin LTD). The aluminum powder was mixed with earlier-prepared 0.4 wt.% cornstarch solution in distilled water. Cornstarch is a binder in this solution. The content of Al powder in final dispersion was 80 wt.%. The dispersion was mixed by using an ultrasonic bath to make it homogeneous. This method consists of 3 steps: (1) coating a polyurethane (PU) sponge with a powder slurry; (2) burning the PU and binder out; (3) final sintering of the foam. The schematic illustration is below in Figure 3. An open-cell polyurethane foam cylinder (Polinazell PPI 20. supplier United

Service Company LLC, Chicago, IL, USA) with a height of 3 mm and a diameter of 19 mm was used as a template.

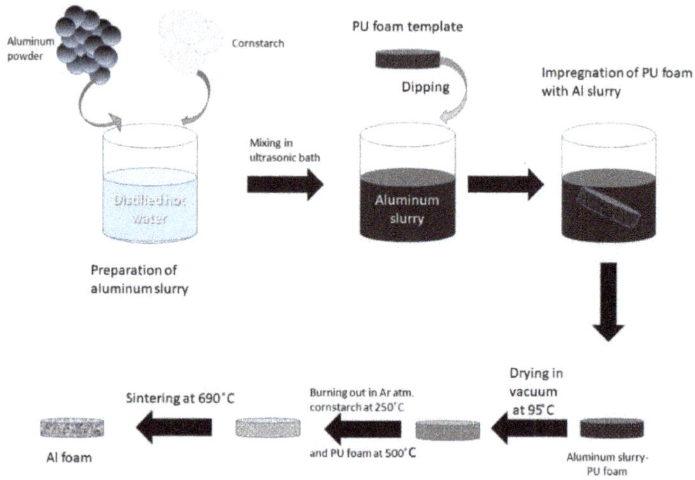

Figure 3. Schematic illustration of replication technique.

As a first step, the PU foam was drowned into previously mentioned dispersion, soaked for 5 min, and then the excess was removed by mechanical wringing. After this step, the dispersion-coated template was dried for 17 h in a vacuum at a temperature of 95 °C. Further, the sample was placed in a tube electric furnace to burn out the PU foam and binder. The final thermal processing step was carried out also in a tube electric furnace. All burning and thermal processes were conducted in the Argon atmosphere, the flow rate was 40 mL/min. The conditions of processing are listed below in Table 1. Due to the cooling effect of Argon flow, the sintering temperature is higher than the melting point of aluminum.

Table 1. Conditions of binder and PU burning and sintering of aluminum foam sample.

Conditions	Binder Burning	PU Burning	Sintering
Initial temperature, °C	20	470	520
Final temperature, °C	470	520	690
Heating time, min	60	25	20
Holding time, min	0	0	480

2.4. Characterisation

The morphology, microstructure, and chemical composition of the samples were studied using Mira 3 Tescan scanning electron microscope with an EDX Oxford Instruments X-max 80 energy dispersive detector for X-ray spectroscopy. X-ray phase analysis was carried out on a Bruker Advance D8 diffractometer in the range of angles from 25° to 85° with a step of 0.02° and an exposure of 1.5 s at each step. The wavelength was 1.5406 Å. The lattice parameters were measured by the Le Bail method using software TOPASS. The particle size (PS) distribution of the powders was studied using a Fritsch Analysette 22 NanoTec plus laser diffraction unit. To calculate the particle size distribution, the Fraunhofer model was used. The electrical conductivity was measured using an AKIP-2101 voltmeter, a GPD-74303S current source, and a FLUKE-289 multimeter. Mechanical compression tests of the specimens were carried out on a Zwick/Roell Z100 testing machine.

3. Results and Discussion

3.1. Aluminum Foam Formation

In Table 2 different conditions of sintering are demonstrated. First, it is worth noting the fact that at a temperature of 600 °C and a pressure of 19 MPa and above, aluminum powder melts and flows out of the mold. At 500 °C, and pressure 38 MPa and below, spherical non-deformed particles are observed, which indicates worse sintering. however, at pressure above 38 MPa the powder starts to melt. The best results of sintering were obtained at temperature 550 °C and pressure 38 MPa.

Table 2. Different conditions of spark plasma sintering method.

Pressure/ Temperature	19 MPa	29 MPa	38 MPa	48 MPa	57 MPa
500 °C	Not sintered	Not sintered	Not sintered	Melted	Melted
550 °C	Not sintered	Partly sintered	Sintered sample	Melted	Melted
600 °C	Melted	Melted	Melted	Melted	Melted

In Figure 4 the obtained samples of aluminum foam are presented. Figure 4a is the SPS sample, and Figure 4b is the replication sample. At first glance, it seems that the SPS sample has more open porosity than the sample fabricated by replication technique; in the next subsections the physicochemical and mechanical properties are considered in more detail. The shape and size of the samples were selected on the basis of further research as current collectors in lithium-ion batteries. The diameter and height of the samples should be no more than 20 mm and 2 mm, respectively, so that they can be placed in the CR2032 case for further testing.

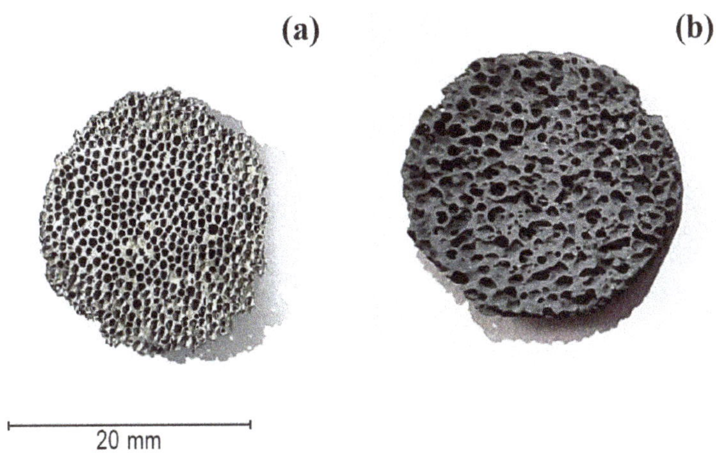

Figure 4. Obtained open-cell aluminum foam by SPS (**a**), by replication method (**b**).

3.2. Morphology and Microstructure Analysis

Figure 5 shows the surface morphology of a sample obtained by the method of spark plasma sintering. The sample has a cellular structure with open porosity. It is apparent that the material has a homogeneous structure with a pore diameter of about 600 μm. It is possible to control the pore diameter of fabricating foam by choosing different fractions of space holding material—sodium chloride.

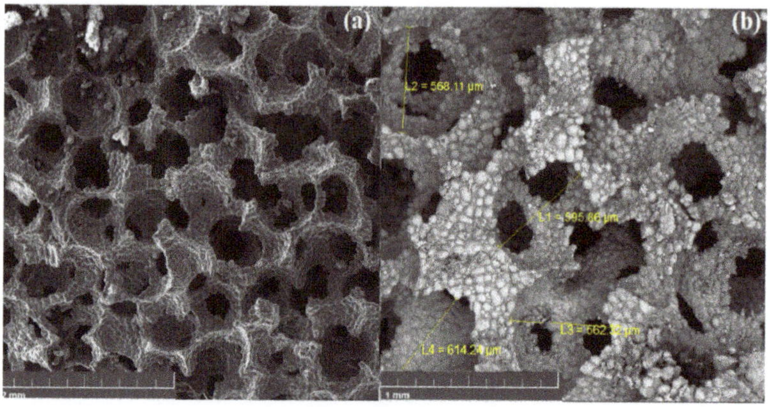

Figure 5. SEM image of foam obtained by SPS method: (**a**) general view, (**b**) pore size investigation.

Figure 6 shows images of the original polyurethane sponge and a sample of aluminum foam obtained by the replication method. The sample has open porosity. The pore diameter is in a wide range (300–1000 microns) and it is comparable to the pore size of the PU sponge. Small pores can be seen on the replication sample due to the incomplete repetition of the template relief, which is justified due to the high viscosity of the suspension.

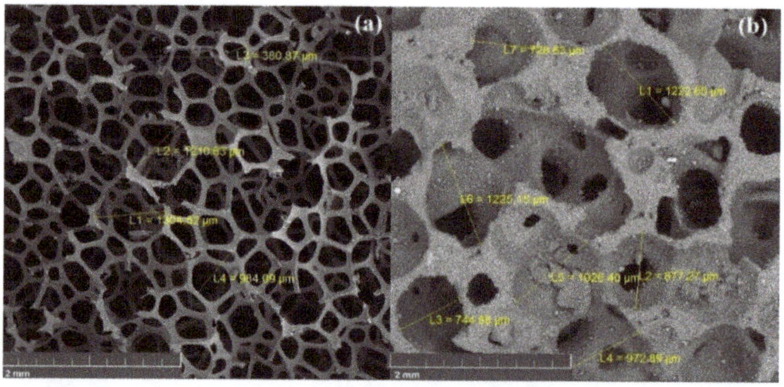

Figure 6. SEM image of PU sponge (**a**) and obtained foam by replication technique (**b**).

From EDX spectra (Figure 7b) it can be concluded that spreading of alloy elements in the SPS sample after sintering is uniform, and the amount of sodium and chlorine is less than 1 wt.%; the presence of sulfur is due to its impurity in the composition of the salt. The results of the replication sample EDX analysis of Figure 7d represents that the chemical composition after thermal processing is characterized by 70.8 wt.% aluminum, and a large amount of the oxygen is recorded as well. However, this method gives a large error when measuring the number of light elements.

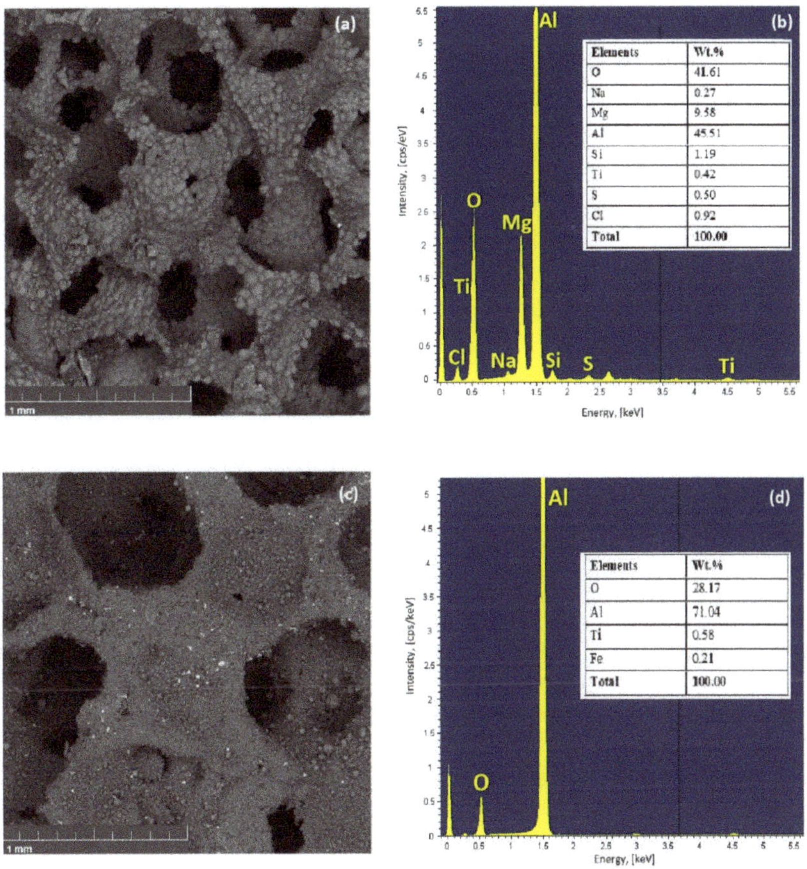

Figure 7. SEM image (**a**,**c**) and corresponding energy-dispersive X-ray spectroscopy spectra (**b**,**d**) of SPS sample (**a**,**b**) and replication sample (**c**,**d**).

Figure 8 shows the diffraction patterns of Al alloy powder (a), Al powder (b), the sample obtained by the SPS (c), and by the replication method (d). The main peaks of all diffraction patterns are indexed by the cubic Al phase (PDF 03-065-2869) with the space group Fm-3m. The unit cell parameters obtained by the Rietveld method are presented in Table 3. In the sample obtained by the replication method, the presence of the γ-Al_2O_3 phase is observed.

Table 3. Unit cell parameters were obtained by the Le Bail method.

Sample	Al Alloy Powder	Al Powder	SPS	Replication
a, Å	4.062	4.049	4.062	4.051

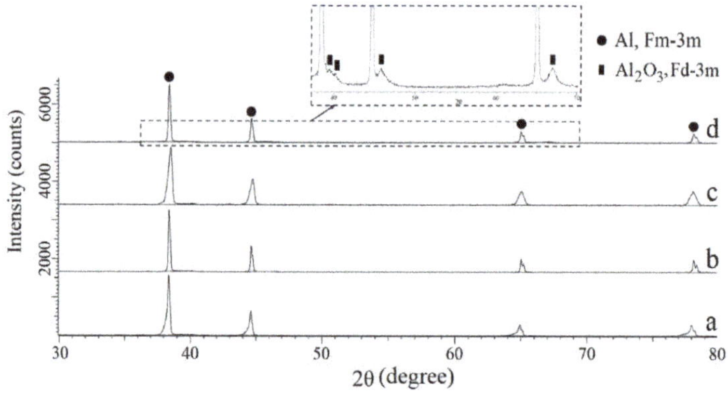

Figure 8. Diffraction patterns of Al alloy powder (**a**), Al powder (**b**), SPS sample (**c**), and replication sample (**d**).

3.3. Porosity

The density of the aluminum foams was determined by hydrostatic weighing (Archimedes method). As a first step, the samples were weighed in air, then they were coated with paraffin to prevent the fluid from entering inside the foam pores. After this step, the samples were weighed in air and further in ethanol, the density of which amounts to 0.785 g/cm^3 at 25 °C. The relationship between the porosity P_r (%) of the aluminum foam and the relative density (ρ_f/ρ_s) of the foam are presented in Equation (1), where ρ_f and ρ_b represent densities of the foam sample and bulk aluminum (about 2.7 g/cm^3), respectively. The density of the aluminum foam was determined by hydrostatic weighing. The measurements of samples densities and porosity are presented in Table 4.

$$P_r = \left(1 - \frac{\rho_f}{\rho_b}\right) \times 100 \quad (1)$$

Table 4. The measurements of density and porosity of aluminum foam.

Sample	Dry Weight, g	Dry Weight with Paraffin, g	Underwater Weight, g	Density, g/cm^3	Porosity%
SPS	0.278	0.621	0.125	0.45	83%
Replication	0.328	0.783	0.164	0.42	85%

3.4. Electrical Conductivity

The electrical conductivity of samples was determined by measuring the resistance using the 4-wire testing method. Electrodes for measuring the potential difference were installed directly on the sample with a constant distance between electrodes (10 mm) for each measurement. The accuracy of measuring the potential difference with the used device was 0.01%.

To determine the resistance of the samples, a direct current was applied. The accuracy of the applied current was ±0.5%. To clarify the measured results, the current was measured with an ammeter with an error of no more than ±0.2%. The potential of aluminum foil and samples was measured. Aluminum foil is a frequently used material as a cathode current collector in lithium-ion batteries. In this test, it was used as a reference to compare with. The diameter of the foil was 20 mm, the thickness, 14 μm. For each sample, three measurements were taken at different locations. The resistance of the samples for each measurement was calculated using Ohm's law equation. The results of the measurements are presented in Table 5.

Table 5. The resistance measurement of samples by using the 4-wire testing method.

Sample (Measurement)	Applied Current, A	Measured Potential Difference, V	Calculated Resistance, Ohm
Aluminum foil 1	0.149	0.002	0.013
Aluminum foil 2	0.151	0.002	0.013
Aluminum foil 3	0.148	0.001	0.007
SPS 1	0.153	0.001	0.006
SPS 2	0.157	0.002	0.012
SPS 3	0.156	0.001	0.006
Replication 1	0.151	47.291	313.185
Replication 2	0.155	49.517	319.464
Replication 3	0.148	46.536	314.432

The key factors affecting the electrical conductivity of the samples are the degree of sintering and the amount of aluminum oxide on the surface of the samples. The determination of the electrical conductivity of the samples showed results consistent with the X-ray analysis—the sample obtained by replication technique has the worst electrical conductivity, which can be explained by the formation of aluminum oxide film on the sample surface and the surface of grains. In addition, the calculated resistances of aluminum foil and SPS sample gave us an understanding that the SPS sample is the closest in electrical conductivity to aluminum foil. The obtained results demonstrate that the SPS sample obtained due to its properties is better for application as a three-dimensional current collector than the sample obtained by replication.

3.5. Mechanical Properties

The compression stress–strain curves of the obtained samples are shown in Figure 9. The dimensions of samples were: diameter—19 mm, height—2 mm. Three samples of each obtaining method were taken for the compressive test. For both samples, coefficient of variation was calculated. The average coefficient of variation for SPS samples was 12%, and for replication samples was 13%.

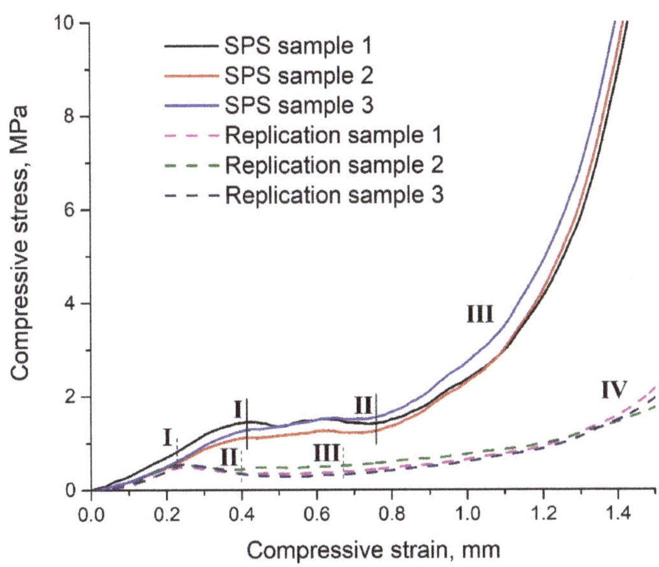

Figure 9. Compression strain curves for aluminum foams: SPS sample, replication sample.

The stress–strain curves of the SPS sample are characterized by three distinct regions (designate by solid lines). The first region is an elastic region characterized by a linear elastic region at very low deformation without peak stress. The second area is the plateau area on which pores collapse (destruction of the bridges), which may indicate the growing deformation without increasing pressure. Fluctuations of the curve in this area can be associated with the uneven distribution of pores in the sample. The third area is the area of compaction, where a sharp increase in compressive stress occurs.

The stress-strain curve of the replication sample is characterized by four regions (designate by dash lines). The first region is the region of elastic deformation; relative to the SPS sample, the region of elastic deformation ends at a lower pressure. In the second region, the pressure decreases with the continued growth of deformation, which may indicate not plastic compression of the bridges, but their brittle fracture. This behavior is demonstrated by all samples. The brittle fracture occurs due to the presence of aluminum oxide in the sample. The third region is less pronounced than for the SPS sample due to the smaller amount of aluminum in the sample. In the fourth region, the same process occurs as in region three for the SPS sample.

For the samples, the adsorbed energies were also compared during mechanical strength tests. Due to the small thickness of the sample, the tests were carried out without using an extensometer, which would not allow the calculation of the nominal adsorbed energy. However, it is admissible to compare the adsorbed energies of the two samples. The calculation of the adsorbed energies was carried out according to the method presented in the article [38]. The ratio of the adsorbed energies of the samples (a)/(b) was 5.24. This value shows that, upon compression, sample (a) absorbed 5.24 times more energy, which also indirectly indicates the metallic structure of sample (a) and the nonmetallic structure of sample (b).

3.6. *Properties Comparison of Samples*

Babcsán et al. have used the foam in their studies, manufactured by the batch casting process. It is possible to obtain aluminum foam with closed porosity by this method. The main difference between our method and theirs is that we pursued the goal of obtaining open-cell aluminum foam. Additionally, our research is distinguished from the Babcsán research since no additives were used in the foam production. In their method, calcium and titanium hydride were used as thickener and foaming agent, respectively. Using our method, as a result, we obtained a sample with a minimum content of impurities. Impurities, which were found by EDX analysis, also were present in the initial materials.

The comparison of samples is presented in Table 6. The low conductivity of the SPS sample can be explained by incomplete sintering of powder during the formation of the sample. The replication sample result was previously explained: the large amount of aluminum oxide has influence on the conductivity.

Table 6. Properties comparison of open-cell aluminum foams.

Property	Babcsán, N et al. Sample [39]	SPS Sample	Replication Sample
Porosity, %	83	83	85
Density, g/m^3	0.46	0.45	0.42
Electrical conductivity, S/m	3.43×10^6	1.39×10^4	3.18×10^{-1}

4. Conclusions

Open-cell aluminum foams were successfully obtained by two various methods: spark plasma sintering and replication on polyurethane template. The porosity of final samples was 83% and 85%, respectively. According to the investigations carried out on the microstructure and morphology, electrical conductivity, and mechanical properties of obtained samples, we can conclude that the sample obtained by the replication method has more aluminum oxide film on its surface than the SPS sample. The consequence of this fact is the lower electric conductivity of the replication sample. The electrical conductivity of the SPS sample is comparable to the aluminum foil. This fact allows us to assume that the electrochemical behavior of the SPS sample will be similar to the foil as a part of lithium-ion batteries. Mechanical tests of the foams also demonstrated the presence of a brittle phase in the replication sample. The SPS sample showed metallic properties and determined sufficient mechanical strength. Due to investigated properties of the SPS sample, it is evident that the spark plasma sintering method is the most promising way to produce open-cell aluminum foam for further use in lithium-ion batteries.

The future research plan is to find the optimal balance of temperature, pressure and sintering time in order to improve the processing mode of the SPS method. Additionally, the further improvement of the samples' characteristics is possible by using fluxes, which limit the oxide film growth and improve the sintering of the aluminum particles. The next step is to assemble mock-ups of CR2032 batteries to study the influence of a new type of current collector on batteries' behavior using cyclic voltammetry and electrochemical impedance spectroscopy methods.

Author Contributions: Conceptualization, P.N. and K.P.; methodology, A.K. (Alexandra Kosenko), A.K. (Artem Kim), and K.P.; software, A.K. (Alexandra Kosenko) and K.P.; validation, A.K. (Alexandra Kosenko) and K.P.; formal analysis, A.K. (Alexandra Kosenko), A.K. (Artem Kim), and K.P.; investigation, A.K. (Alexandra Kosenko) and K.P.; resources, A.A.P. and P.N.; data curation, A.K. (Alexandra Kosenko); writing—original draft preparation, A.K. (Alexandra Kosenko); writing—review and editing, A.K. (Alexandra Kosenko) and K.P.; visualization, A.K. (Alexandra Kosenko) and K.P.; supervision, K.P., P.N., and A.A.P.; project administration, P.N.; funding acquisition, A.A.P. All authors have read and agreed to the published version of the manuscript.

Funding: The research is funded by the Ministry of Science and Higher Education of the Russian Federation as part of the World-Class Research Center program: Advanced Digital Technologies (contract No. 075-15-2020-934 dated 17 November 2020).

Institutional Review Board Statement: Not applicable.

Informed Consent Statement: Not applicable.

Data Availability Statement: The data presented in this study are available on request from the corresponding author.

Conflicts of Interest: The authors declare no conflict of interest.

References

1. Banhart, J. Manufacture, characterisation and application of cellular metals and metal foams. *Prog. Mater. Sci.* **2001**, *46*, 559–632. [CrossRef]
2. Singh, S.; Bhatnagar, N. A survey of fabrication and application of metallic foams (1925–2017). *J. Porous Mater.* **2018**, *25*, 537–554. [CrossRef]
3. Kim, S.; Lee, C.-W. A Review on Manufacturing and Application of Open-cell Metal Foam. *Procedia Mater. Sci.* **2014**, *4*, 305–309. [CrossRef]
4. García-Moreno, F. Commercial Applications of Metal Foams: Their Properties and Production. *Materials* **2016**, *9*, 85. [CrossRef]
5. Gancarczyk, A.; Sindera, K.; Iwaniszyn, M.; Piątek, M.; Macek, W.; Jodłowski, P.J.; Wroński, S.; Sitarz, M.; Łojewska, J.; Kołodziej, A. Metal Foams as Novel Catalyst Support in Environmental Processes. *Catalysts* **2019**, *9*, 587. [CrossRef]
6. Nawaz, A.; Rani, S. Fabrication and evaluation of percent porosity and density reduction of aluminium alloy foam. *Mater. Today Proc.* **2021**, *47*, 6025–6029. [CrossRef]
7. Yang, X.; Li, Z.; Qin, J.; Wu, M.; Liu, J.; Guo, Y.; Ma, Y.; Feng, F. Preparation of Ni-Fe alloy foam for oxygen evolution reaction. *J. Fuel Chem. Technol.* **2021**, *49*, 827–834. [CrossRef]
8. Nakamura, N.; Yokoshima, T.; Nara, H.; Mikuriya, H.; Shiosaki, A.; Ahn, S.; Momma, T.; Osaka, T. High-rate and high sulfur-loaded lithium-sulfur batteries with a polypyrrole-coated sulfur cathode on a 3D aluminum foam current collector. *Mater. Lett.* **2021**, *285*, 129115. [CrossRef]
9. Fritsch, M.; Standke, G.; Heubner, C.; Langklotz, U.; Michaelis, A. 3D-cathode design with foam-like aluminum current collector for high energy density lithium-ion batteries. *J. Energy Storage* **2018**, *16*, 125–132. [CrossRef]
10. Zhang, H.; Xu, Z.; Shi, B.; Ding, F.; Liu, X.; Wu, H.; Shi, C.; Zhao, N. Enhanced Cyclability of Cr_8O_{21} Cathode for PEO-Based All-Solid-State Lithium-Ion Batteries by Atomic Layer Deposition of Al_2O_3. *Materials* **2021**, *14*, 5380. [CrossRef]

11. Zhu, P.; Gastol, D.; Marshall, J.; Sommerville, R.; Goodship, V.; Kendrick, E. A review of current collectors for lithium-ion batteries. *J. Power Sources* **2021**, *485*, 229321. [CrossRef]
12. Patel, P.; Bhingole, P.P.; Makwana, D. Manufacturing, characterization and applications of lightweight metallic foams for structural applications: Review. *Mater. Today Proc.* **2018**, *5*, 20391–20402. [CrossRef]
13. Tripathi, O.; Singh, D.P.; Dwivedi, V.K.; Agarwal, M. A focused review on aluminum metallic foam: Processing, properties, and applications. *Mater. Today Proc.* **2021**, *47*, 6622–6627. [CrossRef]
14. Zaragoza, G.; Goodall, R. Metal Foams with Graded Pore Size for Heat Transfer Applications. *Adv. Eng. Mater.* **2013**, *15*, 123–128. [CrossRef]
15. Quadbeck, P.; Kümmel, K.; Hauser, R.; Standke, G.; Adler, J.; Stephani, G.; Kieback, B. Structural and Material Design of Open-Cell Powder Metallurgical Foams. *Adv. Eng. Mater.* **2011**, *13*, 1024–1030. [CrossRef]
16. Yang, X.; Hu, Q.; Du, J.; Song, H.; Zou, T.; Sha, J.; He, C.; Zhao, N. Compression fatigue properties of open-cell aluminum foams fabricated by space-holder method. *Int. J. Fatigue* **2019**, *121*, 272–280. [CrossRef]
17. Banhart, J. Aluminum Foams: On the Road to Real Applications. *MRS Bull.* **2003**, *28*, 290–295. [CrossRef]
18. Almonti, D.; Mingione, E.; Tagliaferri, V.; Ucciardello, N. Design and analysis of compound structures integrated with bio-based phase change materials and lattices obtained through additive manufacturing. *Int. J. Adv. Manuf. Technol.* **2021**. [CrossRef]
19. Almonti, D.; Ucciardello, N. Design and Thermal Comparison of Random Structures Realized by Indirect Additive Manufacturing. *Materials* **2019**, *12*, 2261. [CrossRef]
20. Al-Ethari, H.; Haleem, A.H.; Ali, O.I. Optimization of Manufacturing Parameters Affecting on Characterization of Porous Sintered Tin-Bronze Alloy. *Int. J. Eng. Technol.* **2019**, *8*, 299–307.
21. Baloyo, J.M. Open-cell porous metals for thermal management applications: Fluid flow and heat transfer. *Mater. Sci. Technol.* **2017**, *33*, 265–276. [CrossRef]
22. Wen, C.E.; Mabuchi, M.; Yamada, Y.; Shimojima, K.; Chino, Y.; Hosokawa, H.; Asahina, T. Processing of fine-grained aluminum foam by spark plasma sintering. *J. Mater. Sci. Lett.* **2003**, *22*, 1407–1409. [CrossRef]
23. Soni, B.; Biswas, S. Development of Al Foams by a Low-cost Salt Replication Method for Industrial Applications. *Mater. Today Proc.* **2015**, *2*, 1886–1891. [CrossRef]
24. Attia, M.A.A.; Garroni, S.; Chiriu, D.; Ricci, C.; Delogu, F.; Orrù, R.; Cao, G. X-ray induced color change on dense yttria samples obtained by spark plasma sintering. *Chem. Phys. Lett.* **2015**, *618*, 108–113. [CrossRef]
25. Khosravani, M.R.; Reinicke, T. On the Use of X-ray Computed Tomography in Assessment of 3D-Printed Components. *J. Nondestruct. Eval.* **2020**, *39*, 75. [CrossRef]
26. Zaman, E.; Keleş, Ö. Open Cell Aluminum Foams Produced by Polymer Impregnation Method. *Acta Phys. Pol. A* **2014**, *125*, 445–448. [CrossRef]
27. Zhai, W.; Yu, X.; Song, X.; Ang, L.Y.L.; Cui, F.; Lee, H.P.; Li, T. Microstructure-based experimental and numerical investigations on the sound absorption property of open-cell metallic foams manufactured by a template replication technique. *Mater. Des.* **2018**, *137*, 108–116. [CrossRef]
28. Bowen, C.R.; Thomas, T. Macro-porous Ti2AlC MAX-phase ceramics by the foam replication method. *Ceram. Int.* **2015**, *41*, 12178–12185. [CrossRef]
29. Bracconi, M.; Ambrosetti, M.; Okafor, O.; Sans, V.; Zhang, X.; Ou, X.; Da Fonte, C.P.; Fan, X.; Maestri, M.; Groppi, G.; et al. Investigation of pressure drop in 3D replicated open-cell foams: Coupling CFD with experimental data on additively manufactured foams. *Chem. Eng. J.* **2019**, *377*, 120123. [CrossRef]
30. Fabrizio, Q.; Boschetto, A.; Rovatti, L.; Santo, L. Replication casting of open-cell AlSi7Mg0.3 foams. *Mater. Lett.* **2011**, *65*, 2558–2561. [CrossRef]
31. Combaz, E.; Bacciarini, C.; Charvet, R.; Dufour, W.; Mortensen, A. Multiaxial yield behaviour of Al replicated foam. *J. Mech. Phys. Solids* **2011**, *59*, 1777–1793. [CrossRef]
32. Li, J.P.; Li, S.H.; Van Blitterswijk, C.A.; de Groot, K. A novel porous Ti6Al4V: Characterization and cell attachment. *J. Biomed. Mater. Res. Part A* **2005**, *73*, 223–233. [CrossRef] [PubMed]
33. Manonukul, A.; Srikudvien, P.; Tange, M.; Puncreobutr, C. Geometry anisotropy and mechanical property isotropy in titanium foam fabricated by replica impregnation method. *Mater. Sci. Eng. A* **2016**, *655*, 388–395. [CrossRef]
34. Trunov, M.A.; Umbrajkar, S.M.; Schoenitz, M.; Mang, J.T.; Dreizin, E.L. Oxidation and Melting of Aluminum Nanopowders. *J. Phys. Chem. B* **2006**, *110*, 13094–13099. [CrossRef]
35. Kwon, H.; Kawasaki, A. Effect of Spark Plasma Sintering in Fabricating Carbon Nanotube Reinforced Aluminum Matrix Composite Materials. In *Advances in Composite Materials for Medicine and Nanotechnology*; InTech, Advanced Materials Processing, Empa-Swiss Federal Laboratories for Materials Science and Technology: Thun, Switzerland; Department of Materials Processing, Tohoku University: Sendai, Japan, 2011.
36. Tzeng, Y.-C.; Chen, R.-Y.; Lee, S.-L. Nondestructive tests on the effect of Mg content on the corrosion and mechanical properties of 5000 series aluminum alloys. *Mater. Chem. Phys.* **2021**, *259*, 124202. [CrossRef]
37. Batool, S.A.; Ahmad, A.; Wadood, A.; Mateen, A.; Hussain, S.W. Development of Lightweight Aluminum-Titanium Alloys for Aerospace Applications. *Key Eng. Mater.* **2018**, *778*, 22–27. [CrossRef]

38. Li, D.; Liao, W.; Dai, N.; Xie, Y.M. Comparison of Mechanical Properties and Energy Absorption of Sheet-Based and Strut-Based Gyroid Cellular Structures with Graded Densities. *Materials* **2019**, *12*, 2183. [CrossRef]
39. Babcsán, N.; Mészáros, I.; Hegman, N. Thermal and electrical conductivity measurements on aluminum foams. *Materwiss. Werksttech.* **2003**, *34*, 391–394. [CrossRef]

materials

MDPI

Article

Multi-Steps Magnetic Flux Entrance/Exit at Thermomagnetic Avalanches in the Plates of Hard Superconductors

Viktor Chabanenko [1,*], Adam Nabiałek [2] and Roman Puźniak [2]

[1] O. Galkin Donetsk Institute for Physics and Engineering, National Academy of Science, Pr. Nauki 46, 03028 Kyiv, Ukraine

[2] Institute of Physics, Polish Academy of Sciences, Aleja Lotników 32/46, PL-02668 Warsaw, Poland; nabia@ifpan.edu.pl (A.N.); puzni@ifpan.edu.pl (R.P.)

* Correspondence: vikchabanenko@gmail.com; Tel.: +380-67-457-9862

Abstract: Avalanche cascades of magnetic flux have been detected at thermomagnetic instability of the critical state in the plates of Nb-Ti alloy. It was found that, the magnetic flux Φ enters conventional superconductor in screening regime and leaves in trapping regime in the form of a multistage "stairways", with the structure dependent on the magnetic field strength and magnetic history, with approximately equal successive portions $\Delta\Phi$ in temporal $\Phi(t)$ dependence, and with the width depending almost linearly on the plate thickness. The steady generation of cascades was observed for the full remagnetization cycle in the field of 2–4 T. The structure of inductive signal becomes complex already in the field of 0–2 T and it was shown, on the base of Fourier analysis, that, the avalanche flux dynamic produces, in this field range, multiple harmonics of the electric field. The physical reason of complex spectrum of the low-field avalanche dynamics can be associated with rough structure of moving flux front and with inhomogeneous relief of induction. It was established that the initiation of cascades occurs mainly in the central part of the lateral surface. The mechanism of cascades generation seems to be connected to the "resonator's properties" of the plates.

Keywords: hard superconductors type II; thermomagnetic instability; avalanche flux dynamics; screening and trapping regimes; cascades of flux

Citation: Chabanenko, V.; Nabiałek, A.; Puźniak, R. Multi-Steps Magnetic Flux Entrance/Exit at Thermomagnetic Avalanches in the Plates of Hard Superconductors. *Materials* **2022**, *15*, 2037. https://doi.org/10.3390/ma15062037

Academic Editors: Sophie Tingry, Marc Cretin and Zhenghua Tang

Received: 12 February 2022
Accepted: 8 March 2022
Published: 10 March 2022

1. Introduction

Magnetic properties of hard type-II superconductors are usually described in the frames of Bean critical state model [1,2]. In this model, the superconductor volume is screened by a current of the critical density, j_c. In increasing external magnetic field, magnetic flux enters into the superconductor's volume as a series of tiny flux jumps of different scales [3]. These jumps allow to relax the critical state and were studied experimentally [4,5].

In the terms of thermodynamics, the critical state is metastable. At certain conditions, small fluctuation of an external magnetic field, temperature, or tiny magnetization fluctuations in superconductor may lead to appearance of a catastrophic thermo-magnetic avalanche [6,7]. Magnetic flux jumps and accompanying heat release are the phenomena commonly observed both in conventional and in high temperature superconductors [8]. During the thermo-magnetic avalanche, the magnetic flux enters/exits into/from superconductors abruptly and magnetic moment decreases sharply. The temperature range, in which the critical state is unstable, is determined by the material parameters and can reach the temperatures as high as ten Kelvin. Low temperatures, where the critical current density increases significantly, are especially interesting for practical application. However, in this case, the instability of the critical state increases strongly too, leading to a giant flux avalanches.

An experimental study of avalanche flux dynamics, in particular, visualization of magnetic field penetration patterns, began more than fifty years ago [9–13]. Direct observation with high speed cinematography of the flux distribution and its change in time over the entire surface of a superconducting sample made it possible to study the kinetic of flux jumps. Particular attention was paid to the study of the flux front velocity, both in pure metals and in alloys [10–14]. It was found that the flux front velocity for conventional superconductors is of an order of tens of meters per second [11,14]. At the same time, the structure of magnetic flux avalanche spot in disks was found to be very diverse: from circular spots [12] to flux fingers and "dendrite-like" structures ("irregular" jumps [13]) in inhomogeneous bulk superconductors. Here, "irregular" jumps are the jumps leading to a complex, irregular distribution of the magnetic flux in the superconductor. The analysis of the visualization patterns of the flux dynamics, presented in the above papers, allowed, within the framework of Bean's concept, to understand the mechanism of local inversion of the surface self-field in the volume of conventional NbTi superconductor [15,16] and the complete inversion of the magnetic moment in the $YBa_2Cu_3O_{7-\delta}$ single crystal [17].

Avalanche flux dynamics in thin films, with penetration field depth greater than their thickness, attracts special attention because of several phenomena, recently discovered by means of magneto-optic, including dendritic flux patterns [18–24]. Such avalanches consist of sharp impulses of the magnetic flux rushing into the specimen. The "branches" of dendritic structures repel each other as they grow. Numerical simulations of the vortex avalanches in superconducting films confirmed a phenomenon of magnetic flux fragmentation [25,26], occurring when the hot spot of the thermomagnetic avalanche reflects from inhomogeneities or the boundary, on which magnetization currents either vanish or change direction. As a result, thermomagnetic avalanche produce complex flux patterns similar to dendrite.

Variation of Nb film geometry extended diversity of magnetic flux structures that penetrate superconductor during avalanches. One of such structures is referred to as "huge compact avalanches" [27] and it is different from dendrites. A recent study found an unambiguous relationship between the size and shape distributions of avalanches and thermomagnetic conditions of instability development [28]. It was established that the transition from a curved "finger-like" structure of avalanche at low fields to a dendritic one at higher fields allows dividing the regimes of dynamical process, where either thermal or magnetic diffusivity prevail.

In addition to the variety of avalanche structures, the study of the dynamic properties of the flux led to an unexpected result: the dendrite front propagation reaches the velocity up to 160 km/s [18,22]. An impressive experimental fact, discovered recently, is the rate of nanoportions of the flux, the Abrikosov vortex, penetrating into the superconducting Pb film [29,30] or into the Nb-C superconductor [31], being up to tens of kilometers per second and exceeding the theoretical estimates of pair breaking speed limit for superconducting condensate. Moreover, as it was shown, calculating the electromagnetic response of a nonequilibrium superconducting state, much lower velocities of the superconducting condensate lead to the appearance of multiple harmonics of the electric field [32], which means that the condensate dynamics is a highly nonlinear process. These results indicate that the study of the dynamic properties of magnetic flux helps to uncover unknown properties of the superconducting state. However, limited attention only was paid to the effective mass of Abrikosov's vortex [33–35], which can manifest itself, for example, in the experiments with circular dichroism of far-infrared light [36].

The purpose of current research was to establish the regularities of dynamic processes, occurring in a superconductor, when the path length of the finger-shaped avalanche is significantly limited by the size of the superconductor, i.e., it was supposed to implement the conditions of the "resonator" for a directed electromagnetic shot in the plates of the most popular hard superconductor. We present, here, an evidence of a multi-step "stairway" structure of magnetic flux dynamics at the thermomagnetic avalanches in the volume of the plates of conventional NbTi superconductor, being the number one material in the world in the terms of wide application. The width of the step, depending on the thickness of the plate, was determined in inductive measurements. The phenomenon of multi-steps magnetic flux dynamics was observed, both at the flux avalanche entry (screening mode) and at its exit (trapping mode). Spectral analysis of the structure of inductive signal allowed us to characterize the avalanche flux dynamics in the entire region of instability of the critical state, and to establish the region of magnetic fields where steady state generation of cascades is realized. The experimental observation of multi-step "stairway" structure of magnetic flux dynamic at avalanches is reported here for the first time.

The experimental results presented here demonstrate the new dynamical properties, appearing during thermomagnetic avalanches, in the plates of type II superconductors. It was found that, the magnetic flux enters conventional superconductor in screening regime and leaves in trapping regime in the form of a multistage "stairways" (cascades), with the structure dependent on the magnetic field strength and magnetic history, with approximately equal successive portions in temporal dependence, and with the width depending almost linearly on the plate thickness. The mechanism of cascades generation seems to be connected to the "resonator's properties" of the plates.

2. Experiments and Materials

Time resolved measurements were performed with the aid of different type of inductive sensors, where a voltage (U_{coil} (t)~$d\Phi/dt$, t—time) caused by the avalanche flux jump was induced. Figure 1a shows flux jump at T = 1.7 K in Nb disc, frozen in the form of "fingers". Figure 1b–d shows the arrangement of pickup coils on superconducting plate for the avalanche registration: in the whole plate's volume (b), on the three height levels (c), and in the magnetic stray field (L_{ext}) (d). Time-dependent voltage U_{coil} (t) was registered with NI DAQ 6115S (National Instruments, Austin, TX, USA) (data acquisition board) simultaneously in four channels with time resolution of ~10^{-7} s. External magnetic field (H_{ext}) applied in direction parallel to the plate was swept with the rate of 0.6 T/min and local magnetic induction (B_{surf}) at the center of the sample surface was measured with Hall sensor. The magnetization M of superconducting plate was determined by the difference between local and external magnetic field induction, $\mu_0 M = B_{surf} - \mu_0 H_{ext}$. The specimen was immersed in liquid helium bath and the temperature down to 2 K was reached in pumped helium.

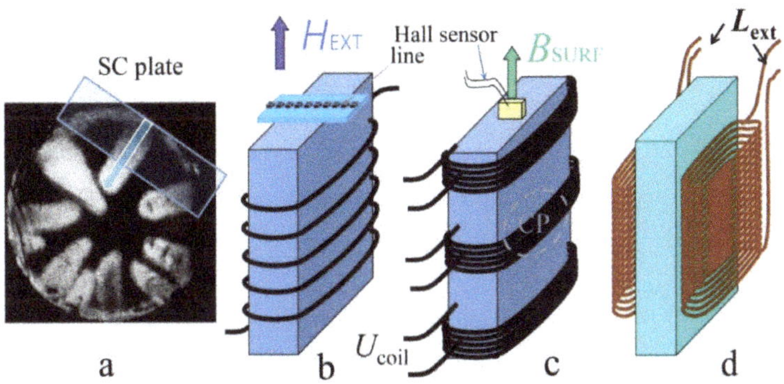

Figure 1. (**a**) Flux jump frozen in the form of "fingers"; Nb disc, T = 1.7 K (Figure 1a from Ref. [13]). The arrangement of pickup coils on superconducting plate for the avalanche flux registration: (**b**) in the whole plate's volume, (**c**) on three height levels, and (**d**) in magnetic stray field (L_{ext}). Hall sensor line and probe for magnetic induction (B_{surf}) measurement are presented in (**b**,**c**); U_{coil} (t)—voltage on the pickup coil; CP is the central part, where the avalanches are mainly triggered.

Initial plate with the size of 20 × 14 × 7 mm^3 has been cut from extruded cylindrical rod of NbTi 50 at% alloy with diameter of 15 mm. Hot extrusion of Nb-Ti 50% alloy was carried out according to the standard technology [37] through a deformation of matrix at a temperature of 750 °C along the route from Ø 50 mm → Ø 15 mm (Ø—diameter of the rod) with a draw value $R = S_{\text{before}}/S_{\text{after}} \approx 11$, where S_{before} and S_{after} are the areas of the sample sections before and after deformation, respectively. The side surfaces of the plate were ground with diamond (corundum) powder in order to remove the layers of the material strongly deformed during cutting. Such procedure was repeated at each stepwise thinning of the plate to the required thickness d. The pickup coils were close-wound with 0.1–0.2 mm diameter of copper wire and consisted of 15–100 turns.

3. Results and Discussion

The salient features of the manifestation of thermomagnetic avalanches in the plates at 2 K are their realization in the form of a multistage "stairway" manner of magnetic flux dynamics (Figure 2a,b). One example of the voltage $U_{\text{coil}}(t)$ signal inherent to a cascade induced on a pick up coil at the avalanche in trapping regime is presented in Figure 2a (2nd quadrant of hysteresis loop on Figure 2e, Jump 34). The signal consists of a series of regularly spaced impulses (portioned avalanches) and shows a multi-step "stairway" structure of magnetic flux dynamic [$\Phi(t)$], received as a result of integration of voltage $U_{\text{coil}}(t)$ curve (Figure 2b below Figure 2a). Thus, the magnetic flux enters (in screening regime) or leaves superconductor (in trapping regime) in "steps-like" manner (Figure 2b). Such well-structured cascades are observed at the temperature of 2 K for plates of different thicknesses with d = 2.7–6 mm in a certain region of the magnetic field, marked with rectangles and arrows on the hysteresis loops $M(H_{\text{ext}})$ (Figure 2d–f, right column). Remagnetization loop at 4.2 K for the plate with thickness of 6 mm—for comparison with $M(H_{\text{ext}})$ curve at 2 K—is shown in Figure 2c.

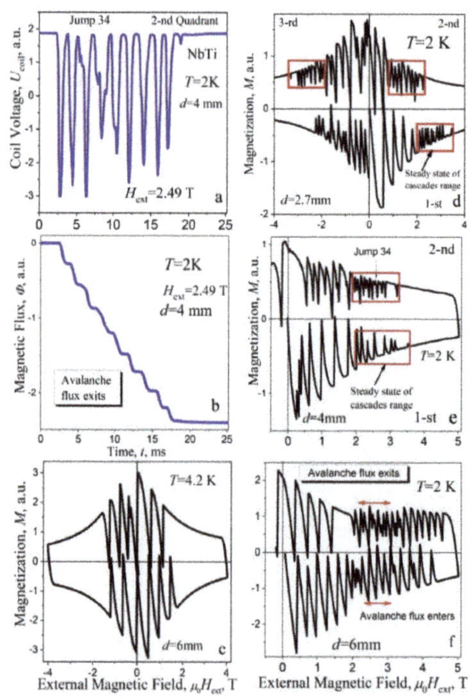

Figure 2. (**a**) The voltage $U_{\text{coil}}(t)$ at the flux cascade in a trapping regime [Jump 34 on hysteresis loop (**e**)]; (**b**) time-integrated voltage $U_{\text{coil}}(t)$—a multi-step "stairway" structure of magnetic flux $\Phi(t)$); (**d**–**f**) remagnetization curves $M(H_{\text{ext}})$ for the plates with the different thickness: (**d**) d = 2.7 mm, (**e**) 4 mm, (**f**) 6 mm at the temperature of 2 K; the ranges of cascades are shown by the rectangles and arrows on hysteresis loops. (**c**) Hysteresis loop at 4.2 K, d = 6 mm.

Figure 3 presents a panorama of the voltage impulses structures $U_{coil}(t)$ at 2 K, induced on a pick up coil at the flux cascades, originating for the plates with various thickness, i.e., with thickness of 2.7 mm (a), 3.1 mm (b), 4 mm (c), and 6 mm (d). Left side of each figure contains cascades in screening mode (1st quadrant) and the right one in trapped mode (2nd quadrant). The most regular step-like structures were found for the plates with diameter of 2.7 mm, 3.1 mm, and 4 mm. Nevertheless, the step-like structure is also visible in a narrow range of magnetic fields for the plate with a diameter of 6 mm (Figure 3d). The maximal number of steps in the cascades was found for magnetic field in the range of 2–4 T.

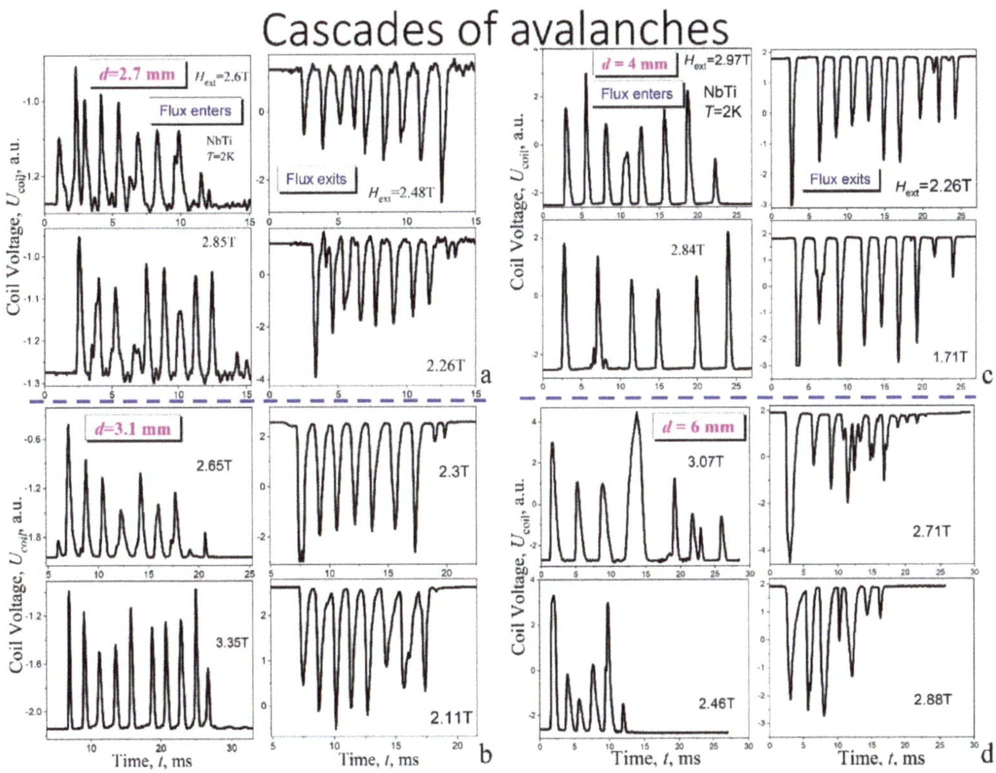

Figure 3. A panorama of the voltage impulses $U_{coil}(t)$ at the flux cascades for the plates with various thickness: 2.7 mm (**a**), 3.1 mm (**b**), 4 mm (**c**), and 6 mm (**d**). Each panel presents cascades recorded in four various fields and in the left column there are cascades in screening mode (1st quadrant of hysteresis loops) while in the right one—in trapped mode (2nd quadrant).

3.1. Evolution of the Structure of Cascades at the Change of Magnetic Field

Screening regime. Time evolution of the voltage impulses $U_{coil}(t)$ at T = 2 K at the avalanches with increasing magnetic field in screening regime, measured after cooling in zero magnetic field (ZFC—zero field cooling), for the 1st quadrant in the plate with d = 2.7 mm, is shown in Figure 4a. Amplitude of the signals presented was normalized to the maximal peak value $U^{max}{}_{coil}$ for each cascade. The absolute value of voltage impulses decreases exponentially with increasing external magnetic field, as it is presented in Figure 4b. The structure of the avalanche is relatively simple for the first and for the last jump: only one single peak of voltage was found, as it is seen in Figure 4a.

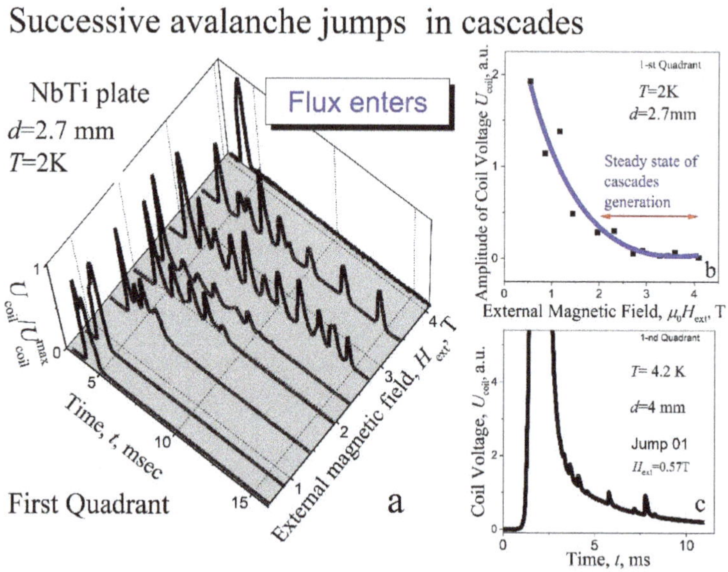

Figure 4. (**a**) Evolution of the normalized voltage impulses structure $U_{coil}(t)/U^{max}{}_{coil}$ at the avalanche during the increase of magnetic field; (**b**) the amplitude of voltage impulse vs magnetic field and exponential fitting to the experimental data: d = 2.7 mm, T = 2 K; (**c**) the voltage impulse at first flux jump: T = 4.2 K, $\mu_0 H_{ext}$ = 0.57 T, d = 4 mm. Screening regime; ZFC.

However, for the first avalanche already, the process of excitation or triggering of cascades of avalanches is observed (Figure 5a), despite that, the "oscillations" $U(t)$ associated with this process are still weakly expressed. With increasing magnetic field, the voltage impulse expands into a shape of a crown with numerous peaks at the fourth avalanche (Figure 5b), and the step structure of the flux becomes apparent. In stronger fields, the "crown" is transformed into a structure of clearly separated, isolated avalanche impulses (Figures 3a and 4a). It means that, the whole avalanche, occurring in the period of several tens of milliseconds, exhibits a multi-step structure, i.e., the magnetic flux enters (leaves) in approximately equal, successive portions $\Delta\Phi$, the number of which can reach the value of ten (Figures 2b and 3). The range of magnetic field between 2 and 4 T may be defined as a range of steady state of cascades generation. It should be noted that duration of entering of each individual portion of the flow $\Delta\Phi$ in the cascade is the same as in the single avalanches that are observed at the beginning and at the end of the region of instability of the critical state (Figure 4a). The nucleation of cascades is shifted into stronger magnetic fields with thickening the plate to d = 3.1 mm (Figure 5c).

It should be noted that with an increase in temperature to 4.2 K, the signal from the excitations of the cascades is practically not observed on the avalanche pulse (Figure 4c, d = 4 mm). At the same time, one can see, on the trailing part of the signal, a certain sequence of small peaks of the flux entry.

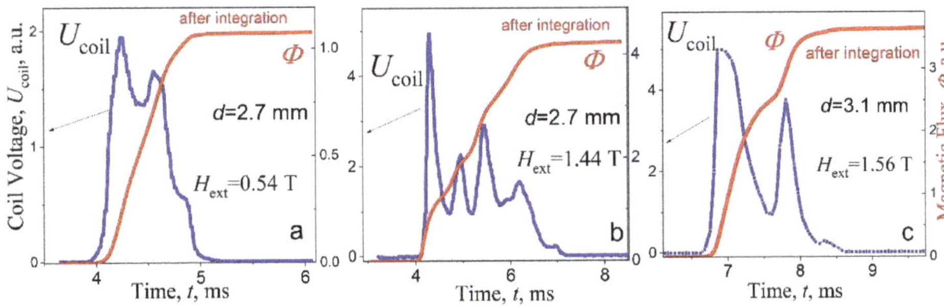

Figure 5. (**a**–**c**) The voltage impulses $U_{coil}(t)$ (left ordinate) and magnetic flux $\Phi(t)$ steps (right ordinate) at the avalanches: (**a**) first jump ($\mu_0 H_{ext}$ = 0.54 T), (**b**) fourth jump ($\mu_0 H_{ext}$ = 1.44 T) for the plate with thickness d = 2.7 mm; (**c**) fourth jump ($\mu_0 H_{ext}$ = 1.56 T), d = 3.1 mm; ZFC, screening regime (1st quadrant), T = 2 K.

3.2. The Flux Step-Like Structure in Avalanche Cascades

Periodicity analysis. To study the patterns of stepped structure of the flux $\Phi(t)$ (Figure 2b) under the influence of external factors, its derivative $U_{coil}(t)$ should be analyzed, which allows to consider changes in the repetition period T of individual avalanche pulses (flux portions) in a cascade, or to analyze their characteristic frequencies F (pulses) of their repetition. For simple harmonic signals, this is, in principle the same, since the relationship between these quantities is trivial: $T = 1/F$. In the case of the studies performed here, it is not always easy to visually determine the period from the signal structure. This is especially true for the region of relatively weak magnetic fields −1.5 T–0–+1.5 T in the second and third quadrants.

Let us first consider the periods T_c between peaks of voltage $U_{coil}(t)$ in cascades for different forms of studied signal. In Figure 6a,c and Figure 7a,c, some characteristic and their detailed analysis, for 2.7 mm and 3.1 mm samples, respectively, are shown, with the voltage impulse $U_{coil}(t)$ (left ordinate) inherent to a cascade at the avalanche, and a multi-step "stairway" structure of magnetic flux $\Phi(t)$ (right ordinate) in a screening (a) and a trapping (c) regime. Two types of signals have been selected here: in the first case (a) avalanche impulses overlap slightly (the end of the "crown", Figure 4a), in the second—Figure 6c—avalanche impulse are spaced (separated) in time. The latter case is typical for avalanches in strong magnetic field, near the boundaries of the region of instability of the critical state.

The periods T_c between jumps in cascade vs number of impulses n (or steps) are shown in the main frame of Figure 6b,d, left ordinate. As follows from presented data, the screening and trapping regimes T_c increase over time in both cases. The experimental points are scattered around the straight lines, constructed with least squares method. The values of the frequency pulse $F_T = 1/T_{av}$ in cascades (T_{av} is the average period) are given below the lines. Importantly, a similar period is observed in three quadrants for the plates of different thicknesses, in the range of steady state of cascades generation.

The increase in the T_c period with time can be associated with an increase in the dissipation of the induction-current system in the critical state as the flux enters the superconductor. Such increase is characteristic for oscillatory with an increase in the attenuation. Structured cascades with a uniquely visually identified frequency rarely appear during magnetization reversal in the range of 0–±1.5 T, in the second and third quadrants. Here, avalanches exhibit, mainly, a complex structure. An example of the signals for a plate with a thickness of d = 3.1 mm, where it is difficult to visually assess the main period of T_c and the corresponding frequency, is presented in Figure 7c. Here, spectral analysis helps to establish the frequency components of the spectrum and, if necessary, to determine the appropriate periods.

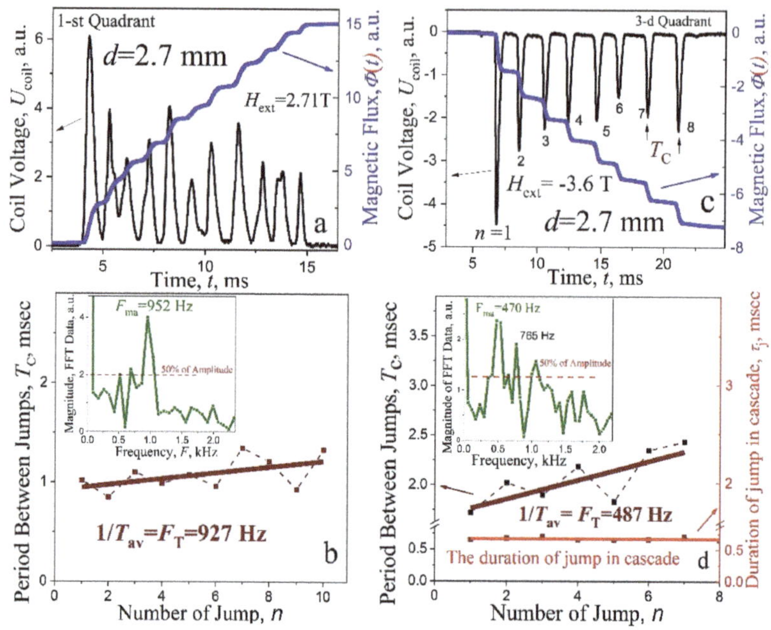

Figure 6. (**a**,**c**) The voltage impulse $U_{\text{coil}}(t)$ at the avalanche in screening (**a**) and in trapping (**c**) regime (left ordinate) and multi-step "stairway" structure of magnetic flux $\Phi(t)$ (right ordinate); (**b**,**d**) main frame—periods T_c between jumps in cascade vs number of jumps n. The duration of separate flux impulses in the cascade is given by bottom line in (**d**). Data for the magnitude in fast Fourier transformation of signal $U_{\text{coil}}(t)$ are presented in the inserts (**b**,**d**); d = 2.7 mm, T = 2 K.

Spectral analysis. With an increase in the magnetic field, the structure of avalanche pulses passes smoothly from a single pulse with small oscillations (Figure 5a) to almost periodic oscillations limited in time (Figure 3). Proper analysis of the data requires information on the amplitude and frequency of the signal spectrum components, i.e., on the amplitude-frequency spectrum and thus, fast Fourier transformation (FFT) from the Origin program (OriginLab Corporation, Northampton, MA, USA) was utilized in the analysis of the $U_{\text{coil}}(t)$ spectra in the entire region of instability of the critical state. The characteristic F_{FFT} frequencies in the signal spectrum, determined with this package and compared with the values of the pulse repetition frequency F_{T}, obtained directly from the average period T_{av}, exhibited its good efficiency (see, the inset in Figure 6b,d and Figure 7b,d—the amplitude spectra of avalanche signals with characteristic F_{FFT} frequencies). The values of the pulse frequency gained from the average period F_{T} for plates with a thickness of d = 2.7 mm and 3.1 mm, are presented in Figures 6 and 7, respectively. The spectrum component with the maximum amplitude presented in Figure 6b corresponds to the frequency F_{ma} = 952 Hz, which is in good agreement with the frequency value estimated from the average period: F_{T} = 927 Hz. The spectrum shown in the inset in Figure 6d is more complex. Here, there is a fundamental frequency F_{ma} = 470 Hz with a maximum amplitude, which practically coincides with the frequency F_{T} = 487 Hz. In addition, the spectrum contains two more components, the amplitude of which exceeds the amplitude level of 50% of the fundamental harmonic. A good agreement of spectral analysis is presented in Figure 7: F_{ma} = 724 Hz and F_{T} = 730 Hz (b) and F_{ma} = 551 Hz and F_{T} = 564 Hz (d).

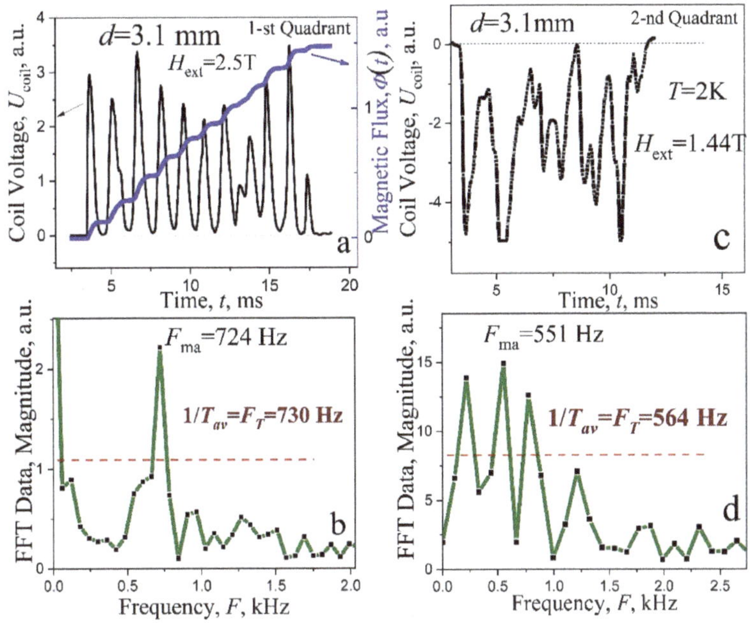

Figure 7. (**a**,**c**) The voltage impulse $U_{coil}(t)$ at the avalanche in trapping (**a**) and screening (**c**) regime (left ordinate) and multi-step "stairway" structure of magnetic flux $\Phi(t)$ [right ordinate in (**a**)]; (**b**,**d**) magnitude data from fast Fourier transformation of signal $U_{coil}(t)$; d = 3.1 mm, T = 2 K.

It should be noted that the signal $U_{coil}(t)$ presented in Figure 7c exhibits a spectrum with three component frequencies (Figure 7d), with similar amplitude values. This is definitely due to the strong beats in the signal. Apparently, the presence of such frequency components, with close amplitude values, leads to "jumps" in the experimental dependences of the T_c periods vs number n (Figure 6b,d) around the approximating straight lines. The data indicate that the harmonic F_{ma}, with the maximum amplitude in the signal spectrum, is suitable for analyzing the effect of external influences and parameters of a superconducting plate on the phenomenon of multi-steps magnetic flux dynamics in the entire region of instability of the critical state.

The results of studying signals from avalanches for three quadrants in a plate with a thickness of 3.1 mm are shown in Figure 8 and the features of the hysteresis loop $M(H_{ext})$ associated with the excitation of cascades are shown in Figure 8a. A stepwise decrease in the amplitudes of magnetization jumps ΔM is clearly expressed in the region of cascades for three quadrants of hysteresis loop and for all thicknesses (Figure 2d,e), which is a result of the excitation of cascades. This is evidenced by magnetic field dependence of the avalanche duration $\Delta t(H)$, shown in Figure 8b (main frame) and by their structures in the region of the amplitude step in field dependence of magnetization. The avalanches in this place [inserts into (b), jump 5 and 6] undergo threshold changes. During the sixth avalanche, the origin of multistage nature of the flux dynamics sharply increases, while the duration of the avalanche suddenly increases by three times, and the incoming flux is divided into six portions spaced in time (jump no. 6). The explanation of the step in magnetization is quite simple. Step-like behavior is associated with a change in the localization of the avalanche spot of the incoming flux and with Hall sensor method of induction measurements, where a local field change at the center of the surface is registered. During avalanche no. 5, the magnetic spot was located mainly in the central part, under the sensor. In the case of cascades (avalanche 6), the magnetic flux was dispersed like a comb of six teeth along the entire side of the plate. In this case, the level of induction under the sensor becomes significantly smaller.

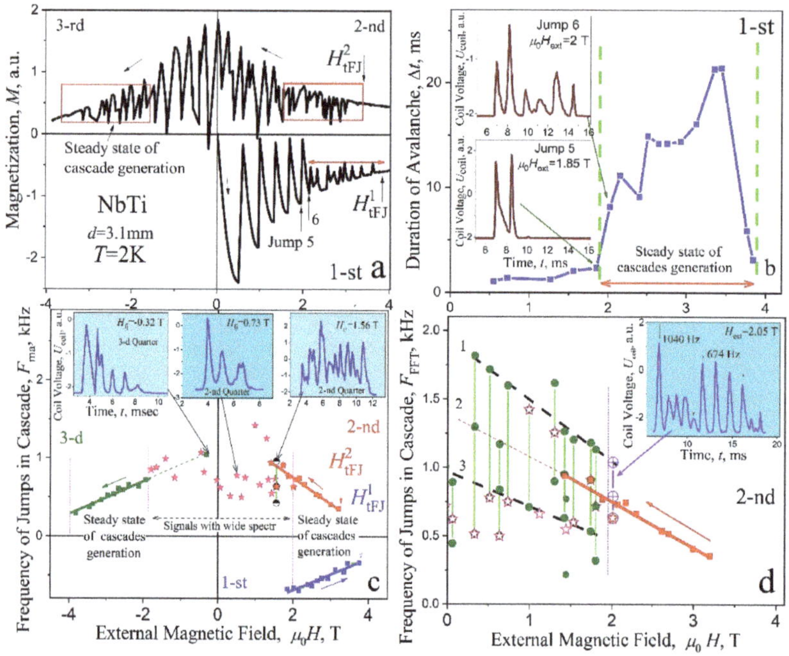

Figure 8. (**a**) Remagnetization curve $M(H_{ext})$ of superconducting plate; the ranges of steady state cascades generation are shown by the rectangles and arrows. (**b**) Main frame—the duration of avalanches $\Delta t(H_{ext})$ in screening mode; insert—the structure of jump 5 and jump 6. (**c**) Main frame—frequencies F_{ma} with the maximal amplitude in specters of signals vs. magnetic field for different quadrants; inserts—the voltage impulses induced at avalanches in fields $\mu_0 H_{ext}$ = 1.56 T and 0.73 T (2nd quadrant) and −0.32 T (3rd quadrant). (**d**) Main frame: the frequencies F_{ma} marked by asterisk (☆) and frequencies of components with the amplitude closest in the value vs magnetic field marked by full circles (•); insert—the structure of signal in field $\mu_0 H_{ext}$ = 2.01 T with two characteristic pronounced frequencies, 1040 Hz and 674 Hz, marked by crossed open circles (⊕), 2nd quadrant. d = 3.1 mm, T = 2 K.

Here, it is appropriate to emphasize practical implication of the importance of phenomenon of cascades at avalanche in the volume of superconductors: the incoming avalanche flux is dispersed in time and becomes non-localized in high field region. One of the main problem challenges in superconducting applications is to avoid thermomagnetic breakdown of critical state (quench) and as a result, the sudden disappearance of beneficial properties. To avoid creation of conditions for non-localized place of quench is one of the important practical issues solved by technologists today [38]. The discovery of a multi-steps magnetic flux entrance/exit at thermomagnetic avalanches in the plates of hard superconductors could make a difference in the case of applications of electric motors or generators built with superconducting elements in the form of plates. In the plates with certain thicknesses, this phenomenon reduces the risk of failure at maximum loads.

The evolution of the avalanche structure in the first quadrant (shielding mode) was previously analyzed in detail, using the example of a plate with d = 2.7 mm (Figures 4a and 5). Now, we will consider the properties of cascades in the second and third quadrants. The instability of the critical state is observed here in a field range from +4T to −4T (Figure 8a,c). The frequency values F_{ma} for the components with the maximum amplitude in the signal spectra are shown in Figure 8c (main frame). The F_{ma} frequencies, in the range of high magnetic field ±2–±4 T (steady state cascades generation), are marked by full squares (■). Here, the frequency decreases linearly with an increase of magnetic field for all quadrants. Such behavior of the F_{ma} may be interpreted as a result of increasing of dynamical dissipation of critical state and by decreasing of the avalanche start velocity as a result of the critical current density $j_c(H_{ext})$ decreasing.

The stable cascades with a small number of flux steps $\Delta\Phi$ appear in the field region −2 T–+2 T, as it is shown, for example, in the left and in the center inserts at the top of Figure 8c. The data in the main panel, described by asterisks (☆), indicate the frequencies F_{ma}. The signal $U_{coil}(t)$ exhibits here a complex structure with a wide spectrum (Figure 7c,d) and it seems that, in such a case, determination of the fundamental harmonic F_{ma} (asterisks in Figure 8b) is very difficult, if it is possible at all. Therefore, an attempt was made to present, for each signal $U_{coil}(t)$, the frequencies of all spectrum components with amplitudes higher than 50% of the maximal value. The result for the second quadrant is shown in Figure 8d. Here, the frequencies F_{ma}, described by asterisks (☆), as in Figure 8c, are shown, together with frequencies of the harmonics with slightly lower amplitudes, described by the full circles (•). In the field region $H < 2$ T, one can visually distinguish between certain lines, designated with the numbers 1, 2, and 3, on which most of the experimental points lie on the array. Field dependence for lines 1 and 3 is similar to that of line 2, which is a continuation of the straight line $F_c(H_{ext})$ from the field range of steady state generation of cascades. In terminology taken from other branches of physics, the lines 1, 2, and 3 can be described as follows: Frequency "level" 2 (line 2) for $F_{ma}(H_{ext})$ corresponds to the dynamic state of the flux with steady state generation of cascades and it splits in a magnetic fields below 2 T into two "levels", corresponding to the states of lines 1 and 3. Such behavior can be expressed as a certain degeneration of the "frequency levels" on lines 1 and 3, which occurs in a magnetic field $\mu_0 H_{ext} > 2$ T.

We managed to observe, for one of the cascades in the transition region (inset in Figure 8d), a signal, occurring at the boundary in $H_{ext} = 2.01$ T, where two frequencies marked by crossed open circles (⊕) are clearly visually present: the upper $F = 1040$ Hz belongs to the straight line 1 and the lower frequency $F = 674$ Hz, which falls on line 2 for frequencies corresponding to stable cascades. Analysis of the avalanche flux dynamics in the second quadrant indicates the presence, in the region of weak fields ($\mu_0 H_{ext} < 2$ T), of some effective mechanism, leading to the appearance of electromagnetic radiation of a complex spectral composition. In a thinner plate with $d = 2.7$ mm, the field dependences of the frequency F_{ma}, which characterize the avalanche flux dynamics look similar—Figure 9a. In the range of steady state generation of cascades, the frequency dependences $F_{ma}(H_{ext})$ for the first and third quadrants practically coincide. The data for the second quadrant also exhibit a significant "scatter-rocking" of the experimental values of $F_{ma}(H_{ext})$ and are lowered, in the same field range, in comparison with those in the first quadrant. Two measurements are presented for the second quadrant [marked by two different kind of stars (☆)] in order to show the repeatability and rocking range of experimental data in complex spectra.

The amount of magnetic fluxes in avalanches significantly differs between the first and the second quadrant (Figure 9b): in the same magnetic field much more flux enters in the shielding mode than exits in the trapping mode. Moreover, in the latter case, experimental values for the exiting fluxes are extensively scattered, as do the values of the frequencies in the dependence $F_{ma}(H_{ext})$. It should be noted that the amount of magnetic fluxes decreases exponentially with increasing external magnetic field for the 1st quadrant (Figure 9b) as well as the absolute values of voltage impulse (Figure 4b). Figure 9c shows the period $T_{ma} = 1/F_{ma}$ between jumps in cascade as a function of magnetic field for various quadrants. It becomes clear here, the pulse frequency in the inductive signal depends linearly on magnetic field.

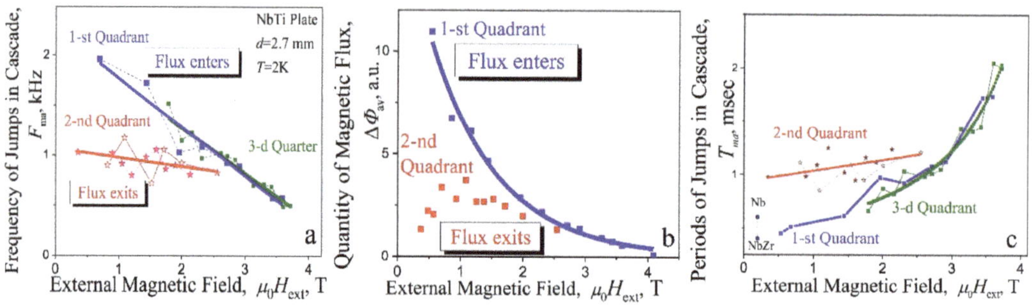

Figure 9. (**a**) The frequencies F_{ma} with the maximal amplitude in specters of signals vs magnetic field for different quadrants. (**b**) Quantity of the magnetic flux enters (1st quadrant) and exits (2nd quadrant) at avalanches vs magnetic field; (**c**) the period $T_{ma} = 1/F_{ma}$ between jumps in cascade vs magnetic field for different quadrants; d = 2.7 mm, T = 2 K. Symbol (•)—the time during which the avalanche flux front travels a distance of 3 mm in Nb and NbZr at 1.6 K and at magnetic field $\mu_0 H_{ext}$ = 0.2 T [14].

The role of plate thickness. Magnetic field dependence of the jump frequency F_{ma} for cascades in screening regime (1st quadrant) for the plates with various thickness d is shown in Figure 10. It can be seen from these data that the region of the magnetic field, where cascades are stably realized, rises up along the frequency axis when the plate becomes thinner. At the same time, the range of magnetic fields and frequencies of cascades is significantly expanded. Insert in Figure 10 shows the dependence of period $1/F_{ma}$ between jumps in cascades as a function of plate thickness along vertical lines at $\mu_0 H_{ext}$ = 2.5 and 3 T in the main frame. The period between avalanches inside cascades increases almost linearly with increasing plate thickness. It may indicate on the functional role of the plate thickness in triggering of cascades.

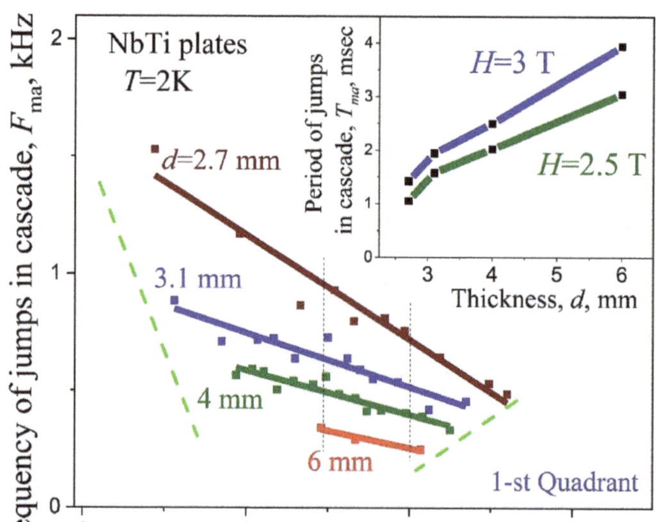

Figure 10. Frequency F_{ma} vs. H_{ext} for plates with different thickness d; screening regime (1st quadrant). Insert—the value of period $T_{ma} = 1/F_{ma}$ along vertical lines at the magnetic field $\mu_0 H_{ext}$ = 2.5 and 3 T on main frame vs. thickness d of plate.

3.3. Triggering, Propagation and Localization of the Avalanche Flux Cascades

The place of avalanche origin. An avalanche appears, usually, in a small volume, almost at a point compared to the size of the plate, and a bundle of vortices trapped at pinning centers can stimulate an avalanche. The process of magnetic field penetration constantly accompanies jumping of the bundles [3,4]. Simultaneous registration of the signal at different levels along the height of the plate (Figure 1c) opened possibility to find the level where the signal appears first and to estimate the propagation velocity of the disturbance along the magnetic induction line. The analysis of the time delay Δt_{delay}, between signals (more than 60 signals) from the avalanche beginning, allowed us to fix the most probable place of their origin. Statistics showed that the avalanches originate in 68% near the central region (CP, Figure 1c) of the lateral surface of the plate, and in 32% only—at its edges. This result agrees with calculation of field penetration into the plate [39]. In the central part, the screening is maximal, the magnetic field pressure is the highest, and, accordingly, the initiation of instability is most likely here. An attempt, to find side surface, where an avalanche originates (and to feel from which side a separate jump of the flux enters), using two external inductive sensors on different sides of the plate (L_{ext}, Figure 1d), while recording signals, was unsuccessful. It means that, the delay (or accelerates) of the avalanche onset on one of the lateral surfaces and the asymmetry of scattering field dynamics on opposite sides of the plate, during the stepwise entry of the avalanches, were not detectable. A beautiful shaped cascade when the signals from two sensors were added and its almost complete compensation to the noise level, when they were subtracted, was observed. It should be noted, here, that, in wide coils, covering a significant part of the side surface of the plate, the stages looked to be more structured. This may be due to the fact that the origin of the cascades travels along the vertical or along the lateral surface and their appearance is somewhat different in different sections. Most probable speed of propagation of the disturbance along the direction of the magnetic field, $V_{||}$, was determined on the basis of statistical data analysis and it was found that, it corresponds to the maximum velocity distribution function. The velocity reaches $V_{||max} = 0.5\, h/\Delta t_{delay} = 286 \pm 20$ m/s, where h is the plate height.

Spatial-temporal cascades behavior. In order to establish the location of the parts of the incoming flux on the central section of the plate, the pickup coils for the avalanche flux registration were placed on superconducting plate (Figure 11a) in the whole plate's width (W) − (L), in the $1/3W$ ($L_{1/3}$) and $2/3W$ ($L_{2/3}$). In order to cover individual sections of the plate, a hole with a diameter of 0.7 mm was made for the placement of inductive sensors $L_{1/3}$ and $L_{2/3}$. Self-consistent cascade for the plate with d = 3.1 mm, recorded at T = 2 K in $\mu_0 H_{ext}$ = 3.35 T simultaneously by three coils L, $L_{1/3}$, and $L_{2/3}$, is shown in Figure 11b. Full signal registered by coil L is divided proportionally between the coils $L_{1/3}$ and $L_{2/3}$, confirming spatial separation of impulses in the cascades along the lateral surface of the central section. Schematic view of possible spatial localization of successive flux avalanches (steps) in the form of fingers is shown on the cross-section of a plate orthogonally to magnetic field (Figure 11c). The numbers given in Figure 11b,c indicate the sequence of avalanches occurrence. The individual flux pulses are shown here in the form of "fingers" since it is known [13] that the shape of avalanche spots transforms at low temperatures from round at T = 4.2 K to "finger-like" at 1.8 K. "Fingers" of avalanches in the scheme (Figure 11c) cross the plate all the way to the opposite side, rather than stopping at its middle, which is in agreement with the results of experimental studies of the distribution of surface induction—signal registered by an array of the Hall probes (Figure 1b) before and after appearance of thermomagnetic avalanches in increasing and decreasing external magnetic field [15]. Here, local induction inversion as a result of a thermomagnetic avalanche over the entire thickness (d = 4 mm) of the studied NbTi plate occurred already at T = 4.2 K (Figure 3b,c from Ref. [15]), which may indicate that the avalanche front crosses the entire plate. However, an unambiguous answer can only be given by direct observation of the flux front using magneto-optics.

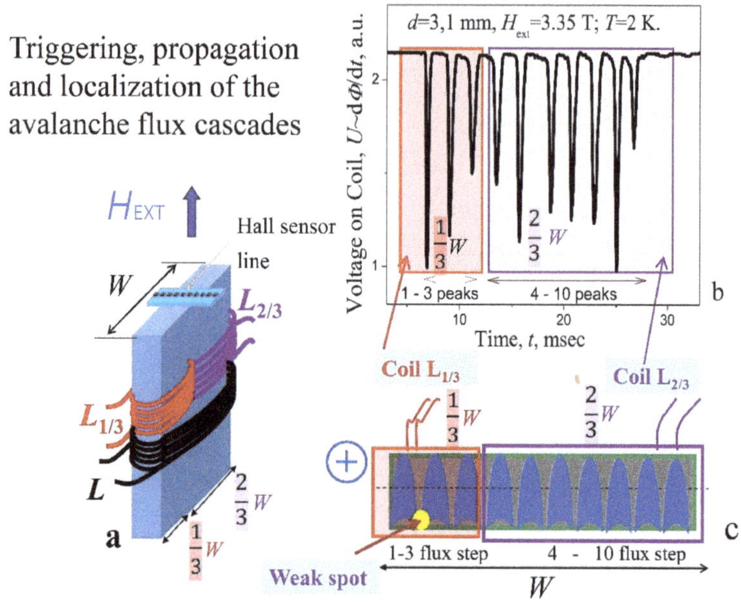

Figure 11. (**a**) The arrangement of pickup coils for the avalanche flux registration in the whole plate's width (W) – (L), in the $1/3W$ ($L_{1/3}$) and in $2/3W$ ($L_{2/3}$); (**b**) one of the cascades recorded simultaneously by three coils L, $L_{1/3}$, and $L_{2/3}$; d = 3.1 mm, H = 3.35 T, T = 2 K. Full signal registered by coil L is divided proportionally between the coils $L_{1/3}$ and $L_{2/3}$, confirming spatial separation of impulses in cascades. (**c**) A schematic cross-section of a plate oriented orthogonally to magnetic field and spatial localization of successive flux avalanches. Weak spots appear in the concave corner, near the finger's gate.

Simulation of avalanches movement in superconducting films [25,26] indicates that the avalanche front tends to displace the superconducting current from the surface to the opposite side of the sample. It means that the superconducting current is pushed out of the hot zone of the avalanche finger into the cold zone of the superconductor [40]. In shielding mode (diamagnetic induction), an appearance of "paramagnetic" profile is expected if the flux front manages to cross the middle of the sample, which takes place in the simulation. This is exactly what was observed in the experiment (Figure 3b from Ref. [15]).

3.4. Possible Mechanism of Cascades Triggering

The linear dependence of the cascades period on plate thickness suggests that the period of step structure of magnetic flux avalanche dynamics $\Phi(t)$ in cascades may be controlled by plate thickness as well as the limitation of path length of the front of an individual avalanche finger in the cascade, introduced by plate thickness, can lead to the appearance of a certain sequence of excitation in the induction-current system of the critical state of the superconductor.

Let us consider a possible scenario of cascades triggering. When the critical induction drop ΔB_{FJ} is reached, the initial flux step (finger) in cascades can be stimulated in the superconductor critical state layer by small jumps of the flux bundles. Penetrated first finger's flux redistributes current in the sample, and thus creates weak spots for penetration of the consequent avalanche finger, etc., [40]. Weak spots appear in the concave corner (Figure 26 from Ref. [41]), near the finger's gate (Figure 11c), where the current lines looming (become thicken) occur. Subsequent avalanches in the cascade can trigger a weak spot by a locally amplified pulse of a magnetic or electric field [42] and, accordingly, a current ΔJ due to deceleration of the previous finger of the flux reaches the opposite side of the plate. The plate in this way can show its "resonator" properties.

The expansion of the range of magnetic fields, where cascades stably arise, with decreasing plate thickness, as well as the disappearance of this phenomenon for plates with $d > 6$ mm (Figure 10), testifies in favor of the important role of finger's-type flux reflection from the opposite side (resonator properties of the plate) in the process of self-excitation.

Let us try to estimate the time it takes for the flux front to cross our plate in order to compare it with the period of multistage "stairway" structures. The dynamics of the avalanche front in conventional superconductors Nb and NbZr at a temperature of $T = 1.4$ K and in a magnetic field of 0.2 T was studied by the magneto-optical method [14] and time dependence of path traveled by the avalanche front is shown here. The magnetic flux front runs a distance of 3 mm in 0.4 ms for NbZr and in 0.6 ms for Nb, respectively, which is plotted in Figure 9c, presenting the data on the pulse periodicity in cascades for a 2.7-mm thick plate, corresponding well to the values of the periods of cascades. Hence, we can conclude that the plate, showing its "resonator" properties, can serve as a metronome that sets the period of magnetic flux dynamics in plates and lead to a multi-step "stairway" structure of flux $\Phi(t)$. Existence of avalanches in the form of a cascade of almost equidistant events can be unambiguously confirmed by simulating it only. However, it is quite plausible since penetrated flux redistributes current in the sample, and thus creates weak spots for penetration of the consequent avalanches, etc., [40].

3.5. Impact of Magnetic Prehistory

The change in the profile of magnetic field in the plate, with the change from increasing to decreasing field in the second quadrant, leads to a displacement of the boundary of critical state $H^2{}_{\text{tfj}}$ stability region into weaker fields (Figure 8b), as compared to the first quadrant ($H^1{}_{\text{tfj}}$). The physical reason for this is a decrease in the critical current density due to the trapped flux, since the trapped flux increases, in a given external magnetic field, the average effective field in superconductor, as compared to that in shielding regime [43]. The profile of magnetic field induction can affect the spectral characteristics of the avalanche dynamics of the front flux. The magnitude of induction gradient in superconductor sets the impulse to the flux in the avalanche at the "start". The subsequent movement of the avalanche front in the plate passes along the induction relief formed before this, with the exception of the first avalanche in the screening mode. A schematic representation of Bean's magnetic induction profile $B(x)$ in a plate corresponding to situation before first flux jumps in screening regime and for trapping regime in strong magnetic field (field ramping backward) is shown in Figure 12 [left side of (a) and (b), respectively]. The result of avalanche flux entry/exits process is presented on the right side of Figure 12a,b. Here, a cross-section of a plate perpendicular to magnetic field is shown.

Screening regime. Avalanches appear above the field of the first instability $H_{\text{ext}} > H_{1\text{fj}}$ (Figure 12a) and the regime of cascades generation appears at the first avalanche, as it was depicted in Figure 5a. The critical current is maximal here, the pressure is high, and the flux front speed is high. Thus, a large amount of magnetic flux enters the region of the Meissner state of the superconductor (Figure 12a, right side). The central strong avalanche appeared, apparently, in the form of a finger. However, when its propagation was limited by the second side of the plate, the width of the finger under the pressure of the flow of vortices began to increase in the orthogonal direction, occupying a significant part of the plate. According to the signal structure (Figure 5a), subsequent flux fingers in the cascade were weaker. As the magnetic field increases, the critical current density decreases. Therefore, in the next cascades, the amount of flux that entered at the first pulse decreases, and to fill the plate with the flux, the number of subsequent avalanches in the cascade increases.

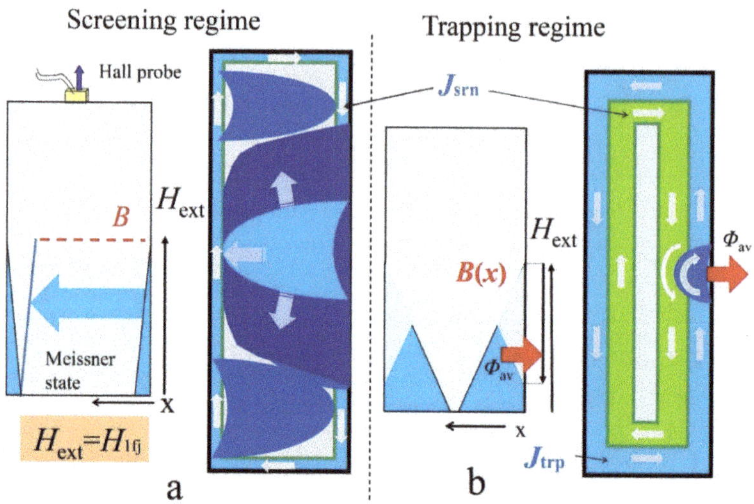

Figure 12. (**a**,**b**) Left side—a magnetic induction $B(x)$ of Bean's profile into plate before the 1st flux jumps in screening regime (**a**) and before the cascade in trapping regime appears (**b**); right side is cross section of plate perpendicular to magnetic field direction: a schematic representation of the flux avalanche entry process, corresponding to the 1st flux jump (**a**) and to the cascade creation (**b**).

Trapping regime. The induction-current structure of the critical state in the flux trapping mode is fundamentally different from that in the screening mode. With shielding, only one critical current direction is present (Figure 12a, right side). In the second quadrant, due to a change from increase to decrease of external field, a second current loop in the opposite direction appears in the current structure (Figure 12b, right side). In this case, a boundary arises that separates these counter currents. In Bean model it is a straight line. In reality, due to the inhomogeneity of the pinning in the superconductor or the stochasticity of the vortex dynamics, this boundary is a jagged line [44–47]. The dynamics of this jagged (zigzag) current boundary can lead to turbulent formations along the front and circular currents (vortices). Moreover, these vortex formations can also keep the opposite direction of the circulation of currents. Therefore, the destruction of the critical state and the breaking the flux off a dome-like profile of induction can be accompanied by a dynamic "mixture" of oppositely directed electromagnetic fields, which leads to a more complex spectral composition of the induction signal than in the screening mode. This electromagnetic mixture, superimposed on the "resonator manifestations" of the plate, can strongly influence the triggering of first finger and steady state of cascades generation. As a consequence, a scatter-swing of experimental points in field dependences of the $T_c(H_{ext})$ periods (Figure 6b,d) and frequencies in the $F_{FFT}(H_{ext})$ spectra (Figure 8c,d) may appear.

The possible physical reasons of the complicated signal spectrum, accompanying avalanche flux dynamics, can be associated with the structure roughness of the moving flux front and inhomogeneous relief of induction. The electromagnetic radiation, accompanying the avalanche, is formed as a result of the acceleration and deceleration of individual sections of the front of moving flux, during interaction with fixed vortex bundles, followed by their separation from the pinning centers. A non-monotonic increase in the amount of flux entering the superconductor may occur and a moving flux can sweep away pinned flux hills (vortex bundles), like a tsunami, as simulation of thermomagnetic avalanches demonstrates [40]. In this case, the magnitude of the pinning inhomogeneity determines the level of electromagnetic "noise", accompanying the flux dynamics. The magnitude of inhomogeneities can decrease, with an increase in magnetic field, due to a decrease in the critical current density. In the third quadrant, when the direction of the external field changes, regions of the superconductor arise, where the magnetic field induction has the opposite direction of the field lines compared to the trapped flux. In the fields of 0–−2T, this can be strongly reflected in the excitation conditions and, accordingly, in the spectrum of avalanche pulses. In stronger fields $\mu_0 H_{ext} \geq -2$ T, the induction of external field direction only remains in the plate and its distribution is similar to that in the first quadrant [43], leading to the coincidence of frequency dependences $F_{ma}(H_{ext})$ of cascades (Figure 9a) for two quadrants.

4. Conclusions

The cascades of magnetic flux avalanches at the remagnetization of superconducting bulk plates, with the magnetic flux Φ entering (screening mode) or leaving (trapping regimes) in the form of a multistage "stairway" manner with approximately equal successive portions $\Delta\Phi$, were discovered. The width of steps in temporal dependence $\Phi(t)$ was found to increase almost linearly with an increase of the plate thickness. Cascades in the *screening mode* after ZFC are realized in the form of fairly stable flux steps in the entire range of the instability of critical state. In this case, the number of flux steps gradually increases, with an increase in the field, reaching ten in the field range of 2–4 T. When the magnetic field is ramping backward (flux *trapping mode*) and then its orientation is changed to the opposite, the structure of the cascades becomes much more complicated.

Spectral analysis of the inductive signal allowed us to characterize the avalanche flux dynamics in the second and third quadrants in the entire region of critical state instability. It was found that, in the magnetic field range of ±2–±4 T, an avalanche is realized in a form of multistage "stairway" manner (steady state of cascades generation). In the 0–2 T region, stable cascades with small quantity of steps were rarely observed. Here, the signal spectrum consists of several spectral components with approximately equal amplitudes. Such a spectrum can arise as a result of an avalanche rearrangement of the current configuration of the critical state under the action of magnetic flux front. In the second quadrant, there are two powerful current circuits with the opposite direction of circulation. These are the shielding current and flux trapping current circuits.

The possible physical reasons of the complicated signal spectrum accompanying avalanche flux dynamics can be associated with rough structure of moving flux front and inhomogeneous relief of induction, formed as a result of previous avalanche processes. In the third quadrant, when the direction of the external field changes, the regions of superconductor arise, where the induction has the opposite direction of the field lines compared to the orientation of the trapped flux. This, in the fields of 0–−2 T, is strongly reflected in the conditions of excitation of cascades and, accordingly, in the spectrum of avalanche pulses. In stronger fields $\mu_0 H_{ext} \geq 2$ T, the induction of external field direction only remains in the plate. In this region of fields, the distributions of induction in the plate in the third and first quadrants behave in the same way. Similarity in behavior is characteristic of the periodicity of cascades.

The use of various sensors allowed us to establish that, the initiation of cascades occurs mainly in the central part of the lateral surface of the plate. The speed of propagation of an avalanche front along the magnetic field line in a superconductor was experimentally determined and possible mechanism of cascades generation, related to the "resonator's properties" of plate, was suggested.

Author Contributions: V.C. conceived an experiment. V.C. and A.N. performed the magnetic measurements and analyzed the data. V.C. proposed a mechanism for cascades triggering. V.C., R.P. and A.N. wrote the paper. All authors have read and agreed to the published version of the manuscript.

Funding: This research received no external funding.

Institutional Review Board Statement: Not applicable.

Informed Consent Statement: Not applicable.

Data Availability Statement: The data that support the findings of this study are available from the corresponding author upon reasonable request.

Acknowledgments: We thank I. S. Aronson for fruitful discussions of the results, for the proposed possible mechanism for the occurrence of self-consistent cascades, and for the simulation of some avalanche processes. Thanks are due to S. Vasylyev for his valuable help in taking part of the data acquisition that are reported here.

Conflicts of Interest: The authors declare no conflict of interest.

Abbreviations

Φ	magnetic flux
t	time
U_{coil}	voltage on the pickup coil (inductive sensors)
M	magnetization
B_{surf}	induction of the magnetic field
H_{ext}	external magnetic field
H_{1fj}	the field of the first instability
d	thickness of the superconducting plate
W	width of the plate
T_c	period of individual avalanche pulses in cascade
T_{av}	the average period
F_{FFT}	frequencies of the Fourier spectrum
F_{ma}	frequency of the spectrum component with the maximum amplitude
F_T	pulse repetition frequency
j_c	critical current density
$V_{\vert\vert}$	speed of propagation of the disturbance along the direction of the magnetic field
FFT	Fast Fourier transformation
ZFC	zero field cooling

References

1. Bean, C.P. Magnetization of hard superconductors. *Phys. Rev. Lett.* **1962**, *8*, 250–253. [CrossRef]
2. Schwartz, P.S.; Bean, C.P. A model for magnetic instabilities in hard superconductors: The adiabatic critical state. *J. Appl. Phys.* **1968**, *39*, 4991–4998. [CrossRef]
3. Altshuler, E.; Johansen, T.H. Colloquium: Experiments in vortex avalanches. *Rev. Mod. Phys.* **2004**, *76*, 471–487. [CrossRef]
4. Nowak, E.R.; Taylor, O.W.; Liu, L.; Jaeger, H.M.; Selinder, T.I. Magnetic flux instabilities in superconducting niobium rings: Tuning the avalanche behavior. *Phys. Rev. B* **1997**, *55*, 11702–11705. [CrossRef]
5. Field, S.; Witt, J.; Nori, F.; Ling, W.S. Superconducting vortex avalanches. *Phys. Rev. Lett.* **1995**, *74*, 1206–1209. [CrossRef] [PubMed]
6. Wipf, S.L. Magnetic instabilities in type-II superconductors. *Phys. Rev.* **1967**, *161*, 404–416. [CrossRef]
7. Mints, R.G.; Rakhmanov, A.L. Critical state stability in type-II superconductors and superconducting-normal-metal composites. *Rev. Mod. Phys.* **1981**, *53*, 551–592. [CrossRef]
8. Wipf, S.L. Review of stability in high temperature superconductors with emphasis on flux jumping. *Cryogenics* **1991**, *31*, 936–948. [CrossRef]
9. De Sorbo, W.; Healy, W.A. The intermediate state of some superconductors. *Cryogenics* **1964**, *4*, 257–323. [CrossRef]
10. Goodman, B.B.; Wertheimer, M. Observation of the speed of flux jumps in superconducting niobium. *Phys. Lett.* **1965**, *18*, 236–238. [CrossRef]
11. Keyston, J.; Wertheimer, M. A new method for measuring the speed of flux jumps in type-II superconductors. *Cryogenics* **1966**, *6*, 341–343. [CrossRef]
12. Goodman, B.B.; Lacaze, A.; Wertheimer, M.R. Un appareil pour l'etude de la cinetique des sauts de flux des supraconducteurs de la deuxiem espece. *C. R. Acad. Sci. Paris Ser. A* **1966**, *262*, 12.
13. Wertheimer, M.R.; Gilchrist, J.G. Flux jumps in type-II superconductors. *J. Phys. Chem. Solids* **1967**, *28*, 2509–2524. [CrossRef]
14. Harrison, R.B.; Wright, S.L.; Wertheimer, M.R. Kinetics of flux jumps in type-II superconductors. *Phys Rev. B* **1973**, *7*, 1864–1866. [CrossRef]
15. Nabiałek, A.; Chabanenko, V.V.; Vasylyev, S.; Rusakov, V.F.; Shushmakova, G.; Szymczak, H. The reversal of the local magnetic field profile at the surface of superconducting sample caused by the thermomagnetic avalanche. *J. Low Temp. Phys.* **2009**, *154*, 55–67. [CrossRef]
16. Chabanenko, V.V.; Kuchuk, E.I.; Rusakov, V.F.; Abaloszewa, I.; Nabiałek, A.; Pérez-Rodríguez, F. Critical state transformation in hard superconductors resulting from thermomagnetic avalanches. *Low Temp. Phys.* **2016**, *42*, 239–257. [CrossRef]

17. Chabanenko, V.V.; Nabiałek, A.; Puźniak, R.; Kuchuk, O.I.; Chumak, O.M.; Pérez-Rodríguez, F.; Garcia-Vazquez, V.; Pal, U.; Cortés-Maldonado, R.; Jun, Q.; et al. Magnetic moment inversion at giant flux jump: Dynamical property of critical state in type-II superconductors. *Sci. Rep.* **2019**, *9*, 6233. [CrossRef] [PubMed]
18. Leiderer, P.; Boneberg, J.; Brüll, P.; Bujok, V.; Herminghaus, S. Nucleation and growth of a flux instability in superconducting $YBa_2Cu_3O_{7-x}$ films. *Phys. Rev. Lett.* **1993**, *71*, 2646–2649. [CrossRef] [PubMed]
19. Durán, C.A.; Gammel, P.L.; Miller, R.E.; Bishop, D.J. Observation of magnetic-field penetration via dendritic growth in superconducting niobium films. *Phys. Rev. B* **1995**, *52*, 75–78. [CrossRef]
20. Johansen, T.H.; Baziljevich, M.; Shantsev, D.V.; Goa, P.E.; Galperin, Y.M.; Kang, W.N.; Kim, H.J.; Choi, E.M.; Kim, M.-S.; Lee, S.I. Dendritic magnetic instability in superconducting MgB_2 films. *Europhys. Lett.* **2002**, *59*, 599–605. [CrossRef]
21. Qviller, A.J.; Qureishy, T.; Xu, Y.; Suo, H.; Mozhaev, P.B.; Hansen, J.B.; Vestgården, J.I.; Johansen, T.H.; Mikheenko, P. Scaling behavior of quasi-one-dimensional vortex avalanches in superconducting films. *Sci. Rep.* **2020**, *10*, 5641. [CrossRef] [PubMed]
22. Bolz, U.; Biehler, B.; Schmidt, D.; Runge, B.-U.; Leiderer, P. Dynamics of the dendritic flux instability in $YBa_2Cu_3O_{7-\delta}$ films. *Europhys. Lett.* **2003**, *64*, 517–523. [CrossRef]
23. Biehler, B.; Runge, B.-U.; Leiderer, P.; Mints, R.G. Ultrafast magnetic flux dendrite propagation into thin superconducting films. *Phys. Rev. B* **2005**, *72*, 024532. [CrossRef]
24. Vestgården, J.I.; Johansen, T.H.; Galperin, Y.M. Nucleation and propagation of thermomagnetic avalanches in thin-film superconductors. *Low Temp. Phys.* **2018**, *44*, 460–476. [CrossRef]
25. Aranson, I.S.; Gurevich, A.; Vinokur, V. Vortex avalanche and magnetic flux fragmentation in superconductors. *Phys. Rev. Lett.* **2001**, *87*, 067003. [CrossRef] [PubMed]
26. Aranson, I.S.; Gurevich, A.; Welling, M.S.; Wijngaarden, R.J.; Vlasko-Vlasov, V.K.; Vinokur, V.M.; Welp, U. Dendritic flux avalanches and nonlocal electrodynamics in thin superconducting films. *Phys. Rev. Lett.* **2005**, *94*, 037002. [CrossRef] [PubMed]
27. Welling, M.S.; Westerwaal, R.J.; Lohstroh, W.; Wijngaarden, R.J. Huge compact flux avalanches in superconducting Nb thin films. *Phys. C* **2004**, *411*, 11–17. [CrossRef]
28. Alvarez, S.B.; Brisbois, J.; Melinte, S.; Kramer, R.B.G.; Silhanek, A.V. Statistics of thermomagnetic breakdown in Nb superconducting films. *Sci. Rep.* **2019**, *9*, 3659. [CrossRef] [PubMed]
29. Embon, L.; Anahory, Y.; Suhov, A.; Halbertal, D.; Cuppens, J.; Yakovenko, A.; Uril, A.; Myasoedov, Y.; Rappaport, M.L.; Huber, M.E.; et al. Probing dynamics and pinning of single vortices in superconductors at nanometer scales. *Sci. Rep.* **2019**, *5*, 7598. [CrossRef] [PubMed]
30. Embon, L.; Anahory, Y.; Jelić, Ž.L.; Lachman, E.O.; Myasoedov, Y.; Huber, M.E.; Mikitik, G.P.; Silhanek, A.V.; Milošević, M.V.; Gurevich, A.; et al. Imaging of super-fast dynamics and flow instabilities of superconducting vortices. *Nat. Commun.* **2017**, *8*, 85. [CrossRef]
31. Dobrovolskiy, O.V.; Vodolazov, D.Y.; Porrati, F.; Sachser, R.; Bevz, V.M.; Mikhailov, M.Y.; Chumak, A.V.; Huth, M. Ultra-fast vortex motion in a direct-write Nb-C superconductor. *Nat. Commun.* **2020**, *11*, 3291. [CrossRef] [PubMed]
32. Sheikhzada, A.; Gurevich, A. Dynamic pair-breaking current, critical superfluid velocity and nonlinear electromagnetic response of nonequilibrium superconductors. *Phys. Rev. B* **2020**, *102*, 104507. [CrossRef]
33. Kopnin, N.B. Frequency singularities of the dissipation in the mixed state of pure type-II superconductors at low temperatures. *JETP Lett.* **1978**, *27*, 390–393.
34. Volovik, G.E. Comment on vortex mass and quantum tunneling of vortices. *JETP Lett.* **1997**, *65*, 217–223. [CrossRef]
35. Rusakov, V.F.; Chabanenko, V.V.; Nabiałek, A.; Chumak, O.M. Oscillations of a single Abrikosov vortex in hard type-II superconductors. *Low Temp. Phys.* **2017**, *43*, 670–682. [CrossRef]
36. Tesař, R.; Šindler, M.; Kadlec, C.; Skrbek, L.; Lipavský, P.; Koláček, J. Mass of Abrikosov vortex in high-temperature superconductor $YBa_2Cu_3O_{7-\delta}$. *Sci. Rep.* **2021**, *11*, 21708. [CrossRef] [PubMed]
37. Cooley, L.D.; Lee, P.J.; Larbalestier, D.C. Conductor processing of low-T_c materials: The alloy Nb-Ti. In *Handbook of Superconducting Materials, Volume I: Superconductivity, Materials and Processes*; Chapter B.3.3.2; Cardwell, D.A., Ginley, D.S., Eds.; Institute of Physics Publishing: Bristol, UK, 2003; Volume 603.
38. Shen, T.M.; Bosque, E.; Davis, D.; Jiang, J.Y.; White, M.; Zhang, K.; Higley, H.; Turqueti, M.; Huang, Y.B.; Miao, H.P.; et al. Stable, predictable and trainingfree operation of superconducting Bi-2212 Rutherford cable racetrack coils at the wire current density of 1000 A/mm^2. *Sci. Rep.* **2019**, *9*, 10170. [CrossRef]
39. Brandt, E.H. Superconductors of finite thickness in a perpendicular magnetic field: Strips and slabs. *Phys. Rev. B* **1996**, *54*, 4246–4264. [CrossRef]
40. Aronson, I.; PennState University, State College, PA, USA. Private communication, 2022.
41. Jooss, C.; Albrecht, J.; Kuhn, H.; Leonhardt, S.; Kronmuller, H. Magneto-optical studies of current distributions in high-T_c superconductors. *Rep. Prog. Phys.* **2022**, *65*, 651–788. [CrossRef]
42. Olsen, Å.A.F.; Johansen, T.H.; Shantsev, D.; Choi, E.-M.; Lee, H.-S.; Kim, H.J.; Lee, S.-I. Flux dendrites of opposite polarity in superconducting MgB_2 rings observed with magneto-optical imaging. *Phys. Rev. B* **2006**, *74*, 064506. [CrossRef]
43. Chabanenko, V.V.; D'yachenko, A.I.; Zalutskii, M.V.; Rusakov, V.F.; Szymczak, H.; Piechota, S.; Nabiałek, A. Magnetothermal instabilities in type II superconductors—The influence of magnetic prehistory. *J. Appl. Phys.* **2000**, *88*, 5875–5883. [CrossRef]
44. Surdeanu, R.; Wijngaarden, R.J.; Visser, E.; Huijbregtse, J.M.; Rector, J.; Dam, B.; Griessen, R. Kinetic roughening of penetrating flux fronts in high-T_c thin film superconductors. *Phys. Rev. Lett.* **1999**, *83*, 2054–2057. [CrossRef]

45. Vlasko-Vlasov, V.K.; Welp, U.; Metlushko, V.; Crabtree, G.W. Experimental test of the self-organized criticality of vortices in superconductors. *Phys. Rev. B* **2004**, *69*, 140504. [CrossRef]
46. Romero-Salazar, C.; Chabanenko, V.V.; Kuchuk, E.I.; Abaloszewa, I.; Nabialek, A.; Perez-Rodriguez, F. Obtaining a rough flux front in type-II superconductors using a critical state model. *Acta Phys. Pol. A* **2016**, *130*, 645–648. [CrossRef]
47. Romero-Salazar, C.; Hernández-Flores, O.A.; Chumak, O.M.; Pérez-Rodríguez, F.; Chabanenko, V.V. Emulating rough flux patterns in type-II superconducting cylinders using the elliptic critical-state model. *J. Appl. Phys.* **2017**, *122*, 143904. [CrossRef]

materials

Article

Computation of the Electrical Resistance of a Low Current Multi-Spot Contact

Gideon Gwanzuwang Dankat and Laurentiu Marius Dumitran *

Laboratory of Electrical Materials, Faculty of Electrical Engineering, University Politehnica of Bucharest, Splaiul Independentei 313, 060042 Bucharest, Romania; gdankat@elmat.pub.ro
* Correspondence: dumitran@elmat.pub.ro

Abstract: In high complexity electrical systems such as those used in the automotive industries, electric connectors play an important role. The automotive industry is gradually shifting its attention to electric cars, which means more electrical connectors for sensors and data collection. A fault in connectors for sensors used in a vehicle can cause drastic damage to capital equipment and, in the worst case, the loss of life. The studies of faults or degradation of electrical contacts are essential for safety in vehicles and various industries. Although such faults can be due to numerous factors (such as dust, humidity, mechanical vibration, etc.) and some yet to be discovered, high contact resistance is the main factor causing erratic behavior of electrical contacts. This paper presents a study on the computation of electrical contact resistance of two metal conductors (in the form of a disk) with analytical relations and a numerical computation model based on the finite element method (FEM) in COMSOL Multiphysics. The contact spots were considered to have a higher electrical resistivity value (ρ_{cs}) than those of the two metal conductors (ρ_{Cu}). Studies such as the one in view that is carried out on a microscopic level are often difficult to investigate experimentally. Therefore, with the help of a simplified numerical model, the consequences of the degradation of electrical contacts are investigated. To validate the FEM model, the numerical results were compared to those obtained from analytical models.

Keywords: electrical contact; contact resistance; electrical connector; numerical analysis; FEM; COMSOL Multiphysics

Citation: Dankat, G.G.; Dumitran, L.M. Computation of the Electrical Resistance of a Low Current Multi-Spot Contact. *Materials* **2022**, *15*, 2056. https://doi.org/10.3390/ma15062056

Academic Editors: Marc Cretin, Sophie Tingry and Zhenghua Tang

Received: 2 February 2022
Accepted: 8 March 2022
Published: 10 March 2022

1. Introduction

Electric contacts are one of the essential components in electronics and automotive systems. Estimating the contact resistance determines the accuracy of connectors. In machinery assembly, evaluating the reliability and tightness of contact can be done using the contact resistance of metals contact surface. For better conductivity and lower resistance, the point of contact between two bodies must have a large area with fewer impurities [1].

The growing number of electronic devices that equip electric vehicles calls for an increase in the number of electric connectors to be adapted in the vehicles' electrical system. The automotive market demands optimum safety requirements and software-enabled features at a low price; to attain this, automotive manufacturers are trying to miniaturize electrical contacts to reduce weight by manufacturing them with high-performing alloys such as CuNiSi (copper: balance; 0.8–1.8% nickel; 0.01–0.05% phosphorus; and 0.15–0.35% silicon). Additionally, aluminum or copper alloys (CuMg, CuAg) are used for smaller wire cross-sections to reduce weight [2]. An important aspect of electrical contact technology is contact housing. They are usually made of polymer and perform functions such as electrical shielding, insulation, and protection of all connector components. Contact between two conductors occurs at discrete spots owing to the rough nature of the surfaces (asperities). Therefore, the real point of contact is a smaller portion of the apparent contact area [3].

Holm [4] in the 1930s presented a well-known theory describing contact systems and contact resistance. Based on several hypotheses, Holm's model was established for the contact resistance of a single contact spot without considering the effect of surface films. Greenwood [5] presented a more elaborate interpretation of Holm's model; he derived a formula for calculating contact resistance based on the number of a set of circular contact spots and the distance between them. Similar to Holm's model, the influence of interface films was ignored.

Over time, Holm and Greenwood's theory has been further developed and serves as a starting point for researchers in designing quality and reliable electrical contacts. Boyer [6] generalized the Greenwood formula in his analytical solution; he introduced the influence of interface films. Nakamura and Minowa [7–10] computed numerically the contact resistance using the boundary element method (BEM) and finite element method (FEM) of a contact model consisting of two cubic electrodes that communicate through square spots [8,9]. Nakamura also analyzed the contact resistance of regular and irregular conducting spots and their behavior [10]. Timsit [11] examined the electrical conduction through small constrictions and the dependence of electrical resistance on the shape and dimensions of the contact spots [12]. Shujuan et al. [13] built a contact resistance model based on a rough surface contact model by surface profile measurement and statistical analysis. Ren et al. [14] and Lim et al. [15] both analyzed contact resistance by applying a combination of experimental, numerical, and analytical methods. Malucci [16,17] established several models, considering the influence of interface films to predict the performance of degraded contacts. Furthermore, Fukuyama et al. [18] evaluated the contact resistance of a simplified contact sample and compared the result to the contact resistance calculated using Holm and Greenwood equations.

Although few studies of contact resistance focus on comparing numerical simulation and analytical solution, this paper presents a simplified contact model to evaluate the contact resistance by analytical approaches (using Holm equation and Greenwood equation) and by numerical simulation (using FEM in COMSOL Multiphysics). The model consists of two metallic disks coming in contact through multiple regular circular spots (a-spots) that have a higher electrical resistivity compared to the two metallic disks ($\rho_{cs} < \rho_{Cu}$) in view to simulate the contact aging. Though analytical models, in theory, are considered to be the exact solution, they are derived based on multiple assumptions. Despite all the assumptions considered in the analytical models, the present study aims to explore the limits of numerical calculation for the contact resistance in the case of a simplified model. Moreover, this paper examines the influence of the size of the contact spots on the value of contact resistance.

Aging of Electrical Contacts

Electrical contacts undergo different stresses during operation (i.e., mechanical wear due to induced vibrations, atmospheric conditions, corrosion, etc.). Swingler et al. [19] classified these stresses on automotive connectors into two significant groups. One is external stresses due to the environmental conditions, while the other is internal stresses created by the vehicle. Studies on contact resistance calculations and degradation of electrical contacts can be very tasking and complex because of the contact interface's coarse nature and asperities that comprise numerous spots constricting the flow of electrons and the numerous degrading factors manifesting at the interface.

Environmental factors such as temperature, air pollution, humidity, condensation, etc., affect electrical connectors because they promote the rate of corrosion (Figure 1); Corrosion is much more severe at the interface of connectors; it gradually reduces the area of contact at the interface, which increases resistance over time and eventually leads to electrical failure of the connectors.

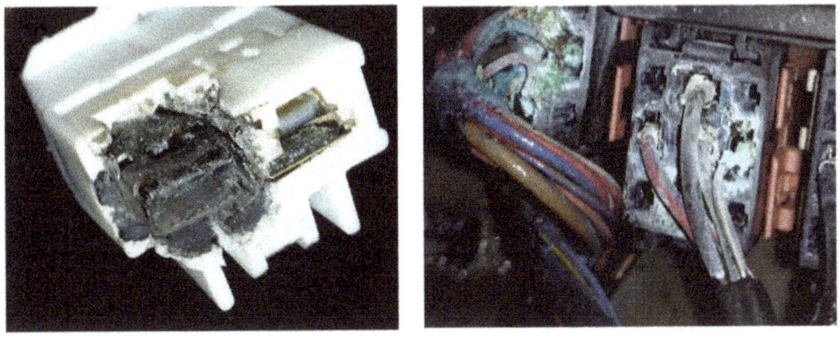

(a) (b)

Figure 1. Examples of corroded connectors: (**a**) corroded Ac blower wire harness connector in an Acura TL [20]; (**b**) corroded wiper control relays in a BMW 7 series [21].

On the other hand, fretting due to vibration, dust, the passage of current, temperature, etc., also contributes and accelerates the degradation process of electrical contacts. Swingler et al. [22] reported that in vehicles, the connectors situated close to the exhaust system can possess high-temperature regimes such as 85, 105, 125, and 155 °C. Additionally, Jedrzejczyk [23] experimented on CuSn/Ni/Sn contact material system (Figure 2); he reported that during the process of degradation, electrical resistance depends on the nonoxidized area of contact at the interface. i.e., when all the areas of contact are fully covered by oxidized films, then the electrical resistance value increases above a normal threshold value.

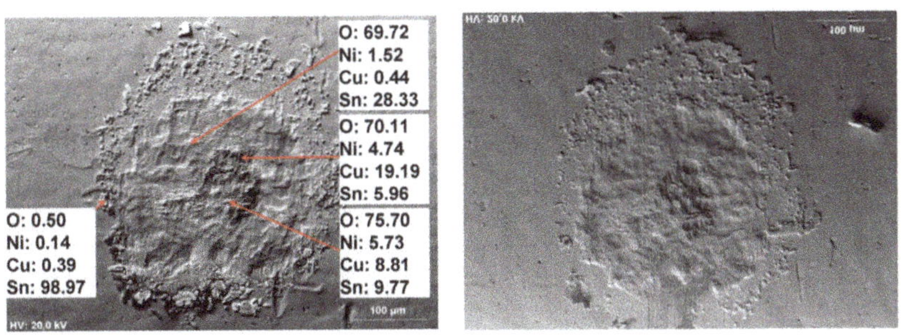

(a) (b)

Figure 2. Fretting scars as seen in an SEM micrograph obtained after 2600 fretting cycles in a CuSn/Ni/Sn contact material system: (**a**) lower sample (**b**) upper sample [23].

2. Analytical Models for Contact Resistance

Numerous studies on contact resistance and electrical contacts have made mention of Greenwood's and Holm's analytical model for contact resistance calculation [6–18]. The research reported in this paper used Holm's and Greenwood's models to calculate the contact resistance.

2.1. Holm's Analytical Model for Contact Resistance

In Holm's analytical model for the calculation of contact resistance [4], two cylinders (C_1 and C_2) in contact were considered (Figure 3).

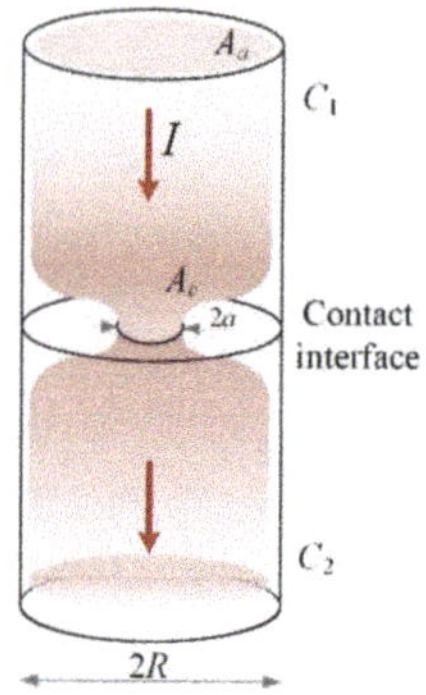

Figure 3. Holm contact resistance model: cylinders in contact showing the apparent area of contact (*Aa*), the real area of contact (*Ac*), and the radius *R* with a circular constriction of the current lines of radius *a*.

Though both cylinders were assumed to be clean and free of any impurity, they come in contact through a smaller portion of the apparent area (*Aa*). This smaller portion is the actual point of contact (*Ac*). Constriction resistance occurs when the current is restricted from its normal flow to pass through (*Ac*). The voltage between both cylinders can be measured as the current flows through and subsequently the resistance between both surfaces. Holm's analytical model calculates the total constriction resistance of one circular spot between two electrically conducting cylinders. It has the expression:

$$R_c = \rho \cdot (2a)^{-1}, \tag{1}$$

where ρ is the resistivity of the conductor and a is the radius of the constriction.

In situations where there is a difference in material properties (when the bodies in contact have different resistivities), Equation (1) becomes:

$$R_c = (\rho_1 + \rho_2) \cdot (4a)^{-1} \tag{2}$$

The assumptions in Holm's analytical model are:

- No presence of oxide films or impurities at the interface of the metallic cylinders;
- There is no axial deviation in the direction of current flow;
- The metallic cylinders in contact have infinite dimensions to the current flow.

2.2. Greenwood's Analytical Model for Contact Resistance

In 1966, J.A. Greenwood published a paper [5] in which he made a detailed interpretation of Holm's analytical model. He considered a set of circular spots (multiple a-spots) within a single cluster located at the interface between two electrodes (Figure 4). The metallic electrodes come in contact through the circular spots with no interface film between them. He obtained the formula for calculating the constriction resistance based on multiple spots within a single cluster (3) by treating the current flow between the metallic electrodes similar to that of an electrostatic charge distribution problem.

$$R_{cG} = \rho \cdot (2\textstyle\sum a_i)^{-1} + \rho \cdot \pi^{-1} \left(\sum_{i \neq j} (a_i a_j) \cdot (s_{ij})^{-1}\right) / (\textstyle\sum a_i)^2, \tag{3}$$

where ρ is the resistivity, s_{ij} is the distance between the centers of the ith and jth spot, a_i and a_j is the radius of the ith and jth spot. The first term $\rho \cdot (2\sum a_i)^{-1}$ is the resistance of all the spots in parallel, the second term $\rho \cdot \pi^{-1} (\sum_{i \neq j} (a_i a_j) \cdot (s_{ij})^{-1}) / (\sum a_i)^2$ is the resistance due to the interaction between all the spots.

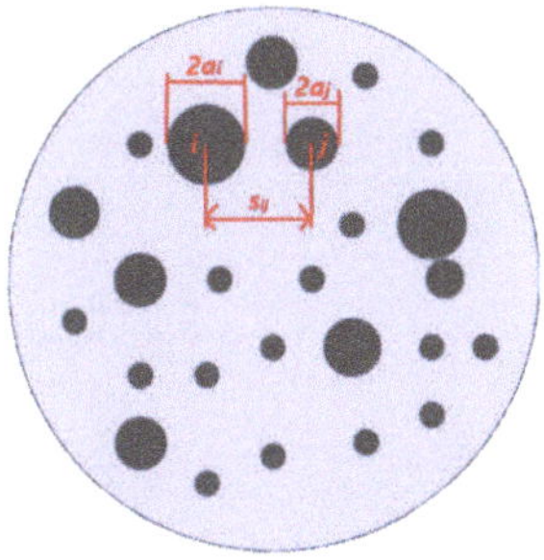

Figure 4. Greenwood's contact resistance model showing multiple spots (a_i—radius of the ith spot; a_j—radius of the jth spot; s_{ij} the distance between the ith and jth spot).

In cases where all the contact spots have the same size, Greenwood further approximated (3) and derived:

$$R_{cG1} = \rho \cdot (2\sum a_i)^{-1} + \rho \cdot (\pi n^2)^{-1} \sum_{i \neq j} (s_{ij})^{-1} \quad (4)$$

Furthermore, the hypothesis used in Holm's model also holds in Greenwood's analytical model.

3. Numerical Model

As shown above, the calculation of contact resistance in the case of a discontinuous interface (spot contact) can be done analytically. However, extensive multiphysics studies in which the electrical problem is coupled with thermal and/or mechanical (contact pressure) problems can only be solved numerically. Thus, in this paper, a numerical model is proposed that allows the calculation of the contact spot resistance, but can be further improved and completed later for the study of the thermal regime or the consideration of other parameters that can influence the value of the contact resistance. Therefore, to validate the proposed numerical model, a simple geometry very similar to that treated by Greenwood and consisting of two copper disks communicating through multiple contact spots (Figures 5 and 6) was considered. At the same time, such a model is indicative of the contact between two metal electrodes through a set of circular spots. To attain the simplification of the model, we considered circular spots that are equal in terms of area and relatively evenly distributed across the contact interface. At the terminals, a low direct current (200 mA) of density J is injected and the electric field distribution is calculated using the finite element method (FEM) in COMSOL Multiphysics software 5.6.

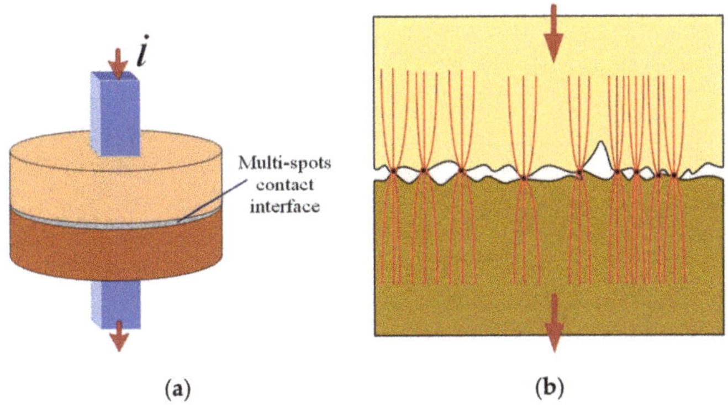

Figure 5. Schematic view of the contact configuration: (**a**) copper disks in contact; (**b**) constriction of the current lines flowing through the multiple contact spots.

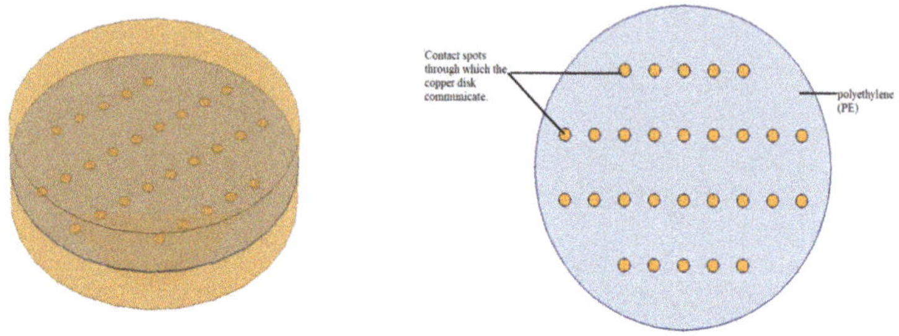

(a) (b)

Figure 6. Geometrical model of the problem investigated: (**a**) two copper disks in contact; (**b**) contact spots and the insulating layer of polyethylene (PE) at the interface.

3.1. Geometrical Model

The two copper disks model investigated come in contact through multiple spots (28 identical circular spots). Both copper disks have a radius $r = 5$ mm and a thickness $h = 1$ mm, while the homogenous contact spots have a radius $a = 0.2$ mm. Apart from the contact spots, the apparent contact area has an insulating layer of polyethylene 30 μm thick; it restricts the current to flow only through the multiple contact spots (Figure 6).

3.2. Mathematical Model

In the COMSOL Multiphysics software, the numerical analysis of the investigated problem was calculated in the stationary electrokinetic regime. It consists of an electromagnetic problem that calculates the distribution of a constant electric current of density $\boldsymbol{J}$ flowing through the multiple contact spots. The essential equations dictating the problem are the electromagnetic induction law (5), the electric conduction law (6), the electric charge conservation law (7), and the electric field strength, which was evaluated as a function of the electric potential V (Equation (8)).

$$\mathrm{rot}\,\boldsymbol{E} = 0, \tag{5}$$

$$\boldsymbol{J} = \sigma \cdot \boldsymbol{E}, \tag{6}$$

$$\mathrm{div}\,\boldsymbol{J} = 0, \tag{7}$$

$$\boldsymbol{E} = -\mathrm{grad}V, \tag{8}$$

where $\boldsymbol{E}$ is the electric field (V/m) and $\sigma = 1/\rho$ is the electric conductivity (S/m).

Boundary Conditions

The boundary conditions applied in this study are:

- Continuity (9): this signifies that the normal components of the injected current flowing through the copper disks are continuous and conserved across the interior boundaries of both disks:

$$\boldsymbol{n} \cdot (\boldsymbol{J}_1 - \boldsymbol{J}_2) = 0 \tag{9}$$

- Insulation (10): this specifies that no current flows across the boundary. It applies to all surfaces except the contact areas:

$$\boldsymbol{n} \cdot \boldsymbol{J} = 0 \tag{10}$$

The discretization of the computational domain shown in Figure 7 has a total of 2,938,081 domain elements.

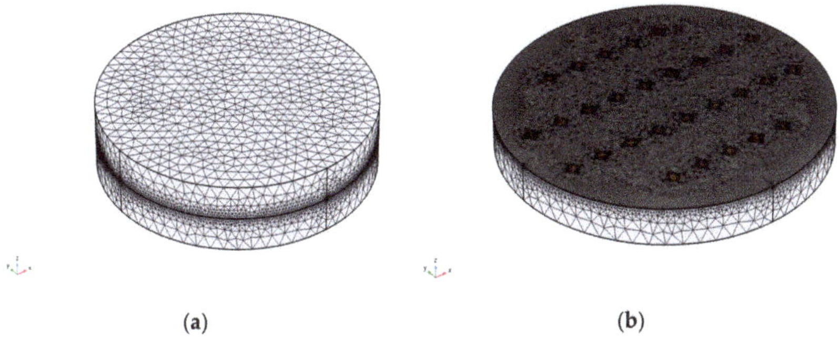

Figure 7. Discretization of the computational domain: (a) two copper disks in contact; (b) contact spots and the insulating layer of polyethylene (PE) at the interface.

4. Results

For an injected current of 200 mA, Figure 8 shows the current lines as they become constricted to flow through the multiple contact spots from one medium of the copper disk to the other.

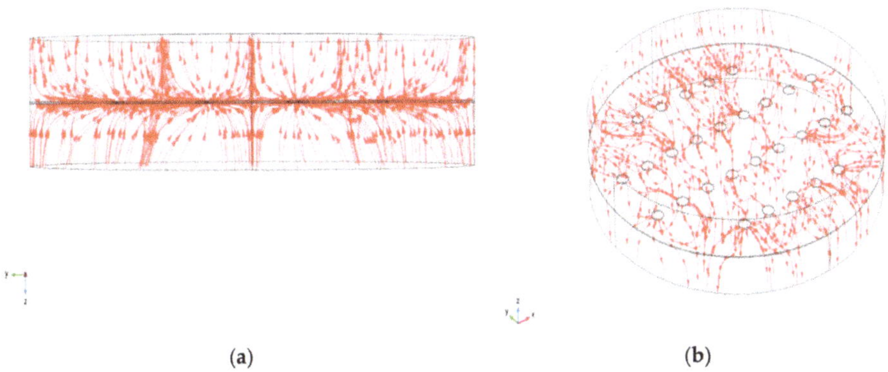

Figure 8. The current line constricted at the contact spots: (a) two-dimensional view and (b) three-dimensional view.

Figure 9 shows the voltage computed for an injected current of 200 mA, the resistivity of the copper disks ($\rho_{Cu} = 1.72 \cdot 10^{-8}$ Ω·m), the resistivity of the contact spots ($\rho_{Cs} = 1.72 \cdot 10^{-2}$ Ω·m), and the resistivity of the polyethylene ($\rho_{PE} = 10^{+17}$ Ω·m). The result shows that for a contact spot radius of 0.1 mm, the maximum voltage is 0.12 V.

After conducting a parametric study on the radius of the contact spots from $a = 0.1$ mm to $a = 0.5$ mm, Figure 10 shows the voltage drop computed for each contact spot radius. As expected, the voltage decrease as the contact spot radius increases. The results show that when the contact spot radius is 0.1 mm, the voltage is 0.12 V; for a radius of 0.3 mm, the voltage is 12.8 mV; and when the radius is 0.5 mm, the voltage becomes 4.6 mV.

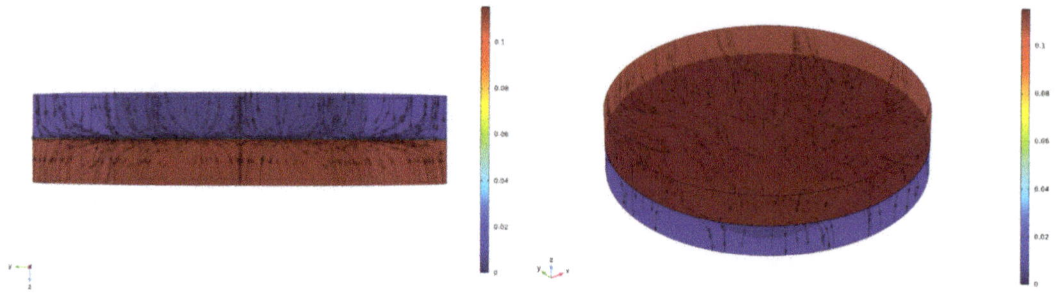

(a) (b)

Figure 9. Computed voltage for contact spot radius of 0.1 mm: (**a**) two-dimensional view and (**b**) three-dimensional view.

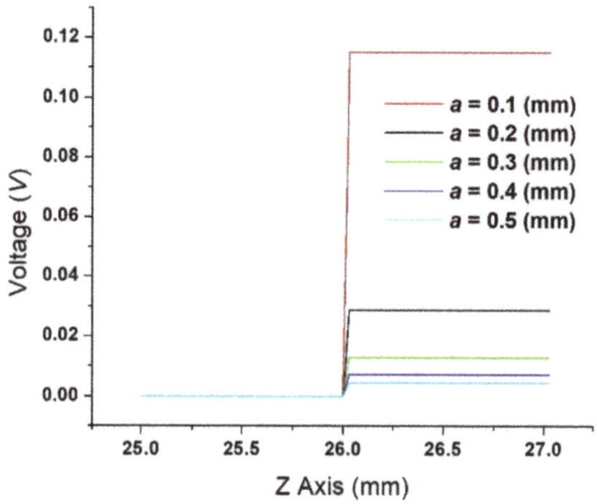

Figure 10. Voltage-drop for different values of contact spot radius a = (0.1 mm/0.5 mm).

Figure 11 shows the contact resistance evaluated in the case where for resistivity of the copper disks is the same as that of the contact spots $\rho_{Cu} = \rho_{Cs} = 1.72 \cdot 10^{-8}$ Ω·m. The result shows that the contact resistance decreases from $3.25 \cdot 10^{-6}$ to $4.34 \cdot 10^{-7}$ Ω (numerical), $3.07 \cdot 10^{-6}$ to $1.02 \cdot 10^{-6}$ Ω (Holms), and $4.76 \cdot 10^{-6}$ to $2.71 \cdot 10^{-6}$ Ω (Greenwood) when the contact spot radius increases from 0.1 to 0.3 mm. When the contact spot radius increases to 0.5 mm, the contact resistance decreases to $3.47 \cdot 10^{-7}$ Ω (numerical), $6.14 \cdot 10^{-7}$ Ω (Holm), and, respectively, $2.3 \cdot 10^{-6}$ Ω (Greenwood).

Figure 12 shows the contact resistance calculated for resistivity of the copper disks $\rho_{Cu} = 1.72 \cdot 10^{-8}$ Ω·m, with the resistivity of the contact spots $\rho_{Cs} = 1.72 \cdot 10^{-6}$ Ω·m (a) and $\rho_{Cs} = 1.72 \cdot 10^{-4}$ Ω·m (b). The result from Figure 12a shows that the contact resistance decreases from $6.03 \cdot 10^{-5}$ to $7.03 \cdot 10^{-6}$ Ω (numerical), $3.07 \cdot 10^{-4}$ to $1.02 \cdot 10^{-4}$ Ω (Holms), and $4.76 \cdot 10^{-4}$ to $2.71 \cdot 10^{-4}$ Ω (Greenwood) when the contact spot radius increases from 0.1 to 0.3 mm. When the contact spot radius increases to 0.5 mm, the contact resistance decreases to $2.58 \cdot 10^{-6}$ Ω (numerical), $6.14 \cdot 10^{-5}$ Ω (Holm), and $2.30 \cdot 10^{-4}$ Ω (Greenwood).

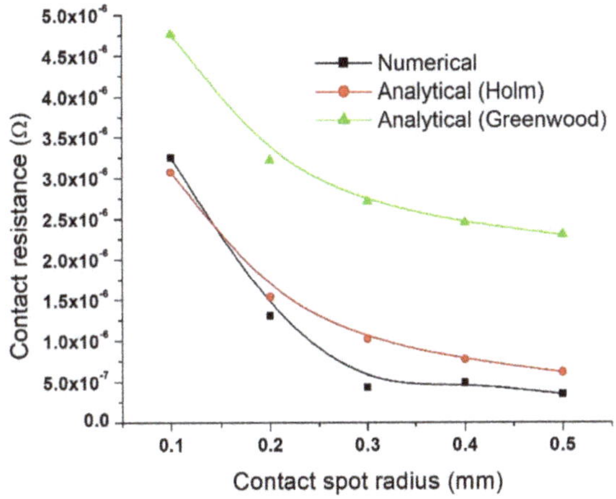

Figure 11. Contact resistance showing values for Holm and Greenwood (analytical) and numerical analysis calculated for ($\rho_{Cu} = \rho_{Cs} = 1.72{\cdot}10^{-8}$ Ω·m).

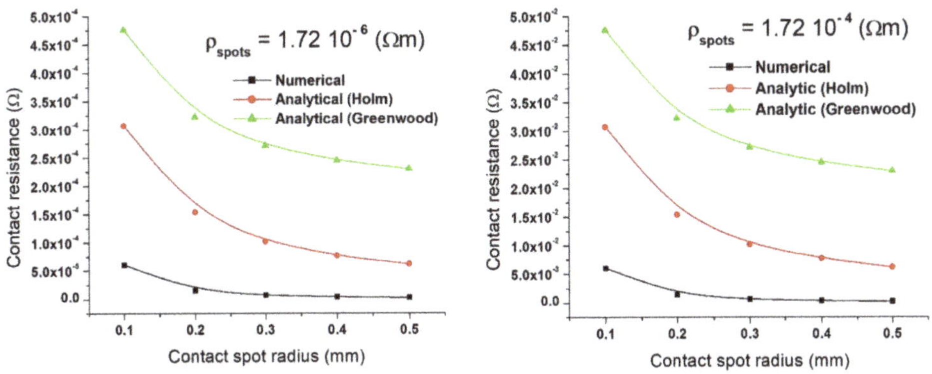

(a) (b)

Figure 12. Contact resistance showing values for Holm and Greenwood (analytical) and numerical analysis calculated for $\rho_{Cu} = 1.72{\cdot}10^{-8}$ Ω·m with **(a)** $\rho_{Cs} = 1.72{\cdot}10^{-6}$ Ω·m and **(b)** $\rho_{Cs} = 1.72{\cdot}10^{-4}$ Ω·m.

In Figure 12b, when the contact spot increases from 0.1 to 0.3 mm, the contact resistance decreases from $6{\cdot}10^{-3}$ to $6.4{\cdot}10^{-4}$ Ω (numerical), $3.07{\cdot}10^{-2}$ to $1.02{\cdot}10^{-2}$ Ω (Holm), and $4.76{\cdot}10^{-2}$ to $2.71{\cdot}10^{-2}$ Ω (Greenwood). At a contact spot radius of 0.5 mm, the contact resistance becomes $2.30{\cdot}10^{-4}$ Ω (numerical), $6.14{\cdot}10^{-3}$ Ω (Holm), and, respectively, $2.30{\cdot}10^{-2}$ Ω (Greenwood).

Figure 13 shows the contact resistance obtained for resistivity of the copper disks $\rho_{Cu} = 1.72{\cdot}10^{-8}$ Ω·m with a resistivity of the contact spots $\rho_{Cs} = 1.72{\cdot}10^{+2}$ Ω·m (a) and $\rho_{Cs} = 1.72{\cdot}10^{+4}$ Ω·m (b). In Figure 13a, the result shows that for an increase in contact radius from 0.1 to 0.3 mm, there is a decrease in the value of the contact resistance from $5.25{\cdot}10^{+3}$ to $6.39{\cdot}10^{+2}$ Ω (numerical), $3.07{\cdot}10^{+4}$ to $1.02{\cdot}10^{+4}$ Ω (Holms), and $4.76{\cdot}10^{+4}$ to $2.71{\cdot}10^{+4}$ Ω (Greenwood). For a contact radius of 0.5 mm, the contact resistance becomes $2.30{\cdot}10^{+2}$ Ω (numerical), $6.14{\cdot}10^{+3}$ Ω (Holm), and, respectively, $2.30{\cdot}10^{+4}$ Ω (Greenwood).

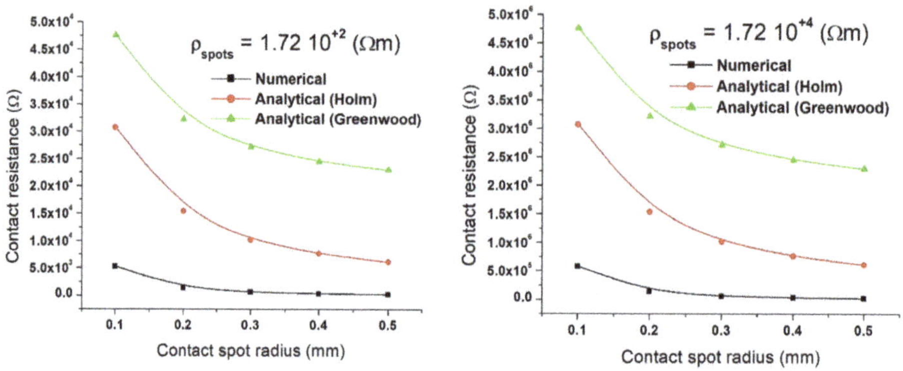

(a) (b)

Figure 13. Contact resistance showing values for Holm and Greenwood models and numerical analysis calculated for $\rho_{Cu} = 1.72{\cdot}10^{-8}$ Ω·m, $\rho_{Cs} = 1.72{\cdot}10^{+2}$ Ω·m (**a**), and $\rho_{Cs} = 1.72{\cdot}10^{+4}$ Ω·m (**b**).

The result in Figure 13b indicates that when the contact radius increases from 0.1 to 0.3 mm, the contact resistance value decreases from $5.75{\cdot}10^{+5}$ to $6.4{\cdot}10^{+4}$ Ω (numerical), $3.07{\cdot}10^{+6}$ to $1.02{\cdot}10^{+6}$ Ω (Holms), and $4.76{\cdot}10^{+6}$ to $2.71{\cdot}10^{+6}$ Ω (Greenwood). For a contact radius of 0.5 mm, the contact resistance becomes $2.3{\cdot}10^{+4}$ Ω (numerical), $6.14{\cdot}10^{+5}$ Ω (Holm), and, respectively, $2.30{\cdot}10^{+6}$ Ω (Greenwood).

Figure 14 presents the contact resistance values obtained only by numerical analysis in a case where the resistivity of the copper disks remains $\rho_{Cu} = 1.72{\cdot}10^{-8}$ Ω·m and the 28 contact spots all have resistivity of $\rho_{Cs} = 1.72{\cdot}10^{-6}$ Ω·m (■), $\rho_{Cs} = 1.72{\cdot}10^{-4}$ Ω·m (•), $\rho_{Cs} = 1.72{\cdot}10^{-2}$ Ω·m (▲), and when the 28 contact spots have different resistivities in the range from $1.72{\cdot}10^{-6}$ to $1.72{\cdot}10^{+6}$ Ω·m (▼). The resistivity was imposed evenly, i.e., (4 contact spots $\rho_{Cs1} = 1.72{\cdot}10^{-6}$ Ω·m, 4 contact spots $\rho_{Cs2} = 1.72{\cdot}10^{-4}$ Ω·m, 4 contact spots $\rho_{Cs3} = 1.72{\cdot}10^{-2}$ Ω·m, 4 contact spots $\rho_{Cs4} = 1.72$ Ω·m, 4 contact spots $\rho_{Cs5} = 1.72{\cdot}10^{+2}$ Ω·m, 4 contact spots $\rho_{Cs6} = 1.72{\cdot}10^{+4}$ Ω·m, and, respectively, 4 contact spots $\rho_{Cs7} = 1.72{\cdot}10^{+6}$ Ω·m). The results show that the contact resistance value varies between 0.6 Ω ($\rho_{Cs} = 1.72{\cdot}10^{-2}$ Ω·m at $a = 0.1$ mm) and $2.58{\cdot}10^{-6}$ Ω ($\rho_{Cs} = 1.72{\cdot}10^{-2}$ Ω·m at $a = 0.5$ mm).

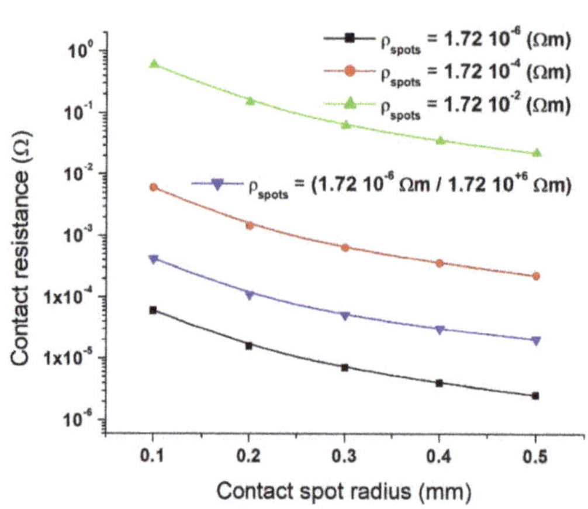

Figure 14. Contact resistance showing values for numerical model calculated for $\rho_{Cu} = 1.72{\cdot}10^{-8}$ Ω·m with $\rho_{Cs} = 1.72{\cdot}10^{-6}$ Ω·m; $\rho_{Cs} = 1.72{\cdot}10^{-4}$ Ω·m; $\rho_{Cs} = 1.72{\cdot}10^{-2}$ Ω·m, and, respectively, $\rho_{Cs} = (1.72{\cdot}10^{-6}$ Ω·m$/1.72{\cdot}10^{+6}$ Ω·m).

5. Discussion

As seen in Figures 11–13, it is clear that the contact resistance decreases as the contact spot radius increases in all cases; a similar outcome was reported by Fukuyama et al. [18]. He experimented on a configuration consisting of samples with SiO_2 insulator layers that have gold (Au) contact spots distributed on the circumference of the electrodes. The numerical and analytical results obtained in this study are consistent with the results presented in [18] in terms of the variation of the contact resistance as a function of the contact spots area. However, comparing the numerical calculation of contact resistance to that of the analytical calculation results (Equations (1) and (4)), it is clear that the numerical results are below the two analytical results. Concerning the data presented in [18], it can be noted that the experimental results were between the two analytical calculations (Holm and Greenwood). Additionally, from the point of view of the relationship between contact resistance and the area of contact, similar variation results were reported by Shujuan et al. [13], Ren et al. [14], Lim et al. [15], and Su et al. [24].

In Figure 14, the contact spots with a high resistivity showed a high value of contact resistance and vice versa, as expected. In practice, this is typical of aged electrical contacts where on the one hand, the parts at the interface with a large concentration of contaminants possess high resistivity to the flow of current and thus, have high contact resistance. On the other hand, the parts with fewer/or no contaminants have low contact resistance. Shibata et al. [25] performed a detailed analysis of the contact distribution of fretting corrosion on samples with low phosphorus–bronze alloy covered with tin plating. He reported that the contact resistance distribution results of fretting corrosion wear as observed by scanning electron microscopy (SEM) and energy dispersive X-ray (EDX) analysis indicated that the parts with few oxide deposit layers showed low contact resistance, while those with a large oxide deposit layer showed high contact resistance. It must be noted that both Holm and Greenwood have some drawbacks. In reality, the actual connector consists of multiple small contact areas (a large number of a-spots), but Holm assumed these a-spots to be equivalent to one solid circular contact. On the other hand, Greenwood's model is suitable for multiple spots within a cluster, and it depends on the distances between the set of the circular spots within that cluster. Though this is an upgrade from Holm's model, it will be difficult to calculate the constriction resistance when the contact consists of more than one cluster [26].

6. Conclusions

This paper presents a study on contact resistance of a model of two metallic disks made of copper that communicate via 28 homogenous circular spots. In the first aspect of the study, the contact resistance calculation was done using Holm's and Greenwood's analytical model, and in the second aspect, using a numerical model in COMSOL Multiphysics. The study was carried out keeping the number of contact spots and the electrical resistivity of the copper disks constant and changing the radius and resistivity of the contact spots as a possible aging consequence, between 0.1 and 0.5 mm and $1.72{\cdot}10^{-6}$ to $1.72{\cdot}10^{+4}$ $\Omega{\cdot}$m, respectively. As expected, the resulting values of contact resistance in all cases indicate that the contact resistance decreases as the contact spot radius increases; it validates that the contact resistance decreases as the effective contact area increases. This means that the interface contains few weaknesses and the contact pressure between the two conductors takes appropriate values. Additionally, the results show that the value of contact resistance obtained for both analytical models is higher with at least an order of magnitude than that of the value obtained by numerical analysis in all cases.

Furthermore, without considering the model used or the contact spot radius, as the resistivity of the contact spot increases in all cases (Holm, Greenwood, and numerical), the contact resistance value obtained increases by two orders of magnitude. In the future, detailed work on the electrothermal coupling of the model will be done to help understand the thermal regime and its effect on contacts.

Author Contributions: Conceptualization, L.M.D.; methodology, G.G.D.; software, G.G.D.; validation, L.M.D. and G.G.D.; formal analysis, G.G.D. and L.M.D. All authors have read and agreed to the published version of the manuscript.

Funding: This research received no external funding.

Institutional Review Board Statement: Not applicable.

Informed Consent Statement: Not applicable.

Data Availability Statement: Not applicable.

Conflicts of Interest: The authors declare no conflict of interest.

References

1. Liu, M. *A New Method for Measuring Contact Resistance*; Beijing Orient Institute of Measurement & Test Chinese Academy of Space Technology: Bejing, China, 2005.
2. Connectors: Technologies and Trends. Available online: https://www.zvei.org/fileadmin/user_upload/Presse_und_Medien/Publikationen/2016/November/Connectors_Technologies-and-Trends_engl/2016-11_Imagebroschuere_Steckverbinder_engl.pdf (accessed on 9 March 2022).
3. Dankat, G.G.; Dobre, A.A.; Dumitran, L.M. Influence of Ageing on Electrothermal Condition of Low Current Contact. In Proceedings of the 2021 12th International Symposium on Advanced Topics in Electrical Engineering (ATEE), Bucharest, Romania, 25–27 March 2021; pp. 1–6.
4. Holm, R. *Electric Contacts, Theory and Application*, 4th ed.; Springer: Berlin/Heidelberg, Germany, 1967.
5. Greenwood, J.A. Constriction resistance and the real area of contact. *Br. J. Appl. Phys.* **1966**, *17*, 1621–1632. [CrossRef]
6. Boyer, L. Contact resistance calculations: Generalizations of Greenwood's formula including interface films. *IEEE Trans. Compon. Packag. Technol.* **2001**, *24*, 50–58. [CrossRef]
7. Nakamura, M. Constriction resistance of conducting spots by the boundary element method. *IEEE Trans. Compon. Hybrids Manuf. Technol.* **1993**, *16*, 339–343. [CrossRef]
8. Nakamura, M.; Minowa, I. Film resistance and constriction effect of current in a contact interface. *IEEE Trans. Compon. Hybrids Manuf. Technol.* **1989**, *12*, 109–113. [CrossRef]
9. Nakamura, M.; Minowa, I. Computer simulation for the conductance of a contact interface. *IEEE Trans. Compon. Hybrids Manuf. Technol.* **1986**, *9*, 150–155. [CrossRef]
10. Nakamura, M. Computer simulation for the constriction resistance depending on the form of conducting spots. *IEEE Trans. Compon. Packag. Technol* **1995**, *18*, 382–383. [CrossRef]
11. Timsit, R.S. Electrical conduction through small contact spots. In Proceedings of the 50th IEEE Holm Conference on Electrical Contacts and the 22nd International Conference on Electrical Contacts, Seattle, WA, USA, 20–23 September 2004; pp. 184–191.
12. Timsit, R.S. Electrical contact resistance: Properties of stationary interfaces. *IEEE Trans. Compon. Packag. Technol.* **1999**, *22*, 85–98. [CrossRef]
13. Shujuan, W.; Fang, H.; Bonan, S.; Guofu, Z. Method for calculation of contact resistance and finite element simulation of contact temperature rise based on rough surface contact model. In Proceedings of the 26th International Conference on Electrical Contacts (ICEC 2012), Beijing, China, 14–17 May 2012; pp. 317–321.
14. Ren, W.; Zhang, C.; Sun, X. Electrical Contact Resistance of Contact Bodies with Cambered Surface. *IEEE Access* **2020**, *8*, 93857–93867. [CrossRef]
15. Lim, J.; Kim, H.; Kim, J.K.; Park, S.J.; Lee, T.H.; Yoon, S.W. Numerical and Experimental Analysis of Potential Causes Degrading Contact Resistances and Forces of Sensor Connectors for Vehicles. *IEEE Access* **2019**, *7*, 126530–126538. [CrossRef]
16. Malucci, R.D. Multispot model of contacts based on surface features. In Proceedings of the 36th IEEE Holm Conference on Electrical Contacts, Montreal, QC, Canada, 20–24 August 1990; pp. 625–634.
17. Malucci, R.D. Making contacts to aged surfaces. In Proceedings of the 38th IEEE Holm Conference on Electrical Contacts, Philadelphia, PA, USA, 18–21 October 1992; pp. 237–248.
18. Fukuyama, Y.; Sakamoto, N.; Kaneko, N.; Kondo, T.; Toyoizumi, J.; Yudate, T. The effect of the distribution of a-spots in the peripheral part of an apparent contact point on the constriction resistance. In Proceedings of the 2017 IEEE Holm Conference on Electrical Contacts, Denver, CO, USA, 10–13 September 2017; pp. 302–305. [CrossRef]
19. Swingler, J.; McBride, J.W. Fretting and the reliability of multi-contact connector terminals. *IEEE Trans. Compon. Package Technol.* **2002**, *25*, 670–676. [CrossRef]
20. Acura TL Corroded/Burnt AC Blower Wire Harness Connector. Available online: https://joshsworld.com/cars/corroded-burnt-ac-blower-connector/ (accessed on 9 March 2022).
21. BMW 7 Series. Available online: https://widnesautoelectrical.co.uk/bmw-7-series/ (accessed on 9 March 2022).
22. Swingler, J.; McBride, J.W.; Maul, C. The degradation of road-tested automotive connectors. *IEEE Trans. Compon. Package Technol.* **2000**, *23*, 157–164. [CrossRef]

23. Jedrzejczyk, P. Analyse and Quantification of Electrical Contacts Endurance under Fretting Loadings. Ph.D. Thesis, Politechnika Lodzka, Lodz, Poland, 2010.
24. Su, B.; Qu, J. Electrical contact resistance of silver with different coatings. In Proceedings of the International Symposium on Advanced Packaging Materials: Processes, Properties, and Interfaces, Irvine, CA, USA, 16–18 March 2005; pp. 75–78. [CrossRef]
25. Shibata, Y.; Oohira, S.; Masui, S.; Sawada, S.; Iida, K.; Tamai, T.; Hattori, Y. Detailed analysis of contact resistance of fretting corrosion track for the tin-plated contacts. In Proceedings of the 26th International Conference on Electrical Contacts (ICEC 2012), Beijing, China, 14–17 May 2012; pp. 228–232. [CrossRef]
26. Fukuyama, Y.; Sakamoto, N.; Kaneko, N.-H.; Kondo, T.; Onuma, M. Constriction resistance of physical simulated electrical contacts with nanofabrication. In Proceedings of the 2014 IEEE 60th Holm Conference on Electrical Contacts (Holm), New Orleans, LA, USA, 12–15 October 2014; pp. 1–5. [CrossRef]

materials

MDPI

Article

Molecular Dynamics Simulation of Sintering Densification of Multi-Scale Silver Layer

Peijie Liang [1], Zhiliang Pan [1], Liang Tang [1], Guoqi Zhang [1,2], Daoguo Yang [1], Siliang He [1,3,*] and Haidong Yan [1,4,*]

1 Guangxi Key Laboratory of Manufacturing System & Advanced Manufacturing Technology, School of Mechanical and Electrical Engineering, Guilin University of Electronic Technology, Guilin 541004, China; liangpj2019@163.com (P.L.); zpan@guet.edu.cn (Z.P.); tangliang@guet.edu.cn (L.T.); g.q.zhang@tudelft.nl (G.Z.); daoguo_yang@163.com (D.Y.)

2 Delft Institute of Microsystems and Nanoelectronics (Dimes), Delft University of Technology, Mekelweg 6, 2628 CD Delft, The Netherlands

3 Institute of Semiconductors, Guangdong Academy of Sciences, Guangzhou 510650, China

4 Hang Zhou Global Scientific and Technological Innovation Center, Zhejiang University, Hangzhou 311200, China

* Correspondence: siliang_he@guet.edu.cn (S.H.); haidong_yan@zju.edu.cn (H.Y.)

Abstract: Based on molecular dynamics (MD), in this study, a model was established to simulate the initial coating morphology of silver paste by using a random algorithm, and the effects of different sizes of particles on sintering porosity were also analyzed. The MD result reveals that compared with the sintering process using large-scale silver particles, the sintering process using multi-scale silver particles would enhance the densification under the same sintering conditions, which authenticates the feasibility of adding small silver particles to large-scale silver particles in theory. In addition, to further verify the feasibility of the multi-scale sintering, a semi in-situ observation was prepared for a sintering experiment using micro-nano multi-scale silver paste. The feasibility of multi-scale silver sintering is proved by theoretical and experimental means, which can provide a meaningful reference for optimizing the sintering process and the preparation of silver paste for die-attach in powering electronics industry. In addition, it is hoped that better progress can be made on this basis in the future.

Keywords: molecular dynamics; initial coating morphology; random algorithm; semi in-situ observation

Citation: Liang, P.; Pan, Z.; Tang, L.; Zhang, G.; Yang, D.; He, S.; Yan, H. Molecular Dynamics Simulation of Sintering Densification of Multi-Scale Silver Layer. *Materials* **2022**, *15*, 2232. https://doi.org/10.3390/ma15062232

Academic Editors: Marc Cretin, Zhenghua Tang and Sophie Tingry

Received: 28 January 2022
Accepted: 15 March 2022
Published: 17 March 2022

1. Introduction

Sintering using silver particles is considered to be one of the ideal bonding techniques for the wide bandgap semiconductors, e.g., silicon carbide (SiC) and gallium nitride (GaN) devices, because this technique is capble of forming a high-melting-point sintered silver joint with sintering temperatures lower than 300 °C [1–4].

Nanosilver paste is considered an ideal bonding material since it can sinter at a temperature lower than 250 °C without pressure and operate at a temperature higher than 400 °C [5–8]. However, the difficulty of preparation and preservation of nanosilver paste results in a high cost in the sintering process. With the continuous improvement of power performance requirements of power electronic equipment, realizing the maximum commercial value of sintered silver is no longer limited to optimizing the sintering process, sintered joint performance, and energy consumption. Rather, optimizing the formula of silver paste is also very important There has been a rapid increase in the demand for power electronics devices because of the strong growth of electric vehicles. In order to maximize commercial value, while considering the sintering behaviors, mechanical properties of joint, energy consumption, and safety of silver particles, researchers are focusing on optimizing the silver particle formula in silver particles paste to achieve compromise solutions. Figure 1a shows the scanning electron microscope (SEM) image of silver nanoparticles after sintering without pressure. Sintering neckings formed between the nano particles can be clearly observed due to the easy diffusion between the adjacent particles [9]. However, the rapid diffusion of silver nanoparticles formed a sintered layer with high porosity during the sintering process. Figure 1b shows the SEM image of silver micron particles sintered under the same sintering conditions as Figure 1a. Compared with Figure 1a, the necking between micron particles is few and inconspicuous. It is difficult to form diffusion necks in the pressureless sintering process due to the low surface energy of large particles. Therefore, applying 10–30 MPa or even higher assist-pressure during the sintering process was used to obtain a reliable micron silver sintered joint [10,11].

Figure 1. The SEM image of Ag sintered layer with (**a**) nano-silver and (**b**) micro-silver by Kazohiko Suganuma et al., adapted with permission from Ref. [9], 2022 Chuantong Chen; (**c**,**d**) are the sintering morphology of multi-scale silver at different sintering temperatures by H D. Yan et al., adapted with permission from Ref. [12]. 2022 Haidong Yan; (**e**) super large silver particles and micron silver particles formed silver neck after sintering.

Figure 1a,b showed the SEM of different particles after pressureless sintering at 250 °C for 60 min by Suganuma's team from Osaka University. It demonstrates the advantages and disadvantages of different scale granular silver sintered solder joints, which are verified by using the experimental method of controlling variables. Based on this study, a pressureless sintering process was proposed by adding the silver nano particles in the silver micro paste to improve the sintering driving force and reduce the sintering temperature, as shown in Figure 1c,d [12]. In our previous study, we were surprised to find that, after pressureless sintering at 250 °C for 60 min, a particle with a diameter of 3.6 μm can form a sintering necking with a silver particle with a diameter of 27.6 μm, as shown in Figure 1e. These results prove the feasibility of low cost nano-micron multi-scale silver paste which prepared by appropriate addition of nanoparticles in low-cost micron silver paste as the potential bonding materials for sintering. The reliability of multi-scale silver sintering as a bonding layer has been clarified in many previous research works [12–15]. However, in past demonstrations of multi-scale silver sintering, most of the experiments are based on the densification analysis at the macro level, and the densification analysis at the micro molecular level is not explained. The initial dynamic diffusion evolution of sintering using multi-scale silver particles, which is important to reveal the sintering mechanism, is not clear and has thus far been difficult to observe. In order to further reveal the micro mechanism of mixed scale particle sintering, we use molecular dynamics simulation to reveal the difference of diffusion mechanism between mixed particles and large particles.

MD is a multidisciplinary technology that relies on Newtonian mechanics and integrates physics, chemistry, and mathematics to simulate the molecular system's motion, which is also widely used in studying the diffusion mechanism of silver sintering [16]. In this study to simulate more authenticity, the sintering of different sizes of particles for MD analysis was prepared with a random algorithm. It is interesting to note that using the random algorithm could well simulate the randomness of particles stacking and arrangement in the silver paste layer, which will be described in the next chapter. After simulation, to further verify the feasibility of the multi-scale sintering, a semi in-situ observation was prepared for a sintering experiment using micro-nano multi-scale silver paste.

2. Method

2.1. Methodology of Simulation

2.1.1. Molecular Dynamics Interatomic Potential

Sintering technology forms a dense joint based on the particles' diffusion below the melting temperature, according to the embedded atom method (EAM) [17]. Using the Large-Scale Atomic Molecular Massively Parallel Simulator (LAMMPS), this study combined the EAM to Simulate the diffusion between particles with different scales at different temperatures. The simulation atom-style was selected as atomic because sintering bonding is a diffusion process between particles. The timestep unit is related to the selected atom-style; In this study, the timestep size is set to the default values 0.001 ps to steadily solve Newton's equation of motion.

The molecular dynamics model is composed of atoms arranged in a specific order, and it could cause geometric distortion if the simulation's model size is too small. In order to reduce the model distortion, we had established the minimum particle radius of 10 Å in this study. The three style particles with radii of 10 Å, 30 Å, and 50 Å were designed by the "equal interval" method. As in past studies [18,19], the model's crystal orientation could affect the simulation results. To make the simulation more closely fit the actual situation, the particles in the simulation box as described in the next section were set at different crystal orientations by random placement.

2.1.2. Random Algorithm

Matlab is a matrix-based language that can provide a most natural expression of computational mathematics and provides researchers with the required algorithm programming for designing and analyzing the data [20].

In order to simulate a random mix like the particles of the silver paste coating on the substrate in the real application, Matlab used mathematical analysis and computation to design a simulation box with the random algorithm. The circles with different radius are randomly arranged in proportion according to the random function in the box. However, it is difficult to avoid intersecting and overlapping circles because generating random circles generates random dots and draws circles. Thus, in this step, we design the circles with a proposed ratio by the random function and remove the circles which intersect or overlap, and, finally, output the random circle's position and size which had been retained.

In this study, we had designed a Half-two-dimensional (H2D) box of 500 × 6 × 300 Å along *X*, *Y*, and *Z* directions with the periodic boundary to simulation the densification process of one cross-section of the silver layer to reduce simulation redundancy, shorten the running time, and achieve a more precise understanding of the densification process In order to understand the effect of multiscale on densification, the following three groups of different mass ratio groups are set in this study as shown in Table 1.

Table 1. Grouping with different mass ratios.

Group	10 Å	30 Å	50 Å
A	0	0	1
B	0	1	1
C	1	1	1

However, the mass ratio in the grouping must transform into the corresponding particle ratio before modeling. In the H2D simulation in this study, the conversion of mass ratio and particle ratio is associated with cross-sectional particle area, and the around ratios of the corresponding particles are shown in Table 2.

Table 2. Particle ratio corresponding to different mass ratio (base on H2D).

Group	10 Å	30 Å	50 Å
A	0	0	1
B	0	3	1
C	25	3	1

The particle ratio was substituted into the random equation in MatLab, and we obtained the random arrangement of various circles shown in Figure 2. Position coordinates of the circle generated in the random function are extracted and substituted into the MD model. In order to calculate the porosity easily and make the simulation the silver paste coating on the substrate, we designed the upper and lower silver plates in the *Z* direction.

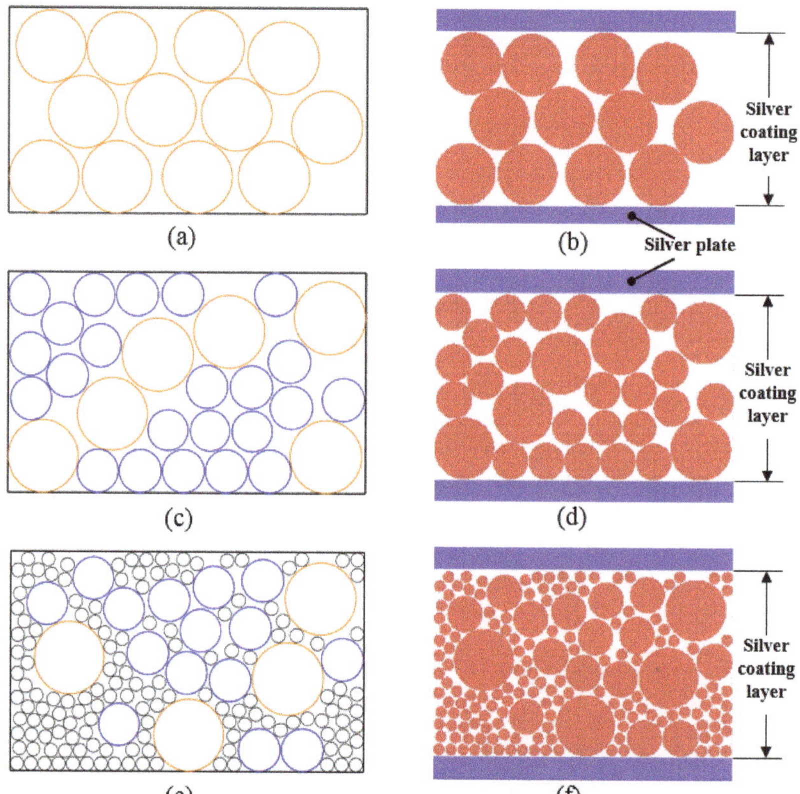

Figure 2. The circles' random placement by MatLab with diameter ratio from Table 2 with (**a**) group A, (**c**) group B, and (**e**) group C. (**b**,**d**,**f**) were the MD model based on randomly placed data to (**a**,**c**,**e**), respectively.

2.2. Semi In-Situ Observation Sintering of Multi-Scale Silver

In-situ sintering observation is a method to observe the diffusion bonding process of silver particles by scanning electron microscope without resampling during the sintering, which observes the morphology change of the particles better than the traditional subsection sampling method [21,22].

In this study, we used a scanning electron microscope equipped with a protochips single tilt heating holder with a maximum heating temperature of 1200 °C to observe the multi-scale silver paste sintering semi in-situ.

During the sintering, in order to avoid electron beam radiation, only open the gun valve when sampling the sample [23]. We referred to the micron silver sintering profile by K. S. Siow [24], and we set the sintering temperature to 250 °C for 10 min on the pressureless setting.

3. Result and Discussion

3.1. Sintering Densification of Silver Particles with Different Proportions

Figure 3 shows the morphology of the sintering densification process of groups A, B, and C sintering at 300 °C. Comparing the sintering processes of A and B, it is obvious to find that group B with small particles is better than group A in the reduction of porosity, the reduction of which is more significant for group C with smaller silver particles based on group B. In the initial stage, driven by thermal energy, the particles in the sintering layer begin to "move" and "attract" the particles around them, and begin to gather to prepare for the start of sintering; which also called the first stage of sintering. With the continuous sintering, the grain boundary energy of particles begins to weaken and the particles begin to diffuse each other, which is a longer process (10–1000 ps). However, with the subsequent sintering, the change of sintering pore decrease becomes weaker, which is more clearly reflected in the change curve of porosity of groups in Figure 4. Multiscale group C not only shows lower porosity than group A and B after sintering but also reduces the area of pore due to its complementarity in the gap of large particles. In the actual production process, the large pore will lead to uneven heat dissipation, resulting in heat concentration and local thermal resistance increasing, which will also affect the reliability of packaging.

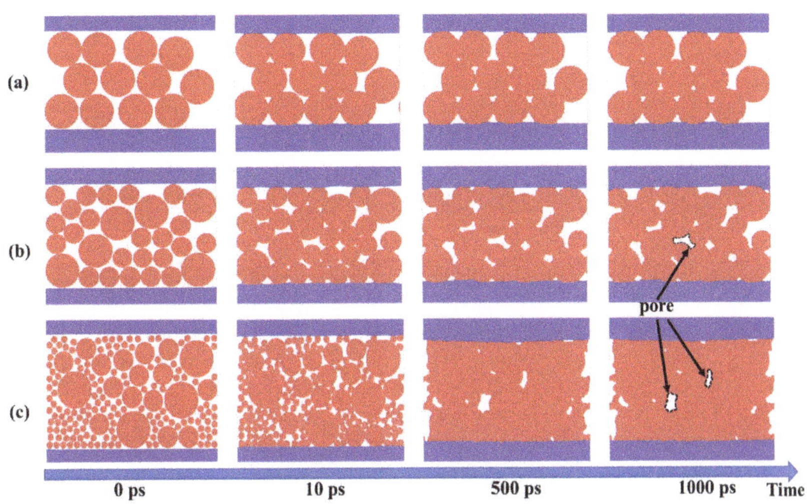

Figure 3. Morphology of sintering densification process of (**a**) group A, (**b**) B, and (**c**) C at 0 ps, 10 ps, 500 ps, and 1000 ps.

Figure 4a shows the changing profiles of different particle ratios' porosity and Figure 4b shows the change rate of porosity. The initial stage is the early stage of sintering, so this stage has the maximum driving force in the whole pressureless sintering process. It shows that the porosity decreases rapidly which is shown in Figure 4a, and the diffusion rate is the fastest in the whole sintering, as shown in Figure 4b. With the continuous sintering, the porosity changes gradually from the initial sharp change to stability. We also found that the larger the particle, the earlier it tends to be stable; see Figure 4b. It is also found from Figure 3 that the larger pores which are found at 500 ps hardly change after 1000 ps, while the smaller pores will be eliminated or further reduced. This is because the bonding completed early greatly weakens the surface energy in the system, which leads to the application temperature being insufficient to provide sufficient diffusion driving force. Porosity slight shrinkage leads to the diffusion rate slowing down.

Combined with the curves from Figure 4, we intuitively find that the porosity and porosity change rate of the simulation group with small particles is better than without the small particles. The relative surface energy is inversely proportional to the particle size, and adding small particles to large particles can improve the sintering driving force. The relationship between particle size and corresponding drive force follows [16]:

$$\sigma = \gamma(\frac{1}{R_1} + \frac{1}{R_2}) \tag{1}$$

where σ_i is the sintering driving force, σ_i is the material surface energy, and R_1 and R_2 are the radii of two particles that diffuse each other. It can be seen from the graph in Figure 4b that the maximum sintering diffusion change rate in the initial stage of mixed particle sintering group C can reach 0.67, which is twice that of group A with only large particles reaching 0.32.

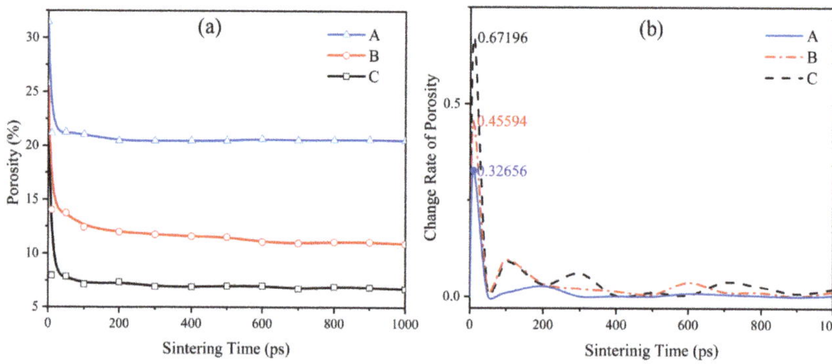

Figure 4. The profile of (**a**) porosity change with sintering time and corresponding (**b**) change rate of porosity.

From the above results, with the same sintering temperature and time, multi-scale silver is better than large particle sintering in sintering porosity. Moreover, the small particles in multi-scale silver act as a "filler", which also greatly reduces the average volume of pores. In order to further understand the diffusion mechanism of silver particles with different sizes in the sintering process, different silver particles' change of necking was simulated and analyzed. This will be described in the next section.

3.2. Evolution of Sintering Silver Neck Size

As shown in Figure 5 the bonding of the following three different size particles pair was studied. The particles pair in the model set different crystal orientations to avoid the contingency of the same crystal orientation. In this chapter, we studied the necking of different particles' radii. We cut 474, 13,240 and 61,370 silver atoms from a large silver FCC supercell to form silver particle pairs with radii of 10 Å, 30 Å and 50 Å respectively as shown in Figure 5, the initial distance of 2 Å between particle pair. As shown in Figure 4, the porosity in the sintering process decreases rapidly within the first 200 ps, and then the densification slows down. Therefore, in this chapter, the simulation mainly takes the first 300 ps for study.

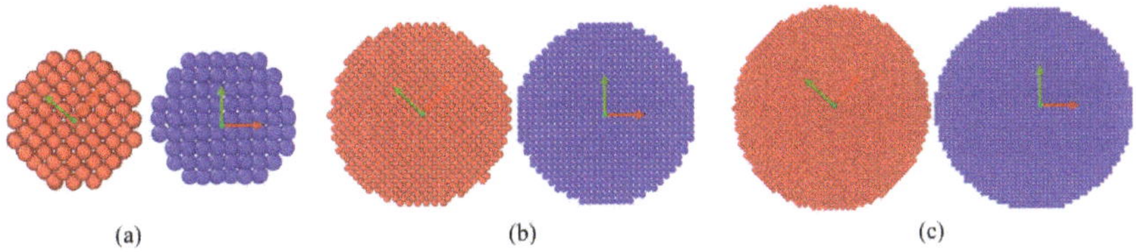

Figure 5. The Silver particles pair with different radius: (**a**) 10 Å, (**b**) 30 Å, and (**c**) 50 Å.

Figure 6 shows the simulated value and the simulated value's fitting curve of x/r (x is the radius of diffusion neck and r in the radius diffusion particle.) of different particles' radii during the sintering. As shown in Figure 6, the necking of particles with radius of 10 Å grows and diffuses rapidly, and the x/r reaches 1 at about 8 ps (the necking was completed). Figure 7 shows the profile of total potential energy during the necking of particles with a radius of 10 Å. Two particles approach and begin to diffuse under the driving force provided by thermal energy. The relative potential energy of the system decreases to provide diffusion kinetic energy. When the diffusion necking is completed, the system's kinetic energy begins to be constant, the relative potential energy stops decreasing and remains stable. As the thermal energy in the system continues to be maintained, some atoms will cross the potential barrier and further diffuse, and potential energy will also change, as shown in the potential energy fluctuation at 250 ps in Figure 7.

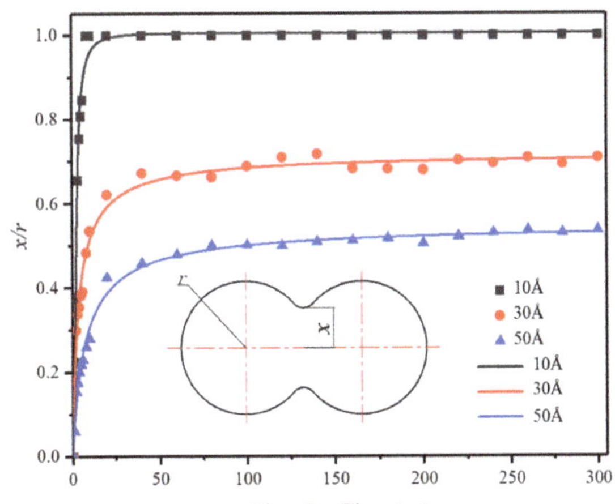

Figure 6. The measurement data and fitting curve of x/r with the time of silver particles pair with different radius sintered at 300 °C. The fitted curve conforms to the growth of the logarithmic function, which is consistent with the rapid change to slow the change of measured data.

However, under the same simulation conditions, the growth rate of particles with radii of 30 Å and 50 Å is slower, and the x/r with the radius of 30 Å tends to be stable when it increases to about 0.7, and the radius of 50 Å tends to be stable when it increases to about 0.52. This difference further indicates that small particles can diffuse more easily than large ones at the same temperature This difference further indicated that small particles are more easily diffused than large particles at the same temperature. The process of necking size accords with the diffusion formula [25]:

$$(\frac{x}{r})^n = \frac{Bt}{(2r)^m} \tag{2}$$

where t is the sintering time, n and m are constants dependent on the specific transport mechanism, and B is a term made of material and geometric constants. Combined with

Equation (2) and the trend of measured data in Figure 6, the radii of 30 Å and 50 Å will continue to increase until the maximum values are reached if the simulation time or sintering temperature increases.

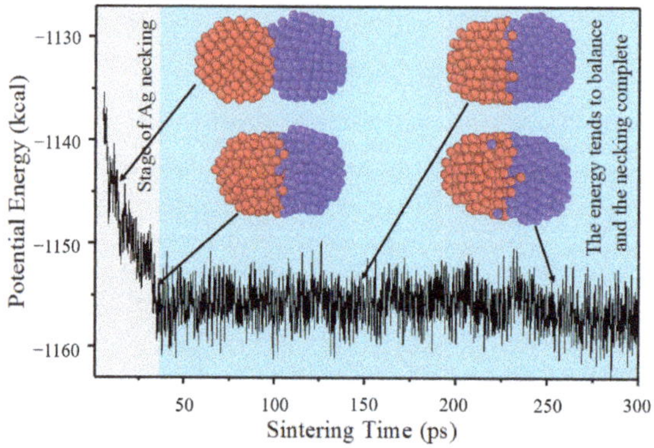

Figure 7. Potential energy with time during sintering of particle pair with a radius of 10 Å.

Small silver particles were doped into large silver particles to shorten the sintering bonding time and increase the sintering neck of large particles in this study, as shown in Figure 8. Figure 8a shows the sectional of 50 Å particles dope with 10 Å particles. In order to observe the diffusion process, the particles were divided into different groups and colored, respectively. As driven by thermal energy, the small particles approach two large particles and diffuses at 1 ps. With the sintering continuing, small particles diffuse into and fill the bonding gap of two large particles, and the diameter of the bonding neck reached 67.8 Å at 10 ps. However, under the same conditions, the diffusion diameter of only large particles is only 28.69 Å. With continued sintering, small particles continue to "climb" along with two large particles, further forming bonds with large particles. At the end of sintering, the neck of the sintering group with small particles shrinks to 66.46 Å, which is larger than 53.66 Å sintering with only large pair particles.

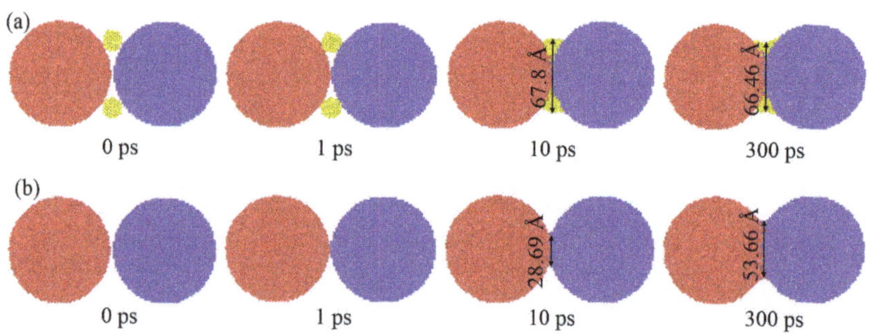

Figure 8. MD profile of particle pair with the radius of 50 Å; (**a**) doping the silver particles with radius of 10 Å, and (**b**) control group without any doping.

Small particles can "close" two large particles far away to form necking to increase the density of sintering, as shown in Figure 9. The initial distance of the two large particles in Figure 9a,b is 10 Å and the difference between them is that Figure 9a had added some small particles in the middle and the lower part between two large particles. Figure 9a shows that large particles pair get close together and form necking due to the pulling of small particles' diffusion at the bottom. This is because small particles provide a certain pulling force for large particles to bond. Driven by this pulling force, the particles can close together and form necking. This can also be confirmed from the particle pair in Figure 9b which, without small particles, always keep the distance during the simulation process.

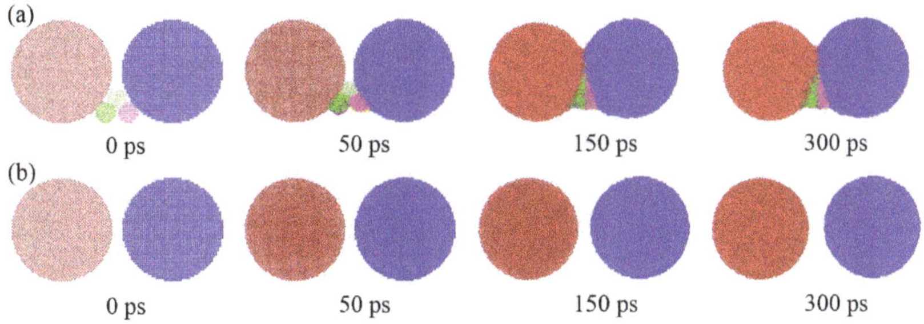

Figure 9. MD profile of a particle pair with radius 50 Å; (**a**) Add small particles in the middle and lower part between large particle pair and (**b**) control group without any doping.

From the above results, whether from the simulation results of silver layer sintering or silver neck diffusion analysis, the evidence points to the superior characteristics of multi-scale silver sintering. From the perspective of molecular dynamics, the small silver particles of multi-scale silver do have a good densification effect on the sintering of large particles. However, this micro analysis does lack some support in practical application. Considering this, in order to further verify this feasibility, we will carry out semi in-situ sintering of multi-scale silver sintering in the next chapter.

3.3. Result of Semi In-Situ Observation

From Figure 10, the diffusion bonding process of Ag particles during the sintering by semi in-situ SEM is clearly shown. From the in-situ sintering diagram, particle pair B which is closed in the initial stage begins to form an obvious diffusion bonding neck when the sintering is carried out for 6 mi, as shown in Figure 10b. In the subsequent sintering, the diffusion silver neck of the B particle pair continues to grow. At the same time, nanosilver particles around the particle pair continue to diffuse to the large particles, also resulting in the also growth of the large particle pair while bonding. The bonding reaction between micron particles and nanoparticles in the A particle pair increase the feasibility of multi-scale sintered silver. For the A particle pair, we can see that the two particles are not in contact before sintering. However, after 8 min sintering, the particle pairs are driven by the nanoparticles and form a necking connection. Also as the particle pair which initial distance was 2.06 μm is shows in Figure 10a, and shorten to 0.6 μm after sintered 8 min. At the same time, the radius of the particle pair also increases from the initial 0.53 μm to 1.14 μm. This sintering change is consistent with the simulation results in Figure 9, which shows that the small silver particles of multi-scale silver paste can provide good sintering driving force in large particles, which can provide better density.

In this study, the micro silver paste with nano-silver is sintered for 8 min without pressure to form a densification bonding, which is much shorter than the traditional pure micro silver paste. Multiscale silver sintering is shortened not only the sintering time but also reduces the sintering temperature, which is also one of our future research directions.

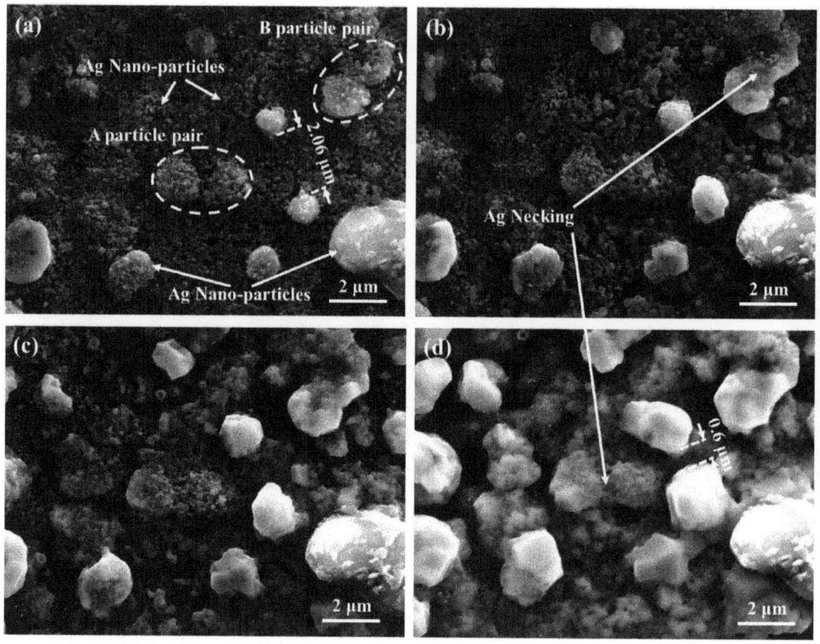

Figure 10. The SEM image of In-Situ multiscale silver sintering at different time steps (**a**) dried, (**b**) after sintering 6 min, (**c**) after sintering 7 min, and (**d**) after sintering 8 min.

4. Conclusions

Using molecular dynamics, combined with MatLab random arrangement,he sintering densification mechanism of silver with different mixed particle ratios has been studied, and the neck growth mechanism of a silver particle pair in the sintering process was analyzed. Combined with the validation experiment, the following conclusions can be summarized:

1. With the same conditions such as sintering temperature and time, adding small silver particles to large silver particles is beneficial to reduce porosity.
2. During the initial sintering stage, the x/r of the small silver particle pair is faster than the large silver particle pair due to higher surface diffusion under heating. Moreover, the x/r of a small particle pair is higher than that of a large particle pair at the same temperature due to the high surface diffusion energy. Using this mechanism improves the radius of the bonding neck and shortens the sintering time by adding small silver particles to large particles.
3. In the multiscale silver sintering experiment, nano-silver as the medium of micron silver bonding could effectively shorten the sintering time and reduce the effective sintering temperature to improve the packaging yield.

Author Contributions: Conceptualization, P.L. and H.Y.; methodology, G.Z., H.Y., D.Y., S.H. and Z.P.; software, P.L., Z.P. and L.T.; formal analysis, P.L., H.Y. and Z.P.; experiment, H.Y., L.T. and S.H.; data curation, P.L., H.Y., S.H., L.T. and; writing, P.L., S.H. and H.Y.; supervision, H.Y., S.H. and D.Y.; project administration, H.Y., S.H. and D.Y.; funding acquisition, H.Y., S.H. and Z.P. All authors have read and agreed to the published version of the manuscript.

Funding: Project supported by National Natural Science Foundation of China (51967005, and 12172096); Science and Technology Planning Project of Guangxi Province (GuiKe AD20297022, and Guike AD20159081); Guangxi Key Laboratory of Manufacturing System & Advanced Manufacturing Technology (Grant No. 20-065-40-003Z, and Grant No. 19-050-44-006Z).

Data Availability Statement: Not applicable.

Conflicts of Interest: The authors declare that there is no conflict of interest regarding the publication of this article.

References

1. Liang, P.; Yan, H.; Li, W.; Yang, D. Void Eliminating Process of Sintered-Silver Die Attachment in Anaerobic-Sintering Atmospheres. In Proceedings of the 21st International Conference on Electronic Packaging Technology (ICEPT), Guangzhou, China, 12–15 August 2020.
2. Zhang, H.; Wang, W.; Bai, H.; Zou, G.; Liu, L.; Peng, P.; Guo, W. Microstructural and mechanical evolution of silver sintering die attach for SiC power devices during high temperature applications. *J. Alloys Compd.* **2018**, *774*, 487–494.
3. Ren, H.; Zou, G.; Jia, Q.; Deng, Z.; Du, C.; Wang, W.; Liu, L. Thermal stress reduction strategy for high-temperature power electronics with Ag sintering. *Microelectron. Reliab.* **2021**, *1247*, 114379. [CrossRef]
4. Yan, H.; Mei, Y.H.; Li, X.; Ma, C.; Lu, G.Q. A Multichip Phase-Leg IGBT Module Using Nanosilver Paste by Pressureless Sintering in Formic Acid Atmosphere. *IEEE Trans. Electron Dev.* **2018**, *65*, 4499–4505. [CrossRef]
5. Chua, S.T.; Siow, K.S. Microstructural studies and bonding strength of pressureless sintered nano-silver joints on silver, direct bond copper (DBC) and copper substrates aged at 300 °C. *J. Alloys Compd.* **2016**, *687*, 486–498. [CrossRef]
6. Gao, R.; Shen, Y.A.; Li, J.; He, S.; Nishikawa, H. Mechanical and microstructural enhancements of Ag microparticle-sintered joint by ultrasonic vibration. *J. Mater. Sci. Mater. Electron.* **2020**, *31*, 21711–21722. [CrossRef]
7. Liu, W.; An, R.; Wang, C.; Zheng, Z.; Tian, Y.; Xu, R.; Wang, Z. Recent Progress in Rapid Sintering of Nanosilver for Electronics Applications. *Micromachines* **2018**, *934*, 346. [CrossRef]
8. Wang, J.; Chen, S.; Zhang, L.; Zhao, X.; Duan, F.; Chen, H. Brief Review of Nanosilver Sintering: Manufacturing and Reliability. *J. Electron. Mater.* **2021**, *50*, 5483–5498. [CrossRef]
9. Chen, C.T.; Suganuma, K. Microstructure and mechanical properties of sintered Ag particles with flake and spherical shape from nano to micro size. *Mater. Des.* **2019**, *162*, 311–321. [CrossRef]
10. Yan, H.; Liang, P.; Mei, Y.; Feng, Z. Brief review of silver sinter-bonding processing for packaging high-temperature power devices. *Chin. J. Electr. Eng.* **2020**, *6*, 25–34. [CrossRef]
11. Liu, Y.; Zhang, H.; Wang, L. Effect of sintering pressure on the porosity and the shear strength of the pressure-assisted silver sintering bonding. *IEEE Trans. Device Mater. Reliab.* **2018**, *18*, 240–246. [CrossRef]
12. Yan, H.; Mei, Y.H.; Wang, M.; Li, X.; Lu, G.Q. Pressureless sintering multi-scale Ag paste by a commercial vacuum reflowing furnace for massive production of power modules. *J. Mater. Sci. Mater. Electron.* **2019**, *30*, 9634–9641. [CrossRef]
13. McCoppin, J.; Reitz, T.L.; Miller, R.; Vijwani, H.; Mukhopadhyay, S.; Young, D. Low Temperature Consolidation of Micro/Nanosilver Die-Attach Preforms. *J. Electron. Mater.* **2014** *43*, 3379–3388. [CrossRef]
14. Yang, Y.; Li, Z.; Yang, S.; Li, Y.; Huang, J. Multiscale simulation study of laser sintering of inkjet-printed silver nanoparticle inks. *Int. J. Heat Mass Transf.* **2020** *159*, 120110. [CrossRef]
15. Lin, Y.C.; Liu, X.; Chou, K.W.; Tsai, E.H.; Zhao, C.; Holler, M.; Diaz, A.; Petrash, S.; Chen-Wiegart, Y.C. Unveiling 3D Morphology of Multiscale Micro-Nanosilver Sintering for Advanced Electronics Manufacturing by Ptychographic X-ray Nanotomography. *Adv. Eng. Mater.* **2020** *22*, 1901250.
16. Badar, M.; Shamsi, S.; Ahmed, J.A. *Molecular Dynamics Simulations: Concept, Methods, and Applications: Molecular Dynamics Simulations: Concept, Methods, and Applications*; Springer Nature: Washington, DC, USA, 2020; pp. 1–32.
17. Daw, M.S.; Foiles, S.M.; Baskes, M.I. The embedded-atom method: A review of theory and applications. *Mater. Sci. Rep.* **1993**, *9*, 251–310. [CrossRef]
18. Kwak, J.H.; Ko, M.S.; Kim, K.H. The Effect of Crystal Orientation on Damping Capacity in Rolled AZ31 Magnesium Alloy. *Mater. Sci. Forum* **2018**, *941*, 1127–1131.
19. Han, Q.N.; Rui, S.; Qiu, W. Effect of crystal orientation on the indentation behaviour of Ni-based single crystal superalloy. *Mater. Sci. Eng.* **2018**, *773*, 138893. [CrossRef]
20. Grigoriu, M. Models for space-time random functions . *Probabilistic Eng. Mech.* **2016**, *43*, 5–19. [CrossRef]
21. Wang, J.; Zeng, Z.; Weinberger, R.C.; Zhang, Z.; Zhang, Z.; Mao, S.X. In situ atomic-scale observation of twinning-dominated deformation in nanoscale body-centred cubic tungsten. *Nat. Mater.* **2015**, *14*, 594–600. [CrossRef]
22. Yuan, W.T.; Zhang, D.W.; Ou, Y. Direct In Situ TEM Visualization and Insight into the Facet-Dependent Sintering Behaviors of Gold on TiO_2. *Angew. Chem.* **2018**, *130*, 17069–17073.
23. Li, G.; Fang, K.; Chen, Y.; Ou, Y.; Mao, S.; Yuan, W.; Wang, Y.; Yang, H.; Zhang, Z. Unveiling the Gas-dependent Sintering Behavior of Au-TiO_2 Catalysts via Environmental Transmission Electron Microscopy. *J. Catal.* **2020**, *388*, 84–90. [CrossRef]
24. Siow, K.S.; Chua, S.T. Thermal ageing studies of sintered micron-silver (Ag) joint as a lead-free bonding material. *Met. Mater. Int.* **2020**, *26*, 1404–1414. [CrossRef]
25. Tan, S.; Zhang, X.; Zhao, Z.; Wu, Z. Driving force evolution in solid-state sintering with coupling multiphysical fields. *Ceram. Int.* **2020**, *46*, 11584–11592.

materials

Article

Improving Energy Harvesting from Bridge Vibration Excited by Moving Vehicles with a Bi-Stable Harvester

Zhiyong Zhou [1,2], Haiwei Zhang [1], Weiyang Qin [3], Pei Zhu [1,*] and Wenfeng Du [1]

1 School of Civil Engineering and Architecture, Henan University, Kaifeng 475004, China; 10160091@vip.henu.edu.cn (Z.Z.); hwzhang@henu.edu.cn (H.Z.); 10160016@vip.henu.edu.cn (W.D.)
2 School of Power and Energy, Northwestern Polytechnical University, Xi'an 710072, China
3 Department of Engineering Mechanics, Northwestern Polytechnical University, Xi'an 710072, China; qinweiyang@nwpu.edu.cn
* Correspondence: 10160096@vip.henu.edu.cn

Abstract: To monitor the health status of the bridge, many sensors are needed to be mounted on it. Converting bridge vibration energy to electrical energy is considered as a potential solution to the problem of providing reliable electric power to these sensors. The objective of this work is to present an operable strategy for improving the electric energy output of a piezoelectric energy harvester installed on a bridge by introducing bi-stable characteristics. A bi-stable harvester is proposed. By adjusting the tip and fixed magnets, different types of potential energy can be generated, and then the harvester can exhibit the linear, mono-stable and bi-stable characteristics. In the bi-stable state, the harvester triggers snap-through motions easily and generates large outputs. The corresponding prototype was fabricated, and the experiment was carried out to validate the advantage of the bi-stable energy harvester. The experiment results show that the bi-stable energy harvester outperforms the classical linear harvester over the whole range of vehicle speed. As the vehicle speed exceeds a critical one, the snap-through motion will happen, which is beneficial to enhancing the electricity output. This conceptual design may provide guidance for promoting the performance of bridge energy harvesting.

Keywords: energy harvesting; bridge vibration; bi-stable energy harvester; linear energy harvester; snap-through motion

Citation: Zhou, Z.; Zhang, H.; Qin, W.; Zhu, P.; Du, W. Improving Energy Harvesting from Bridge Vibration Excited by Moving Vehicles with a Bi-Stable Harvester. *Materials* **2022**, *15*, 2237. https://doi.org/10.3390/ma15062237

Academic Editors: Marc Cretin, Sophie Tingry and Zhenghua Tang

Received: 16 February 2022
Accepted: 16 March 2022
Published: 17 March 2022

1. Introduction

Bridges are becoming a more and more important part of the modern transportation system [1]. It is necessary to monitor the health of bridges at all times to ensure their safety. Although deploying wireless sensors in the detection region is an ideal scenario to monitor the state of the bridge, providing cheap and reliable power to the sensors is a well-known difficulty [2]. At present, the conventional electrochemical battery is still the primary power source of wireless sensors, but replacing the battery costs a lot, especially in dangerous locations [3]. Moreover, the heavy metals in wasted batteries pose a threat to the health and living environment of human beings. Energy harvesting from vibrations is considered a great potential energy solution to powering wireless sensors. In practice, there exists abundant vibration energy when vehicles pass through the bridge. This vibration energy results from passing vehicles and is not affected by the weather, geographical conditions and other factors, so it emerges as a promising solution for the power of bridge health monitoring systems [4]. Moreover, the piezoelectric energy harvester is usually simple in structure and easy to be manufactured. Thus, it can be produced in a large amount, so the vibration-based energy piezoelectric harvester has received more and more attention [5].

In recent decades, there are many studies focusing on converting the energy of bridge vibrations into electric energy. Ali et al. [6] investigated a linear energy harvester for the highway bridge under the excitation of a moving vehicle. The highest efficiency of energy harvesting was sensitively affected by the resonance frequency. Peigney and Siegert [7] investigated energy harvesting from a concrete highway bridge by a cantilever energy harvester. The results showed that the electric energy output could reach 0.03 mW. Zhang et al. [8] used a cantilever beam energy harvester to harvest energy from four bridges excited by truck models. The results showed that poorer road conditions on bridges were helpful for harvesting vibration energy. Karimi et al. [9] considered inertial effects on energy harvesting from a bridge with a heavy vehicle traveling on it. Zhang et al. [10] analyzed the output voltages of the piezoelectric energy harvester for nine different conditions. The vehicle speed had a significant effect on the energy output. Since the bridge response is the excitation source of the linear piezoelectric energy harvester (LPEH), the resonant frequency of LPEH must be tuned to the fundamental frequency of the bridge to obtain a high harvesting efficiency. Thus, some significant efforts have been devoted to studying the vibration response of a bridge and estimating the fundamental frequency and stiffness parameters. Aloisio et al. [11] conducted a modal analysis for a multi-span concrete bridge to estimate the compressive strength of concrete specimens with seven spans. Bonopera et al. [12] predicted the short-term dynamics of the uncracked prestressed concrete members, which was also valuable for designing new concrete bridges. However, it should be noted that the vibration energy of a bridge has a broadband frequency range since the speeds and weights of vehicles crossing the bridge change with time. Thus, the LPEH is inefficient in harvesting the broadband random vibration energy.

In recent years, some researchers tried utilizing nonlinearity to improve the operating bandwidth. The investigations on nonlinear energy harvesters with mono-stability [13], bi-stability [14], tri-stability [15] and quad-stability [16] have been carried out. For example, Stanton et al. [17] studied the hardening and softening response of the nonlinear mono-stable piezoelectric energy harvester. Sebald et al. [18] conducted a theoretical analysis of the nonlinear Duffing oscillator. The simulation results showed that the highest solution could obtain a huge gain in output power. Erturk and Inman [19] introduced a bi-stable piezoelectric energy harvester (BPEH) to improve the performance of power generation, which had a significant advantage over the LPEH. Arrieta et al. [20] introduced bi-stable characteristics in the cantilever composite to enhance the energy harvesting effectiveness under low-amplitude input levels. Masana et al. [21] designed an axially loaded BPEH, which could produce large output voltages. Lan et al. [22] added a small magnet in BPEH to improve the energy harvesting ability. Pellegrini [23] introduced a guideline to classify the BPEHs according to the feasibility of fabrication. Zhou et al. [24] proposed a broadband tri-stable piezoelectric energy harvester to further improve the energy harvesting ability. The results demonstrated that the tri-stable piezoelectric energy harvester could be superior to the bi-stable one, which was more suitable to the low input acceleration level. Litak et al. [25] investigated multiple solutions of a tri-stable energy harvester. It was found that the initial condition played an important role in the formation of the basins of attraction. Zhou et al. [26] designed a quad-stable energy harvester to improve the energy harvesting efficiency. The results proved the energy harvester designed in a quad-stable state made it much easier to attain snap-through motion, and its energy harvesting efficiency was significantly improved. Therefore, introducing nonlinear multi-stability is an effective way for efficient piezoelectric energy harvesting from the excitation in the shaking table test. Recently, Zhou et al. [27] exploited the mono-stable characteristics generated by the magnetic interaction to harvest bridge vibration energy, which brought about a noticeable improvement in the harvesting performance. However, to the authors' best knowledge, the concept of harvesting energy from bridge vibrations by bi-stable characteristics has not been reported so far. In bridge vibration energy harvesting, the bi-stable characteristics may help improve the harvesting performance significantly.

In this paper, the bi-stable characteristics generated by the magnetic interaction were exploited to enhance the performance of energy harvesting from a bridge under the excitation of moving vehicles. When the excitation triggers the BPEH to execute snap-through motion and oscillates in a high-energy orbit, it can generate a large output. In order to study the dynamic characteristics of the BPEH, the potential energy, restoring force and stiffness are analyzed. Then, the output voltages and dynamic strains of LPEH and BPEH are measured in experiments and compared, as a vehicle model is designed to run on the bridge with different speeds. Finally, some important findings and conclusions are drawn.

2. Energy Harvester for Bridge Vibration Induced by a Moving Vehicle

The schematic sketch of the BPEH is shown in Figure 1. The BPEH is composed of a substrate partly covered by a piezoelectric patch or two, a tip magnet and two fixed magnets. A tip magnet is placed at the free end of the substrate, while two fixed magnets are mounted on the frame and positioned near the tip magnet. The tip magnet and fixed magnets have opposite polarities and face each other to produce a nonlinear magnetic attraction force. For clarity, we define two distances: one is the gap distance between two fixed magnets (d_g); the other is the separation distance between the tip and the fixed magnets (d). Both of them can be adjusted to make the BPEH own two stable equilibrium positions. When the BPEH is excited by bridge vibrations, the piezoelectric patch will deflect and generate voltage. There is load resistance in the circuit. The BPEH can jump between the two stable equilibrium positions and execute snap-through motions, and then it will generate high voltage outputs. If the two fixed magnets are removed, the system degenerates to a classical LPEH. For the LPEH, as a linear system, it usually owns a narrow resonant bandwidth. Because the acceleration spectrum of bridge vibration usually exhibits a random characteristic and distributes over a wide range, the energy harvesting efficiency of LPEH under this excitation will be far from satisfactory.

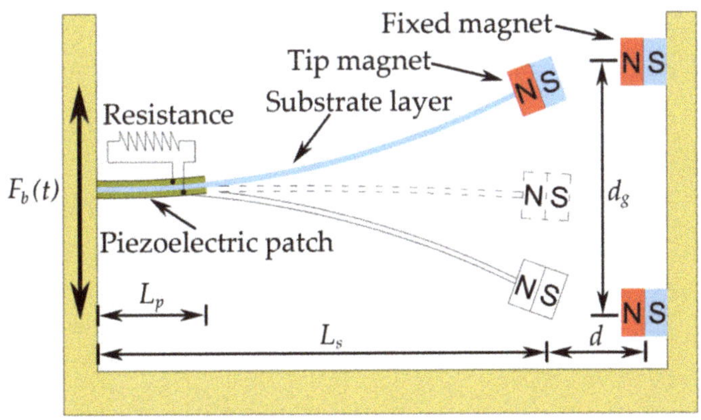

Figure 1. Schematic of the BPEH.

Figure 2 shows the schematic of harvesting vibration energy from a bridge with a moving vehicle. As the vehicle runs on the bridge, the bridge is under the excitation of a moving load, so it will oscillate. In analysis, for simplicity, the vehicle can be modeled as a mass traveling on an elastic beam with length L_b and thickness T_b [9]. When the vehicle moves at a speed of v and passes the bridge, the bridge's oscillation will make the BPEH vibrate and produce snap-through motions.

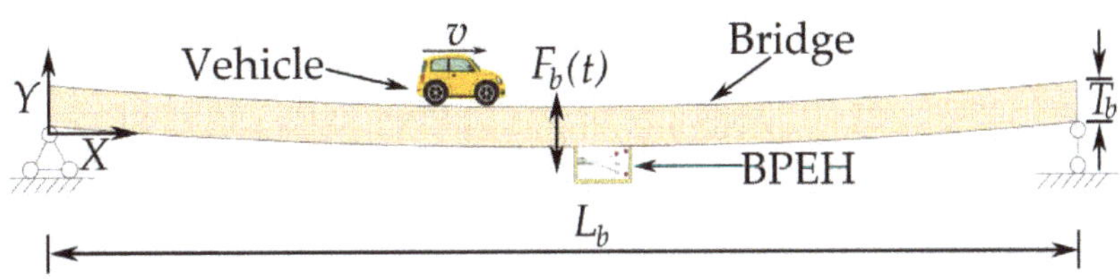

Figure 2. Energy harvesting from a bridge traversed by a vehicle.

3. Modeling Potential Energy, Restoring Force and Stiffness

As we know, the LPEH has a linear restoring force and constant stiffness. It could exhibit a satisfactory work efficiency only when the frequency of the excitation source coincides with the resonance frequency. In fact, the vibration energy in the realistic environment usually distributes over a broadband frequency range, so the LPEH is inefficient in harvesting vibration energy. In contrast, the BPEH may overcome this defect.

To show the characteristics of the BPEH, its potential energy, restoring force and stiffness are presented. The total potential energy of the system includes three parts: the elastic potential energy of the substrate and piezoelectric patch, the gravity potential energy and the magnetic potential energy.

The substrate can be modeled as a Euler–Bernoulli beam, whose strain is proportional to the second spatial derivative of deflection. Thus, the elastic potential energy of the piezoelectric beam can be given by [28]

$$U_b = \frac{1}{2}E_s I_s \int_0^{L_s} \left(\frac{\partial^2 y(x,t)}{\partial x^2}\right)^2 dx + E_p I_p \int_0^{L_p} \left(\frac{\partial^2 y(x,t)}{\partial x^2}\right)^2 dx \tag{1}$$

where E_s and E_p are the Young's modulus of the substrate and that of the piezoelectric patch, respectively; I_s and I_p are the inertial moments of the substrate and the piezoelectric patch, respectively; L_s and L_p are the lengths of the substrate and the piezoelectric patch, respectively; $y(x,t)$ is the displacement of the piezoelectric beam.

The magnetic potential energy of the system comes from the magnetic attraction between the tip and fixed magnets, and the point dipole approximation is employed to model the magnets. Then, the magnetic potential energy considered in the BPEH can be given by [29]:

$$U_m = -\frac{\mu_0 a_1 a_2}{2\pi}\left\{\left(y(L_s,t) - \frac{d_g}{2}\right)^2 + d^2\right\}^{-\frac{3}{2}} - \frac{\mu_0 a_1 a_3}{2\pi}\left\{\left(y(L_s,t) + \frac{d_g}{2}\right)^2 + d^2\right\}^{-\frac{3}{2}} \tag{2}$$

where μ_0 is the magnetic permeability constant; a_1 and a_2 (a_3) are the effective magnetic moments of the tip and fixed magnets, respectively; d is the distance between the tip magnet and the fixed magnets in the horizontal direction (as shown in Figure 1); d_g is the vertical distance between the two fixed magnets.

The gravitation potential energies of the substrate, piezoelectric patch and tip magnet can be given by the following equation:

$$U_g = -m_t g y(L_s,t) - \rho_s h_s b_s g \int_0^{L_s} y(x,t)dx - 2\rho_p h_p b_p g \int_0^{L_p} y(x,t)dx \tag{3}$$

where m_t is the mass of the tip magnet; g is the gravitational acceleration; ρ_s and ρ_p are the densities of the substrate and piezoelectric patch, respectively; h_s and h_p are the thicknesses of the substrate and piezoelectric patch, respectively; b_s and b_p are the widths of the substrate and piezoelectric patch, respectively.

Thus, the total potential energy of BPEH can be given as follows

$$U_t = U_b + U_m + U_g \tag{4}$$

The dynamic characteristics of the energy harvester can be inferred from its potential energy shape. Figure 3 shows the potential energy shapes of the harvester for different separation distances (d). The corresponding parameters are listed in Table 1. As is clear from Figure 3, when d is quite large (e.g., d = 90 mm), the harvester acts like a linear system since the magnetic force generated by the magnetic attractive effect is weak. Then, as d decreases (e.g., d = 17 mm), it turns into a mono-stable one due to the introduction of the magnetic force. It can be found that the curve of the potential energy function has a potential well and is asymmetric due to the gravitation effect. When d is less than a critical value, the attractive magnetic force will be greater than the resultant force of the elastic force and gravitational force. Thus, a potential barrier and two asymmetric potential energy wells are formed in the potential energy diagram. The asymmetry will become more obvious with the decrease in d. Subsequently, as d decreases further (e.g., d = 14 mm), the depths of two potential energy wells will increase, forming two deeper potential wells. For the bi-stable system, its snap-through motion may be excited to happen, which can make the piezoelectric beam have a large deflection and thus generate a relatively large voltage output.

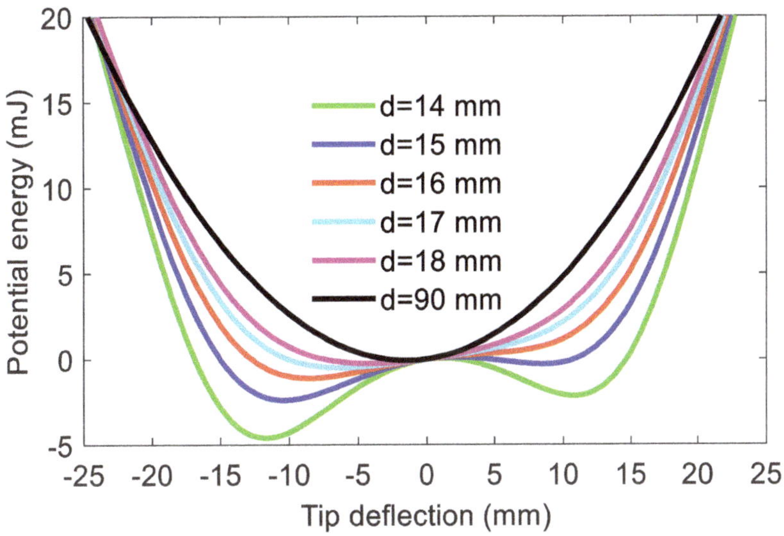

Figure 3. Potential energy shapes of the harvester for various separation distances.

Table 1. Parameters used for the analysis of potential energy function.

Parameters	Value
Substrate properties	
Length L_s	190 mm
Thickness h_s	1 mm
Width b_s	10 mm
Inertial moment I_s	0.833 mm^4
Young modulus E_s	205 GPa
density ρ_s	7.8 g/cm^3
Piezoelectric patch properties	
Length L_p	5 mm
Thickness h_p	0.25 mm
Width b_p	5 mm
Inertial moment I_p	0.49 mm^4
Young modulus E_p	56 GPa
density ρ_s	2.7 g/cm^3
Magnet properties	
Effective magnetic moment $a_1\,(a_2, a_3)$	0.218 Am^2
Mass m_t	3.7 g
Gap distance between two fixed magnets d_g	45 mm
Permeability constant of magnet μ_0	4×10^{-7} NA^{-2}

To further study the characteristics of the BPEH, Figure 4 shows the restoring force and stiffness of the BPEH with respect to different separation distances (d = 14, 15, 16, 17, 18 or 90 mm). As d decreases, the equilibrium position corresponding to the zero value of the restoring force gradually shifts toward the negative direction due to the gravitation effect, as shown in Figure 4a. When the attractive magnetic and elastic forces get to the same order of magnitude, the coupling effect becomes stronger. Now, the gravity effect plays an increasingly important role. As the separation distance decreases, it will influence the equilibrium position of tip deflection, which corresponds to the point of zero restoring force. When there appear three zero points in the restoring force curve, the system exhibits bi-stable behavior. In particular, by adjusting d, the restoring force and stiffness shown in Figure 4 can be reduced to a very small value and even become negative. The restoring force in the bi-stable state is extremely low and exhibits a negative value near zero tip deflection, which is expected to give rise to a large deflection and generate a high output voltage under weak vibration of the bridge.

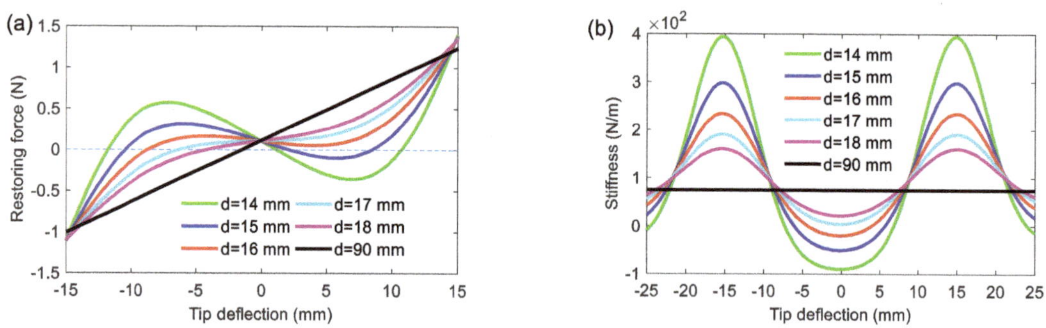

Figure 4. (**a**) Restoring force and (**b**) stiffness of the BPEH for various separation distances.

4. Experimental Setup

To validate the energy harvesting efficiency of the BPEH, corresponding experimental tests were carried out. The experimental setup is shown in Figure 5. The vehicle is represented by a steel ball (1003 kg). The model bridge used for experimental tests is made from an acrylic sheet (1300 mm × 80 mm × 8 mm), as shown in Figure 5a. Two guardrails are attached to the bridge to prevent the ball from falling off the bridge. Because the first mode plays an important role in bridge vibration, the first modal displacement is a determinant element for the installation of the harvester. For collecting high electric energy, the BPEH is installed under the midpoint of the bridge to harvest the vibration energy. As for the excitation, a steel ball is released from different heights and rolls down the acceleration section (inclined track). Then, it passes the bridge, as shown in Figure 5b. The prototype of the BPEH is fabricated, as shown in Figure 6. The two stable equilibrium positions of the BPEH are shown in Figure 7. The substrate is made of stainless steel and has dimensions of 190 mm × 10 mm × 1 mm. The piezoelectric patch has the dimensions of 5 mm × 5 mm × 0.25 mm, which is connected to a high-precision resistance box. A strain sensor (120–5 AA) is bonded on the substrate to measure the dynamic strain. As for the magnets, three permanent magnets have the same dimensions of 10 mm × 10 mm × 5 mm. If the two fixed magnets are removed, the BPEH will degenerate to a classical LPEH (as shown in Figure 8). In all experimental tests, the LPEH and BPEH are put in the same experimental conditions to make a direct comparison. A data acquisition device (DH5922N, Donghua, JingJiang, China) is adopted to measure the output voltage and dynamic strain. The schematic diagram of the experimental procedure is illustrated in Figure 9.

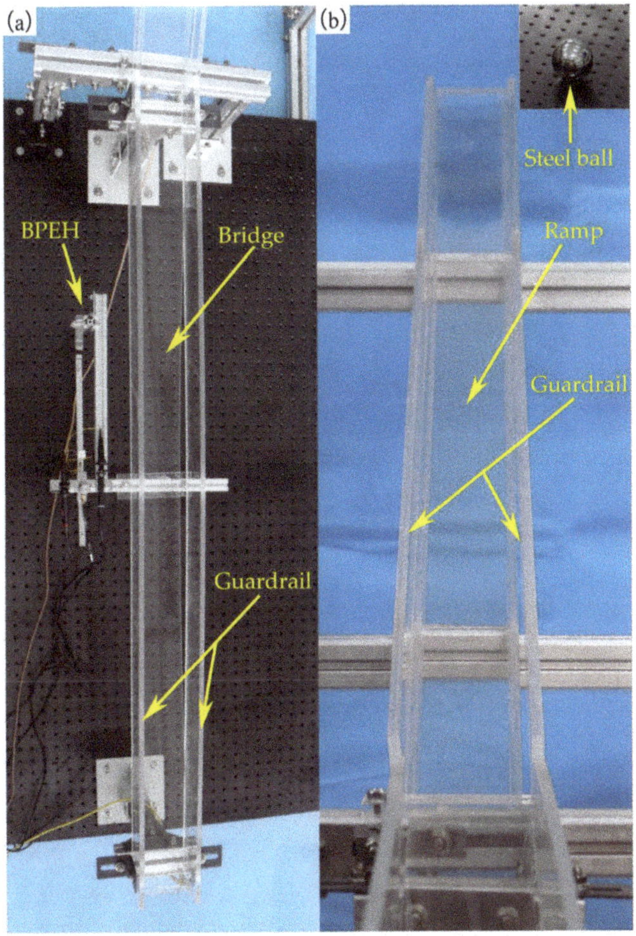

Figure 5. The experiment setup: (**a**) test section and (**b**) acceleration section.

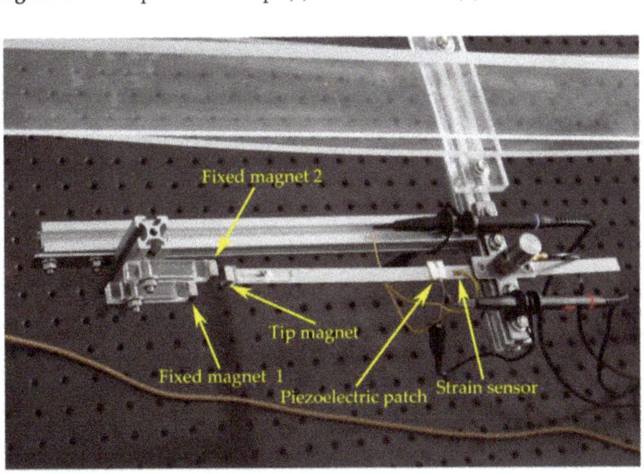

Figure 6. The prototype of the BPEH.

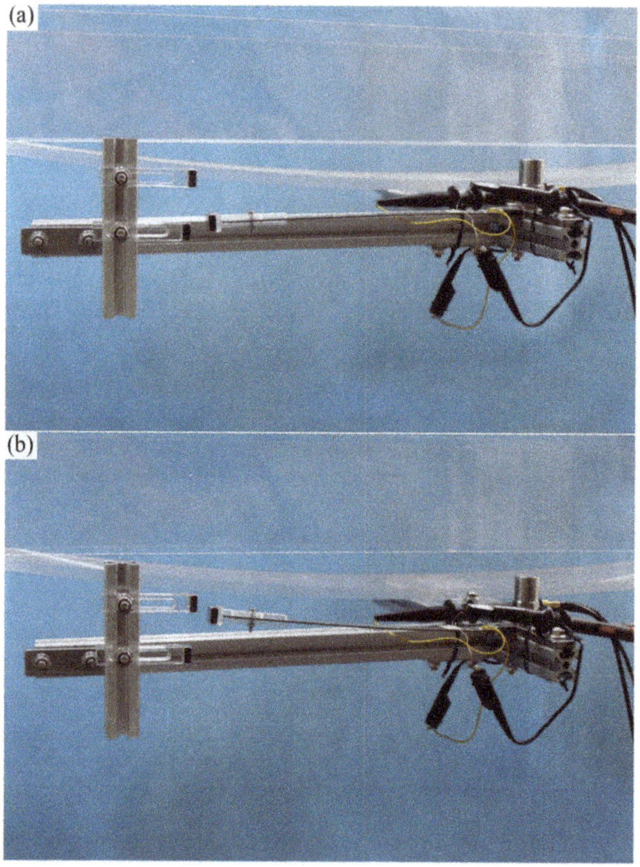

Figure 7. The two stable equilibrium positions of the BPEH: (**a**) stable equilibrium position 1 and (**b**) stable equilibrium position 2.

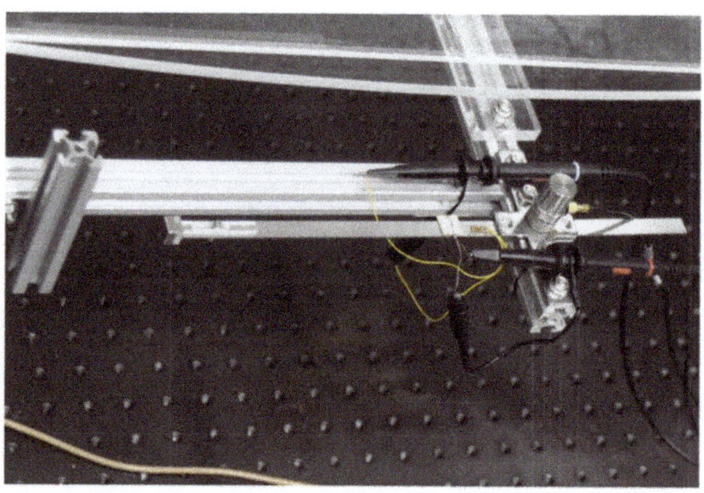

Figure 8. The prototype of the LPEH.

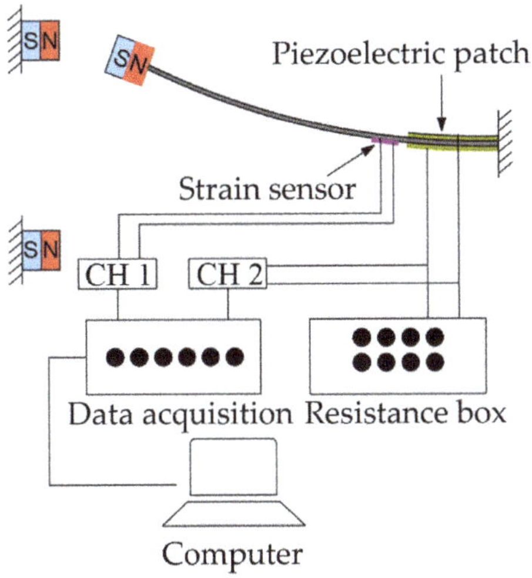

Figure 9. Schematic diagram of experimental procedure.

5. Results and Discussions

The dynamic behavior and electrical output are selected as the key indicators in evaluating the dynamic behavior and output performance. The steel ball is released from a titled track, which is named the acceleration section. In the acceleration section, for the ball, different positions could produce different initial speeds to pass the bridge. In the experiment, eleven positions are chosen to obtain different initial speeds. The running interval of the steel ball for measurement is set to be from 0.52 s to 1.18 s, which is consistent with the actual interval of a vehicle traveling through a bridge. Figure 10 shows the variance of strain and the power density (the load resistance is about 0.9 MΩ) of LPEH and BPEH for different moving speeds. It is clear from Figure 10 that, for BPEH, the variances of strain increase greatly at the speed of 1.68 m/s, resulting in a significant increase in the output power density. This phenomenon indicates that the BPEH is triggered to execute the snap-through motion and oscillates in a high-energy orbit. Thus, the BPEH can significantly improve the energy harvesting performance compared to the LPEH. For example, at the speed of 2.07 m/s, the BPEH's maximum power density can reach 430 W/m^3, whereas the LPEH's maximum RMS voltage is only 261 W/m^3.

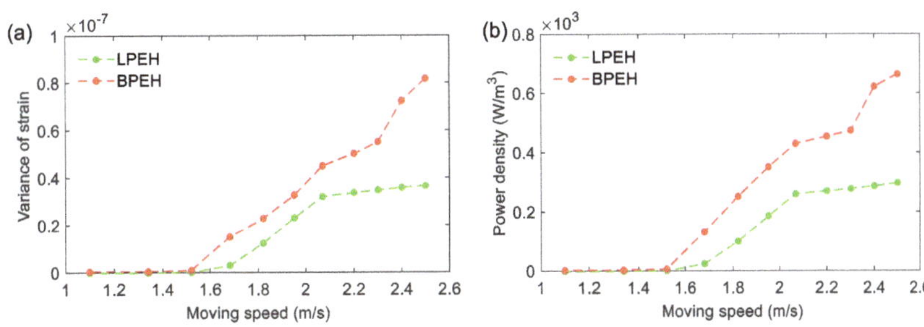

Figure 10. (**a**) Variance of strain and (**b**) power density versus moving speed for LPEH and BPEH.

To reveal the vibration response of the LPEH and BPEH, Figure 11 illustrates the time histories of strain for six moving speeds (v = 1.10, 1.52, 1.82, 2.07, 2.30 and 2.50 m/s). In the time response, the tag "On bridge" indicates the instant when the vehicle enters the bridge, while the tag "Leaving bridge" indicates the instant when the vehicle gets off the bridge. At a very low moving speed v = 1.10 m/s or 1.52 m/s, the BPEH's oscillation cannot cross the potential barrier and thus is trapped in a potential well. Thus, both BPEH and LPEH have a small vibration amplitude now. When the moving speed increases to 1.82 m/s (Figure 11c), the BPEH produces a large-amplitude vibration, for it begins to realize jumps between two stable equilibrium positions. Now the maximum strain of the BPEH can reach 4.5×10^{-4}, whereas the maximum strain of the LPEH is only 3.4×10^{-4}. The amplitude of BPEH is larger than that of LPEH, and the vibration period of BPEH is longer as well. During the running period on the bridge, for the BPEH, the snap-through motion keeps happening. Finally, as the vehicle gets off the bridge, the response will fall into a potential well and oscillate in it. As the moving speed increases to 2.07 m/s, the vibration amplitude of BPEH will have a significant increase, whose maximum strain will increase up to 9.5×10^{-4} (Figure 11c). If the moving speed increases further to a high one (e.g., 2.30 m/s or 2.50 m/s), the frequent snap-through motion will happen in the BPEH, resulting in a much larger amplitude than the LPEH, as shown in Figure 11e,f. For example, at 2.30 m/s, the maximum strain of BPEH can reach 1.16×10^{-3}, nearly two times that of the LPEH (6.8×10^{-4}). Thus, through comparing the time histories of the strain of two systems, it can be concluded that the BPEH is preferable to the LPEH in producing large deformations and giving large outputs for the same running vehicle excitations.

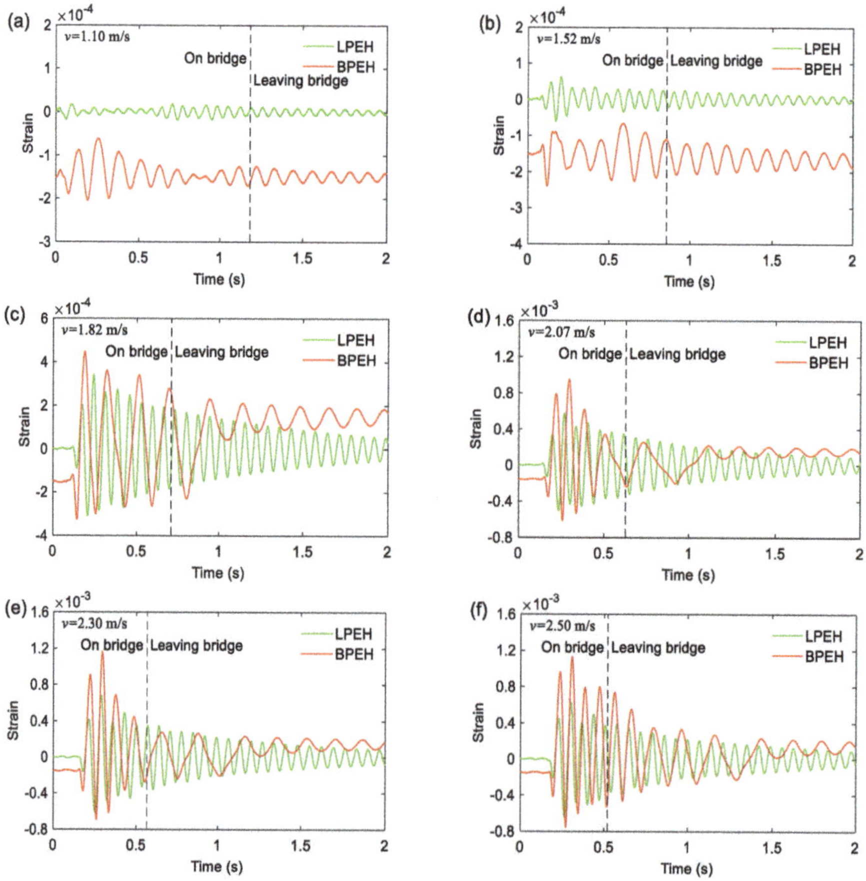

Figure 11. Strain responses of the LPEH and BPEH for the vehicle traveling on and leaving the bridge: (**a**) v = 1.10 m/s; (**b**) v = 1.52 m/s; (**c**) v = 1.82 m/s; (**d**) v = 2.07 m/s; (**e**) v = 2.30 m/s; and (**f**) v = 2.50 m/s.

Figure 12 shows the open-circuit output voltages of LPEH and BPEH for different moving speeds. It is well known that the PZT output is proportional to its deformation or strain. As is clear from Figure 12, the BPEH is superior to the LPEH in open-circuit voltage output at every speed. Specifically, at the low moving speed, e.g., v = 1.10 m/s, as Figure 12a shows, although the oscillation of BPEH is trapped in a potential well, the open-circuit voltage output of BPEH is still larger than that of LPEH due to the tailored potential energy shape. Then, when the moving speed increases to 1.82 m/s (Figure 12c), the BPEH's maximum open-circuit output voltage has a large increase and reaches up to 4.09 V, for the snap-through motion happens now, whereas the LPEH's maximum open-circuit voltage is only 2.44 V. Next, when the moving vehicle runs at a speed of 2.07 m/s, the open-circuit output voltage is further improved and is up to 6.72 V. However, if the moving speed increases continuously, the open-circuit output voltage of BPEH will not increase greatly, even will decrease. For example, as the moving speed reaches 2.30 m/s, the peak voltage of BPEH is 7.73 V. Then, as the moving speed increases a little to 2.50 m/s, the peak voltage is 7.56 V, a little smaller than that at 2.30 m/s. This is due to the protective effect of the fixed magnets. If the cantilever is excited to execute a very large vibration, the cantilever is likely to be damaged rapidly. Therefore, the attractive force between the tip and fixed magnets will take effect and constrain the excessively large vibration to protect the BPEH.

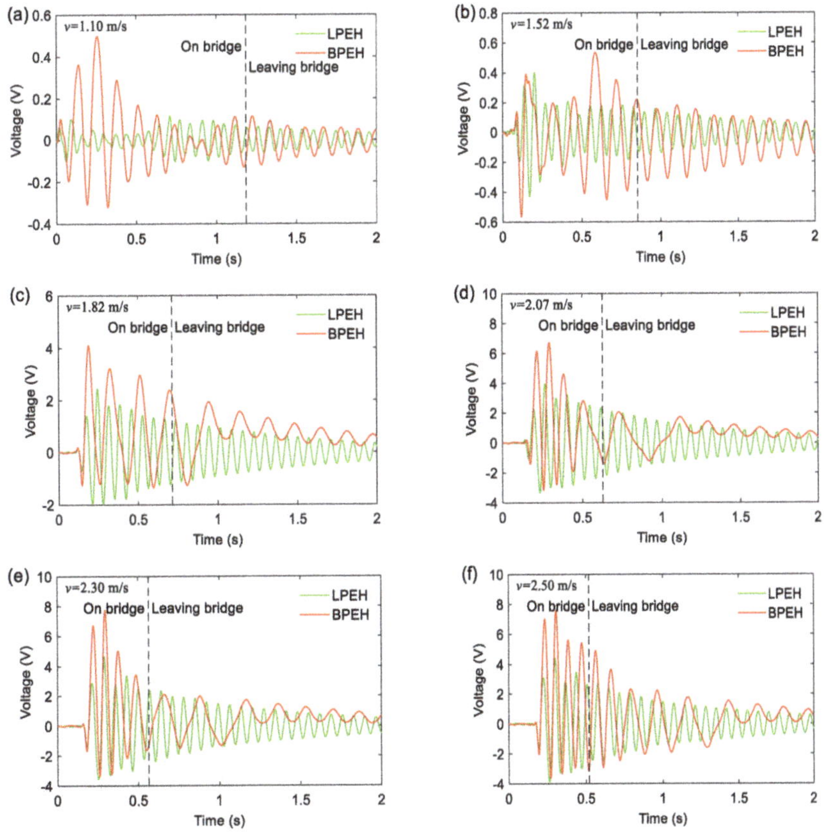

Figure 12. Open-circuit output voltages of the LPEH and BPEH for the vehicle traveling on and leaving bridge: (**a**) v = 1.10 m/s; (**b**) v = 1.52 m/s; (**c**) v = 1.82 m/s; (**d**) v = 2.07 m/s; (**e**) v = 2.30 m/s; and (**f**) v = 2.50 m/s.

In order to show the characteristics of the open-circuit voltage response of LPEH and BPEH, corresponding frequency spectra for different moving speeds are shown in Figure 13. It can be found that the peak of the LPEH is located near 14.3 Hz for any speed of the moving vehicle, implying that it is the system's resonance frequency. However, for the BPEH, the distribution of frequency spectra is different depending on the moving speed. At a relatively low moving speed (e.g., v = 1.10 m/s), the peak of the spectrum of BPEH output is located at 8.5 Hz, the BPEH's amplitude is much larger than that of the LPEH. When the moving speed increases to 1.82 m/s, the main peak shifts to 5.2 Hz, and the frequency energy of BPEH distributes over a wide range of 0–11.4 Hz. This is because the BPEH executes snap-through motions and exhibits a strong nonlinear characteristic. As the moving speed continues to increase, the spectrum of BPEH distributes more widely in a frequency range of 0~17.2 Hz (as shown in Figure 13d,f). Therefore, it can be concluded that the BPEH has a wider bandwidth and exhibits strong nonlinearity, which could help harvest more energy from bridge vibration.

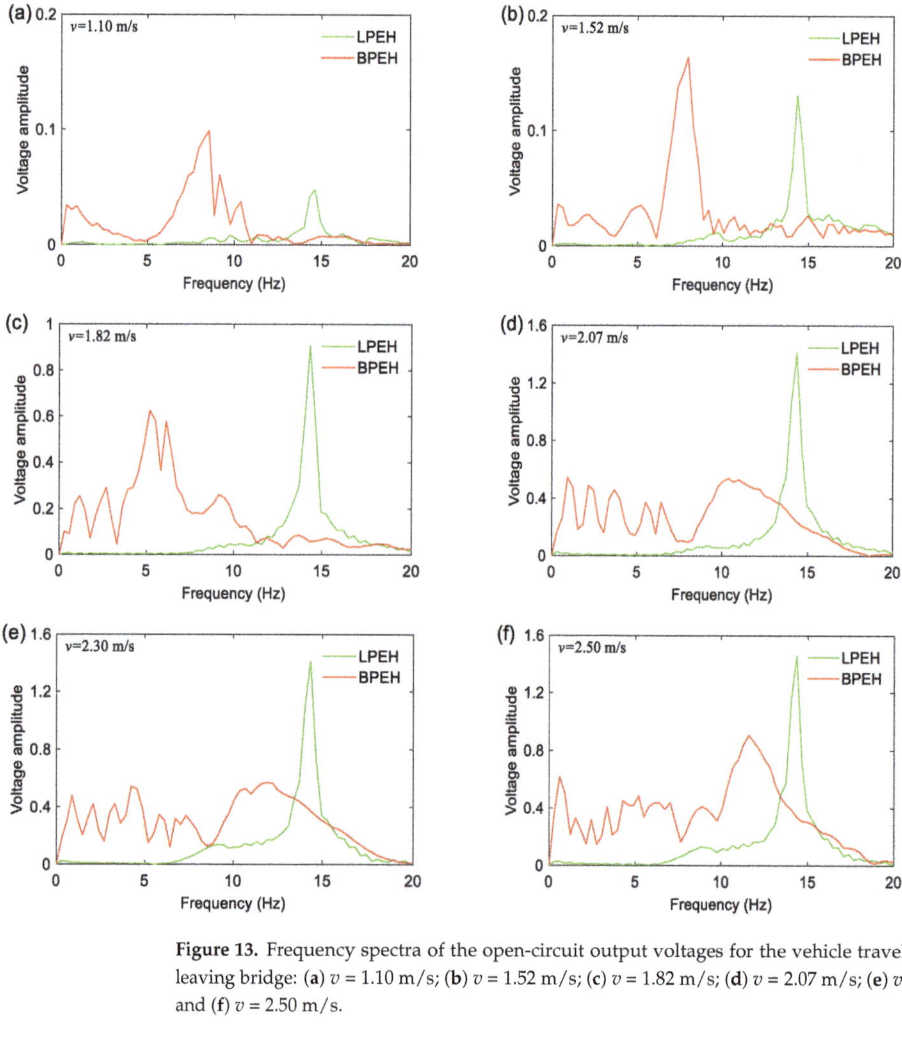

Figure 13. Frequency spectra of the open-circuit output voltages for the vehicle traveling on and leaving bridge: (**a**) v = 1.10 m/s; (**b**) v = 1.52 m/s; (**c**) v = 1.82 m/s; (**d**) v = 2.07 m/s; (**e**) v = 2.30 m/s; and (**f**) v = 2.50 m/s.

6. Conclusions

In this work, a bi-stable PEH was developed by introducing a tip and two fixed magnets. The bi-stable characteristic of this BPEH is generated by the magnetic attractive interaction, which helps enhance the energy harvesting from bridge vibrations. For the BPEH, corresponding theoretical analysis and simulations were carried out, and the validation experiment was conducted. The results show that the BPEH can significantly improve the energy harvesting performance by realizing snap-through motions. Compared to the LPEH, the BPEH has a wide frequency spectrum, which varies with the moving speed. The voltage response of the BPEH outperforms that of the LPEH over the whole range of moving speeds. The BPEH could significantly increase the deflection under the same excitation, which is beneficial for harvesting more energy from bridge vibration induced by traveling vehicles. Compared to the classical designs, the BPEH is simple in structure and can be easily adjusted in size to obtain a satisfactory harvesting performance. This design may lead to the realization of an efficient and reliable energy harvester. However, there remain some aspects to be studied in future work. For example, the separation and gap distances can be optimized so as to make the BPEH realize snap-through motions more easily. In practice, multiple vehicles will run on the bridge simultaneously, the BPEH's response and output under this excitation should be studied.

Author Contributions: Conceptualization, P.Z.; methodology, W.Q.; software, H.Z.; formal analysis, Z.Z. and W.D.; data curation, H.Z.; writing—original draft preparation, P.Z.; writing—review and editing, Z.Z.; supervision, W.Q.; All authors have read and agreed to the published version of the manuscript.

Funding: This work was supported by the National Natural Science Foundation of China (grant # 52005155), the China Postdoctoral Science Foundation (grant # 2020M673470), the Key Scientific Research Project of Colleges and Universities in Henan Province (grant # 20A130001) and the Key Research Development and Promotion Project in Henan Province (grant # 202102310249, 212102310952).

Institutional Review Board Statement: Not applicable.

Informed Consent Statement: Not applicable.

Data Availability Statement: The data presented in this study are available on request from the corresponding author.

Conflicts of Interest: The authors declare no conflict of interest.

References

1. Kwiatkowski, J.; Anigacz, W.; Beben, D. Comparison of Non-Destructive Techniques for Technological Bridge Deflection Testing. *Materials* **2020**, *13*, 1908. [CrossRef] [PubMed]
2. Sofi, A.; Regita, J.J.; Rane, B.; Lau, H.H. Structural health monitoring using wireless smart sensor network–An overview. *Mech. Syst. Signal Process.* **2022**, *163*, 108113. [CrossRef]
3. Zhang, Z.W.; Tang, L.H.; Xiang, H.J. Piezoelectric Energy Harvesting from Bridge Vibrations Using Different Models for Moving Vehicles. *J. Aerosp. Eng.* **2019**, *32*, 04018141. [CrossRef]
4. Wu, N.; Bao, B.; Wang, Q. Review on engineering structural designs for efficient piezoelectric energy harvesting to obtain high power output. *Eng. Struct.* **2021**, *235*, 112068. [CrossRef]
5. Zhou, Z.; Qin, W.; Du, W.; Zhu, P.; Liu, Q. Improving energy harvesting from random excitation by nonlinear flexible bi-stable energy harvester with a variable potential energy function. *Mech. Syst. Signal Process.* **2019**, *115*, 162–172. [CrossRef]
6. Ali, S.F.; Friswell, M.I.; Adhikari, S. Analysis of energy harvesters for highway bridges. *J. Intel. Mat. Syst. Str.* **2011**, *22*, 1929–1938. [CrossRef]
7. Peigney, M.; Siggert, D. Piezoelectric energy harvesting from traffic-induced bridge vibrations. *Smart Mater. Struct.* **2013**, *22*, 095019. [CrossRef]
8. Zhang, Y.; Cai, S.C.S.; Deng, L. Piezoelectric-based energy harvesting in bridge systems. *J. Intel. Mat. Syst. Str.* **2014**, *25*, 1414–1428. [CrossRef]
9. Karimi, M.; Karimi, A.H.; Tikani, R.; Ziaei-Rad, S. Experimental and theoretical investigations on piezoelectric-based energy harvesting from bridge vibrations under travelling vehicles. *Int. J. Mech. Sci.* **2016**, *119*, 1–11. [CrossRef]
10. Zhang, Z.W.; Xiang, H.J.; Shi, Z.F.; Zhan, J.W. Experimental investigation on piezoelectric energy harvesting from vehicle-bridge coupling vibration. *Energy Convers. Manag.* **2018**, *163*, 169–179. [CrossRef]

11. Aloisio, A.; Pasca, D.P.; Di Battista, L.; Rosso, M.M.; Cucuzza, R.; Marano, G.C.; Alaggio, R. Indirect assessment of concrete resistance from FE model updating and Young's modulus estimation of a multi-span PSC viaduct: Experimental tests and validation. *Structures* **2022**, *37*, 686–697. [CrossRef]
12. Bonopera, M.; Liao, W.C.; Perceka, W. Experimental-theoretical investigation of the short-term vibration response of uncracked prestressed concrete members under long-age conditions. *Structures* **2022**, *35*, 260–273. [CrossRef]
13. Fan, K.Q.; Tan, Q.X.; Zhang, Y.W.; Liu, S.; Cai, M.; Zhu, Y. A monostable piezoelectric energy harvester for broadband low-lever excitations. *Appl. Phys. Lett.* **2018**, *112*, 123901. [CrossRef]
14. Huguet, T.; Badel, A.; Lallart, M. Parametric analysis for optimized piezoelectric bistable vibration energy harvesters. *Smart Mater. Struct.* **2019**, *28*, 115009. [CrossRef]
15. Yang, T.; Cao, Q.J. Dynamics and performance evaluation of a novel tristable hybrid energy harvester for ultra-low level vibration resources. *Int. J. Mech. Sci.* **2019**, *156*, 123–136. [CrossRef]
16. Mei, X.T.; Zhou, S.X.; Yang, Z.C.; Kaizuka, T.; Nakano, K. Enhancing energy harvesting in low-frequency rotational motion by a quad-stable energy harvester with time-varying potential wells. *Mech. Syst. Signal Process.* **2020**, *148*, 107167. [CrossRef]
17. Stanton, S.C.; Mcgehee, C.C.; Mann, B.P. Reversible hysteresis for broadband magnetopiezoelastic energy harvesting. *Appl. Phys. Lett.* **2009**, *95*, 174103. [CrossRef]
18. Sebald, G.; Kuwano, H.; Guyomar, D.; Ducharne, B. Simulation of a Duffing oscillator for broadband piezoelectric energy harvesting. *Smart Mater. Struct.* **2011**, *20*, 075022. [CrossRef]
19. Erturk, A.; Inman, D.J. Broadband piezoelectric power generation on high-energy orbits of the bistable Duffing oscillator with electromechanical coupling. *J. Sound Vib.* **2011**, *330*, 2339–2353. [CrossRef]
20. Arrieta, A.F.; Delpero, T.; Bergamini, A.E.; Ermanni, P. Broadband vibration energy harvesting based on cantilevered piezoelectric bi-stable composites. *Appl. Phys. Lett.* **2013**, *102*, 173904. [CrossRef]
21. Masana, R.; Daqaq, M.F. Relative performance of a vibratory energy harvester in mono and bi-stable potentials. *J. Sound Vib.* **2011**, *330*, 6036–6052. [CrossRef]
22. Lan, C.B.; Qin, W.Y. Enhancing ability of harvesting energy from random vibration by decreasing the potential barrier of bistable harvester. *Mech. Syst. Signal Process.* **2017**, *85*, 71–81. [CrossRef]
23. Pellegrini, S.P.; Tolou, N.; Schenk, M.; Herder, J.L. Bistable vibration energy harvesters: A review. *J. Intell. Mater. Syst. Struct.* **2013**, *24*, 1303–1312. [CrossRef]
24. Zhou, S.; Cao, J.; Inman, D.J.; Lin, J.; Liu, S.; Wang, Z. Broadband tristable energy harvester: Modeling and experiment verification. *Appl. Energy* **2014**, *133*, 33–39. [CrossRef]
25. Litak, G.; Margielewicz, J.; Gaska, D.; Wolszczak, P.; Zhou, S.X. Multiple Solutions of the Tristable Energy Harvester. *Energies* **2021**, *14*, 1284. [CrossRef]
26. Zhou, Z.Y.; Qin, W.Y.; Zhu, P. Improve efficiency of harvesting random energy by snap-through in a quad-stable harvester. *Sens. Actuat. A-Phys.* **2016**, *243*, 151–158. [CrossRef]
27. Zhou, Z.Y.; Zhang, H.W.; Qin, W.Y.; Zhu, P.; Wang, P.; Du, W.F. Harvesting Energy from Bridge Vibration by Piezoelectric Structure with Magnets Tailoring Potential Energy. *Materials* **2022**, *15*, 33. [CrossRef]
28. Zhu, Y.; Zu, J.; Su, W. Broadband energy harvesting through a piezoelectric beam subjected to dynamic compressive loading. *Smart Mater. Struct.* **2013**, *22*, 045007. [CrossRef]
29. Tang, L.; Yang, Y.; Soh, C.K. Improving functionality of vibration energy harvesters using magnets. *J. Intell. Mater. Syst. Struct.* **2012**, *23*, 1433–1449. [CrossRef]

materials

MDPI

Article

Synthesis and Characterization of $Sn/SnO_2/C$ Nano-Composite Structure: High-Performance Negative Electrode for Lithium-Ion Batteries

Jaffer Saddique [1], Honglie Shen [1,*], Jiawei Ge [1], Xiaomin Huo [1], Nasir Rahman [2], Muhammad Mushtaq [3], Khaled Althubeiti [4] and Hamza Al-Shehri [5]

[1] Jiangsu Key Laboratory of Materials and Technology for Energy Conversion, College of Materials Science and Technology, Nanjing University of Aeronautics and Astronautics, Nanjing 210016, China; jafferphy@gmail.com (J.S.); 18551755923@sina.cn (J.G.); huoxm0928@163.com (X.H.)
[2] Department of Physics, University of Lakki Marwat, Lakki Marwat 28420, Khyber Pakhtunkhwa, Pakistan; nasir@ulm.edu.pk
[3] Key Laboratory of Advanced Functional Materials of Education Ministry of China, College of Materials Science and Engineering, Beijing University of Technology, Beijing 100124, China; mushtaqphy009@yahoo.com
[4] Department of Chemistry, College of Science, Taif University, P.O. Box 11099, Taif 21944, Saudi Arabia; k.athubeiti@tu.edu.sa
[5] Chemistry Division, King Khaled Military Academy, SANG, Riyadh 11495, Saudi Arabia; h.s.alshehri@outlook.com
* Correspondence: hlshen@nuaa.edu.cn; Tel./Fax: +86-25-52112626

Abstract: Tin oxide (SnO_2) and tin-based composites along with carbon have attracted significant interest as negative electrodes for lithium-ion batteries (LIBs). However, tin-based composite electrodes have some critical drawbacks, such as high volume expansion, low capacity at high current density due to low ionic conductivity, and poor cycle stability. Moreover, complex preparation methods and high-cost carbon coating procedures are considered main challenges in the commercialization of tin-based electrodes for LIBs. In this study, we prepared a $Sn/SnO_2/C$ nano-composite structure by employing a low-cost hydrothermal method, where Sn nanoparticles were oxidized in glucose and carboxymethyl cellulose CMC was introduced into the solution. Scanning electron microscope (SEM) and transmission electron microscope revealed the irregular structure of Sn nanoparticles and SnO_2 phases in the conductive carbon matrix. The as-prepared $Sn/SnO_2/C$ nano-composite showed high first-cycle reversible discharge capacity (2248 $mAhg^{-1}$) at 100 mAg^{-1} with a first coulombic efficiency of 70%, and also displayed 474.64 $mAhg^{-1}$ at the relatively high current density of about 500 mAg^{-1} after 100 cycles. A low-cost $Sn/SnO_2/C$ nano-composite with significant electrochemical performance could be the next generation of high-performance negative electrodes for LIBs.

Keywords: $Sn/SnO_2/C$ composite anode material; lithium-ion battery (LIBs); energy storage; synthesis; electrochemical performance

Citation: Saddique, J.; Shen, H.; Ge, J.; Huo, X.; Rahman, N.; Mushtaq, M.; Althubeiti, K.; Al-Shehri, H. Synthesis and Characterization of $Sn/SnO_2/C$ Nano-Composite Structure: High-Performance Negative Electrode for Lithium-Ion Batteries. *Materials* **2022**, *15*, 2475. https://doi.org/10.3390/ma15072475

Academic Editors: Marc Cretin, Sophie Tingry and Zhenghua Tang

Received: 26 January 2022
Accepted: 15 March 2022
Published: 27 March 2022

1. Introduction

Rechargeable lithium-ion batteries (LIBs) have been used during the last few decades as the main power source of choice with tremendous applications in many fields, such as portable optoelectronics devices, mobile phones, laptops and cameras, etc. [1–4]. However, given the high demands and rapid development of electric vehicles, lithium-ion batteries are now urgently required, especially those with high energy density, a long cycle life, and fast charging capacity [5–7]. A lot of expectations are on LIBs in terms of stable performance to meet the high demands of the current electronic market. However, the conventional anode material graphite (372 $mAhg^{-1}$) used commercially is incapable of fulfilling the requirement of the current demands of power due to its low theoretical capacity [8–10].

According to the current demands, it a priority to explore new-generation electrode materials with excellent electrochemical performances, including high lithiation capacity, long cycle lifespan, low cost, and that is eco-friendly, to boost up the overall performances of LIBs. In this context, much work has been done to introduce high-performance electrode materials for LIBs. Among these, metallic tin is a non-toxic, environmentally friendly, low-cost and highly abundantly available metal with outstanding specific theoretical capacity (994 $mAhg^{-1}$) for LIBs [11,12]. However, metallic tin electrodes have some serious drawbacks concerning the dramatic volume expansion >300% during the lithiation/delithiation process resulting in capacity fading, rapid capacity decay, and low coulombic efficiency, which always limits their potential applications [13,14]. Tin oxide SnO_x, specifically SnO_2 and SnO, have attracted attention as a potential next-generation anode owing to their high theoretical capacity of 1494 $mAhg^{-1}$ and 1273 $mAhg^{-1}$, respectively [15,16]. In the literature, different strategies have been applied to obtain different structures of tin oxide in order to improve the electrochemical performance [17,18]. By controlling and manipulating important parameters of the SnO_2, such as the size of the electrode nanostructure and the confinement of the active material in a carbonaceous matrix to prevent the agglomeration of the nanostructures upon cycling, it is possible to increase the amount of lithium-ion reversibility during conversion reactions [19–21]. During lithiation, the reaction mechanism of SnO_2 can be described as two stages: the (1) conversion reaction and (2) alloying reaction, which are given as follows:

In the conversion stage, SnO_2 is converted to form Li_2O and the elemental Sn as expressed by the following equation:

$$SnO_2 + 4Li^+ + 4e^- \leftrightarrow 2Li_2O + Sn \quad (1)$$

which contributes to a capacity of 711 $mAhg^{-1}$.

In the second stage, the alloying reaction taking place can be expressed as the following equation:

$$Sn + xLi^+ + 4e^- \leftrightarrow LixSn\ (0 < x < 4.4) \quad (2)$$

to which a further 783 $mAhg^{-1}$ can be added over its theoretical capacity, and it can reach as high as 1494 $mAhg^{-1}$ [22,23]. Besides these, SnO_2 has some drawbacks when used as a negative electrode for LIBs, such as high volume expansion during lithiation, poor conductivity, and large first-cycle capacity, which lead to electrode pulverization and electrical exfoliation of the active material from the current collector and further hinder their application, resulting in poor cycle performance, rapid capacity decay, and low coulombic efficiency [24–26]. Different morphology and nanostructures, and the introduction of Sn-based anode into the carbon matrix are helpful to overcome all these obstacles while utilizing and introducing a negative electrode material for LIBs with such high theoretical capacity [27,28]. Herein, an $Sn/SnO_2/C$ nano-composite anode, prepared by facile hydrothermal method, is expected to be able to enhance the performance of LIBs, especially in its real capacity with stable cyclability and coloumbic efficiency. Structural characterizations confirm the existence of Sn and SnO_2 in the carbon conductive framework, ensuring $Sn/SnO_2/C$-based LIBs that possess considerable enhancement in terms of cycle stability as well as other enhanced performances.

2. Experimental

2.1. *Preparation of $Sn/SnO_2/C$ in Carbon Matrix*

A typical hydrothermal method was used for the preparation of $Sn/SnO_2/C$ nano-composite materials. In detail, the desired molar ratio of tin chloride dehydrate $SnCl_2 \cdot 2H_2O$ (99.9%, Aladdin) and Glucose $C_6H_{12}O_6$ (99%, Aladdin Nanjing Chemical Reagent Co., LTO) of about 0.02 mole (4.51 g) and 0.02 mole (3.6 g), respectively, were mixed in 50 mL deionized water. The solution was stirred for 15 min and then 4 g (weight %) carboxymethyl cellulose CMC was added slowly and gradually into the solution and continually stirred for more than 2 h in order to get a homogenous solution milky in color. All the reagents were of the analytical grade and used without any further purification. The solution was then transferred to a 100 mL Teflon-lined autoclave and heated to 200 °C and kept at this temperature for 24 h to get the SnO_x nanoparticles. The solution was cooled naturally to room temperature and was then centrifuged at 9000 r/min several times. Then, the solution was washed thoroughly by using de-ionized water and ethanol to obtain more pure products. Finally, the products were transferred to a vacuum oven and kept at a temperature of 90 °C overnight to dry them. To get the $Sn/SnO_2/C$, the resultant products were ground with the help of a pestle and mortar and then annealed at 800 °C at the rate of 5 °C/min for 4 h under argon atmosphere. The used carboxymethyl cellulose CMC polymer was converted into carbon after high-temperature treatment by following the annealing process. The experimental steps were exercised repeatedly to make it more accurate.

2.2. Materials Characterization

The crystalline structure and phase confirmation of the as-synthesized $Sn/SnO_2/C$ material was characterized via room temperature powder X-ray diffractometer (XRD) in the range of 10–80° on Broker D8 with Cu Kalpha radiation. Scanning Electron Microscopy (SEM model FEI, Quanta 650, HITACHI, Japan) was carried out to analyze the morphological structure of the as-synthesized material. Energy dispersive spectroscopy (EDS) along with SEM was performed to confirm the elemental distribution mapping of the elements contained by the prepared materials. Transmission electron microscopy (TEM) along with selected area electron diffraction (SAED) and high-resolution transmission microscopy (HRTEM, Tokyo, Japan) was performed on EFI Tecnai G2s-Twin instrument with electron gun operating at 200 KV to further confirm the atomic distribution and size of the synthesized $Sn/SnO_2/C$ material. X-ray photoelectron spectroscopy (XPS, PHI-5000 VersaProbe Ulvac–Phi Thermo Fischer Scientific Multilab 2000 spectrometer with an Al Kα radiation monochromator at 1486.6 eV) was carried out to confirm the composition and oxidation state of the prepared $Sn/SnO_2/C$ material.

2.3. Electrochemical Measurement

In order to investigate the electrochemical performance of the synthesized $Sn/SnO_2/C$ material, working electrodes were prepared by a slurry of 70% active materials, 20% carbon black (conductive agent), and 10% polyvinyl difluoride (PVDF) used as a binder in N-methy1-2-pyrrolidone (NMP). The mixture was then stirred with a magnetic stirrer for 2 h to obtain a homogenous solution. After that, the slurry was bladed on copper foil (current collector) and then transferred to a vacuum oven at 80 °C overnight. The prepared dry electrode was cut into a 12 mm round shape with an active mass of about 1.4 mg, and was pressed at 20 MPa. The coin cells (CR-2032) were assembled inside the glovebox (water and oxygen content < 1), where lithium metal foil was used as a reference and counter electrode. The propylene was used as a separator and 1M $LiPF_6$ in EC:DEC = 1:1 (volume ratio) was used as an electrolyte. The galvanostatic charge/discharge curves were tested using a Land automatic batteries tester (LAND-CT2001A, Wuhan, China) in the potential range of 0–3 V as well at different current densities. The cyclic voltammetry (CV) was tested at a scan rate of 0.1 V at the electrochemical workstation (CHI660D, Shanghai, China). The electrochemical impedance spectroscopy (EIS) tests were also performed in the frequency ranging from 0.01 Hz to 1 MHz to check the ionic conductivity of the prepared electrodes.

3. Results

We first prepared a $Sn/SnO_2/C$ nano-composite by using the facile hydrothermal method along with high-temperature post-treatment of about 800 °C. In order to investigate the crystalline and phase confirmation of the prepared materials, X-ray diffraction Broker D8 with Cu Kalpha radiation was performed. Figure 1 shows the XRD pattern of the synthesized $Sn/SnO_2/C$. The XRD pattern shows the prominent diffraction peaks of tin (JCPDS No 90-08570) with high intensity, where some of the diffraction peaks of the SnO_2 phase structure were also observed and matched well with the SnO_2(JCPDS No 14-1445), which had comparatively very low intensity. The diffraction peaks belonging to the SnO_2 crystal structure can be observed in the XRD pattern. The observed diffraction peaks have very low intensity when compared with the prominent phase structure of tin. The X-ray diffraction pattern shows the existence of both structures in the composite, where the low intensity diffraction peaks suggest that the SnO_2 phase shows their existence in the synthesized composite material in a low ratio, or the high-intensity peaks depressed their execution in the pattern.

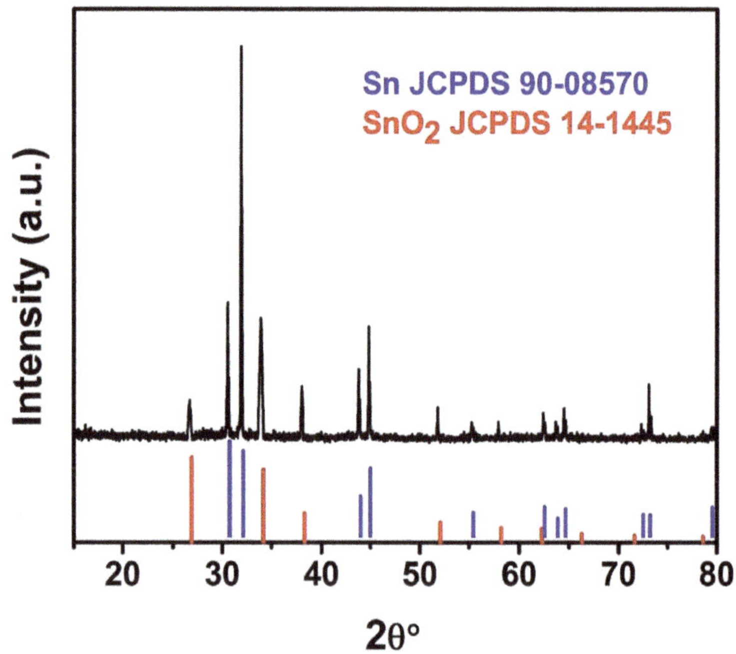

Figure 1. XRD pattern of $Sn/SnO_2/C$.

To investigate and characterize the surface and morphological characteristics of the prepared $Sn/SnO_2/C$ nano-composite, scanning electron microscope (SEM) was performed and the results are presented in Figure 2. Figure 2a contains the typical low-magnification images of the $Sn/SnO_2/C$ nano-composite. It also shows particles irregular in shape as well as size, which may be due to the two different phases generated during the synthesis process. No obvious agglomerates were observed for the tin particles in the pattern. From the high-magnification images presented in Figure 2b, it can be seen clearly that the irregularity in terms of size and shape is due to the transformation of glucose and carboxymethyl cellulose (CMC) into carbon matrix during carbonization and the high-temperature (800 °C) treatment. The observed irregularity of the particles in the presented Figure 2a,b, suggests that the prepared $Sn/SnO_2/C$ nano-composites were inter-connected with each other, which created short pathways and more chances for the penetration of electrolytes and more Lithium-ion diffusion. Energy dispersive spectroscopy (EDS) along with SEM was performed to confirm the elemental mapping for the $Sn/SnO_2/C$ nano-composite and the results are shown in Figure 2c–f. It can be seen in the figure that tin agglomerates are distributed in the carbon matrix; the existence of oxygen mapping in Figure 2f can also be observed. No other contaminations were observed for other particles, which confirms that Sn and SnO_2 particles were distributed in the carbon matrix and further clarifies the even distribution of the Sn in the carbon matrix.

Figure 2. (**a**,**b**) SEM images; and (**c**–**f**) EDS and the corresponding elemental mapping Sn, C and O of as-prepared $Sn/SnO_2/C$ nano-composite.

Transmission electron microscope (TEM) analysis along with high-resolution transmission electron microscope (HRTEM) and selected area electron diffraction (SAED) analysis was performed to investigate and obtain more structural information of the as-prepared $Sn/SnO_2/C$ nano-composite anode, as shown in Figure 3. The presented TEM image in Figure 3a shows the even distribution of Sn and SnO_2 nanoparticles in the carbon matrix. The SAED pattern in Figure 3b shows different lattice fringes, where in fact no obvious patterns can be differentiated easily for any specific phases, which further confirms and indicates the formation of a complex composite structure. A high-resolution TEM was used to reveal the lattice fringes of the prepared $Sn/SnO_2/C$ nano-composite as shown in Figure 3c–e. In the HRTEM image of Figure 3c, two different crystalline domains are embedded in the carbon matrix and are enlarged in the selected area of the image and are further presented in Figure 3d,e to confirm information about their lattices. In Figure 3d, the crystalline structure of Sn with its (200) planes can be identified. Figure 3e contains a SnO_2 crystalline structure with its (110) planes. The inter-atomic spacing distance of 0.293 nm [29] and 0.334 nm [30] corresponding to (200) and (110) for both Sn and SnO_2 match well with the XRD results. The observed lattice fringes of the SnO_2 crystal structure in Figure 3e confirm their existence in the carbon matrix as well in the composite. The corresponding TEM and HRTEM images confirm the $Sn/SnO_2/C$ nano-composite phase and distinguish two crystalline structures embedded in the carbon matrix. As the literature reports, Sn itself as an anode suffers from high volume expansion, whereas SnO_2 tends to agglomerate into large particles, both of which induce the fading of the performance of the battery. So, fabricating a composite structure has advantages such as controlling the volume expansion of Sn and depressing the agglomerates of the SnO_2, which is expected to improve the reaction efficiency of the $Sn/SnO_2/C$ composite structure.

The chemical composition and oxidation states of the as-prepared $Sn/SnO_2/C$ nano-composite was further scrutinized via X-ray photo electron spectroscopy (XPS), as shown in Figure 4a–d. In the low-resolution survey spectra depicted in Figure 4a, the typical peaks of carbon, tin, and oxygen can be identified, indicating the purity of the synthesized $Sn/SnO_2/C$ nano-composite samples. As shown in Figure 4b, two strong characteristic peaks belonging to $Sn3d_{5/2}$ and $Sn3d_{3/2}$ of $Sn/SnO_2/C$ can be observed at 487.2 and 496.6 eV [31]. These characteristic bands of $Sn/SnO_2/C$ confirmed the oxidation of Sn into Sn^{+4} states [32]. Figure 4c represents the high-resolution spectra of Cs1 at 286.1 eV, which further confirm the existence of carbon in the composite sample. The major peak of carbon in the spectra indicates the existence of C species in the Sn/SnO_2/C nano-composite.

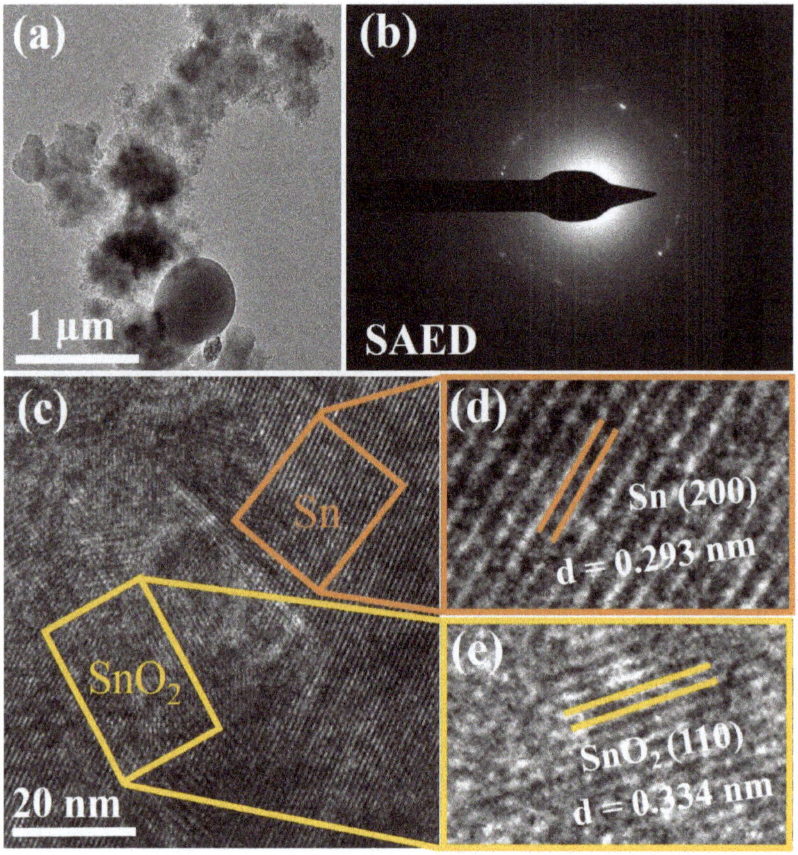

Figure 3. (**a**) TEM image of the Sn/C composite; (**b**) SAED pattern of the composite; and (**c**–**e**) HRTEM and enlarged image of the Sn/SnO_2/C nano-composite.

Figure 4d shows the high-resolution spectra of Os1 with maximum intensity at 533.6 eV attributed to the C–O bonding functional group, which further contributes to the reversibility of LiO_2 during the cycling process [33].

In order to explore the electrochemical performance of the prepared Sn/SnO_2/C nano-composite, half-coin cells were assembled in a glovebox where metallic Li foils were used as counter and reference electrodes, and the corresponding synthesized Sn/SnO_2/C nano-composite materials were used as a negative electrode. The measured electrochemical performance of the Sn/SnO_2/C nano-composite is depicted in Figure 5a–c. The cyclic voltammetry (CV) results were tested in the potential window ranging from 0.01 to 3 V (vs. Li^+/Li) at a scanning rate of 0.1 mV s^{-1} for the initial five cycles for LIBs and are shown in Figure 5a. As observed in Figure 5a, in the first cycle there were six cathodic peaks appearing at different voltages and are identified at 0.338 V, 0.76 V, 1.5 V, 1.7 V, 1.8 V and 2.8 V, respectively. All the observed peaks in the first cycle disappeared, which was attributed to the solid electrolyte interphase (SEI) layer forming on the surface of the active electrodes. Moreover, all the observed oxidation and reduction peaks in the subsequent cycles are distinct and overlap each other, which indicates a stable electrochemical performance. The characteristic electrochemical reduction and oxidation peaks reflect the electrochemical behavior of the Sn and SnO_2 anodes in one electrode system for LIBs.

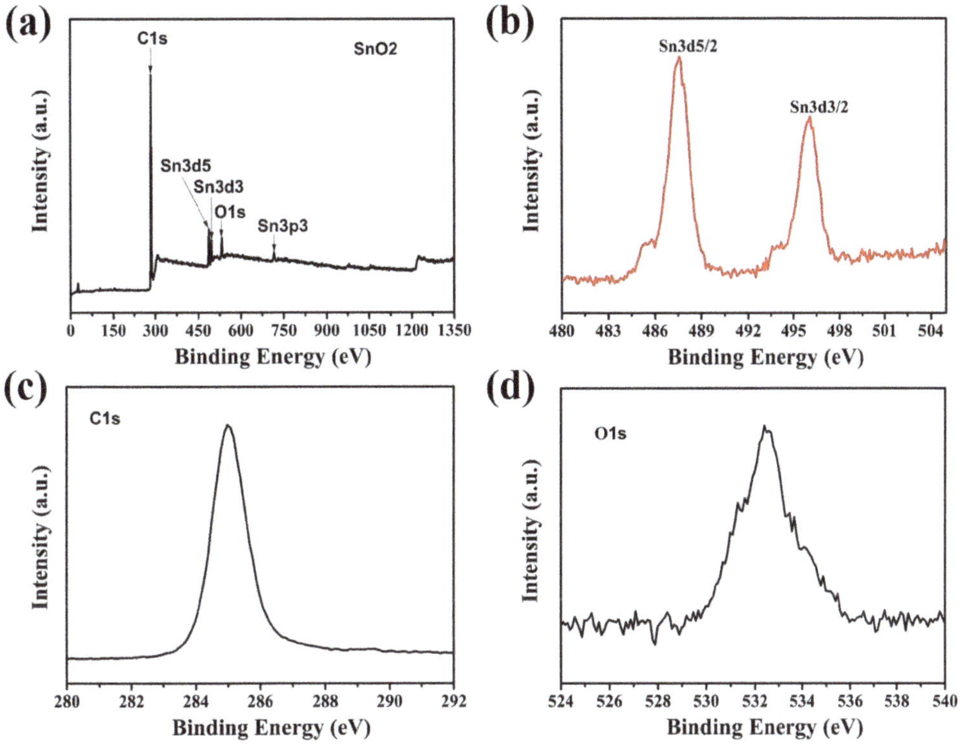

Figure 4. XPS spectra of $Sn/SnO_2/C$: (**a**) survey scan; (**b**) Sn3d spectra; (**c**) C1s spectra; and (**d**) O1s spectra.

The measured electrochemical impedance (EIS) of the $Sn/SnO_2/C$ nano-composite for LIBs is shown in Figure 5b. In general, a Nyquist plot contains a semicircle and a straight line representing charge transfer in high frequency and lower frequency regions, respectively. The depicted EIS results in Figure 5b show lower resistance with a high rate diffusion of Li-ion in the $Sn/SnO_2/C$ nano-composite's negative electrode for LIBs. The inset in Figure 5b represents the EIS results measured for pure Sn as a negative electrode for LIBs, which show high conductivity in a low-frequency range. The absence of the high-frequency region may be due to the ESI layer formation/ decomposition in the pure Sn electrode system which further enhanced the diffusion rate of Li-ion in the $Sn/SnO_2/C$ nano-composite electrode for LIBs. Figure 5c represents the cycle performance of the $Sn/SnO_2/C$ nano-composite electrode and pure tin electrode for LIBs measured at 100 and 500 mAg^{-1} current densities in a potential range 0–3 V for 100 cycles. The first 5 cycles were measured at 100 mAg^{-1} and the remaining 90 cycles were measured at 500 mAg^{-1}. The initial discharge capacity of the $Sn/SnO_2/C$ nano-composite was 2248 $mAhg^{-1}$ with coulombic efficiency 70% of first charge/discharge capacity and a 99% capacity was maintained for the remaining cycles, even at high current density. In Figure 5c, we can also see the cycle performance of the pure tin electrode measured at same current densities in order to compare the cycle performance of both electrodes. Here, we can see the capacity drops after a few cycles at high current density; high capacity decay may be due to the high volume expansion or because of the low ionic conductivity of the tin electrode for LIBs. On the other hand, the discharge capacity decreased to 1685 $mAhg^{-1}$ in the second cycle for the $Sn/SnO_2/C$ nano-composite electrode and then maintained up to 95% for the remaining cycles. The first high discharge capacity may be attributed to the SEI layer on the surface of the electrode which further decomposed in subsequent cycles as a result of a large amount of capacity decay. It is believed that such a high capacity with excellent coulumbic efficiency may be attributed to the combination of Sn and SnO_2 in the conductive carbon matrix [34]. The $Sn/SnO_2/C$ nano-composite shows a high reversible capacity of about 489 $mAhg^{-1}$ at the 100th cycle. As a result, the $Sn/SnO_2/C$ nano-composite showed enhanced electrochemical performance for LIBs and could be a promising candidate as a negative electrode for future prospects.

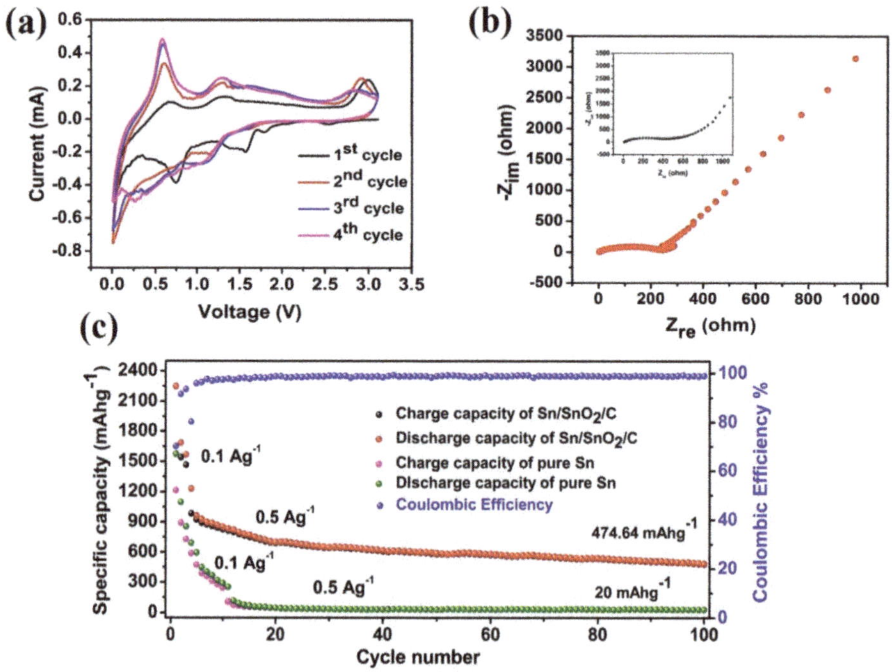

Figure 5. Electrochemical performance of $Sn/SnO_2/C$ nano-composite for LIBs: (**a**) initial five cyclic voltammetry curves scanned from 0.01 to 3 V at a rate of 0.01 mV s^{-1}; (**b**) EIS of $Sn/SnO_2/C$ nano-composite; and (**c**) cycle performance of Sn/C electrode acquired at two different current densities of about 100 mA g^{-1} and 500 mAg^{-1}.

4. Conclusions

In this work, a $Sn/SnO_2/C$ nano-composite was synthesized via a hydrothermal method as an anode for LIBs. The basic structural characterization performed using SEM and TEM revealed the existence and homogenous distribution of the Sn nano-particles and SnO_2 in the carbon matrix, which significantly enhanced the ionic conductivity of the electrode and buffered the volume expansion during repeated lithiation/delithiation processes. Moreover, the synthesized $Sn/SnO_2/C$ nano-composite showed high initial first-cycle discharge capacity (2248 $mAhg^{-1}$) at 100 mAg^{-1} with a first-cycle coulombic efficiency of 70% and also displayed 489 $mAhg^{-1}$ at a relatively high current density of about 500 mAg^{-1} after 100 cycles. As a result, high electrochemical active and crystalline nanoparticles embedded in the carbon matrix anode were achieved. The improved electrochemical performance of the prepared $Sn/SnO_2/C$ nano-composite enables it to be a promising anode for next-generation LIBs.

Author Contributions: Conceptualization and formal analysis, J.S., J.G.; investigation, J.S. and X.H.; Methodology, J.G. and X.H.; writing-origional draft preparation, J.S.; editing, N.R. and M.M.; writing review, J.S., M.M.; formal analysis, N.R. and H.A.-S.; Investigation, H.A.-S. and K.A.; project administration, K.A. and H.S.; funding acquisition, K.A. and H.S.; Supervision, H.S. All authors have read and agreed to the published version of the manuscript.

Funding: The authors gratefully acknowledge the National Natural Science Foundation of China (Grant No. 61774084), the Special Scientific Innovation Found of Sihong County (H201901) and the open project of Key Laboratory of Materials Preparation and Protection for Harsh Environment, Ministry of Industry and Information Technology (XCA20013-3). In addition, the authors extend their appreciation to Taif University Researchers Supporting Project number (TURSP-2020/241), Taif University, Taif, Saudi Arabia.

Institutional Review Board Statement: Not applicable.

Informed Consent Statement: Not applicable.

Data Availability Statement: Data sharing is not applicable for this article.

Conflicts of Interest: The authors declare no conflict of interest.

References

1. Etacheri, V.; Marom, R.; Elazari, R.; Salitra, G.; Aurbach, D. Challenges in the development of advanced Li-ion batteries: A review. *Energy Environ.Sci.* **2011**, *4*, 3243–3262. [CrossRef]
2. Yu, Y.; Chen, C.-H.; Shi, Y. A Tin-Based Amorphous Oxide Composite with a Porous, Spherical, Multideck-Cage Morphology as a Highly Reversible Anode Material for Lithium-Ion Batteries. *Adv. Mater.* **2007**, *19*, 993–997. [CrossRef]
3. Poizot, P.; Laruelle, S.; Grugeon, S.; Dupont, L.; Tarascon, J.M. Nano-sized transition-metal oxides as negative-electrode materials for lithium-ion batteries. *Nature* **2000**, *407*, 496–499. [CrossRef] [PubMed]
4. Zhang, W.-J. A review of the electrochemical performance of alloy anodes for lithium-ion batteries. *J. Power Sources* **2011**, *196*, 13–24. [CrossRef]
5. Ali, S.; Jaffer, S.; Maitlo, I.; Shehzad, F.K.; Wang, Q.; Ali, S.; Akram, M.Y.; He, Y.; Nie, J. Photo cured 3D porous silica-carbon (SiO_2–C) membrane as anode material for high performance rechargeable Li-ion batteries. *J. Alloys Compd.* **2020**, *812*, 152127. [CrossRef]
6. Tarascon, J.M.; Armand, M. Issues and challenges facing rechargeable lithium batteries. *Nature* **2001**, *414*, 359–367. [CrossRef] [PubMed]
7. Sun, X.; Si, W.; Liu, X.; Deng, J.; Xi, L.; Liu, L.; Yan, C.; Schmidt, O.G. Multifunctional Ni/NiO hybrid nanomembranes as anode materials for high-rate Li-ion batteries. *Nano Energy* **2014**, *9*, 168–175. [CrossRef]
8. Sun, Y.; Wu, Q.; Shi, G. Graphene based new energy materials. *Energy Environ. Sci.* **2011**, *4*, 1113–1132. [CrossRef]
9. Lian, P.; Dong, Y.; Wu, Z.-S.; Zheng, S.; Wang, X.; Sen, W.; Sun, C.; Qin, J.; Shi, X.; Bao, X. Alkalized Ti_3C_2 MXene nanoribbons with expanded interlayer spacing for high-capacity sodium and potassium ion batteries. *Nano Energy* **2017**, *40*, 1–8. [CrossRef]
10. Yuan, C.; Wu, H.B.; Xie, Y.; Lou, X.W. Mixed Transition-Metal Oxides: Design, Synthesis, and Energy-Related Applications. *Angew. Chem. Int. Ed.* **2014**, *53*, 1488–1504. [CrossRef]
11. Yin, L.; Chai, S.; Huang, J.; Kong, X.; Wang, J.; Bai, P. Influence of solvent on the structure and electrochemical performances of Sn-based anode for lithium-ion battery. *Ceram. Int.* **2017**, *43*, 12667–12674. [CrossRef]
12. Du, Z.; Zhang, S.; Jiang, T.; Bai, Z. Preparation and characterization of three-dimensional tin thin-film anode with good cycle performance. *Electrochim. Acta* **2010**, *55*, 3537–3541. [CrossRef]
13. Hassoun, J.; Derrien, G.; Panero, S.; Scrosati, B. A Nanostructured Sn–C Composite Lithium Battery Electrode with Unique Stability and High Electrochemical Performance. *Adv.Mat.* **2008**, *20*, 3169–3175. [CrossRef]
14. Ao, X.; Jiang, J.; Ruan, Y.; Li, Z.; Zhang, Y.; Sun, J.; Wang, C. Honeycomb-inspired design of ultrafine SnO_2@C nanospheres embedded in carbon film as anode materials for high performance lithium- and sodium-ion battery. *J. Power Sources* **2017**, *359*, 340–348. [CrossRef]
15. Chen, X.; Guo, J.; Gerasopoulos, K.; Langrock, A.; Brown, A.; Ghodssi, R.; Culver, J.N.; Wang, C. 3D tin anodes prepared by electrodeposition on a virus scaffold. *J. Power Sources* **2012**, *211*, 129–132. [CrossRef]
16. Ahn, D.; Xiao, X.; Li, Y.; Sachdev, A.K.; Park, H.W.; Yu, A.; Chen, Z. Applying functionalized carbon nanotubes to enhance electrochemical performances of tin oxide composite electrodes for Li-ion battery. *J. Power Sources* **2012**, *212*, 66–72. [CrossRef]
17. Hu, R.; Sun, W.; Liu, H.; Zeng, M.; Zhu, M. The fast filling of nano-SnO_2 in CNTs by vacuum absorption: A new approach to realize cyclic durable anodes for lithium ion batteries. *Nanoscale* **2013**, *5*, 11971–11979. [CrossRef]
18. Jiang, P.; Jing, J.; Wang, Y.; Li, H.; He, X.; Chen, Y.; Liu, W. Facilely transforming bulk materials to SnO/pristine graphene 2D-2D heterostructures for stable and fast lithium storage. *J. Alloys Compd.* **2020**, *812*, 152114. [CrossRef]
19. Kim, H.; Kim, M.-C.; Kim, S.-b.; Kim, Y.-S.; Choi, J.-H.; Park, K.-W. Porous SnO_2 nanostructure with a high specific surface area for improved electrochemical performance. *RSC Adv.* **2020**, *10*, 10519–10525. [CrossRef]

20. Zhou, X.; Wan, L.-J.; Guo, Y.-G. Binding SnO_2 Nanocrystals in Nitrogen-Doped Graphene Sheets as Anode Materials for Lithium-Ion Batteries. *Adv. Mater.* **2013**, *25*, 2152–2157. [CrossRef]
21. Ma, B.; Lu, B.; Luo, J.; Deng, X.; Wu, Z.; Wang, X. The hollow mesoporous silicon nanobox dually encapsulated by SnO_2/C as anode material of lithium ion battery. *Electrochim. Acta* **2018**, *288*, 61–70. [CrossRef]
22. Lee, W.H.; Kim, H. Oxidized iridium nanodendrites as catalysts for oxygen evolution reactions. *Catal. Commun.* **2011**, *12*, 408–411. [CrossRef]
23. Saikia, D.; Deka, J.R.; Chou, C.-J.; Kao, H.-M.; Yang, Y.-C. 3D interpenetrating cubic mesoporous carbon supported nanosized SnO2 as an efficient anode for high performance lithium-ion batteries. *J. Alloys Compd.* **2019**, *791*, 892–904. [CrossRef]
24. Saddique, J.; Zhang, X.; Wu, T.; Su, H.; Liu, S.; Zhang, D.; Zhang, Y.; Yu, H. Sn4P3-induced crystalline/amorphous composite structures for enhanced sodium-ion battery anodes. *J. Mater. Sci. Technol.* **2020**, *55*, 73–80. [CrossRef]
25. Chang, L.; Yi, Z.; Wang, Z.; Wang, L.; Cheng, Y. Ultrathin SnO2 nanosheets anchored on graphene with improved electrochemical kinetics for reversible lithium and sodium storage. *Appl. Surf. Sci.* **2019**, *484*, 646–654. [CrossRef]
26. Zhang, J.; Ma, Z.; Jiang, W.; Zou, Y.; Wang, Y.; Lu, C. Sandwich-like CNTs@SnO_2/SnO/Sn anodes on three-dimensional Ni foam substrate for lithium ion batteries. *J. Electroanal. Chem.* **2016**, *767*, 49–55. [CrossRef]
27. Wei, W.; Du, P.; Liu, D.; Wang, H.; Liu, P. Facile mass production of nanoporous SnO2 nanosheets as anode materials for high performance lithium-ion batteries. *J. Colloid Interface Sci.* **2017**, *503*, 205–213. [CrossRef]
28. Mao, Y.; Tian, Q.; Chen, F.; Chen, J.; Yang, L. Improving the lithium storage performance of SnO_2 nanoparticles by in-situ embedding into a porous carbon framework. *J. Alloys Compd.* **2019**, 803. [CrossRef]
29. Liu, D.; Wei, Z.; Liu, L.; Pan, H.; Duan, X.; Xia, L.; Zhong, B.; Wang, H.; Jia, D.; Zhou, Y.; et al. Ultrafine SnO_2 anchored in ordered mesoporous carbon framework for lithium storage with high capacity and rate capability. *Chem. Eng. J.* **2021**, *406*, 126710. [CrossRef]
30. Cheng, Y.; Yi, Z.; Wang, C.; Wu, Y.; Wang, L. Controllable fabrication of C/Sn and C/SnO/Sn composites as anode materials for high-performance lithium-ion batteries. *Chem. Eng. J.* **2017**, *330*, 1035–1043. [CrossRef]
31. Zhou, G.; Wang, D.-W.; Yin, L.-C.; Li, N.; Li, F.; Cheng, H.-M. Oxygen Bridges between NiO Nanosheets and Graphene for Improvement of Lithium Storage. *ACS Nano* **2012**, *6*, 3214–3223. [CrossRef] [PubMed]
32. Reddy, M.J.K.; Ryu, S.H.; Shanmugharaj, A.M. Synthesis of SnO_2 pillared carbon using long chain alkylamine grafted graphene oxide: An efficient anode material for lithium ion batteries. *Nanoscale* **2016**, *8*, 471–482. [CrossRef] [PubMed]
33. Zhou, X.; Dai, Z.; Liu, S.; Bao, J.; Guo, Y.-G. Ultra-Uniform SnOx/Carbon Nanohybrids toward Advanced Lithium-Ion Battery Anodes. *Adv. Mater.* **2014**, *26*, 3943–3949. [CrossRef] [PubMed]
34. Hu, Y.; Yang, Q.-R.; Ma, J.; Chou, S.-L.; Zhu, M.; Li, Y. Sn/SnO2@C composite nanofibers as advanced anode for lithium-ion batteries. *Electrochim. Acta* **2015**, *186*, 271–276. [CrossRef]

materials MDPI

Article

New n-p Junction Floating Gate to Enhance the Operation Performance of a Semiconductor Memory Device

Yi-Yueh Chen *, Su-Jien Lin and Shou-Yi Chang

Department of Materials Science and Engineering, National Tsing Hua University, Hsinchu 30013, Taiwan; sjlin@mx.nthu.edu.tw (S.-J.L.); changsy@mx.nthu.edu.tw (S.-Y.C.)
* Correspondence: yychen@mxic.com.tw; Tel.: +886-3-578-6688

Abstract: To lower the charge leakage of a floating gate device and improve the operation performance of memory devices toward a smaller structure size and a higher component capability, two new types of floating gates composed of pn-type polysilicon or np-type polysilicon were developed in this study. Their microstructure and elemental compositions were investigated, and the sheet resistance, threshold voltages and erasing voltages were measured. The experimental results and charge simulation indicated that, by forming an n-p junction in the floating gate, the sheet resistance was increased, and the charge leakage was reduced because of the formation of a carrier depletion zone at the junction interface serving as an intrinsic potential barrier. Additionally, the threshold voltage and erasing voltage of the np-type floating gate were elevated, suggesting that the performance of the floating gate in the operation of memory devices can be effectively improved without the application of new materials or changes to the physical structure.

Keywords: semiconductor device; memory cell; floating gate; n-p junction; charge leakage

Citation: Chen, Y.-Y.; Lin, S.-J.; Chang, S.-Y. New n-p Junction Floating Gate to Enhance the Operation Performance of a Semiconductor Memory Device. *Materials* **2022**, *15*, 3640. https://doi.org/10.3390/ma15103640

Academic Editors: Zhenghua Tang, Marc Cretin and Sophie Tingry

Received: 1 April 2022
Accepted: 16 May 2022
Published: 19 May 2022
Corrected: 28 September 2022

Publisher's Note: MDPI stays neutral with regard to jurisdictional claims in published maps and institutional affiliations.

1. Introduction

Memory devices, one of the most typical and popularly used electronic devices, generally comprise a plurality of gate structures, which include a control gate and a floating gate [1,2]. The floating gate is a conductive layer normally fabricated from polysilicon that is positioned between the control gate and a silicon substrate [1,2]. The floating gate is not attached to any electrodes or power sources and is generally surrounded by an insulation material [1,2]. The operation of the memory cells is dependent upon the charge stored in the floating gate at the threshold voltage to represent information in these memory devices [3,4]. The performance of the memory cells is determined by the programming speed, which is dominated by the speed of the erasing and writing operations [1,2]. The speed is basically limited by the rate at which electrons can be pumped into (writing) and out of (erasing) the devices without causing damage to the device [1,5–7]. Typically, writing and erasing operations must be capable of operating within 1 ms at a specified applied voltage [1,6,8–11].

Aiming at a higher capability but a smaller chip size, the semiconductor industry has been increasingly driven towards smaller and more numerous electronic devices, including memory cells [2,12,13]. To reduce the size and accordingly increase the number of such devices, while simultaneously maintaining or even improving their respective capabilities, the size of components and the distance between such components need to be reduced [2,14,15]. However, as the cell size is reduced, some other issues arise that prevent a further reduction in size [15,16]. One of these issues is that charge leakage from the floating gate may increase, thereby deteriorating the performance of the devices as the individual layers of the gate structures are made smaller and placed closer to each other [15]. In particular, the tunneling oxide will be more seriously damaged with more programming and erasing sequences, resulting in more charge leakage [15]. In order to overcome the issue of charge leakage, many device structures have been proposed, e.g., SONOS, BE-SONOS,

TAHOS and 3D FLASH [6,17–20]. The 3D NAND FLASH structure was proposed as a solution when 2D NAND FLASH reached the scaling limit of a 15 nm process node [21]. Furthermore, the ReRAM [8,22], PCRAM [23,24], FeRAM [25,26] and MRAM [27,28] devices have also attracted much attention in the past two decades as promising candidates for the next generation of nonvolatile memory cells with improved performance. However, new semiconductor devices with markedly shrunken gate structures and reduced charge leakage that do not sacrifice their performance or suffer from environmental contamination are still elusive.

Hence, in this study, two new floating gate structures, including a p-n junction and an n-p junction, were designed, investigated and processed on 300 mm wafers. In these new designs, no extra new material needs to be employed, no new process needs to be developed and no contamination risk needs to be considered when the devices are processed at the semiconductor manufacturing factory. By forming an n-p junction instead of a p-n junction in the first conductive layer (the floating gate), the charge leakage across the second dielectric layer (the inter-polysilicon dielectric layer) may be reduced. This n-p junction interface is anticipated to provide an intrinsic potential barrier to inhibit the leakage path, successfully reducing the charge leakage and enlarging the programming and erasing window. Additionally, upon the reduction of the charge leakage across the second dielectric layer, the second dielectric layer can be made thinner and/or even be completely removed from wrapping the first conductive layer. The gate structure can thereby be made more compact, allowing a smaller semiconductor device without sacrificing the performance of the device.

2. Materials and Methods

2.1. Device Fabrication

NAND FLASH memory devices with two new floating gate structures were fabricated on p-type 300 mm silicon (Si) wafers with n^+ junctions. As shown in Figure 1, the memory devices comprise the Si substrate, the first dielectric layer (tunneling oxide, denoted as TUN OX) disposed along the substrate, and the first conductive layer (floating gate, FG) disposed along the first dielectric layer (Figure 1a,c, schematically illustrated from the X- and Y-direction cross-sections, respectively). The second dielectric layer (inter-polysilicon dielectric, IPD) is disposed along the sidewall of the first conductive layer, and the second conductive layer (control gate, CG, such as n-type polysilicon) is afterwards deposited. Two new types of the first conductive layer, i.e., the floating gate, were proposed, including the pn-type (a bottom "p^+" region followed by a top "n^+" region) polysilicon and the np-type (bottom "n^+" followed by top "p^+") polysilicon, for which a high-temperature chemical vapor deposition (CVD) boron-doping polysilicon process and a high-temperature furnace phosphorous-doping polysilicon process were applied at 500 °C, in sequence or vice versa. The thickness ratio of the bottom-to-top regions of the pn-type or np-type polysilicon was designed to be around 1:3. For comparison, a conventional floating gate (the control split) was also prepared, with single n^+ polysilicon as the first conductive layer. The concentration of dopants in the n-type and p-type polysilicon was around 1×10^{19} cm^{-3} and 1×10^{21} cm^{-3}, respectively.

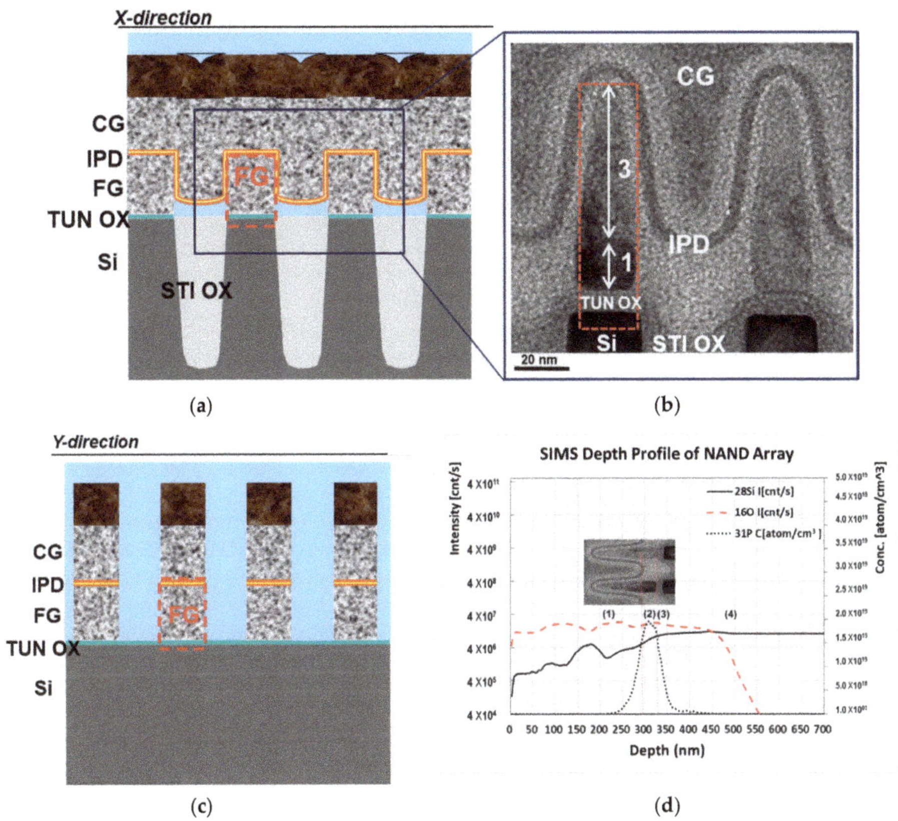

Figure 1. (**a**) Schematic illustration and (**b**) cross-sectional TEM image of memory cells around floating gates with np-type polysilicon from the X-directional view; (**c**) schematic illustration of memory cells from the Y-directional view; (**d**) SIMS depth profile of elemental distribution along the floating gate.

2.2. Characterization and Measurement

Thin foils (cross-sectional) of the memory devices around the floating gates were cut by using a focused ion beam system (USA, FIB, FEI Expida1265) and milled with an ultralow current, and the microstructure was observed by using a transmission electron microscope (Netherlands, TEM, FEI Osiris). The depth profile of elemental compositions along the floating gates for understanding the distribution of dopants was determined by using a secondary ion mass spectrometer (France, SIMS, AMETEK ims-6f). The sheet resistance (R_s) of the floating gates, programing threshold voltage (V_{th}) and erasing voltage (V_{er}) were measured by using a WAT system (USA, Keysight, 4082F). The charge simulation of the floating gates was performed by the TCAD (Technology Computer-Aided Design).

3. Results and Discussion

3.1. Microstructure and Chemical Composition

Figure 1b shows the cross-sectional TEM microstructure of the memory cells around the floating gates with np-type polysilicon from the X-directional view. Clearly, the tunneling oxide layer is disposed between the floating gates and the substrate, and the inter-polysilicon dielectric layer is uniformly deposited on the floating gates. The image contrast indicates two regions in the floating gates: the bright region at the top and the dark region at the bottom, and the thickness ratio of the bottom-to-top regions is roughly estimated to be around 1:3. As further illustrated in Figure 1d, the SIMS depth profile along the floating gate confirms four regions of elemental distribution along the floating gate, from top to bottom: (1) the top p^+ polysilicon for a thickness of about 60 nm, with a silicon element; (2) the bottom n^+ polysilicon for 20 nm, with silicon and a high concentration of phosphorous dopants; (3) the tunneling oxide and (4) the silicon substrate. It was noted that in region (1), boron dopants were not present due to the improper collection condition of light-ionized boron signals from the uneven film structure of the sample instead of a planar/blanket one. However, the gradually dropping intensity of silicon might reveal the existence of other elements that were very likely boron.

3.2. Sheet Resistance and Charge

Figure 2 presents the cumulative probability plot and box plot of sheet resistance for three different floating gates, including the control split and the new pn-type and np-type floating gates. Clearly, the sheet resistance of the new floating gates was higher than that of the control split, (i.e., about 1.7 times for the pn-type floating gates and 2.1 times for the np-type floating gates), which was plausibly caused by the formation of a depletion zone and the narrowed channels for current flow. When a forward bias was applied to the np-type floating gate, or a reverse bias was applied to the pn-type floating gate, a depletion zone of carriers would be formed at the n-p or p-n junction interface [28–32], leading to an open circuit at the bottom region of the floating gates. Current flow was therefore allowed only through the top p^+ or n^+ polysilicon paths, respectively, and the narrowed channel would thus result in increased resistance, particularly for the np-type floating gates, as the mobility of holes in the p^+ polysilicon path was lower than that of electrons in the n^+ polysilicon path [29,30].

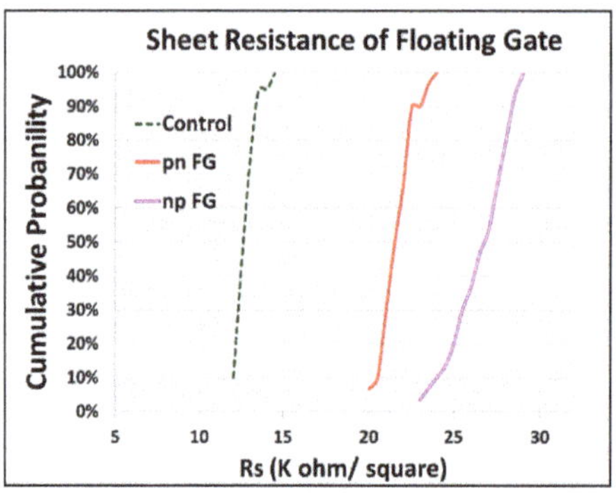

(a)

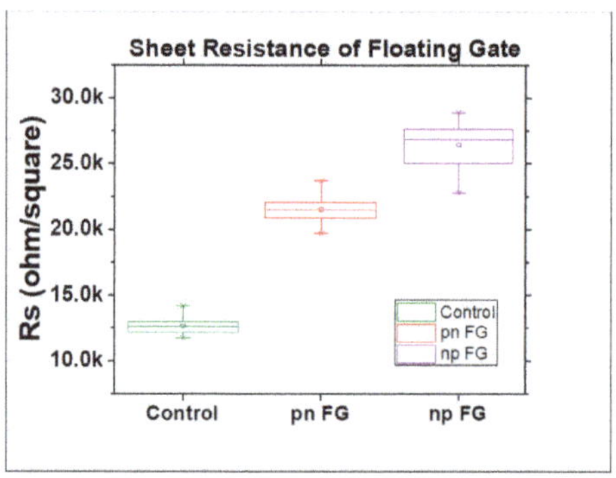

(b)

Figure 2. (a) Cumulative probability plot of sheet resistance of floating gates (30 data points for pn FG and np FG, and 20 data points for the control split). (b) Box plot of sheet resistance (center line: median of the data; top line: Q3, the upper quartile; bottom line: Q1, the lower quartile).

As illustrated in the band diagrams of the neutral and charged states of these three floating gates in Figure 3a,c, different band structures are expected. For the conventional n-type polysilicon floating gate (the control split, Figure 3a) at a programing voltage (positive bias, ΔV) applied to the control gate, the energy band near the control gate will bend downward for ΔV to form a channel near the tunneling oxide for carriers to tunnel through the tunneling oxide into the floating gate for programming [1,2]. The charge in the floating gate depends on the gate coupling ratio (GCR) to influence the efficiency of the device programming [13,33]. In comparison, for the pn-type polysilicon floating gate (Figure 3b) and the np-type polysilicon floating gate (Figure 3c) at a thermal equilibrium state, the Fermi level (E_f) is close to the valence band in the p^+ region (conduction by holes) and close to the conduction band in the n^+ region (conduction by electrons). At a constant Fermi level, the distributions of carriers as well as the energy levels of the conduction band (E_c) and valence band (E_v) are thus different in the p^+ and n^+ regions at the neutral state, and a depletion zone (a thin region with very few carriers) of high electrical resistance will accordingly be formed at the p-n or n-p junction interface [29–32].

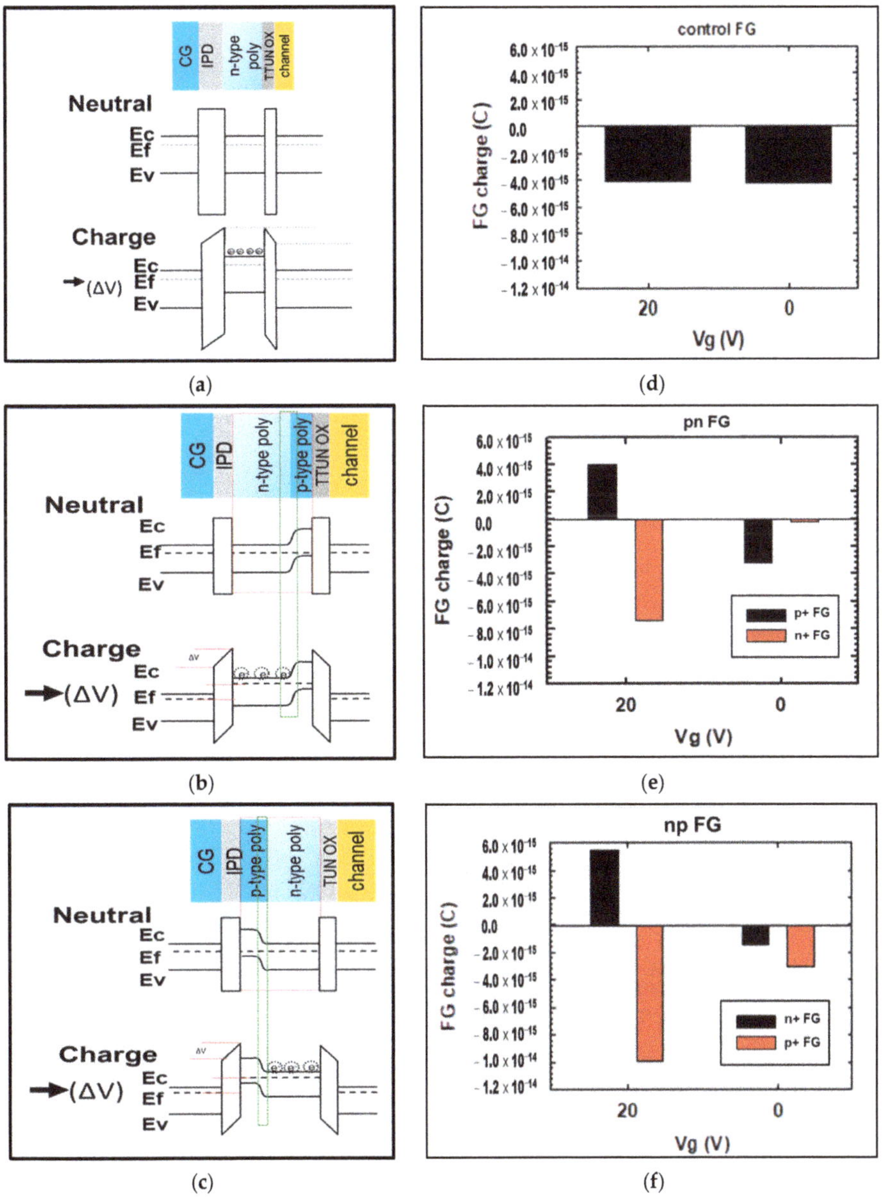

Figure 3. Band diagrams of the natural and charge states of (**a**) conventional n-type polysilicon floating gate (the control split), (**b**) pn-type polysilicon floating gate, and (**c**) np-type polysilicon floating gate. Charge simulations for (**d**) conventional n-type floating gate (the control split), (**e**) pn-type floating gate, and (**f**) np-type floating gate (V_g 20 V: programming state, V_g 0 V: retention state).

When a programming voltage ΔV is applied to the control gate, the energy band bends downward, and the carriers will tunnel through the tunneling oxide into either the pn- or the np-type floating gate in the same way as the conventional floating gate. However, owing to the different space charge distributions in the p^+ and n^+ regions, electrons stay mainly in the n^+ region [29,34]. The carriers (electrons) into the pn-type floating gate will induce a reverse bias in the p-n junction to cause the expansion of the depletion zone and the shrinkage of the top n^+ region for carrier storage, therefore reducing the total stored charge. On the contrary, in the np-type floating gate case, a forward bias will lead to the contraction of the depletion zone and the extension of the bottom n^+ region for carrier storage, which in turn increases the total stored charge. In addition, because the bottom n^+ region is close to the tunneling oxide channel and has a low energy barrier for programming, and the top p^+ region is adjacent to the inter-polysilicon dielectric and has a high energy barrier, the charge leakage of the control gate is expected to be inhibited, which aids in improving the programming efficiency and elevating the programming threshold voltage (V_{th}) of the np-type floating gate as investigated below.

Furthermore, the charge simulations given in Figure 3d–f confirm the aforementioned assumption regarding charging in the three different floating gates. When the voltage applied to the control gate (V_g) is switched from 20 V (the programming state) to 0 V (the retention state), as expected, there is no change in the amount or distribution of charge in the conventional floating gate (the control split, Figure 3d), since the n-type floating gate is simply composed of a single material (n^+ polysilicon). However, the charge is obviously redistributed, and a part of the charge is lost in the pn-type and np-type floating gates when the gate voltage V_g is switched. Clearly, at V_g = 20 V, the charge in the n^+ or p^+ region of the np-type floating gate is larger than that of the pn-type floating gate. At V_g = 0 V, in addition to the fact that more charge in the n^+ region of the np-type floating gate is retained, a portion of charge in the p^+ region is retained as well, suggesting that this n-p junction design in the floating gate will benefit the retention of charge, particularly because the p^+ region is much farther away from the tunneling oxide, making it less likely that a charge leakage will occur.

3.3. Threshold Voltage and Erasing Voltage

Two other important factors dominating the programming (writing) window and performance of memory devices include the threshold voltage (V_{th}, the gate voltage required to create strong inversion under the gate when the floating gate contains the electrons [35]) and the erasing voltage (V_{er}, the voltage required for removing the stored charge (electrons) in the floating gates [36]). When the gate voltage is below the threshold voltage, this device is no longer in strong inversion. This region of device operation is called the "cutoff", which corresponds to a logical "0" stored in the cell [37]. A higher threshold voltage yields a wider programming window and thereafter benefits more precise control over the read operation state of the devices. For example, two states with programming threshold voltages of 4 V and 2 V define a memory window, ΔV, of 2 V, which is clearly better than a window of 1.5 V attained in the case where the programming threshold voltages of the 0 and 1 states are, respectively, 3 V and 1.5 V. On the other hand, a higher erasing voltage is conducive for a more stable state and more effective retention of the stored charge in the memory devices. However, a higher programming threshold voltage may also cause a more serious impact on the tunneling oxide and induce larger current leakage to lower the erasing voltage.

As mentioned above and presented below in Figure 4, the new types of floating gates, in particular the np-type, are observed to effectively improve the performance of the memory devices fabricated without the application of any new materials or changes to their physical structure. As clearly seen in the cumulative probability plot and box plot, the programming threshold voltage of the np-type floating gate was as high as about 1.2 times that of the conventional one (the control split) and much higher than that of the pn-type one (Figure 4a,b), while the erasing voltage of the np-type floating gate was close to that of the conventional one and also higher than that of the pn-type one, both suggesting the better performance of the np-type floating gate in controlling the operation of the memory devices (Figure 4c,d). The erasing voltages of the floating gates that we showed in Figure 4 were actually measured with a deliberately designed test key to check the floating gate state after the charges of the floating gate were cleaned up by applying a high voltage on the substrate. The lower the $|V_{er}|$, the easier it is for the cell to be turned on, which typically corresponds to a logical "1" stored in the cell. The degradations of the programming and erasing operation (after 3000 cycles) were also investigated to understand the performance of the different types of floating gates, as given in Figure 5. It was clear that the np-type floating gate showed a much better performance than the pn-type one and had the same performance as the control split, indicating no extra current leakage from the tunneling oxide even at a higher threshold voltage.

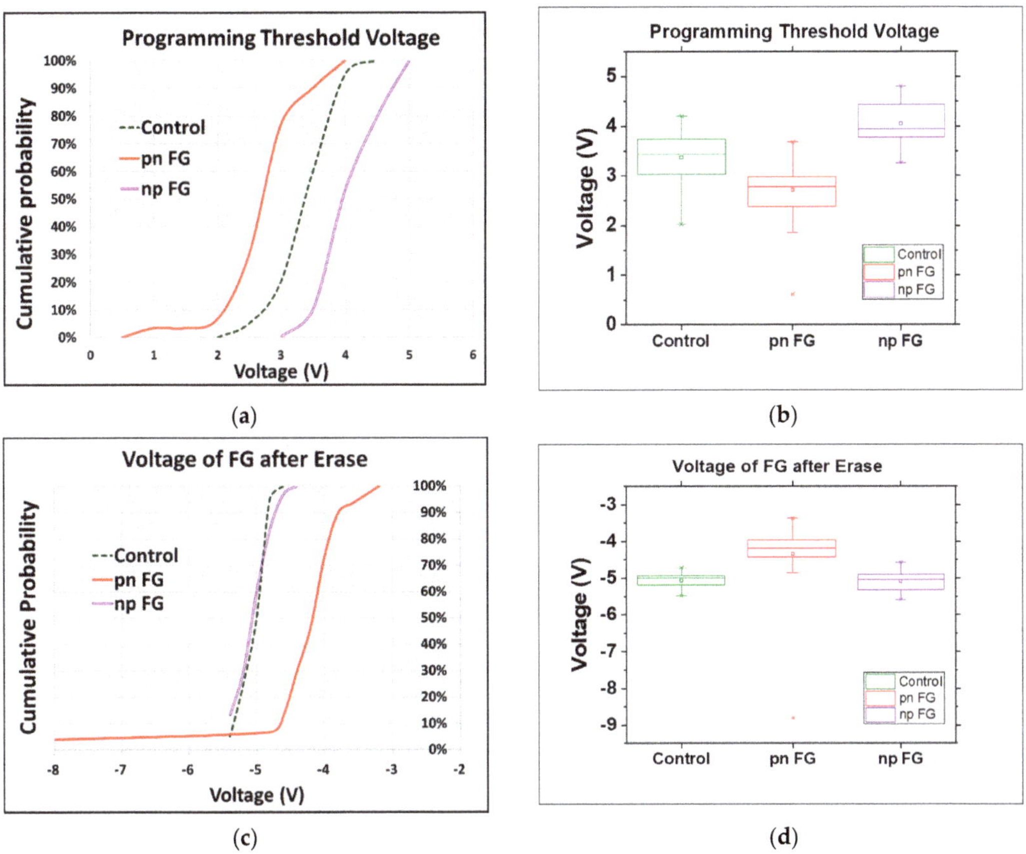

Figure 4. (**a**) Cumulative probability plot of programming threshold voltage of floating gates. (**b**) Box plot of programming threshold voltage. (**c**) Cumulative probability plot of erasing voltage of floating gates. (**d**) Box plot of erasing voltage (center line: median of the data; top line: Q3, the upper quartile; bottom line: Q1, the lower quartile).

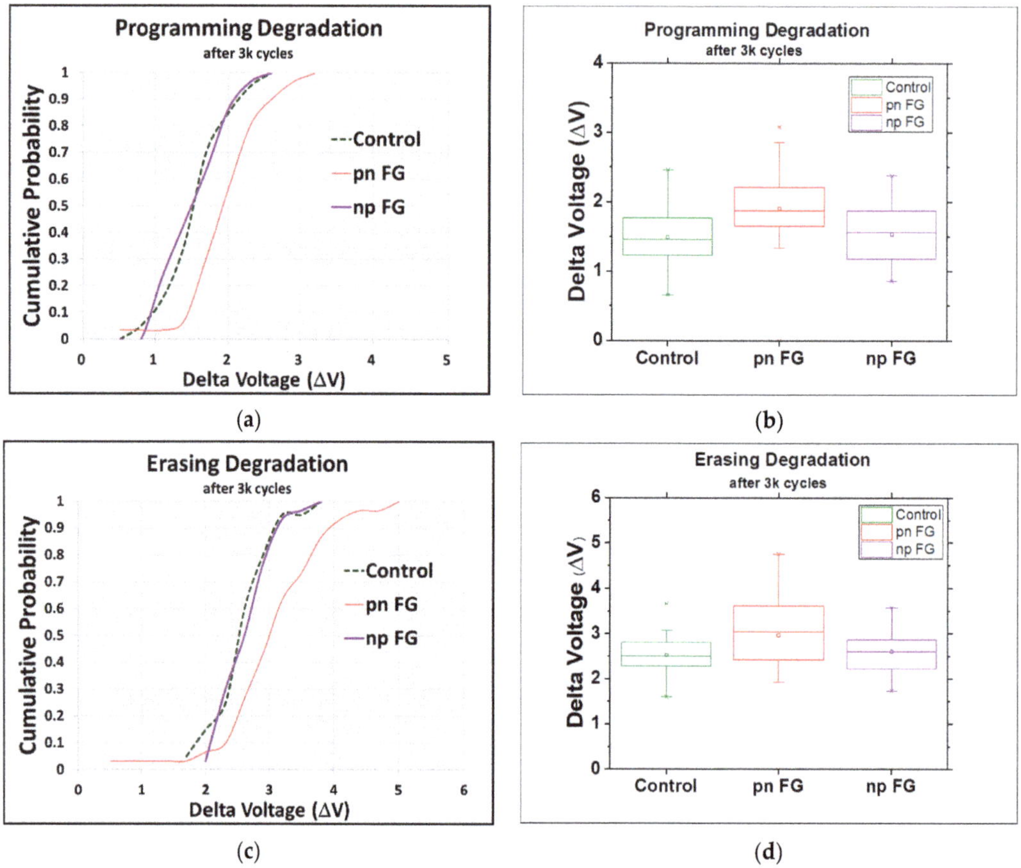

Figure 5. (**a**) Cumulative probability plot of degradation of programming voltage of floating gates. (**b**) Box plot of degradation of programming voltage. (**c**) Cumulative probability plot of degradation of erasing voltage of floating gates. (**d**) Box plot of degradation of erasing voltage (all after 3000 cycles; center line: median of the data; top line: Q3, the upper quartile; bottom line: Q1, the lower quartile).

4. Conclusions

In summary, a new np-type floating gate with n-p junction polysilicon (bottom "n^+" followed by top "p^+" with a thickness ratio of 1:3) was developed in this study to reduce the charge leakage and improve the operation performance of memory devices. A depletion zone of carriers was formed at the n-p junction interface, leading to a narrowed channel and thus an increased sheet resistance that was 2.1 times that of the conventional floating gate. The relatively high charge storage and retention in the np-type floating gate is able to inhibit the charge leakage, owing to the high energy barrier at the n-p junction interface. Moreover, the programming threshold voltage difference between the 0 and 1 states (i.e., the memory window) of the np-type floating gate was effectively elevated by 1.2 times, while the erasing voltage and its degradation were close to that of the conventional one, indicative of no extra current leakage even at a higher programming threshold voltage and the better operation performance of the memory devices.

Author Contributions: Conceptualization, Y.-Y.C.; methodology, Y.-Y.C.; formal analysis, Y.-Y.C.; investigation, Y.-Y.C.; resources, Y.-Y.C.; data curation, Y.-Y.C.; writing—original draft, Y.-Y.C. and S.-Y.C.; writing—review & editing, Y.-Y.C. and S.-Y.C.; supervision, S.-J.L. and S.-Y.C.; project administration, Y.-Y.C.; funding acquisition, S.-J.L. All authors have read and agreed to the published version of the manuscript.

Funding: This research was funded by Macronix International Co., Ltd.

Institutional Review Board Statement: Not Applicable.

Informed Consent Statement: Not Applicable.

Data Availability Statement: The data presented in this study are available on request from corresponding author.

Acknowledgments: The authors gratefully acknowledge Macronix International Co., Ltd. for their support and technical discussions about this work.

Conflicts of Interest: The authors declare no conflict of interest.

References

1. Kahng, D.; Sze, S.M. A Floating Gate and It's Application to Memory Devices. *Bell Syst. Tech. J.* **1967**, *46*, 1283. [CrossRef]
2. Sze, S.M. Evolution of nonvolatile semiconductor memory: From floating-gate concept to single-electron memory cell. In *Future Trends in Microelectronics*; Luryi, S., Xu, J., Zaslavsky, A., Eds.; John Wiley & Sons: Hoboken, NJ, USA, 1999; pp. 291–303.
3. Yau, L.D. A simple theory to predict the threshold voltage of short-channel IGFET's. *Solid State Electron.* **1974**, *17*, 1059–1063. [CrossRef]
4. Himeno, T.; Matsukawa, N.; Hazama, H.; Sakui, K.; Oshikiri, M.; Masuda, K.; Kanda, K.; Itoh, Y.; Miyamoto, J. A new technique for measuring threshold voltage distribution in flash EEPROM devices. In Proceedings of the International Conference on Microelectronic Test Structures, Cork, Ireland, 8–11 April 2002; IEEE: New York, NY, USA, 2002. [CrossRef]
5. Lee, C.H.; Yang, I.C.; Cheng, C.H.; Chong, L.H.; Chen, K.F.; Huang, J.S.; Ku, S.H.; Zous, N.K.; Huang, I.J.; Han, T.T.; et al. Junction optimization for Reliability issues in floating gate NAND flash cells. In Proceedings of the 2011 International Reliability Physics Symposium, Monterey, MA, USA, 24 January 2011; IEEE: New York, NY, USA, 2011. [CrossRef]
6. Lue, H.T.; Wang, S.Y.; Lai, E.K.; Hsieh, K.Y.; Liu, R.; Lu, C.Y. A BE-SONOS (bandgap engineered SONOS) NAND for post-floating gate era flash memory. In Proceedings of the 2007 International Symposium on VLSI Technology, Systems and Applications (VLSI-TSA), Hsinchu, Taiwan, 23–25 April 2007; IEEE: New York, NY, USA, 2007.
7. Jin, Y.; Lee, B. A comprehensive survey of issues in solid state drives. *Adv. Comput.* **2019**, *114*, 1–69.
8. Zahoor, F.; Azni Zulkifli, T.Z.; Khanday, F.A. Resistive random access memory (RRAM): An overview of materials, switching mechanism, performance, multilevel cell (MLC) storage, modeling, and applications. *Nanoscale Res. Lett.* **2020**, *15*, 90. [CrossRef] [PubMed]
9. Meena, J.S.; Sze, S.M.; Chand, U.; Tseng, T.-Y. Overview of emerging nonvolatile memory technologies. *Nanoscale Res. Lett.* **2014**, *9*, 526. [CrossRef] [PubMed]
10. Aritome, S. Advanced flash memory technology and trends for file storage application. In Proceedings of the International Electron Devices Meeting 2000, San Francisco, CA, USA, 10–13 December 2000; IEEE: New York, NY, USA, 2002; pp. 763–766.
11. Waser, R. *Nanoelectronics and Information Technology: Advanced Electronic Materials and Novel Devices*, 2nd ed.; Wiley-VCH: Weinheim, Berlin, Germany, 2012; Chapter IV; p. 548.
12. Afshari, K. Nonvolatile Memory with Multi-Stack Nanocrystals as Floating Gates. In *2007 REU Research Accomplishments*; National Nanotechnology Infrastructure Network: Ithaca, NY, USA, 2007; pp. 38–39.
13. Blomme, P.; Rosmeulen, M.; Cacciato, A.; Kostermans, M.; Vrancken, C.; Van Aerde, S.; Schram, T.; Debusschere, I.; Jurczak, M.; Van Houdt, J. Novel dual layer floating gate structure as enabler of fully planar flash memory. In Proceedings of the 2010 Symposium on VLSI Technology, Honolulu, HI, USA, 15–17 June 2010; IEEE: New York, NY, USA, 2010; pp. 129–130.
14. Soin, N. Dependency of threshold voltage on floating gate and inter-polysilicon dielectric thickness for nonvolatile memory devices. In Proceedings of the 2010 IEEE International Conference on Semiconductor Electronics (ICSE2010), Malacca, Malaysia, 28–30 June 2010; IEEE: New York, NY, USA, 2010; pp. 83–87.
15. Lee, M.C.; Wong, H.Y. The Impact of Tunnel Oxide Nitridation to Reliability Performance of Charge Storage Non-Volatile Memory Devices. *J. Nanosci. Nanotechnol.* **2014**, *14*, 1508–1520. [CrossRef] [PubMed]
16. Ohba, R.; Sugiyama, N.; Uchida, K.; Koga, J.; Toriumi, A. Nonvolatile Si quantum memory with self-aligned doubly-stacked dots. *IEEE Trans. Electron Devices* **2002**, *49*, 1392–1398. [CrossRef]
17. Lin, Y.H.; Yang, Y.Y. Work Function Adjustment by Using Dipole Engineering for TaN-Al2O3-Si3N4-HfSiOx-Silicon Nonvolatile Memory. *Materials* **2015**, *8*, 5112–5120. [CrossRef] [PubMed]
18. Spassov, D.; Paskaleva, A.; Guziewicz, E.; Davidović, V.; Stanković, S.; Djorić-Veljković, S.; Ivanov, T.; Stanchev, T.; Stojadinović, N. Radiation Tolerance and Charge Trapping Enhancement of ALD HfO2/Al2O3 Nanolaminated Dielectrics. *Materials* **2021**, *14*, 849. [CrossRef] [PubMed]

19. Yoon, J.H. Fabrication of Sn@Al2O3 Core-shell Nanoparticles for Stable Nonvolatile Memory Applications. *Materials* **2019**, *12*, 3111. [CrossRef] [PubMed]
20. Wang, B.; Gao, B.; Wu, H.; Qian, H. A drain leakage phenomenon in poly silicon channel 3D NAND flash caused by conductive paths along grain boundaries. *Microelectron. Eng.* **2018**, *192*, 66–69. [CrossRef]
21. Goda, A. Recent Progress on 3D NAND Flash Technologies. *Electronics* **2021**, *10*, 3156. [CrossRef]
22. Kim, H.-J.; Kim, D.-W.; Lee, W.-Y.; Kim, K.; Lee, S.-H.; Bae, J.-H.; Kang, I.-M.; Kim, K.; Jang, J. Flexible Sol-Gel-Processed Y2O3 RRAM Devices Obtained via UV/Ozone-Assisted Photochemical Annealing Process. *Materials* **2022**, *15*, 1899. [CrossRef] [PubMed]
23. Le Gallo, M.; Sebastian, A. An overview of phase-change memory device physics. *J. Phys. D Appl. Phys.* **2020**, *53*, 213002. [CrossRef]
24. Hatayama, S.; Sutou, Y.; Shindo, S.; Saito, Y.; Song, Y.-H.; Ando, D.; Koike, J. Inverse Resistance Change Cr2Ge2Te6-Based PCRAM Enabling Ultralow-Energy Amorphization. *ACS Appl. Mater. Interfaces* **2018**, *10*, 2725–2734. [CrossRef] [PubMed]
25. Fan, Z.; Chen, J.; Wang, J. Ferroelectric HfO2-based materials for next-generation ferroelectric memories. *J. Adv. Dielectr.* **2016**, *6*, 1630003. [CrossRef]
26. Park, M.H.; Lee, Y.H.; Mikolajick, T.; Schroeder, U.; Hwang, C.S. Review and perspective on ferroelectric HfO_2-based thin films for memory applications. *MRS Commun.* **2018**, *8*, 795–808. [CrossRef]
27. Guo, Z.; Yin, J.; Bai, Y.; Zhu, D.; Shi, K.; Wang, G.; Cao, K.; Zhao, W. Spintronics for Energy- Efficient Computing: An Overview and Outlook. *Proc. IEEE* **2021**, *109*, 1398–1417. [CrossRef]
28. Nehra, V. High-Performance Computing-in-Memory Architecture Using STT-/SOT-Based Series Triple-Level Cell MRAM. *IEEE Trans. Magn.* **2021**, *57*, 1–12. [CrossRef]
29. Shockley, W. *Electrons and Holes in Semiconductors*; Van Nostrand: Princeton, NJ, USA, 1950.
30. Grove, A.S. *Physics and Technology of Semiconductor Devices*; Wiley: New York, NY, USA, 1967.
31. Nicollian, E.H.; Brews, J.R. *MOS Physical and Technology*; Wiley: New York, NY, USA, 1982.
32. Saatci, A.E.; Özdemir, O.; Kutlu, K. Conduction Mechanism Analysis of Inversion Current in MOS Tunnel Diodes. *Mater. Sci. Appl.* **2013**, *04*, 794–801. [CrossRef]
33. Hoefflinger, B. *ITRS: The International Technology Roadmap for Semiconductors*, 2007th ed.; Springer: Berlin/Heidelberg, Germany, 2011.
34. Kim, G.; Sun, M.-C.; Kim, S.W.; Kim, H.W.; Kim, J.H.; Park, E.H.; Kim, H.; Park, B.-G.; Byung-Gook, P. Novel MOSFET structure using p-n junction gate for ultra-low subthreshold-swing. In Proceedings of the 2011 International Semiconductor Device Research Symposium (ISDRS), Baltimore, MD, USA, 7–9 December 20; IEEE: New York, NY, USA, 2011; pp. 1–2. [CrossRef]
35. Misra, V.; Ozturk, M.C. *Electrical Engineering Handbook*; Elsevier: Amsterdam, The Netherlands, 2004; Chapter 3.2.4.
36. Micheloni, R.; Crippa, L. *Advances in Non-volatile Memory and Storage Technology*; Woodhead Publishing: Sawston, Cambridge, UK, 2019; Chapter 3.3.4.
37. Boukhobza, J.; Olivier, P. *Flash Memory Integration*; Elsevier: Amsterdam, The Netherlands, 2017; Chapter 2.1.2.

materials

MDPI

Correction

Correction: Chen et al. New n-p Junction Floating Gate to Enhance the Operation Performance of a Semiconductor Memory Device. *Materials* 2022, *15*, 3640

Yi-Yueh Chen *, Su-Jien Lin and Shou-Yi Chang

Department of Materials Science and Engineering, National Tsing Hua University, Hsinchu 30013, Taiwan
* Correspondence: yychen@mxic.com.tw; Tel.: +886-3-578-6688

Citation: Chen, Y.-Y.; Lin, S.-J.; Chang, S.-Y. Correction: Chen et al. New n-p Junction Floating Gate to Enhance the Operation Performance of a Semiconductor Memory Device. *Materials* 2022, *15*, 3640. *Materials* **2022**, *15*, 6738. https://doi.org/10.3390/ma15196738

Received: 2 August 2022
Accepted: 9 August 2022
Published: 28 September 2022

The authors would like to make corrections to a recently published paper [1].

Removal of Authors

In consideration of the contributions to this work, the following authors are removed from the publication: Feng-Ming Lee, Yu-Yu Lin, Chih-Hsiung Lee, Wei-Chen Chen, Che-Kai Shu, and Chih-Yuan Lu. Their support is acknowledged as indicated in the Acknowledgements. The corrected authors are Yi-Yueh Chen, Su-Jien Lin, and Shou-Yi Chang.

Removal of Affiliations

According to the change in authorship, the corrected affiliation appears below:

Department of Materials Science and Engineering, National Tsing Hua University, Hsinchu 30013, Taiwan; sjlin@mx.nthu.edu.tw (S.-J.L.); changsy@mx.nthu.edu.tw (S.-Y.C.).

Update of Author Contributions

According to the change in authorship, the corrected Author Contributions statement appears below:

Conceptualization, Y.-Y.C.; methodology, Y.-Y.C.; formal analysis, Y.-Y.C.; investigation, Y.-Y.C.; resources, Y.-Y.C.; data curation, Y.-Y.C.; writing—original draft, Y.-Y.C. and S.-Y.C.; writing—review & editing, Y.-Y.C. and S.-Y.C.; supervision, S.-J.L. and S.-Y.C.; project administration, Y.-Y.C.; funding acquisition, S.-J.L. All authors have read and agreed to the published version of the manuscript.

Change of Acknowledgements

The corrected Acknowledgments statement appears below:

The authors gratefully acknowledge Macronix International Co., Ltd. for their support and technical discussions about this work.

The authors apologize for any inconvenience caused and state that the scientific conclusions are unaffected. This correction was approved by the Academic Editor. The original publication has also been updated.

Reference

1. Chen, Y.-Y.; Lin, S.-J.; Chang, S.-Y. New n-p Junction Floating Gate to Enhance the Operation Performance of a Semiconductor Memory Device. *Materials* **2022**, *15*, 3640. [CrossRef] [PubMed]

materials

Article

Insight into a Nitrogen-Doping Mechanism in a Hard-Carbon-Microsphere Anode Material for the Long-Term Cycling of Potassium-Ion Batteries

Changdong Chen [1,2,†], Kai Zhao [3,†], Ming La [2,*] and Chenghao Yang [1,*]

1 School of Enviornment and Energy, South China University of Technology, Guangzhou 510006, China; lmccd5613@163.com
2 College of Chemistry and Environmental Engineering, Pingdingshan University, Pingdingshan 467000, China
3 College of Information Engineering, Pingdingshan University, Pingdingshan 467000, China; zhaokai@pdsu.edu.cn
* Correspondence: laming82@126.com (M.L.); esyangc@scut.edu.cn (C.Y.)
† These authors contributed equally to this work.

Abstract: To investigate the alternatives to lithium-ion batteries, potassium-ion batteries have attracted considerable interest due to the cost-efficiency of potassium resources and the relatively lower standard redox potential of K^+/K. Among various alternative anode materials, hard carbon has the advantages of extensive resources, low cost, and environmental protection. In the present study, we synthesize a nitrogen-doping hard-carbon-microsphere (N-SHC) material as an anode for potassium-ion batteries. N-SHC delivers a high reversible capacity of 248 mAh g^{-1} and a promoted rate performance (93 mAh g^{-1} at 2 A g^{-1}). Additionally, the nitrogen-doping N-SHC material also exhibits superior cycling long-term stability, where the N-SHC electrode maintains a high reversible capacity at 200 mAh g^{-1} with a capacity retention of 81% after 600 cycles. DFT calculations assess the change in K ions' absorption energy and diffusion barriers at different N-doping effects. Compared with an original hard-carbon material, pyridinic-N and pyrrolic-N defects introduced by N-doping display a positive effect on both K ions' absorption and diffusion.

Keywords: potassium-ion batteries; hard-carbon microspheres; nitrogen-doping; energy storage mechanisms

Citation: Chen, C.; Zhao, K.; La, M.; Yang, C. Insight into a Nitrogen-Doping Mechanism in a Hard-Carbon-Microsphere Anode Material for the Long-Term Cycling of Potassium-Ion Batteries. *Materials* **2022**, *15*, 4249. https://doi.org/10.3390/ma15124249

Academic Editors: Zhenghua Tang, Marc Cretin and Sophie Tingry

Received: 9 May 2022
Accepted: 10 June 2022
Published: 15 June 2022

1. Introduction

In recent years, energy storage devices represented by lithium-ion batteries (LIBs) have been widely used in various fields to accelerate the goal of carbon neutrality [1–3]. However, limited lithium resources cannot satisfy the burgeoning demand of LIBs [4]. Thus, it is particularly important to develop new energy storage devices. As one of the alkali metal-ions, potassium-ion is rich in resources and has similar physicochemical properties to lithium-ions; furthermore, potassium-ion batteries (PIBs) also deliver a reaction mechanism similar to LIBs, which belongs to the "rocking chair batteries" [5,6]. In comparison to sodium (−2.71 V vs. standard hydrogen electrode), the standard potential of K^+/K (−2.93 V vs. SHE) is relatively close to that of Li^+/Li (−3.04 V vs. SHE), which suggests that PIBs have the potential of working in a broad voltage window and thus delivering a high energy density [7,8]. Additionally, in PIB electrolytes, due to the weaker Lewis acidity and smaller de-solvation activation energy of K ions than those of Na and Li, K ions show a smaller Stokes radius and interfacial reaction resistance in electrolytes, which display a more rapid conductivity and better interfacial reaction kinetics of K ions [9]. Due to those advantages, PIBs have attracted considerable interest as an alternative candidate for energy storage systems (ESSs).

To date, the design of anode materials is the key to developing PIBs. In consideration of the similar development of anode materials in LIBs, several anode materials have attracted much attention, mainly including carbon-based materials [10], metal oxides [11,12], and alloy-type materials [13,14]. Although potassium-ions intercalated into graphite carbon layers can form a thermodynamically stable intercalation compound, its larger radius (0.138 nm) hinders the rapid intercalation behavior between the carbon layers and shows unsatisfactory specific capacity [15,16]. Recently, hard carbon with extended carbon-layer spacing has received increasing attention from researchers and has been demonstrated as an anode for PIBs [17]. Compared to graphite, the arrangement of graphite-like microcrystallines in hard carbon is disordered, and there are only a few regularly stacked microcrystalline regions. Hard carbon mainly consists of a sp2-hydridized poly-hexagonal carbon-ring structure [18]. The contained intrinsic defects in hard carbon induce the dislocation and deformation of carbon-ring structures, leading to a long-range disordering structure with the large and irregular d-spacing [19]. Those expanded graphitic lattices of hard carbons could be more suitable for K ions' (de)intercalation. At the same time, hard carbon has the advantages of extensive resources, low cost, and being environmentally friendly, which attracts much interest concerning its role as a promising anode for PIBs [20,21]. However, pure hard carbon also suffers from a low specific capacity and poor rate capability.

To solve this limitation, extensive ideas have been presented; heteroatom-doping, in particular, is the usual modified strategy [22,23]. Nitrogen-doping is one of the most popular doping strategies, which can not only provide additional active defects in hard carbon, but also improve conductivity by tuning the intrinsic electronic state [24]. Urea is used as a dopant, and Deng et al. [25] designed an N-doped three-dimensional biomass-derived porous carbon from bagasse. The experimental results show that the introduction of nitrogen not only promotes the change in morphology, but also accelerates the transport of potassium-ions and electrons. Even following 400 cycles at a current density of 200 mA g^{-1}, the reversible capacity of the modified derived carbon can maintain 100.4 mAh g^{-1} without significant capacity decay. In addition, Zhou et al. [26] synthesized an N-doped carbon hollow turbostratic tube by a amidation reaction and heating treatment process. Benefiting from the doping modification of the N atom, the discharge capacity of the carbon hollow turbostratic tube reaches 397 mAh g^{-1} at 0.1 A g^{-1}, and the capacity of 212 mAh g^{-1} can be maintained, even when the current density is increased to 2 A g^{-1}. Furthermore, when the scan rate of cyclic voltammetry (CV) is 0.6 mV s^{-1}, the electrode has 71% capacitance contribution, corresponding to 46.3% of the battery, implying that capacitively controlled processes dominate at high rates. Wang et al. [27] synthesized the edge-enrich N-doped graphitic carbon via pyrolyzing carbon nitride for the PIB anode. Owing to the high N-doping level, the edge-enrich N-doped graphitic carbon could deliver a high capacity of 266 mAh g^{-1} and a remarkable rate performance of 228.9 mAh g^{-1} at 2 A g^{-1}. Moreover, an ultra-long lifespan is displayed in edge-enriched N-doped graphitic carbon, which maintains 188.9 mAh g^{-1} even after 2200 cycles.

The above reports all confirm that N-doping can significantly improve the electrochemical performance of hard carbon, but there are many types of N-doping configurations, which are generally divided into pyridinic-N, pyrrolic-N, and graphitic-N, and their respective effects are rarely reported [28,29]. In a previous report, it was suggested that the graphitic-N site in hard carbon could enhance the electronic conductivity of carbon materials, and pyrrolic-N and graphitic-N could be capable of providing more active sites for K ions' adsorption [30]. However, a systematic evaluation of the effects of K-ion diffusion after N-doping is still lacking. In this work, N-doped hard-carbon microspheres were obtained by using glucose and melamine as the precursors, combined with the hydrothermal reaction and high-temperature heat-treatment process. The differences in the electrochemical performance before and after doping are compared in detail. Furthermore, we employ the Density Functional Theory (DFT) calculation to assess the effect on K^+ absorption and, subsequently, K-ion diffusion of different N-doping effects, which confirms the positive role of pyridinic-N and pyrrolic-N in enhancing the electrochemical properties of hard-carbon microspheres.

2. Experimental Section

Material synthesis and characterization: the hard-carbon microspheres were synthesized via the hydrothermal reaction. A total of 6.4 g glucose (99% purity, Macklin) was dissolved in 45 mL of deionized water and the solution was poured into an autoclave. After being heated at 200 °C for 5 h, the black powders were collected by filtration and drying at 80 °C. To obtain the N-SHC, the obtained powder was mixed with melamine (99% purity, Macklin), and subsequently annealed at 900 °C for 5 h in an argon atmosphere. For the original SHC sample, the black powders collected by filtration were heat treated directly at 900 °C for 5 h in the argon atmosphere without melamine. The structural features of the obtained materials were determined by powder X-ray diffraction (XRD) recorded by CuKα radiation (λ = 1.5418 Å) in the scan range of 10–70°. Field-emission scanning electron microscopy (SEM, Hitachi S-4800) and transmission electron microscopy (TEM, JEOL-JEM-2100) were employed to characterize the morphologies and microstructures. The surface chemical states were analyzed using X-ray photoelectron spectroscope (XPS, PHI500 Versaprobe-II).

Electrochemical measurements: the electrochemical properties were evaluated in the 2032 coin-type cells. The working electrodes were prepared by coating the pulp suspension, which consisted of 80% active material, 10% super P, and 10% carboxymethyl cellulose (CMC) on copper (Cu) foil. After drying at 80 °C in a vacuum oven overnight, the copper foil was cut into 13 mm electrodes and assembled in an Ar gas-filled glovebox. For the assemblage of coin-type cells, the employed electrolyte was 0.8 M KPF6 in ethylene carbonate and propylene carbonate (EC:PC = 1:1), glass fiber (Whatman) was used as the separator, and metallic potassium served as a counter electrode. The galvanostatic charge-discharge tests were conducted in the battery measurement system (Land, CT2001), and the cyclic voltammogram (CV) and electrochemical impedance spectroscopy (EIS) in the frequency range of 0.01–100 kHz were performed in the electrochemical work station (CHI660E).

Computational Methods: The modeling and simulation were based on the spin-polarized density functional theory (DFT), which employed the Vienna Ab initio Simulation Package (VASP). The simulation adopted the Perdew–Burke–Ernzerhof test for the exchange-correlation potential and projector-augmented wave (PAW) with a setting cut-off energy of 400 eV. The Brillouin zone was modeled as a structure of 3 × 3 × 2. The diffusion barriers of K ions between the adjacent carbon layer were calculated by using the CI-NEB [31].

3. Results and Discussion

The morphologies of as-prepared N-SHC and SHC samples were investigated by SEM and TEM. As show in Figure 1b, the obtained N-SHC sample consists of uniform spherical hard-carbon particles with diameters of around 3 μm. Additionally, the energy-dispersive X-ray spectroscopy (EDS) mapping (inserted in Figure 1b) of N-SHC displays a homogeneous distribution of the N element on the hard-carbon sphere. The TEM was employed to observe the microstructure further. As depicted in Figure 1c, the average interlayer distance of N-SHC is 0.376 nm, which is greater than the original SHC sample (0.349 nm in Figure S1). The expanded interlayer space of N-SHC should contribute to the nitrogen groups introduced by N-doping, which provides a buffer for the structural change during the insertion–extraction process of K^+ ions.

Figure 1d represents the XRD patterns of N-SHC and SHC. Broad diffraction peaks appear at around 25° in both N-SHC and SHC samples, which can be indexed on the (002) plane [32]. Additionally, the weak diffraction peak at about 43° suggests the relatively low degree of graphitization. Compared to SHC (24.4°), the (002) peak of N-SHC shifts to a lower 2θ angle (23.6°), indicating the enlarged interspaces of N-SHC, which is in accordance with the analysis of TEM. Additionally, Figure 1e displays the Raman spectra of N-SHC and SHC. The peaks centered at around 1330 and 1580 cm^{-1} are ascribed to the disorder/defect-induced D band and in-plane vibrational G band, respectively [33]. The intensity ratios (I_D/I_G) of D and G peaks have been established to assess the degree of graphitization and the amount of disorder defects [34]. It can be observed that the I_D/I_G value of SHC is 1.09 and the I_D/I_G ratio of N-SHC is calculated to be 1.31, revealing the higher degree of disorder defects induced by N-doping. Moreover, there appears to be a slight blue shift in the G band in N-SHC (1585.2 cm^{-1}), which results from increased carrier concentration via N-doping [35]. Figure 1f represents the high-resolution X-ray photoelectron spectroscopy (XPS) of the N element in N-SHC. This can be deconvoluted as the peaks in the N 1s spectrum of N-SHC: the peak at 401.8 eV is indexed to graphitic-N; peaks at 400.8 and 398.6 eV correspond to pyrrolic-N and pyridinic-N, respectively [36]. Accordingly, the nitrogen concentration in the N-SHC sample was determined as 6.3% by the XPS analyses. Moreover, the proportion of various N-defects were calculated via convolution and the results are inserted in Figure 1f. Previous research has mentioned the high chemical activity of pyrrolic-N and pyridinic-N. These high relative concentrations of pyrrolic-N and pyridinic-N in N-SHC could lead to better chemical properties for K^+ ions' storage.

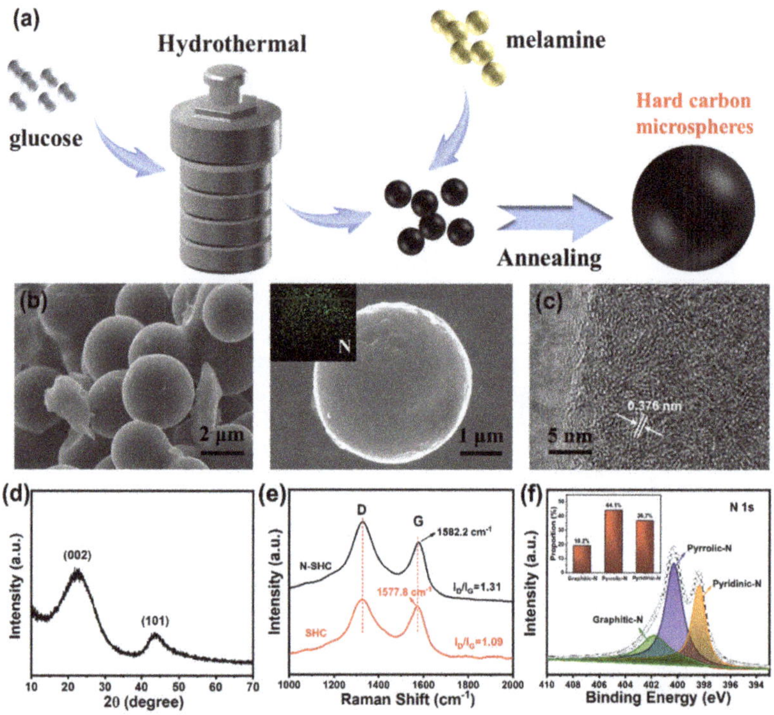

Figure 1. The (**a**) schematic of the synthesis process, (**b**) SEM images and EDS element mapping of as-prepared N-SHC, (**c**) TEM image, (**d**) XRD pattern, (**e**) Raman spectra and (**f**) high-resolution XPS spectra of nitrogen in N-SHC.

The K-ion storage properties of as-prepared N-SHC and SHC were investigated by the galvanostatic test, using potassium metal as the counter electrode. Figure 2a demonstrates the typical cyclic voltammetry (CV) curve for the N-SHC material at the potential window of 0.01–3.0 V. There were two cathodic peaks centered at ≈0.5 and 0.75 V that could be observed in the first cycle, but disappeared in the following cycles. The cathodic peaks that disappeared could probably be attributed to the formation of a solid electrolyte interface (SEI) layer. The anodic peak at about 0.55 V and corresponding sharp cathodic peak at 0.12 V should be ascribed to the de-intercalation and intercalation processes of K^+ ions forming into carbon, respectively. The overlap at CV curves between second and third cycles revealed a stable cycling capability. However, an obvious difference was noticed when comparing the first cathodic process with the second one, which suggested a relatively low initial coulombic efficiency (ICE) of N-SHC. The ICE of the anode material is defined as the ratio of charge capacity and discharge capacity in the first cycle. This can also be confirmed by the charge/discharge profiles. Figure 2b shows the potassiation/depotassiation profiles of the initial three cycles for N-SHC at 1 C (where 1 C is defined as 200 mA g^{-1} in this work). The first potassiation capacity of N-SHC is 466 mAh g^{-1} with the depotassiation capacity of 248 mAh g^{-1}, which reveals an ICE of 53%. The ICE of the N-SHC electrode is similar to SHC (51%), whose initial discharge/charge capacities are 446/228 mAh g^{-1} (Figure S2). The promotion of the reversible capacity of N-SHC (248 mAh g^{-1}) compared to SHC (228 mAh g^{-1}) can be noted. This high capacity of N-SHC may be attributed to the enhanced adsorption effect, which is triggered by N-doping in N-SHC sample. In the first cycle of both the N-SHC and SHC samples, three quasi-plateau regions and a high-potential sloping region can be observed, and the plateau at ≈0.8 V versus K^+/K disappears in the following potassiation process. This irreversible reaction may be related to the formation of the solid electrolyte interface (SEI) and should be responsible for the low ICE [37].

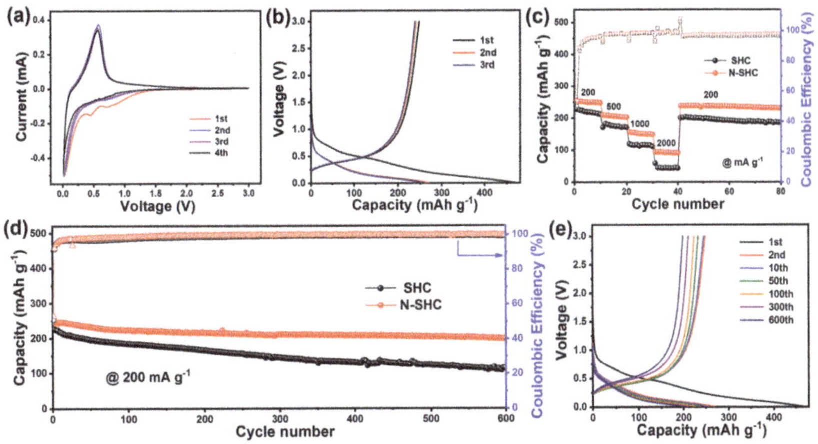

Figure 2. (**a**) CV curves and (**b**) charge-discharge profiles of the N-SHC electrode. Compared (**c**) rate performance and (**d**) long-term cycle properties at 200 mA g^{-1} for N-SHC and SHC electrodes. (**e**) Corresponding charge-discharge profiles of N-SHC at 200 mA g^{-1} after various cycles.

The rate performances of N-SHC and SHC samples were evaluated at various current densities. It was interesting to note that the rate properties of N-SHC were improved, where the N-SHC electrode delivered charge capacities of 251, 206, 152, and 93 mAh g^{-1} at the increased current densities of 200, 500, 1000, and 2000 mA g^{-1}, respectively, as shown in Figure 2c. Compared to the SHC sample, both samples displayed a similar rate capacity retention at a low current rate of 500 mAh g^{-1}. However, as the current densities increased to 1000 and 2000 mA g^{-1}, the capacity of SHC declined more rapidly than the N-SHC electrode. At the rates of 1000 and 2000 mA g^{-1}, N-SHC presented 152 and 93 mAh g^{-1}, respectively, while the reversible charge capacities of pristine SHC were 116 and 43 mAh g^{-1} at the same current densities. When the current density returned to 200 mA g^{-1}, the reversible charge capacity of N-SHC was recovered to 245 mAh g^{-1} following the high-current-rate test, while SHC returned to 200 mAh g^{-1}. This indicates the stabler structure of the N-SHC material compared to SHC. The better rate performance of N-SHC may be due to N-doping leading to a lower diffusion barrier for K^+-ion transfer.

The long-term cycling properties of N-SHC and SHC are evaluated at 200 mA g^{-1}. As shown in Figure 2d, the SHC electrode delivers an initial charge capacity of 228 mAh g^{-1}. Additionally, the reversible capacity of SHC rapidly decayed to 178 mAh g^{-1} after 100 cycles at 200 mA g^{-1}, corresponding to a capacity retention of 78%. Compared to SHC, the N-

SHC shows a better cycle performance. After cycling at 200 mA g^{-1} for 100 cycles, N-SHC presented a higher capacity of 224 mAh g^{-1} with a capacity retention of 90%. When cycled further, at 600 cycles, the superiority of long-term cycling properties of N-SHC were featured more obviously. After 600 cycles at 200 mA g^{-1}, the capacity retention of SHC was no more than 50%. Surprisingly, the N-SHC electrode maintained a high reversible capacity at 200 mAh g^{-1} after 600 cycles, which displayed a superior capacity retention of 81%. Furthermore, Figure 2e and Figure S3 exhibit the charge/discharge profiles after different cycles of N-SHC and SHC samples, respectively. It can be observed that, compared to the SHC electrode, N-SHC shows a slight voltage polarization. This capacity fading and voltage polarization may be a result of the pulverization of SHC particles, where the spherical particle of the N-doping N-SHC sample reveals better structural stability. The structural stability of the N-SHC sample also can be verified by electrochemical impedance spectroscopy (EIS). Figure S4 represents the Nyquist plots of N-SHC and SHC electrodes after various cycles. The Nyquist plots consist of a semicircle and a straight line. The interception of the semicircle corresponds to the contact ohmage, the semicircle in the high-frequency regions is relevant to the charge transfer resistance, and the straight line in the low-frequency region is associated with the Warburg impedance, which reflect ion transportation in the electrode [38]. It can be fitting that the initial charge transfer resistance (R_{ct}) of the SHC electrode is 4868 Ω, and continues decline to 1797 Ω after 100 cycles, which should be attributed to the activization of the electrode. Then, R_{ct} begins to increase and, after 300 cycles, the R_{ct} of SHC increases to 3921 Ω, while N-SHC delivers the R_{ct} of 2982, 684, and 1139 Ω at the original rate and after 100 and 300 cycles, respectively. For comparison, the lower-charge transfer resistance and minor change in the N-SHC electrode after cycling indicates the improved conductivity and better cyclability of N-doping N-SHC.

To obtain further insight into the dynamic characteristic of the potassium-ion-storage behavior involved in the N-SHC electrode, we employed CV analysis at various scan rates from 0.1 to 1 mV s^{-1}. Figure 3a shows the CV test result of N-SHC. Additionally, the electrochemical-reaction behavior was assessed by the relationship of the peak-current (i) response with the scan rate (v), according to Equations (1) and (2):

$$i = a \cdot v^b \tag{1}$$

$$\log(i) = b \cdot \log(v) + \log(a) \tag{2}$$

where i is the peak current, v is the scan rate, and the calculated a and b are adjustable parameters [39]. In particular, the storage behavior can be reflected by parameter b; when b is close to 0.5, the K^+ ion's storage behavior is dominated by the diffusion process; and when b approaches 1.0, the capacitive process predominates. The b-value could be determined by the slope of the plotted log(i)−log(v) curve [40]. In this case, as displayed in Figures 3b and S5, the b-values for N-SHC and SHC are calculated as 0.608 and 0.527, respectively, indicating that the diffusion process predominates in the K^+-ion storage behavior. Furthermore, the capacitive contribution fractions could be distinguished based on Equation (3):

$$i = k_1 \cdot v^{\frac{1}{2}} + k_2 \cdot v \tag{3}$$

where $k_1 \cdot v^{1/2}$ is related to the diffusion-controlled reaction, $k_2 \cdot v$ represents a surface-driven capacitive process, and k_1 and k_2 are constants [41]. Figure 3c shows the calculated capacitive contribution fractions of the N-SHC electrode at different scan rates, and the capacitive contribution ratio reaches 36.1% at 1 mV s^{-1}. Compared to pristine SHC (Figure S5), the surface-driven capacitive behavior of N-SHC is promoted, which should be ascribed to the generation of active defects on the surface induced by N-doping and enhances the adsorption of K^+ ions.

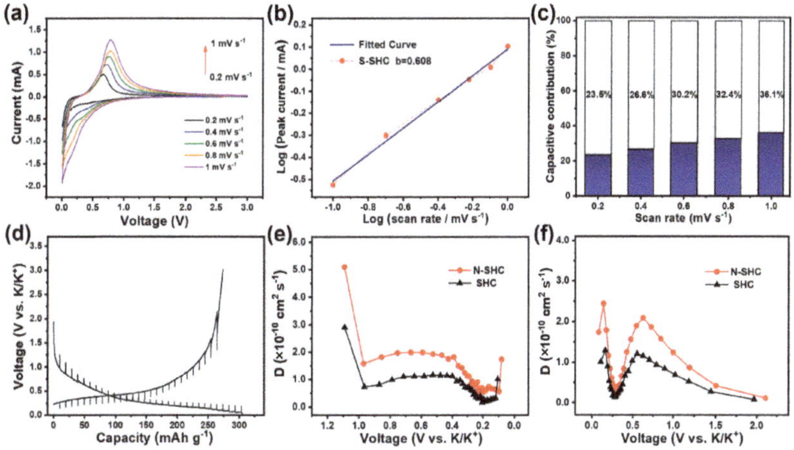

Figure 3. (**a**) CV curves of N-SHC at scan rates from 0.2 to 1 mV s^{-1}, corresponding (**b**) relationship between log (peak current) versus log (scan rate) and (**c**) calculated normalized percentages of capacitive capacities' contributions at various scan rates. (**d**) GITT profiles and diffusion coefficients in (**e**) discharge and (**f**) charge processes.

Furthermore, the galvanostatic intermittent titration technique (GITT) test was employed to evaluate the diffusion ability of K^+ ions in electrodes. Figure 4d exhibits the GITT profiles with the test condition of a pulse current density of 20 mA g^{-1} for 30 min and 2 h rest intervals. The corresponding ionic diffusion coefficients are determined based on Fick's second law as the following equation:

$$D = \frac{4}{\pi \cdot \tau} \cdot \left(\frac{m_B \cdot V_m}{M_B \cdot S} \right)^2 \cdot \left(\frac{\Delta E_S}{\Delta E_\tau} \right)^2 \quad \left(\tau \ll L^2/D \right) \tag{4}$$

Here, the labels D, m_B, V_M, M_B, and S correspond to the K^+-ions diffusion coefficient, the active mass of electrode, molar volume for carbon, molar mass for carbon, and active surface area, respectively [42]. Additionally, the detailed parameters in the GITT test, on which the diffusion coefficient calculations are based, are illustrated in Figure S6, where τ represents the duration of the pulse current and ΔE_τ and ΔE_s represent the voltage variations during the pulse current and adjacent rest steps. Figure 3e displays the diffusion coefficients of N-SHC and the SHC electrode during the insertion process, where diffusion coefficients first decline until ≈0.2 V, and then begin to recover. The corresponding diffusion coefficients during de-potassiation are represented in Figure 3f. It is interesting to note that the diffusion coefficients first decrease, then recover until they attain a voltage value of 0.55 V, and then decline again. Additionally, an increase in the K^+-ion diffusion coefficient of N-SHC can be observed during both the charge and discharge processes. We attribute this faster K^+-ion migration in N-SHC to the defect introduced by N-doping, which reduces the diffusion barriers of K^+ ions.

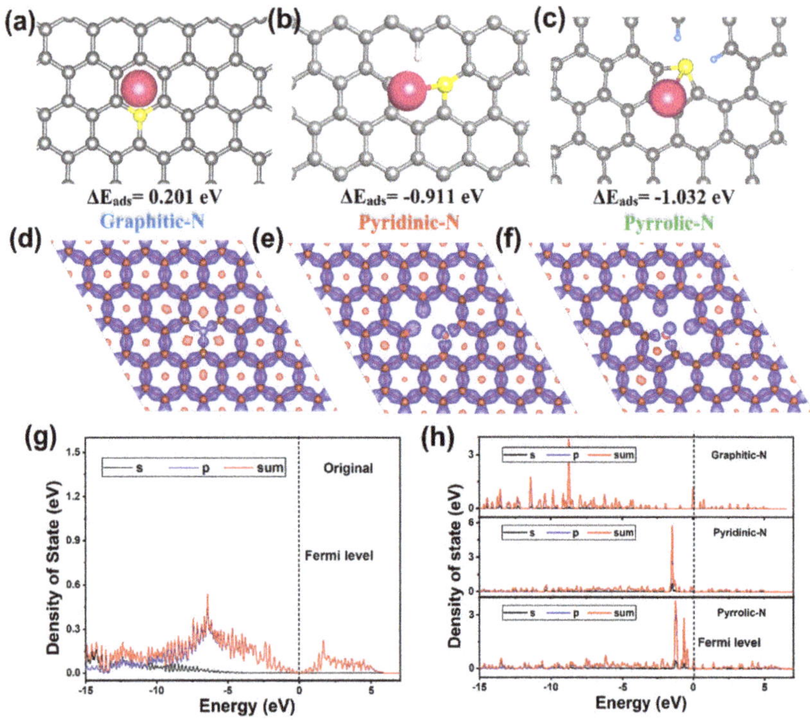

Figure 4. The DFT illustrations of adsorption energy of K ions (set adsorption energy at original hard carbon as 0 eV) and differential charge densities in (**a**,**d**) graphitic-N, (**b**,**e**) pyrindic-N, and (**c**,**f**) pyrrolic-N defects. The density states of (**g**) original hard carbon and (**h**) various defects of N−doping hard−carbon materials.

To interpret the reaction mechanism of K^+ ions' storage behavior, the differential charge density and adsorption abilities of different N-doping defects were investigated. Figure 4a–c simulate the K^+ ions' adsorption energy for three kinds of N-doping-surface defects on hard carbon (where the adsorption energy on original hard carbon was set as 0 eV). In contrast, the K^+ ions' adsorption energies on various introduced N-doping-surface defect sites of graphitic-N, pyrindic-N, and pyrrolic-N were calculated to be 0.201 eV (Figure 4a), −0.911 eV (Figure 4b), and −1.032 eV (Figure 4c), respectively. The DFT calculation exhibited the relatively low adsorption energy of graphitic-N, which indicated a less attractive tendency to K^+ ions in this kind of N-doping defect. Higher K^+-ion adsorption energies can be observed at the pyrindic-N and pyrrolic-N active sites, which reveals a stronger driving force and tendency for K^+ ions to be caught by those defects. This stronger adsorption of surface sites induced by N-doping is beneficial to enhancing the capacitive behavior of hard carbon, especially showing a positive effect on improving the reversible capacity of the N-SHC sample. The distinguishable adsorption energies between the graphitic-N site and pyridinic-N and pyrrolic-N sites could be verified by the differential charge density around N-doping defects. Figure 4d–e show the differential charge densities around graphitic-N, pyridinic-N, and pyrrolic-N sites, respectively. Heterogeneous electron distribution and electron vacancy appear around the pyridinic-N and pyrrolic-N defects, which cause the electron cloud of K^+ ions to be inclined to be combined with pyridinic-N and pyrrolic-N defects. The electron-rich structure that shows a negative effect on K^+-ion adsorption at the graphitic-N site should be responsible for the relatively low adsorption energy of graphitic-N [43]. Additionally, the density of states (DOSs) of original hard carbon and different N-doping sites were investigated (Figure 4g,h). It can be observed that, compared to the original one (Figure 4g), the DOSs around the Fermi level increase as we introduce all three kinds of N-doping defects into the original hard carbon. This increases in DOS around the Fermi level suggest the promotion of electronic conductivity and a higher affinity with K^+ [44].

To further solidify the improvement of diffusion kinetics in N-doping hard-carbon materials, we assessed the K^+ ions' migration barriers in the situation of K^+ ions across N-doping defects (Figure 5). Original hard-carbon material displays the largest K^+-ion migration barrier of 104 meV and lowest diffusion thermodynamics of −3.6 meV, which should be responsible for the sluggish rate capacity of the SHC electrode. In comparison, when the K^+ ions passed through the hard-carbon structures with graphitic-N, pyridinic-N, and pyrrolic-N defects, the migration barrier of K^+ ions was simulated to be 43.8, 7.3, and 6.8 meV, respectively. Obvious decreases in the K^+-ion migration barriers induced by N-doping appeared, particularly in the pyridinic-N and pyrrolic-N defects. This promoted the diffusion of K^+ ions between adjacent C layers more easily. Additionally, the diffusion thermodynamics across graphitic-N, pyridinic-N, and pyrrolic-N defects were calculated as −51.1, −35.4, and −98.4 meV, which revealed a stronger driving force for K^+-ion migration. Those simulation results could explain the better rate capacity delivered by N-doping the N-SHC electrode.

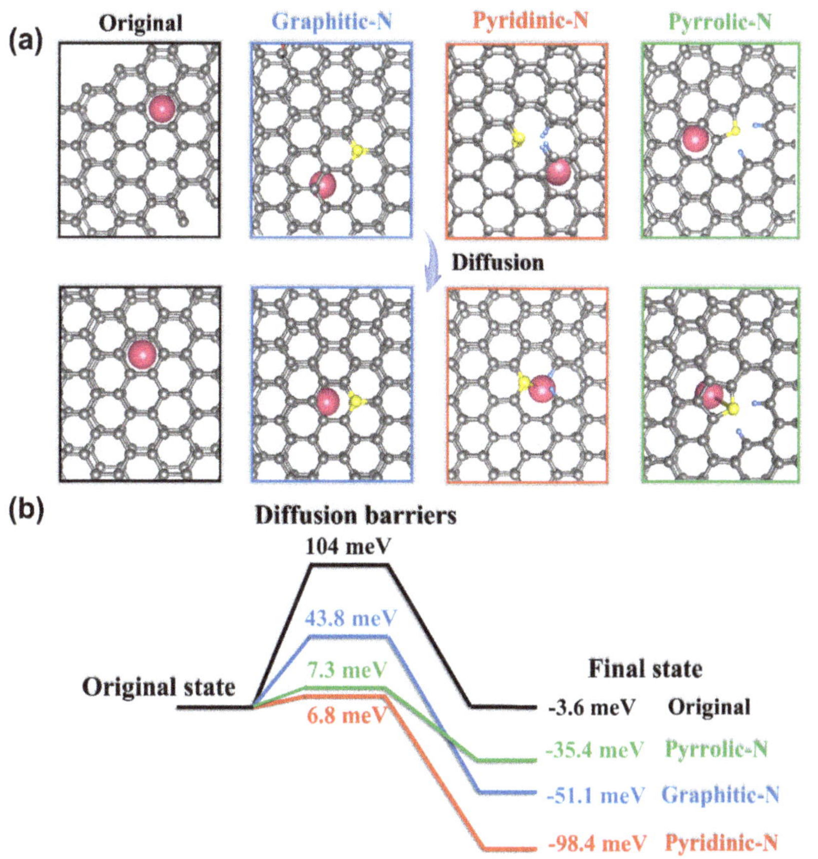

Figure 5. (**a**) DFT illustrations of K-ion diffusion, and (**b**) simulated diffusion barriers and thermodynamics across original hard carbon, graphitic-N, pyrindic-N, and pyrrolic-N defects.

4. Conclusions

In summary, we synthesized nitrogen-doping hard-carbon microspheres of N-SHC material as an anode for PIBs. In PIBs, N-SHC delivers a high reversible capacity of 248 mAh g^{-1} and a promoted rate performance (251, 206, 152, and 93 mAh g^{-1} at 200, 500, 1000, and 2000 mA g^{-1}, respectively). Moreover, nitrogen-doping N-SHC material also exhibits superior cycling long-term stability, where the N-SHC electrode maintains a high reversible capacity at 200 mAh g^{-1} with a capacity retention of 81% after 600 cycles. DFT calculations assessed the change of K ions' absorption energy and K ions' diffusion barriers at different N-doping effects. Compared with original hard-carbon material, pyridinic-N and pyrrolic-N defects introduced by N-doping displayed the positive effects of both K-ion absorption and diffusion. Based on the experimental observations and theoretical calculations, the superior electrochemical properties of N-SHC are ascribed to the enhancement of K-ion absorption and migration induced by N-doping.

Supplementary Materials: The following supporting information can be downloaded at: https://www.mdpi.com/article/10.3390/ma15124249/s1. Figure S1: (a) SEM and (b) TEM images of SHC; Figure S2: XRD pattern of SHC; Figure S3: (a) The initial charge-discharge profiles of SHC. And (b) Charge-discharge profiles of SHC at various cycles; Figure S4: EIS Nyquist plots of (a) SHC and (b) N-SHC after various cycles; Figure S5: (a) The initial CV curves of SHC at the scan rate of 0.1 mV s^{-1}. (b) CV curves of SHC at scan rates from 0.2 to 1 mV s^{-1}, corresponding (c) Relationship between log (peak current) versus log (scan rate) and (d) calculated normalized percentages of capacitive capacities contribution at various scan rates; Figure S6: The detailed GITT profiles of N-SHC; Figure S7: The GITT profiles of SHC; Table S1 Comparison of the electrochemical performance of N-doping hard carbons. References [45–48] are cited in the supplementary materials.

Author Contributions: Conceptualization, M.L. and C.Y.; methodology, K.Z. and C.C.; software, K.Z.; formal analysis, K.Z. and C.C.; data curation, K.Z.; writing—original draft preparation, K.Z. and C.C.; writing—review and editing, M.L. and C.Y.; supervision, M.L. All authors have read and agreed to the published version of the manuscript.

Funding: This research was funded by the National Natural Science Foundation of China (Nos. 51922042 and 51872098), the Young Key Teacher Foundation of Henan Province (2020GGJS227), and the Science and Technology Development Program of Henan province (No. 222102320057).

Institutional Review Board Statement: Not applicable.

Informed Consent Statement: Not applicable.

Data Availability Statement: The data presented in this study are available upon request from the corresponding author.

Conflicts of Interest: The authors declare no conflict of interest.

References

1. Xie, J.; Lu, Y.-C. A retrospective on lithium-ion batteries. *Nat. Commun.* **2020**, *11*, 2499. [CrossRef] [PubMed]
2. Yang, Y.; Okonkwo, E.G.; Huang, G.; Xu, S.; Sun, W.; He, Y. On the sustainability of lithium ion battery industry-A review and perspective. *Energy Storage Mater.* **2021**, *36*, 186–212. [CrossRef]
3. Peçenek, H.; Dokan, F.K.; Onses, M.S.; Yılmaz, E.; Sahmetlioglu, E. Outstanding supercapacitor performance with intertwined flower-like NiO/MnO_2/CNT electrodes. *Mater. Res. Bull.* **2022**, *149*, 111745. [CrossRef]
4. Opra, D.P.; Neumoin, A.I.; Sinebryukhov, S.L.; Podgorbunsky, A.B.; Kuryavyi, V.G.; Mayorov, V.Y.; Ustinov, A.Y.; Gnedenkov, S.V. Moss-like Hierarchical Architecture Self-Assembled by Ultrathin $Na_2Ti_3O_7$ Nanotubes: Synthesis, Electrical Conductivity, and Electrochemical Performance in Sodium-Ion Batteries. *Nanomaterials* **2022**, *12*, 1905. [CrossRef] [PubMed]
5. Sun, J.; Lee, H.-W.; Pasta, M.; Yuan, H.; Zheng, G.; Sun, Y.; Li, Y.; Cui, Y. A phosphorene-graphene hybrid material as a high-capacity anode for sodium-ion batteries. *Nat. Nanotechnol.* **2015**, *10*, 980–985. [CrossRef] [PubMed]
6. Xiao, N.; Rooney, R.T.; Gewirth, A.A.; Wu, Y. The Long-Term Stability of KO_2 in K-O_2 Batteries. *Angew. Chem. Int. Ed.* **2018**, *57*, 1227. [CrossRef] [PubMed]
7. Rajagopalan, R.; Tang, Y.; Ji, X.; Jia, C.; Wang, H. Advancements and Challenges in Potassium Ion Batteries: A Comprehensive Review. *Adv. Funct. Mater.* **2020**, *30*, 1909486. [CrossRef]
8. Komaba, S.; Hasegawa, T.; Dahbi, M.; Kubota, K. Potassium intercalation into graphite to realize high-voltage/high-power potassium-ion batteries and potassium-ion capacitor. *Electrochem. Commun.* **2015**, *60*, 172. [CrossRef]

9. Okoshi, M.; Yamada, Y.; Komaba, S.; Yamada, A.; Nakai, H. Theoretical analysis of interactions between potassium ions and organic electrolyte solvents: A comparison with lithium, sodium, and magnesium ions. *J. Electrochem. Soc.* **2017**, *164*, A54. [CrossRef]
10. Peçenek, H.; Yetiman, S.; Dokan, F.K.; Onses, M.S.; Yılmaz, E.; Sahmetlioglu, E. Effects of carbon nanomaterials and MXene addition on the performance of nitrogen doped MnO_2 based supercapacitors. *Ceram. Int.* **2022**, *48*, 7253. [CrossRef]
11. Sultana, I.; Rahman, M.; Mateti, S.; Ahmadabadi, V.; Glushenkov, A.M.; Chen, Y. K-ion and Na-ion storage performances of Co_3O_4–Fe_2O_3 nanoparticle-decorated super P carbon black prepared by a ball milling process. *Nanoscale* **2017**, *9*, 3646. [CrossRef]
12. Yetiman, S.; Peçenek, H.; Dokan, F.K.; Onses, M.S.; Yılmaz, E.; Sahmetlioglu, E. Microwave-assisted fabrication of high-performance supercapacitors based on electrodes composed of cobalt oxide decorated with reduced graphene oxide and carbon dots. *J. Energy Storage* **2022**, *49*, 104103. [CrossRef]
13. McCulloch, W.D.; Ren, X.; Yu, M.; Huang, Z.; Wu, Y. Potassium-ion oxygen battery based on a high capacity antimony anode. *ACS Appl. Mater. Interfaces* **2015**, *7*, 26158. [CrossRef]
14. Qian, J.; Wu, X.; Cao, Y.; Ai, X.; Yang, H. High capacity and rate capability of amorphous phosphorus for sodium ion batteries. *Angew. Chem. Int. Ed.* **2013**, *52*, 4633. [CrossRef]
15. Min, X.; Xiao, J.; Fang, M.; Wang, W.; Zhao, Y.; Liu, Y.; Abdelkader, A.M.; Xi, K.; Kumarb, R.V.; Huang, Z. Potassium-ion batteries: Outlook on present and future technologies. *Energy Environ. Sci.* **2021**, *14*, 2186–2243. [CrossRef]
16. Zhu, Z.; Zhong, W.; Zhang, Y.; Dong, P.; Sun, S.; Zhang, Y.; Li, X. Elucidating electrochemical intercalation mechanisms of biomass-derived hard carbon in sodium-/potassium-ion batteries. *Carbon Energy* **2021**, *3*, 541–553. [CrossRef]
17. Ju, Z.C.; Zhang, S.; Xing, Z.; Zhuang, Q.C.; Qiang, Y.H.; Qian, Y.T. Direct synthesis of few-layer F-doped graphene foam and its lithium/potassium storage properties. *ACS Appl. Mater. Interfaces* **2016**, *8*, 20682. [CrossRef]
18. Cho, S.Y.; Yun, Y.S.; Jang, D.; Jeon, J.W.; Kim, B.H.; Lee, S.; Jin, H.-J. Ultra strong pyroprotein fibres with long-range ordering. *Nat. Commun.* **2017**, *8*, 74. [CrossRef]
19. Oberlin, A. Carbonization and graphitization. *Carbon* **1984**, *22*, 521. [CrossRef]
20. Chen, M.; Wang, W.; Liang, X.; Gong, S.; Liu, J.; Wang, Q.; Guo, S.; Yang, H. Sulfur/Oxygen Codoped Porous Hard Carbon Microspheres for High-Performance Potassium-Ion Batteries. *Adv. Energy Mater.* **2018**, *8*, 1800171. [CrossRef]
21. Zhang, K.; He, Q.; Xiong, F.; Zhou, J.; Zhao, Y.; Mai, L.; Zhang, L. Active sites enriched hard carbon porous nanobelts for stable and high-capacity potassium-ion storage. *Nano Energy* **2020**, *77*, 105018. [CrossRef]
22. Cui, R.; Xu, B.; Dong, H.; Yang, C.; Jiang, Q. N/O Dual-Doped Environment-Friendly Hard Carbon as Advanced Anode for Potassium-Ion Batteries. *Adv. Sci.* **2020**, *7*, 1902547. [CrossRef] [PubMed]
23. Liu, Y.; Dai, H.; Wu, L.; Zhou, W.; He, L.; Wang, W.; Yan, W.; Huang, Q.; Fu, L.; Wu, Y. A Large Scalable and Low-Cost Sulfur/Nitrogen Dual-Doped Hard Carbon as the Negative Electrode Material for High-Performance Potassium-Ion Batteries. *Adv. Energy Mater.* **2019**, *9*, 1901379. [CrossRef]
24. Li, Y.; Chen, M.; Liu, B.; Zhang, Y.; Liang, X.; Xia, X. Heteroatom Doping: An Effective Way to Boost Sodium Ion Storage. *Adv. Energy Mater.* **2020**, *10*, 2000927. [CrossRef]
25. Deng, Q.; Liu, H.; Zhou, Y.; Luo, Z.; Wang, Y.; Zhao, Z.; Yang, R. N-doped three-dimensional porous carbon materials derived from bagasse biomass as an anode material for K-ion batteries. *J. Electroanal. Chem.* **2021**, *899*, 115668. [CrossRef]
26. Zhou, Y.; Zhu, Y.; Xu, B.; Zhang, X. Nitrogen-doped porous carbon with complicated architecture and superior K+ storage performance. *Sustain. Energy Fuels* **2021**, *5*, 396–400. [CrossRef]
27. Wang, B.; Gu, L.; Yuan, F.; Zhang, D.; Sun, H.; Wang, J.; Wang, Q.; Wang, H.; Li, Z. Edge-enrich N-doped graphitic carbon: Boosting rate capability and cyclability for potassium ion battery. *Chem. Eng. J.* **2022**, *432*, 134321. [CrossRef]
28. Schutjajew, K.; Pampel, J.; Zhang, W.; Antonietti, M.; Oschatz, M. Influence of Pore Architecture and Chemical Structure on the Sodium Storage in Nitrogen-Doped Hard Carbons. *Small* **2021**, *17*, 2006767. [CrossRef]
29. Agrawal, A.; Janakiraman, S.; Biswas, K.; Venimadhav, A.; Srivastava, S.K.; Ghosh, S. Understanding the improved electrochemical performance of nitrogen-doped hard carbons as an anode for sodium ion battery. *Electrochim. Acta* **2019**, *317*, 164–172. [CrossRef]
30. Li, Y.; Yang, C.; Zheng, F.; Ou, X.; Pan, Q.; Liu, Y.; Wang, G. High pyridine N-doped porous carbon derived from metal-organic frameworks for boosting potassium-ion storage. *J. Mater. Chem. A* **2018**, *6*, 17959. [CrossRef]
31. Henkelman, G.; Jónsson, H. A climbing image nudged elastic band method for finding saddle points and minimum energy paths. *J. Chem. Phys.* **2000**, *113*, 9901–9904. [CrossRef]
32. Sun, J.; Sun, Y.; Oh, J.; Gu, Q.; Zheng, W.; Goh, M.; Zeng, K.; Cheng, Y.; Lu, L. Insight into the structure-capacity relationship in biomass derived carbon for high-performance sodium-ion batteries. *J. Energy Chem.* **2021**, *62*, 497–504. [CrossRef]
33. Xu, Y.; Zhang, C.; Zhou, M.; Fu, Q.; Zhao, C.; Wu, M.; Lei, Y. Highly nitrogen doped carbon nanofibers with superior rate capability and cyclability for potassium ion batteries. *Nat. Commun.* **2018**, *9*, 1720. [CrossRef]
34. Xu, J.; Wang, M.; Wickramaratne, N.P.; Jaroniec, M.; Dou, S.; Dai, L. High-Performance Sodium Ion Batteries Based on a 3D Anode from Nitrogen-Doped Graphene Foams. *Adv. Mater.* **2015**, *27*, 2042–2048. [CrossRef]
35. Share, K.; Cohn, A.P.; Carter, R.; Rogers, B.; Pint, C.L. Role of Nitrogen-Doped Graphene for Improved High-Capacity Potassium Ion Battery Anodes. *ACS Nano* **2016**, *10*, 9738–9744. [CrossRef]
36. Wang, X.; Weng, Q.; Liu, X.; Wang, X.; Tang, D.M.; Tian, W.; Zhang, C.; Yi, W.; Liu, D.; Bando, Y. Atomistic Origins of High Rate Capability and Capacity of N-Doped Graphene for Lithium Storage. *Nano Lett.* **2014**, *14*, 1164. [CrossRef]

37. Qian, Y.; Li, Y.; Yi, Z.; Zhou, J.; Pan, Z.; Tian, J.; Wang, Y.; Sun, S.; Lin, N.; Qian, Y. Revealing the Double-Edged Behaviors of Heteroatom Sulfur in Carbonaceous Materials for Balancing K-Storage Capacity and Stability. *Adv. Funct. Mater.* **2021**, *31*, 2006875. [CrossRef]
38. Yan, H.; Huang, X.; Hong, L.; Chen, L. Electrochemical study on $LiCoO_2$ synthesized by microwave energy. *Solid State Ion.* **1998**, *11*, 113–115. [CrossRef]
39. Qiu, Z.; Zhao, K.; Liu, J.; Xia, S. Nitrogen-doped mesoporous carbon as an anode material for high performance potassium-ion batteries. *Electrochim. Acta* **2020**, *340*, 135947. [CrossRef]
40. Li, P.; Hwang, J.; Sun, Y. Highly wrinkled carbon tubes as an advanced anode for K-ion full batteries. *J. Mater. Chem. A* **2019**, *7*, 20675–20682. [CrossRef]
41. Alvin, S.; Cahyadi, H.S.; Hwang, J.; Chang, W.; Kwak, S.K.; Kim, J. Revealing the Intercalation Mechanisms of Lithium, Sodium, and Potassium in Hard Carbon. *Adv. Energy Mater.* **2020**, *10*, 2000283. [CrossRef]
42. Zhong, W.; Liu, X.; Cheng, Q.; Tan, T.; Huang, Q.; Deng, Q.; Hu, J.; Yang, C. Suppressing the interlayer-gliding of layered P3-type $K_{0.5}Mn_{0.7}Co_{0.2}Fe_{0.1}O_2$ cathode materials on electrochemical potassium-ion storage. *Appl. Phys. Rev.* **2021**, *8*, 031412. [CrossRef]
43. Ma, C.; Shao, X.; Cao, D. Nitrogen-doped graphenenanosheets as anode materials for lithium ion batteries: A first-principles study. *J. Mater. Chem.* **2012**, *22*, 8911. [CrossRef]
44. Li, D.; Ren, X.; Ai, Q.; Sun, Q.; Zhu, L.; Liu, Y.; Liang, Z.; Peng, R.; Si, P.; Lou, J.; et al. Facile Fabrication of Nitrogen-Doped Porous Carbon as Superior Anode Material for Potassium-Ion Batteries. *Adv. Energy Mater.* **2018**, *8*, 1802386. [CrossRef]
45. Zhu, L.F.; Zhang, Z.; Zhang, H.; Wang, Y.; Luo, J.; Yu, J.; Qu, Y.; Yang, Z. Tunable 2D tremella-derived carbon nanosheets with enhanced pseudocapacitance behavior for ultrafast potassium-ion storage. *Sci. China Technol. Sci.* **2021**, *64*, 2047–2056. [CrossRef]
46. Wanga, X.; Zhao, J.; Yao, D.; Xu, Y.; Xu, P.; Chen, Y.; Chen, Y.; Zhu, K.; Cheng, K.; Ye, K.; et al. Bio-derived hierarchically porous heteroatoms doped-carbon as anode for high performance potassium-ion batteries. *J. Electroanal. Chem.* **2020**, *871*, 114272. [CrossRef]
47. Ma, L.; Li, J.; Li, Z.; Ji, Y.; Mai, W.; Wang, H. Ultra-stable potassium ion storage of nitrogen-doped carbon nanofiber derived from bacterial cellulose. *Nanomaterials* **2021**, *11*, 1130. [CrossRef]
48. Luo, H.; Chen, M.; Cao, J.; Zhang, M.; Tan, S.; Wang, L.; Zhong, J.; Deng, H.; Zhu, J.; Lu, B. Cocoon silk-derived, hierarchically porous carbon as anode for highly robust potassium-ion hybrid capacitors. *Nano-Micro Lett.* **2020**, *12*, 113. [CrossRef]

materials

Review

Review of the Electrospinning Process and the Electro-Conversion of 5-Hydroxymethylfurfural (HMF) into Added-Value Chemicals

Maximilien Coronas, Yaovi Holade * and David Cornu *

Institut Européen des Membranes, IEM UMR 5635, Univ Montpellier, ENSCM, CNRS, 34090 Montpellier, France; maximilien.coronas@enscm.fr
* Correspondence: yaovi.holade@enscm.fr (Y.H.); david.cornu@enscm.fr (D.C.)

Abstract: Electrochemical converters (electrolyzers, fuel cells, and batteries) have gained prominence during the last decade for the unavoidable energy transition and the sustainable synthesis of platform chemicals. One of the key elements of these systems is the electrode material on which the electrochemical reactions occur, and therefore its design will impact their performance. This review focuses on the electrospinning method by examining a number of features of experimental conditions. Electrospinning is a fiber-spinning technology used to produce three-dimensional and ultrafine fibers with tunable diameters and lengths. The thermal treatment and the different analyses are discussed to understand the changes in the polymer to create usable electrode materials. Electrospun fibers have unique properties such as high surface area, high porosity, tunable surface properties, and low cost, among others. Furthermore, a little introduction to the 5-hydroxymethylfurfural (HMF) electrooxidation coupled to H_2 production was included to show the benefit of upgrading biomass derivates in electrolyzers. Indeed, environmental and geopolitical constraints lead to shifts towards organic/inorganic electrosynthesis, which allows for one to dispense with polluting, toxic and expensive reagents. The electrooxidation of HMF instead of water (OER, oxygen evolution reaction) in an electrolyzer can be elegantly controlled to electro-synthesize added-value organic chemicals while lowering the required energy for CO_2-free H_2 production.

Keywords: electrospinning; gold; polyacrylonitrile; electrosynthesis; thermal treatment; hydrogen; 5-hydroxymethylfurfural

Citation: Coronas, M.; Holade, Y.; Cornu, D. Review of the Electrospinning Process and the Electro-Conversion of 5-Hydroxymethylfurfural (HMF) into Added-Value Chemicals. *Materials* **2022**, *15*, 4336. https://doi.org/10.3390/ma15124336

Academic Editors: Zhenghua Tang, Marc Cretin and Sophie Tingry

Received: 29 April 2022
Accepted: 17 June 2022
Published: 19 June 2022

1. Introduction

Twenty years ago, Professor Peter William Atkins, one of the best chemistry educators of our era, considered that we need to broaden our view of what electrochemistry is; it is not just about electrode potentials and electrolysis, this is half of modern chemistry; electrochemistry—in the broadest sense—will be one of the great achievements in chemistry in the next millennium, and we need prepare people for it [1]. Indeed, for centuries, fossil fuel resources have played an essential role in the industry, including chemical and petrochemical manufacturing, making them central to us nowadays. Due to the current environmental situation, where CO_2 levels have risen rapidly, there is an increasing demand for energy, and there is a decrease in fossil fuel feedstock, a lot of research has been dedicated to the development of alternative processes [2–4].

Current developments are oriented towards producing less waste by making organic chemicals in a "greener" fashion from biomass derivates (for example, avoiding the use of organic solvents, working under room temperature, having better control of coproducts, etc.) [2–6]. For instance, several candidates such as 2,5-diformylfuran (DFF), 5-hydroxymethyl-2-furancarboxylic acid (HFCA), 5-formyl-2-furancarboxylic acid (FFCA), and 2,5-furandicarboxylic acid (FDCA) have been generated by the oxidation of 5-hydroxymethylfurfural (HMF) under different conditions [2,7–9]. HMF derives from cellulose and hemicellulose, the two primary constituents of lignocellulosic biomass (60–75%), and has the potential to be converted into platform biofuels [10,11]. Specifically, the biosourced compound HMF is an intermediate for the synthesis of bio-renewable FDCA, which is the monomer for the polyethylene furanoate (PEF) biopolymer materials of industrial relevance as it is a green alternative to polyethylene terephthalate (PET) [10,12,13]. On the other hand, by producing these value-added chemicals from an electro-oxidation process in an electrolyzer, the electro-reduction step could be used to create another value-added chemical. For example, the electrocatalytic reduction of water (in alkaline media) or protons (in acidic media) is considered to be an ideal process for generating H_2 without by-products and does not lead to the emission of polluting species, unlike the current methods, which produce a large amount of CO_2 [3,14,15]. By using this coproduction method, the use of fossil fuels for energy production can be gradually replaced by generating H_2 from renewable resources (H_2O) and for the chemical industry by using the abundant biomass. However, for this electrochemical biomass-to-hydrogen technology, further efforts are needed to overcome the kinetics and/or the selectivity of both anode and cathode half-cell reactions. This will reduce the price of H_2 production and make the technology competitive with current methods based on the thermal decomposition of fossil fuels [16–18].

To be used in electrochemical reactors, the catalytic materials (ideally in the form of nanoparticles in order to reduce the total amount and to regulate the kinetics) must be immobilized on an electrically conducting support. Therefore, the need for an increased reaction yield and selectivity by minimizing the waste production resulted in the development of novel catalytic supports. Atomic layer deposition (ALD) has presented the possibility to design supports for catalytic reactions; however, this method can be expensive and time consuming, and it is quite impossible to dope the interior of the fibers by metallic species [19–23]. Moreover, electrospinning is a versatile technique for generating ultrathin fibers that can lead to catalytic supports. Metal particles can be added on the surface and inside the fibers. Different types of spinning methods have been explored in the past decades, for example, wet spinning, dry spinning, melt spinning, and gel spinning. However, the fibers could not reach the sub-micrometer scale. Then, electrospray and electrospinning were discovered by adding voltage in order to create ultra-thin continuous fibers, leading to the possibility of reaching the sub-micrometer scale [5,24–29].

Different polymers have been tested in electrospinning, such as polyvinylpyrrolidone (PVP) [30–32], polylactic acid (PLA) [33], polyacrylonitrile (PAN) [34,35], poly(ethylene oxide) (PEO) [31], etc. [5,24–26,36], and the results are promising. However, further investigations are needed now to fabricate advanced conducting materials that could lead to electrode materials for the electrosynthesis of value-added chemicals. A recent advancement in the method has shown that it is possible to add metallic species into the polymeric fiber matrix and then thermally convert the electrospun mat into nanometer sized catalysts with metallic species inside and outside each microfiber. To perform such a synthesis for an electrically conducting material (to act as a support in electrochemical applications) and an electrocatalytically active material (to act as active site in heterogeneous electrocatalysis), a suitable procedure of thermal treatment is needed. To achieve that from a raw PAN-based polymer modified by metallic cations, two steps need to be followed: stabilization and the carbonization [37]. Their tight control enables one to fabricate a free-standing catalytic support for the implementation in electrochemical fields. The details are explained later in this review. Currently, the main precursor of the electrospun carbon fiber (CF, also referred to as carbon nanofiber (CNF) even if the diameter of the fiber is mostly higher than 100 nm) is the PAN; 90% of carbon fibers is made from PAN [5,35,38].

The aim of this review is to summarize the latest developments regarding the use of electrospinning to design advanced electrocatalytic materials. We will present the different steps starting from PAN solutions, the parameters of the electrospinning process, the impact of metallic species, and the thermal treatment. Then, the electrocatalysis of the HMF will be discussed along with the electrocatalytic characterizations.

2. Synthesis of PAN Fibers and Gold Nanoparticles

2.1. Formulation of a Suitable Electrospinning Solution

PAN is the most used polymer for the preparation of CNFs [5,35,38]. It offers the possibility to add different metals [32,39,40] or molecules [27,41–43] to the fibers during the synthesis process. Like the other polymers (PVP, PLA, PEO, etc.), PAN is commercially available. The commonly used solvents to dissolve PAN are dimethyl sulfoxide (DMSO) and dimethylformamide (DMF). Although DMF has side effects (it is carcinogenic, flammable, and harmful), it is widely used to solubilize the polymer. DMSO and DMF could both be used for the electrospinning process because they provide the two desired criteria: the ability to solubilize the polymer and evaporate during the electrospinning process [34]. As summarized by Pastoriza-Santos and Liz-Marzan, under appropriate conditions, DMF can also act as a reducing agent and lead to metal nanoparticles, mainly Au and Ag [30]. Different studies have reported the preparation of PAN solutions in DMF with or without adding metals for the electrospinning such as Zhang et al. [35], Holade et al. [39], and Both Engel et al. [34]. The formulation of a suitable electrospinning solution is very critical for the electrospinning itself. It is important to ensure that the solution is as homogeneous as possible and that it does not change significantly during the electrospinning process, which can take several hours. The majority of the studies used PAN with a molecular weight of 150,000 g/mol [44]. This makes it easier to reproduce the experiments and to have a similar background of information, which is not the case for the other parameters that will be reviewed below. The key parameter is the percentage of PAN in the electrospinning solution to avoid having a mixture that is neither too liquid nor too viscous and therefore difficult for electrospinning. Moreover, the addition of an inorganic or organic compound can impact the stability of the mixture, as well as the electrospinning conditions. The reported amount ranges from 3 wt.% [45] to 14 wt.% [46], while the majority of the studies are done for 10–11 wt.% of PAN [34,39,40,47–52]. The underlying parameter is the true volume of PAN solution loaded in the syringe (see Figure 1a). The reported values are 7.2 mL [39], 4 mL [34], and 2 mL [53]. We note that not all of the prepared volume of the solution is always used; it depends on the type of desired material. This key parameter is not specified in the majority of studies, which compromises the repeatability and replicability of the results because the total amount of electrospun mat logically depends on the used volume for a fixed concentration.

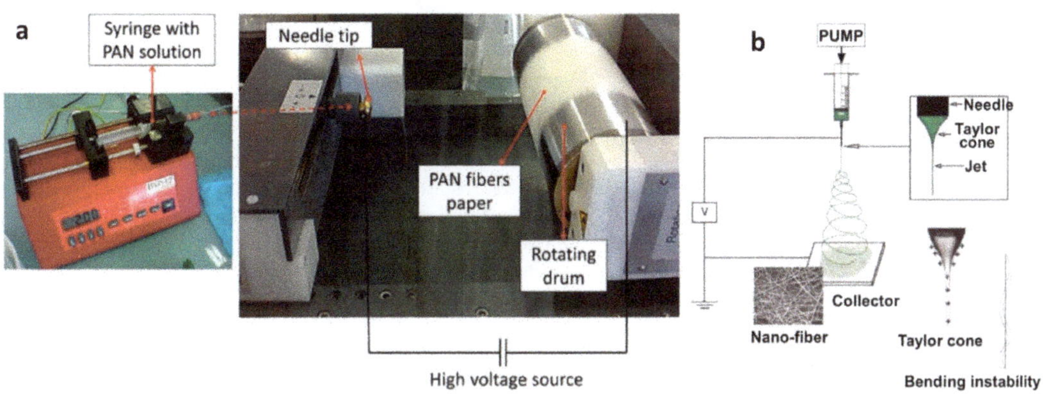

Figure 1. (**a**) Pictures of an electrospinning setup: reprinted and adapted from ref. [54]; Copyright 2015, American Chemical Society, and courtesy provided by Dr. Adriana Both Engel. (**b**) Illustration of the working principle of electrospinning: reprinted and adapted with permission from ref. [55]; Copyright 2015, Elsevier Inc.

Even if the formulation duration, or the temperature, are not always fully disclosed in all studies, the duration is 12 h at 60 °C [56], 3–6 h at 70 °C [34,39], and 2 h at 80 °C [47]

(50 °C without an indication of the used duration [46]). These differences between the used conditions to prepare the electrospinning solution may come from the intended end-use, the addition of other chemical species (for example, metallic species), or the type of solvent for the dissolution of PAN. After ending the formulation and preparation of the solution, the mixture is loaded in a syringe (see Figure 1a) and the electrospinning process can properly start.

2.2. Electrospinning Process

Electrospinning is routinely conducted at room temperature [57] and is recognized as an efficient technique for the fabrication of 1D nanofibers [17,58]. The diameter of the constituting fibers ranges from micrometers (e.g., 10–100 μm) to sub-microns (and even nanometers) [32,58]. It is a cost-effective technology that generates non-woven fibers with a high ratio of surface-area-to-volume and tunable porosity. As shown in Figure 1b, the electrospinning involves an electro-hydrodynamic process, whereby a liquid droplet is electrified to generate a jet, followed by stretching and elongation to generate fibers [55]. An electrostatic force should be created in order to produce polymer filaments, as shown in the Figure 1 [58,59]. Furthermore, a DC voltage in the range of several tens of kVs is necessary to run the electrospinning. There are basically three components needed to fulfill the process: a high-voltage supplier, a capillary tube with a pipette or needle of small diameter, and a metal collecting screen, as shown in Figure 1 [5,57,58]. Over the years, more than 200 polymers have been electrospun for various applications, and more than 50 different polymers have been successfully electrospun into ultra-fine fibers, as represented in Table 1 [55,57,58].

The electrospinning process is relatively simple. Before reaching the electrically conducting collector, the jet solution evaporates or solidifies and is collected as an interconnected network of fibers. Mutual charge repulsion and the contraction of the surface charges of the counter electrode cause a force directly opposite to the surface tension. When the applied electrostatic forces overcome the fluid surface tension, the electrified fluid forms a jet out of the capillary tip towards a grounded collecting screen. At a high rotation speed up to thousands of rpm (revolutions per minute), electrospun nanofibers can be oriented circumferentially [5,32,58–60]. To summarize, the process consists of three stages: (i) jet initiation and the extension of the jet along a straight line; (ii) the growth of whipping instability and the further elongation of the jet, which may or may not be accompanied with the jet branching and/or splitting; and (iii) the solidification of the jet into fibers.

During an electrospinning process, there are few parameters that need to be controlled to ensure that the experiments meet the three criteria of being highly repeatable, reproducible, and replicable. Electrospun fibers and the control of their diameters are largely determined by the processing parameters, including [5,32,57–60]: (i) solution parameters such as viscosity, elasticity, conductivity, surface tension, and polymer-solvent affinity; (ii) process parameters such as hydrostatic pressure in the capillary tube, electric potential at the capillary tip, the gap (the distance between the tip and the collector), and the feed rate; and (iii) ambient parameters such as solution temperature, humidity, and air velocity in the electrospinning chamber.

Furthermore, a dopant such as metal or alloy could be added in the electrospun solution. If the dopant increases the conductivity of the polymer solution, it will make the nanofiber thinner [32]. Each of these parameters can affect the electrospinning process—by harming the fibers' morphology, for example. Hence, by a proper control of these parameters, one can fabricate electrospun fibers with desired morphologies and diameters. Further information about the impact of the different parameters can be found in ref. [60].

Table 1. Examples of electrospun (bio)polymer materials and the used solvents (in alphabetical order). Reprinted and adapted with permission from ref. [55]; Copyright 2015, Elsevier Inc.

Materials	Solvents
Acrylonitrile-butadiene-styrene (ABS)	N,N-Dimethyl formamide (DMF) or tetrahydrofuran (THF)
Cellulose	Ethylene diamine
Cellulose acetate	Dimethylacetamide (DMAc) and acetone or acetic acid
Ethyl-cyanoethyl cellulose [(E-CE)C]	THF
Chitosan and chitin	1,1,1,3,3,3-hexafluoro-2-propanol (HFIP)
Dextran	Water, dimethyl sulfoxide (DMSO)/water, DMSO/DMF
Gelatine	2,2,2-Trifluoroethanol
Nylon	Formic acid
Poly(2-acrylamido-2-methyl-1-propane sulfonic acid) (PMAPS)	Ethanol/water
Polyacrylonitrile (PAN)	DMF
Polyalkyl methacrylate (PMMA)	Toluene/DMF
Polycarbonate	THF/DMF
Poly(ethylene oxide) (PEO)	Water, ethanol, DMF
Polyethylene terephthalate (PET)	Trifluoroacetic acid (TFA) and dichloromethane (DCM)
Polylactic based polymers	Chloroform, HFIP, and DCM
Pol(ε-caprolacone) based polymers	Acetone, acetone/THF, chloroform/DMF, DCM/methanol, chloroform/methanol, and THF/acetone
Poly(3-hydroxybutyrate-co-3-hydroxyvalerate) (PHBV)	2,2,2-Trifluoroethanol
Polyphosphazenes	Chloroform
Polystyrene	1,2 Dichloroethane, DMF, ethylacetate, methylethylketone (MEK), and THF
Bisphenol-A Polysulfone	DMAc/acetone
Polyurethane (PU)	THF/DMF
Polyvinyl alcohol (PVA)	Water
Polyvinyl chloride (PVC)	DMF, DMF/THF
Poly(vinylidene fluoride) (PVDF)	DMF/THF
Poly(vinyl pyrrolidone)	Ethanol, DCM, and DMF
Silk	Hexafluoroacetone (HFA), HFIP, and formic acid

2.3. Formation and Synthesis of Nanoparticles

There are different ways to synthesize NPs, for example, the constant potential electrolysis [61], the annealing of a salt precursor coated on a surface [5], the use of a reducing agent [39,62], ultra-violet (UV) light [46,47], the phase transfer [16,63], the sol-gel method [40], or directly by electrospinning. Additionally, different methods can be combined as reported by Anka et al. [47] and Sawada et al. [46]. In their studies, they formed AuNPs by electrospinning under UV light, reactions of photo-polymerization, and photo-crosslinking. UV, microwave, thermal treatment requires extra energy to produce the AuNPs. A study made by Both Engel et al. [34] showed that it is possible to create metallic particles and specifically nanostructured gold particles on carbon fibers by the electrospinning process and prevent impurities that are usually present on the surface of AuNPs electrodes prepared by chemical vapor deposition (CVD). By the method of Both Engel et al. [34], small AuNPs are formed inside the fibers and play a role in the physical properties of the carbon electrodes, whereas the large ones are on the surfaces of the fibers to act as active sites during the electrocatalytic reactions.

The electrospinning process allows a one-pot synthesis to have the support and metallic particles [5,17,32–34,64,65] and does not impact the morphology and the size of particles created during the process [32]. Different metals were tested with PAN in DMF (see Table 2). Figure 2 shows the scanning electron microscopy (SEM) and transmission electron microscopy (TEM) images and different metals on the fibers. This method enabled one to engineer a three-dimensional support with different metals, leading to a high surface area support that can act as a scaffold for enzymes immobilization [41] or catalytic reactions [65]. For Figure 2, electrospun PAN nanofibers were prepared, and the nanoparticles were then loaded onto the carbonized electrospun mat to yield supported high-entropy-alloy nanoparticles (HEA-NPs) [65].

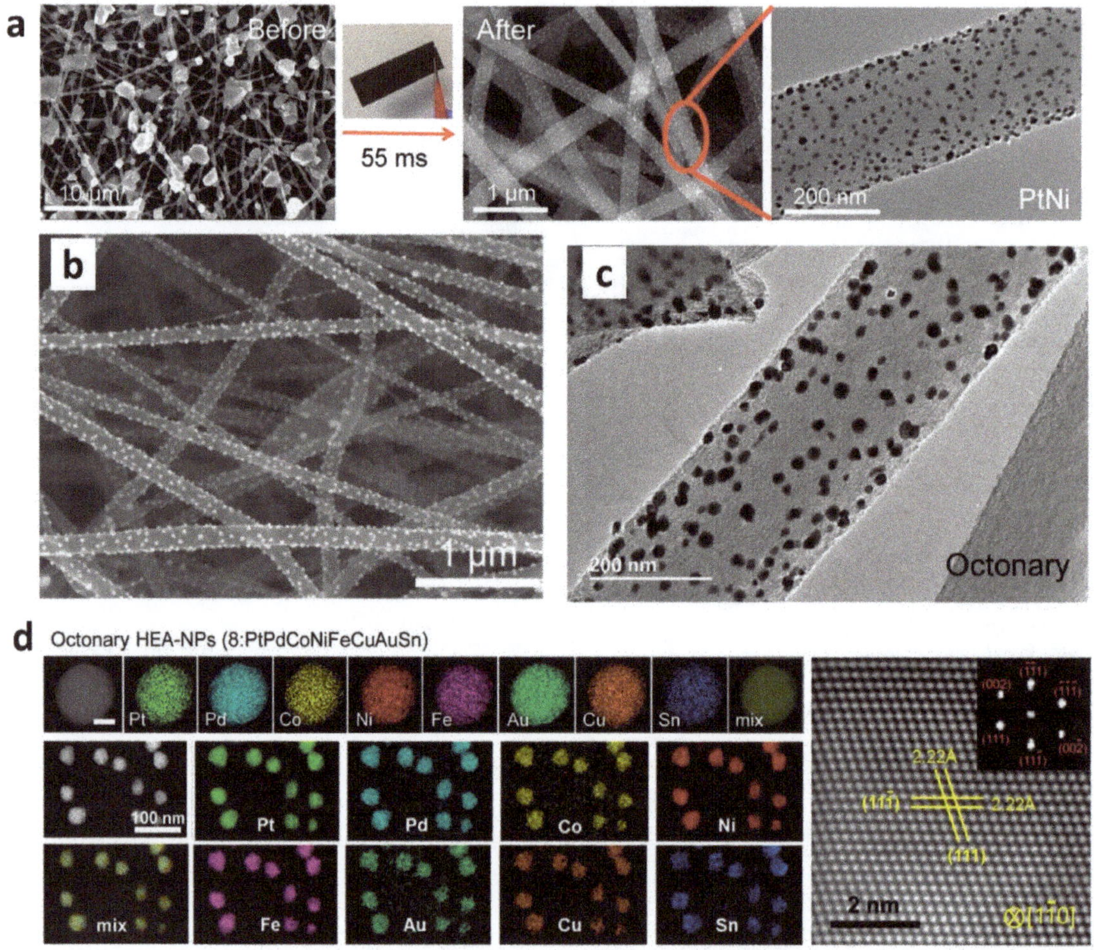

Figure 2. Nanoparticles loading onto carbon nanofibers derived from electrospun PAN nanofibers. (**a**) Illustration of carbothermal shock (CTS) method after electrospinning to synthesize high-entropy-alloy nanoparticles (HEA-NPs): SEM images of microsized precursor salt particles before and after CTS. (**b–d**) SEM, TEM, and HAADF-STEM mapping images of octonary alloy nanoparticles. Reprinted and adapted with permission from ref. [65]; Copyright 2018, AAAS.

Two analyses are commonly performed to get the information about the fiber quality. By SEM, it is possible to get the information on the size of the particles and the fibers. The coupling to energy-dispersive X-ray spectroscopy (EDX) enables one to get the qualitative data on the nature of chemical species. To get quantitative data (bulk analysis), complementary elementary analysis by inductively coupled plasma (ICP-OES or ICP-MS) for heavier elements and CHNS-O (the determination of C, H, N, S, and O contents) should be done. For thinner samples, transmission electron microscopy (TEM) can be used to reach a nanometer scale information and provide more detail about the sample. SEM is more used than TEM because the samples made by electrospinning could be too thick for TEM analysis, because SEM is more representative of the sample (not a nm information), and because it is less expensive and less time consuming. Some direct analysis by TEM can be carried out [31], but most of the time, the materials are ground into a powder similar to the preparation for the electrocatalytic tests [66–70].

Table 2. Different types of materials prepared as fibers with PAN via electrospinning.

Material	Solvent	Reference
Ni/N,S-doped carbon	DMF	[48]
CoSe/N-doped carbon	DMF	[71]
Fe_3O_4/N-doped carbon	DMF	[49]
WS_2/N-doped carbon	DMF	[50]
$MnCo_2O_4$@N-doped carbon	DMF	[51]
Co_3O_4-CNFs (CNF: carbon nanofibers)	DMF	[72]
Au@CFs (CFs: carbon fibers)	DMF	[34]
Casein/PAN	DMSO	[42]
Li/CNFs	DMF	[35]
NCNFs (nitrogen-doped electrospun carbon nanofibers)	DMF	[45]
PAN NFM (polyacrylonitrile electrospun nanofibrous membrane)	DMF	[41]

To explain the formation of AuNPs during the electrospinning process without extra energy, it was argued that DMF can act as a reducing agent under suitable conditions to convert Au(+III) into Au(0) [30,34]. However, the ability of DMF to reduce metallic cation into the zero oxidation state depends on the nature of the involved metal ($AuCl_4^-$ ions into Au^0 vs. Ag^+ into Ag) [30]. Given that DMSO medium is not known to act as a reducing agent and that the AuNPs are formed in the presence of DMF or DMSO, it is argued that PAN polymer can also act as a reducing agent to reduce the Au(+III) salt under mild conditions (70 °C) [34]. However, there is a limitation to this method. The quantity of metal salt added in solution should not exceed a certain percentage depending on the targeted metal because it becomes more difficult to create fiber during electrospinning (the formulated solution is not so homogeneous, see the previous section).

2.4. Process of Electrospinning for PAN

A thorough literature analysis shows a certain difference in the used experimental parameters to spin PAN fibers, but in the majority, they are in the same range. One common thing is that the grounded counter electrode is connected to an aluminum foil as a collector for the experiment. For the electrospinning process, the applied voltage ranges from 11 kV [47,52,64] to 80 kV [42]; it should be noted that this parameter depends on the needle-to-collector distance and the composition of the electrospinning solution (see Section 2.1). The most commonly used voltage is 20 kV [34,35,39–41,46,53,56] for an average needle-to-collector distance of 15 cm (that is, 100 kV/m as an estimated electric field), whereby the majority of the studies are done for 10–11 wt.% of PAN in DMF [34,39,40,47–52]. Salles et al. reported a voltage of 4.5 kV when the distance between the needle tip and the metallic target was 10 cm, that is, an estimated electric field of 45 kV/m [73]. By assuming that it is almost impossible to perform electrospinning below 10 kV for 10–15 cm as the needle-to-collector distance, it is therefore likely that the measurement of this voltage value was underestimated. Another parameter of key importance is the distance between the needle and the collector, which is kept constant through the experiment. The lowest values are 6–10 cm (voltage of 4.5 kV [73] or 12 kV [47]), the highest values are 20–25 cm (voltage of 80 kV [42] or 20–25 kV [56]), and the most used value is 15 cm [34,35,39–41,46,53] (voltage of 20 kV). The presence of additives can also lead to a change in the conditions. For example, according to Both Engel et al. [34], 10 mL DMF with 1 g PAN is performed under 20 kV for a tip-to-collector distance of 14 cm (flow rate of 2.4 mL/h at 2000 rpm) while in the presence of 0.5 g $HAuCl_4$, and the voltage is increased to 22 kV for a tip-to-collector distance of 8 cm (the drum speed reduced to 1000 rpm for a flow rate of 2 mL/h) because of the change in the solution properties. For the flow rate, the widely used value ranges from

0.075–0.1 mL/h [47,64,73] to 2.4 mL/h [34,39], with an average of 1.5 ± 0.5 mL/h [35,41,46,53,56]. For the needle, the commonly used diameter is 800 μm, and diameters of 1.2 mm [35] and 0.5 mm [64] were also reported. For the drum collector, the rotational speed is 1000 rpm [34] or 2000 rpm [34,39]; unfortunately, many works [41,53] do not specify the value, which compromises the reproducibility of the experiments.

The previous points highlight different experimental conditions for the synthesis of PAN-based electrospun mats. The difference might be directed by the intended application of the fabricated materials. The parameters for one system are not necessarily the same for others (see above). For a given synthesis, the weight percent of the PAN polymer and the addition of metallic salts or organic molecules could impact the optimized parameters shown before. To make sure that high-quality fibers will be collected, a further optimization might be necessary. For example, considering "PAN" and "PAN with gold", the parameters are different in order to create the mat of fibers [34]: the voltage, needle-to-collector distance, the flow rate, and the collector rotational speed are different. It should be kept in mind that not only operational parameters have an impact on the fiber synthesis but also environmental parameters such as temperature, relative humidity, etc., and, to our knowledge, only one study [42] among all those cited above has provided this key information. For example, electrospinning between summer and winter periods is subject to a temperature difference of up to 30 °C. As a result, the quality of the collected material would not be the same if the parameters are left unchanged, e.g., the humidity level will impact the viscosity. The issue of research reproducibility [74] is inherently linked to the non-disclosure of all experimental conditions, which is very problematic and may arise from the complexity of the studied materials or from a lack of rigor [75–78]. In an editorial entitled "The Experimental Section: The Key to Longevity of Your Research", Buriak and Korgel summarized that "one of the greatest compliments anyone can give your published work is to reproduce it and build upon it" [76]. The two identified reasons of irreproducibility in scientific reports are the sheer carelessness and the misguided attempt at obtaining or maintaining a competitive advantage by intentionally withholding critical experimental details; these lead to short-term gains but very profound long-term losses as future papers are viewed with a skeptical eye or simply ignored [76].

Once the electrospinning is properly designed and done, at the end of the process, a mat composed of micro/nanofibers is obtained, as shown in SEM images and the photos of Figure 3 in which the fibers' diameter ranges from 1 to 5 μm [34]. The yellowish color is obvious for the tissues collected after electrospinning of a solution containing $HAuCl_4$. This mat could then undergo thermal or chemical treatment depending on the intended use, which is the purpose of the next section.

Figure 3. SEM images of Au@PAN fibers (insets are the tissues collected after electrospinning, yellowish for the presence of gold (III) salt) obtained with: (**A**) regular secondary electrons detector and (**B**) back-scattered electrons detector [34]. Reprinted and adapted with permission from ref. [34]; Copyright 2016, Wiley-VCH Verlag GmbH and Co. KGaA, Weinheim.

2.5. Thermal Treatment

Thermal treatment is needed to create CNFs with enhanced electrical conductivity, which is important for electrochemical applications. The step is also needed to consolidate the electrospun mat. For that, two steps are needed. The first step is the stabilization process to make a reticulation of the polymer. After that, the second stage is the carbonization process to make the carbon fibers with a graphitic structure. For the PAN-based materials, extensive studies have been done and the parameters are the same as for the first step. This step is routinely done under air atmosphere at a temperature between 220 °C and 300 °C for a dwell time of 1–3 h [17,34,35,37,39,52,53] in order to trigger cyclization, the dehydrogenation and oxidation process [35,38,52,56,79], as shown in Figure 4, which is the evolution of the structure of the PAN molecule during the thermal treatment.

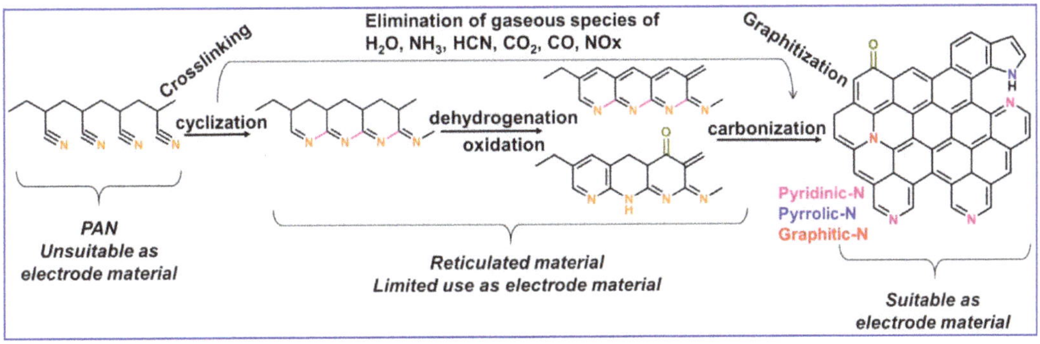

Figure 4. Sketch of the thermal treatment starting from polyacrylonitrile-based material to the fabrication of an electrode material.

By increasing the temperature under air atmosphere, the color of the PAN fibers goes from white to black, as shown in Figure 5, and is assigned to the formation of a ladder ring structure [35]. The weight loss also increases from 5 wt.% at 235 °C to 15 wt.% at 300 °C [17], as shown in the thermogravimetric analysis (TGA) of Figure 6a. At the same time, the shrinking of the PAN mat diminishes, depending on the temperature and dwelling time. TGA is a thermal analysis technique that consists of measuring the variation in mass of a sample as a function of time, for a given temperature or temperature profile. It is used to determine the degradation temperatures, the moisture absorbed by the material, and the amount of organic and inorganic compounds in a material. Here, it is used to determine the best temperature for the stabilization process to get the full cyclization of the PAN molecule and the weight loss after thermal treatments. It could be combined with differential scanning calorimetry (DSC). This analysis measures the difference in heat exchange between the sample and a reference and allows one to determine the phase transitions, for instance, the glass transition temperature and the melting or crystallization temperature. For PAN, it is possible to determine the best stabilization condition that corresponds to the main peak in the DSC curve; the position ranges from 250 to 290 °C. This step corresponds to the shrinking of the PAN.

After this first step, the carbonization process could be done at a temperature of 600–1400 °C under either argon or nitrogen gas [52,80–82]. This step should be done in an inert atmosphere to: (i) avoid the total loss of weight by the complete mineralization and (ii) create an electrically conductive carbon material for use in electrocatalytic tasks. Figure 6a clearly shows that if the second step is done in an O_2 atmosphere, only CO_2 will be created, and structures such as Figure 6b,c will not be obtained. This step permits one to increase the C/N ratio and, at the same time, to create an electrically conductive graphitic-like structure. The latter depends on the temperature and the dwell time. As shown in Table 3 by X-ray photoelectron spectroscopy (XPS) analysis, when the temperature increases, the C/N ratio increases (4.8 and 7.6 at 573 K (300 °C) and 1073 K (800 °C, respectively) [80], which impacts the electrical properties of the material [52,80,81]. The type of nitrogen species also changes: the main species are pyridinic-N (N-6), pyrrolic-N (N-5), oxidized-N (N-X), and quaternary-N (N-Q: the exact nature is subject to keen debate) [80].

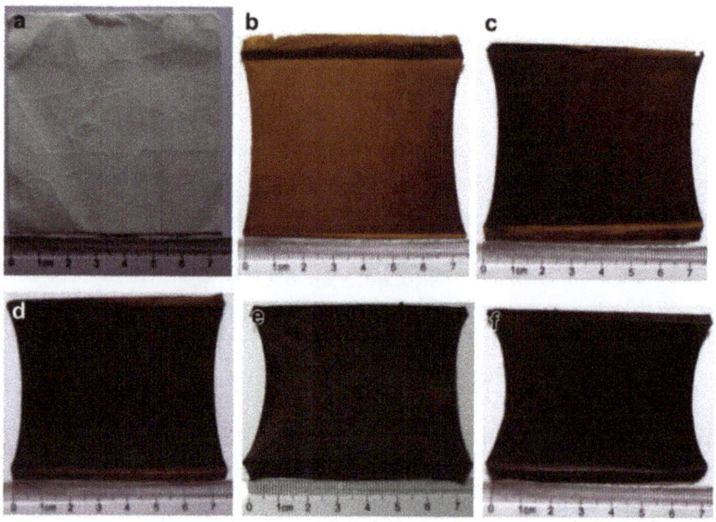

Figure 5. Pictures showing the color change of electrospun PAN mats during stabilization: (**a**) pristine and (**b**–**f**) stabilized (heating rate of 2 °C min^{-1} in a constant air flow, and then maintained at the target temperature): (**b**) 250 °C for 1 h, (**c**) 265 °C for 1 h, (**d**) 280 °C for 1 h; (**e**) 280 °C for 2 h, and (**f**) 280 °C for 3 h. Reprinted and adapted with permission from ref. [52]; Copyrights 2012, Elsevier Ltd.

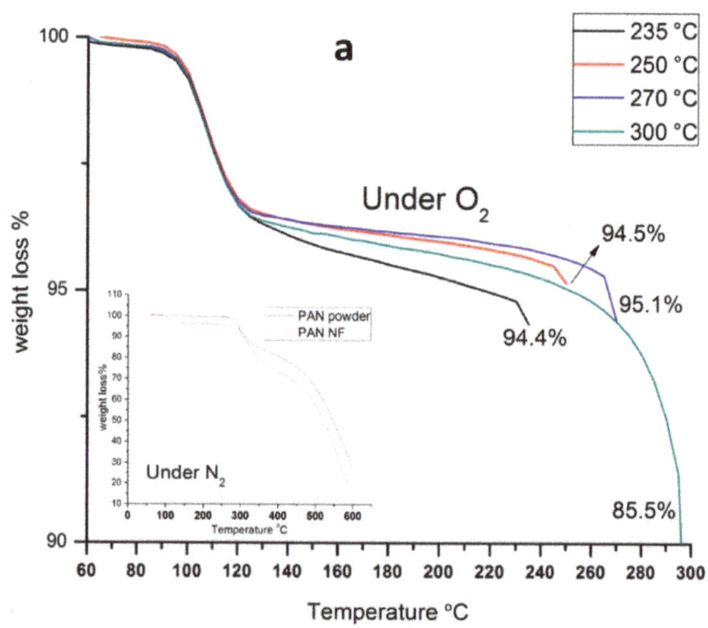

Figure 6. *Cont.*

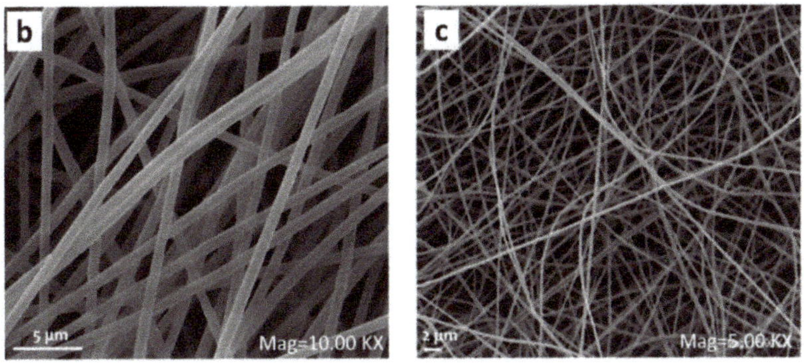

Figure 6. (**a**) The TGA curves of aligned PAN nanofibers (5 °C min^{-1}, air) until the targeted oxidation temperature (235, 250, 270, and 300 °C) for a dwell of 6 h (the inset represents the TGA under inert nitrogen atmosphere). SEM images of aligned PAN nanofibers (stabilization at 250 °C for 3 h in air atmosphere and carbonization at 900 °C for 1 h under nitrogen atmosphere) produced with: (**b**) rotating collector and (**c**) with fixed collector. Reprinted and adapted with permission from ref. [53]; Copyrights 2017, Gergin et al.; licensee Beilstein-Institut.

For PAN electrospun mats with gold, Both Engel et al. [34] have substantially increased the duration of the carbonization step from 1 to 10 h at 1000 °C to study the impact on the formation of nanostructured gold particles inside and outside the graphitized carbon microfibers. SEM images of the sample annealed for 10 h are presented in Figure 7. The main observation is the formation of two particle-sized populations, which was approximately the same result upon annealing for 1 h. The thermal treatment seems to not affect the particles' size or morphology; only the polymer changes into CNFs, and likely the nitrogen content too [35]. The most widely used temperature for the carbonization of PAN-based materials ranges from 550 to 1200 °C for 1 h [34,35,39,42,53]. Zhang et al. [35] showed that the crystallites' size increases while the nitrogen content decreases when the carbonization temperature goes from 550 to 950 °C under nitrogen atmosphere; 950 °C (1 h) leads to the optimal electrochemical performance.

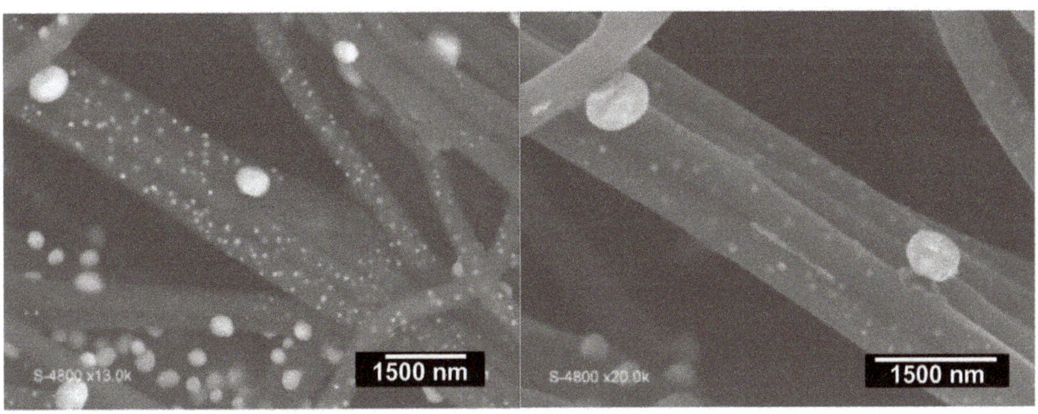

Figure 7. SEM images of Au@CFs at different magnifications after stabilization under air (2 °C min^{-1} until 250 °C for 2 h stay) and calcination under high-purity nitrogen atmosphere (2 °C min^{-1} until 1000 °C for 10 h stay). Reprinted and adapted with permission from ref. [34]; Copyrights 2016, Wiley-VCH Verlag GmbH and Co. KGaA, Weinheim.

Table 3. XPS data from calcined PAN at different temperatures (heating rate of 1 °C/min, dwell time of 1 h, and Ar atmosphere). Reprinted and adapted with permission from ref. [80]; Copyright 1995, Elsevier Science Ltd.

Precursor	Pyrolysis Temperature	Atomic Percentage			Percentage of Nitrogen Functionalities (from XPS N1s)			
		C	N	O	Pyridinic-N (N-6)	Pyrrolic-N (N-5)	Quaternary-N (N-Q)	Oxidized-N (N-X)
PAN	573 K (300 °C)	79.5	16.2	4.3	100	0	0	0
PAN	773 K (500 °C)	78.4	15.5	6.1	69	19	8	4
PAN	1073 K (800 °C)	85.4	11.2	3.5	40	29	23	9

2.6. Other Characterizations of the Mat Fibers

Other characterizations are possible for the PAN mats before and after any thermal treatment. The Fourier Transform Infrared Spectroscopy (FTIRS) gives information about the structure of the molecule, that is, the type of organic functions (alcohol, amine, etc.). Figure 8a,b shows the FTIR spectra for different types of PAN-based materials. For PAN, the bands at 2940, 2250, and 1450 cm^{-1} correspond to the stretching of C–H, C≡N and CH_2, respectively [35,40,52,56,79]. After stabilization, their intensities decrease and an additional peak appears at 810 cm^{-1}, which is assigned to the vibration of =C–H. The new peak comes from the aromatic ring that is created during the cyclization reaction (see Figure 4) [35,52]. Indeed, after carbonization under N_2 gas, the C≡N and CH_2 bands must disappear because of crosslinks between the PAN molecules. Therefore, any presence of a peak at 2250 cm^{-1} suggests that the carbonization step is not yet completed [56]. It was reported that at 500 °C, C=N and C–N bands at 1374 and 1284 cm^{-1}, respectively, are strengthened due to the conversion of C≡N, while at 950 °C, the C=N vibration band disappears (other reaction products can be detected depending on the applied temperature) [35,52,56,79].

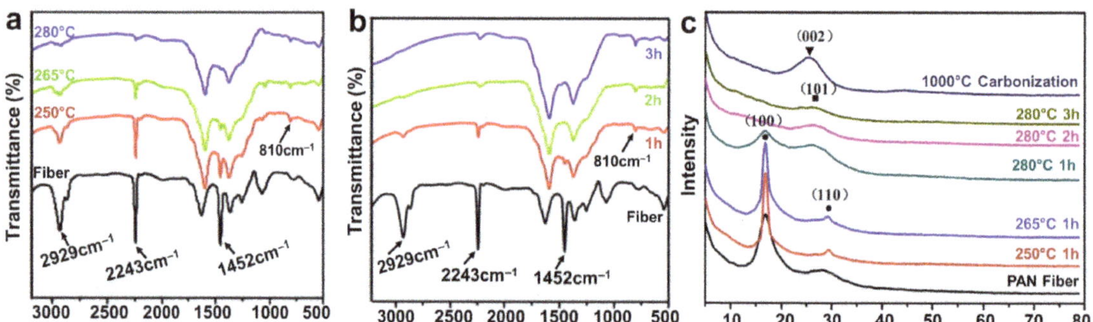

Figure 8. (**a**,**b**) FTIR spectra of stabilized PAN nanofibers (heating rate of 2 °C min^{-1} in a constant air flow): (**a**) temperature effect (holding time: 1 h); (**b**) time effect (holding temperature: 280 °C). (**c**) XRD patterns of the PAN precursor nanofibers, the stabilized (under air), and the carbonized PAN nanofibers (2 °C min^{-1} until 1000 °C for 1 h stay under N_2 atmosphere). Reprinted and adapted with permission from ref. [52]; Copyrights 2012, Elsevier Ltd.

X-ray diffraction (XRD) can be used for the crystallinity characterization of the polymer and the metallic particles. Figure 8c shows XRD patterns for different electrospun PAN-based materials. We note the presence of a strong diffraction peak centered at 17° and a weak diffraction peak centered at 29°, which correspond to the X-ray diffraction of (100) and (110) of PAN nanofibers, respectively [52,64,79]. By FTIRS, it was observed that the aromatic growth is the major process during the carbonization at a high temperature, and it triggers the formation of a graphite-like structure by eliminating the nitrogen-containing groups. The decrease in nitrogen content could be confirmed by XPS [35]. When the carbonization temperature increases, the intensity of the peaks corresponding to the graphitic carbon increase while that at 17° disappears [35]. In Figure 8c, at 1000 °C, the carbonized PAN material shows a single diffraction peak centered at 25.6° for the (002) crystallographic plane of graphite [35,52]. For the gold species within the electrospun mats (with and without carbonization), Figure 9 shows the XRD patterns. Figure 9a confirms the role of the solvent and PAN on the formation of gold particles. The peaks situated at 38.2°, 44.5°, 65.6°, and 78.6° correspond to, respectively, (111), (200), (220), and (311) planes of face-centered cubic structure of gold [34,47,64,83]. Leveraging on the evolution of the intensity and the width of the (200) and (220) peaks, it was argued that metallic gold particles are created during the electrospinning process and the thermal treatment changes the crystallinity [34].

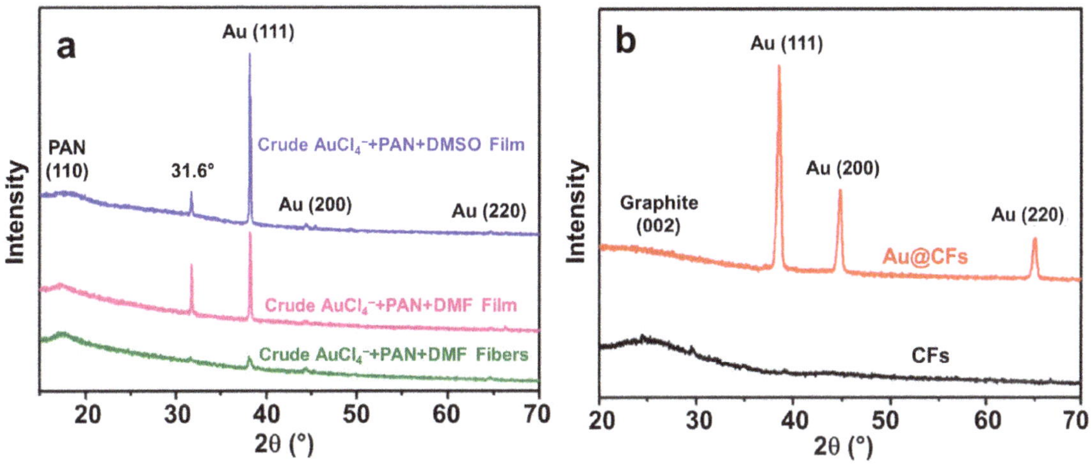

Figure 9. (**a**) XRD of crude fibers obtained from the solution containing $AuCl_4^-$+PAN+DMF (green curve), crude film from the same solution (pink curve), and crude film from the solution prepared in DMSO $AuCl_4^-$ + PAN + DMSO (blue curve). (**b**) XRD of carbonized fibers with no Au (black curve) and Au@CFs (red curve), after stabilization under air (2 °C min^{-1} until 250 °C for 2 h stay) and calcination under high-purity nitrogen atmosphere (2 °C min^{-1} until 1000 °C for 1 h stay). Reprinted and adapted with permission from ref. [34]; Copyrights 2016, Wiley-VCH Verlag GmbH and Co. KGaA, Weinheim.

3. HMF Electrooxidation for Paired Electrosynthesis of Valuable Chemicals

In the previous Section 2, we focused on the electrospinning method to obtain materials that can be used as electrocatalysts. The current Section 3 focuses on a summary review of the electrocatalysis of HMF. Ideally, we would like to review the performance of the materials from Section 2; however, as there are no data available, to our knowledge, on the use of electrocatalysts derived from electrospinning for the electroconversion of HMF, we will review other systems to provide some inspirational ideas.

To reduce the H_2 price by an electrochemical process, few solutions are reported. One of them is to reduce the overpotential of the oxygen evolution reaction (OER) by the electrooxidation of organic molecules to decrease the quantity of energy used to form H_2 at the cathode of an electrolyzer [6,12,84–88]. It has been showed that by oxidizing at the anode a biomass derivate, it is possible to reduce the required voltage compared to the electrooxidation of water molecules [89]. The techno-economic analysis of renewable energy-powered biomass electrolysis where organics are oxidized at the anode in lieu of water has shown the promise of H_2 production at the cathode with the possible co-generation of valuable organics at the anode [11,85,89,90]. For example, Paul Kenis' group reported lower electricity consumption by up to 53% compared to conventional water electrolysis [91]. Figure 10a shows that glucose and glycerol have a much lower oxidation potential than the oxidation of water, so less energy will be required to produce H_2 by such an electrolysis [6]. We note that the cell voltage of an electrolyzer is $U_{cell} = E_{(anode)} - E_{(cathode)}$. Consequently, the potential of the anode where the electrooxidation is happening must be as small as possible for a given cathode material. As shown in Figure 10b, HMF electrooxidation is less energy-demanding than OER [12], and its electrocatalysis can be regulated to produce value-added chemicals to replace petroleum-based polymers [10,12,13].

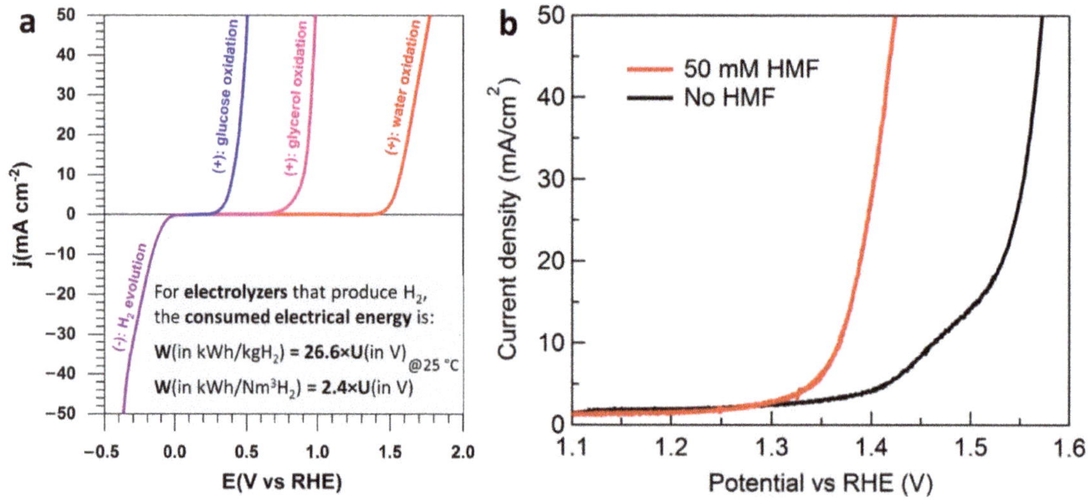

Figure 10. (**a**) Typical half-cell electrolysis polarization curves illustrating the working principle of solid alkaline membrane electrolysis cell [6]. (**b**) LSVs of Co−P/CF (1 M KOH, 2 mV s^{-1}) in the absence (black) and presence of 50 mM HMF (red). (**a**) Reprinted and adapted with permission from ref. [6], Copyright 2020, The Royal Society of Chemistry 2020. (**b**) Reprinted and adapted with permission from ref. [12], Copyright 2016, American Chemical Society.

The linear sweep voltammetry (LSV) of Figure 10b shows that the presence of 50 mM HMF in 1 M KOH is accompanied by a drastic shift in the potential needed to drive more current density (and thus more conversion of the reactants into products, the second law of Faraday). To reach 50 mA cm^{-2}, a potential of ca. 1.42 V vs. RHE (reversible hydrogen electrode) is needed in the presence of HMF, which is 150 mV smaller than that of OER (1.57 V vs. RHE). The goal of using organic molecules at the anodic compartment of a H_2 production electrolyzer is not to perform the complete electrooxidation into CO_2 as it seemingly violates the original intention of reduction the carbon footprint [6,92]. Of course, the CO_2 from the full degradation of organic molecules could be sent back to the cathodic compartment; this is only valid for CO_2 electrolyzers [91,93,94], it is not a universal strategy. Here, Figure 11 highlights the dreamed scenarios of the electrocatalytic oxidation of HMF, whereby selective electrooxidation could be a solution to produce useful organic molecules. Hence, in addition to the significant decrease of the total energy input (see Figure 10a), this collateral gain at the anodic compartment contributes to the decrease of the cost of H_2 production (Figure 12a). The two possible pathways without any carbon–carbon bonds cleavage are the aldehyde route (Pathway A) and the alcohol route (Pathway B), which can change depending on the used electrode material (Figure 12). We note that HMF can be conventionally oxidized by using several methods: (i) in hard conditions with organic solvents at high pressure of O_2 and temperature [7,95,96], (ii) by an enzymatic method [97], and (iii) by an electrocatalytic method using synthetized metal-based electrodes [2,13,16,34,98–101]. The promise of the breakthrough electroconversion of HMF by electrocatalysis relies on the use of mild conditions (temperature, pressure), only water as an oxygen source, solid electrode materials, and electrons as the triggers (from any renewable electricity source, solar, wind, etc.). Compared to catalysis by oxidants such as O_2 at elevated temperature and pressure, this set of specifiers contributes to making electrocatalysis a more sustainable process and renders electrochemistry environmentally useful as a "green tool" [102–104].

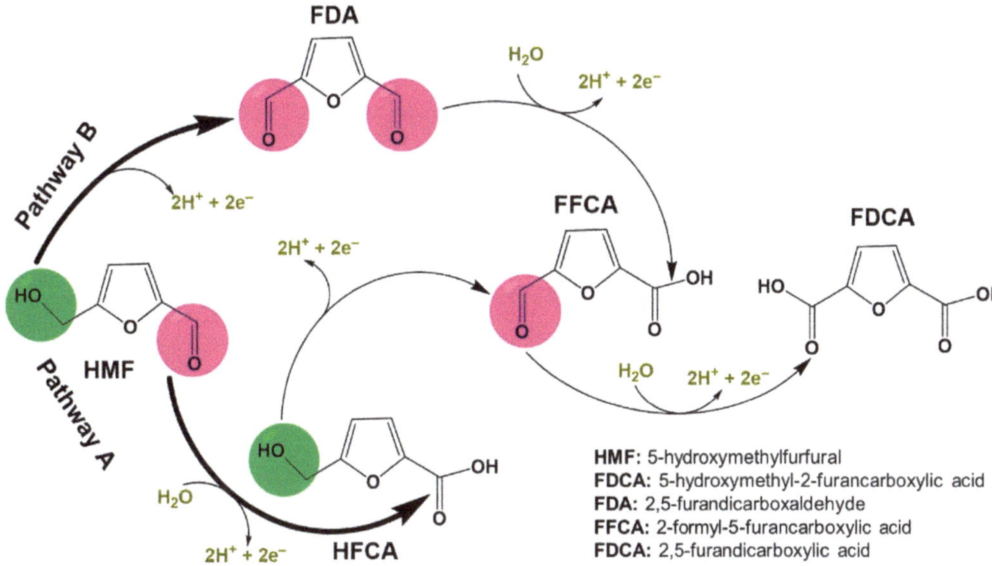

Figure 11. Scenarios of the electrocatalytic oxidation of HMF and the resulting products without C-C bond cleavage (written in acidic media).

For Pd-Au materials, the required cell voltage (Figure 12b), the reaction paths (Figure 12c,f), and the yield (Figure 12d,e) depend on the exposed active sites at the electrocatalyst surface. Under chronoamperometry condition (0.82 V vs. RHE in 1 M KOH + 5 mM HMF), the conversion of HMF into FDCA is more important at bimetallic Au-Pd in comparison to the monometallic Pd and Au materials [16]. The selectivity improvement can be explained by the synergistic actions between Au and Pd, driven by the coupling of electronic and geometric effects. It can also be observed that Au outperforms Pd for the oxidation of HMF into HFCA (Figure 12e). It is well-known that Au is more active for the oxidation of the aldehyde group, whereas Pt and Pd catalysts are more active for the oxidation of the alcohol groups [98]. When they are combined, the electrooxidation could start at low potentials [13,16,98]. For the cost reduction, it is possible to use non-noble metals. For example, HMF was successfully converted into DFF with 78% isolated yield and 100% selectivity by an environmentally friendly organic electrosynthesis using graphite as an anode and stainless steel as a cathode [2]. You et al. [87] reported 3D hierarchically porous nickel-based electrocatalyst obtained by the electrodeposition to reduce the cell voltage by 220 mV at 50 mA cm^{-2} when compared to conventional water splitting.

The above examples show the variety of opportunities to electrooxidize HMF into different components. Other options exist, starting from fructose because of the possible stability issues of HMF in aqueous alkaline solutions where the aldehyde function could be converted into carboxylic acid by a nucleophilic attack by OH^-. Control trials are currently missing to fully know the impact of such a hypothesis on the whole mechanism. Indeed, the electrooxidation of any organic molecule is a set of proton-coupled electron (PCET) steps so that the pH has a key impact on the reactivity that is expected to be maximal at pH close to the compound's p*K*a [105–110]. Given the HMF's p*K*a of 12.8, working in an alkaline electrolyte to maximize electrocatalytic activity exposes HMF to a nucleophilic attack by OH^-, which will be amplified as the reaction time increases. Therefore, to provide high yield and selectivity, optimization has to be made. For example, the pH impacts the yield/selectivity, the metal has its own selectivity properties, and the temperature or duration of the electrolysis could impact the stability/yield [2,16,96,97]. Furthermore, the technical report of IUPAC indicates that "the simultaneous transfer of more than one electron is highly improbable" [111]. Thus, the electrocatalysts for HMF electroconversion must withstand a cascade of PCET steps, which will induce many reaction intermediates and energy barriers of larger overpotentials [106]. Each of the discussed parameters should be properly optimized, and theoretical calculations (DFT, MD, etc.) can help to better guide the operation of electrocatalytic materials [6,112].

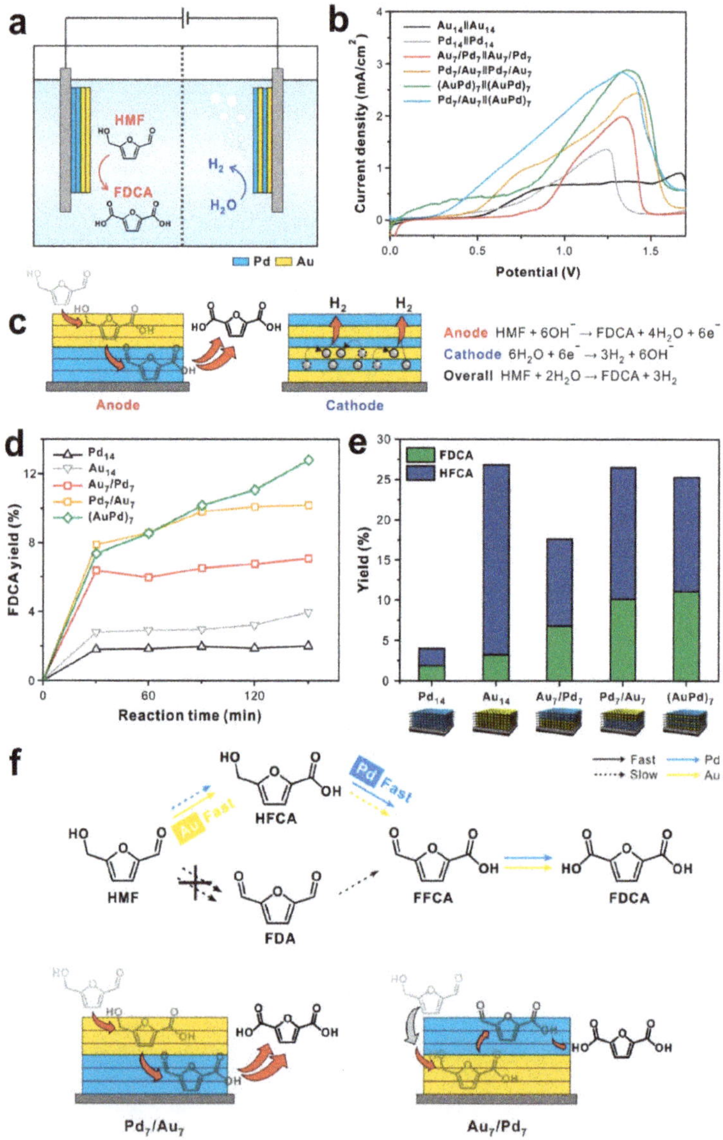

Figure 12. (**a**) Sketch of the two-electrode electrolysis cell made of (+)Pd$_7$/Au$_7$||(AuPd)$_7$(−) for the simultaneous generation of FDCA from the anode (+) and H_2 from the cathode (−). (**b**) LSV

(2 mV s^{-1}) in 1 M KOH + 10 mM HMF. (**c**) Pd_7/Au_7 multilayer anode and the $(AuPd)_7$ multilayer cathode with the corresponding electron transfer processes. Bulk electrolysis at 0.82 V vs. RHE in 1 M KOH + 5 mM HMF for 2 h at different anode materials. (**d**) Yield of FDCA and (**e**) comparison between HFCA and FDCA. (**f**) Pathways of HMF electrooxidation. Reprinted and adapted with permission from ref. [16], Copyright 2020, American Chemical Society.

4. Conclusions and Perspective

This review discussed the advances of electrospinning by giving an overview of the experimental conditions for reproducible methodologies. The polymer solution formulation, the spinning, the thermal treatment, and the different analyses have been critically reviewed to understand the chemical and structural changes in the polyacrylonitrile (PAN) molecule and how it can be exploited to engineer high-performance electrode materials for electrocatalysis. During the formulation of the polymer solution for the electrospinning process, it is possible to introduce metallic or organic species to change the final materials' structure and target different electrocatalytic properties. However, the addition of such a chemical species modifies the physico-chemical properties of the electrospinning solution, as well as the operation parameters. The latter are not rationalized as the methodologies change from one report to another one, which calls for further efforts to structure the field and understand the impact on the electrospinning and the calcination steps.

A focused analysis was conducted for the electroconversion of 5-hydroxymethylfurfural (HMF) into added-value chemicals during CO_2-free and low energy consumption H_2 production in electrolyzers. For HMF electrooxidation, further studies should be conducted to obtain free-standing electrocatalysts from electrospinning and target a strong anchoring of metals to support. Thermodynamically, it is true that the electrooxidation of HMF instead of water (OER, oxygen evolution reaction) in the anodic compartment of an electrolyzer for H_2 production is accompanied by a significant saving of the consumed electrical energy. Nonetheless, this is purely conceptual at the moment because the recorded current densities of a few tens of mA/cm^2 are too far from the industrial needs of at least 0.5 A/cm^2 [77,113]: the current density-dependent H_2 rate at 25 °C is ca. D (L/h per cm^2) = 0.45 × j (A/cm^2). In addition, the field of organic electrosynthesis coupled to the production of decarbonized H_2 should go beyond the hundreds of reports being published each year by simple three-electrode setups. In fact, there is no guarantee that the implementation in real two-electrode and zero-gap electrolyzers will meet the promise. Theoreticians and experimenters are invited to join their forces in order to significantly increase the current densities while preserving the selectivity that is often obtained at the expense of the current density. In terms of the bigger picture, any developed electrocatalyst for the selective electro-conversion of HMF can be a widespread solution in low-energy input power-to-X technologies where hydrogen evolution (HER), CO_2 reduction (CO2RR), and N_2 reduction (N2RR) reactions are considered to enable the electrosynthesis of high value-added fuels and/or chemicals with a significantly reduced environmental footprint. For those interested in electrolysis, we end this review by extending an invitation to a recent Perspective by Siegmund et al. [77] on "Crossing the Valley of Death: From Fundamental to Applied Research in Electrolysis".

Author Contributions: Conceptualization, M.C., Y.H. and D.C.; methodology, M.C., Y.H. and D.C.; project administration, Y.H. and D.C.; writing—original draft, M.C.; and writing—review and editing, Y.H. and D.C. All authors have read and agreed to the published version of the manuscript.

Funding: This work has been partially funded by the CNRS Energy unit (Cellule Energie) through the project "PEPS19-ELECTROFUEL".

Institutional Review Board Statement: Not applicable.

Informed Consent Statement: Not applicable.

Data Availability Statement: Not applicable.

Conflicts of Interest: The authors declare no conflict of interest.

References

1. Bond, A.M. *Broadening Electrochemical Horizons: Principles and Illustration of Voltammetric and Related Techniques*; Oxford University Press Inc.: New York, NY, USA, 2002; p. 532.
2. Kisszekelyi, P.; Hardian, R.; Vovusha, H.; Chen, B.L.; Zeng, X.H.; Schwingenschlogl, U.; Kupai, J.; Szekely, G. Selective Electrocatalytic Oxidation of Biomass-Derived 5-Hydroxymethylfurfural to 2,5-Diformylfuran: From Mechanistic Investigations to Catalyst Recovery. *Chemsuschem* **2020**, *13*, 3127–3136. [CrossRef] [PubMed]
3. Luna, P.D.; Hahn, C.; Higgins, D.; Jaffer, S.A.; Jaramillo, T.F.; Sargent, E.H. What would it take for renewably powered electrosynthesis to displace petrochemical processes? *Science* **2019**, *364*, eaav3506. [CrossRef]
4. Montoya, J.H.; Seitz, L.C.; Chakthranont, P.; Vojvodic, A.; Jaramillo, T.F.; Norskov, J.K. Materials for solar fuels and chemicals. *Nat. Mater.* **2017**, *16*, 70–81. [CrossRef] [PubMed]
5. Xue, J.; Wu, T.; Dai, Y.; Xia, Y. Electrospinning and Electrospun Nanofibers: Methods, Materials, and Applicaations. *Chem. Rev.* **2019**, *119*, 5298–5415. [CrossRef]
6. Holade, Y.; Tuleushova, N.; Tingry, S.; Servat, K.; Napporn, T.W.; Guesmi, H.; Cornu, D.; Kokoh, K.B. Recent advances in the electrooxidation of biomass-based organic molecules for energy, chemicals and hydrogen production. *Catal. Sci. Technol.* **2020**, *10*, 3071–3112. [CrossRef]
7. Dai, J. Synthesis of 2,5-diformylfuran from renewable carbohydrates and its applications: A review. *Green Energy Environ.* **2021**, *6*, 22–32. [CrossRef]
8. Kwon, Y.; Schouten, K.J.P.; van der Waal, J.C.; de Jong, E.; Koper, M.T.M. Electrocatalytic Conversion of Furanic Compounds. *ACS Catal.* **2016**, *6*, 6704–6717. [CrossRef]
9. Rosatella, A.A.; Simeonov, S.P.; Frade, R.F.M.; Afonso, C.A.M. 5-Hydroxymethylfurfural (HMF) as a building block platform: Biological properties, synthesis and synthetic applications. *Green Chem.* **2011**, *13*, 754–793. [CrossRef]
10. Patel, P.; Schwartz, D.; Wang, X.; Lin, R.; Ajao, O.; Seifitokaldani, A. Technoeconomic and Life-Cycle Assessment for Electrocatalytic Production of Furandicarboxylic Acid. *ACS Sustain. Chem. Eng.* **2022**, *10*, 4206–4217. [CrossRef]
11. Lucas, F.W.S.; Grim, R.G.; Tacey, S.A.; Downes, C.A.; Hasse, J.; Roman, A.M.; Farberow, C.A.; Schaidle, J.A.; Holewinski, A. Electrochemical Routes for the Valorization of Biomass-Derived Feedstocks: From Chemistry to Application. *ACS Energy Lett.* **2021**, *6*, 1205–1270. [CrossRef]
12. Jiang, N.; You, B.; Boonstra, R.; Terrero Rodriguez, I.M.; Sun, Y. Integrating Electrocatalytic 5-Hydroxymethylfurfural Oxidation and Hydrogen Production via Co–P-Derived Electrocatalysts. *ACS Energy Lett.* **2016**, *1*, 386–390. [CrossRef]
13. Chadderdon, D.J.; Xin, L.; Qi, J.; Qiu, Y.; Krishna, P.; More, K.L.; Li, W. Electrocatalytic oxidation of 5-hydroxymethylfurfural to 2,5-furandicarboxylic acid on supported Au and Pd bimetallic nanoparticles. *Green Chem.* **2014**, *16*, 3778–3786. [CrossRef]
14. Ermakova, M.A.; Ermakov, D.Y.; Kuvshinov, G.G. Effective catalysts for direct cracking of methane to produce hydrogen and filamentous carbon: Part I. *Nickel catalysts. Appl. Catal. A Gen.* **2000**, *201*, 61–70. [CrossRef]
15. Inocêncio, C.V.M.; Holade, Y.; Morais, C.; Kokoh, K.B.; Napporn, T.W. Electrochemical hydrogen generation technology: Challenges in electrodes materials for a sustainable energy. *Electrochem. Sci. Adv.* **2022**, e2100206. [CrossRef]
16. Park, M.; Gu, M.; Kim, B.-S. Tailorable Electrocatalytic 5-Hydroxymethylfurfural Oxidation and H2 Production: Architecture–Performance Relationship in Bifunctional Multilayer Electrodes. *ACS Nano* **2020**, *14*, 6812–6822. [CrossRef]
17. Zhang, L.; Zhao, H.; Xu, S.; Liu, Q.; Li, T.; Luo, Y.; Gao, S.; Shi, X.; Asiri, A.M.; Sun, X. Recent Advances in 1D Electrospun Nanocatalysts for Electrochemical Water Splitting. *Small Struct.* **2021**, *2*, 2000048. [CrossRef]
18. Stamenkovic, V.R.; Strmcnik, D.; Lopes, P.P.; Markovic, N.M. Energy and fuels from electrochemical interfaces. *Nat. Mater.* **2017**, *16*, 57–69. [CrossRef]
19. Assaud, L.; Brazeau, N.; Barr, M.K.S.; Hanbucken, M.; Ntais, S.; Baranova, E.A.; Santinacci, L. Atomic Layer Deposition of Pd Nanoparticles on TiO_2 Nanotubes for Ethanol Electrooxidation: Synthesis and Electrochemical Properties. *ACS Appl. Mater. Interfaces* **2015**, *7*, 24533–24542. [CrossRef]
20. Hajar, Y.; Di Palma, V.; Kyriakou, V.; Verheijen, M.A.; Baranova, E.A.; Vernoux, P.; Kessels, W.M.M.; Creatore, M.; van de Sanden, M.C.M.; Tsampas, M.N. Atomic layer deposition of highly dispersed Pt nanoparticles on a high surface area electrode backbone for electrochemical promotion of catalysis. *Electrochem. Commun.* **2017**, *84*, 40–44. [CrossRef]
21. Weber, M.; Tuleushova, N.; Zgheib, J.; Lamboux, C.; Iatsunskyi, I.; Coy, E.; Flaud, V.; Tingry, S.; Cornu, D.; Miele, P.; et al. Enhanced electrocatalytic performance triggered by atomically bridged boron nitride between palladium nanoparticles and carbon fibers in gas-diffusion electrodes. *Appl. Catal. B Environ.* **2019**, *257*, 117917. [CrossRef]
22. Weber, M.; Collot, P.; El Gaddari, H.; Tingry, S.; Bechelany, M.; Holade, Y. Enhanced Catalytic Glycerol Oxidation Activity Enabled by Activated-Carbon-Supported Palladium Catalysts Prepared through Atomic Layer Deposition. *ChemElectroChem* **2018**, *5*, 743–747. [CrossRef]
23. Zhang, R.; Wei, H.; Si, W.; Ou, G.; Zhao, C.; Song, M.; Zhang, C.; Wu, H. Enhanced Electrocatalytic Activity for Water Splitting on NiO/Ni/Carbon Fiber Paper. *Materials* **2017**, *10*, 15. [CrossRef] [PubMed]
24. Cavaliere, S.; Subianto, S.; Savych, I.; Jones, D.J.; Roziere, J. Electrospinning: Designed architectures for energy conversion and storage devices. *Energy Environ. Sci.* **2011**, *4*, 4761–4785. [CrossRef]

25. Sood, R.; Cavaliere, S.; Jones, D.J.; Rozière, J. Electrospun nanofibre composite polymer electrolyte fuel cell and electrolysis membranes. *Nano Energy* **2016**, *26*, 729–745. [CrossRef]
26. Xue, J.; Xie, J.; Liu, W.; Xia, Y. Electrospun Nanofibers: New Concepts, Materials, and Applications. *Acc. Chem. Res.* **2017**, *50*, 1976–1987. [CrossRef]
27. Rodríguez-Tobías, H.; Morales, G.; Grande, D. Comprehensive review on electrospinning techniques as versatile approaches toward antimicrobial biopolymeric composite fibers. *Mater. Sci. Eng. C* **2019**, *101*, 306–322. [CrossRef]
28. Saleh, A.; Marhuenda, E.; Fabre, C.; Hassani, Z.; Weille, J.D.; Boukhaddaoui, H.; Guelfi, S.; Maldonado, I.L.; Hugnot, J.-P.; Duffau, H.; et al. A novel 3D nanofibre scaffold conserves the plasticity of glioblastoma stem cell invasion by regulating galectin-3 and integrin-β1 expression. *Sci. Rep.* **2019**, *9*, 14612. [CrossRef]
29. Marhuenda, E.; Fabre, C.; Zhang, C.; Martin-Fernandez, M.; Iskratsch, T.; Saleh, A.; Bauchet, L.; Cambedouzou, J.; Hugnot, J.-P.; Duffau, H.; et al. Glioma stem cells invasive phenotype at optimal stiffness is driven by MGAT5 dependent mechanosensing. *J. Exp. Clin. Cancer Res.* **2021**, *40*, 139. [CrossRef]
30. Pastoriza-Santos, I.; Liz-Marzán, L.M. N,N-Dimethylformamide as a Reaction Medium for Metal Nanoparticle Synthesis. *Adv. Funct. Mater.* **2009**, *19*, 679–688. [CrossRef]
31. Saquing, C.D.; Manasco, J.L.; Khan, S.A. Electrospun Nanoparticle–Nanofiber Composites via a One-Step Synthesis. *Small* **2009**, *5*, 944–951. [CrossRef]
32. Wang, Y.; Li, Y.; Sun, G.; Zhang, G.; Liu, H.; Du, J.; Yang, S.; Bai, J.; Yang, Q. Fabrication of Au/PVP nanofiber composites by electrospinning. *J. Appl. Polym. Sci.* **2007**, *105*, 3618–3622. [CrossRef]
33. Casasola, R.; Thomas, N.L.; Trybala, A.; Georgiadou, S. Electrospun poly lactic acid (PLA) fibres: Effect of different solvent systems on fibre morphology and diameter. *Polymer* **2014**, *55*, 4728–4737. [CrossRef]
34. Both Engel, A.; Bechelany, M.; Fontaine, O.; Cherifi, A.; Cornu, D.; Tingry, S. One-Pot Route to Gold Nanoparticles Embedded in Electrospun Carbon Fibers as an Efficient Catalyst Material for Hybrid Alkaline Glucose Biofuel Cells. *ChemElectroChem* **2016**, *3*, 629–637. [CrossRef]
35. Zhang, B.; Yu, Y.; Xu, Z.-L.; Abouali, S.; Akbari, M.; He, Y.-B.; Kang, F.; Kim, J.-K. Correlation Between Atomic Structure and Electrochemical Performance of Anodes Made from Electrospun Carbon Nanofiber Films. *Adv. Energy Mater.* **2014**, *4*, 1301448. [CrossRef]
36. Thavasi, V.; Singh, G.; Ramakrishna, S. Electrospun nanofibers in energy and environmental applications. *Energy Environ. Sci.* **2008**, *1*, 205–221. [CrossRef]
37. Xue, T.J.; McKinney, M.A.; Wilkie, C.A. The thermal degradation of polyacrylonitrile. *Polym. Degrad. Stab.* **1997**, *58*, 193–202. [CrossRef]
38. Badii, K.; Church, J.S.; Golkarnarenji, G.; Naebe, M.; Khayyam, H. Chemical structure based prediction of PAN and oxidized PAN fiber density through a non-linear mathematical model. *Polym. Degrad. Stab.* **2016**, *131*, 53–61. [CrossRef]
39. Holade, Y.; Both Engel, A.; Tingry, S.; Cherifi, A.; Cornu, D.; Servat, K.; Napporn, T.W.; Kokoh, K.B. Insights on Hybrid Glucose Biofuel Cell Based on Bilirubin Oxidase Cathode and Gold-Based Nanomaterials Anode. *ChemElectroChem* **2014**, *1*, 1976–1987. [CrossRef]
40. Aslan, S.; Bal Altuntaş, D.; Koçak, Ç.; Kara Subaşat, H. Electrochemical Evaluation of Titanium (IV) Oxide/Polyacrylonitrile Electrospun Discharged Battery Coals as Supercapacitor Electrodes. *Electroanalysis* **2021**, *33*, 120–128. [CrossRef]
41. Li, S.-F.; Chen, J.-P.; Wu, W.-T. Electrospun polyacrylonitrile nanofibrous membranes for lipase immobilization. *J. Mol. Catal. B Enz.* **2007**, *47*, 117–124. [CrossRef]
42. Diestelhorst, E.; Mance, F.; Mamun, A.; Ehrmann, A. Chemical and Morphological Modification of PAN Nanofibrous Mats with Addition of Casein after Electrospinning, Stabilisation and Carbonisation. *Tekstilec* **2020**, *63*, 38–49. [CrossRef]
43. Rodríguez-Tobías, H.; Morales, G.; Ledezma, A.; Romero, J.; Saldívar, R.; Langlois, V.; Renard, E.; Grande, D. Electrospinning and electrospraying techniques for designing novel antibacterial poly(3-hydroxybutyrate)/zinc oxide nanofibrous composites. *J. Mater. Sci.* **2016**, *51*, 8593–8609. [CrossRef]
44. Che, A.-F.; Germain, V.; Cretin, M.; Cornu, D.; Innocent, C.; Tingry, S. Fabrication of free-standing electrospun carbon nanofibers as efficient electrode materials for bioelectrocatalysis. *New J. Chem.* **2011**, *35*, 2848–2853. [CrossRef]
45. Wang, S.; Dai, C.; Li, J.; Zhao, L.; Ren, Z.; Ren, Y.; Qiu, Y.; Yu, J. The effect of different nitrogen sources on the electrocatalytic properties of nitrogen-doped electrospun carbon nanofibers for the oxygen reduction reaction. *Int. J. Hydrogen Energy* **2015**, *40*, 4673–4682. [CrossRef]
46. Sawada, K.; Sakai, S.; Taya, M. Polyacrylonitrile-based electrospun nanofibers carrying gold nanoparticles in situ formed by photochemical assembly. *J. Mater. Sci.* **2014**, *49*, 4595–4600. [CrossRef]
47. Anka, F.H.; Perera, S.D.; Ratanatawanate, C.; Balkus, K.J. Polyacrylonitrile gold nanoparticle composite electrospun fibers prepared by in situ photoreduction. *Mater. Lett.* **2012**, *75*, 12–15. [CrossRef]
48. Liu, D.; Li, W.; Li, L.; Ling, H.; You, T. Facile preparation of Ni nanowire embedded nitrogen and sulfur dual-doped carbon nanofibers and its superior catalytic activity toward urea oxidation. *J. Colloid Int. Sci.* **2018**, *529*, 337–344. [CrossRef]
49. Guo, L.; Sun, H.; Qin, C.; Li, W.; Wang, F.; Song, W.; Du, J.; Zhong, F.; Ding, Y. Flexible Fe_3O_4 nanoparticles/N-doped carbon nanofibers hybrid film as binder-free anode materials for lithium-ion batteries. *Appl. Surf. Sci.* **2018**, *459*, 263–270. [CrossRef]
50. Yu, X.; Pei, C.; Chen, W.; Feng, L. 2 dimensional WS2 tailored nitrogen-doped carbon nanofiber as a highly pseudocapacitive anode material for lithium-ion battery. *Electrochim. Acta* **2018**, *272*, 119–126. [CrossRef]

51. Cai, N.; Fu, J.; Chan, V.; Liu, M.; Chen, W.; Wang, J.; Zeng, H.; Yu, F. $MnCo_2O_4$@nitrogen-doped carbon nanofiber composites with meso-microporous structure for high-performance symmetric supercapacitors. *J. Alloys Compd.* **2019**, *782*, 251–262. [CrossRef]
52. Wu, M.; Wang, Q.; Li, K.; Wu, Y.; Liu, H. Optimization of stabilization conditions for electrospun polyacrylonitrile nanofibers. *Polym. Degrad. Stab.* **2012**, *97*, 1511–1519. [CrossRef]
53. Gergin, İ.; Ismar, E.; Sarac, A.S. Oxidative stabilization of polyacrylonitrile nanofibers and carbon nanofibers containing graphene oxide (GO): A spectroscopic and electrochemical study. *Beilstein J. Nanotechnol.* **2017**, *8*, 1616–1628. [CrossRef] [PubMed]
54. Both Engel, A.; Holade, Y.; Tingry, S.; Cherifi, A.; Cornu, D.; Servat, K.; Napporn, T.W.; Kokoh, K.B. Electrospun Carbon Fibers: Promising Electrode Material for Abiotic and Enzymatic Catalysis. *J. Phys. Chem. C* **2015**, *119*, 16724–16733. [CrossRef]
55. Chronakis, I.S. Chapter 22—Micro- and Nano-fibers by Electrospinning Technology: Processing, Properties, and Applications. In *Micromanufacturing Engineering and Technology*, 2nd ed.; Qin, Y., Ed.; Elsevier Inc.: Boston, MA, USA, 2015; pp. 513–548. [CrossRef]
56. Khan, W.S. Chemical and thermal investigations of electrospun polyacrylonitrile nanofibers incorporated with various nanoscale inclusions. *J. Therm. Eng.* **2017**, *3*, 1375–1390. [CrossRef]
57. Bhardwaj, N.; Kundu, S.C. Electrospinning: A fascinating fiber fabrication technique. *Biotechnol. Adv.* **2010**, *28*, 325–347. [CrossRef]
58. Huang, Z.-M.; Zhang, Y.Z.; Kotaki, M.; Ramakrishna, S. A review on polymer nanofibers by electrospinning and their applications in nanocomposites. *Compos. Sci. Technol.* **2003**, *63*, 2223–2253. [CrossRef]
59. Lei, Y.; Wang, Q.; Peng, S.; Ramakrishna, S.; Zhang, D.; Zhou, K. Electrospun Inorganic Nanofibers for Oxygen Electrocatalysis: Design, Fabrication, and Progress. *Adv. Energy Mater.* **2020**, *10*, 1902115. [CrossRef]
60. Li, Z.; Wang, C. Effects of Working Parameters on Electrospinning. In *One-Dimensional Nanostructures: Electrospinning Technique and Unique Nanofibers*; Springer: Berlin/Heidelberg, Germany, 2013; pp. 15–28.
61. Gotti, G.; Fajerwerg, K.; Evrard, D.; Gros, P. Electrodeposited gold nanoparticles on glassy carbon: Correlation between nanoparticles characteristics and oxygen reduction kinetics in neutral media. *Electrochim. Acta* **2014**, *128*, 412–419. [CrossRef]
62. Holade, Y.; Hickey, D.P.; Minteer, S.D. Halide-regulated growth of electrocatalytic metal nanoparticles directly onto a carbon paper electrode. *J. Mater. Chem. A* **2016**, *4*, 17154–17162. [CrossRef]
63. Gittins, D.I.; Caruso, F. Spontaneous Phase Transfer of Nanoparticulate Metals from Organic to Aqueous Media. *Angew. Chem. Int. Ed.* **2001**, *40*, 3001–3004. [CrossRef]
64. Shamsabadi, A.S.; Ranjbar, M.; Tavanai, H.; Farnood, A. Electrospinning of gold nanoparticles incorporated PAN nanofibers via in-situ laser ablation of gold in electrospinning solution. *Mater. Res. Express* **2019**, *6*, 055051. [CrossRef]
65. Yao, Y.; Huang, Z.; Xie, P.; Lacey, S.D.; Jacob, R.J.; Xie, H.; Chen, F.; Nie, A.; Pu, T.; Rehwoldt, M.; et al. Carbothermal shock synthesis of high-entropy-alloy nanoparticles. *Science* **2018**, *359*, 1489–1494. [CrossRef] [PubMed]
66. Ju, Y.-W.; Yoo, S.; Kim, C.; Kim, S.; Jeon, I.-Y.; Shin, J.; Baek, J.-B.; Kim, G. Fe@N-Graphene Nanoplatelet-Embedded Carbon Nanofibers as Efficient Electrocatalysts for Oxygen Reduction Reaction. *Adv. Sci.* **2016**, *3*, 1500205. [CrossRef] [PubMed]
67. Li, M.; Wang, H.; Zhu, W.; Li, W.; Wang, C.; Lu, X. RuNi Nanoparticles Embedded in N-Doped Carbon Nanofibers as a Robust Bifunctional Catalyst for Efficient Overall Water Splitting. *Adv. Sci.* **2020**, *7*, 1901833. [CrossRef]
68. Wang, X.; Li, Y.; Jin, T.; Meng, J.; Jiao, L.; Zhu, M.; Chen, J. Electrospun Thin-Walled $CuCo_2O_4$@C Nanotubes as Bifunctional Oxygen Electrocatalysts for Rechargeable Zn–Air Batteries. *Nano Lett.* **2017**, *17*, 7989–7994. [CrossRef]
69. Zhao, Y.; Zhang, J.; Li, K.; Ao, Z.; Wang, C.; Liu, H.; Sun, K.; Wang, G. Electrospun cobalt embedded porous nitrogen doped carbon nanofibers as an efficient catalyst for water splitting. *J. Mater. Chem. A* **2016**, *4*, 12818–12824. [CrossRef]
70. Mooste, M.; Kibena-Põldsepp, E.; Vassiljeva, V.; Kikas, A.; Käärik, M.; Kozlova, J.; Kisand, V.; Külaviir, M.; Cavaliere, S.; Leis, J.; et al. Electrospun Polyacrylonitrile-Derived Co or Fe Containing Nanofibre Catalysts for Oxygen Reduction Reaction at the Alkaline Membrane Fuel Cell Cathode. *ChemCatChem* **2020**, *12*, 4568–4581. [CrossRef]
71. Liu, J.; Liang, J.; Wang, C.; Ma, J. Electrospun CoSe@N-doped carbon nanofibers with highly capacitive Li storage. *J. Energy Chem.* **2019**, *33*, 160–166. [CrossRef]
72. Zhang, F.; Yuan, C.; Zhu, J.; Wang, J.; Zhang, X.; Lou, X.W. Flexible Films Derived from Electrospun Carbon Nanofibers Incorporated with Co_3O_4 Hollow Nanoparticles as Self-Supported Electrodes for Electrochemical Capacitors. *Adv. Funct. Mater.* **2013**, *23*, 3909–3915. [CrossRef]
73. Salles, V.; Bernard, S.; Brioude, A.; Cornu, D.; Miele, P. A new class of boron nitride fibers with tunable properties by combining an electrospinning process and the polymer-derived ceramics route. *Nanoscale* **2010**, *2*, 215–217. [CrossRef]
74. Allison, D.B.; Shiffrin, R.M.; Stodden, V. Reproducibility of research: Issues and proposed remedies. *Proc. Natl. Acad. Sci. USA* **2018**, *115*, 2561–2562. [CrossRef] [PubMed]
75. Akbashev, A.R. Electrocatalysis Goes Nuts. *ACS Catal.* **2022**, *12*, 4296–4301. [CrossRef]
76. Buriak, J.M.; Korgel, B. The Experimental Section: The Key to Longevity of Your Research. *Chem. Mater.* **2014**, *26*, 1765–1766. [CrossRef]
77. Siegmund, D.; Metz, S.; Peinecke, V.; Warner, T.E.; Cremers, C.; Grevé, A.; Smolinka, T.; Segets, D.; Apfel, U.-P. Crossing the Valley of Death: From Fundamental to Applied Research in Electrolysis. *JACS Au* **2021**, *1*, 527–535. [CrossRef]
78. Wang, Q.; Guesmi, H.; Tingry, S.; Cornu, D.; Holade, Y.; Minteer, S.D. Unveiling the Pitfalls of Comparing Oxygen Reduction Reaction Kinetic Data for Pd-Based Electrocatalysts without the Experimental Conditions of the Current–Potential Curves. *ACS Energy Lett.* **2022**, *7*, 952–957. [CrossRef]
79. Moafi, H.F.; Fallah Shojaie, A.; Ali Zanjanchi, M. Photoactive polyacrylonitrile fibers coated by nano-sized titanium dioxide: Synthesis, characterization, thermal investigation. *J. Chil. Chem. Soc.* **2011**, *56*, 610–615. [CrossRef]

80. Pels, J.R.; Kapteijn, F.; Moulijn, J.A.; Zhu, Q.; Thomas, K.M. Evolution of nitrogen functionalities in carbonaceous materials during pyrolysis. *Carbon* **1995**, *33*, 1641–1653. [CrossRef]
81. Both Engel, A.; Cherifi, A.; Tingry, S.; Cornu, D.; Peigney, A.; Laurent, C. Enhanced performance of electrospun carbon fibers modified with carbon nanotubes: Promising electrodes for enzymatic biofuel cells. *Nanotechnology* **2013**, *24*, 245402. [CrossRef]
82. Zhou, Z.; Lai, C.; Zhang, L.; Qian, Y.; Hou, H.; Reneker, D.H.; Fong, H. Development of carbon nanofibers from aligned electrospun polyacrylonitrile nanofiber bundles and characterization of their microstructural, electrical, and mechanical properties. *Polymer* **2009**, *50*, 2999–3006. [CrossRef]
83. Zhao, G.; Liu, G. Electrochemical Deposition of Gold Nanoparticles on Reduced Graphene Oxide by Fast Scan Cyclic Voltammetry for the Sensitive Determination of As(III). *Nanomaterials* **2019**, *9*, 41. [CrossRef]
84. Miao, J.; Teng, X.; Zhang, R.; Guo, P.; Chen, Y.; Zhou, X.; Wang, H.; Sun, X.; Zhang, L. "Carbohydrate-Universal" electrolyzer for energy-saving hydrogen production with Co_3FePx@NF as bifunctional electrocatalysts. *Appl. Catal. B Environ.* **2020**, *263*, 118109. [CrossRef]
85. Rafaïdeen, T.; Baranton, S.; Coutanceau, C. Highly efficient and selective electrooxidation of glucose and xylose in alkaline medium at carbon supported alloyed PdAu nanocatalysts. *Appl. Catal. B Environ.* **2019**, *243*, 641–656. [CrossRef]
86. Caravaca, A.; Sapountzi, F.M.; de Lucas-Consuegra, A.; Molina-Mora, C.; Dorado, F.; Valverde, J.L. Electrochemical reforming of ethanol–water solutions for pure H2 production in a PEM electrolysis cell. *Int. J. Hydrogen Energy* **2012**, *37*, 9504–9513. [CrossRef]
87. You, B.; Liu, X.; Liu, X.; Sun, Y.J. Efficient H-2 Evolution Coupled with Oxidative Refining of Alcohols via A Hierarchically Porous Nickel Bifunctional Electrocatalyst. *ACS Catal.* **2017**, *7*, 4564–4570. [CrossRef]
88. Zhao, H.; Lu, D.; Wang, J.; Tu, W.; Wu, D.; Koh, S.W.; Gao, P.; Xu, Z.J.; Deng, S.; Zhou, Y.; et al. Raw biomass electroreforming coupled to green hydrogen generation. *Nat. Commun.* **2021**, *12*, 2008. [CrossRef]
89. Lepage, T.; Kammoun, M.; Schmetz, Q.; Richel, A. Biomass-to-hydrogen: A review of main routes production, processes evaluation and techno-economical assessment. *Biomass Bioenergy* **2021**, *144*, 105920. [CrossRef]
90. Khan, M.A.; Al-Attas, T.A.; Yasri, N.G.; Zhao, H.; Larter, S.; Hu, J.; Kibria, M.G. Techno-economic analysis of a solar-powered biomass electrolysis pathway for coproduction of hydrogen and value-added chemicals. *Sustain. Energy Fuels* **2020**, *4*, 5568–5577. [CrossRef]
91. Verma, S.; Lu, S.; Kenis, P.J.A. Co-electrolysis of CO_2 and glycerol as a pathway to carbon chemicals with improved technoeconomics due to low electricity consumption. *Nat. Energy* **2019**, *4*, 466–474. [CrossRef]
92. Holade, Y.; Servat, K.; Tingry, S.; Napporn, T.W.; Remita, H.; Cornu, D.; Kokoh, K.B. Advances in Electrocatalysis for Energy Conversion and Synthesis of Organic Molecules. *ChemPhysChem* **2017**, *18*, 2573–2605. [CrossRef]
93. Houache, M.S.E.; Safari, R.; Nwabara, U.O.; Rafaïdeen, T.; Botton, G.A.; Kenis, P.J.A.; Baranton, S.; Coutanceau, C.; Baranova, E.A. Selective Electrooxidation of Glycerol to Formic Acid over Carbon Supported Ni1–xMx (M = Bi, Pd, and Au) Nanocatalysts and Coelectrolysis of CO_2. *ACS Appl. Energy Mater.* **2020**, *3*, 8725–8738. [CrossRef]
94. Bardow, A.; Wessling, M. Converting two wastes to value. *Nat. Energy* **2019**, *4*, 440–441. [CrossRef]
95. Boisen, A.; Christensen, T.B.; Fu, W.; Gorbanev, Y.Y.; Hansen, T.S.; Jensen, J.S.; Klitgaard, S.K.; Pedersen, S.; Riisager, A.; Ståhlberg, T.; et al. Process integration for the conversion of glucose to 2,5-furandicarboxylic acid. *Chem. Eng. Res. Des.* **2009**, *87*, 1318–1327. [CrossRef]
96. Nocito, F.; Ventura, M.; Aresta, M.; Dibenedetto, A. Selective Oxidation of 5-(Hydroxymethyl)furfural to DFF Using Water as Solvent and Oxygen as Oxidant with Earth-Crust-Abundant Mixed Oxides. *ACS Omega* **2018**, *3*, 18724–18729. [CrossRef] [PubMed]
97. Zhang, C.; Chang, X.; Zhu, L.; Xing, Q.; You, S.; Qi, W.; Su, R.; He, Z. Highly efficient and selective production of FFCA from CotA-TJ102 laccase-catalyzed oxidation of 5-HMF. *Int. J. Biol. Macromol.* **2019**, *128*, 132–139. [CrossRef]
98. Chen, C.; Wang, L.; Zhu, B.; Zhou, Z.; El-Hout, S.I.; Yang, J.; Zhang, J. 2,5-Furandicarboxylic acid production via catalytic oxidation of 5-hydroxymethylfurfural: Catalysts, processes and reaction mechanism. *J. Energy Chem.* **2021**, *54*, 528–554. [CrossRef]
99. Kang, M.J.; Yu, H.J.; Kim, H.S.; Cha, H.G. Deep eutectic solvent stabilised Co–P films for electrocatalytic oxidation of 5-hydroxymethylfurfural into 2,5-furandicarboxylic acid. *New J. Chem.* **2020**, *44*, 14239–14245. [CrossRef]
100. Chadderdon, X.H.; Chadderdon, D.J.; Pfennig, T.; Shanks, B.H.; Li, W. Paired electrocatalytic hydrogenation and oxidation of 5-(hydroxymethyl)furfural for efficient production of biomass-derived monomers. *Green Chem.* **2019**, *21*, 6210–6219. [CrossRef]
101. Villa, A.; Schiavoni, M.; Campisi, S.; Veith, G.M.; Prati, L. Pd-modified Au on Carbon as an Effective and Durable Catalyst for the Direct Oxidation of HMF to 2,5-Furandicarboxylic Acid. *ChemSusChem* **2013**, *6*, 609–612. [CrossRef]
102. Minteer, S.D.; Baran, P. Electrifying Synthesis: Recent Advances in the Methods, Materials, and Techniques for Organic Electrosynthesis. *Acc. Chem. Res.* **2020**, *53*, 545–546. [CrossRef]
103. Dey, A.; Gunnoe, T.B.; Stamenkovic, V.R. Organic Electrosynthesis: When Is It Electrocatalysis? *ACS Catal.* **2020**, *10*, 13156–13158. [CrossRef]
104. Frontana-Uribe, B.A.; Little, R.D.; Ibanez, J.G.; Palma, A.; Vasquez-Medrano, R. Organic electrosynthesis: A promising green methodology in organic chemistry. *Green Chem.* **2010**, *12*, 2099–2119. [CrossRef]
105. Lai, S.C.S.; Kleijn, S.E.F.; Öztürk, F.T.Z.; van Rees Vellinga, V.C.; Koning, J.; Rodriguez, P.; Koper, M.T.M. Effects of electrolyte pH and composition on the ethanol electro-oxidation reaction. *Catal. Today* **2010**, *154*, 92–104. [CrossRef]
106. Koper, M.T.M. Theory of multiple proton-electron transfer reactions and its implications for electrocatalysis. *Chem. Sci.* **2013**, *4*, 2710–2723. [CrossRef]

107. Joo, J.; Uchida, T.; Cuesta, A.; Koper, M.T.M.; Osawa, M. Importance of Acid–Base Equilibrium in Electrocatalytic Oxidation of Formic Acid on Platinum. *J. Am. Chem. Soc.* **2013**, *135*, 9991–9994. [CrossRef] [PubMed]
108. Kwon, Y.; Lai, S.C.S.; Rodriguez, P.; Koper, M.T.M. Electrocatalytic Oxidation of Alcohols on Gold in Alkaline Media: Base or Gold Catalysis? *J. Am. Chem. Soc.* **2011**, *133*, 6914–6917. [CrossRef]
109. Zope, B.N.; Hibbitts, D.D.; Neurock, M.; Davis, R.J. Reactivity of the Gold/Water Interface during Selective Oxidation Catalysis. *Science* **2010**, *330*, 74–78. [CrossRef]
110. Ketchie, W.C.; Murayama, M.; Davis, R.J. Promotional effect of hydroxyl on the aqueous phase oxidation of carbon monoxide and glycerol over supported Au catalysts. *Top. Catal.* **2007**, *44*, 307–317. [CrossRef]
111. Guidelli, R.; Compton, R.G.; Feliu, J.M.; Gileadi, E.; Lipkowski, J.; Schmickler, W.; Trasatti, S. Defining the transfer coefficient in electrochemistry: An assessment (IUPAC Technical Report). *Pure Appl. Chem.* **2014**, *86*, 245–258. [CrossRef]
112. Seh, Z.W.; Kibsgaard, J.; Dickens, C.F.; Chorkendorff, I.; Nørskov, J.K.; Jaramillo, T.F. Combining theory and experiment in electrocatalysis: Insights into materials design. *Science* **2017**, *355*, eaad4998. [CrossRef]
113. Burdyny, T.; Smith, W.A. CO_2 reduction on gas-diffusion electrodes and why catalytic performance must be assessed at commercially-relevant conditions. *Energy Environ. Sci.* **2019**, *12*, 1442–1453. [CrossRef]

materials

MDPI

Article

Large Electrocaloric Responsivity and Energy Storage Response in the Lead-Free $Ba(Ge_xTi_{1-x})O_3$ Ceramics

Bouchra Asbani [1,2], Yaovi Gagou [1,*], Said Ben Moumen [3], Jean-Luc Dellis [1], Abdelilah Lahmar [1], M'Barek Amjoud [3], Daoud Mezzane [3], Mimoun El Marssi [1], Brigita Rozic [4] and Zdravko Kutnjak [4]

[1] Laboratoire de Physique de La Matière Condensée (LPMC), Université de Picardie, Jules Verne, 33 rue Saint-Leu, CEDEX 1, 80039 Amiens, France; asbani.bouchra@gmail.com (B.A.); dellisjeanluc@gmail.com (J.-L.D.); lahmar2@yahoo.fr (A.L.); mimoun.elmarssi@u-picardie.fr (M.E.M.)

[2] Unit of Dynamics and Structure of Molecular Materials—UDSMM (EA 4476), MREI-1, Université du Littoral Côte d'Opale, 59140 Dunkerque, France

[3] IMED-Lab, Department of Applied Physics, Faculty of Sciences and Techniques, Cadi Ayyad University, P.O. Box 549, Marrakesh 40000, Morocco; said.benmoumen@yahoo.fr (S.B.M.); m.amjoud@uca.ac.ma (M.A.); daoudmezzane@gmail.com (D.M.)

[4] Jozef Stefan Institute, Jamova Cesta 39, 1000 Ljubljana, Slovenia; brigita.rozic@ijs.si (B.R.); zdravko.kutnjak@ijs.si (Z.K.)

* Correspondence: yaovi.gagou@u-picardie.fr

Abstract: Ferroelectric property that induces electrocaloric effect was investigated in $Ba(Ge_xTi_{1-x})O_3$ ceramics, known as BTGx. X-ray diffraction analysis shows pure perovskite phases in tetragonal symmetry compatible with the *P4mm* (No. 99) space group. Dielectric permittivity exhibits first-order ferroelectric-paraelectric phase transition, confirmed by specific heat measurements, similar to that observed in $BaTiO_3$ (BTO) crystal. Curie temperature varies weakly as a function of Ge-content. Using the direct and indirect method, we confirmed that the adiabatic temperature change ΔT reached its higher value of 0.9 K under 8 kV/cm for the composition BTG6, corresponding to an electrocaloric responsivity $\Delta T/\Delta E$ of 1.13×10^{-6} K.m/V. Such electrocaloric responsivity significantly exceeds those obtained so far in other barium titanate-based lead-free electrocaloric ceramic materials. Energy storage investigations show promising results: stored energy density of ~17 mJ/cm^3 and an energy efficiency of ~88% in the composition BTG5. These results classify the studied materials as candidates for cooling devices and energy storage applications.

Keywords: BGT; ceramics; ferroelectric; electrocaloric; energy storage

Citation: Asbani, B.; Gagou, Y.; Ben Moumen, S.; Dellis, J.-L.; Lahmar, A.; Amjoud, M.; Mezzane, D.; El Marssi, M.; Rozic, B.; Kutnjak, Z. Large Electrocaloric Responsivity and Energy Storage Response in the Lead-Free $Ba(Ge_xTi_{1-x})O_3$ Ceramics. *Materials* **2022**, *15*, 5227. https://doi.org/10.3390/ma15155227

Academic Editors: Marc Cretin, Zhenghua Tang and Sophie Tingry

Received: 28 April 2022
Accepted: 22 July 2022
Published: 28 July 2022

1. Introduction

Electrocaloric effect (ECE) is similar to magnetocaloric effect (MCE), in which the temperature change of the material is achieved by the electrical and magnetic field control, respectively [1–3]. Therefore, ECE occurs in switchable dipolar materials, where the application of electric field leads to change in the reversible polarization and consequently in the entropy [4]. Briefly, the applied electric bias induces an adiabatic increase of temperature in the ferroelectric material. The field removal causes similarly adiabatic cooling in a reversible process known as ECE [5–7].

Nowadays, ECE attracts scientists' attention because of its promising frigorific applications and microelectronic cooling devices [8–10]. Various contributions were recently reported on ferroelectric thin films, ceramic and polymers [1,7,11,12].

Prototype refrigerators using ceramic ECE have been experimented with by a few research groups [13–15]. The low ECE in these materials prevent them from being used in efficient calorimetric applications. Therefore, high performance solid-state cooling devices could be realized only using materials with large ECE, for example, in electronic elements cooling of microelectronic devices, sensors, textile, etc.

Moreover, electrocaloric (EC) coolers based on eco-friendly materials stimulate a resurgent focus on the development of high-performance lead-free barium titanate $BaTiO_3$ (BTO)-based materials [12–15] due to the cancerous nature and neurotoxicity of Lead (Pb) [16,17].

BTO is one of the most frequently used ceramic materials in electronic devices due to its outstanding dielectric properties [18]. In particular, remarkable tailoring of functional characteristics can be achieved by substitution with foreign ions on the A, B, or in both sites of the ABO_3 perovskite structure. Isovalent substitution ions such as Zr^{4+}, Sn^{4+} and Ce^{4+} onto the Ti^{4+} site gave rise to enhanced material properties (e.g., dielectric, pyro- and piezoelectric), high thermal stability in a large temperature range [2,4]. Otherwise, substitution can induce also a decrease in polarisation [2], tunability [19] and hysteresis loops [20]. These substitutions can also induce structural-phase transitions or polymorphic phases overlapping for specific compositions, and sometimes a ferroelectric relaxor by increasing the dopant concentration [21]. Substitutions with smaller radius ions such as Ge are far less investigated and apparently do not modify the structural transition temperatures: the room temperature tetragonality and the ferroelectric character [22]. There are only a few reports concerning the role of Ge insertion, which has mainly been used as sintering aids to promote better densification of BTO-based ceramics at a lower sintering temperature [23]. Chu et al. [24] reported that germanates are used in industrial applications to produce heterophasic ceramic bodies. Köferstein et al. [25] have investigated the influence of $BaGeO_3$ (BGO) on the sintering behaviour and properties of fine- and coarse-grained BTO powder density. GeO_2 can be used as a sintering aid to reduce the sintering temperature of BTO ceramics below 1000 °C. Guha and Kolar [26] studied the $BaTiO_3-BaGeO_3$ (BTO-BGO) system, and they determined a eutectic composition of 68 mol% $BaGeO_3$ with a melting temperature of approximately 1120 ± 5 °C. The authors did not observe any shifting of the cubic to the tetragonal phase transition temperature in BTO by the addition of GeO_2, in agreement with the investigations by Plessner and West [27]. These authors noticed a reduction in the sharpness of the permittivity maximum. In contrast, Pulvari [28] and Baxter [29] found a slight decrease in the Curie temperature with the addition of GeO_2, and determined the limit of solubility of some doping elements in BTO. Consequently, the knowledge about the effect of sintering additive materials on the dielectric properties of BTO-based ceramics is essential for potential technological applications.

Therefore, ECE is evidenced in BTO-doped ferroelectric material using indirect or direct methods [30]. The indirect method used in our case is based on thermal P-E hysteresis loop data analysis using Maxwell relations [31]. In addition, the results were confirmed by direct method using a modified high-resolution calorimeter [32].

The literature indicated that the EC responsivity $\Delta T/\Delta E$ is the most suitable coefficient to highlight the electrocaloric effect [33–35]. This is supported by the aim of producing high ECE under optimal applied electric field. The electrocaloric effect in $Na_{0.5}Bi_{0.5}TiO_3$–$BaTiO_3$ (NBT-BTO) [30] and BTO bulk ceramics had recently been investigated by Bai et al. [7], who reported a maximum electrocaloric responsivity of 8×10^{-7} K.m/V. Wang et al. [36] obtained a $\Delta T/\Delta E$ value of 1.5×10^{-7} K.m/V in the $Ba_{0.98}Ca_{0.02}(Zr_{0.085}Ti_{0.915})O_3$ (BCZT) ceramics. Liu et al. [37] reported that ECE responsivity in $Ba_{0.65}Sr_{0.35}TiO_3$ (SBT) ceramics prepared by spark plasma sintering was enhanced up to 2.33×10^{-7} K.m/V. In BTO single crystal, Novak et al. [6] evidenced a $\Delta T/\Delta E$ value of 5×10^{-7} K.m/V by high-resolution EC measurements in the narrowed region of transition. We reported previously a large ECE in a much broader temperature range extended in 50 K a maximum $\Delta T/\Delta E$ of 3.4×10^{-7} K.m/V in $Ba_{0.8}Ca_{0.2}TiO_3$ Zr-doped, lead-free ferroelectric ceramics at a relatively small electric field of 7.95 kV/cm [34]. Previous results demonstrate a significant challenge to achieving high EC adiabatic temperature changes in the lead-free bulk materials with the optimal applied electric field and the best doping cation for higher EC responsivity. It is worth noting that the optimal electric field minimises material fatigue and Joule heating effect. Operating at low voltage could also be a suitable way to reduce the technology cost and ensure low energy consumption.

In the present paper, we focus our investigation on the ECE and energy storage performance in BTGx materials (x = 0.02, 0.03, 0.05, 0.06 and 0.09). We study the influence of germanium (Ge) doping in BT matrix on the structural, electrical, ferroelectric, electrocaloric and stored energy properties in this system. It should be noted that ECE results obtained using indirect methods were found to be in good agreement with data directly measured using a high-resolution calorimeter, and the energy stored shows promising results.

2. Experimental Section

BTGx ceramics were elaborated using a conventional powder processing technique and starting from high purity raw materials $BaCO_3$ (99%), TiO_2 (99.8%) and GeO_2 (99.8%), which were mixed in the desired stoichiometry. The mixture powder was grounded in ethanol medium for an hour in agate mortar, followed by a thermal treatment. The obtained powder of each Ge-content was calcined at 900 °C for 4 h. Heating rate was 5 °C/min. This thermal process had been repeated three times for grain size reduction and for accurate grain reactions to achieve a homogenous state with single-phase powder.

The symmetry phases were confirmed by X-ray powder diffraction (XRD) data analysis which was recorded with detector step of 0.0199° and waiting time of 10 s using Brucker D8 with λ_{Cu} = 1.5406 Å in θ-2θ Bragg–Bretano geometry configuration. The structural resolution was carried out using the Rietveld method implemented in the FullProf software [38]. The finely granulated powder was compacted under a hydraulic press at 250 MPa pressure to obtain circular pellet discs of 6 mm diameter and 0.4 mm thickness. The obtained pellets were then placed into an alumina crucible and sintered at 1100 °C for 1 h. The obtained sintered ceramic samples were crack-free, and their density was found in the range 5.03 to 5.80 g/cm^3. These densities measured based on Archimedes' method, to represent 93 to 96% of their theoretical value. Silver paste was used to form electrodes covering both faces of the pellets to form a plane capacitor shape for electrical measurements. Measurements can be started after 30 min heating of electrodes at 300 °C constant temperature. Then, P-E hysteresis loops were registered as a function of temperature using AixACCT TF1000 apparatus, and dielectric permittivity was measured for several frequencies (100 Hz–1 MHz) as a function of temperature using Solartron SI 1200 Impedancemeter. Temperature controller is Linkam THMS600 with heating and cooling rate of 5 °C/min. Temperature module offered 0.01 °C in accuracy. All the thermal measurements have been performed under air. The Impedance measurements were performed every 2.00 °C $\pm$ 0.01 °C, under oscillator source of 50 mV applied on ~0.5 mm sample thickness. Sample heat capacity was deduced from heat flow measurements performed using Netzsch DSC 204F1 apparatus. SEM images were recorded under 5 kV electrons acceleration source and a working distance in the range 10–15 mm. The Energy Dispersive X-ray (EDX) analysis had been performed using the FEI Quanta 200F apparatus.

3. Structural Studies

3.1. X-ray Studies

X-ray diffraction patterns of BTGx ceramics (x = 0.02, 0.03, 0.05, 0.06, 0.09) performed at room temperature were shown in Figure 1. Small displacement of the rays to high angles can be observed in this figure versus Ge-content, which results in the contraction of lattice parameters. No significant structural change was observed at first look based on the diffraction lines. This means Ge-insertion in BTO matrix induces weak distortion in the crystal lattice parameters, and the tetragonal symmetry was observed for all the compositions. A zoom conducted on the tetragonal symmetry lines (202) and (220) showed a weak displacement of ~0.5°, as presented in the inset of Figure 1. The structural resolution was then conducted for all the elaborated compositions satisfactorily. Beginning from profiles adjustment, the calculations led to rapid convergence. By way of the Rietveld method using FullProf software, the atomic positions, thermal isotropic agitation factors, occupation rates and scale factor were adjusted at room temperature to minimize the reliability factors in a coherent way, based on BTO tetragonal matrix in accordance with the *P4mm* (No. 99) space group and JCPDS N° 05-0626. We observed globally a small decrease of lattice parameters leading to a small decrease in lattice volume. This behaviour can be expected, since the ionic radius [39] of Ge^{4+} (0.530 Å) in substitution of Ti^{4+} ion in the octahedral site is slightly smaller than Ti^{4+} ionic radius (0.605 Å), which could lead to a decrease of volume if the substitution occurred. It is worth recalling that a systematic microstructural analysis and the phase equilibrium diagram for the system $BaTiO_3$-$BaGeO_3$ (BTO-BGO) was reported by Guha et al. [26], in which the solubility limit of BGO in BTO was determined around 1.8 mol% but no electrical characterization was performed. In the present case, structural resolution leads to several observations: (i) small octahedra distortion is observed, (ii) volume decreases weakly, (iii) Ge doping atoms in the octahedral site have weak concentration and consequently do not impact greatly the structural symmetry.

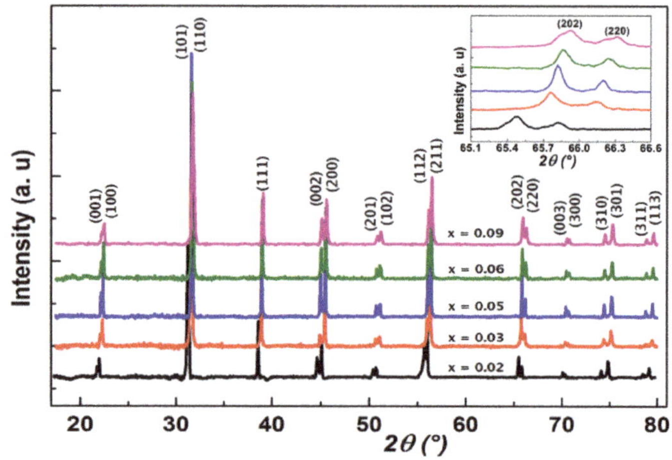

Figure 1. Room temperature X-ray patterns of elaborated compound BTGx (x = 0.02, 0.03, 0.05, 0.06, 0.09). The inset indicates the zoom of evolution of (202) and (220) lines versus Ge-content.

The refined lattice parameters (a = b and c), the unit cell volume, atomic positions and the reliability factors are summarized in Table 1, and the curves of evolution of lattice parameter and volume versus Ge-content are presented in Figure 2. Moreover, considering the ionic radius of chemical elements and Ge-content, we calculated the Goldschmidt tolerance factor to characterize the perovskite structural stability. The obtained values gathered in Table 1, demonstrate progressive stabilization to pseudo-cubic perovskite structure while Ge-content increases.

3.2. Raman Spectroscopy

To confirm the structural analysis, we performed Raman spectroscopy measurements at room temperature. The spectra are plotted in Figure 3 and show the BTO known active modes in its tetragonal symmetry (C_{4v}) for all the compositions as reported in the literature [40–43]. The active modes are covered in the frequency range 100–800 cm^{-1} [41,44]. Three E(TO) modes appear around 167, 304 and 518 cm^{-1}, which were often observed at 170, 306 and 520 cm^{-1} in pure BTO ceramics [44]. Moreover, the modes 260, 471 and the somewhat broader 720 cm^{-1} constitute the A_1 modes in this structure. An extra A_1 mode appears clearly at 800 cm^{-1} for composition x = 0.09. This mode is sensitive to Ge-rate, and very close observation indicates that it occurs earlier at lower concentration with weak intensity and evolves to be observable at x = 0.09. It was previously reported [45] that this mode is sensitive only to B-site occupation in perovskite matrix and move to high frequency as a function of BTO-doping rate and was affected to the A1(TO) mode. This confirms that Ge is inserted in the B-site of the perovskite structure [45]. Therefore, we can conclude that no significant displacement or variation of the modes is observed with Ge-content until x = 0.09, where the A1(TO) mode appears. This can be explained by the weak Ge-doping, which corroborates the X-ray analysis.

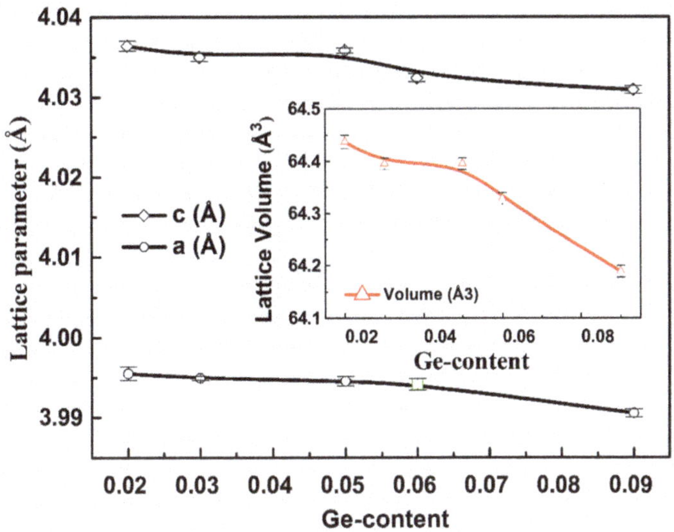

Figure 2. Lattice parameters (a and c) and volume variation of BTGx (x = 0.02, 0.03, 0.05, 0.06, 0.09) versus Ge-content. These parameters were obtained from Rietveld structural resolution using FullProf software.

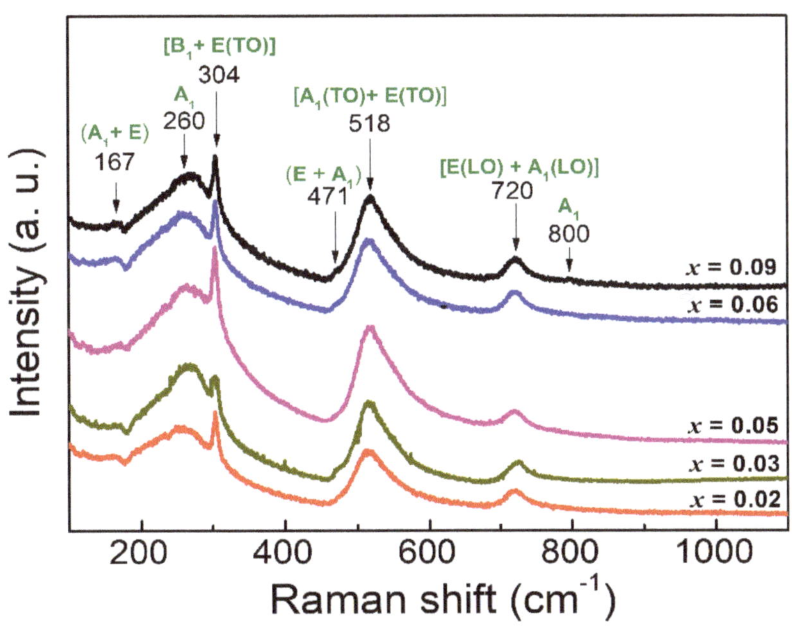

Figure 3. Room temperature Raman spectra for the studied BTGx (x = 0.02, 0.03, 0.05, 0.06, 0.09) compositions. The A_1(TO) mode at 800 cm^{-1} confirms Ge-insertion in the octahedral site in the perovskite.

3.3. Microstructure Analysis

We present in Figure 4 the SEM micrographs of the five elaborated BTGx (x = 0.02, 0.03, 0.05, 0.06, 0.09) ceramics. These images show a significant variation in grain size depending on the composition. We can observe the increase of grain size with Ge-content, except from the particular grain size obtained for the composition BTG6. The grains seem to bathe in a lacquer, assuming the relative compacity of the structure with Ge-content. This behaviour can explain the effect of Ge-content around the limit of solubility of BGO in BTO matrix. The grain bonding defects disappear at the interfaces, leading to acceptable density of the ceramics with Ge-content. Moreover, the density of the ceramics was calculated based on the Archimedes method, and the values range between 93–96% of their theoretical values, as reported in Table 1. A typical morphology was observed in the BTG6 ceramic, which also exhibits particular behaviour in X-ray analysis and other results. Chemical analysis based on Energy Dispersive X-Ray (EDX) analysis confirmed the compositions, as we can observe at bottom right part of Figure 4 for the composition BTG3. Experimental relative weight values are close to theoretical ones, which are: 58, 19, 20, and 0.9% mass percentage, for the chemical element Ba, Ti, O and Ge, respectively (see inset of Figure 4f). Furthermore, the density values range in 5.035 to 5.801 g/cm^3 and the grain sizes determined using ImageJ software range between 3.6 to 18.7 μm. The values for each composition are depicted in Table 1.

Table 1. Atomic positions and Rietveld reliability factors for BTGx (x = 0.02, 0.03, 0.05, 0.06, 0.09) compositions.

Atoms	*Atomic Positions*	*Compositions*				
		0.02	*0.03*	*0.05*	*0.06*	*0.09*
Ba	*x*	0	0	0	0	0
	y	0	0	0	0	0
	z	−0.0422 (4)	0.1827 (5)	−0.0293 (8)	0.3342 (4)	0.1724 (2)
	Occ.	1	1	1	1	1
	B_{iso}	1.664	0.795	0.586	1.074	1.581
Ti	*x*	0.5	0.5	0.5	0.5	0.5
	y	0.5	0.5	0.5	0.5	0.5
	z	0.4798 (5)	0.6677 (3)	0.4874 (5)	0.8681 (3)	0.7117 (2)
	Occ.	0.98	0.97	0.95	0.94	0.91
	B_{iso}	2.1	1.272	1.218	1.433	1.874
Ge	*x*	0.5	0.5	0.5	0.5	0.5
	y	0.5	0.5	0.5	0.5	0.5
	z	0.4798 (5)	0.6677 (3)	0.4874 (5)	0.8681 (3)	0.7117 (2)
	Occ.	0.02	0.03	0.05	0.06	0.09
	B_{iso}	2.1	1.272	1.218	1.433	1.874
O1	*x*	0.5	0.5	0.5	0.5	0.5
	y	0.5	0.5	0.5	0.5	0.5
	z	−0.0481 (3)	0.1413 (8)	−0.0068 (2)	0.2753 (7)	0.0853 (1)
	Occ.	1	1	0.9421	1	1
	B_{iso}	3.331	2.675	1	2.018	3.798

Table 1. *Cont.*

Atoms	Atomic Positions	Compositions				
		0.02	*0.03*	*0.05*	*0.06*	*0.09*
O2	*x*	0.5	0.5	0.5	0.5	0.5
	y	0	0	0	0	0
	z	0.5064 (2)	0.7333 (5)	0.5447 (6)	0.7720 (2)	0.7306 (7)
	Occ.	1	1	0.98	1	1
	B_{iso}	1.974	0.222	2.982	0.438	3.034
Symmetry Group		*P4mm*	*P4mm*	*P4mm*	*P4mm*	*P4mm*
Unit cell parameters	*a*(Å)	3.9955 (9)	3.9949 (3)	3.9945 (6)	3.9941 (7)	3.9905 (5)
	c(Å)	4.0364 (6)	4.0350 (5)	4.0358 (3)	4.0324 (5)	4.0309 (5)
Unit cell Volume V ($Å^3$)		64.4371 (7)	64.3954 (8)	64.3953 (5)	64.3282 (1)	64.1884 (2)
Nb Refined param.		45	33	37	38	34
Grain size (μm)		3.63	5.45	10.90	8.32	18.77
Theoretical Density (g/cm^3)		6.067	6.052	6.045	6.041	6.049
Calculated Density (g/cm^3)		5.801	5.785	5.645	5.035	5.648
Relative Density (%)		96	95	93	83	93
χ^2		2.85	2.1	2.14	2.5	3.72
Rwp		5.57	3.84	3.07	3.18	4.62
R_P		4.23	2.71	2.21	2.14	3.19
R_{Bragg}		4.64	6.54	5.26	3.37	3.5
$Ti\text{-}O_1$ *(c-axis)*		1.8278 (7)	1.7198 (8)	1.9947 (7)	2.0302 (3)	1.8998 (6)
$Ti\text{-}O_2$ *(c-axis)*		2.2086 (6)	2.3151 (0)	2.0410 (2)	2.0023 (4)	2.1311 (3)
Goldschmidt factor		0.9929	0.9932	0.9940	0.9944	0.9956
Curie C ($\times 10^5$ *K*)		3.78	3.42	3.87	3.40	3.44
T_0 *(K)* $\pm$ *1.00 K*		372.11	353.79	359.13	348.63	354.40
T_C *(K)* $\pm$ *0.12 K*		400.83	399.40	400.10	399.73	399.53

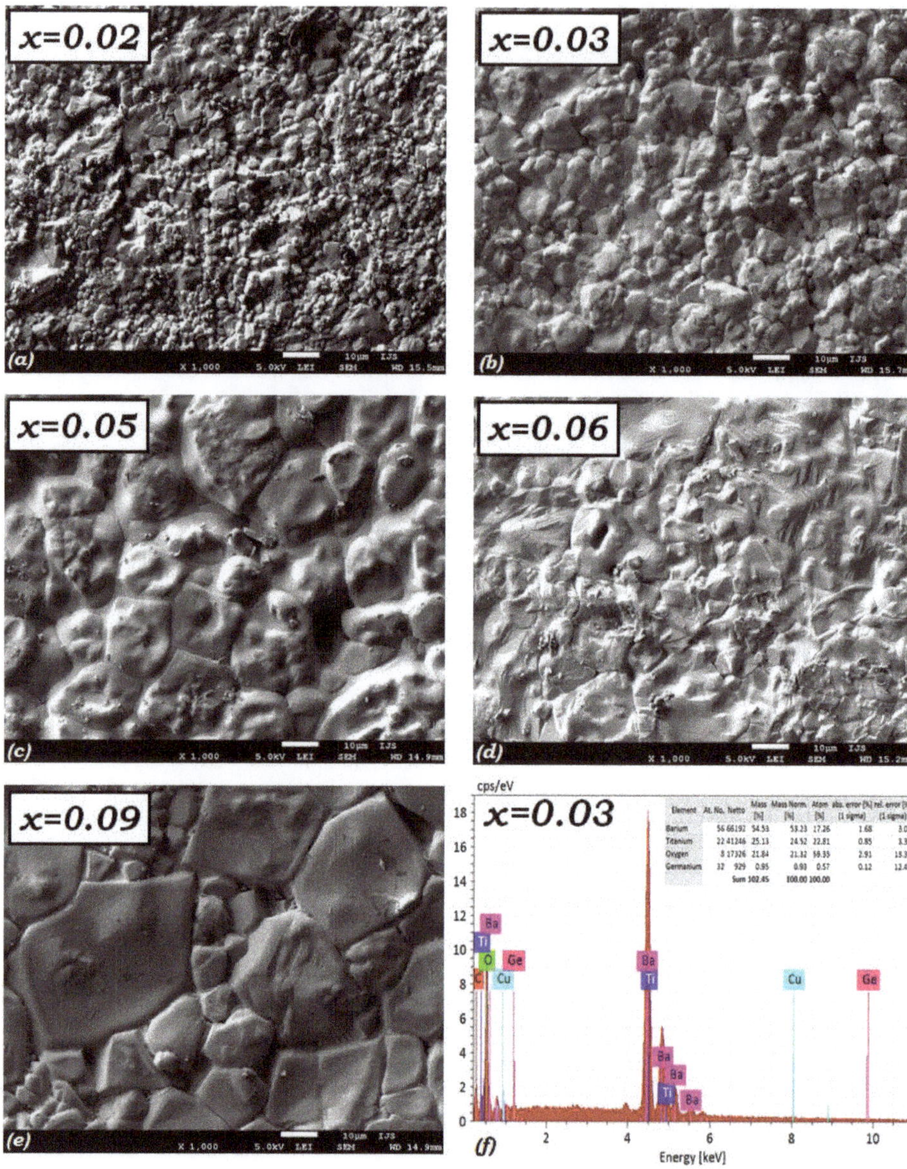

Figure 4. (**a–e**) SEM micrographs of BTGx (x = 0.02, 0.03, 0.05, 0.06, 0.09) ceramics and (**f**) EDX analysis of the composition BTG3 showing mass percentage of containing elements in agreement with the theoretical values.

4. Dielectric Measurements

Figure 5 displays the temperature dependence of the dielectric permittivity measured at different frequencies for all the ceramics. The Ferroelectric-to-Paraelectric (FE-PE) structural phase transition is marked by the dielectric permittivity jump at Curie temperature (T_C). The symmetry changes from tetragonal *P4mm* (No. 99) to cubic *Pm-3m* (No. 221) space groups in the paraelectric phase. The real part of relative dielectric permittivity of the BGTx ceramics reaches at T_C a high value of 10,140 for the composition BTG2 and then decreases to 5918 for BTG9, with a very weak variation of the Curie temperature between the studied compounds. Moreover, no relaxor effect was observed in this system, which can be considered as classical ferroelectric material. Horchidan et al. [21] reported that for small Ge additions to BTO $x \leq 0.10$, for which the perovskite tetragonal phase is predominant, the dielectric properties are quite similar to ones of BTO ceramics, with all the structural phase transitions in the same temperature range and a small shift of the Curie temperature to higher with Ge-content. This seems to be in contradiction with the present results. T_C decreases weakly when Ge-content increases. Furthermore, the structural phase transition is of first-order type, as evidenced by the drastic dielectric permittivity jump and the thermal hysteresis observed for all the compositions. Curie temperature remains almost constant in-between temperature variation $\Delta T_C = 1$ K for these studied Ge-contents. Plessner et al. [27] reported a similar result on electrical measurement, showing no significant variation of Curie temperature with Ge-rate. We confirm this result by the specific heat variation depicted from DSC Signal measurements that evidenced, for all the compositions, an asymmetric peak versus temperature, in favour of first order type phase transition as shown later in this work. Although the Curie temperature varies almost weakly, for all the samples we observed variation of the Curie constant, reflecting the mechanism of the different dynamics of phase transitions in these samples, especially in BTG2 and BTG9. The Curie–Weiss temperature T_0 in this system presents its low value for the composition BTG6, as indicated in Table 1.

We also report in Table 1 the Curie constant values and the Curie temperatures obtained from dielectric permittivity measurements. The higher gap value (~51 K) between T_C and T_0 is observed for the composition BTG6. This composition constitutes that in which metastable transformation exists in a large range temperature, one of the reasons for the better electrocaloric adiabatic temperature variation, but not beneficial to energy storage.

Samples globally presented a real permittivity thermal hysteresis of about 2 K, as showed in Figure 6 for the composition BTG3 recorded at 1 kHz. The inset of this figure also highlights a global weak loss factor less than 4.1%.

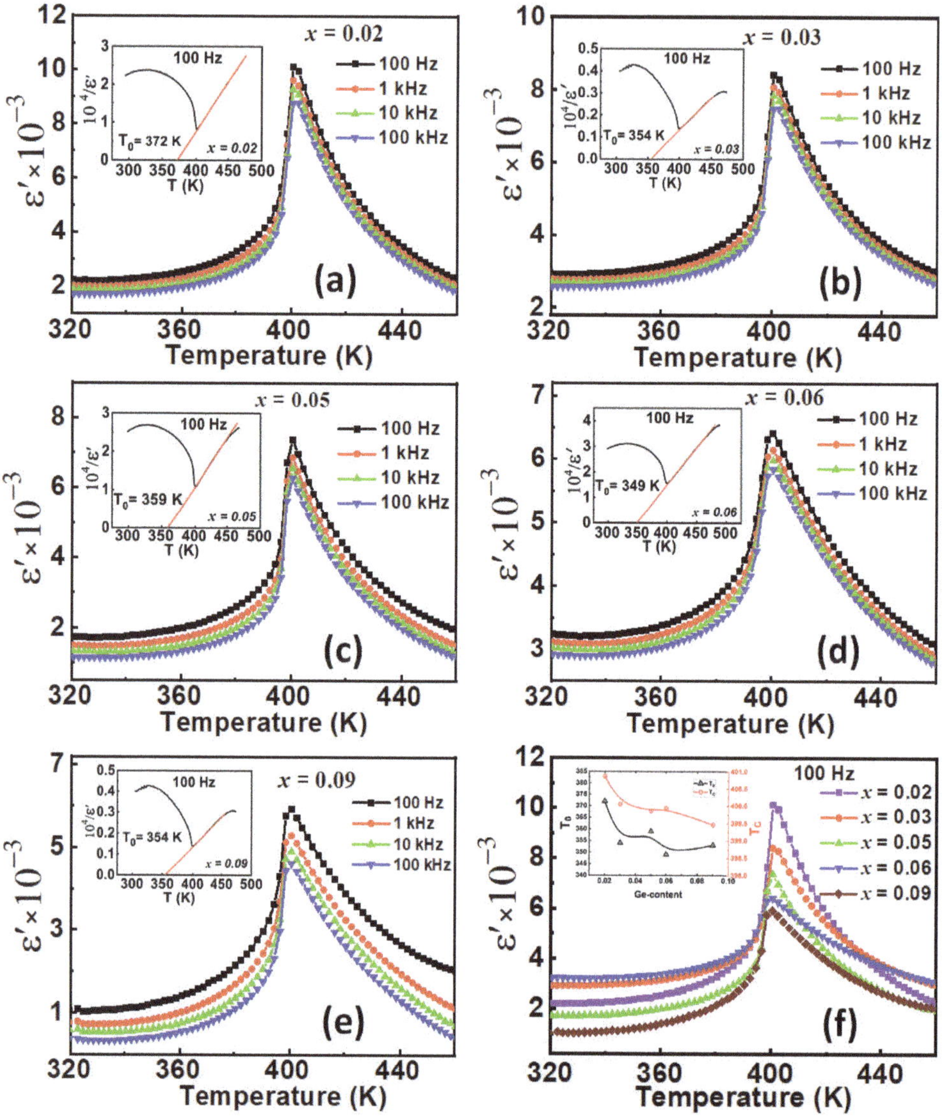

Figure 5. (**a**–**e**) Temperature dependence of dielectric permittivity in BGTx (x = 0.02, 0.03, 0.05, 0.06, 0.09) ceramics in the frequency range 10^2–10^5 Hz. Insets represent the Curie–Weiss plot of each composition. (**f**) Dielectric permittivity at 100 Hz for all the BTGx compositions. The inset shows T_C and T_0 variation as a function of Ge-content.

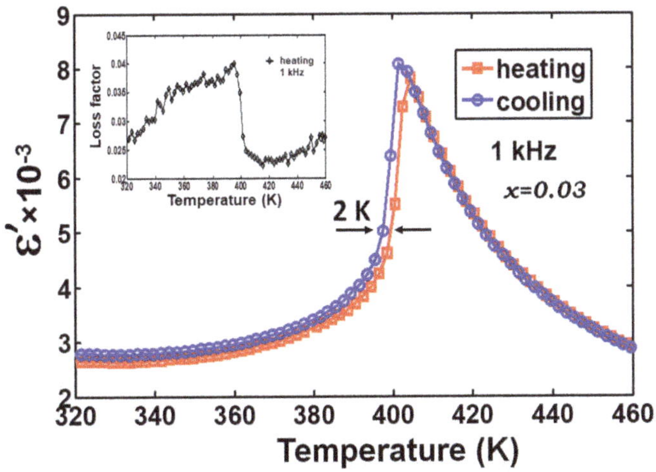

Figure 6. Thermal hysteresis loop of real permittivity for BTG3 compound, recorded at 1 kHz, showing a discard of ΔT_C = 2.00 K between heat and cool data. Inset the dielectric shows Loss factor.

5. Ferroelectricity and Electrocaloric Effect

5.1. Ferroelectric Properties

P-E hysteresis loops were recorded on cooling in the temperature range from 473 to 273 K to minimize the polarization inaccuracy induced by fatigue during heating. Figure 7 presents the P-E hysteresis loops variation recorded at 5 Hz in the BTGx samples as a function of temperature. These curves show a ferroelectric character for $T < T_C$, confirmed by the non-linear behaviour, and then evolves towards a paraelectric phase for $T > T_C$, characterized by the linear curve. In the insets of Figure 7, we present the polarization variation versus temperature under three selected applied electric fields. We remark on the decrease of polarization as a function of temperature followed by an abrupt drop at Tc, for all the compositions (Figure 7a–e), which was particularly important for the compound BTG6. A particularly rapid jump was observed at T_C for the composition BTG6, and therefore, the highest value of the pyroelectric coefficient *dP/dT*, favouring a significant electrocaloric effect, was observed for this composition.

In Figure 7f, we present comparative P-E hysteresis loops for all the compositions at room temperature (T = 303 K). The lowest coercive field Ec value is observed for BTG6. The inset of Figure 7f shows the evolution of remnant polarization versus Ge-content that exhibits a maximum at the composition BTG3, before decreasing to a low value when Ge-content increases. This behaviour is attributed to dipole reordering and domain wall motion mechanism versus Ge-content.

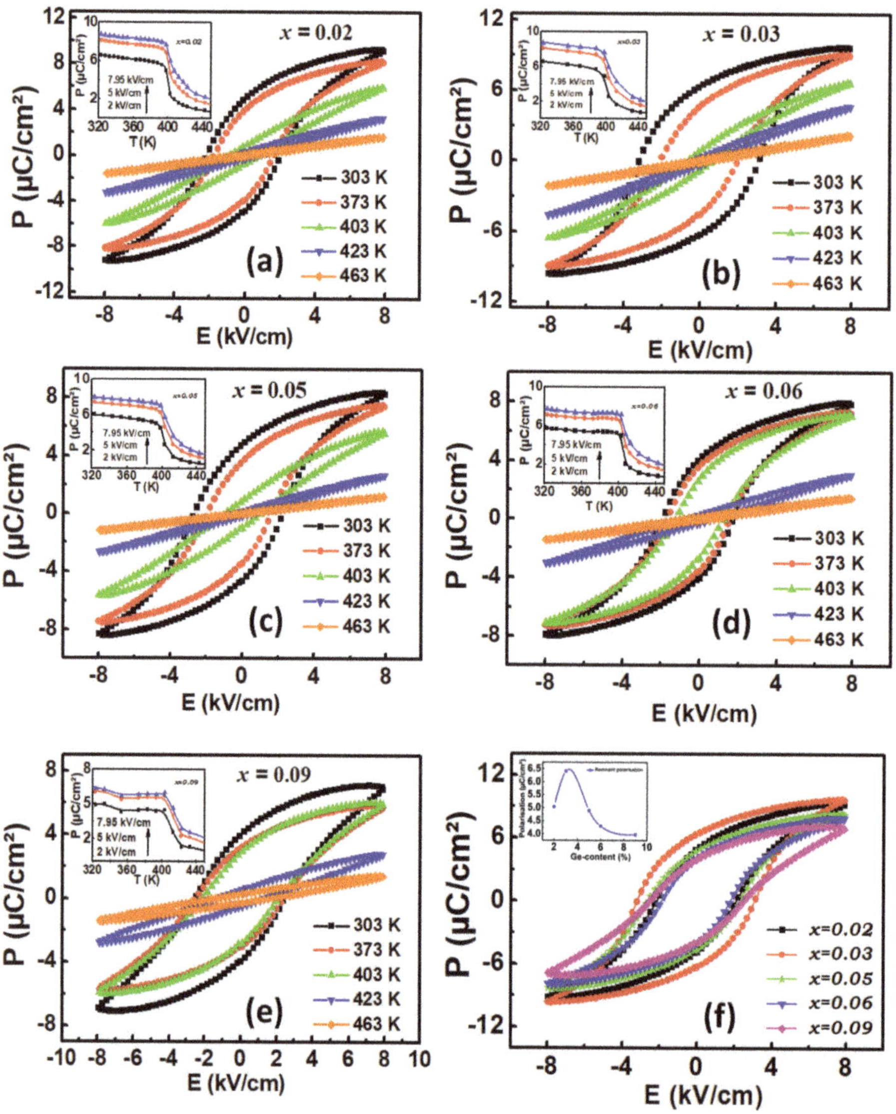

Figure 7. (**a–e**) P-E hysteresis loops of BTGx ceramics (x = 0.02, 0.03, 0.05, 0.06, 0.09) as a function of temperature and extracted maximum polarization (insets) variation versus temperature at three applied fields. (**f**) Comparative hysteresis loops at T = 303 K and remnant polarization (inset) versus Ge-content.

5.2. Indirect Electrocaloric Effect

The electrocaloric effect was then investigated by an indirect method deduced from P-E hysteresis measurements. We extracted the upper branches from the temperature-dependent P-E hysteresis loops to calculate the variation of polarization versus temperature $P(T)$. The pyroelectric coefficient $\partial P/\partial T$ is then calculated from fourth-order polynomial fits of the $P(T)$ data, and the adiabatic electrocaloric temperature change (ΔT) was deduced from this analysis according to the equation:

$$\Delta T = -\frac{1}{\rho}\int_{E_1}^{E_2} \frac{T}{c_p}\left(\frac{\partial P}{\partial T}\right)_E dE \quad (1)$$

where ρ is the density of each studied material, E_1 and E_2 are the starting and the final applied electric fields, respectively, and *Cp* is the specific heat capacity of each studied material. Figure 8a–e show the electrocaloric adiabatic temperature change as a function of temperature for all studied Ge-doped compounds at three selected applied electric fields: 2.00, 5.00 and 7.95 kV/cm. The absolute maximum of each obtained ΔT curve occurs at FE-PE phase transition temperature. The insets show heat capacity *Cp* deduced from heat flow measurement versus temperature, which were adjusted to their background polynomial fits. Note that electrocaloric effect depends mainly on excess of specific heat at phase transition. The higher value of ΔT was evidenced for the composition BTG6, which reached the value of 0.8 K. This high value is expected at this particular composition, since a drastic drop was observed in the polarisation at T_C and also due to its particular crystallinity. A broad anomaly in ΔT was observed for BTG9, as shown in Figure 8e, attributed to the limit of Ge substituting in the BTO matrix that induced conductivity which appears in the less saturated P-E hysteresis or the broad specific heat curve in the inset of Figure 8e. The evolution of maximal variation of electrocaloric responsivity as a function of Ge-content is plotted on Figure 8f.

In the BGTx system, the calculation of electrocaloric responsivity leads to a high value of $\Delta T/\Delta E = 1.01$ K.m/V at 400 K in BTG6. This electrocaloric responsivity was obtained just under an applied electric field value of 7.95 kV/cm. To our knowledge, this value is one of the higher values of electrocaloric responsivity reported in lead-free barium-based oxides, making the BTGx system a candidate for refrigeration devices.

5.3. Direct Electrocaloric Measurement

Direct electrocaloric measurements have been performed using a modified high-resolution calorimeter on the composition BGT6, which exhibits the highest ECE response in the case of indirect method. As presented in Figure 9, the adiabatic temperature variation in this compound reaches $\Delta T = 0.9$ K under an applied electric field of 8 kV/cm. This value is in an excellent agreement to that obtained by indirect calculation of the adiabatic temperature variation in this compound, whose value was 0.8 K under 7.95 kV/cm. Sharp ECE and dielectric peaks demonstrate the first order character of the ferroelectric transition in BTGx. The latent heat enhancement can explain the relatively large ECE obtained at small field changes at the FE-PE transition, similar to that observed in BTO [6].

Indeed, large ECE responsivity $\Delta T/\Delta E = 1.13 \times 10^{-6}$ K.m/V was obtained by direct measurements, putting this compound into the category of promising materials for refrigeration applications.

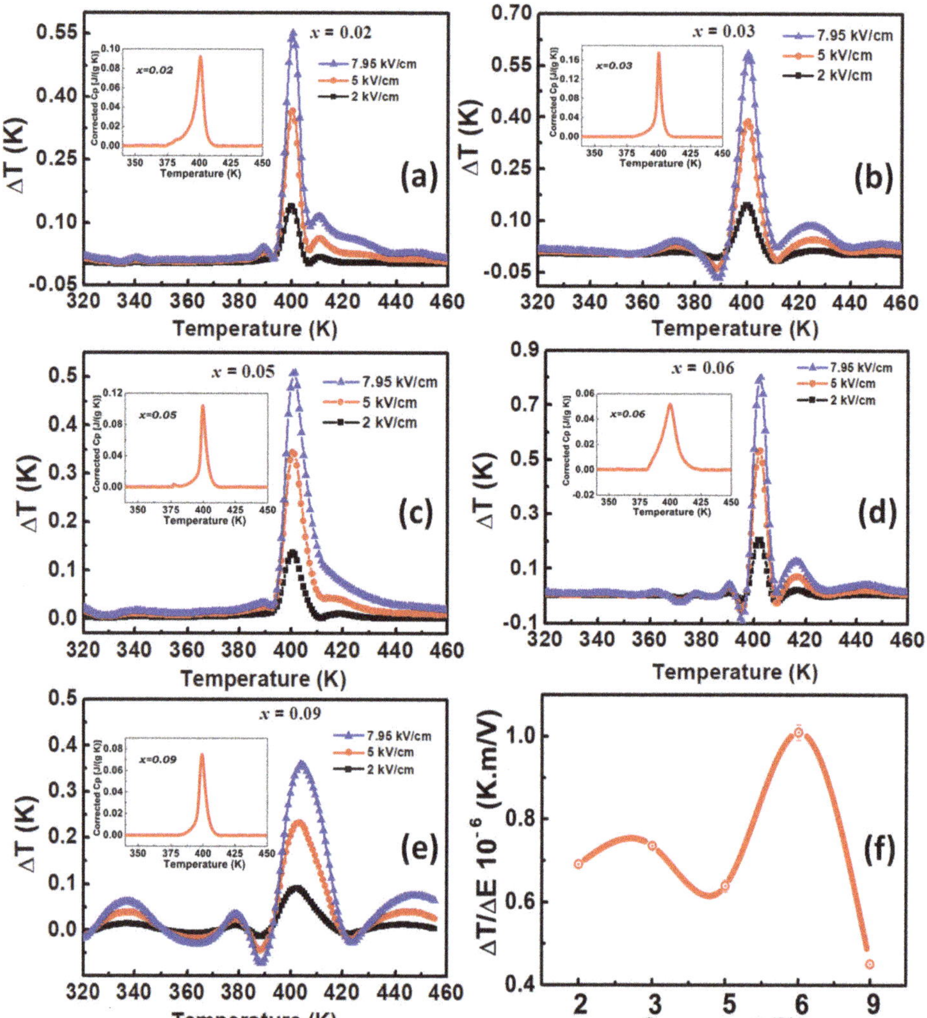

Figure 8. (**a**–**e**) Electrocaloric adiabatic temperature change (ΔT) as a function of temperature at three different applied electric fields (2.00, 5.00 and 7.95 kV/cm). The insets show corrected heat capacity Cp deduced from heat flow measurement versus temperature. (**f**) Electrocaloric responsivity as a function of Ge-content.

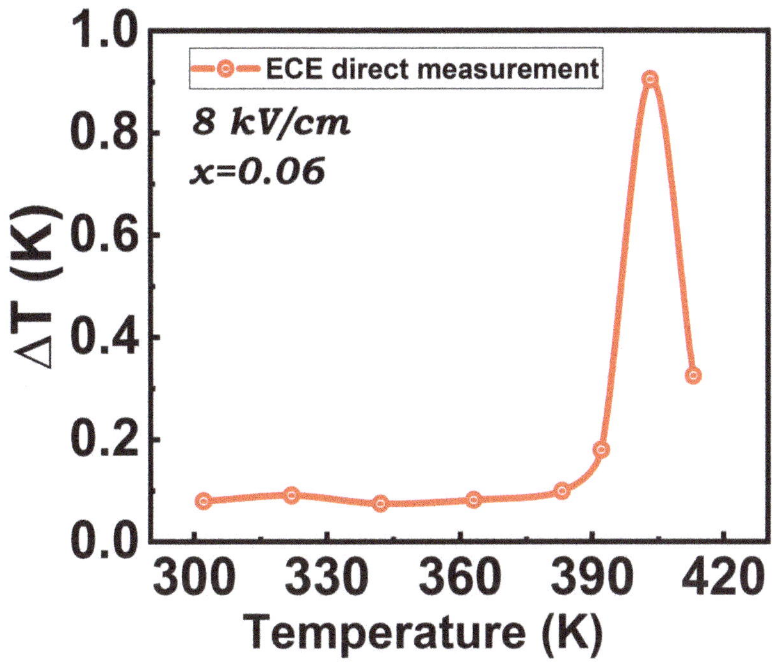

Figure 9. Adiabatic electrocaloric temperature variation obtained from composition BTG6 by the direct method under applied electric field of 8 kV/cm.

6. Energy Storage Investigations

Electrostatic energy storage studies have been investigated. The charged energy density (W_{ch}), the loss energy density (W_{loss}), energy storage density (W_{rec}) and energy storage efficiency (η) were calculated from P-E hysteresis data. These physical quantities can be expressed by the following equations, respectively:

$$W_{ch} = \int_0^{P_m} E\,dP, \tag{2}$$

$$W_{rec} = \int_{P_r}^{P_m} E\,dP, \tag{3}$$

$$W_{loss} = W_{ch} - W_{rec}, \tag{4}$$

$$\eta = \frac{W_{rec}}{W_{ch}} \times 100, \tag{5}$$

where P_m, P_r and E denote maximum polarization, remnant polarization and electric field strength, respectively. W_{loss} represents the difference of the energy brought during the charge and that during the discharge process, equivalent to W_{rec} and η is energy storage efficiency coefficient [46,47].

In Figure 10a–c, we plot energy loss density, energy stored density, and energy storage efficiency coefficient versus temperature for all the studied compounds.

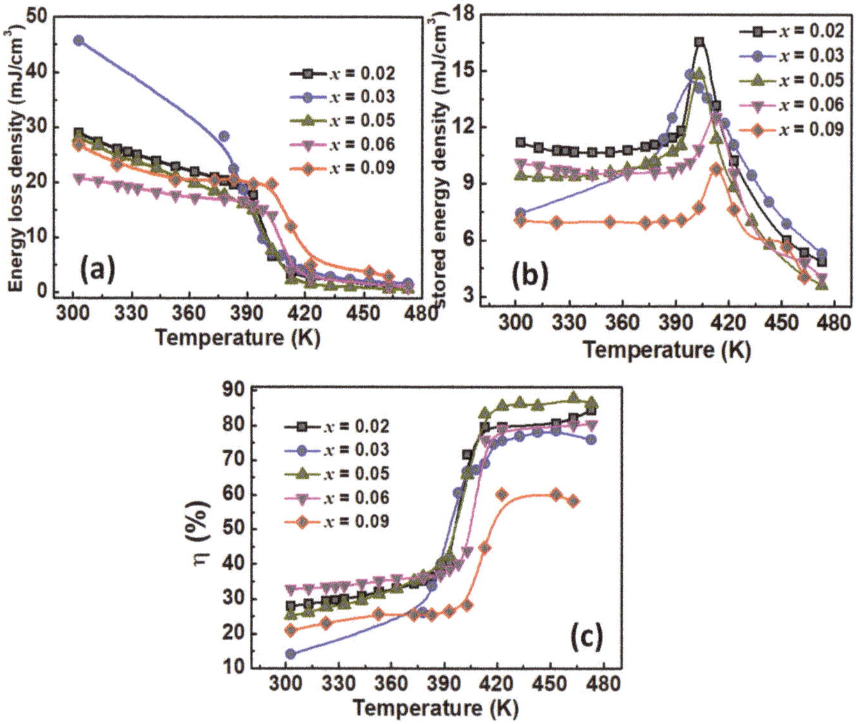

Figure 10. (**a**) Energy lost, (**b**) energy stored and (**c**) energy storage efficiency for all the BTGx compositions (x = 0.02, 0.03, 0.05, 0.06, 0.09). Energy loss decreases globally versus temperature for all the compositions. Maximal energy stored is obtained for BTG2. Higher energy storage efficiency is observed in BTG5.

For all the compositions, energy loss density value plotted in Figure 10a decreases and presents a step shape at the Curie point, dropping even more in the paraelectric phase. At the same time, the energy storage density (Figure 10b) shows a λ-shape curve with a maximum around the Curie point, where energy storage appears to be at its maximum. As the energy loss seems to be minimal in the paraelectric phase, this is favourable to a high energy storage efficiency, as shown in Figure 10c in all the studied Ge-content BTGx ceramics.

As expected, the composition BTG6 shows the lower energy lost in accordance with its density and its dielectric and ferroelectric responses. On the other hand, higher energy loss was observed in the ferroelectric phase for the composition BTG3. The sample BTG2, on the other hand, presents a higher energy stored value, and this energy decreases when Ge-content increases. This behaviour is different to the electrocaloric responsivity behaviour, which showed a maximum for BTG6. This result shows the decorrelation between the electrocaloric effect and energy storage mechanism. The former depends on pyroelectric coefficient jump and domain walls dynamic, while the latter depends on the ceramic density and polarization value. Furthermore, as shown in Figure 10b, the higher energy storage density 16.65 mJ/cm^3 was observed for the composition BTG2 at the Curie temperature. This global small value can be attributed to samples density, the shape of PE-hysteresis loops and maximal polarisation value, since higher values are usually obtained from slimmer PE-hysteresis loops, similarly to relaxor-type ferroelectrics [48,49]. However, the energy storage efficiency of BTGx samples was observed in the range 55 to 88% in the paraelectric phase. The maximum value 87.67% is observed for BTG5. Only the composition BTG9 shows somewhat smaller energy storage efficiency; this Ge concentration approaching limit of solubility. Nevertheless, these results place this family compound in stable energy storable compounds at relatively high temperatures due to the energy storage efficiency.

7. Conclusions

High ECE was evidenced in the lead-free BTGx system (x = 0.02, 0.03, 0.05, 0.06, 0.09). These compounds exhibit classical ferroelectric behaviour confirmed from P-E hysteresis and FE-PE phase transition confirmed by dielectric and heat capacity measurements. The substitution of titanium (Ti) by germanium (Ge) ions in BTO matrix in the octahedral sites was confirmed by structural analysis based on X-ray diffraction patterns, Raman spectroscopy and SEM images. All compositions are of pure perovskite tetragonal structure and acceptable compactness ceramics. However, Ge-doping did not induce structural symmetry change but the decrease of lattice parameters and volume. Electrical and heat capacity measurements show first-order-type phase transition for all the BTGx compounds, the T_C value varies weakly and a thermal hysteresis of about 2.00 K is observed. ECE responsivity was calculated for all the compositions from the indirect method that reveals large values, especially for the compound BTG6 ($\Delta T/\Delta E$ = 1.01 K.m/V at 400 K), in good agreement with direct EC measurements result of $\Delta T/\Delta E$ of 1.13×10^{-6} K.m/V under 8 kV/cm applied electric field. Energy storage investigations show moderate energy stored of 16.65 mJ/cm^3 and energy storage efficiency of 87.97%. These results make the lead-free BTGx system a promising alternative candidate for refrigeration and energy storage materials.

Author Contributions: Conceptualization, Y.G., D.M. and B.A.; methodology, D.M. and A.L.; software, Y.G. and S.B.M., validation, Z.K. and M.E.M.; formal analysis, J.-L.D.; investigation, B.R. and J.-L.D.; resources, M.A. and A.L.; data curation, Y.G. and B.A.; writing—original draft preparation, B.A. and M.A.; writing—review and editing, Z.K. and Y.G.; visualization, S.B.M. and A.L.; supervision, M.E.M.; project administration, Y.G.; funding acquisition, Y.G. All authors have read and agreed to the published version of the manuscript.

Funding: This work was supported financially by The Ministry for Europe and Foreign Affairs (MAEDI) via PHC Carlos Finlay project No 47075NF and University of Picardie Jules Verne via S2R Action 3.

Institutional Review Board Statement: Not applicable.

Informed Consent Statement: Not applicable.

Data Availability Statement: Not applicable.

Conflicts of Interest: The authors declare no conflict of interest.

References

1. Moya, X.; Kar-Narayan, S.; Mathur, N.D. Caloric materials near ferroic phase transitions. *Nat. Mater.* **2014**, *13*, 439–450. [CrossRef] [PubMed]
2. Kaddoussi, H.; Gagou, Y.; Lahmar, A.; Allouche, B.; Dellis, J.L.; Courty, M.; Khemakhem, H.; El Marssi, M. Ferroelectric phase changes and electrocaloric effects in $Ba(Zr_{0.1}Ti_{0.9})_{1-x}SnxO_3$ ceramics solid solution. *J. Mater. Sci.* **2016**, *51*, 3454–3462. [CrossRef]
3. Ben Moumen, S.; Gagou, Y.; Chettab, M.; Mezzane, D.; Amjoud, M.; Fourcade, S.; Hajji, L.; Kutnjak, Z.; El Marssi, M.; El Amraoui, Y.; et al. Synthesis of $La_{0.5}Ca_{0.5-x}\square_xMnO_3$ nanocrystalline manganites by sucrose assisted auto combustion route and study of their structural, magnetic and magnetocaloric properties. *J. Mater. Sci. Mater. Electron.* **2019**, *30*, 20459–20470. [CrossRef]
4. Bai, Y.; Han, X.; Qiao, L. Optimized electrocaloric refrigeration capacity in lead-free $(1-x)BaZr_{0.2}Ti_{0.8}O_3$-$xBa_{0.7}Ca_{0.3}TiO_3$ ceramics. *Appl. Phys. Lett.* **2013**, *102*, 252904. [CrossRef]
5. Chukka, R.; Vandrangi, S.; Shannigrahi, S.; Chen, L. An electrocaloric device demonstrator for solid-state cooling. *Eur. Lett.* **2013**, *103*, 47011. [CrossRef]
6. Novak, N.; Kutnjak, Z.; Pirc, R. High-resolution electrocaloric and heat capacity measurements in barium titanate. *Eur. Lett.* **2013**, *103*, 47001. [CrossRef]
7. Bai, Y.; Zheng, G.; Shi, S. Direct measurement of giant electrocaloric effect in $BaTiO_3$ multilayer thick film structure beyond theoretical prediction. *Appl. Phys. Lett.* **2010**, *96*, 192902. [CrossRef]

8. Aprea, C.; Greco, A.; Maiorino, A.; Masselli, C. Electrocaloric refrigeration: An innovative, emerging, eco-friendly refrigeration technique. *J. Physics Conf. Ser.* **2017**, *796*, 12019. [CrossRef]
9. Kutnjak, Z.; Rožič, B.; Pirc, R. Electrocaloric Effect: Theory, Measurements, and Applications. In *Wiley Encyclopedia of Electrical and Electronics Engineering*; John Wiley & Sons, Inc.: Hoboken, NJ, USA, 2015; pp. 1–19. [CrossRef]
10. Correia, T.; Zhang, Q. (Eds.) *Electrocaloric Materials: New Generation of Coolers*, 1st ed.; Springer: Berlin/Heidelberg, Germany, 2014. [CrossRef]
11. Suchaneck, G.; Gerlach, G. Lead-free Relaxor Ferroelectrics for Electrocaloric Cooling. *Mater. Today: Proc.* **2016**, *3*, 622–631. [CrossRef]
12. Lu, S.-G.; Zhang, Q.M.; Kutnjak, Z. The electrocaloric effect (ECE) in ferroelectric polymer films. In *Thin Film Growth*; Elsevier: Amsterdam, The Netherlands, 2011; pp. 364–383. [CrossRef]
13. Mischenko, A.S.; Zhang, Q.; Scott, J.F.; Whatmore, R.W.; Mathur, N.D. Giant Electrocaloric Effect in Thin-Film $PbZr_{0.95}Ti_{0.05}O_3$. *Science* **2006**, *311*, 1270–1271. [CrossRef]
14. Cao, H.-X.; Li, Z.-Y. Electrocaloric effect in $BaTiO_3$ thin films. *J. Appl. Phys.* **2009**, *106*, 94104. [CrossRef]
15. Valant, M. Electrocaloric materials for future solid-state refrigeration technologies. *Prog. Mater. Sci.* **2012**, *57*, 980–1009. [CrossRef]
16. Lidsky, T.I.; Schneider, J.S. Lead neurotoxicity in children: Basic mechanisms and clinical correlates. *Brain* **2003**, *126*, 5–19. [CrossRef] [PubMed]
17. Meng, Y.; Tang, C.; Yu, J.; Meng, S.; Zhang, W. Exposure to lead increases the risk of meningioma and brain cancer: A meta-analysis. *J. Trace Elem. Med. Biol.* **2020**, *60*, 126474. [CrossRef]
18. Li, W.-B.; Zhou, D.; Xu, R.; Wang, D.-W.; Su, J.-Z.; Pang, L.-X.; Liu, W.-F.; Chen, G.-H. $BaTiO_3$-Based Multilayers with Outstanding Energy Storage Performance for High Temperature Capacitor Applications. *ACS Appl. Energy Mater.* **2019**, *2*, 5499–5506. [CrossRef]
19. Mahmoud, A.; Moen, S.; Gerges, M.K. Enhanced Tunability Properties of Pure (Ba,Sr)TiO_3 Lead free Ferroelectric by Polar Nano-region Contributions. *Res. Sq.* **2021**. *preprint*. [CrossRef]
20. Asbani, B.; Gagou, Y.; Trček, M.; Dellis, J.-L.; Amjoud, M.; Lahmar, A.; Mezzane, D.; Kutnjak, Z.; El Marssi, M. Dielectric permittivity enhancement and large electrocaloric effect in the lead free $(Ba_{0.8}Ca_{0.2})_{1-x}La_{2x/3}TiO_3$ ferroelectric ceramics. *J. Alloys Compd.* **2018**, *730*, 501–508. [CrossRef]
21. Horchidan, N.; Curecheriu, L.; Ciomaga, C.E.; Lupu, N.; Mitoseriu, L. Preparation and Functional Properties of $BaTiO_3$–$BaGeO_3$ Ceramics. *IEEE Trans. Ultrason. Ferroelectr. Freq. Control* **2021**, *68*, 279–287. [CrossRef] [PubMed]
22. Garbarz-Glos, B.; Lisińska-Czekaj, A.; Czekaj, D.; Bąk, W. Effect of Semiconductor Element Substitution on the Electric Properties of Barium Titanate Ceramics. *Arch. Met. Mater.* **2016**, *61*, 887–890. [CrossRef]
23. Köferstein, R.; Ebbinghaus, S.G. $BaGeO_3$ as sintering additive for $BaTiO_3$–$MgFe_2O_4$ composite ceramics. *RSC Adv.* **2015**, *5*, 71491–71499. [CrossRef]
24. Chu, M.S.-H.; Bultitude, J.; Hood, C.; Nimmo, K.L.; Rand, M. Temperature Stable Dielectric. EU Patent No. EP 0,731,066,A1, 1996.
25. Köferstein, R.; Jäger, L.; Zenkner, M.; Müller, T.; Abicht, H.-P. Shrinkage mechanism and phase evolution of fine-grain $BaTiO_3$ powder compacts containing 10 mol% $BaGeO_3$ prepared via a precursor route. *Mater. Chem. Phys.* **2008**, *112*, 531–535. [CrossRef]
26. Guha, J.P.; Kolar, D. Phase equilibria in the system $BaTiO_3$-$BaGeO_3$. *J. Mater. Sci.* **1972**, *7*, 1192–1196. [CrossRef]
27. Plessner, K.W.; West, R. Replacement of Ti in $BaTiO_3$ Ceramic by Si and Ge. *Proc. Phys. Soc. Sect. B* **1955**, *68*, 1150–1152. [CrossRef]
28. Pulvari, C.F. Effect of Impurities on Electrical Solid-state Properties of Barium Titanate. *J. Am. Ceram. Soc.* **1959**, *42*, 355–363. [CrossRef]
29. Baxter, P.; Hellicar, N.J.; Lewis, B. Effect of Additives of Limited Solid Solubility on Ferroelectric Properties of Barium Titanate Ceramics. *J. Am. Ceram. Soc.* **1959**, *42*, 465–470. [CrossRef]
30. Bai, Y.; Zheng, G.-P.; Shi, S.-Q. Abnormal electrocaloric effect of $Na_{0.5}Bi_{0.5}TiO_3$–$BaTiO_3$ lead-free ferroelectric ceramics above room temperature. *Mater. Res. Bull.* **2011**, *46*, 1866–1869. [CrossRef]
31. Liu, H.; Yang, X. Theoretical prediction of electrocaloric effect based on non-linear behaviors of dielectric permittivity under temperature and electric fields. *AIP Adv.* **2015**, *5*, 117134. [CrossRef]
32. Sanlialp, M.; Luo, Z.; Shvartsman, V.V.; Wei, X.; Liu, Y.; Dkhil, B.; Lupascu, D.C. Direct measurement of electrocaloric effect in lead-free $Ba(Sn_xTi_{1-x})O_3$ ceramics. *Appl. Phys. Lett.* **2017**, *111*, 173903. [CrossRef]
33. Singh, G.; Bhaumik, I.; Ganesamoorthy, S.; Bhatt, R.; Karnal, A.K.; Tiwari, V.S.; Gupta, P.K. Electro-caloric effect in $0.45BaZr_{0.2}Ti_{0.8}O_3$-$0.55Ba_{0.7}Ca_{0.3}TiO_3$ single crystal. *Appl. Phys. Lett.* **2013**, *102*, 82902. [CrossRef]
34. Asbani, B.; Dellis, J.-L.; Lahmar, A.; Courty, M.; Amjoud, M.; Gagou, Y.; Djellab, K.; Mezzane, D.; Kutnjak, Z.; El Marssi, M. Lead-free $Ba_{0.8}Ca_{0.2}(Zr_xTi_{1-x})O_3$ ceramics with large electrocaloric effect. *Appl. Phys. Lett.* **2015**, *106*, 42902. [CrossRef]
35. Kaddoussi, H.; Lahmar, A.; Gagou, Y.; Dellis, J.-L.; Khemakhem, H.; El Marssi, M. Electro-caloric effect in lead-free ferroelectric Ba1−Ca $(Zr_{0.1}Ti_{0.9})_{0.925}Sn_{0.075}O_3$ ceramics. *Ceram. Int.* **2015**, *41*, 15103–15110. [CrossRef]
36. Wang, J.; Yang, T.; Chen, S.; Li, G.; Zhang, Q.; Yao, X. Nonadiabatic direct measurement electrocaloric effect in lead-free Ba,Ca(Zr,Ti)O_3 ceramics. *J. Alloy. Compd.* **2013**, *550*, 561–563. [CrossRef]
37. Liu, X.Q.; Chen, T.T.; Wu, Y.J.; Chen, X.M. Enhanced Electrocaloric Effects in Spark Plasma-Sintered $Ba_{0.65}Sr_{0.35}TiO_3$-Based Ceramics at Room Temperature. *J. Am. Ceram. Soc.* **2013**, *96*, 1021–1023. [CrossRef]
38. Rietveld, H.M. A profile refinement method for nuclear and magnetic structures. *J. Appl. Crystallogr.* **1969**, *2*, 65–71. [CrossRef]

39. Shannon, R.D. Revised effective ionic radii and systematic studies of interatomic distances in halides and chalcogenides. *Acta Crystallogr. Sect. A* **1976**, *32*, 751–767. [CrossRef]
40. Parsons, J.; Rimai, L. Raman spectrum of $BaTiO_3$. *Solid State Commun.* **1967**, *5*, 423–427. [CrossRef]
41. Pinczuk, A.; Taylor, W.; Burstein, E.; Lefkowitz, I. The Raman spectrum of $BaTiO_3$. *Solid State Commun.* **1967**, *5*, 429–433. [CrossRef]
42. Venkateswaran, U.D.; Naik, V.M.; Naik, R.R. High-pressure Raman studies of polycrystalline $BaTiO_3$. *Phys. Rev. B* **1998**, *58*, 14256–14260. [CrossRef]
43. DiDomenico, M.; Wemple, S.H.; Porto, S.P.S.; Bauman, R.P. Raman Spectrum of Single-Domain $BaTiO_3$. *Phys. Rev.* **1968**, *174*, 522–530. [CrossRef]
44. Wei, A.N.; Liu, T.-H.; Wang, C.-H.; Diao, C.-L.; Luo, N.-N.; Liu, Y.; Qi, Z.-M.; Shao, T.; Wang, Y.-Y.; Jiao, H.; et al. Assignment for Vibrational Spectra of $BaTiO_3$ Ferroelectric Ceramic Based on the First-Principles Calculation. *Acta Physico-Chim. Sin.* **2015**, *31*, 1059–1068. [CrossRef]
45. Pokorný, J.; Pasha, U.M.; Ben, L.; Thakur, O.P.; Sinclair, D.C.; Reaney, I.M. Use of Raman spectroscopy to determine the site occupancy of dopants in $BaTiO_3$. *J. Appl. Phys.* **2011**, *109*, 114110. [CrossRef]
46. Zhao, N.; Fan, H.; Li, C.; Huang, F.; Cao, J.; Li, Z. Enhanced energy storage density and efficiency in Sm3+-doped $((Bi_{0.5}Na_{0.5})_{0.7}(Sr_{0.7}Bi_{0.2})_{0.3}))TiO_3$ ceramics. *J. Mater. Sci. Mater. Electron.* **2021**, *32*, 24930–24938. [CrossRef]
47. Mezzourh, H.; Belkhadir, S.; Mezzane, D.; Amjoud, M.; Choukri, E.; Lahmar, A.; Gagou, Y.; Kutnjak, Z.; El Marssi, M. Enhancing the dielectric, electrocaloric and energy storage properties of lead-free $Ba_{0.85}Ca_{0.15}Zr_{0.1}Ti_{0.9}O_3$ ceramics prepared via sol-gel process. *Phys. B Condens. Matter.* **2021**, *603*, 412760. [CrossRef]
48. Hanani, Z.; Mezzane, D.; Amjoud, M.; Razumnaya, A.G.; Fourcade, S.; Gagou, Y.; Hoummada, K.; El Marssi, M.; Gouné, M. Phase transitions, energy storage performances and electrocaloric effect of the lead-free $Ba_{0.85}Ca_{0.15}Zr_{0.10}Ti_{0.90}O_3$ ceramic relaxor. *J. Mater. Sci. Mater. Electron.* **2019**, *30*, 6430–6438. [CrossRef]
49. Veerapandiyan, V.; Benes, F.; Gindel, T.; DeLuca, M. Strategies to Improve the Energy Storage Properties of Perovskite Lead-Free Relaxor Ferroelectrics: A Review. *Materials* **2020**, *13*, 5742. [CrossRef]

membranes

MDPI

Article

Application and Analysis of Bipolar Membrane Electrodialysis for LiOH Production at High Electrolyte Concentrations: Current Scope and Challenges

Alonso González [1], Mario Grágeda [1,*], Adrián Quispe [1], Svetlana Ushak [1], Philippe Sistat [2] and Marc Cretin [2]

1 Departamento de Ingeniería Química y Procesos de Minerales and Center for Advanced Study of Lithium and Industrial Minerals (CELiMIN), Universidad de Antofagasta, Campus Coloso, Av Universidad de Antofagasta, Antofagasta 02800, Chile; alonso.gonzalez@celimin.com (A.G.); adrian.quispe.huayta@ua.cl (A.Q.); svetlana.ushak@uantof.cl (S.U.)

2 Institut Européen des Membranes, IEM, UMR-5635, Université de Montpellier, ENSCM, CNRS, Place Eugène Bataillon, CEDEX 5, 34095 Montpellier, France; philippe.sistat@umontpellier.fr (P.S.); marc.cretin@umontpellier.fr (M.C.)

* Correspondence: mario.grageda@uantof.cl; Tel.: +56-552637513

Abstract: The objective of this work was to evaluate obtaining LiOH directly from brines with high LiCl concentrations using bipolar membrane electrodialysis by the analysis of Li^+ ion transport phenomena. For this purpose, Neosepta BP and Fumasep FBM bipolar membranes were characterized by linear sweep voltammetry, and the Li^+ transport number in cation-exchange membranes was determined. In addition, a laboratory-scale reactor was designed, constructed, and tested to develop experimental LiOH production tests. The selected LiCl concentration range, based on productive process concentrations for Salar de Atacama (Chile), was between 14 and 34 wt%. Concentration and current density effects on LiOH production, current efficiency, and specific electricity consumption were evaluated. The highest current efficiency obtained was 0.77 at initial concentrations of LiOH 0.5 wt% and LiCl 14 wt%. On the other hand, a concentrated LiOH solution (between 3.34 wt% and 4.35 wt%, with a solution purity between 96.0% and 95.4%, respectively) was obtained. The results of this work show the feasibility of LiOH production from concentrated brines by means of bipolar membrane electrodialysis, bringing the implementation of this technology closer to LiOH production on a larger scale. Moreover, being an electrochemical process, this could be driven by Solar PV, taking advantage of the high solar radiation conditions in the Atacama Desert in Chile.

Keywords: lithium hydroxide; bipolar membrane electrodialysis; high concentration; lithium brine; sustainable process

Citation: González, A.; Grágeda, M.; Quispe, A.; Ushak, S.; Sistat, P.; Cretin, M. Application and Analysis of Bipolar Membrane Electrodialysis for LiOH Production at High Electrolyte Concentrations: Current Scope and Challenges. *Membranes* **2021**, *11*, 575. https://doi.org/10.3390/membranes11080575

Academic Editor: Tuti Mariana Lim

Received: 7 July 2021
Accepted: 27 July 2021
Published: 29 July 2021

1. Introduction

In recent years, lithium has become a mineral of great interest worldwide. Its demand has increased due to its use in lithium-ion batteries for electric vehicles and consumer electronics, which has been boosted in some countries by energy policies that promote clean energy usage. Currently, lithium hydroxide (LiOH) shows high projections in the production of battery cathodes [1,2]. By 2030, lithium consumption in electric vehicle batteries is expected to account for 80% of aggregate lithium consumption. By the same year, lithium hydroxide is projected to account for 57% of lithium compound demand, compared to 24% in 2019 [3]. Some key advantages of cathodes produced from lithium hydroxide over other chemical compounds are better power density, longer life cycle, and improved safety characteristics [4–7].

Conventional industrial processes for obtaining lithium salts from brines include several stages and unit operations. Lithium is extracted by pumping natural brines to evaporation ponds, where solar radiation produces energy to evaporate brine water, precipitating salts in a sequential process lasting between 12 and 15 months, until a concentration of 5.5–6.0 wt% lithium is reached [8]. Lithium-concentrated brine is then used as a raw material to obtain lithium carbonate through solvent extraction, purification, chemical reaction, filtration, and classification processes [9]. Some of the lithium carbonate obtained is technical grade (99.0% purity), and is subsequently used as a raw material to obtain lithium hydroxide by the reaction $Li_2CO_3 + Ca(OH)_2 \rightarrow 2LiOH + CaCO_3$. As a product of this reaction, an aqueous solution is obtained that reaches approximately 3 wt% LiOH concentration, which is subjected to an evaporation and crystallization process to obtain lithium hydroxide monohydrate. At the same time, calcium carbonate is obtained as waste, which is separated by settling tanks and filters. Conventional processing from concentrated brine to lithium hydroxide monohydrate crystals requires several pieces of equipment and, therefore, a large installation space, as well as the addition of chemical reagents (e.g., Na_2CO_3, CaO, among others), producing liquid waste (mother liquor) and solid wastes (e.g., $Mg(OH)_2$, $CaCO_3$, among others) [10].

Ion-exchange-membrane-based electrochemical processes are an alternative for the production of alkaline products, avoiding the use of chemical reagents and reducing waste production. Thais is the case of the chloralkali process for NaOH production. In this process, a cationic membrane is used between two electrodes and, by applying an electric current, water reduction at the cathode is achieved to generate NaOH and H_2 gas through the semi-reaction $2H_2O + 2e^- \rightarrow H_2 + 2OH^-$. Gas generation is an electrochemical phenomenon that is best avoided for certain applications where it is not used and simply discarded. This can be achieved by using bipolar membranes, which are also less expensive compared to an electrode for OH^- ion generation. A bipolar membrane (BPM) consists of a cationic layer, an anionic layer, and an intermediate zone that allows water catalysis of dissociation into OH^- and H^+ ions without gas generation [11,12]. Thus, production of acids and bases from a salt solution is possible. Bipolar membrane electrodialysis (BMED) is a technology that has been studied in recent years for various applications, such as desalination [13–16] and sodium hydroxide production [17–19], as well as boron recovery [20] and acid recovery [21,22].

In the case of lithium, its recovery from waste streams has been studied using conventional electrodialysis [23] and bipolar membrane electrodialysis [21,24–26], generally recovering lithium in dilute LiOH solutions. Few works have been reported whose main focus is the production of lithium hydroxide by electromembrane processes [27]. They usually study membranes' interaction with aqueous solutions more dilute than those used in the lithium hydroxide production industry. Among them, the work of Melnikov et al. [28] stands out, where from a LiCl solution with organic solvents, the authors produced LiOH by bipolar membrane electrodialysis, reaching a LiOH concentration of 0.3 M with a specific energy consumption of 6.6 $kWh \cdot kg^{-1}$ and a current efficiency of 0.6. Grageda et al. [29] analyzed a membrane electrodialysis process to produce high-purity lithium hydroxide, determining the effects of concentration, current density, and temperature on process energy performance and product purity; the authors reported a specific energy consumption of 7.25 $kWh \cdot kg^{-1}$ of LiOH. On the other hand, Ryabtsev et al. [30] used membrane electrolysis to obtain lithium hydroxide from lithium carbonate treated with sulfuric acid, taking advantage of the high solubility of Li_2SO_4, and managing to obtain a high average LiOH concentration of 45 $g \cdot L^{-1}$, similar to that obtained in the conventional LiOH process. Another interesting work is the one carried out by Jiang et al. [31], in which LiOH with a purity of 95% was obtained from aqueous solutions of Li_2CO_3 (0.18 M), with a specific electricity consumption of 6.66 $kWh \cdot kg^{-1}$. BMED has also been studied for hydrochloric acid production, reporting specific energy consumptions between 4.4 $kWh \cdot kg^{-1}$ and 8.3 $kWh \cdot kg^{-1}$ of HCl [32,33].

In recent years, interest in the application of bipolar membranes has increased due to their technical, environmental, and economic advantages for the production of acids and bases, compared to their conventional production processes [27]. Today, this technology is applied for research and some pilot projects [34,35]. Electrodialysis processes are affected by different variables, such as current density, electrolyte concentration, flow rate, osmotic pressure, and ion activity. These parameters affect electrochemical equilibrium between membrane-separated solutions by influencing ion transport across the membrane. During a concentration process with ion-exchange membranes, transport rate varies according to changes in concentration and membrane stability. On the other hand, co-ion flux, osmosis, and electro-osmosis affect process performance. For the application of LiOH in lithium battery materials, high purity is required, which to date has not been achieved using BMED. Therefore, there is a need to understand the transport mechanism of Li^+ ions across membranes. This work analyzes the feasibility of BMED for application in lithium hydroxide production from concentrated brines, with the aim of achieving high-purity lithium hydroxide concentrations for applications in the battery industry. Thus, our aim is to reduce the energy consumption of the overall production process by achieving a smaller gap between final LiOH concentration and saturation concentration, reducing energy expenditure in the evaporation, crystallization, and lithium hydroxide production stages that currently exist. This work presents the results of the feasibility and scope of LiOH production by BMED using high concentrations, identifying the main phenomena affecting lithium transport and LiOH formation throughout the concentration process with BMED. The main novelty of this study is the use of high-concentrated aqueous solutions, similar to those obtained during the industrial lithium concentration process in the Salar de Atacama. Moreover, experimental process performance data are presented, establishing the current scope of BMED technology in industrial LiOH production. In addition, the energy supply for this process could be photovoltaic solar energy [33], given the high solar irradiation levels in the Salar de Atacama, making this an even more sustainable process from energy point of view.

2. Materials and Methods

2.1. Materials

Cationic (Neosepta CMX and CMB) and anionic (Neosepta AMX, manufactured by Astom Corporation, Tokyo, Japan) ion-exchange membranes were used in this work. The bipolar membranes used were Neosepta BP (Astom Corporation, Tokyo, Japan) and Fumasep FBM (FuMA-Tech GmbH, Bietigheim-Bissingen, Germany). The technical specifications of the membranes are shown in

Tables 1 and 2. For the preparation of aqueous solutions, LiCl (Winkler ltda, Santiago, Chile), LiOH, HCl, Na_2SO_4 (Merck, Darmstadt, Germany), and deionized water were used as reagents. The presence of other elements such as K, Mg, and Ca present in Salar de Atacama brine was not considered, as this study focuses on transport phenomena associated with lithium and the effect of high concentrations. Therefore, a purified brine free of Mg and Ca—as can be obtained by chemical precipitation and ion-exchange processes—was considered for experimental development [36]. The presence of these elements should be reduced to the minimum possible in order to avoid membrane poisoning [37].

Table 1. Monopolar membranes' technical specifications [38–41].

Characteristics	AMX	CMX	CMB
Type	Anion	Cation	Cation
Electrical resistance ($\Omega\cdot cm^2$)	2.0–3.5 (0.5 M NaCl)	2.5–3.5 (0.5 M NaCl)	4.5
Thickness of wet membrane (mm)	0.14–0.18	0.17–0.19	0.21
Burst pressure	4.5–5.5 (kg/cm^2)	5–6 (kg/cm^2)	≥0.40 MPa
Exchange capacity ($meq\cdot g^{-1}$)	1.4–1.7	1.5–1.8	2.4–2.7
Water content (%)	25–30	25–30	37–42
Thermal stability (°C)	40	40	60

Table 2. Bipolar membranes' technical specifications.

Characteristics	Fumasep FBM [42,43]	Neosepta BP [27,38]
Type	Bipolar	Bipolar
Thickness (mm)	0.18–0.20 (wet) 0.13–0.16 (dry)	0.22 (wet)
Burst pressure	-	>0.40 MPa
Thermal stability (°C)	40	-
Water dissociation voltage at 100 $mA\cdot cm^{-2}$	<1.2 [1]	1.2 [2]
Water dissociation efficiency at 100 $mA\cdot cm^{-2}$	>98%	>98%

[1] At 100 $mA\cdot cm^{-2}$ in 0.5 M NaCl at 25 °C; [2] Potential difference measured between silver/silver chloride electrodes. 1 N NaOH/1 N HCl 10 $A\cdot dm^{-2}$ 30 °C.

2.2. Measurement of Water Uptake and Thickness of Membranes

In order to ascertain the influence of different operating solutions on membranes' water uptake and thickness, these were equilibrated in their respective operating solutions. Thus, for cation-exchange membranes, LiOH (0.5–5.0 wt%) and LiCl (14–34 wt%) solutions were used, while for bipolar membranes, LiOH and HCl (0.5–5.0 wt%) solutions were used, in which membranes were equilibrated for at least 24 h. Then, the electrolytes on the membranes' surface were carefully cleaned, thickness was measured using a high-accuracy digital thickness gauge, and the wet membranes' mass was recorded. Subsequently, membranes were dried in an oven at 70 °C until they showed zero mass variation. Water uptake (W_u) was calculated using Equation (1), from dry mass (W_d) and wet mass (W_w) values, in grams. At least two replicates were performed for each measurement.

$$W_u = \frac{W_w - W_d}{W_d} \times 100 \quad (1)$$

2.3. Determination of Lithium Transport Number

In order to determine the influence of current density and initial lithium concentration on transport in cation-exchange membranes, tests were performed to determine the lithium transport number. For this purpose, the Hittorf method was used according to the scheme in Figure 1, in which a three-compartment cell separated by two cation-exchange membranes with an effective membrane area of 7 cm^2 was used. Two different cation-exchange membranes were used: Neosepta CMX and CMB. In the cathodic compartment, lithium from a LiCl solution reacted with OH^- anions and was concentrated in LiOH form within 60 min. To reduce the concentration variation of LiCl solution, a larger volume of LiCl (100 mL) was used compared to the LiOH solution receiving lithium ions (20 mL). A 2.5 wt% LiOH solution (g/100 g solution) was used as an anolyte under the assumption that lithium ions are transported to the LiCl compartment at the same rate at which they migrate to the cathode, contributing to keep its concentration unchanged. After 60 min, the LiOH solution in the cathode compartment was extracted and analyzed by ion chromatography to determine the difference in Li concentration and calculate its Li ion molar flux (J_{Li}) across the cation-exchange membrane using Equation (2):

$$J_{Li} = \frac{C_{Li,0}\cdot V_{LiOH,0} - C_{Li,f}\cdot V_{LiOH,f}}{A\cdot t} \quad (2)$$

where C_{Li} is the molar concentration of lithium in the LiOH solution ($mol\cdot L^{-1}$), V_{LiOH} is the volume of the LiOH solution (L), A is the effective membrane area (m^2), and t is the time (s). Subscripts 0 and f refer to initial and final time, respectively.

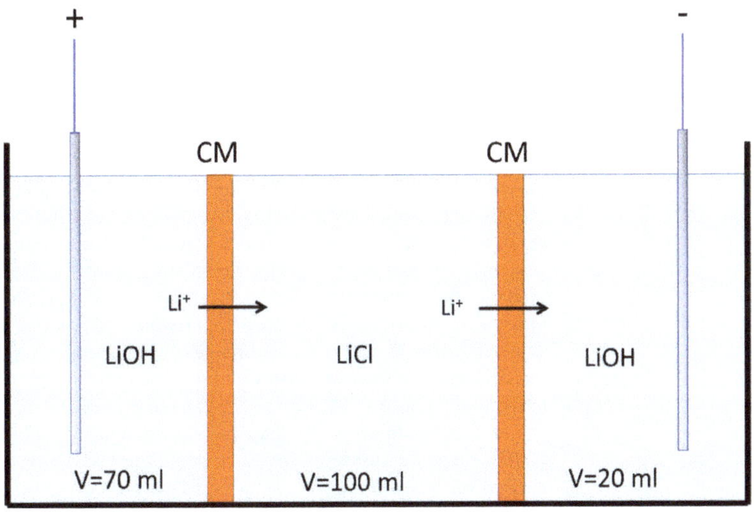

Figure 1. Hittorf method experimental scheme for the determination of lithium ion transport number (CM: Cation exchange membrane).

Once molar flux (J_{Li}) was determined, the lithium ion transport number t_{Li} was calculated according to Equation (3):

$$t_{Li} = \frac{F \cdot z_{Li} \cdot J_{Li}}{I} \tag{3}$$

where F is the Faraday constant (96,485 $A \cdot s \cdot mol^{-1}$), z_{Li} is the lithium-ion valence ($z_{Li} = 1$), J_{Li} is the lithium-ion molar flux ($mol \cdot m^{-2} \cdot s^{-1}$), and I is the electric current (A).

The different concentration conditions used are presented in Table 3. The effects of current densities (300, 700, and 1100 $A \cdot m^{-2}$) were also studied. All tests were performed at room temperature (20–22 °C). The pH of the LiCl solution was measured before and after the experiment.

Table 3. Initial LiCl and LiOH concentrations in solution for lithium ion transport number determination.

Configuration	1	2	3	4	5	6
LiCl (wt%)	14	14	14	14	25	25
LiOH (wt%)	0.5	2.5	5.0	8.0	0.5	5.0

Electrolyte concentration selection corresponded to ranges established based on concentrated lithium brine obtained from solar evaporation ponds used in productive processes at the Salar de Atacama, where a brine rich in lithium up to 5.5–6.0 wt% was obtained, equivalent to approximately 33.6–36.6 wt% LiCl. In this work, concentrated LiCl solutions between 14 wt% and 34 wt% were considered. On the other hand, we performed experiments with a LiOH concentration range between 0.5 wt% and 8.0 wt%, covering higher concentrations than those obtained industrially (approximately 3.0 wt%) without reaching LiOH solution saturation concentration.

2.4. Linear Sweep Voltammetry (LSV)

In the characterization of bipolar membranes, the linear sweep voltammetry (LSV) technique allows the determination of the degree of salt leakage that occurs through them, and of the limitations of water diffusion towards bipolar membrane reactive interface [44]. In the case of cation-exchange membranes, LSV can be used to determine the limiting current density caused by the polarization concentration effect at membrane surface. Both phenomena affect the efficiency of the electrodialysis process.

To determine current–voltage curves and estimate membranes' apparent electrical resistance, a six-compartment cell was used, according to electrochemical techniques indicated by Balster et al. [44]. To measure potential differences in the membranes under study, Haber–Luggin capillaries filled with 3 M KCl solution were used, where 99.99% silver wires connected to a voltmeter (Tenma 72-1016, Springboro, OH, USA) were placed. The experimental cell used had an effective membrane area of 4 cm^2. The flow velocity in each compartment was between 1.0 and 1.4 cm·s^{-1}. The three cases considered are presented in Figure 2. The salt leakage and performance of bipolar membranes under production conditions were determined according to configurations (a) and (b), respectively. Cation-exchange membranes were tested according to configuration (c). The solution concentrations and membranes used in each test are presented in Table 4.

LSV tests were performed using a power supply (Tenma 72-2550, Springboro, OH 45066, USA); the step voltage increase was 0.15 V, and the corresponding current value was measured after 30–60 s after voltage stabilization was observed. For higher current values, more noise was observed in the measurements, so the step increase was 0.30 V with a waiting time between 60 and 90 s in order to reach voltage stabilization.

Table 4. Solution concentrations and membranes used in LSV tests.

Configuration (a): Bipolar Membrane's Salt Leakage						
Test	1	2	3	4		
Bipolar membrane	Fumasep FBM	Neosepta BP	Fumasep FBM	Neosepta BP		
LiCl solution (wt%)	14	14	25	25		
Configuration (b): Bipolar Membrane's Apparent Electrical Resistance						
Test	1	2	3	4	5	6
Bipolar membrane	Fumasep FBM	Fumasep FBM	Fumasep FBM	Neosepta BP	Neosepta BP	Neosepta BP
LiOH solution (wt%)	0.5	2.5	5.0	0.5	2.5	5.0
HCl solution (wt%)	0.5	3.5	7.8	0.5	3.5	7.8
Configuration (c): Cation-Exchange Membrane's Apparent Electrical Resistance						
Test	1	2	3	4	5	6
Membrane	CMX	CMX	CMX	CMB	CMB	CMB
LiCl solution (wt%)	14	14	14	14	14	14
LiOH solution (wt%)	0.5	2.5	5.0	0.5	2.5	5.0

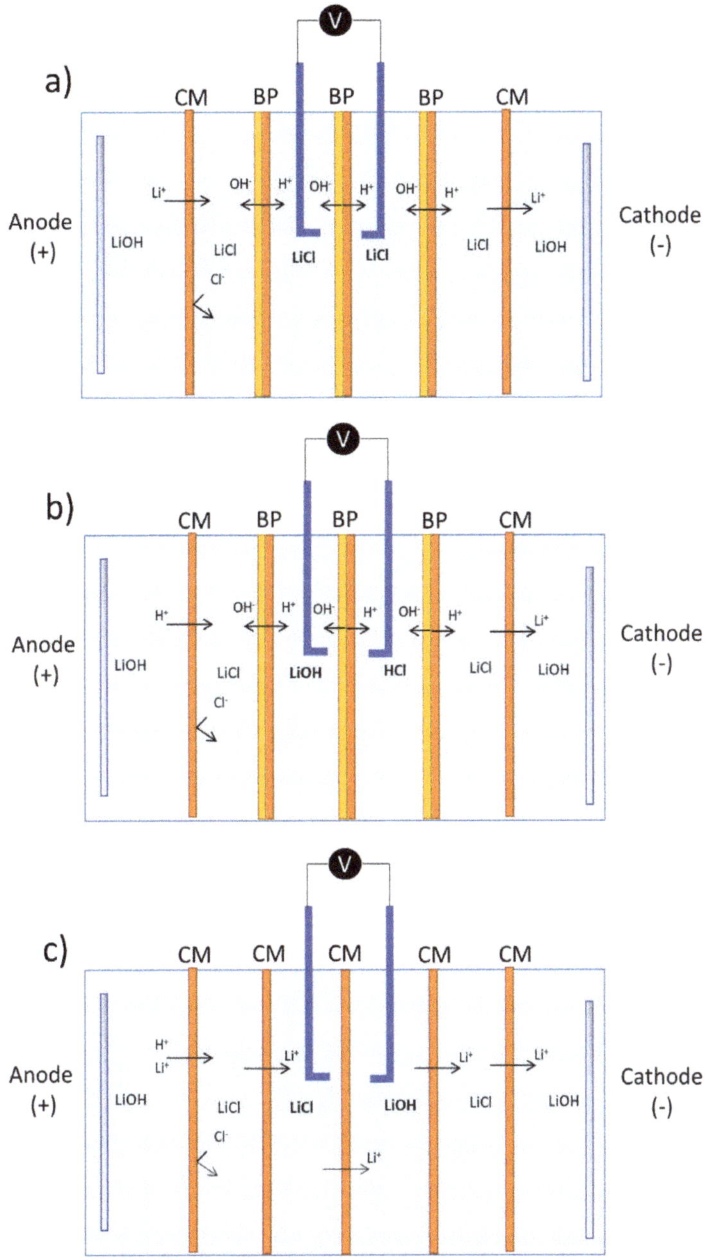

Figure 2. LSV tests' cell configuration. (**a**) Limiting current density measurement as related to salt leakage in bipolar membranes (BP). (**b**) Bipolar membranes' apparent electrical resistance measurements as related to operating conditions. (**c**) Cation-exchange membranes' (CM) concentration polarization determination in the range under study.

2.5. Bipolar Membrane Electrodialysis System

2.5.1. Stack Design and Construction

To perform experimental tests on LiOH production, an electrodialysis stack with bipolar membranes was designed and constructed. The stack was constructed from PTFE, with an effective membrane area of 27.5 cm^2 (55 mm × 50 mm). The design was a "filter press" type where membranes and separators could be configured according to different flow compartments (see Figure 3). The separators were made of EPDM, with a thickness of 1.6 mm.

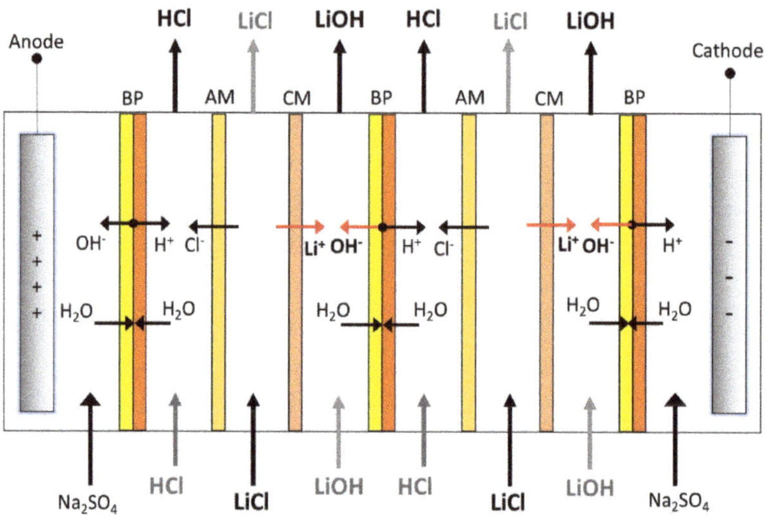

Figure 3. LiOH production by bipolar membrane electrodialysis; configuration of two three-compartment cells. BP: Bipolar membrane, AM: Anion exchange membrane, CM: Cation exchange membrane.

The electrodialysis stack's repetitive unit corresponded to a three-compartment cell, consisting of a HCl compartment, an anion-exchange membrane, a LiCl compartment, a cation-exchange membrane, and a LiOH compartment. These were stacked between two bipolar membranes. In this way, the stack could be configured with several three-compartment cells, among which different electrolytes were distributed. The stack was fed with five flow streams corresponding to LiCl, LiOH, HCl, and two electrode solution streams. The main feed to the stack was the LiCl solution, from which lithium ions migrated into the LiOH solution and chloride ions migrated into the HCl stream. As the electrode solution, 0.5 M Na_2SO_4 was used; its purpose was to provide conductivity in the electrode compartment, and simultaneously prevent chlorine gas (Cl_2) formation in the anode compartment.

2.5.2. LiOH Production Experimental Tests Using BMED

To test the LiOH production range, six long-running tests were performed. The system was configured in batch mode, recirculating different feed solutions and measuring concentration variation over time (see Figure 4).

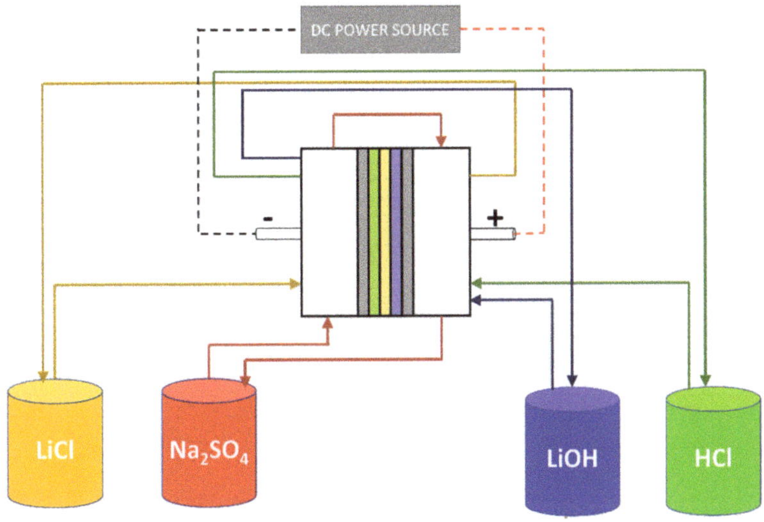

Figure 4. LiOH production system by BMED.

Flow rate in each compartment was set between 1.0 and 1.5 cm·s^{-1} using peristaltic pumps (Watson-Marlow 520SN/R2, Falmouth, UK), and a DC power supply (GW Instek GPR-1810HD, New Taipei, Taiwan) was used to set the electric current.

Experimental long-running tests of LiOH production were performed according to the operating conditions shown in Table 5. Tests 1 and 2 compare the effects of two different cation-exchange membranes (CMX and CMB), while Tests 3 and 4 compare the effects of different bipolar membranes (Neosepta BP and Fumasep FBM). On the other hand, 14 wt% and 25 wt% LiCl concentration effects can be compared by Tests 1 and 3 at 1000 A·m^{-2}, while Tests 5 and 6 compare 14 wt% and 34 wt% LiCl concentrations at a current density of 500 A·m^{-2}. In the latter case, to obtain comparable initial and final LiOH concentration ranges, the number of compartments was increased to four three-compartment cells. Equal initial LiOH and HCl concentrations of 0.5 wt% were used in all tests. An initial LiCl concentration between 14 and 34 wt% was considered. Neosepta AMX was used as an anionic membrane. The cation-exchange membranes and bipolar membranes were previously conditioned in a 0.5 M LiOH solution for 24 h, while the anion-exchange membrane was conditioned in 0.5 M LiCl solution.

Table 5. Operating conditions of long-running tests of LiOH production according to different bipolar membranes, initial LiCl concentration, and current density.

Test	1	2	3	4	5	6
Cation membrane	CMX	CMB	CMX	CMX	CMX	CMX
Bipolar membrane	Neosepta BP	Neosepta BP	Neosepta BP	Fumasep FBM	Fumasep FBM	Fumasep FBM
Current density (A·m^{-2})	1000	1000	1000	1000	500	500
LiCl initial concentration (wt%)	14	14	25	25	14	34
Time (min)	360	360	360	360	360	440
Number of three-compartment cells	2	2	2	2	4	4

2.5.3. SEC and ϕ Calculation

Concentrations of different elements—such as Li^+, OH^-, and Cl^-—in the LiOH solution were measured at the end of each experiment. Lithium concentration was determined by atomic absorption spectrophotometry, chloride ion concentration was determined by an argentometric method, and hydroxide content was analyzed by acid–base volumetry.

For each test, specific electrical consumption (SEC) was calculated according to Equation (4):

$$SEC = \frac{I \cdot U \cdot t}{m} \quad (4)$$

where I is the electrical current (A), U is the average voltage (V), t is the process time (h), and m is the mass of LiOH produced (kg). On the other hand, current efficiency (ϕ) was calculated according to Equation (5):

$$\phi = \frac{z \cdot F \cdot m}{N \cdot I \cdot t \cdot M} \quad (5)$$

where F is the Faraday constant (96,485 $A \cdot s \cdot mol^{-1}$), m is the mass of LiOH produced (g), z is the valence number, N is the number of LiOH compartments, I is the electrical current (A), t is the process time (s), and M is the molar mass.

Final LiOH solution purity was calculated according to Equation (6):

$$P = \frac{m_{LiOH}}{m_{TDS}} \times 100 \quad (6)$$

where m_{LiOH} is the LiOH mass in solution, and m_{TDS} is the mass of total dissolved salts (g).

3. Results

3.1. Water Uptake and Membrane Thickness

The CMX and CMB cation-exchange membranes' water uptake and thickness were measured after equilibrium with LiCl and LiOH solutions, which they were in contact with during the BMED LiOH production process. Meanwhile, the same parameters were determined for the bipolar membranes Fumasep FBM and Neosepta BP in equilibrium with LiOH and HCl solutions at different concentrations. The results of water uptake and membrane thickness are presented in Tables 6 and 7, respectively. In the cation-exchange membranes, water uptake in the membrane decreased with LiCl concentration. For a LiCl concentration of 34 wt%, the CMX membrane shows a 49.1% lower water uptake compared to a LiCl concentration of 14 wt%, while the CMB membrane shows a 38.3% lower water uptake for the same comparison. In contact with LiOH solutions, water uptake in the CMX membrane tended to decrease with concentration, while for the CMB membrane, water uptake increased for a 2.5 wt% LiOH solution and then decreased for a 5.0 wt% LiOH solution. The latter can be attributed to the high hydration shell associated with the lithium ion [45], which by increasing its concentration in the membrane increases water uptake in the membrane. Results suggest that between a concentration of 2.5 wt% and 5.0 wt%, water uptake would reach a maximum, which would then decrease with concentration due to osmotic deswelling [46]. Different behaviors observed in the CMX and CMB membranes can be attributed to the density of their polymeric structures [27]. In all cases, the CMB membrane presented a higher water uptake than the CMX membrane.

Table 6. Water uptake measured in cation-exchange and bipolar membranes.

Water Uptake (%)									
Solution	**HCl 0.5 wt%**	**HCl 2.5 wt%**	**HCl 5.0 wt%**	**LiOH 0.5 wt%**	**LiOH 2.5 wt%**	**LiOH 5.0 wt%**	**LiCl 14 wt%**	**LiCl 25 wt%**	**LiCl 34 wt%**
CMX	-	-	-	31.2 ± 0.6	30.5 ± 0.4	29.5 ± 0.2	27.1 ± 0.0	20.3 ± 0.6	13.8 ± 0.2
CMB	-	-	-	35.9 ± 0.7	38.0 ± 0.8	37.1 ± 0.8	30.6 ± 0.5	26.5 ± 0.8	18.9 ± 0.7
Fumasep FBM	37.5 ± 0.3	48.2 ± 0.2	40.6 ± 0.3	52.1 ± 0.6	55.2 ± 0.0	53.1 ± 0.0	-	-	-
Neosepta BP	31.1 ± 0.7	30.9 ± 0.3	30.4 ± 0.1	33.8 ± 0.4	35.9 ± 0.2	30.4 ± 0.6	-	-	-

Table 7. Wet membrane thickness measured in cation-exchange and bipolar membranes.

	Thickness of Wet Membrane (μm)								
Solution	**HCl 0.5 wt%**	**HCl 2.5 wt%**	**HCl 5.0 wt%**	**LiOH 0.5 wt%**	**LiOH 2.5 wt%**	**LiOH 5.0 wt%**	**LiCl 14 wt%**	**LiCl 25 wt%**	**LiCl 34 wt%**
CMX	-	-	-	172 ± 1	170 ± 1	168 ± 1	167 ± 1	165 ± 1	164 ± 1
CMB	-	-	-	198 ± 2	198 ± 2	198 ± 1	196 ± 0	196 ± 1	195 ± 1
Fumasep FBM	176 ± 2	171 ± 1	171 ± 1	189 ± 1	191 ± 2	192 ± 1	-	-	-
Neosepta BP	229 ± 2	229 ± 1	230 ± 1	263 ± 1	266 ± 1	265 ± 2	-	-	-

Bipolar membranes in contact with LiOH solutions presented a behavior similar to that observed for the CMB membrane. A higher water uptake was observed for a 2.5 wt% LiOH solution, which then decreased for a 5.0 wt% LiOH concentration. When they were equilibrated with HCl solutions, the same behavior was observed for the Fumasep FBM membrane as concentration increased, while the Neosepta BP membrane showed a slight tendency to decrease its water uptake. These differences would be determined by the ion-exchange layers' specific characteristics in the respective membranes. The Fumasep FBM membrane showed a higher water uptake compared to the Neosepta BP membrane in all cases.

With respect to membrane thickness, the CMX membrane shows a clear tendency to decrease in thickness with concentration, which is consistent with the water uptake results. In contrast, the CMB membrane presents less variability in thickness for different LiOH concentrations, and an average thickness 1.2% lower for LiCl solutions. When comparing bipolar membranes, the Neosepta BP membrane shows a higher thickness than the Fumasep FBM membrane. Regarding different solutions, the Neosepta BP and Fumasep FBM membranes, when equilibrated with LiOH solutions, presented 15.4% and 10.4% higher thickness, respectively, compared to HCl solutions. This could be attributed to a greater lithium ion effect on the cation-exchange layer due to its higher hydrated radius compared to other monovalent cations [45].

3.2. Lithium Transport Number

The influence of current densities (300, 700, and 1100 $A{\cdot}m^{-2}$) and initial LiOH concentrations (0.5, 2.5, 5.0, and 8.0 wt%) on lithium transport number at 14 wt% LiCl for CMX and CMB was studied; the results are summarized in Figure 5.

3.2.1. Influence of Current Density

Current density shows influence on lithium transport. As can be seen in Figure 5, the lithium transport number decreases at higher current densities. This can be best observed for initial concentrations of 0.5 wt% and 2.5 wt% LiOH. In the CMX membrane, comparing LiOH concentrations at 0.5 wt% and 2.5 wt%, it was observed that when increasing current density from 300 to 1100 $A{\cdot}m^{-2}$, the lithium transport number decreased by 13% and 30%, respectively. In the case of the CMB membrane for the same comparison, the lithium transport number decrease was 11% and 35%, respectively. At higher concentrations of LiOH (5.0 wt% and 8.0 wt%), although transport number variation followed the trend of decrease with LiOH concentration, it failed to clearly differentiate the effect of current density. However, at 1100 $A{\cdot}m^{-2}$, it is appreciated that CMB membrane use showed a greater reduction in the lithium transport number—49% lower than that obtained with the CMX membrane. On the other hand, the results obtained for 8.0 wt% LiOH for both membranes (see Figure 5) indicate that a higher current density could counteract the effect of decreasing lithium transport number with concentration. This might be related to a lithium flux mostly associated to migration transport over diffusion transport, and could be explained qualitatively by the Nernst–Planck equation, where a higher electric field increases migration transport, while diffusive transport depends on the concentration difference on both sides of the membrane. Thus, high LiCl concentrations promote diffusive transport of lithium ions, whose effect on total transport decreases at higher current densities.

A problem observed when applying the Hittorf method at high concentrations was that when working with a 20-mL LiOH sample volume, the sensitivity of calculating molar flux and lithium-ion transport number increased due to the low final volume and high concentration, according to Equations (1) and (2).

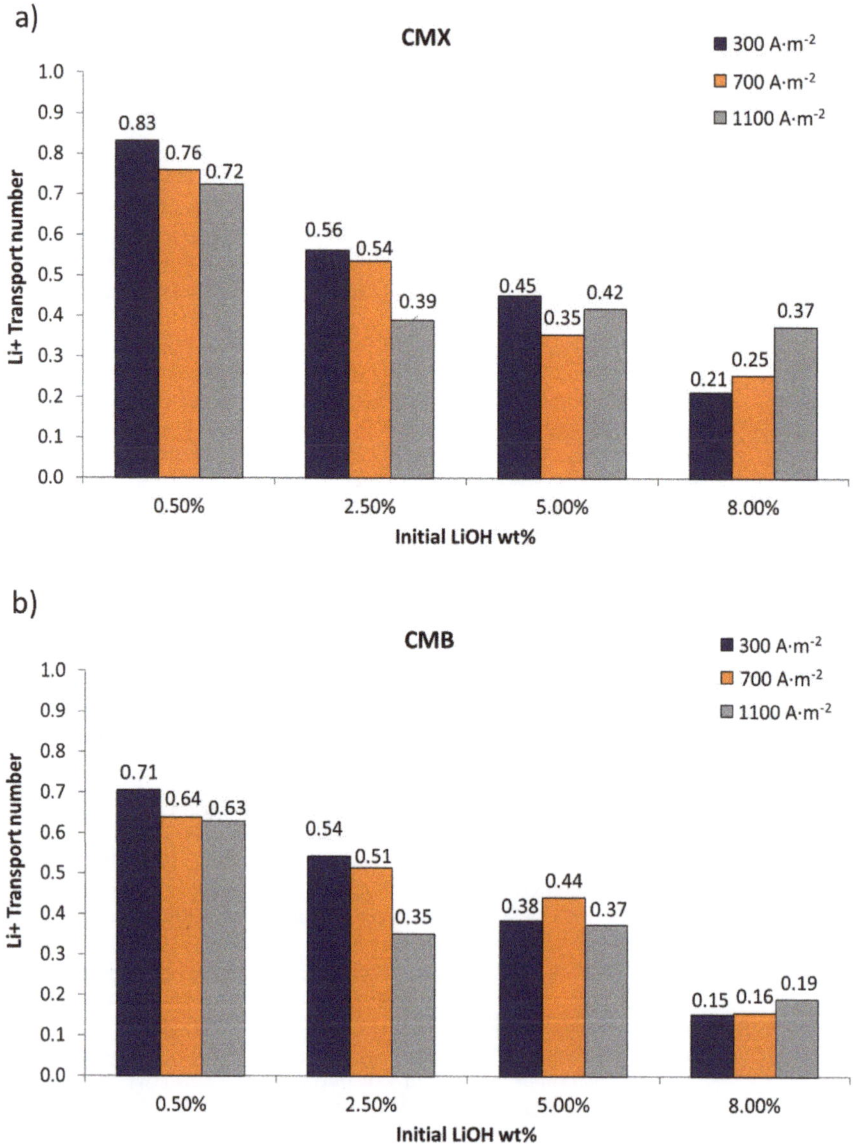

Figure 5. Lithium transport number in cation-exchange membranes according to LiOH concentration and current density (14 wt% LiCl): (**a**) CMX membrane; (**b**) CMB membrane.

3.2.2. Influence of Initial LiOH Concentration

The lithium transport number decreases with LiOH concentration. For the CMX and CMB membranes, at 0.5 wt% LiOH concentration, the lithium transport number was in the range of 0.83–0.72 and 0.71–0.63, respectively. That is, more than 60% of the electric current was used for lithium transport through the cationic membranes. For the CMX membrane, when concentrating the LiOH solution from 0.5 wt% to 2.5, 5.0, and 8.0 wt%, the average decrease in the lithium transport number was 35%, 47%, and 64%, respectively. Meanwhile, in the CMB membrane, for the same concentrations, the average decrease was 29%, 40%, and 75%, respectively. The best result was 0.83 with the CMX membrane at an initial LiOH concentration of 0.5 wt%.

One of the possible causes of the decrease in the lithium transport number with LiOH concentration is the decrease in water content in the cationic membrane [37]. Due to the increase in concentration, membranes tend to lose water; this can be attributed to the accumulation of osmotic pressure within them [37,46]. According to the work varied out by Izquierdo-Gil et al. [47], comparing water content and salt transport in cationic Nafion membranes of different thickness, it was reported that water uptake increases with membrane thickness and decreases with cation size. In the present work, although the CMB membrane was thicker and showed higher water uptake than the CMX membrane, it had a lower lithium transport number—a condition that can be attributed to other characteristics of the CMB membrane, such as fixed charge density and polymeric matrix cross-linking [48]. Water uptake into the membrane allows ionic binding between two sides of the membrane, making electrical conductivity possible [47,49]. Increasing the concentration of electrolytes in contact with the membrane causes an increase in the concentration of counterions in the membrane, which in turn generates osmotic deswelling, further increasing its ion concentration [46]. Eventually, water loss and high counterion concentration can reduce the membrane's fixed charge density. This has been explained thermodynamically by Kamcev et al. [46,50] by means of the counterion condensation model. Such a situation occurs when counterions close to fixed charges at a separation of less than the Bjerrum length do not have sufficient thermal energy to free themselves from the interaction of electrostatic forces, and tend to remain in this region, generating a shielding in fixed charges. Counterions trapped in this region are considered "condensed" and, therefore, immobile under a concentration gradient. However, counterions, under an electric field, exhibit higher diffusion coefficients than those that are not condensed, due to the shorter distance they must move in the membrane structure [51,52]. Although counterions' condensation favors their transport, high concentration of these counterions implies a reduction in Donnan potential, and causes an increase in co-ion sorption in the membrane [52]. For the application of membranes in LiOH production with concentrated solutions, it is highly probable that counterion condensation occurs, which under an electric field would favor lithium migration through the cation-exchange membrane. However, this may increase OH^- co-ion concentration, which, due to its high mobility (the highest among all anions) promotes undesired OH^- transport, resulting in a decrease in the lithium transport number. The use of membranes with higher fixed charge density would present higher ionic conductivity, facilitating the transport of lithium ions across the membrane [51], as long as OH^- ion leakage is reduced.

In this work, membranes' water uptake tended to decrease with LiOH concentration. However, this variation does not capture the behavior of the lithium transport number for all cases. This lithium transport number reduction is better explained by the presence of OH^-, and its interaction with the cation-exchange membrane. On the other hand, the decrease in the observed lithium transport number can be explained by the fact that the salt transport rate across the membrane decreases with decreasing concentration difference on both sides of the membrane, causing the permeation rate to decrease [47]—as occurs in the case of this work, where the LiOH concentration increased and the concentration difference with the LiCl solution decreased.

3.2.3. pH Variation

During experiments, it was observed that the LiCl solution's measured pH increased in all cases from pH 7–8 to a basic pH between 12–13. This increase can be attributed to the undesired migration of OH^- anions across the cation-exchange membrane due to the membrane not being 100% permselective, and also to the high OH^- ion mobility associated with the Grotthuss mechanism, involving proton hopping by ionization of water molecules. It has been reported that OH^- ions' migration through the cation-exchange membranes reduces current efficiency in membrane processes for sodium hydroxide production [40], which in bipolar membrane electrodialysis can be manifested by an increase in the feed salt solution's pH [53]. In this work, a higher pH increase was observed in the LiCl feed solution when working with more concentrated LiOH solutions. Hence, the lithium transport number decrease can be mainly attributed to electrical current transport by OH^- leakage through the cation-exchange membrane.

3.2.4. Influence of LiCl Concentration

To obtain a better understanding of the effects of LiCl concentration and concentration difference between electrolytes on each side of the membrane, additional experiments were performed for 0.5 wt% and 5.0 wt% LiOH at 25 wt% LiCl (Tests 5 and 6 in Table 3); The corresponding results are presented in Figure 6. When comparing the effects of LiCl concentration (see Figures 5 and 6), results show that for the CMX membrane at 0.5 wt% LiOH concentration, the transport number decreased with LiCl concentration. Thus, when using a LiCl solution at 14 wt%, a value of 0.72 was obtained, while with a LiCl solution at 25 wt%, the transport number was 0.66. The same behavior was not obtained for the CMB membrane, where the transport number was 0.63–0.65 for the same comparison. This can be attributed to the fact that the CMB membrane has an average thickness 18.9% higher compared to the CMX membrane, with 50–60% higher exchange capacity and 13.3–37.5% higher water uptake (see Tables 1, 6 and 7). In addition, the higher electrical resistance of the membrane would be associated with a denser polymeric structure [27]. These characteristics would allow the reduction of the influence of high LiCl concentrations on the lithium transport number. At higher LiOH concentrations, for the CMX membrane it was measured that at a concentration of 5.0 wt% LiOH, the lithium transport number decreased from 0.42 to 0.36 when using LiCl solutions of 14 wt% and 25 wt%, respectively. In the case of the CMB membrane, the lithium transport number was 0.37 and 0.44 for the same comparison.

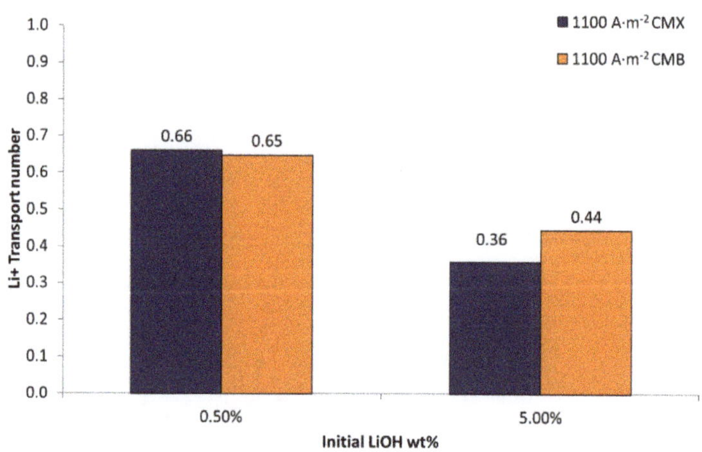

Figure 6. Lithium transport number in cation membranes (CMX and CMB) according to LiOH concentration and current density. LiCl 25 wt%.

In the particular case of this work, the transport number decrease can be attributed to the fact that high concentrations of LiCl and LiOH promote higher concentration of Li^+ in the membrane as a counterion. This, along with the principle of electroneutrality and Donnan exclusion, causes increase in the concentration of Cl^- and OH^- co-ions in the membrane [48]. According to the Nernst–Planck flux equation applied to the membrane, the latter would promote more OH^- ion leakage across the cation-exchange membrane due to increased concentration and current density [54,55]. In this work, this is manifested by a greater influence of the LiOH solution on the lithium transport number compared to the LiCl solution, which can be attributed to the presence of OH^- as a co-ion in the membrane.

Regarding the membranes used, the CMX membrane presents a higher transport number for lithium ions than the CMB membrane, which can be attributed to its lower thickness and electrical resistance. On the other hand, at high LiOH concentrations (8.0 wt%), an increase in current density increases the lithium transport number—especially for the CMX membrane.

3.3. Current–Voltage Curves

To determine the performance of bipolar and ion-exchange membranes, linear sweep voltammetry was performed as presented in Figures 7–9.

3.3.1. Salt Leakage through Bipolar Membranes

In the BMED process for LiOH production, Li^+ leakage into the HCl solution and Cl^- leakage into the LiOH solution through the bipolar membrane can occur.

Figure 7 presents current–voltage curves measured for the determination of salt transport through bipolar membranes at two different concentrations: LiCl 14 wt% and 25 wt%. Results are presented for the Fumasep FBM and Neosepta BP membranes. At a concentration of LiCl 14 wt%, a plateau in the curve was clearly observed for both membranes, indicating a limiting current density associated with the transport of salts through the membrane. At 25 wt% LiCl concentration, the plateau was less defined and presented a slope, which is probably due to a significant transport of co-ions—which carry electric current—through the membrane [44]. When using 14 wt% LiCl as an electrolyte, the measured limiting current densities were 176 $A{\cdot}m^{-2}$ and 77 $A{\cdot}m^{-2}$ for the Fumasep FBM and Neosepta BP membranes, respectively. In contrast, for a 25 wt% LiCl electrolyte, the measured limiting current densities were approximately 334 $A{\cdot}m^{-2}$ and 79 $A{\cdot}m^{-2}$, respectively. Limiting current densities for the Neosepta BP membrane were lower, indicating that it possesses characteristics more suitable for operating at high electrolyte concentrations (see Table 2). Studies have been reported that indicate a decrease in salt leakage through bipolar membranes by increasing the thickness of one of the cationic or anionic layers that compose a bipolar membrane [44]. A similar effect can be expected when comparing the Neosepta BP and Fumasep FBM membranes, with the former being thicker.

For production process application, results indicate that for high concentrations (25% LiCl), the use of high current densities could reduce undesired transport of salts (for instance, in the case of the bipolar membrane Fumasep FBM, a current density of 1000 $A{\cdot}m^{-2}$). Thus, electric current through bipolar membranes is transported mostly by H^+ and OH^- ions generated by water dissociation in the bipolar membrane, and to a lower extent by leakage of Li^+ and Cl^- salts.

Balster et al. [44] studied salt leakage through a BP-1 membrane using the same characterization method. They found that the first limiting current density associated with salt transport was 0.61 $mA{\cdot}cm^{-2}$ (6.1 $A{\cdot}m^{-2}$) using a 2 M NaCl solution (approximately 11 wt%). By laminating AMX membranes on the anionic face of the BP-1 membrane (asymmetric bipolar membranes), they achieved a 25–30% decrease in the first current density, at the cost of an increase in electrical resistance by 32% and 84% by the addition of one and two AMX membranes, respectively. As reported in their study, the first limiting current density decreases with the thickness of the anionic layer, meaning reduced salt leakage through the bipolar membrane and, therefore, a base with fewer impurities might be obtained.

In none of the obtained curves (see Figure 7) the second limiting current density—associated with the limited diffusion of water molecules towards the catalytic interface of the bipolar membrane—was observed.

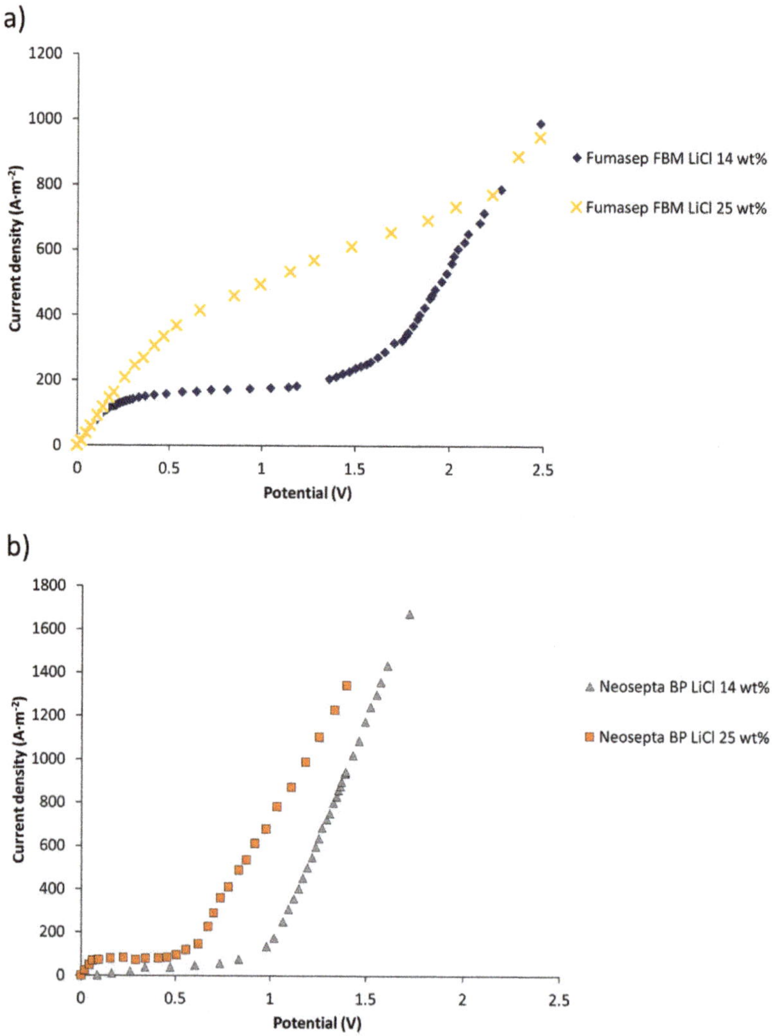

Figure 7. Salt leakage current–potential curves in bipolar membranes: (**a**) Fumasep FBM; (**b**) Neosepta BP.

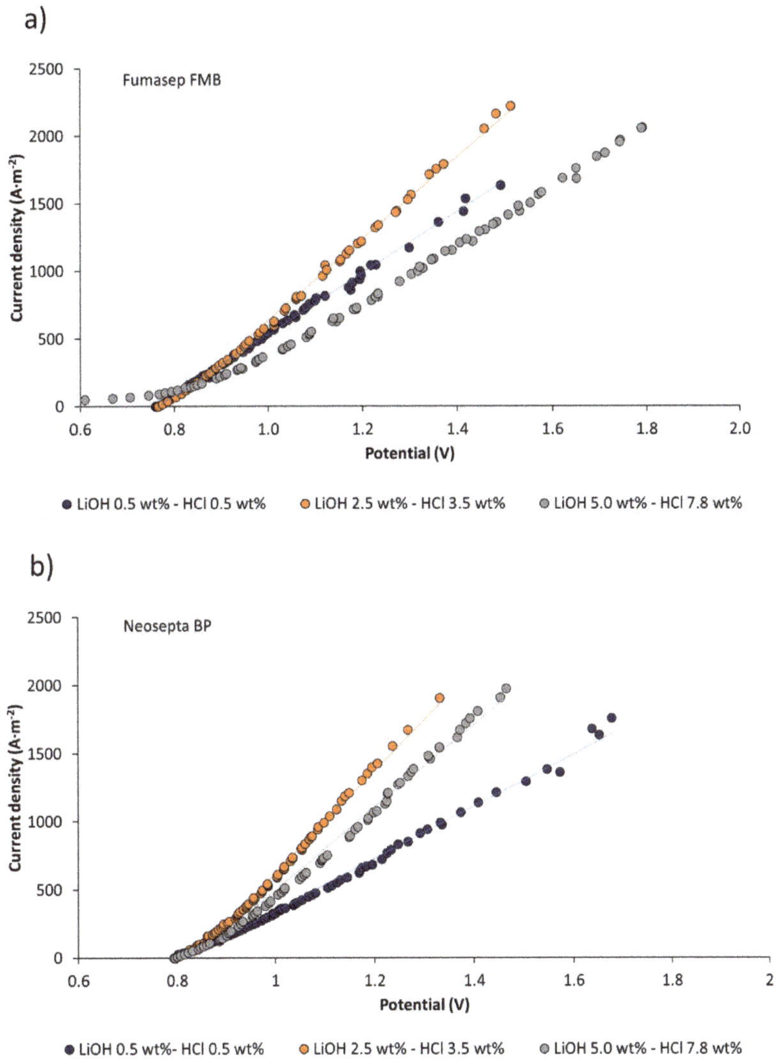

Figure 8. Linear sweep voltammetry of bipolar membranes under production operating conditions: (**a**) Fumasep FBM; (**b**) Neosepta BP.

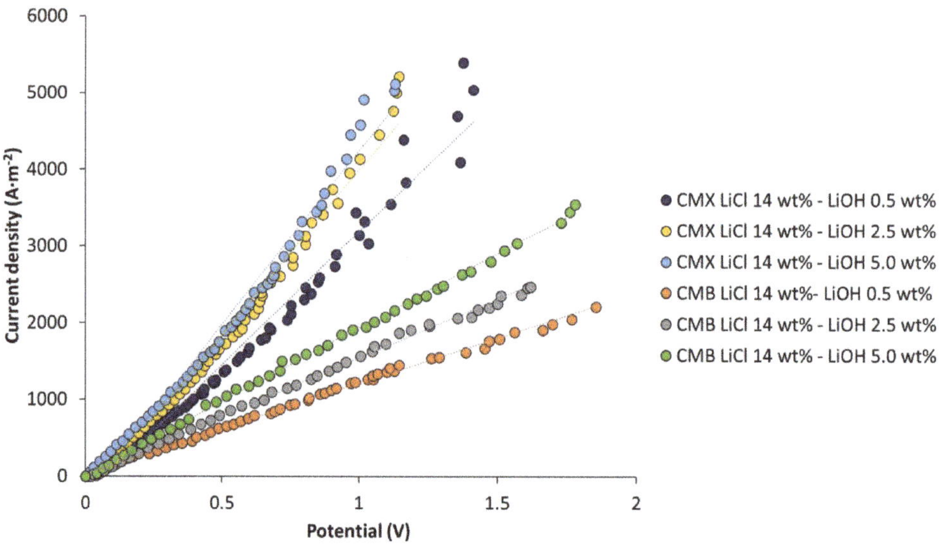

Figure 9. Linear sweep voltammetry in cationic membranes.

3.3.2. Linear Sweep Voltammetry on Bipolar Membranes under Production Conditions

Figure 8 presents current–voltage curves for Fumasep FBM and Neosepta BP membranes at different degrees of LiOH and HCl concentration. The A/V ratio (amperes/voltage) has a linear behavior for measured potential range. When comparing the effects of LiOH and HCl concentration processes, for the same current density, the voltage drop tends to decrease when reaching a concentration of 2.5 wt% LiOH and 3.5 wt% HCl. Then, it presents a tendency to increase at higher concentrations (LiOH 5.0 wt% and HCl 7.8 wt%). For both types of membrane, the highest A/V value is obtained for 2.5 wt% LiOH and 3.5 wt% HCl concentrations. This means lower apparent electrical resistance of the membranes at such concentrations, which can be attributed to increased electrolytic conductivity in the solution on the membrane surface and osmotic deswelling at higher concentrations (LiOH 5.0 wt% and HCl 7.8 wt%), leading to an increase in membrane electrical resistance [56]. When LiOH and HCl concentration increases—up to 5.0 wt% and 7.8 wt%, respectively—A/V ratio decreases and apparent electrical resistance increases, suggesting that there is an optimal concentration value close to 2.5 wt% LiOH and 3.5 wt% HCl, where the membrane voltage drop is lower. Regarding the increase in the electrical resistance of the bipolar membrane at concentrations of 5.0 wt% LiOH and 7.8 wt% HCl, this can be attributed to concentrated electrolyte that is absorbed by the membrane [57]. At high concentrations, the effect of membrane dehydration has been observed [56,57]; at the same time, membrane conductivity decreases due to absorbed electrolyte by the membrane. At 5.0 wt% HCl and 7.8 wt% LiOH concentrations, the Fumasep FBM membrane exhibits higher apparent electrical resistance than the Neosepta BP membrane. This can be attributed to its higher electrolyte absorption compared to the Neosepta FBM membrane (see Table 6). The latter is associated with a lower electric charge density, which is inversely proportional to water uptake [58].

The dissociation voltage of water in bipolar membranes has been reported to be 0.83 V, but it is usually higher in practical applications [59,60]. In the case of linear sweep voltammetry in Figure 8, the dissociation voltage measured for the Fumasep FBM and Neosepta BP membranes was 0.76–0.80 V and 0.82–0.84 V, respectively. Low water dissociation voltages observed for the Fumasep FBM membrane have also been reported by Xu et al. [61], and can be attributed to an exergonic secondary neutralization reaction between the OH^- and H^+ ions, which reduces the electrical energy requirement for water electrolysis. On the other hand, this result can be attributed to leakage current associated with a non-ideal selectivity of the membrane [62].

In the case of the Fumasep FBM membrane at higher concentrations (5.0 wt% LiOH and 7.8 wt% HCl), a small plateau is detected in the graph for potential values below 0.8 V. This can be explained by the increase in salt leakage with concentration. For practical purposes of the production of bases and acids at high concentrations, the bipolar membrane Neosepta BP presents better performance with respect to salt leakage reduction.

3.3.3. Linear Sweep Voltammetry in Cation-Exchange Membranes

Linear sweep voltammetry was performed on the CMX and CMB membranes at different LiOH concentrations. Figure 9 presents a comparative plot of the different current–voltage curves obtained. Current variation with potential was linear for the measurement range used, and limiting current density was not reached in any case. It can be observed from the graph that current densities close to 5000 A·m^{-2} for the CMX membrane, and between 2200 and 3500 A·m^{-2} for the CMB membrane, were reached for a potential range between 0 and 1.8 V. This indicates that the high electrolyte concentrations studied allow high availability of Li^+ ions on the membrane surface to migrate across the membrane without the occurrence of concentration polarization. The highest current densities were reached at higher electrolyte concentrations. For theCMX membrane, the current density at 5.0 wt% LiOH was on average 32% higher than the value at 0.5 wt% LiOH. In the case of the CMB membrane, this difference was 60%. On the other hand, when comparing the membranes at the same concentration, the CMX membrane allowed us to obtain current densities 113–159% higher than those obtained with the CMB membrane. Better energy efficiency was observed for the CMX membrane, as it achieved a higher electric current density at a lower potential difference. This can be attributed to the fact that the CMX membrane has 13–21% less thickness than the CMB membrane (see Table 7), and an electrical resistance 22–24% lower (see Table 1).

For the application of the membranes in LiOH production, the selected concentrations and flux rates (1.0–1.4 cm/s) were adequate for lithium transport through cationic membranes without reaching the limiting current density.

3.4. Long-Running Production Tests of LiOH by BMED

Long-running tests of LiOH production were performed using two different bipolar membranes (Neosepta BP and Fumasep FBM) and cation-exchange membranes (CMX and CMB) (Table 5). The current densities used were 500 and 1000 A·m^{-2}. These current densities were chosen according to the LSV results in order to reduce salt leakage through the bipolar membranes (see Figure 7). The initial LiCl feed concentration was between 14 and 34 wt%. Initial LiOH and HCl concentrations equal to 0.5 wt% were used in all tests. The obtained results are summarized in Table 8.

Table 8. Summary of the main results obtained in long-running tests.

Test	Test 1			Test 2			Test 3			Test 4			Test 5			Test 6		
Cation membrane	CMX			CMB			CMX			CMX			CMX			CMX		
Bipolar membrane	Neosepta BP			Neosepta BP			Neosepta BP			Fumasep FBM			Fumasep FBM			Fumasep FBM		
Current density (A·m^{-2})	1000			1000			1000			1000			500			500		
LiCl initial concentration (wt%)	14			14			25			25			14			34		
Number of unit cells	2			2			2			2			4			4		
LiOH concentration (wt%)	1.98	3.16	4.05	2.11	3.34	4.35	1.93	3.25	4.43	1.81	2.98	3.97	2.05	3.35	4.13	3.10	4.80	5.20
Cl^- concentration in LiOH solution (wt%)	0.10	0.19	0.26	0.05	0.13	0.21	0.06	0.13	0.46	0.07	0.32	0.52	0.18	0.40	0.69	0.48	1.23	1.24
Final purity (wt%)	94.6	93.5	93.4	97.9	96.0	95.4	96.0	95.3	88.6	92.9	86.5	83.8	90.7	87.8	83.6	76.8	66.0	54.7
SEC (kWh·kg^{-1})	6.94	7.72	8.71	7.57	8.58	9.45	8.23	8.58	9.01	8.98	9.46	10.23	5.97	7.14	9.29	6.81	8.92	11.94
CE	0.72	0.65	0.58	0.77	0.66	0.59	0.68	0.64	0.60	0.63	0.58	0.53	0.69	0.58	0.46	0.54	0.39	0.31

3.4.1. Product Purity

Among the results, Figure 10 shows variation in LiOH and Cl^- ion concentrations over time according to different operating conditions. The presence of chloride as an impurity can be attributed to the leakage of Cl^- into the bipolar membrane and undesired transport of this anion across the cation-exchange membrane, due to high LiCl concentration and its effect on co-ion concentration in the membrane [48]. Cl^- molar flux into the LiOH compartment was calculated to be between 0.47 and 1.06 mol·m^{-2}·h^{-1} when using a 14 wt% LiCl concentration, whereas for a 34 wt% LiCl concentration, Cl^- flux was between 1.29 and 2.28 mol·m^{-2}·h^{-1}, evidencing the influence of LiCl concentration on undesired Cl^- transport.

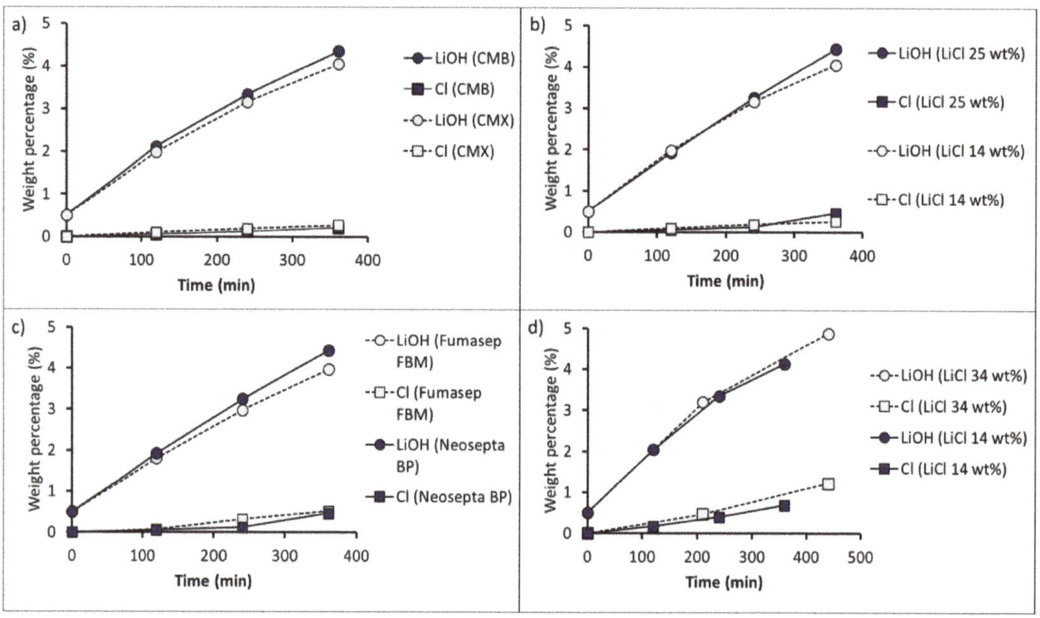

Figure 10. Concentration variation of LiOH and Cl^- in LiOH solution according to different operating conditions: (**a**) comparison of the cation-exchange membranes CMX and CMB (Test 1 and Test 2); (**b**) comparison of 14 and 25 wt% LiCl concentrations at 1000 A·m^{-2} (Test 1 and Test 3); (**c**) comparison of the bipolar membranes Neosepta BP and Fumasep FBM (Test 3 and Test 4); (**d**) comparison of 14 and 34 wt% LiCl concentrations at 500 A·m^{-2} (Test 5 and Test 6). Initial LiOH concentration 0.5 wt%.

In relation to the two cation-exchange membranes used, the CMB membrane (Test 1) presented a LiOH concentration on average 6.6% higher than that obtained with the CMX membrane (Test 2), with a Cl^- content of 0.21 wt% and 0.26 wt%, respectively (see Figure 10a). This means that with the CMB membrane, as the LiOH solution was concentrated from 2.11 wt% to 4.35 wt%, the solution purity decreased from 97.9% to 95.4%, respectively. On the other hand, in the case of the CMX membrane, the increase in LiOH concentration from 1.98 wt% to 4.05 wt% implied a reduction in purity from 94.6 to 93.4%. The latter can be attributed to the fact that the CMX membrane is thinner and shows lower water uptake, presenting less resistance to undesired Cl^- diffusion and OH^- leakage through it.

Figure 10b presents the LiOH concentration results when using different LiCl concentrations (14 wt% and 25 wt%) at 1000 A·m^{-2}. The use of a 25 wt% LiCl concentration allowed us to obtain a final LiOH concentration 9.4% higher compared to the use of a 14 wt% LiCl solution. However, this implies a 76.9% higher Cl^- content (see Test 1 and 3 in

Table 8). After 360 min of processing, the final LiOH purity was 93.4% and 88.6% when using 14 wt% and 25 wt% LiCl solutions, respectively.

Regarding the comparison of bipolar membranes in Tests 3 and 4 (see Figure 10c), a final LiOH concentration of 4.43 wt% and 3.97 wt%. with a Cl^- content of 0.52 wt% and 0.46 wt%, was obtained for the Neosepta BP and Fumasep FBM membranes, respectively. The LiOH solution purity decreased with LiOH concentration. When the Neosepta BP bipolar membrane was used (Test 3), the LiOH solution purity decreased from 96.0% to 88.6% when concentrated from 1.93 wt% to 4.43 wt%. On the other hand, in the case of the Fumasep FBM membrane (Test 4), the solution purity decreased from 92.9% to 83.8% when concentrating the LiOH solution from 1.81 wt% to 3.97 wt%.

In Tests 5 and 6, the use of four three-compartment cells at a current density of 500 $A{\cdot}m^{-2}$ (see Figure 10d) resulted in a final LiOH solution with higher Cl^- content. This can be attributed to an increase in Cl^- diffusion into the cation-exchange membrane as the total membrane area in the stack increased, and the current density of 500 $A{\cdot}m^{-2}$ was not able to mitigate bipolar membrane Cl^- ion leakage. When using a 14 wt% LiCl feed (Test 5), the LiOH solution purity decreased from 90.7% to 83.6% when concentrating from 2.05 wt% to 4.13 wt% LiOH. On the other hand, when a 34 wt% LiCl feed was used (Test 6), Figure 10d shows that, at 210 min of operation, LiOH concentration increased from 0.50 wt% to 3.10 wt%, with a Cl^- concentration of 0.47 wt%. After this point, an accelerating trend in the increase of Cl^- concentration in the LiOH compartment can be observed in the graph, reaching 1.23 wt% after 440 min (corresponding to a purity of 66.0% LiOH in solution). This behavior was accompanied with a decrease in LiOH concentration rate, as observed in a change of slope in the curve, suggesting a greater leakage of Cl^- ions through the bipolar membrane promoted by HCl concentration increase. On the other hand, there was a higher undesired transport of Cl^- ions across the cation-exchange membrane due to a high LiCl concentration (34 wt%), affecting the final product purity.

3.4.2. SEC and Current Efficiency

Figure 11 shows the total specific electricity consumption (SEC) and current efficiency (CE) corresponding to LiOH production in long-running tests.

Energy efficiency according to cation-exchange membrane type was compared based on the results of Tests 1 and 2, and the results are presented in Figure 11a. For Test 1 with the Neosepta BP bipolar membrane and the CMX membrane, when concentrating the LiOH solution from 1.98 wt% to 4.05 wt%, specific electricity consumption (SEC) increased by 25.6% and current efficiency (CE) decreased by 19.9%. On the other hand, in Test 2, when using the CMB membrane, SEC increased by 24.8% and CE decreased by 23.4%. However, the average SEC and CE with the CMB membrane were 10% and 3% higher, respectively, compared to the CMX membrane. The higher SEC obtained with the CMB membrane can be attributed to the higher electrical resistance of this membrane (see Table 1).

Figure 11b compares SEC and CE for the process at 1000 $A{\cdot}m^{-2}$ for initial LiCl solutions of 14 wt% and 25 wt% according to Tests 1 and 3, respectively. For Test 1, with 14 wt% LiCl solution, when concentrating the LiOH solution from 1.98 wt% to 4.05 wt%, SEC increased by 25.6% and CE decreased by 19.9%. Meanwhile, for Test 3, with 25 wt% LiCl solution, when concentrating the LiOH solution from 1.93 wt% to 4.43 wt%, SEC increased by 9.5% and CE decreased by 11.8%. With 14 wt% LiCl solution, a 12.9% lower SEC and 3.6% higher CE was observed up to LiOH concentrations of 3.16–3.25 wt%. After this point, energy efficiency reduction could be attributed to a decrease in the concentration and electrical conductivity of the LiCl solution over time. When initial LiCl solutions of 14 wt% and 25 wt% were used, the final LiCl concentration was 8.71 wt% and 18.16 wt%, respectively.

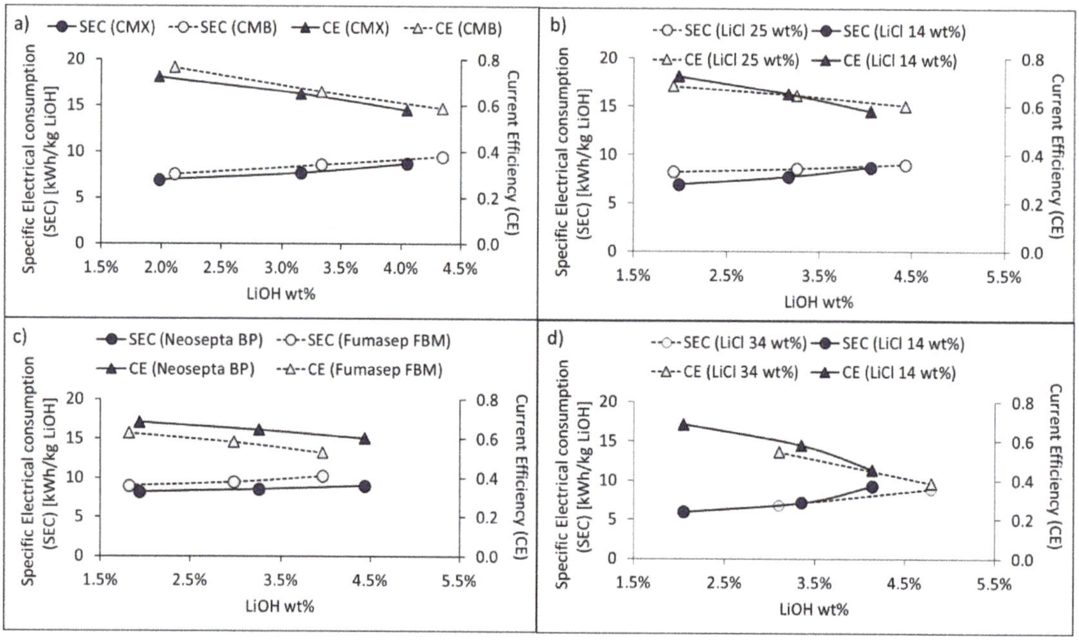

Figure 11. Specific electricity consumption and current efficiency under different operating conditions: (**a**) comparison of the cationic membranes CMX and CMB (Test 1 and Test 2); (**b**) comparison of 14 and 25 wt% LiCl concentrations at 1000 $A \cdot m^{-2}$ (Test 1 and Test 3); (**c**) comparison of the bipolar membranes Neosepta BP and Fumasep FBM (Test 3 and Test 4); (**d**) comparison of 14 and 34 wt% LiCl concentrations at 500 $A \cdot m^{-2}$ (Test 5 and Test 6) Initial LiOH concentration 0.5 wt%.

The influence of the bipolar membranes Neosepta BP and Fumasep FBM influence can be observed when comparing Tests 3 and 4 in Figure 11c. When the Neosepta BP membrane was used, SEC was 10% lower compared to the Fumasep FBM membrane, while CE was 10% higher. The latter results in a better use of electric current in OH^- production by the Neosepta BP membrane, resulting in a final LiOH concentration of 4.43 wt%—11.6% higher when compared to the Fumasep FBM membrane. With the Neosepta BP membrane, when concentrating LiOH solution from 1.93 wt% to 4.43 wt%, SEC increased by 9.5% and CE decreased by 11.8%. On the other hand, when concentrating the LiOH solution from 1.81 wt% to 3.97 wt% with the Fumasep FBM membrane, SEC increased by 13.9% and CE decreased by 15.7%. These results show a higher energy yield when using the Neosepta BP membrane.

The effects of LiCl concentrations of 14 wt% (Test 5) and 34 wt% (Test 6) with four three-compartment cells and a current density of 500 $A \cdot m^{-2}$ are presented in Figure 11d. For Test 5, with 14 wt% LiCl solution, when concentrating the LiOH solution from 2.05 wt% to 4.13 wt%, specific electricity consumption (SEC) increased from 5.97 to 9.29 $kWh \cdot kg^{-1}$, while electrical current efficiency (CE) decreased from 0.69 to 0.46, respectively. On the other hand, for Test 6, Figure 11d shows that by concentrating LiOH from 0.50 wt% to 3.10 wt%, a specific electricity consumption of 6.81 kWh per kg LiOH was obtained, which then increased to 8.92 $kWh \cdot kg^{-1}$ when reaching a LiOH concentration of 3.80 wt%. When using a 14 wt% LiCl solution (Test 5), current efficiency was on average 6.2% higher compared to 34 wt% LiCl solution (Test 6). Nevertheless, after reaching a LiOH concentration of 3.35 wt%, CE decreased in an accelerated way due to a decrease in LiCl concentration.

For a LiCl concentration of 14 wt%, specific electricity consumption was between 12% and 20% lower compared to 34 wt% LiCl. This can be attributed to a decrease in electrolytic conductivity in concentrated aqueous LiCl solutions greater than 7 $mol \cdot kg^{-1}$ [63] (approximately 23 wt%).

The lowest SEC value obtained was 5.97 kWh per LiOH kg in Test 5 after the first 120 min of processing, when reaching a LiOH concentration of 2.05 wt%. This can be attributed to the lower current density used (500 $A \cdot m^{-2}$) and the advantage of increasing the number of membranes in the stack to four LiOH compartments. However, current efficiency (CE) when concentrating the LiOH solution to 3.35 wt% was 0.58, and as LiOH concentration increased to 4.13 wt%, the current efficiency decreased to 0.46. Thus, part of the electric current was consumed in leakage by Cl^- migration across the bipolar membrane, rather than in OH^- production. This is consistent with the results obtained in the LSV tests. This behavior evidences the advantage of using operational current densities higher than the first limiting current density in bipolar membranes.

According to our results, process energy efficiency is affected with increasing LiOH concentration according to OH^- leakage effects and cation membrane deswelling. On the other hand, increasing HCl solution concentration promotes a higher leakage of Cl^- ions into the LiOH compartment, affecting current efficiency in the bipolar membrane in the generation of OH^- ions, which also affects solution purity.

Table 8 presents a summary of results obtained in long-running tests. The best result was achieved in Test 2, where a Neosepta BP membrane, CMB cationic membrane and current density of 1000 $A \cdot m^{-2}$ were used. The possibility was shown of achieving LiOH solutions between 3.34 and 4.35 wt% concentration, with a range of 96.0–95.4% purity. Product purity is therefore inversely proportional to the final concentration obtained. The lowest average specific electricity consumption (SEC) was achieved in Test 1 between 6.94 and 8.71 kWh per kilogram of LiOH. On the other hand, the highest electrical current efficiency (CE) was achieved in Test 2, in the range of 0.77–0.59, with an SEC between 7.57 and 9.45 kWh per kilogram of LiOH.

The CMB membrane presented a high current efficiency despite its higher electrical resistance, which can be attributed to a higher resistance to co-ion leakage. However, it is not recommended to use it with LiCl solutions higher than 14 wt%. This membrane was tested in long-running tests at 25 wt% LiCl concentrations, showing damage to its structure after two hours of processing due to the high osmotic pressure reached, causing its rupture and contamination between LiCl and LiOH solutions. Consequently, these results are not presented.

3.4.3. Specific Electricity Consumption (SEC) Comparison

Specific BMED electricity consumption depends mainly on membranes' characteristics and their interaction with concentrated solutions under a specific electric current density. The SEC results obtained in this work were determined by final LiOH concentration, and are comparable to other membrane processes used for lithium recovery and LiOH production. Specific energy consumptions between 5.43 and 6.20 $kWh \cdot kg^{-1}$ of LiOH from lithium-rich solutions produced from salt lake brine have been reported for processes such as nanofiltration, reverse osmosis, and conventional electrodialysis integrated with BMED, achieving final LiOH concentrations between 0.58 and 1.03 M (approximately 1.37–2.41 wt%), with current efficiencies between 0.36 and 0.44% [64]. Regarding direct BMED application, SEC in the order of 6.60 $kWh \cdot kg^{-1}$ was reported from 0.725–0.730 wt% LiCl solutions containing organic solvents reaching 0.3 M LiOH concentrations (approximately 0.71 wt%) [28], while recently, from LiCl solutions of 100 $g \cdot L^{-1}$ (approximately 9.5 wt%), SECs between 2.78 $kWh \cdot kg^{-1}$ and 7.80 $kWh \cdot kg^{-1}$ of LiOH have been reported, with current efficiencies between 0.7384 and 0.1445 at 640 $A \cdot m^{-2}$ using CMB and BP membranes, achieving LiOH concentrations of 1.08 M and 2.40 M (approximately 2.5 wt% and 5.4 wt% LiOH), respectively [64]. These reports are consistent with the results presented in this work, demonstrating the influence of LiCl concentration and the degree of final LiOH concentration on process performance.

The same technology has been used with different starting solutions. Thus, a 0.5 M Li_2SO_4 feed [65] reports a specific electricity consumption of 7 $kWh \cdot kg^{-1}$, which increases when working with more concentrated solutions. Meanwhile, the use of aqueous Li_2CO_3

solutions up to 0.18 M presents an SEC of up to 20.4 kWh·kg^{-1} of LiOH [31], depending on operating conditions.

On the other hand, LiOH production by membrane electrolysis has reported an SEC of 6.1–14.6 kWh·kg^{-1} LiOH using initial LiOH solutions between 4 and 8 wt% as the initial catholyte [66] and 7.25 kWh·kg^{-1} of LiOH for an initial catholyte of 2.3 wt% [29].

For conventional industrial process of obtaining lithium hydroxide from Li_2CO_3 by chemical reaction with lime slurry, the specific energy consumption is 14.04 kWh·kg^{-1} of LiOH·H_2O (or 24.6 kWh·kg^{-1} of LiOH) [67], of which electricity, fuel, and natural gas consumption represent 17.3%, 10.3%, and 72.4% of the total energy required, respectively. High natural gas consumption can be attributed to the thermal energy required for the evaporation and crystallization stages of the process to obtain lithium hydroxide monohydrate, in which a lithium hydroxide solution of approximately 3 wt% must be evaporated to saturation.

The BMED process of obtaining LiOH does not eliminate the need to evaporate LiOH solution to crystallize LiOH·H_2O; however, it is expected that obtaining LiOH concentrations higher than 3 wt% would contribute to reduce heat requirements in these stages. Thus, if the LiOH solution is evaporated to saturation at 70°C, initial LiOH concentrations of 4 wt% and 5 wt% would allow the thermal energy requirement to be reduced by 9.4% and 18.9%, respectively, compared to using a 3 wt% LiOH solution.

Obtaining LiOH via the chemical reaction of Li_2CO_3 with lime slurry presents a conversion efficiency of 59.0–59.5% after one hour of processing at 60–100 °C [68], which, based on mass and energy balances, allows the estimation of a theoretical SEC of 1.27–2.84 kWh·kg^{-1} of LiOH. This required energy is lower than that obtained in this work by BMED. However, membrane processes could become competitive if process sustainability is considered by reducing the use of chemical reagents, reducing waste generation, and potential coupling with non-conventional renewable energies.

4. Future Challenges

The results in this work present the current scope of obtaining high concentrations of LiOH by BMED, variation of electrical energy consumption with concentration, and salt leakage related to Cl^- ion contamination of LiOH solution. The results of this work suggest that from an initial LiOH concentration of 0.5 wt%, it is possible to obtain concentrated LiOH solutions in the range of 3.34–4.35 wt%, with 96.0–95.4% purity and a specific electricity consumption between 7.57 and 9.45 kWh per kilogram of LiOH. After this point, current efficiency tends to decrease below 0.50, significantly increasing the specific energy consumption of the process. This is associated with high OH^- ion leakage in the cation-exchange membrane [40,53] and salt leakage in the bipolar membrane [44,69], which causes undesired Cl^- transport into the LiOH compartment. In the production of LiOH by membrane electrolysis, energy consumption can be reduced in a concentration range between 40 and 50 g·L^{-1} [30], which is approximately 3.5 wt% and4.3 wt% LiOH, respectively.

For the implementation of a lithium hydroxide production process by electrodialysis with bipolar membranes, there exist limitations related to membrane performance, affecting final product purity and energy efficiency. For the use of lithium hydroxide as a precursor for lithium batteries, work must be done to reduce the transport of impurities in membranes. Both monopolar and bipolar membranes are not 100% permselective. Thus, as LiOH and HCl concentrations increase, LiOH solution purity and process efficiency are affected.

Implementation-related challenges for this technology in LiOH production involve aspects such as the performance of monopolar membranes and bipolar membranes, stack design related to the optimal number of unit cells, membrane lifetime, and technology implementation coupled with renewable energy sources.

In this work, evidence is shown that process efficiency is mainly established by membrane performance. To date, several investigations have been carried out in order to improve cationic membranes' selective transport [70–72]. In the case of bipolar membranes, the salt leakage problem can be addressed by using asymmetric bipolar membranes [44]. However, this improvement in salt leakage reduction leads to higher electrical resistance and, therefore, an increase in electrical power consumption. Based on these investigations, it could be expected that new membranes will emerge in the future, being more suitable for application in the production of concentrated bases. Thus, the final LiOH concentration will be increased with monopolar membranes by higher resistance to OH^- leakage, and with bipolar membranes by reduced salt leakage and higher water dissociation efficiency.

Currently, these difficulties could be addressed by specific operating conditions. For instance, it has been shown in this work that at 14 wt% LiCl concentration, lower specific electricity consumption and higher electric current efficiency are obtained for LiOH production, so if a constant LiCl concentration could be established during LiOH production, this would increase energy efficiency, and would also help to slow down reduction in concentration difference between LiCl and LiOH solutions separated by the cationic membrane which, in turn, would contribute to reducing the rate of lithium transport decrease through this membrane [47]. On the other hand, with respect to Cl^- leakage in the bipolar membrane, an increase in Cl^- leakage rate related to simultaneous concentration of LiOH and HCl was observed. Given the molar mass difference between these salts, the mass concentration rate (wt%) of HCl solution is always higher compared to LiOH. Therefore, for production processes, a low HCl concentration can be set while LiOH concentration increases. In this way, Cl^- leakage into the LiOH compartment through the bipolar membrane would be reduced.

On the other hand, the application of a high current density would reduce the membrane area requirement for a given production, reducing the overall process cost, as membranes are expensive. Moreover, according to linear sweep voltammetry analysis and long-running tests, working at high current densities favors water dissociation over salt leakage [69]. However, current efficiency decreases and, therefore, specific energy consumption increases, in addition to the fact that catalytic interlayer may deteriorate, which would imply a decrease in membrane lifetime and an increase in voltage drop after a long period of operation [73].

In industry, obtaining a high concentration in the final LiOH solution (greater than 3.0 wt%) would bring benefits to the current crystallization process of $LiOH{\cdot}H_2O$ production, since it would reduce the current difference between LiOH concentration and saturation concentration at which crystallization of lithium hydroxide monohydrate begins. In addition, this would encourage the use of green technologies, given the potential use of this technology driven by renewable energy sources. The Atacama Desert in Chile is characterized by high levels of solar radiation—between 6.7 and 10.55 $kWh{\cdot}m^{-2}$ per day [74,75]. The application of membrane technologies for LiOH production—such as membrane electrodialysis [29] and bipolar electrodialysis membranes—is a potential alternative to the LiCl to Li_2CO_3 and subsequent Li_2CO_3 to LiOH conversion steps. Obtaining Li_2CO_3 from lithium brine reports an approximate SEC of 1.31 $kWh{\cdot}kg^{-1}$ of Li_2CO_3 [67], which implies an equivalent SEC of 6.83 $kWh{\cdot}kg^{-1}$ of LiOH (considering obtaining 0.38 kg of LiOH per kg of Li_2CO_3 for a 59% reaction conversion [68]). Therefore, the authors estimate that specific electricity consumptions determined by BMED and other membrane technologies offset those obtained by conventional methods, with the advantage of being potentially driven by clean technologies given the conditions of the Atacama Desert. Future studies should be conducted to quantify the real impact of integrating solar energy into these processes.

The design choice of a high-purity lithium hydroxide production process would be conditioned by investment costs and operating costs. Pilot tests and process simulations would allow the obtaining of key information according to different scenarios and process configurations. The latter constitutes work in progress that the authors would like to present in a follow-up to the present results.

5. Conclusions

Experimental tests were developed to determine and analyze the scope and feasibility of BMED application in high-purity lithium hydroxide production. Among the main results, the influence of high concentrations of LiOH and LiCl solutions on LiOH production energy efficiency and final solution purity were determined and analyzed. The best results for energy efficiency were obtained at low initial LiOH concentrations. A high LiCl concentration (25–34 wt%) was shown to increase electrical resistance and promote Cl^- diffusion into the LiOH solution. On the other hand, increasing the LiOH concentration was shown to cause OH^- ion leakage phenomena in the cation-exchange membrane, reducing its performance. Regarding bipolar membranes, high LiOH and HCl concentrations were associated with accelerated leakage of co-ions, impairing LiOH solution purity. This can currently be addressed with asymmetric bipolar membranes, or by avoiding high HCl concentrations with a high current density that promotes water dissociation over salt leakage.

For the application of electrodialysis with bipolar membranes, the results of this work show that it is possible to obtain a LiOH concentration in a concentration range of 3.34–4.35 wt%, with 96.0–95.4% purity. After this point, electrical power consumption and LiOH contamination with chloride ions tend to increase significantly, affecting solution purity. From a 0.5 wt% LiOH solution and a current density of 1000 $A \cdot m^{-2}$, a specific electrical energy consumption (SEC) of 9.45 $kWh \cdot kg^{-1}$ was determined with a current efficiency (CE) of 0.77–0.59, obtaining a final LiOH concentration of 4.35 wt%. On the other hand, with a current density of 500 $A \cdot m^{-2}$, an SEC of 5.97 $kWh \cdot kg^{-1}$ was obtained with a CE of 0.69, obtaining a LiOH concentration of 2.05 wt%. The highest current efficiency obtained was 0.77 at 0.5 wt% LiOH and 14 wt% LiCl concentrations.

To achieve high LiOH concentrations with higher efficiency, it is necessary to improve the performance of bipolar membranes at high concentrations, and the resistance of cation-exchange membranes to OH^- leakage, supported by suitable operating conditions.

Author Contributions: Conceptualization, M.G. and A.G.; methodology, A.G., P.S., M.C. and M.G.; formal analysis, M.G.; investigation, A.G.; resources, M.G. and M.C.; data curation, A.G. and A.Q.; writing—original draft preparation, A.G., S.U. and M.G.; writing—review and editing, A.G., M.G. and S.U.; supervision, M.G., P.S. and M.C.; funding acquisition, M.G. and S.U. All authors have read and agreed to the published version of the manuscript.

Funding: We give thanks to ANID/FONDAP/15110019, CONICYT/FONDECYT REGULAR N° 1191347. Alonso Gonzalez expresses thanks to CONICYT for funding his postgraduate studies: CONICYT-PFCHA/Doctorado Nacional/2017-21170998.

Institutional Review Board Statement: Not applicable.

Informed Consent Statement: Not applicable.

Data Availability Statement: Not applicable.

Acknowledgments: A.G. and A.Q. acknowledge the infrastructure and support of the Programa de Doctorado en Ingeniería de Procesos de Minerales of the Universidad de Antofagasta.

Conflicts of Interest: The authors declare no conflict of interest.

References

1. Zybert, M.; Ronduda, H.; Szczęsna, A.; Trzeciak, T.; Ostrowski, A.; Żero, E.; Wieczorek, W.; Raróg-Pilecka, W.; Marcinek, M. Different strategies of introduction of lithium ions into nickel-manganese-cobalt carbonate resulting in $LiNi_{0.6}Mn_{0.2}Co_{0.2}O_2$ (NMC622) cathode material for Li-ion batteries. *Solid State Ion.* **2020**, *348*, 115273. [CrossRef]
2. Li, H.; Eksteen, J.; Kuang, G. Recovery of lithium from mineral resources: State-of-the-art and perspectives—A review. *Hydrometallurgy* **2019**, *189*, 105129. [CrossRef]
3. Cochilco.cl. Oferta y Demanda de Litio Hacia el 2030. Available online: https://www.cochilco.cl/Mercado%20de%20Metales/Produccion%20y%20consumo%20de%20litio%20hacia%20el%202030.pdf (accessed on 16 July 2021).
4. Li, H.; Cormier, M.; Zhang, N.; Inglis, J.; Li, J.; Dahn, J. Is cobalt needed in Ni-rich positive electrode materials for lithium ion batteries? *J. Electrochem. Soc.* **2019**, *166*, A429–A439. [CrossRef]

5. Kim, J.; Scheers, J.; Hwang, G.; Zhao, X.; Kang, S.; Johansson, P.; Ahn, J.; Jacobsson, P. Role of lithium precursor in the structure and electrochemical performance of $LiFePO_4$. *Scr. Mater.* **2013**, *69*, 716–719. [CrossRef]
6. Liu, A.; Liu, Y.; Hu, Z.; Gao, G.; Xu, Y.; Lei, L. Electrochemical performance of $LiFePO_4/C$ synthesized by solid state reaction using different lithium and iron sources. *J. Phys. Chem. Solids* **2011**, *72*, 831–835. [CrossRef]
7. Fitch, B.; Yakovleva, M.; Meiere, S. Lithium hydroxide based performance improvements for nickel rich ncm layered cathode material. *ECS Meet. Abstr.* **2016**. [CrossRef]
8. Garrett, D. *Handbook of Lithium and Natural Calcium Chloride*; Elsevier: Amsterdam, The Netherlands, 2007; pp. 1–235.
9. Grágeda, M.; González, A.; Alavia, W.; Ushak, S. Development and optimización of a modified process for producing the battery grade LiOH: Optimization of energy and water consumption. *Energy* **2015**, *89*, 667–677. [CrossRef]
10. Chagnes, A. *Lithium Process Chemistry: Resources, Extraction, Batteries, and Recycling*; Elsevier Science: Amsterdam, The Netherlands, 2015.
11. Pourcelly, G. Electrodialysis with bipolar membranes: Principles, optimization, and applications. *Russ. J. Electrochem.* **2002**, *38*, 919–926. [CrossRef]
12. Koter, S.; Warszawski, A. A new model for characterization of bipolar membrane electrodialysis of brine. *Desalination* **2006**, *198*, 111–123. [CrossRef]
13. Ghyselbrecht, K.; Silva, A.; van der Bruggen, B.; Boussu, K.; Meesschaert, B.; Pinoy, L. Desalination feasibility study of an industrial NaCl stream by bipolar membrane electrodialysis. *J. Environ. Manag.* **2014**, *140*, 69–75. [CrossRef] [PubMed]
14. Reig, M.; Casas, S.; Valderrama, C.; Gibert, O.; Cortina, J. Integration of monopolar and bipolar electrodialysis for valorization of seawater reverse osmosis desalination brines: Production of strong acid and base. *Desalination* **2016**, *398*, 87–97. [CrossRef]
15. Fernandez-Gonzalez, C.; Dominguez-Ramos, A.; Ibañez, R.; Chen, Y.; Irabien, A. Valorization of desalination brines by electrodialysis with bipolar membranes using nanocomposite anion exchange membranes. *Desalination* **2016**, *406*, 16–24. [CrossRef]
16. Chen, B.; Jiang, C.; Wang, Y.; Fu, R.; Liu, Z.; Xu, T. Selectrodialysis with bipolar membrane for the reclamation of concentrated brine from RO plant. *Desalination* **2018**, *442*, 8–15. [CrossRef]
17. Zabolotskii, V.; Sheldeshov, N.; Melnikov, S. Heterogeneous bipolar membranes and their application in electrodialysis. *Desalination* **2014**, *342*, 183–203. [CrossRef]
18. Wei, Y.; Wang, Y.; Zhang, X.; Xu, T. Comparative study on regenerating sodium hydroxide from the spent caustic by bipolar membrane electrodialysis (BMED) and electro-electrodialysis (EED). *Sep. Purif. Technol.* **2013**, *118*, 1–5. [CrossRef]
19. Ye, W.; Huang, J.; Lin, J.; Zhang, X.; Shen, J.; Luis, P.; van der Bruggen, B. Environmental evaluation of bipolar membrane electrodialysis for NaOH production from wastewater: Conditioning NaOH as a CO_2 absorbent. *Sep. Purif. Technol.* **2015**, *144*, 206–214. [CrossRef]
20. Sun, M.; Li, M.; Zhang, X.; Wu, C.; Wu, Y. Graphene oxide modified porous P84 co-polyimide membranes for boron recovery by bipolar membrane electrodialysis process. *Sep. Purif. Technol.* **2020**, *232*, 115963. [CrossRef]
21. Hwang, C.W.; Jeong, M.H.; Kim, Y.J.; Son, W.K.; Kang, K.S.; Lee, C.S.; Hwang, T.S. Process design for lithium recovery using bipolar membrane electrodialysis system. *Sep. Purif. Technol.* **2016**, *166*, 34–40. [CrossRef]
22. Sun, X.; Lu, H.; Wang, J. Recovery of citric acid from fermented liquid by bipolar membrane electrodialysis. *J. Clean. Prod.* **2017**, *143*, 250–256. [CrossRef]
23. Parsa, N.; Moheb, A.; Mehrabani-Zeinabad, A.; Masigol, M. Recovery of lithium ions from sodium-contaminated lithium bromide solution by using electrodialysis process. *Chem. Eng. Res. Des.* **2015**, *98*, 81–88. [CrossRef]
24. Iizuka, A.; Yamashita, Y.; Nagasawa, H.; Yamasaki, A.; Yanagisawa, Y. Separation of lithium and cobalt from waste lithium-ion batteries via bipolar membrane electrodialysis coupled with chelation. *Sep. Purif. Technol.* **2013**, *113*, 33–41. [CrossRef]
25. Bunani, S.; Arda, M.; Kabay, N.; Yoshizuka, K.; Nishihama, S. Effect of process conditions on recovery of lithium and boron from water using bipolar membrane electrodialysis (BMED). *Desalination* **2017**, *416*, 10–15. [CrossRef]
26. Bunani, S.; Yoshizuka, K.; Nishihama, S.; Arda, M.; Kabay, N. Application of bipolar membrane electrodialysis (BMED) for simultaneous separation and recovery of boron and lithium from aqueous solutions. *Desalination* **2017**, *424*, 37–44. [CrossRef]
27. Pärnamäe, R.; Mareev, S.; Nikonenko, V.; Melnikov, S.; Sheldeshov, N.; Zabolotskii, V.; Hamelers, H.; Tedesco, M. Bipolar membranes: A review on principles, latest developments, and applications. *J. Membr. Sci.* **2021**, *617*, 118538. [CrossRef]
28. Melnikov, S.; Sheldeshov, N.; Zabolotsky, V.; Loza, S.; Achoh, A. Pilot scale complex electrodialysis technology for processing a solution of lithium chloride containing organic solvents. *Sep. Purif. Technol.* **2017**, *189*, 74–81. [CrossRef]
29. Grageda, M.; Gonzalez, A.; Quispe, A.; Ushak, S. Analysis of a process for producing battery grade lithium hydroxide by membrane electrodialysis. *Membranes* **2020**, *10*, 198. [CrossRef] [PubMed]
30. Ryabtsev, A.; Nemkov, N.; Kotsupalo, N.; Serikova, L. Preparation of high-purity lithium hydroxide monohydrate from technical-grade lithium carbonate by membrane electrolysis. *Russ. J. Appl. Chem.* **2004**, *77*, 1108–1116. [CrossRef]
31. Jiang, C.; Wang, Y.; Wang, Q.; Feng, H.; Xu, T. Production of lithium hydroxide from lake brines through electro–electrodialysis with bipolar membranes (EEDBM). *Ind. Eng. Chem. Res.* **2014**, *53*, 6103–6112. [CrossRef]
32. Yang, Y.; Gao, X.; Fan, A.; Fu, L.; Gao, C. An innovative beneficial reuse of seawater concentrate using bipolar membrane electrodialysis. *J. Membr. Sci.* **2014**, *449*, 119–126. [CrossRef]
33. Herrero-Gonzalez, M.; Diaz-Guridi, P.; Dominguez-Ramos, A.; Ibañez, R.; Irabien, A. Photovoltaic solar electrodialysis with bipolar membranes. *Desalination* **2018**, *433*, 155–163. [CrossRef]

34. Ghyselbrecht, K.; Huygebaert, M.; van der Bruggen, B.; Ballet, R.; Meesschaert, B.; Pinoy, L. Desalination of an industrial saline water with conventional and bipolar membrane electrodialysis. *Desalination* **2013**, *318*, 9–18. [CrossRef]
35. Jiang, C.; Wang, Q.; Zhang, Y.; Li, Y.; Wang, Y.; Xu, T. Separation of methionine from the mixture with sodium carbonate using bipolar membrane electrodialysis. *J. Membr. Sci.* **2016**, *498*, 48–56. [CrossRef]
36. Grágeda, M.; González, A.; Grágeda, M.; Ushak, S. Purification of brines by chemical precipitation and ion-exchange processes for obtaining battery-grade lithium compounds. *Int. J. Energy Res.* **2018**, *42*, 2386–2399. [CrossRef]
37. Hayes, T.; Severin, B. Electrodialysis of highly concentrated brines: Effects of calcium. *Sep. Purif. Technol.* **2017**, *175*, 443–453. [CrossRef]
38. ASTOM Corporation. Available online: http://www.astom-corp.jp/en/catalog/pdf/Astom_Products_Catalogue.pdf (accessed on 18 July 2021).
39. Kentish, S.; Kloester, E.; Stevens, G.; Scholes, C.; Dumée, L. Electrodialysis in aqueous-organic mixtures. *Sep. Purif. Rev.* **2014**, *44*, 269–282. [CrossRef]
40. Jaime-Ferrer, J.S.; Couallier, E.; Viers, P.; Durand, G.; Rakib, M. Three-compartment bipolar membrane electrodialysis for splitting of sodium formate into formic acid and sodium hydroxide: Role of diffusion of molecular acid. *J. Membr. Sci.* **2008**, *325*, 528–536. [CrossRef]
41. Sarapulova, V.; Shkorkina, I.; Mareev, S.; Pismenskaya, N.; Kononenko, N.; Larchet, C.; Dammak, L.; Nikonenko, V. Transport characteristics of fujifilm ion-exchange membranes as compared to homogeneous membranes AMX and CMX and to heterogeneous membranes MK-40 and MA-41. *Membranes* **2019**, *9*, 84. [CrossRef] [PubMed]
42. Fuelcellstore.com. Available online: https://www.fuelcellstore.com/spec-sheets/fumasep-fbm-pk-technical-specifications.pdf (accessed on 18 July 2021).
43. Products—Fumatech GmbH. Available online: https://www.fumatech.com/EN/Membranes/Water+treatment/Products+fumasep/ (accessed on 21 July 2021).
44. Balster, J.; Sumbharaju, R.; Srikantharajah, S.; Pünt, I.; Stamatialis, D.; Jordan, V.; Wessling, M. Asymmetric bipolar membrane: A tool to improve product purity. *J. Membr. Sci.* **2007**, *287*, 246–256. [CrossRef]
45. Israelachvili, J. Interactions involving polar molecules. *Intermol. Surf. Forces* **2011**, 71–90. [CrossRef]
46. Kamcev, J.; Paul, D.; Freeman, B. Ion activity coefficients in ion exchange polymers: Applicability of manning's counterion condensation theory. *Macromolecules* **2015**, *48*, 8011–8024. [CrossRef]
47. Izquierdo-Gil, M.; Barragán, V.; Villaluenga, J.; Godino, M. Water uptake and salt transport through Nafion cation-exchange membranes with different thicknesses. *Chem. Eng. Sci.* **2012**, *72*, 1–9. [CrossRef]
48. Strathmann, H. *Ion-Exchange Membrane Separation Processes*; Elsevier: Amsterdam, The Netherlands, 2004.
49. Hu, J.; Zhang, H.; Xu, W.; Yuan, Z.; Li, X. Mechanism and transfer behavior of ions in Nafion membranes under alkaline media. *J. Membr. Sci.* **2018**, *566*, 8–14. [CrossRef]
50. Kamcev, J.; Galizia, M.; Benedetti, F.; Jang, E.; Paul, D.; Freeman, B.; Manning, G. Partitioning of mobile ions between ion exchange polymers and aqueous salt solutions: Importance of counter-ion condensation. *Phys. Chem. Chem. Phys.* **2016**, *18*, 6021–6031. [CrossRef]
51. Kamcev, J.; Paul, D.; Manning, G.; Freeman, B. Ion diffusion coefficients in ion exchange membranes: Significance of counterion condensation. *Macromolecules* **2018**, *51*, 5519–5529. [CrossRef]
52. Luo, H.; Agata, W.; Geise, G. Connecting the ion separation factor to the sorption and diffusion selectivity of ion exchange membranes. *Ind. Eng. Chem. Res.* **2020**, *59*, 14189–14206. [CrossRef]
53. Fu, L.; Gao, X.; Yang, Y.; Aiyong, F.; Hao, H.; Gao, C. Preparation of succinic acid using bipolar membrane electrodialysis. *Sep. Purif. Technol.* **2014**, *127*, 212–218. [CrossRef]
54. Gimmi, T.; Alt-Epping, P. Simulating Donnan equilibria based on the Nernst-Planck equation. *Geochim. Cosmochim. Acta* **2018**, *232*, 1–13. [CrossRef]
55. Wang, Y.; Wang, A.; Zhang, X.; Xu, T. Simulation of electrodialysis with bipolar membranes: Estimation of process performance and energy consumption. *Ind. Eng. Chem. Res.* **2011**, *50*, 13911–13921. [CrossRef]
56. Aritomi, T.; van den Boomgaard, T.; Strathmann, H. Current-voltage curve of a bipolar membrane at high current density. *Desalination* **1996**, *104*, 13–18. [CrossRef]
57. Peng, J.; Zawodzinski, T. Describing ion exchange membrane-electrolyte interactions for high electrolyte concentrations used in electrochemical reactors. *J. Membr. Sci.* **2020**, *593*, 117340. [CrossRef]
58. Avci, A.; Rijnaarts, T.; Fontananova, E.; Di Profio, G.; Vankelecom, I.; De Vos, W.; Curcio, E. Sulfonated polyethersulfone based cation exchange membranes for reverse electrodialysis under high salinity gradients. *J. Membr. Sci.* **2020**, *595*, 117585. [CrossRef]
59. Simons, R. Preparation of a high-performance bipolar membrane. *J. Membr. Sci.* **1993**, *78*, 13–23. [CrossRef]
60. Wang, Q.; Wu, B.; Jiang, C.; Wang, Y.; Xu, T. Improving the water dissociation efficiency in a bipolar membrane with amino-functionalized MIL-101. *J. Membr. Sci.* **2017**, *524*, 370–376. [CrossRef]
61. Xu, J.; Amorim, I.; Li, Y.; Li, J.; Yu, Z.; Zhang, B.; Araujo, A.; Zhang, N.; Liu, L. Stable overall water splitting in an asymmetric acid/alkaline electrolyzer comprising a bipolar membrane sandwiched by bifunctional cobalt-nickel phosphide nanowire electrodes. *Carbon Energy* **2020**, *2*, 646–655. [CrossRef]
62. Chabi, S.; Wright, A.; Holdcroft, S.; Freund, M. Transparent bipolar membrane for water splitting applications. *ACS Appl. Mater. Interfaces* **2017**, *9*, 26749–26755. [CrossRef]

63. Wu, X.; Gong, Y.; Xu, S.; Yan, Z.; Zhang, X.; Yang, S. Electrical conductivity of lithium chloride, lithium bromide, and lithium iodide electrolytes in methanol, water, and their binary mixtures. *J. Chem. Eng. Data* **2019**, *64*, 4319–4329. [CrossRef]
64. Zhao, Y.; Wang, H.; Li, Y.; Wang, M.; Xiang, X. An integrated membrane process for preparation of lithium hydroxide from high Mg/Li ratio salt lake brine. *Desalination* **2020**, *493*, 114620. [CrossRef]
65. Chen, X.; Ruan, X.; Kentish, S.; Li, G.; Xu, T.; Chen, G. Production of lithium hydroxide by electrodialysis with bipolar membranes. *Sep. Purif. Technol.* **2021**, *274*, 119026. [CrossRef]
66. Turan, A.; Baloglu, H.; Ünveren, E.; Bulutcu, A. The behaviour of Nafion® 424 membrane in the electrochemical production of lithium hydroxide. *J. Chem. Technol. Biotechnol.* **2015**, *91*, 2529–2538. [CrossRef]
67. Kelly, J.; Wang, M.; Dai, Q.; Winjobi, O. Energy, greenhouse gas, and water life cycle analysis of lithium carbonate and lithium hydroxide monohydrate from brine and ore resources and their use in lithium ion battery cathodes and lithium ion batteries. *Resour. Conserv. Recycl.* **2021**, *174*, 105762. [CrossRef]
68. Yuan, B.; Wang, J.; Cai, W.; Yang, Y.; Yi, M.; Xiang, L. Effects of temperature on conversion of Li_2CO_3 to LiOH in $Ca(OH)_2$ suspension. *Particuology* **2017**, *34*, 97–102. [CrossRef]
69. Wilhelm, F. Optimisation strategies for the preparation of bipolar membranes with reduced salt ion leakage in acid-base electrodialysis. *J. Membr. Sci.* **2001**, *182*, 13–28. [CrossRef]
70. Ji, Z.; Chen, Q.; Yuan, J.; Liu, J.; Zhao, Y.; Feng, W. Preliminary study on recovering lithium from high Mg^{2+}/Li^{+} ratio brines by electrodialysis. *Sep. Purif. Technol.* **2017**, *172*, 168–177. [CrossRef]
71. Nie, X.; Sun, S.; Sun, Z.; Song, X.; Yu, J. Ion-fractionation of lithium ions from magnesium ions by electrodialysis using monovalent selective ion-exchange membranes. *Desalination* **2017**, *403*, 128–135. [CrossRef]
72. Zabolotsky, V.; Utin, S.; Bespalov, A.; Strelkov, V. Modification of asymmetric bipolar membranes by functionalized hyperbranched polymers and their investigation during pH correction of diluted electrolytes solutions by electrodialysis. *J. Membr. Sci.* **2015**, *494*, 188–195. [CrossRef]
73. Peng, F.; Peng, S.; Huang, C.; Xu, T. Modifying bipolar membranes with palygorskite and $FeCl_3$. *J. Membr. Sci.* **2008**, *322*, 122–127. [CrossRef]
74. Gonzalez, A.; Grágeda, M.; Ushak, S. Assessment of pilot-scale water purification module with electrodialysis technology and solar energy. *Appl. Energy* **2017**, *206*, 1643–1652. [CrossRef]
75. Hanel, M.; Escobar, R. Influence of solar energy resource assessment uncertainty in the levelized electricity cost of concentrated solar power plants in Chile. *Renew. Energy* **2013**, *49*, 96–100. [CrossRef]

membranes

MDPI

Article

Highly Proton-Conducting Membranes Based on Poly(arylene ether)s with Densely Sulfonated and Partially Fluorinated Multiphenyl for Fuel Cell Applications

Tzu-Sheng Huang [1], Tung-Li Hsieh [2], Chih-Ching Lai [1], Hsin-Yi Wen [3], Wen-Yao Huang [1,*] and Mei-Ying Chang [1,*]

[1] Department of Photonics, National Sun Yat-Sen University, Kaohsiung 80424, Taiwan; zxp86133@gmail.com (T.-S.H.); jackko61010@gmail.com (C.-C.L.)
[2] General Education Center, Wenzao Ursuline University of Languages, Kaohsiung 80793, Taiwan; tunglihsieh@gmail.com
[3] Department of Green Energy and Environmental Resources, Chang Jung Christian University, Tainan City 71101, Taiwan; hywen@mail.cjcu.edu.tw
* Correspondence: wyhuang@mail.nsysu.edu.tw (W.-Y.H.); mychang01@mail.nsysu.edu.tw (M.-Y.C.)

Abstract: Series of partially fluorinated sulfonated poly(arylene ether)s were synthesized through nucleophilic substitution polycondensation from three types of diols and superhydrophobic tetra-trifluoromethyl-substituted difluoro monomers with postsulfonation to obtain densely sulfonated ionomers. The membranes had similar ion exchange capacities of 2.92 ± 0.20 mmol g^{-1} and favorable mechanical properties (Young's moduli of 1.60–1.83 GPa). The membranes exhibited considerable dimensional stability (43.1–122.3% change in area and 42.1–61.5% change in thickness at 80 °C) and oxidative stability (~55.5%). The proton conductivity of the membranes, higher (174.3–301.8 mS cm^{-1}) than that of Nafion 211 (123.8 mS cm^{-1}), was the percent conducting volume corresponding to the water uptake. The membranes were observed to comprise isolated to tailed ionic clusters of size 15–45 nm and 3–8 nm, respectively, in transmission electron microscopy images. A fuel cell containing one such material exhibited high single-cell performance—a maximum power density of 1.32 W cm^2 and current density of >1600 mA cm^{-2} at 0.6 V. The results indicate that the material is a candidate for proton exchange membranes in fuel cell applications.

Keywords: poly(arylene ether)s; ionomers; proton exchange membranes; fuel cell

Citation: Huang, T.-S.; Hsieh, T.-L.; Lai, C.-C.; Wen, H.-Y.; Huang, W.-Y.; Chang, M.-Y. Highly Proton-Conducting Membranes Based on Poly(arylene ether)s with Densely Sulfonated and Partially Fluorinated Multiphenyl for Fuel Cell Applications. *Membranes* **2021**, *11*, 626. https://doi.org/10.3390/membranes11080626

Received: 28 July 2021
Accepted: 12 August 2021
Published: 15 August 2021

1. Introduction

The fuel cell is a type of energy conversion device, and the prototype fuel cell, which converted chemical energy into electrical energy through an electrochemical mechanism, was produced at the beginning of the 19th century [1–3]. Due to the world's thirst for energy, fuels such as coal, oil, and gas are constantly consumed, and the generation of greenhouse gases has become an environmental concern. Hydrogen energy—a clean energy resource, producing little pollution (no CO_2), small emissions, and low noise—is being developed by countless researchers in this era of the pursuit of green energy [4]. On the other hand, hydrogen is produced following the opposite principle of fuel cells, according to which it is produced from water splitting and used as a carrier for energy storage. Additionally, the water electrolysis unit can be used in combination with a fuel cell unit to store intermittent or overflow energy (ex. solar, wind, waste heat, or nuclear). In addition, fuel cells can be used for energy storage applications, which is how they differ from, or have advantages over, lithium-ion batteries [5] and redox flow batteries [6]. In addition to the basic ion selectivity differences, membrane performance needs to be able to perform well at extreme temperatures (i.e., −40 to 60 °C) in order to ensure environmental sustainability. Correspondingly, they may be affected during cold/hot shock cycling and charge/discharge cycling to maintain good ionic conductivity, mechanical properties, electrochemical and thermal stability [5,6]. However, their future development will depend on functional molecular designs while meeting the key economic and technical drivers of cost, power density, efficiency, and durability.

Among them, proton-exchange membrane (PEM) fuel cell using functional polymer materials as the core has been widely developed. In addition to having high proton conductivity and favorable mechanical properties, PEMs meet various performance requirements—low sensitivity to humidity, low oxidant permeability, and adequate electrochemical, chemical, thermal, and dimensional stability—If a satisfactorily low-cost and functional polymer material is to be manufactured [7–10]. PEMs have four major types—perfluorinated sulfonic acid (PFSA), fluorinated, hydrocarbon (HC)-based, and functional PEMs, all of which are widely researched. PFSA polymer materials (such as the Nafion, Flamion, Aciplex, and Dow series membranes) have excellent performance and are widely used in commercial applications [11]. However, the high pollution and external costs due to perfluorinated compounds have prompted many research teams to investigate alternatives. By contrast, HC-based PEMs are the most environmentally friendly because they are completely free of halogens, but they rarely combine high proton conductivity and favorable mechanical or other crucial properties. Functional sulfonated poly(arylene ether)s have higher thermostability and increased proton conductivity at temperatures above 80 °C and in low humidity. Poly(arylene ether)s have sulfonated derivatives of various forms, such as poly(arylene ether ketone)s [12–15], poly(arylene sulfide)s [16,17], polyimides [18–20], and polybenzimidazoles [21–23], which have been extensively investigated because of their high thermal stability, favorable chemical stability, appropriate mechanical properties, and low production cost, as well as the easy adjustment of their molecular structure [24,25]. Countless efforts have been made to synthesize new polymer architectures with conductive components (moieties); nevertheless, the influence on the morphology at many scales is not clear; therefore, PEMs have high proton conductivity, proper hygroscopicity, and sufficient mechanical properties, but at the same time, are still under study [9].

In general, sulfonated HC-based polymers have a wide diversity of morphology types because hydrophobic and hydrophilic moieties are multiscale distributed by structural effects. These are microphase separations to form proton transport channels. As reported widely, the chemistry and architecture of PEMs have effects on proton conductivity, they can adjust the structural design of polymer composites to improve by the ion exchange capacity and microphase separation [26]. We recently reported a series of poly(arylene ether)s with a high-free-volume multiarylbenzene (MAB) structure in the polymer backbone and a trifluoromethylphenyl side chain; this series has high proton conductivity, mechanical and dimensional stability, and film-forming ability [27–29]. Adamski et al. reported two classes of HC-based PEMs, sulfonated phenylated poly(phenylene) biphenyl and sulfonated poly(arylene ether), with branching multiphenylated structures. They described the relationship between densely sulfonated polymers and water uptake and transport, aiding in the design of next-generation solid polymer electrolytes [30]. Lee et al. synthesized a series of ethynyl-terminated sulfonated–fluorinated poly(arylene ether) random copolymers, including benzenesulfonate, hexafluoroisopropylidene, and perfluorobenzene derivatives; molar volume per charge (MCV), percent conducting volume (PCV), and derivative parameters were introduced for evaluation of the membrane properties and comparison of cross-linked poly(arylene ether)s with Nafion membranes [31]. The hydrophilic and hydrophobic substituents of the hydrophobic polymer backbone strongly influenced the microphase separation. With dense sulfonation, a superhydrophilic domain can be produced, and local halogenation or perfluoroalkane substitution can create a superhydrophobic domain in the structure. Furthermore, the trifluoromethyl-substituted group is a strongly electron-withdrawing group and can deactivate the nucleophilic aromatic substitution (S_NAr) reaction of the sulfonating agent, creating favorable localization for hydrophobicity [13,32,33]. However, compatibility challenges may occur when the polarities of the components are extremely different, regardless of whether the monomers are polymerized through physical blending or preacidification.

Herein, we report a series of novel poly(arylene ether)s synthesized using a multiphenylated difluoro monomer with a trifluoromethylphenyl side chain and three types of multiphenylated bisphenol monomer. Sulfonated poly(arylene ether)s were prepared through treatment with chlorosulfuric acid, with the trifluoromethyl (-CF_3) substituent expected to inhibit the S_NAr reaction in specific moieties, effectively dividing the local, densely sulfonated moieties. After sulfonation, bipolar domains form in the molecules, which become partially fluorinated sulfonated poly(arylene ether)s with highly efficient proton transport. Furthermore, the free volume effect and hydrogen bond cohesion of bisphenol monomers with different degrees of sulfonation provide corresponding water sorption ability and preserve high mechanical stability. The synthesis of sulfonated poly(arylene ether)s was confirmed using proton nuclear magnetic resonance (^{1}H NMR) and Fourier transform infrared (FTIR) spectroscopy. PEMs were prepared through solution casting with dimethyl sulfoxide (DMSO), and their properties—water uptake, dimensional stability, mechanical strength, proton conductivity, morphology, and single-cell performance—were determined.

2. Materials and Methods

2.1. General Methods

2,5-Bis(4-bromophenyl)-3,4-bis(3-(trifluoromethyl)phenyl)cyclopenta-2,4-dienone (DTF-EO), 2′′,3′′,5′′,6′′-Tetraphenyl-[1,1′:4′,1′′:4′′,1′′′:4′′′,1′′′′-quinquephenyl]-4,4′′′′-diol (9B-DO), 4-Bromo-4′-(4-bromophenyl-3′,5′-diphenyl-1,1′;2′1′′-terphenyl (6B-DB), and 3′,6′-bis(4-bromophenyl)-1,1′:2′,1′′-terphenyl (5B-DB) were synthesized, as described previously [27,29]. Toluene and dichloromethane (DCM) were dried over CaH_2 and P_2O_5, respectively, and then freshly distilled under a N_2 atmosphere and deoxygenated through purging with N_2 for 30 min prior to use. Other reagents and solvents were purchased from Alfa Aesar (Haverhill, MA, USA), Aldrich Chemical Co. (St. Louis, MI, USA), Fisher Scientific (Hampton, NH, USA), Merck (Darmstadt, Germany), or Tokyo Chemical Industry Co. (Tokyo, Japan) and used without further purification. All reactions were performed under repurified N_2 atmosphere. The monomer reaction steps are presented in Scheme 1.

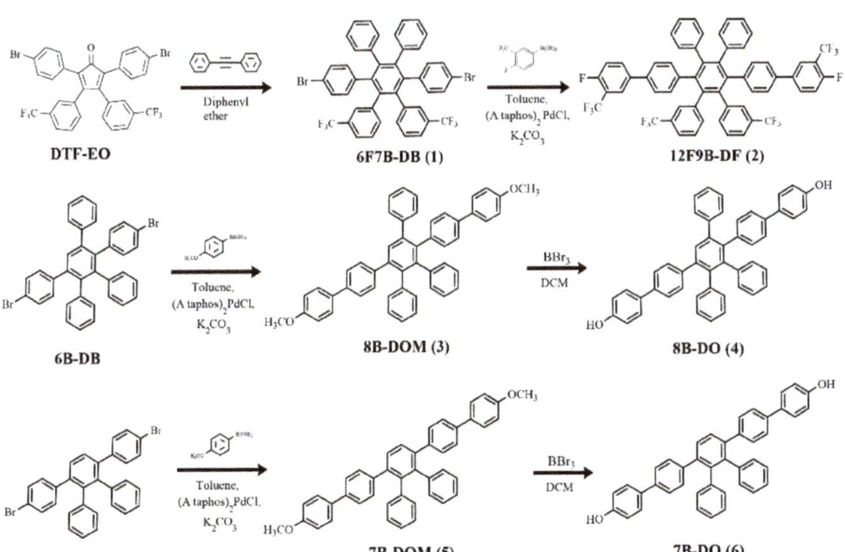

Scheme 1. Synthesis of monomers.

2.2. 3′,6′-Bis(4-bromophenyl)-4′,5′-diphenyl-3,3″-bis(trifluoromethyl)-1,1′:2′,1″-terphenyl (6F7B-DB) (1)

DTF-EO (8.10 g, 11.94 mmol), diphenyl acetylene (2.55 g, 1.20 eq.) diphenyl ether (8.00 g) were placed in a 3-necked round-bottomed flask with condenser and thermometer under N_2 atmosphere at 40 °C. The reaction temperature was then increased to 220 °C for 20 h. After being cooled to room temperature, the crude mixture was poured into methanol and deionized water (DI water), and extraction was subsequently performed using ethyl acetate and DI water; dewatering was then performed using $MgSO_4$. Finally, the resultant product was obtained through reduced pressure distillation and recrystallization from tetrahydrofuran (THF)/n-hexane, producing a crystal in a 54.3% yield. ^{1}H-NMR (500 MHz, $CDCl_3$, 25 °C): d (ppm) 7.15 (d, 1H), 7.09 (s, 1H), 6.96–7.07 (m, 4H), 6.88–6.93 (m, 3H), 6.82–6.95 (m, 2H), 6.64–6.70 (m, 2H). Matrix-assisted laser desorption ionization–tandem time-of-flight mass spectrometry MALDI-TOF/TOF MS (m/z): Calculated for $C_{44}H_{26}Br_2F_6$: 828.48, found 828.30.

2.3. 4,4⁗-Difluoro-3,3⁗-bis(trifluoromethyl)-2″,3″,5″,6″,4,4⁗-difluoro-3,3⁗-bis(trifluoromethyl) -2″,3″,5″,6″-tetra (trifluoromethyl)phenyl-[1,1′:4′,1″:4″,1‴:4‴,1⁗-quinquephenyl] (12F9B-DF) (2)

For the aforementioned reactor, 6F7B-DB (10.00 g, 12.1 mmol) and 4-fluoro-3-(trifluoromethyl) phenylboronic acid (5.00 g, 36.18 mmol) were dissolved in toluene (600 mL) under N_2 atmosphere. Subsequently, 2.0 M K_2CO_3 (aq.) (5.00 g, 28.94 mmol) was added to the reactor. The mixture was boiled, (A-^{ta}Phos)$_2$PdCl$_2$ (catalyst, 100 mg, 0.14 mmol) was added, and the mixture was then stirred for 2 days. After being cooled to room temperature, toluene and saturated saline were used to extract the organic layer. Any impurities in the mixture were removed using activated carbon, and the mixture was filtered using $MgSO_4$. Finally, the crude product was purified through recrystallization from ethyl acetate/n-hexane, producing a crystal in 80.1% yield. ^{1}H-NMR (500 MHz, $CDCl_3$, 25 °C): d (ppm) 7.60 (d, 1H), 7.52–7.56 (m, 1H), 7.10–7.18 (m, 5H), 7.05–7.09 (m, 1H), 7.03 (d, 1 H), 6.98 (d, 1H) 6.82–6.95 (m, 7H). MALDI-TOF/TOF MS (m/z): Calculated for $C_{58}H_{32}F_{14}$: 944.85, found 994.66.

2.4. 4,4⁗-Dimethoxy-2″,3″,5″-triphenyl-1,1′:4′,1″:4″,1‴:4‴,1⁗-quinquephenyl (8B-DMO) (3)

Briefly, 6B-DB (5.00 g, 8.11 mmol) and 4-methoxyphenylboronic acid (3.10 g, 8.11 mmol) were dissolved in toluene (800 mL) under the N_2 atmosphere. Subsequently, 2.0 M K_2CO_3 (aq.) (4.50 g, 32.6 mmol) was added to the mixture. The mixture was boiled, (A-^{ta} Phos)$_2$PdCl$_2$ (catalyst, 100 mg, 0.14 mmol) was added, and the mixture was then stirred for 2 days. Toluene and saturated saline were used to extract the organic layer. Any impurities in the mixture were removed using activated carbon, and the mixture was filtered using $MgSO_4$. Finally, the crude product was purified through recrystallization from ethyl acetate/n-hexane, producing a crystal in 85.1% yield. ^{1}H-NMR (500 MHz, $CDCl_3$, 25 °C): d (ppm) 7.62 (s, 1H), 7.48 (d, 2H), 7.40 (d, 2H), 7.12–7.21 (m, 9H), 7.70 (d, 2H), 6.92–6.96 (m, 5H), 6.85−6.91 (m, 9H), 6.80–6.83 (m, 2H), 3.83 (s, 3H), 3.81 (s, 3H). MALDI-TOF/TOF MS (m/z): Calculated for $C_{50}H_{38}O_2$: 670.84, found 670.37.

2.5. 2″,3″,5″-Triphenyl-[1,1′:4′,1″:4″,1‴:4‴,1⁗-quinquephenyl]-4,4⁗-diol (8B-DO) (4)

All glassware used must be dried rigorously and assembled under the N_2 atmosphere. 8B-DMO (6.00 g, 4.47 mmol) was dissolved in dry DCM (120 mL). Boron tribromide (5.1 mL, 20.4 mmol) was then injected at −78 °C under cryogenic storage dewar, and the solution was stirred for 1 day. DI water was slowly added to stop the reaction. Extraction was performed using ethyl acetate and DI water, and dewatering was performed using $MgSO_4$. Finally, the resultant product was obtained through reduced pressure distillation and recrystallization from THF/n-hexane, producing a crystal in 47.0% yield. ^{1}H-NMR (500 MHz, $CDCl_3$, 25 °C): d (ppm) 7.62 (s, 1H), 7.43 (d, 2H), 7.34–7.37 (m, 4H), 7.21–7.12 (m, 10H), 6.93–6.95 (m, 3H), 6.85–6.90 (m, 9H), 6.80–6.83 (m, 4H). MALDI-TOF/TOF MS (m/z): Calculated for $C_{48}H_{34}O_2$: 642.78, found 642.39.

2.6. *4,4''''-Dimethoxy-2'',3''-diphenyl-1,1':4',1'':4'',1'''4'''1''''-quinquephenyl (7B-DMO) (5)*

Briefly, 5B-DB (6.00 g, 11.1 mmol) and 4-methoxyphenylboronic acid (5.06 g, 33.3 mmol) were dissolved in toluene (700 mL) under the N_2 atmosphere. Subsequently, 2.0 M K_2CO_3 (aq.) (4.50 g, 39.1 mmol) was added to the mixture. The mixture was boiled, (A-^{ta} Phos)$_2$PdCl$_2$ (catalyst, 100 mg, 0.14 mmol) was added, and the mixture was stirred for 2 days. Moreover, toluene and saturated saline were used to extract the organic layer. Any impurities in the mixture were removed using activated carbon, and the mixture was filtered using $MgSO_4$. Finally, the crude product was purified through recrystallization from ethyl acetate/n-hexane, producing a crystal in a 63.9% yield. ^{1}H-NMR (500 MHz, $CDCl_3$, 25 °C): d (ppm) 7.56 (s, 1H), 7.48 (d, 2H), 7.35 (d, 2H), 7.16 (d, 2H), 6.92–6.95 (m, 5H), 6.86–6.83 (m, 2H). MALDI-TOF/TOF MS (m/z): Calculated for $C_{44}H_{34}O_2$: 594.74, found 594.24.

2.7. *2'',3''-Diphenyl-[1,1':4',1'':4'',1''':4''',1''''-quinquephenyl]-4,4''''-diol (7B-DO) (6)*

As regards the aforementioned reactor and cooling bath, 7B-DOM (4.00 g, 6.73 mmol) was dissolved in DCM (150 mL). Boron tribromide (3.90 mL, 15.6 mmol) was then injected at −78 °C, and the solution was stirred for 1 day. DI water was slowly added to stop the reaction. Extraction was performed using ethyl acetate and DI water, and dewatering was performed using $MgSO_4$. Finally, the resultant product was obtained through reduced pressure distillation and recrystallization from THF/n-hexane, producing a crystal in 92.1% yield. ^{1}H-NMR (500 MHz, $CDCl_3$, 25 °C): d (ppm) 7.56 (s, 1H), 7.44 (d, 2H), 7.34 (d, 2H), 7.16 (t, 2H), 6.92–6.96 (m, 3H), 6.83–6.88 (m, 4H). MALDI-TOF/TOF MS (m/z): Calculated for $C_{42}H_{30}O_2$: 566.69, found 566.22.

2.8. *General Procedure for the Synthesis of Polymers*

Polymerization reactions were conducted in a three-necked 100-mL flame-dried flask equipped with a stirring bar and Dean–Stark apparatus fitted with a condenser under N_2 atmosphere. The flask was charged with K_2CO_3 (3.61 mmol), dimethylacetamide (DMAC; 25 mL), dry toluene (15 mL), difluoro monomer 12F9B-DF (1.50 g, 1.51 mmol), and bisphenol monomer (1.51 mmol). Three types of polymers were used: 9B-DO, 8B-DO, and 7B-DO. The polymer reaction steps are presented in Scheme 2.

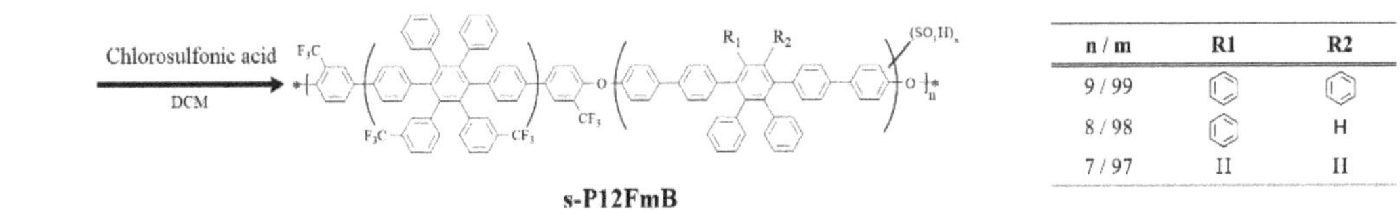

Scheme 2. Synthesis of sulfonated ionomers.

The one-pot reaction was dried through azeotropic distillation at 130 °C for 2 h; then, the reaction temperature was increased to 160 °C for 36 h. After being cooled to room temperature, the crude mixture was poured into methanol and DI water to precipitate a beige fibrous polymer. The polymer was filtered, washed several times with water and hot methanol, and dried in a vacuum at 80 °C for 8 h.

2.9. General Procedure for Sulfonation

To a solution of the polymer (1.20 g) in DCM (75 mL) at room temperature, chlorosulfuric acid in DCM was added dropwise. The reaction mixture was stirred for 24 h and then poured into water. The polymer precipitate was filtered, washed thoroughly with DI water until the pH became neutral, and then dried in a vacuum at 60 °C overnight to obtain the sulfonated polymer. The polymers were sulfonated to different extents by using the aforementioned procedure and adding 4 or 12 mL of a sulfonating agent, respectively. The sulfonated polymer was readily soluble in polar aprotic solvents such as dimethylformamide, DMAC, DMSO, and n-methyl-2-pyrrolidone (NMP). The ionomer reaction steps are presented in Scheme 2.

2.10. Measurements

^{1}H NMR spectroscopy was performed using a 500-MHz Varian Unity Inova 500 spectrometer using $CDCl_3$ or DMSO-d6 as the solvent. MALDI-TOF mass spectra were obtained using a Bruker Daltonics Autoflex III TOF/TOF with 2,5-dihydroxybenzoic acid employed as the matrix. The FTIR spectra of the polymer membranes were obtained using a Bruker VERTEX 70 FTIR spectrometer. Gel permeation chromatography analysis was conducted using a Viscotek 270 Max with a refractive index detector and THF used as the eluent at a flow rate of 1 mL min^{-1}. For calibration, a polystyrene standard (molecular weight between 75 and 117 kDa) was used. Stress–strain curves were obtained for the film specimens (length, 10.0 mm; width, 1.0 mm; and thickness, 30–40 µm) by using the PerkinElmer Pyris Diamond TMA at 25 °C. Membranes with various thicknesses may possess different physical–chemical properties, including mechanical strength, proton conductance (not electrical conductivity), water uptake, and resistance [34]. Accordingly, the thickness was controlled between 30 and 40 µm to match the comparable range of the selected standard, Nafion 211 (25.4 µm). For sufficient ion cluster growth time, it is necessary to have both high-concentration dissolution characteristics and appropriate solvent evaporation rate under film-forming temperature conditions. Therefore, this study uses high-boiling DMSO as the casting solvent.

2.11. Stability

The thermal stability of the polymers was evaluated through thermogravimetric analysis (TGA) performed using a PerkinElmer Pyris 1 from 50 to 800 °C at a heating rate of 10 °C min^{-1} under N_2 atmosphere. Before the analysis, the membranes were dried in the TGA furnace at 130 °C under N_2 for 1 h to remove moisture.

The membranes were weighed and soaked in Fenton's reagent (3% H_2O_2 aqueous solution containing ferrous sulfate at 2.0 ppm) at 80 °C for observation after 24 h. Oxidative stability (OS) was evaluated using the change in weight of the membranes after exposure to Fenton's reagent.

2.12. Water Uptake and Dimensional Stability

The water uptake (WU) of the membranes is gauged by comparing the weights of the dry and the wet membrane samples by means of Equation (1). The dry membrane weight (W_{dry}) is obtained by vacuum drying the sample at 80 °C for 24 h immediately before weighing it. The weight of the corresponding membrane in wet conditions (W_{wet}) is obtained by immersing the membrane sample in DI water at a specified temperature for 24 h, wiping off the surface moisture with filter paper, and then quickly weighing it. The final *WU* is obtained from the average of three experiments.

$$Water\ uptake\ (WU) = \frac{W_{wet} - W_{dry}}{W_{dry}} \times 100\%, \quad (1)$$

The swelling ratio was calculated using the following Equations (2) and (3):

$$In\ plane\ selling\ ratio\ (\Delta T\%) = \frac{T_{wet} - T_{dry}}{T_{dry}} \times 100\% \quad (2)$$

$$Through\ plane\ selling\ ratio\ (\Delta A\%) = \frac{A_{wet} - A_{dry}}{A_{dry}} \times 100\% \quad (3)$$

where A_{wet} and A_{dry} are the areas, and T_{wet} and T_{dry} are the thickness of the wet and dry membranes, respectively.

2.13. Ion Exchange Capacity

The ion exchange capacity (IEC) of the membranes was determined through acid-base titration. Each dried membrane was weighed and then immersed in 1.0 M HCl (aq.) for 24 h to protonate the acid groups; it was then washed thoroughly with DI water to reach a neutral pH. Subsequently, the membrane was immersed in a 1.0 M NaCl (aq.) for 24 h to replace the protons of the sulfonic acid groups with Na+ ions. The solution was titrated using 0.01 M NaOH (aq.), and phenolphthalein was used as an indicator. The IEC (mmol g^{-1}) was calculated from the titration results by using the following Equation (4):

$$\text{IEC (mmol g}^{-1}) = (V_{NaOH} \times M_{NaOH})/W_{dry} \quad (4)$$

where V_{NaOH} and M_{NaOH} are the volume and concentration of the NaOH solution, respectively, and W_{dry} is the weight of the dry membrane.

2.14. Hydrated Molar Volume per Charge

The molar volume per charge (MVC) is an estimate of equivalent volume (cm^3 per ionomer or the mol equivalent of acid groups) based on the summation of molar volume subunits rather than true volume measurements, as shown in Equation (5) [31,35] as follows:

$$MVC = \sum_i n_i V_i \quad (5)$$

where V_i is the volumetric contribution of structural group i, which appears n_i times per charge.

The hydrated molar volume per charge ($MVC_{(wet)}$) was the sum of the molar volumes of the component atoms or water uptake of the polymers, which is given by Equation (6) [31,35] as follows:

$$MVC_{(wet)} = \sum_i n_i V_i + V_{H_2O} \times \lambda \quad (6)$$

where V_{H_2O} is the molar volume of water, 18 ($cm^3\ mol^{-1}$)., and λ is the number of water molecules per charge based, as shown in Equation (7) as follows:

$$\lambda = \frac{[H_2O]}{[SO_3^-]} = \frac{10 \times WU}{M_{H_2O} \times IEC} \quad (7)$$

where M_{H_2O} is the molar mass of water.

2.15. Percent Conducting Volume

Percent conducting volume (PCV), which is the ratio of the water uptake volume to the acid volume concentration, was calculated using the following Equation (8) [31,35]:

$$PCV = \frac{V_{H_2O} \times \lambda}{MVC_{(wet)}} \quad (8)$$

Recently, research teams have employed this equation to better compare the proton conductivity of sulfonated polymers in PEMs in various states of WU [31,35,36].

2.16. Proton Conductivity

The membranes' proton conductivity was measured using a frequency response analyzer (Solartrom 1260) along the in-plane direction over the frequency range 10 MHz to 100 Hz at a voltage amplitude of 100 mV. Conductivity was measured after clamping a 10.0 mm × 5.0 mm sample between the two platinum electrodes of a conducting cell. The test cell was placed in an Espec SH-241 environmental chamber to measure its conductivity at 80 °C and various values of relative humidity (RH). Proton conductivity (σ) was then calculated using the equation $\sigma = L/RA$, where L (cm) is the distance between the electrodes, R (Ω) is the membrane's resistance, and A (cm^2) is the cross-sectional area of the sample.

2.17. Microstructure Analysis

Transmission electron microscopy (TEM) was performed using a JEM-2100 (high-resolution) TEM instrument operated at an accelerating voltage of 200 kV. The acid form of the s-P12FmB-X membranes was dyed and converted into Ag^+ ions through overnight immersion in 1.0 M $AgNO_3$. Subsequently, the membranes were washed thoroughly with DI water and vacuum dried at 80 °C for 12 h. These dyed membranes were placed in an enclosure of epoxy resin and ultramicrotome under cryogenic conditions to obtain samples with a thickness of 30 nm.

2.18. Single-Cell Performance

Catalyst ink was prepared by mixing Pt/C (HiSPEC 4000, Alfa Aesar, Haverhill, MA, USA) with a 5 wt% Nafion D520 binder. The ink was sprayed onto both sides of the membranes. The active surface area was 5 cm^2 with an overall Pt loading of 0.4 mg cm^{-2}. A membrane electrode assembly (MEA) was obtained by sandwiching the s-P12FmB-X membranes between two gas diffusion layers (GDL 24 BC, Hephas Energy Co., Ltd., Hsinchu, Taiwan). An MEA featuring a Nafion 211 membrane was fabricated using the same procedure to serve as a reference. The anode and cathode were supplied with hydrogen at a flow rate of 0.2 L min^{-1} and oxygen at a flow rate of 0.4 L min^{-1}, respectively. The fabricated cell was activated for 4 h with hydrogen and oxygen at 80 °C and 0.5 V.

3. Results and Discussion

3.1. Synthesis and Characterization of the Monomers and Polymers

A novel tetra-trifluoromethyl-substituted hexaarylbenzene (HAB) derivative difluoro monomer was successfully synthesized in three steps, as illustrated in Scheme 1. Using the method discussed in our previous report, DTF-EO was synthesized from 1,3-bis(4-bromophenyl)propan-2-one and 1,2-bis(3-(trifluoromethyl)phenyl)ethane-1,2-dione through aldol condensation. DTF-EO was then reacted with diphenylacetylene through the Diels–Alder reaction to obtain monomer1 (6F7B-DB). To obtain the leaving for the condensation reaction, monomer 2 (12F9B-DF) was prepared from 6F7B-DB and 4-fluoro-3-(trifluoromethyl)phenyl)boronic acid through Suzuki coupling. To obtain the diol monomers, a series of volume-related MAB guides—nB-DB, where m is the number of Benzene Rings and synthesized, as described previously. Then, nB-DMO series were synthesized from three guides and 4-methoxyphenylboronic acid through Suzuki coupling. Finally, diol monomers (nB-DO) were prepared from nB-DMOs through BBr_3 demethylation. All the monomer structures were confirmed through ^1H NMR and MALDI-TOF-MS spectroscopy (Figures S1–S12).

The polymers were synthesized through S_NAr with a difluoro monomer (12F9B-DF) and diol (nB-DO); these polymers are denoted P12F99B, P12F98, and P12F97B, respectively, as shown in Scheme 2. The polymers were readily soluble in common solvents DMAC, DMSO, NMP, THF, and chloroform. The P12FmB series polymers had molecular weights of 75–117 kDa and polydispersity indices of 2.0–2.7 mmol g^{-1}. These results reveal that a short perfluoroalkyl group could contribute sufficient solubility, and the tetra-trifluoromethyl substitution in the phenyl provided a strong electron-withdrawing group to increase the substitution activity of halophenyl enough to obtain high-molar-mass poly(arylene ether)s during polymerization [37,38].

Postsulfonation of the P12FmB polymers was conducted using chlorosulfonic acids of various concentrations to obtain s-P12FmB-X ionomers, where m indicates the composition of phenyl rings and X indicates the IEC. The structures were confirmed through ^{1}H NMR spectroscopy. The NMR signals of ionomers and polymers are presented in Figure 1, Figures S13 and S14, respectively. After polycondensation, the NMR signal of the P12FmB series polymers comprised a single sharp peak (labeled 12) and was shifted to 7.70 ppm because of a change in the chemical environment of the trifluoromethyl aromatic ring at the junction. For instance, inP12F98B, as trifluoromethyl passivates the strong electron-withdrawing and deactivating aromatic ring, the substituent of the sulfonated group is often not in the trifluoromethyl substituted phenyl group. The new signal at 7.59–7.77 ppm is assigned to the proton next to the sulfonated pendant benzene ring; the chemical shift is similar to that reported in our previous work [27,28,39].

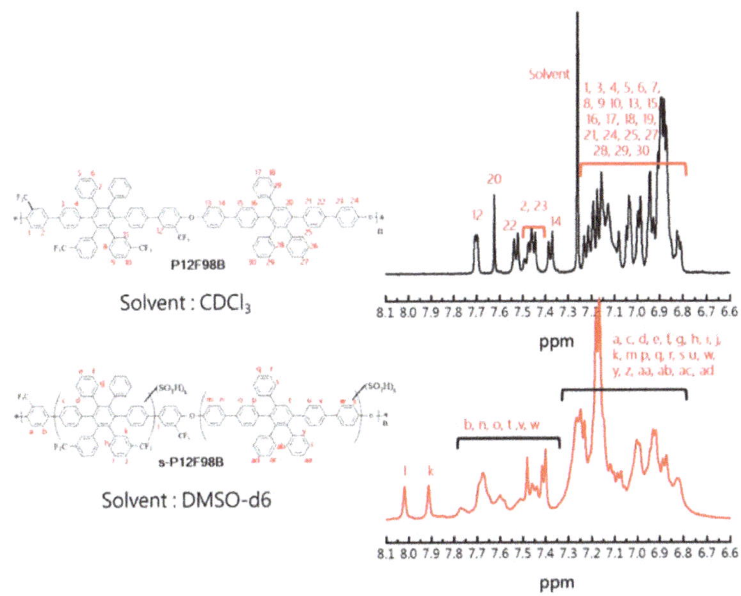

Figure 1. 1H NMR spectrum of P12F99B (**top**) and s-P12F99B (**bottom**) from 6.6 ppm to 8.1 ppm.

To confirm the sulfonation of the polymers, FTIR spectroscopy was performed to determine the structural composition of the untreated polymer and polymers with different degrees of sulfonation. After the sulfonation treatment, a broad absorption band was generated from 3749 to 2425 cm^{-1} (–OH stretching) and was accompanied by changes in the absorption peaks in fingerprint regions, as shown in Figure S15a,b, which reveals a slight shift of the peak corresponding to symmetrical stretching of C=C absorption from 1492 to 1478 cm^{-1}. Larger absorption peaks appear at 1350 and 1010 cm^{-1} and correspond to symmetrical stretching and in-plane bending of the O=S=O functional group [40–42]. The appearance of an enlarged shoulder at 1225 cm^{-1} is ascribed to asymmetric stretching [43]. The peak corresponding to the absorption of aromatic ethers (Ar–O–Ar) on the main chain remains at 1035 cm^{-1} [42,44].

3.2. Thermal Stability

The s-P12FmB-X series arylene ether ionomers obtained through sulfonation were dried to remove moisture and then kept at 120 °C for 30 min before their thermal stability was evaluated to guarantee the high stability of the cell during operation. Compared with the untreated polymers, the s-P12FmB-X series ionomers exhibited less than 2 wt% thermogravimetric loss due to loss of residual moisture before 120 °C. The trend, similar for all of the sulfonated polymers, consisted of a drying stage and two degradation stages. The thermal degradation curve obtained at 200–400 °C was attributed to the decomposition of sulfonic acid groups [45]. The secondary degradation curve obtained above 600 °C approximately corresponded to the degradation of the polymers' main chain. Figure 2 presents the degradation of polymers, and their 5% weight loss temperature ($T_{d5\%}$) is listed in Table 1. All the untreated polymers exhibited excellent thermal stability, with $T_{d5\%}$ higher than 589 °C. This was attributed to the multiple phenyl ionomer because rigid aromatic groups are known to provide strong resistance to thermal degradation [14].

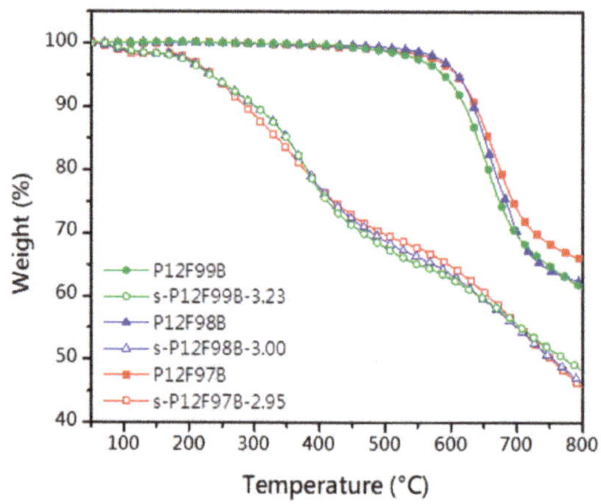

Figure 2. Thermogravimetric analysis curves of the P12FmB polymers and after sulfonation.

Table 1. Thermal stability and mechanical properties of the polymers and ionomers.

Membrane	$T_{d5\%}$ (°C)	Young's Modulus [a] (GPa)	Tensile Strength [a] (MPa)	Elongation at Break [a] (%)
P12F99B	589.0	1.69	86.2	25.6
s-P12F99B-2.89	265.4	0.51	68.3	74.9
s-P12F99B-3.23	230.8	0.45	48.7	57.7
P12F98B	610.1	1.83	80.3	34.9
s-P12F98B-2.61	262.6	0.59	70.7	75.6
s-P12F98B-3.00	232.2	0.53	56.2	66.9
P12F97B	608.9	1.60	97.1	40.3
s-P12F97B-2.84	251.0	1.03	76.6	95.8
s-P12F97B-2.92	234.7	0.73	59.6	82.7
Nafion 211	-	0.28	34.2	101.5

[a] Measured at 25 °C and 40% RH.

3.3. Mechanical Properties

The mechanical properties of the P12FmB (innate form, solid symbol lines in Figure 3) and s-P12FmB-X (acid form, hollow symbol lines in Figure 3) series membranes and of the control sample (PFSA, Nafion 211) were measured; the detailed results are shown in Table 1. The innate-form membranes were discovered to have Young's moduli of 1.60–1.83 GPa, a tensile strength of 80.3–97.1 MPa, elongation at break of 25.6–40.3%, and excellent source characteristics. The toughness of poly(arylene ether)s is derived from the contribution of rigid aromatic groups and their van der Waals forces, giving the full range of polymers similar strength and generally low elongation at break as a more brittle characteristic. The elongation at break P12F97B > P12F98B > P12F97B can be attributed to the architecture effect of the polymer chains. P12F97B has fewer side chains on the main chain, which allows it to stack better and obtain slightly stronger properties. After treatment, the degree of acidification was higher, and the strength was lower, conforming to the characteristics of strongly hydrophilic group implantation [8,46,47]. Moisture acted similar to a plasticizer to enhance the ductility of the membranes (elongation at break, 57.7–95.8%) but combined with the rigid structure of the poly(arylene ether)s to yield sufficient toughness. Although a significant increase in elongation at break was obtained after the sulfonation treatment, a decrease in strength was observed in Table 2. The sustained increase in IEC does not allow for better toughness of the membranes. Consequently, Young's moduli and tensile strength of the acid-form membranes were 0.34–0.57 GPa and 48.7 MPa–82.8 MPa, respectively. This result demonstrated that the mechanical properties of the s-P12F97B-X series of membranes were comparable to those of Nafion 211 membranes (Young's modulus, 0.28 GPa; tensile strength, 34.2 MPa). Regarding the elongation at break, s-P12F97B-2.92 > s-P12F98B-3.00 > s-P12F97B-2.89, which is attributed to the following two points: 1. the enhancement of IEC is accompanied by the enhancement of WU, and the effect of self-wetting increases the plasticizer content and enhances the ductility; 2. the stacking surface, which mainly contributes to van der Waals forces, changes with the substitution of sulfonic acid groups, decreasing van der Waals forces. Since the active zone of sulfonation substitution (diol-monomers, Scheme 2) is also located at the difference of the three polymer structures, the increase of IEC also means that the force dominance changes from van der Waals forces to hydrogen bonding. In the presence of hydrogen bonding and plasticizer (moisture), the ductility is increased and the strength is decreased, allowing for proper toughness. The detailed hydration behavior will be discussed subsequently.

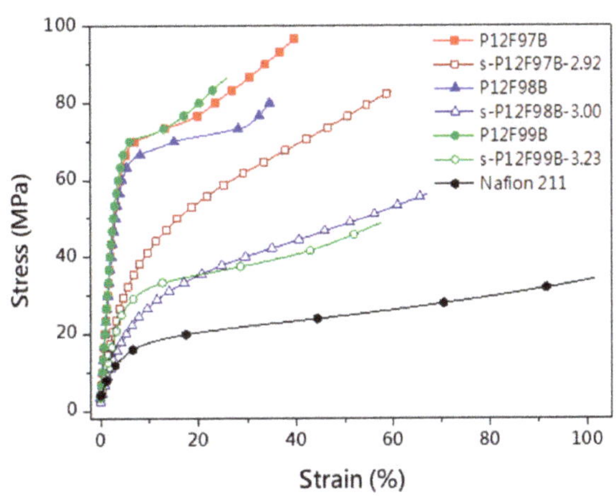

Figure 3. Stress–strain curves of MEAs at 25 °C and 40% RH.

3.4. *Hydration Behavior*

Fuel cells are operated under high humidity, and the water management characteristics of PEMs must not be neglected. Excellent water uptake (WU) for the proton carrier may be necessary, and its powerful influence on fuel cell performance [2,9,48]. However, the hygroscopic swelling of polymers is accompanied by dimensional changes, microphase transitions, and even polymer corrosion [9]. The WU (Figure 4), hydration number, dimensional stability, and oxidative stability (OS) of the s-P12FmB series membranes were investigated, and the results are listed in Table 2. The Ave designation in the s-P12FmB-Ave implies the average characterization of this series of ionomers. Since the

IEC represented by X cannot be mathematically averaged to express its significance, the average characterization and distribution are expressed as s-P12FmB-Ave.

Table 2. IEC, water uptake, swelling, and oxidative stability of the s-P12FmB-X and Nafion 211 membranes.

Membrane	IEC [a]	WU (%)		Λ [b]		ΔA (%)		ΔT [c] (%)		OS (%)
	(mmol g^{-1})	30 °C	80 °C	30 °C	80 °C	30 °C	80 °C	30 °C	80 °C	80 °C
s-P12F99B-2.89	2.89	96.6	114.3	18.6	28.0	64.3	100.1	35.9	61.5	-
s-P12F99B-3.23	3.23	106.7	142.9	18.3	26.0	76.5	122.3	35.5	-	-
s-P12F98B-2.61	2.61	65.4	88.5	13.9	18.8	36.9	52.0	29.7	54.1	55.9
s-P12F98B-3.00	3.00	66.7	103.3	12.3	19.1	33.7	49.4	31.9	56.0	61.2
s-P12F97B-2.84	2.84	40.0	74.6	7.8	14.6	28.6	45.1	26.3	42.1	50.0
s-P12F97B-2.92	2.92	39.3	78.6	7.5	14.9	32.4	43.1	29.1	42.5	55.0
Nafion 211	0.91	17.0	29.0	7.5	22.0	-	-	-	-	88.6

[a] IEC determined by acid-base titration. [b] λ calculation from WU. and IEC. [c] Change in membrane length (ΔL) and thickness (ΔT).

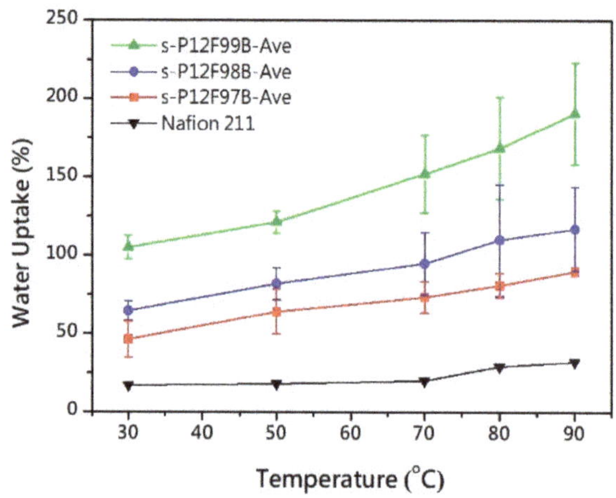

Figure 4. Water uptake as a function of temperature for s-P12FmB-Ave and Nafion 211 membranes.

As expected, the WU of the membranes increased with an increase in temperature, IEC, and the number of arms of the phenyl group substituent. The amount of WU of the s-P12FmB series membranes was 24–67% at 30 °C but increased to 74–143% at 80 °C. The highest WU, for the s-P12F99B-2.89 membrane, was twice that of the others at 80 °C. In addition, the slightly higher IEC of the polymer was attributed to the numerous sulfonated active sites and larger free volume on the structure. Within the range of comparable IECs, s-P12F98B-3.00, which had an asymmetric notch in its structure, had the highest initial WU (60%) at low temperatures, and this uptake was greater than that of Nafion 211. The effective WU properties are expected to provide an increase in the proton conductivity of the film. Typically, water is required and generated during the operation of a PEM fuel cell. As a result, membranes with excessive swelling rates may shrink under repeated cell on/off (charge/discharge cycling) operations, resulting in creep and the risk of catalytic layer peeling. In this study, the drawbacks of obtaining films with acceptable swelling rates need to be verified by subsequent rigorous fuel cell lifetime studies.

In this system, the hydration number (λ), the number of equivalent carriers that participate in carrying ions that are calculated using the experimental value of WU, was equivalent to the number of water molecules per sulfonic acid group. As the result, the s-P12F99B-X series membranes had higher λ (average 27.0) under operating conditions (80 °C; RH, ~99%). This may be because the free space and the distribution of sulfonic acids in the structure are relatively concentrated. On the other hand, the λ of the s-P12F98B-Ave and s-P12F97B-Ave membrane were approximately 19.0 and 14.8, respectively; the λ value decreases as the number of arms decreases. In this case, the equivalent carrying capacity of the s-P12F99B-Ave membrane may be better. Under similar IEC conditions, λ was discovered to increase with an increase in the number of sulfonyl phenyl arms, indicating that the densely hydrophilic concentration affects the hydration behavior.

The swelling of the s-P12FmB series membranes resulted in an approximate 30–120% change in area and 30–60% change in thickness at 80 °C, as illustrated in Figure 5. Notably, the s-P12F97B series membranes exhibited excellent dimensional stability, with an average area change of 43.8 ± 10.3% and thickness change of 42.4 ± 3.6%. The s-P12F99B-3.23 membrane had the largest dimensional change at low temperature because of its lowest trifluoromethyl content and highest water absorption capacity, which were crushing at 80 °C. This result can be attributed to the variable free volume and rigidity of the multiphenylated structure; the sulfonated territory allows water to occupy it [49,50], and the strong hydrophobic phase (–CF_3) improves the solubility and acceptable swelling rate of the membrane [27,51].

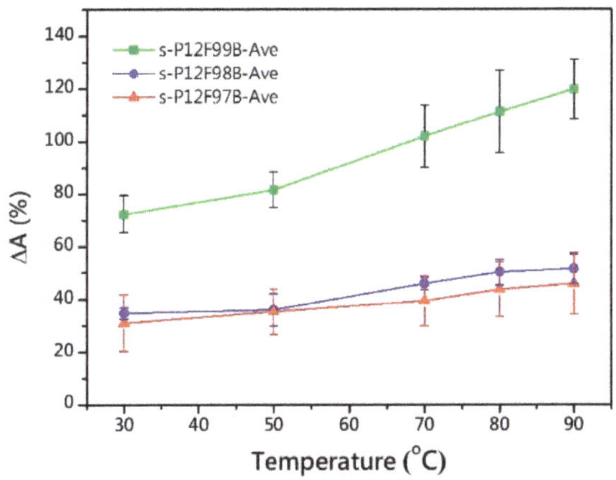

Figure 5. Swelling as a function of temperature for s-P12FmB-Ave.

The resistance of the membranes to oxidation was tested using Fenton's reagent, and all membranes exhibited excellent OS at 80 °C for 1 h; the results are listed in Table 2. After 16 h, continuous swelling was observed, and only the s-P12F99B-3.23 membranes completely degraded. The degradation of the s-P12F99B-Ave film was attributed to its high WU, which caused collapse and dissolution and increased the reaction area. Until 24 h, the asymmetric s-P12F98B-Ave still had higher oxidative stability than s-P12F97B-Ave—58.6 and 52.5 wt%, respectively. Inevitably, the long-term stability of the prepared membranes is not that optimistic or promising, compared to Nafion membranes. It is recommended to introduce functional groups that can effectively form electron-deficient structures in the polymer structure to improve the OS of membranes [37]. Although the trifluoromethyl group can effectively form an electronic trap structure, the OS was expected to increase as the volume ratio of trifluoromethyl molecules in the repeating unit was increased [52,53]. Up to 24 h, the stability of the s-P12F98B-Ave was similar to that of the s-P12F97B-Ave—58.6 and 59.5 wt%, respectively. This implies a threshold degree of intensive sulfonation as a function of the number of arms of phenyl group substituents, and this threshold can be used as a design reference for subsequent research.

3.5. Proton Conductivity

Proton conductivity is a principal indicator in the evaluation of the mass transfer characteristics of an MEA, which usually involve multiple transport mechanisms and hydration behaviors on multiple scales [9,26,32,37,54,55]. This article describes space-related hydration factors such as IEC, λ, $MVC_{(wet)}$, and PCV; the characteristics are summarized in Table 2. The proton conductivity of the s-P12FmB series membranes was measured as a function of RH at 80 °C and compared with that of the iconic standard Nafion membrane (Figure 6). Compared with the Nafion membrane (123.8 mS cm^{-1}) under the same measurement conditions, the s-P12FmB membranes (>173.9 mS cm^{-1}) had superior proton conductivity due to the dense sulfonated structure. The conductivity of the s-P12FmB membranes decreased with a decrease in the RH but was higher than that of the Nafion at all RHs. The proton conductivity of the s-P12FmB-Ave membranes with comparable IECs increased gradually with an increase in the number of phenyl groups in the structure; this increase may be attributed to the higher concentration of local sulfonate groups. Due to the existence of trifluoromethyl groups, the probability of sulfonation of the benzene ring can be reduced [13], and the concentration of localized sulfonate groups on the polymer chain can be further increased. Accordingly, s-P12F99B-2.89 had high proton conductivity (294.1 mS cm^{-1}) at low IEC, whereas s-P12F98B-3.00 and s-P12F97B-2.92 had lower proton conductivity (219.0 and 186.6 mS cm^{-1}, respectively) at high IEC.

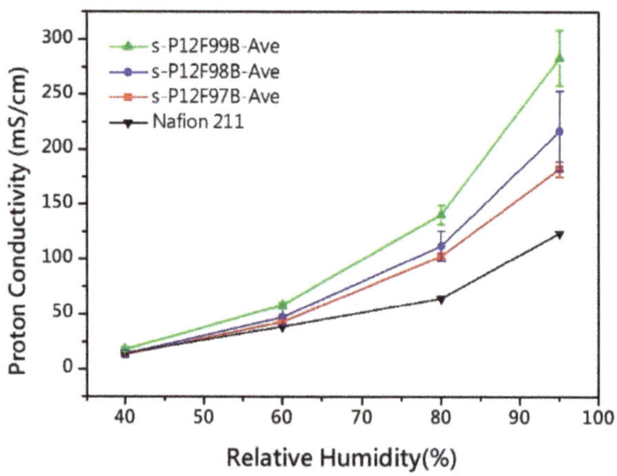

Figure 6. Proton conductivity of the s-P12FmB-Ave and Nafion 211 membranes as a function of relative humidity.

Due to the nature of the monomer components, the molecular volumes of their repeat units were different, and the IEC was introduced; the MVC values of the polymers in the dry state were calculated. To approximate the operating conditions of a fuel cell, the λ under high humidity and temperature was measured to obtain the $MVC_{(wet)}$ in this case. Although the actual volume was not obtained, when saturated, the estimated volume change was assumed to be mainly due to the filling effect caused by the ion clusters absorbing water and swelling under high humidity [9,31,36,52]. Incidentally, we can notice that the conductivity of s-P12F97B-X membranes under low humidity conditions has a contradictory behavior. It does not comply with the rising trend of conductivity with the increase of IEC. However, the results from MVC to $MVC_{(wet)}$ can be observed to be very similar, which means that the equivalent space provided by the concentration of sulfonate [$-SO_3H$] is similar. With the increase of RH, the master control of conductivity returns to IEC and returns to the normal trend at high RH conditions. Accordingly, we believe that the counterintuitive behavior of s-P12F97B-X membranes under low humidity conditions had an acceptable margin of error.

To determine the influence of hydration relative to conductivity, the influence of free volume was focused on the hydration domain (water-rich areas in the membrane), and the hydration ratio of ionomers under fully hydrated conditions was determined as the PCV; the result is listed in Table 3. With the change in the number of side-chain arms, the difference in IEC was insufficient to demonstrate the effect of the hydration zone on conductivity, whereas the PCV clearly differed. Consequently, even if the IEC was different, the PCV exhibited by ionomers with the same number of arms in the system under full hydration was similar. For example, even if the λ value of the s-P12F99 membranes was different, the proportion of domains that were hydrated was similar (PCV $\approx$ 0.71), and the conductivity was also similar. Conversely, as the PCV decreased, the conductivity decreased considerably, such as for s-P12F98B-2.61 and s-P12F98B-3.00. Although s-P12F98B-2.61 had a higher $MCV_{(wet)}$ (562.8 and 543.8 cm^2 [eq. mol.]$^{-1}$, respectively), its PCV was low (0.60 and 0.63, respectively); the ratio of the volume of the hydrophilic phase to that of the hydrated membrane was relatively small; therefore, the conductivity was low (255.1 and 212.6 cm^2 [eq. mol]$^{-1}$, respectively). s-P12FmB series membranes decreased PCV with an increase in the number of arms that were sulfonated. Notably, the membrane had low sensitivity to humidity, and the conductivity remained higher than 100 cm^2 (eq. mol)$^{-1}$ and was comparable with the Nafion membrane (39–65 cm^2 [eq. mol]$^{-1}$) at low humidity (RH, 60–80%). The low-humidity sensitivity is beneficial because humidity often varies during cell operation.

Table 3. Proton conductivity, MVC, and PCV of the s-P12FmB-X membranes at 80 °C.

Membrane	Proton Conductivity at 80 °C (mS/cm)				λ	MVC	MCV (Wet) at 80 °C	PCV
	40%RH	60%RH	80%RH	95%RH		(cm^3 eq.$^{-1}$ mol^{-1})	(cm^3 eq.$^{-1}$ mol^{-1})	at 80 °C
s-P12F99B-2.89	15.8	55.8	134.2	294.1	28.0	151.8	713.5	0.71
s-P12F99B-3.23	19.5	60.2	151.0	301.8	26.0	139.5	658.4	0.71
s-P12F98B-2.61	11.3	42.7	103.3	212.6	18.8	161.6	562.8	0.60
s-P12F98B-3.00	17.2	55.8	128.0	255.1	19.1	145.1	543.8	0.63
s-P12F97B-2.84	14.2	43.8	100.9	173.9	14.6	142.6	458.1	0.57
s-P12F97B-2.92	11.2	40.7	101.6	187.3	14.9	139.7	460.3	0.58
Nafion 211	14.6	38.8	64.5	123.8	14.8	-	-	-

3.6. Microstructure Analysis

The s-P12FmB series membrane had clear phase separation, including main isolated clusters (size, 3–8 nm) and mesoscale leopard-like clusters (15–45 nm), as visualized in the dry state by using TEM (Figure 7). In Figure 7, the shadowy areas represent hydrophilic ionic clusters, whereas the brighter areas represent the hydrophobic polymer matrix. The expansion of the shadowy area implies the expansion of the proton transmission channel. The aforementioned characteristics are observable to varying degrees in the cross-sectional TEM images of the s-P12FmB series membranes. The overall shape of the s-P12F99B-2.89 membrane was gyroid. In the microphase separation, isolated clusters tended to be connected to condensate subregions with low sulfonate density and a diameter of approximately 45 nm, as shown in Figure 7a. In the image of the s-P12F98B-3.00 membrane, the tendency to connect appears to be reduced, forming a metastable state between the gyroid and cylindrical types, as shown in Figure 7e. Under similar IEC, the number of side chains passing through the aromatic group decreased, and the density and morphology of large clusters changed until round isolated clusters formed in the s-P12F97B-2.92 membrane, as revealed in Figure 7f. Thus, the trend in the microphase separation pattern could be successfully changed by controlling the number of aromatic side chains. Predictably, the generated operating groups produced during the cell process form the main path for further growth while being transported.

3.7. Fuel Cell Performance

A single-MEA fuel cell with suitable mechanical characteristics and excellent proton conductivity was fabricated using s-P12FmB-X series and Nafion 211 membranes with overall Pt loading of 0.4 mg cm^{-2} on both sides of the membrane. The fuel cell was examined at 80 °C at full hydration. Figure 8 shows the polarization and power density curves of the MEA containing the s-P12FmB-X and Nafion 211 membranes. The s-P12F97B-2.92 cell exhibited an excellent current density of more than 1670 mA cm^{-2} (at 0.6 V) and power density similar to that of the MEA containing Nafion 211 with identical graphite bipolar plate and other fuel cell operating conditions. As shown in Figure 8, the cell with the s-P12F97B-2.92 membrane demonstrated the highest maximum power density (1.32 W cm^{-2}) and had superior power density to that containing Nafion 211. The fuel cell containing the s-P12F98B-3.00 membrane performed well (current density > 1650 mA cm^{-2} at 0.6 V), but greater ohmic region loss induced power density that was inferior to that for Nafion 211 (>2000 mA cm^{-2}). On the other hand, all s-P12FmB-X series membranes are known to have reached the realm of overhydration (PCV > 0.35) from PCV, which means that the hydrophilic phase may extend to satiation. The original TEM images used to evaluate the ion channels will need to be discussed in terms of their opposites. The gyroid from the ion cluster, which was expected to be beneficial for proton transport, instead has a detrimental effect on the structural support of the membrane. In addition to the macroscopic swelling of the membrane under high temperature and high RH conditions during cell operation, water crossover and fuel permeation may also occur. As a result, the s-P12F99B-2.89 membrane, which has the highest proton conductivity characteristics, exhibited cells performance that were not that expected. The optimization of membrane permeation modulation IEC may be the direction of research to obtain improved cell performance. Encouragingly, this study has been successful in improving cell performance through a slight structural design (different number of phenyl substitution arms and dense sulfonation), which is also expected to provide a reference for subsequent researchers. Accordingly, the high performance of the cell containing the s-P12F97B-2.92 membrane may have been due to its high dimensional stability and sufficient proton conductivity.

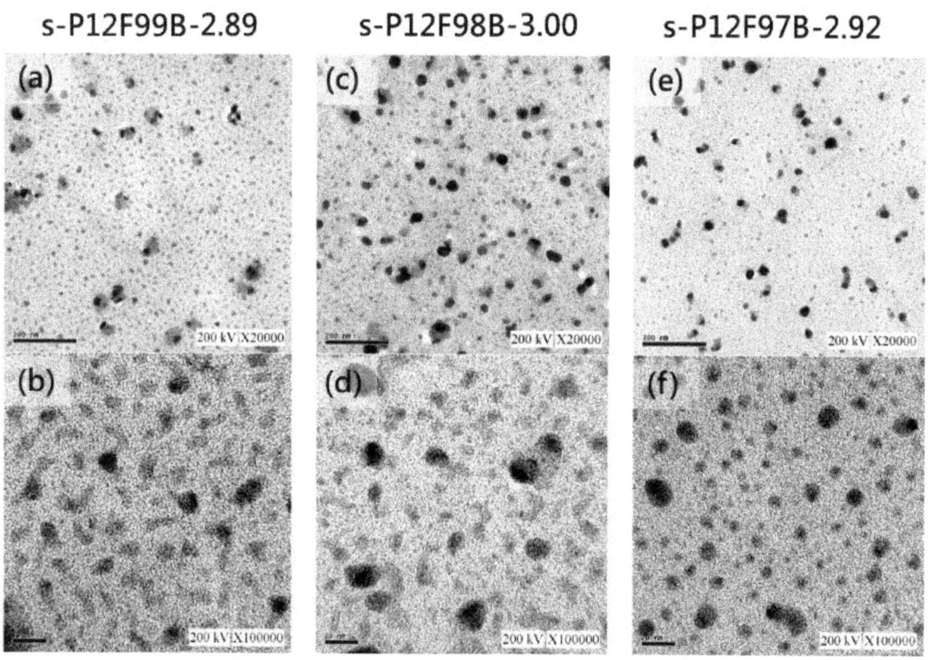

Figure 7. TEM image of the (**a**) s-P12F99B-2.89 membrane at 20k and (**b**) 100k magnification; (**c**) s-P12F98B-3.00 membrane at 20k and (**d**) 100k magnification; (**e**) s-P12F97B-2.92 membrane at 20k and (**f**) 100k magnification.

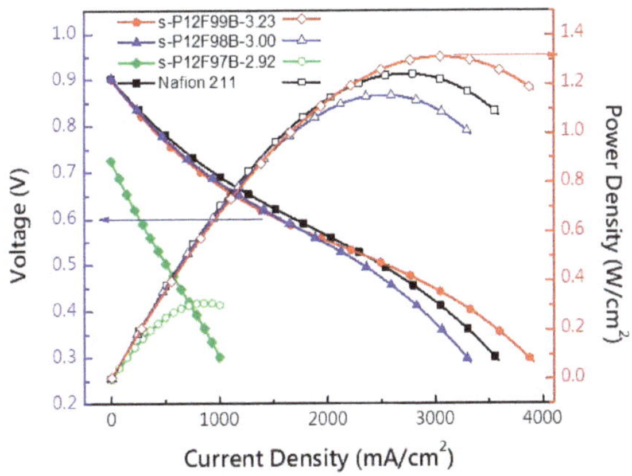

Figure 8. Fuel cell performance of s-P12F99B-2.89, s-P12F98B-3.00, s-P12F97B-2.92 and Nafion 211 membranes at 80 °**C**.

4. Conclusions

In summary, we introduced a series of novel sulfonated poly(arylene ether)s that contain superhydrophobic tetra-trifluoromethyl-substituted HAB and three types of densely sulfonated phenylated MAB structures. The polymers were prepared through nucleophilic substitution polycondensation, and postsulfonation was performed using chlorosulfuric acid. The electron-withdrawing pendant groups of trifluoromethyl activated polymerization and improved solubility to yield high-molecular-weight polymers. The results indicate that trifluoromethyl inhibits the sulfonation activity of local phenyl substitutions and postsulfonation can achieve more precise positioning in polymer while maintaining the distribution of hydrophobic and hydrophilic moieties. Furthermore, all of the membranes exhibited high thermal stability, favorable dimensional stability, and conductivity that was not sensitive to humidity. The fuel cell containing the s-P12F97B-2.92 membrane demonstrated excellent performance with a current density of more than 1670 mA cm^{-2} at 0.6 V and a maximum power density of 1.32 W cm^2 at 80 °C at full hydration, properties that are superior to those of Nafion 211. In addition, the s-P12F97B membranes exhibited tensile stress at a maximum load of 59.6–76.6 MPa and elongation at break of 82.8–95.8%. The combination of high thermal stability, acceptable dimensional stability, high proton conductivity, and excellent single-cell performance makes s-P12F97B-2.92 attractive as a PEM material for fuel cell applications.

Supplementary Materials: The following are available online at https://www.mdpi.com/article/10.3390/membranes11080626/s1, Figure S1: 1H NMR spectrum of 3′,6′-bis(4-bromophenyl)-4′,5′-diphenyl-3,3″-bis (trifluoromethyl)-1,1′:2′,1″-terphenyl (1) (6F7B-DB), Figure S2: ^{1}H NMR spectrum of 4,4′′′′-difluoro-3,3′′′′-bis(trifluoromethyl)-2″,3″,5″,6″,4,4′′′′-difluoro-3,3′′′′-bis(trifluoromethyl)-2″,3″,5″,6″,-Tetra (trifluoromethyl)phenyl-[1,1′:4′,1″:4″,1‴:4‴,1′′′′-quinquephenyl] (2) (12F9B-DF), Figure S3: ^{1}H NMR spectrum of 4,4′′′′-Dimethoxy-2″,3″,5″-triphenyl-1,1′:4′,1″:4″,1‴:4‴,1′′′′-quinquephenyl (3) (8B-DMO), Figure S4: ^{1}H NMR spectrum of 2″,3″,5″-Triphenyl-[1,1′:4′,1″:4″,1‴:4‴,1′′′′-quinquephenyl]-4,4′′′′-diol (4) (8B-DO), Figure S5: ^{1}H NMR spectrum of 4,4′′′′-Dimethoxy-2″,3″-diphenyl-1,1′:4′,1″:4″,1‴:4‴,1‴-quinquephenyl (5) (7B-DMO), Figure S6: ^{1}H NMR spectrum of 2″,3″-Diphenyl-[1,1′:4′,1″:4″,1‴:4‴,1‴-quinquephenyl]-4,4′′′′-diol (6) (7B-DO), Figure S7: MALDI-TOF spectrum of 6F7B-DB, Figure S8: MALDI-TOF spectrum of 12F9B-DF, Figure S9: MALDI-TOF spectrum of 8B-DMO, Figure S10: MALDI-TOF spectrum of 8B-DO, Figure S11: MALDI-TOF spectrum of 7B-DMO, Figure S12: MALDI-TOF spectrum of 7B-DO, Figure S13: ^{1}H NMR spectrum of P12F99B (top) and s-P12F99B (bottom), Figure S14: ^{1}H NMR spectrum of P12F97B (top) and s-P12F97B (bottom), Figure S15: FTIR spectra of the P12FmB polymers and after sulfonation from (a) 650 to 3800 cm^{-1} and (b) 700 to 1700 cm^{-1}.

Author Contributions: T.-S.H. and C.-C.L. conducted the experiment. T.-S.H. wrote the manuscript and with the help of H.-Y.W. and T.-L.H. T.-S.H. generated the figures and tables. M.-Y.C. and W.-Y.H. supervised the project. M.-Y.C. and W.-Y.H. proposed the idea for this project. All authors have read and agreed to the published version of the manuscript.

Funding: This research was funded by the Ministry of Science and Technology of Taiwan, Grant Number MOST 108-2221-E-110-035.

Institutional Review Board Statement: Not applicable.

Informed Consent Statement: Not applicable.

Data Availability Statement: Data are available in a publicly accessible repository.

Acknowledgments: The authors are grateful to the Ministry of Science and Technology, Taiwan (MOST 108-2221-E-110-035), for the financial support of this work. This manuscript was edited by Wallace Academic Editing. We also gratefully acknowledge Brightness & Green Technology Co., Ltd. for the high-performance liquid chromatography measurements.

Conflicts of Interest: The authors declare no conflict of interest.

References

1. Andujar, J.M.; Segura, F. Fuel cells: History and updating. A walk along two centuries. *Renew. Sustain. Energy Rev.* **2009**, *13*, 2309–2322. [CrossRef]
2. Dodds, P.E.; Staffell, I.; Hawkes, A.D.; Li, F.; Grünewald, P.; McDowall, W.; Ekins, P. Hydrogen and fuel cell technologies for heating: A review. *Int. J. Hydrog. Energy* **2015**, *40*, 2065–2083. [CrossRef]
3. Perry, M.L.; Fuller, T.F. A historical perspective of fuel cell technology in the 20th century. *J. Electrochem. Soc.* **2002**, *149*, S59–S67. [CrossRef]
4. Sharaf, O.Z.; Orhan, M.F. An overview of fuel cell technology: Fundamentals and applications. *Renew. Sustain. Energy Rev.* **2014**, *32*, 810–853. [CrossRef]
5. Hou, J.; Yang, M.; Wang, D.; Zhang, J. Fundamentals and challenges of lithium ion batteries at temperatures between −40 and 60 °C. *Adv. Energy Mater.* **2020**, *10*, 1904152. [CrossRef]
6. Zhang, H.; Sun, C. Cost-effective iron-based aqueous redox flow batteries for large-scale energy storage application: A review. *J. Power Sources* **2021**, *493*, 229445. [CrossRef]
7. Schiavone, M.-M.; Lamparelli, D.H.; Zhao, Y.; Zhu, F.; Revay, Z.; Radulescu, A. The Effects of Temperature and Humidity on the Microstructure of Sulfonated Syndiotactic-polystyrene Ionic Membranes. *Membranes* **2020**, *10*, 187. [CrossRef] [PubMed]
8. Kim, A.R.; Vinothkannan, M.; Lee, K.H.; Chu, J.Y.; Ryu, S.K.; Kim, H.G.; Lee, J.-Y.; Lee, H.-K.; Yoo, D.J. Ameliorated Performance of Sulfonated Poly(Arylene Ether Sulfone) Block Copolymers with Increased Hydrophilic Oligomer Ratio in Proton-Exchange Membrane Fuel Cells Operating at 80% Relative Humidity. *Polymers* **2020**, *12*, 1871. [CrossRef] [PubMed]
9. Kusoglu, A.; Weber, A.Z. New Insights into Perfluorinated Sulfonic-Acid Ionomers. *Chem. Rev.* **2017**, *117*, 987–1104. [CrossRef]
10. Kraytsberg, A.; Ein-Eli, Y. Review of Advanced Materials for Proton Exchange Membrane Fuel Cells. *Energy Fuels* **2014**, *28*, 7303–7330. [CrossRef]
11. Mehta, V.; Cooper, J.S. Review and analysis of PEM fuel cell design and manufacturing. *J. Power Sources* **2003**, *114*, 32–53. [CrossRef]
12. Iulianelli, A.; Basile, A. Sulfonated PEEK-based polymers in PEMFC and DMFC applications: A review. *Int. Hydrog. Energy* **2012**, *37*, 15241–15255. [CrossRef]
13. Dhara, M.G.; Banerjee, S. Fluorinated high-performance polymers: Poly(arylene ether)s and aromatic polyimides containing trifluoromethyl groups. *Prog. Polym. Sci.* **2010**, *3*, 1022–1077. [CrossRef]
14. Liu, B.; Robertson, G.P.; Kim, D.-S.; Guiver, M.; Hu, A.W.; Jiang, Z. Aromatic poly(ether ketone)s with pendant sulfonic acid phenyl groups prepared by a mild sulfonation method for proton exchange membranes. *Macromolecules* **2007**, *40*, 1934–1944. [CrossRef]
15. Kaliaguine, S.; Mikhailenko, S.; Wang, K.; Xing, P.; Robertson, G.; Guiver, M. Properties of SPEEK based PEMs for fuel cell application. *Catal. Today* **2003**, *82*, 213–222. [CrossRef]
16. Li, Y.; Zhang, X.; He, G.; Zhang, F. Sulfonated poly(phenylene sulfide) grafted polysulfone proton exchange membrane with improved stability. *Int. J. Hydrog. Energy* **2017**, *42*, 2360–2369. [CrossRef]
17. Weiber, E.A.; Takamuku, S.; Jannasch, P. Highly Proton Conducting Electrolyte Membranes Based on Poly(arylene sulfone)s with Tetrasulfonated Segments. *Macromolecules* **2013**, *46*, 3476–3485. [CrossRef]
18. Yao, Z.; Zhang, Z.; Hu, M.; Hou, J.; Wu, L.; Xu, T. Perylene-based sulfonated aliphatic polyimides for fuel cell applications: Performance enhancement by stacking of polymer chains. *J. Membr. Sci.* **2018**, *547*, 43–50. [CrossRef]
19. Chen, K.C.; Ji, M.D. High Tempreture Fuel Cell Performance and Anisotropy of Carbonyl and Sulfone Groups Co-crosslinked Sulfonated Polyimides Proton Exchange Membranes. *Chem. J. Chin. Univ.* **2016**, *37*, 989–995.
20. Yin, Y.; Fang, J.; Watari, T.; Tanaka, K.; Kita, H.; Okamoto, K.-I. Synthesis and properties of highly sulfonated proton conducting polyimides from bis(3-sulfopropoxy)benzidine diamines. *J. Mater. Chem.* **2004**, *14*, 1062–1070. [CrossRef]

21. Pan, H.; Chen, S.; Jin, M.; Chang, Z.; Pu, H. Preparation and properties of sulfonated polybenzimidazole-polyimide block copolymers as electrolyte membranes. *Ionics* **2018**, *24*, 1629–1638. [CrossRef]
22. Singha, S.; Jana, T.; Modestra, J.A.; Kumar, A.N.; Mohan, S.V. Highly efficient sulfonated polybenzimidazole as a proton exchange membrane for microbial fuel cells. *J. Power Sources* **2016**, *317*, 143–152. [CrossRef]
23. Bai, H.; Ho, W.S.W. New sulfonated polybenzimidazole (SPBI) copolymer-based proton-exchange membranes for fuel cells. *J. Taiwan Inst. Chem. Eng.* **2009**, *40*, 260–267. [CrossRef]
24. Shin, D.W.; Lee, S.Y.; Kang, N.R.; Lee, K.H.; Guiver, M.; Lee, Y.M. Durable Sulfonated Poly(arylene sulfide sulfone nitrile)s Containing Naphthalene Units for Direct Methanol Fuel Cells (DMFCs). *Macromolecules* **2013**, *46*, 3452–3460. [CrossRef]
25. Hickner, M.A.; Ghassemi, H.; Kim, Y.S.; Einsla, B.R.; McGrath, J.E. Alternative polymer systems for proton exchange membranes (PEMs). *Chem. Rev.* **2004**, *104*, 4587–4611. [CrossRef] [PubMed]
26. Shin, D.W.; Guiver, M.D.; Lee, Y.M. Hydrocarbon-Based Polymer Electrolyte Membranes: Importance of Morphology on Ion Transport and Membrane Stability. *Chem. Rev.* **2017**, *117*, 4759–4805. [CrossRef]
27. Huang, Y.-C.; Lee, H.-F.; Tseng, Y.-C.; Lee, C.-C.; Chang, M.-Y.; Huang, W.-Y. Synthesis of novel sulfonated poly(arylene ether)s containing a tetra-trifluoromethyl side chain and multi-phenyl for proton exchange membrane fuel cell application. *RSC Adv.* **2017**, *7*, 33068–33077. [CrossRef]
28. Huang, Y.-C.; Tai, R.-H.; Lee, H.-F.; Wang, P.-H.; Gopal, R.; Lee, C.-C.; Chang, M.-Y.; Huang, W.-Y. Synthesis of Highly Sulfonated Poly(arylene ether) Containing Multiphenyl for Proton Exchange Membrane Materials. *Int. J. Polym. Sci.* **2016**, *2016*, 6545362. [CrossRef]
29. Lee, H.-F.; Wang, P.-H.; Huang, Y.-C.; Su, W.-H.; Gopal, R.; Lee, C.-C.; Holdcroft, S.; Huang, W.-Y. Synthesis and Proton Conductivity of Sulfonated, Multi-phenylated Poly(arylene ether)s. *J. Polym. Sci. Part A Polym. Chem.* **2014**, *52*, 2579–2587. [CrossRef]
30. Adamski, M.; Skalski, T.J.; Schibli, E.M.; Killer, M.; Wu, Y.; Peressin, N.; Frisken, B.J.; Holdcroft, S. Molecular branching as a simple approach to improving polymer electrolyte membranes. *J. Membr. Sci.* **2020**, *595*, 117539. [CrossRef]
31. Lee, K.-S.; Jeong, M.-H.; Lee, J.-P.; Kim, Y.-J.; Lee, J.-S. Synthesis and Characterization of Highly Fluorinated Cross-linked Aromatic Polyethers for Polymer Electrolytes. *Chem. Mater.* **2010**, *22*, 5500–5511. [CrossRef]
32. Zheng, Y.; Li, S.; Weng, Z.; Gao, C. Hyperbranched polymers: Advances from synthesis to applications. *Chem. Soc. Rev.* **2015**, *44*, 4091–4130. [CrossRef] [PubMed]
33. Kim, D.S.; Robertson, G.P.; Kim, Y.S.; Guiver, M.D. Copoly(arylene ether)s Containing Pendant Sulfonic Acid Groups as Proton Exchange Membranes. *Macromolecules* **2009**, *42*, 957–963. [CrossRef]
34. Sun, C.Y.; Zhang, H. Investigation of Nafion series membranes on the performance of iron-chromium redox flow battery. *Int. J. Energy Res.* **2019**, *43*, 8739–8752. [CrossRef]
35. Van Krevelen, D.W.; Te Nijenhuis, K. *Properties of Polymers: Their Correlation with Chemical Structure;* Their Numerical Estimation and Prediction from Additive Group Contributions; Elsevier: Amsterdam, The Netherlands, 2009.
36. Fuessl, A.; Yamamoto, M.; Schneller, A. *Polymer Science: A Comprehensive Reference;* Matyjaszewski, K., Möller, M., Eds.; Elsevier: Amsterdam, The Netherlands, 2012.
37. Mohanty, A.K.; Mistri, E.A.; Banerjee, S.; Komber, H.; Voit, B. Highly Fluorinated Sulfonated Poly(arylene ether sulfone) Copolymers: Synthesis and Evaluation of Proton Exchange Membrane Properties. *Ind. Eng. Chem. Res.* **2013**, *52*, 2772–2783. [CrossRef]
38. Banerjee, S.; Maier, G.; Salunke, A.; Madhra, M. Novel high-T-g, high-strength poly(aryl ether) containing quadrephenyl unit. *J. Appl. Polym. Sci.* **2001**, *82*, 3149–3156. [CrossRef]
39. Lee, H.-F.; Huang, Y.-C.; Wang, P.-H.; Lee, C.-C.; Hung, Y.-S.; Gopal, R.; Holdcroft, S.; Huang, W.-Y. Synthesis of highly sulfonated polyarylene ethers containing alternating aromatic units. *Mater. Today Commun.* **2015**, *3*, 114–121. [CrossRef]
40. Parreno, R.P., Jr.; Liu, Y.-L.; Beltran, A.-B.; Carandang, M.-B. Effect of a direct sulfonation reaction on the functional properties of thermally-crosslinked electrospun polybenzoxazine (PBz) nanofibers. *RSC Adv.* **2020**, *10*, 14198–14207. [CrossRef]
41. Pirali-Hamedani, M.; Mehdipour-Ataei, S. Effect of sulfonation degree on molecular weight, thermal stability, and proton conductivity of poly(arylene ether sulfone)s membrane. *Des. Monomers Polym.* **2017**, *20*, 54–65. [CrossRef]
42. Sun, C.; Negro, E.; Vezzù, K.; Pagot, G.; Cavinato, G.; Nale, A.; Bang, Y.H.; Di Noto, V. Hybrid inorganic-organic proton-conducting membranes based on SPEEK doped with WO3 nanoparticles for application in vanadium redox flow batteries. *Electrochimica Acta* **2019**, *309*, 311–325. [CrossRef]
43. Iulianelli, A.; Clarizia, G.; Gugliuzza, A.; Ebrasu, D.; Bevilacqua, A.; Trotta, F.; Basile, A. Sulfonation of PEEK-WC polymer via chloro-sulfonic acid for potential PEM fuel cell applications. *Int. J. Hydrog. Energy* **2010**, *35*, 12688–12695. [CrossRef]
44. Siengchum, T.; Isenberg, M.; Chuang, S.S.C. Fast pyrolysis of coconut biomass—An FTIR study. *Fuel* **2013**, *105*, 559–565. [CrossRef]
45. Takenaka, M.; Kimura, Y.; Ohara, H. Influence of decomposition temperature of aromatic sulfonic acid catalysts on the molecular weight and thermal stability of poly(L-lactic acid) prepared by melt/solid state polycondenstaion. *Polymer* **2018**, *155*, 218–224. [CrossRef]
46. Mukherjee, R.; Banerjee, S.; Komber, H.; Voit, B. Carboxylic acid functionalized fluorinated sulfonated poly(arylene ether sulfone) copolymers with enhanced oxidative stability. *J. Membr. Sci.* **2016**, *510*, 497–509. [CrossRef]
47. Jo, S.G.; Kim, T.-H.; Yoon, S.J.; Oh, S.-G.; Cha, M.S.; Shin, H.Y.; Ahn, J.M.; Lee, J.Y.; Hong, Y.T. Synthesis and investigation of random-structured ionomers with highly sulfonated multi-phenyl pendants for electrochemical applications. *J. Membr. Sci.* **2016**, *510*, 326–337. [CrossRef]

48. Ijaodola, O.S.; Hassan, Z.E.-; Ogungbemi, E.; Khatib, F.; Wilberforce, T.; Thompson, J.; Olabi, A. Energy efficiency improvements by investigating the water flooding management on proton exchange membrane fuel cell (PEMFC). *Energy* **2019**, *179*, 246–267. [CrossRef]
49. Lee, S.; Ann, J.; Lee, H.; Kim, J.-H.; Yang, T.-H.; Bae, B. Synthesis and characterization of crosslink-free highly sulfonated multi-block poly(arylene ether sulfone) multi-block membranes for fuel cells. *J. Mater. Chem. A* **2015**, *3*, 1833–1836. [CrossRef]
50. Gao, Y.; Robertson, G.P.; Kim, D.-S.; Guiver, M.D.; Mikhailenko, S.D.; Li, X.; Kaliaguine, S. Comparison of PEM properties of copoly(aryl ether ether nitrile)s containing sulfonic acid bonded to naphthalene in structurally different ways. *Macromolecules* **2007**, *40*, 1512–1520. [CrossRef]
51. Voit, B.I.; Lederer, A. Hyperbranched and Highly Branched Polymer Architectures-Synthetic Strategies and Major Characterization Aspects. *Chem. Rev.* **2009**, *109*, 5924–5973. [CrossRef]
52. Kim, S.; Melnyk, A.; Andrienko, D.; Suzuki, Y. Solid-State Electron Affinity Analysis of Amorphous Fluorinated Polymer Electret. *J. Phys. Chem. B* **2020**, *124*, 10507–10513. [CrossRef] [PubMed]
53. Mohanty, A.K.; Mistri, E.A.; Ghosh, A.; Banerjee, S. Synthesis and characterization of novel fluorinated poly(arylene ether sulfone)s containing pendant sulfonic acid groups for proton exchange membrane materials. *J. Membr. Sci.* **2012**, *409*, 145–155. [CrossRef]
54. Nandjou, E.; Poirot-Crouvezier, J.-P.; Chandesris, M.; Blachot, J.-F.; Bonnaud, C.; Bultel, Y. Impact of heat and water management on proton exchange membrane fuel cells degradation in automotive application. *J Power Sources* **2016**, *326*, 182–192. [CrossRef]
55. Zhang, J. *PEM Fuel Cell Electrocatalysts and Catalyst Layers: Fundamentals and Applications*; Springer Science & Business Media: Vancouver, BC, Canada, 2008.

membranes

MDPI

Article

Flexible, Transparent and Highly Conductive Polymer Film Electrodes for All-Solid-State Transparent Supercapacitor Applications

Xin Guan [1], Lujun Pan [2,*] and Zeng Fan [2,*]

[1] International Studies College, National University of Defense Technology, Nanjing 210012, China; xguan1408@163.com

[2] School of Physics, Dalian University of Technology, No. 2 Linggong Road, Ganjingzi District, Dalian 116024, China

* Correspondence: lpan@dlut.edu.cn (L.P.); fanzeng@dlut.edu.cn (Z.F.)

Abstract: Lightweight energy storage devices with high mechanical flexibility, superior electrochemical properties and good optical transparency are highly desired for next-generation smart wearable electronics. The development of high-performance flexible and transparent electrodes for supercapacitor applications is thus attracting great attention. In this work, we successfully developed flexible, transparent and highly conductive film electrodes based on a conducting polymer, poly(3,4-ethylenedioxythiophene):poly(styrenesulfonate) (PEDOT:PSS). The PEDOT:PSS film electrodes were prepared via a simple spin-coating approach followed by a post-treatment with a salt solution. After treatment, the film electrodes achieved a high areal specific capacitance (3.92 mF/cm^2 at 1 mA/cm^2) and long cycling lifetime (capacitance retention >90% after 3000 cycles) with high transmittance (>60% at 550 nm). Owing to their good optoelectronic and electrochemical properties, the as-assembled all-solid-state device for which the PEDOT:PSS film electrodes were utilized as both the active electrode materials and current collectors also exhibited superior energy storage performance over other PEDOT-based flexible and transparent symmetric supercapacitors in the literature. This work provides an effective approach for producing high-performance, flexible and transparent polymer electrodes for supercapacitor applications. The as-obtained polymer film electrodes can also be highly promising for future flexible transparent portable electronics.

Keywords: conducting polymers; electrodes; all-solid-state supercapacitors; transparency; flexibility

Citation: Guan, X.; Pan, L.; Fan, Z. Flexible, Transparent and Highly Conductive Polymer Film Electrodes for All-Solid-State Transparent Supercapacitor Applications. *Membranes* **2021**, *11*, 788. https://doi.org/10.3390/membranes11100788

Academic Editors: Marc Cretin, Sophie Tingry and Zhenghua Tang

Received: 4 September 2021
Accepted: 12 October 2021
Published: 16 October 2021

1. Introduction

As one of the most attractive power sources, supercapacitors that deliver rapid charging/discharging rates, high power densities and long lifecycles have received significant attention over the past few years [1–3]. High-performance supercapacitors have even been regarded as the most vital part for a wide application scope ranging, from miniaturized medical devices to large electric vehicles [4–6]. With the increasing desire for portable and wearable electronics, the supercapacitors in early works, which are commonly bulky, heavy and rigid, have gradually become unsatisfactory. Therefore, research interest toward developing novel energy storage devices that confer features including small size, light weight, good mechanical flexibility and even optical transparency has risen [7–12]. As the key component of a supercapacitor, the property of its electrodes is critical. To meet the aforementioned demands, the electrodes should possess not only superior electrochemical properties, but also easy processability, high mechanical flexibility and good optoelectronic performance [13].

To date, a variety of materials, including metal nanowires [14], conducting polymers [15], carbon-based nanomaterials [16,17] and nanocomposites containing transition metal oxides [18–20], are used to fabricate the electrodes for flexible and transparent supercapacitors. Among these materials, the hybrids of metal networks with conducting polymers have been most extensively investigated. For example, Yang et al. reported the synthesis of nickel oxide/nickel vanadate ($Ni_3V_2O_8$) and polyaniline (PANI) composites via in situ chemical bath deposition [21]. The as-fabricated electrodes exhibited a high transparency of 71.8%, a high specific capacitance of 2565.7 F/g at 5 mV/s and a wide potential window of 1.6 V. Lee et al. developed composite electrodes consisting of poly(3,4-ethylenedioxythiophene): poly(styrenesulfonate) (PEDOT:PSS) and silver nanowires [22]. The electrodes delivered 87% transmittance and capacitance of 113 $\pm$ 18 F/g at 10 mV/s. However, the adhesion between the metal network with either the conducting polymer or the plastic substrate was weak, so the metal network was usually easily delaminated from the substrate [23]. Aiming to improve the mechanical flexibility, He et al. developed flexible and transparent film electrodes, using cellulose nanofibrils and reduced graphene oxide via a layer-by-layer self-assembly method [24]. The areal specific capacitance was 2.25 mF/cm^2 at a current density of 0.01 mA/cm^2. However, the carbon-based nanomaterials commonly exhibit limited optoelectronic performance [25–27].

Owing to the good mechanical and electrical properties, flexible and transparent electrodes made of neat conducting polymers are attracting great attention. PEDOT:PSS is considered the most successful conducting polymer because of its unique advantages, e.g., high electrical conductivity, excellent thermal stability, easy solution processability, high transparency in the visible range and possible pseudo-capacitive effects [28–32]. In particular, compared with the complex processing techniques for developing various inorganic-organic composite electrodes, the PEDOT:PSS film electrodes can be prepared by a simple spin-coating approach, thus allowing a high-efficient fabrication on a large scale. Due to their high electrical conductivity, PEDOT:PSS films can be applied as both the current collectors and active electrodes in flexible and transparent all-solid-state supercapacitors [33]. Lai et al. reported a flexible transparent supercapacitor based on the high-performance PEDOT:PSS films doped with ethylene glycol (EG) [33]. The highly flexible PEDOT:PSS electrodes showed high transparency of >82% and an areal specific capacitance of 4.72 mF/cm^2 at a current density of 0.01 mA/cm^2. The resultant all-solid-state symmetric supercapacitor yielded an areal specific capacitance of 1.18 mF/cm^2 and an energy density of 0.105 μWh/cm^2 at a power density of 36 μW/cm^2. Lee et al. fabricated flexible, lightweight and coplanar transparent supercapacitor electrodes by embedding conducting silver nanofibers into PEDOT:PSS [23]. The highly flexible electrode delivered an areal capacitance of 3.64 mF/cm^2 at 85% transmittance, while the symmetric transparent supercapacitor device exhibited an areal capacitance of 0.91 mF/cm^2 at a scan rate of 5 mV/s and an energy density of 0.05 μWh/cm^2 at a power density of 74.1 μW/cm^2. However, a challenging task still remains for the development of lightweight, flexible and transparent supercapacitors with high power/energy densities.

In addition, a simple and environmental-friendly technique for the electrode fabrication is of practical significance. Previous studies have shown that the optoelectronic properties of the PEDOT:PSS films could be greatly improved via either secondary doping or a post-treatment process with solvents [34], solutions [35], ionic liquids [36], zwitterions [37], acids [29], etc. Among these approaches, the treatment using solutions consisting of common salts and solvents is particularly mild, without causing damage to the plastic substrates, e.g., polyethylene terephthalate (PET). In this regard, the PEDOT:PSS films after such a solution treatment can be promising for use as the electrodes of flexible and transparent supercapacitors.

In this work, we reported a flexible and transparent electrode with high performance based on PEDOT:PSS films. The PEDOT:PSS films were prepared through directly spin-coating the aqueous solution onto PET substrates, while their optoelectronic property was systematically investigated and modulated as a function of the spin coating cycles. The as-prepared PEDOT:PSS films were then post-treated with a solution consisting of a common inorganic salt and organic solvent. The treatment has improved both the sheet conductance and optical transmittance of the as-prepared films. An all-solid-state supercapacitor was assembled, using the treated PEDOT:PSS films as both the electrodes and current collectors.

The electrochemical properties of the PEDOT:PSS electrodes and the all-solid-state device were studied in terms of the areal specific capacitance, cycling durability and mechanical flexibility. In comparison with other PEDOT-based transparent and flexible supercapacitors in literature, our device exhibited a remarkably higher power/energy density. This work offers a simple, efficient and environmental-friendly approach to fabricate high-performance electrodes for future flexible and transparent energy storage devices.

2. Experimental Section

2.1. Materials

The PEDOT:PSS aqueous solution (Clevios PH1000) was purchased from Heraeus (Shanghai, China). The concentration of PEDOT:PSS was 1.3 wt.%, and the weight ratio of PSS to PEDOT was 2.5 in solution. Isopropanol (IPA), zinc chloride ($ZnCl_2$) and *N*, *N*-dimethylformamide (DMF) were purchased from Sigma-Aldrich (Shanghai, China). Phosphoric acid (H_3PO_4) and polyvinyl alcohol (PVA) powder were purchased from Tianjin Kaixin Chem (Tianjin, China) and Evs Chemical Technology Co. Ltd. (Dalian, China), respectively. All the chemicals were used as received. Transparent PET substrates were purchased from Dawan Plastic Electronics Company (Suzhou, China).

2.2. Fabrication of PEDOT:PSS Electrodes

The PEDOT:PSS electrodes was fabricated by spin-coating the PEDOT:PSS aqueous solution on PET substrates of 1.5×1.5 cm^2, which were pre-treated successively by detergent, IPA and plasma cleaning (air plasma, Shenzhen Tonson Tech Automation Equipment Co., Ltd. Shenzhen, China). After spin-coating, the PEDOT:PSS films were dried at 120 °C on a hotplate for 20 min. The PEDOT:PSS electrodes with different thicknesses were prepared by repeating the spin-coating process for 1, 2, 4 and 5 cycles. The as-prepared PEDOT:PSS films were then treated with 0.1 M $ZnCl_2$-DMF solution at 120 °C. After the solvent DMF was evaporated completely, the treated films were cooled down to room temperature, followed by rinsing with deionized (DI) water three times. After that, the films were dried again on the hot plate.

2.3. Fabrication of All-Solid-State Symmetric Supercapacitors

A gel electrolyte was prepared by mixing 1 mL H_3PO_4, 1g PVA powder and 9 mL DI water. The whole mixture was heated to 90 °C under stirring until the blend solution turned into a gel. The all-solid-state supercapacitors were assembled in a sandwich configuration. Typically, two PEDOT:PSS electrodes were coated on the upper and lower sides of the H_3PO_4/PVA gel electrolyte, respectively, and then firmly pressed together. The gel electrolyte served as both the binder and the separator. The effective area of the symmetric supercapacitor was 1.5×1.0 cm^2.

2.4. Characterizations

The optical transmittance of the PEDOT:PSS electrodes was measured on the UV-vis-NIR spectrometer (UV-2600, Shimadzu Co., Ltd. Suzhou, China) in the visible light range (400–700 nm). The surface morphology of the PEDOT:PSS electrodes was observed on an optical microscope (PSM-1000 Motic China Group Co., Ltd. Xiamen, China). The sheet resistances of the film electrodes were measured by the van der Pauw probe technique, using a Keithley 2450 source/m. The electrical contacts were prepared on the four corners of the films, using silver paste.

2.5. Electrochemical Measurements

The electrochemical properties of both the PEDOT:PSS electrodes and all-solid-state devices were investigated using a CHI 760E electrochemical workstation (Chenhua Instrument Co., Shanghai, China). Cyclic voltammetry (CV), galvanostatic charge–discharge (GCD) and electrochemical impedance spectroscopy (EIS) measurements for the PEDOT:PSS electrodes were conducted in a three-electrode configuration. A 1 M sulfuric acid (H_2SO_4)

was used as the electrolyte, while the PEDOT:PSS films, Hg/$HgSO_4$ and platinum (Pt) foil served as the working, reference and counter electrodes, respectively. The EIS measurements were performed in the 10 mHz to 100 kHz frequency range with a potential amplitude of 5 mV. The electrochemical properties of the all-solid-state symmetric supercapacitor were evaluated in a two-electrode configuration by CV, GCD and EIS measurements.

The areal specific capacitances (C_{SC}, mF/cm^2) of the film electrodes and the symmetric supercapacitor were both calculated from their GCD curves according to the following formula:

$$C_{SC} = \frac{I\Delta t}{A\Delta V} \quad (1)$$

where I is the discharge current, Δt is the discharging time based on the scan rate, A is the effective area of the electrode, and ΔV is the potential window. The areal energy density (E, μWh/cm^2) and power density (P, μW/cm^2) were calculated according to the equations

$$E = \frac{1}{2} C_{SC} \Delta V^2 \quad (2)$$

And

$$P = \frac{E}{\Delta t} \quad (3)$$

respectively, where C_{SC} is the specific capacitance of the device.

3. Results and Discussion

Compared with other transparent supercapacitor electrodes composed of inorganic and organic blends, neat polymer film electrodes can possess various advantages, e.g., low-cost, good material uniformity and easy processability. The pristine PEDOT:PSS films directly produced from its aqueous solution usually have a very low electrical conductivity of <1 S/cm [38]. To achieve highly conductive electrodes in a mild way, the as-prepared PEDOT:PSS films were subjected to a post-treatment with 0.1 M $ZnCl_2$-DMF solution. Notably, since the PEDOT:PSS films was prepared via spin-coating from its aqueous solution, and the post-treatment was performed by simple solution drop-casting, the process for producing the transparent and flexible PEDOT:PSS film electrodes is, therefore, all-solution-based, which can enable a scalable fabrication in practice. The mechanism for the electrical conductivity enhancement of PEDOT:PSS by the $ZnCl_2$-DMF treatment is illustrated in Figure 1. During the $ZnCl_2$-DMF treatment, the solvent DMF and the salt ions can both induce charge screening on PEDOT and PSS [39]. The charge screening effect weakens the Coulombic attraction between the PEDOT and PSS chains, thus facilitating a structural transition of the PEDOT:PSS chains from a coiled to a linear conformation. The expansion of the polymer chains consequently results in a remarkably improved electrical performance. Since the solvent and salt ions in the salt solution can play a synergetic effect on altering the polymer conformation, the salt solution treatment turns out to facilitate greater enhancement in the electrical conductivity of PEDOT:PSS than the neat solvent treatment methods, which are commonly used elsewhere [22].

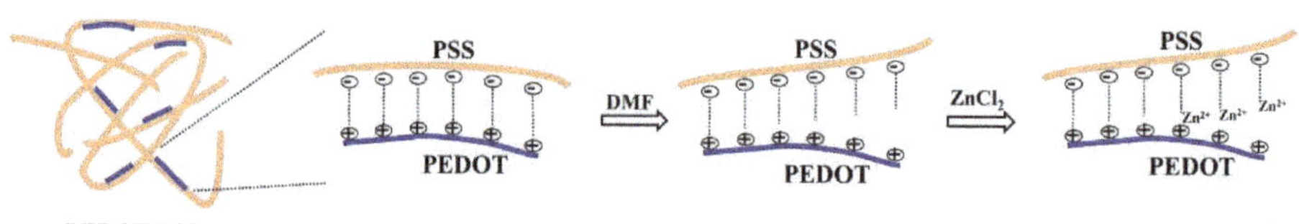

Figure 1. Schematic diagram of the mechanism for conductivity enhancement of PEDOT:PSS by $ZnCl_2$-DMF treatment.

For use as the flexible and transparent supercapacitor electrodes, the optoelectronic performance of the PEDOT:PSS films in terms of the sheet resistance and transmittance were measured, as shown in Figure 2. After the $ZnCl_2$-DMF solution treatment, the sheet resistance of the PEDOT:PSS film prepared via only once spin-coating sharply decreases from 700 to 0.163 kΩ/sq, while its transmittance at 550 nm surprisingly increases from 89.7% to 93.9%, indicating the effectiveness of the treatment process. The sheet resistance and optical transmittance of the PEDOT:PSS film electrodes were further investigated as a function of the spin-coating cycles. As the number of the coating cycles increases from 2 to 5, the film electrode exhibits simultaneous decreases in both the sheet resistance and transparency. Typically, a PEDOT:PSS electrode prepared from 4-cycle spin-coating yields a sheet resistance of as low as 41 Ω/sq, with a transmittance of over 60% at 550 nm. These properties are quite comparable or even better than those of the PANI- and graphene-based counterparts reported in the literature [15,17].

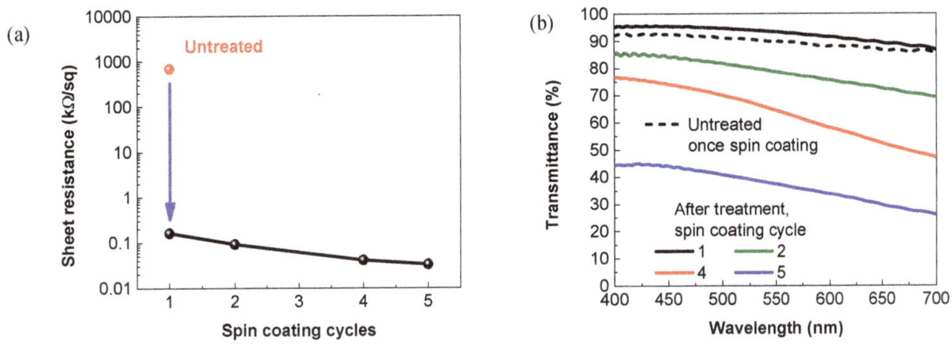

Figure 2. Optoelectronic property of PEDOT:PSS film electrodes. (**a**) Sheet resistances and (**b**) optical transmittance of the PEDOT:PSS film electrodes as a function of the spin-coating cycles.

Figure 3 displays the electrochemical properties of the PEDOT:PSS film electrodes with varying spin-coating cycles. As shown in Figure 3a, all the CV curves are nearly rectangular without redox peaks at a scan rate of 100 mV/s, indicating the typical electrical-double-layer (EDL) capacitive characteristics and good electrical conductivity of the PEDOT:PSS electrodes [40]. As the coating cycle increases from 1 to 5, the area of the closed CV loop accordingly increases. The capacitive behaviors of the PEDOT:PSS electrodes are thus improved by increasing the thickness of the PEDOT:PSS films. At the applied current density of 0.025 mA/cm^2, all their GCD curves exhibit a nearly perfect triangular shape (Figure 3b). The small voltage drop (IR drop) suggests a decreased mass transport resistance and a good charge propagation of ions within the electrodes [41]. According to the GCD curves, the areal specific capacitances of the PEDOT:PSS electrodes prepared via spin-coating of 1, 2, 4 and 5 cycles are calculated to be 1.89, 2.27, 3.92 and 4.82 mF/cm^2, respectively. As the discharge current density increases from 0.025 to 0.1 mA/cm^2, their areal specific capacitances only slightly decrease to 1.52, 1.84, 3.2 and 4.02 mF/cm^2, respectively. The corresponding capacitance retentions are 80.6%, 81.0%, 81.6%, and 83.5%, respectively, indicating the good rate capability of these electrodes (Figure 3c) [42]. The Nyquist plots as a function of the spin-coating cycles are shown in Figure 3d. At the low frequency regions, all the electrodes exhibit a steep slope, further confirming their ideal capacitive behavior [43]. At the high-frequency regions (the inset of Figure 3d), the equivalent series resistances (ESRs), which correspond to the internal resistances, are calculated to be ~3 Ω for all these electrodes.

Our previous work has shown that the post-treatment with organic solutions of inorganic salts (e.g., $ZnCl_2$, $NiCl_2$ and $CuCl_2$) could effectively screen the Coulombic interaction between the positively charged PEDOT and negatively charged PSS. This process not only facilitates the effective PSS depletion from PEDOT:PSS, but also induces the PEDOT conformation change from a benzoid to a quinoid structure. After such a treatment, the PEDOT:PSS films could achieve a remarkable conductivity enhancement by nearly 4 orders of magnitude, from ~0.2 to ~1400 S/cm [39]. The high conductivity of the electrodes contributes to the low ESR. A low ESR has shown to be very important for reducing the IR drop of the electrodes and the assembled devices during the charging/discharging process at high current densities [13,44]. It is worth noting that the ESR of our PEDOT:PSS electrodes is significantly lower than the corresponding values of many transparent electrodes reported previously [16,45–47].

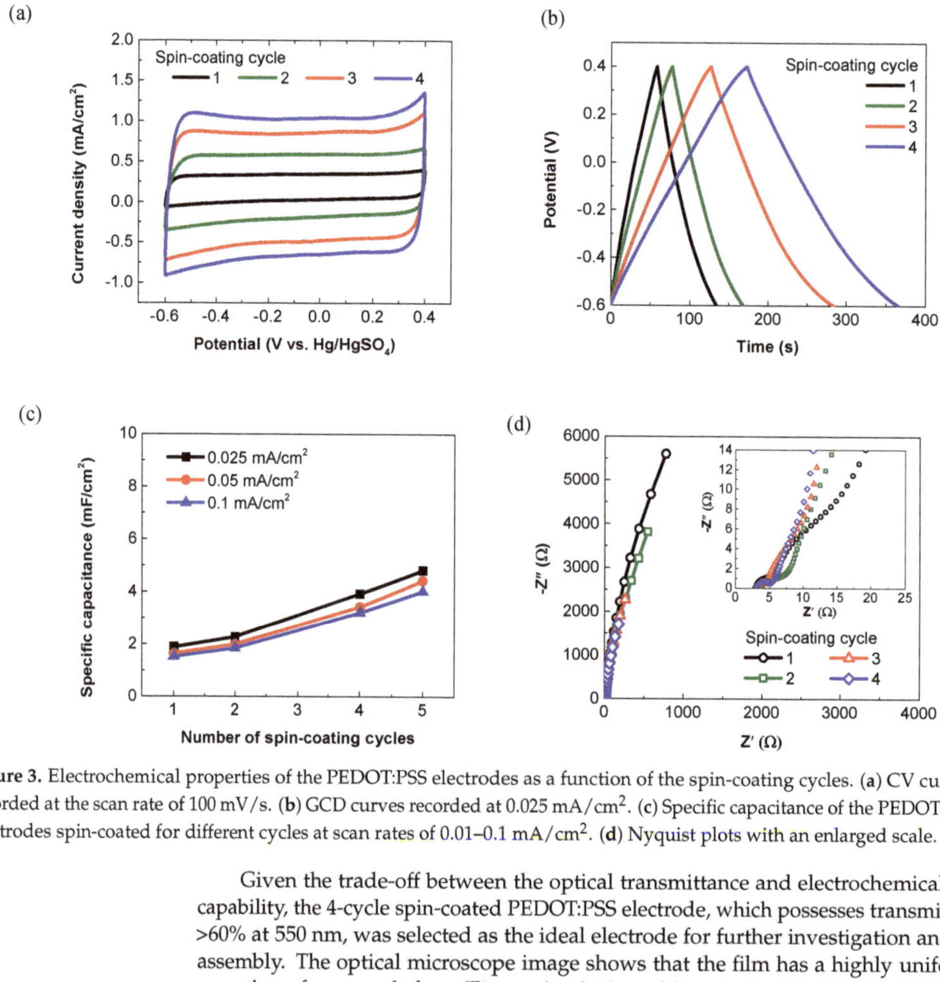

Figure 3. Electrochemical properties of the PEDOT:PSS electrodes as a function of the spin-coating cycles. (**a**) CV curves recorded at the scan rate of 100 mV/s. (**b**) GCD curves recorded at 0.025 mA/cm^2. (**c**) Specific capacitance of the PEDOT:PSS electrodes spin-coated for different cycles at scan rates of 0.01–0.1 mA/cm^2. (**d**) Nyquist plots with an enlarged scale.

Given the trade-off between the optical transmittance and electrochemical storage capability, the 4-cycle spin-coated PEDOT:PSS electrode, which possesses transmittance of >60% at 550 nm, was selected as the ideal electrode for further investigation and device assembly. The optical microscope image shows that the film has a highly uniform and smooth surface morphology (Figure 4), which could be attributed to the good wettability of the PEDOT:PSS aqueous dispersion on the PET substrates. Within a potential window of −0.6–0.4 V, the CV curves of the typical 4-cycle spin-coated PEDOT:PSS electrode at scan rates of 20–500 mV/s are shown in Figure 5a. Notably, the CV curve retains a good rectangular shape, even at a high scan rate of 500 mV/s, thus suggesting the fast electron transfer and ion transfer throughout the electrode [48].

Derived from the GCD curves (Figure 5b), its specific capacitance is calculated to be 6.3 mF/cm^2 at a current density of 0.01 mA/cm^2, and 3.92 mF/cm^2 at a current density of 0.1 mA/cm^2, significantly greater than that of the electrodes prepared from the PEDOT:PSS solution doped with EG and a surfactant [33]. The cyclic stability of the 4-cycle spin-coated PEDOT:PSS electrode was assessed by GCD over 3000 cycles, as shown in Figure 5c. After 3000 charge/discharge cycles, the electrode still exhibits a capacitance retention of >90%. The corresponding GCD curves after 1000 and 3000 cycles almost overlap with the initial one (inset of Figure 5c). These results suggest the excellent cycling performance of the electrode, which is quite comparable with the reported electrochemical stability of other polymer-based counterparts [42,49,50]. The PEDOT:PSS electrode exhibits good mechanical flexibility and stable electrochemical performance at bending states. As the bending angle increases from 0 to 180°, its sheet resistance only slightly increases, i.e., by less than 20% (Figure 6). The CV and cyclic GCD measurements were also performed when the electrode was bent at a bending angle of 90°. For the electrode at normal and bending states, no significant difference could be observed in the capacitance retention and the CV and GCD curves after 500 cycles (Figure 5d), further suggesting its outstanding electrochemical stability and robust flexibility.

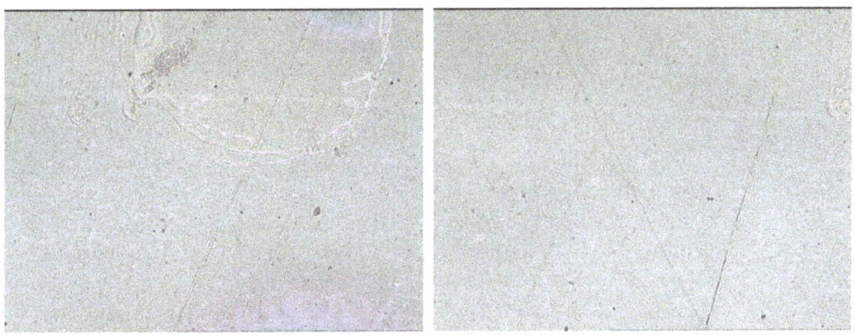

Figure 4. The optical microscopy images of the 4-layer PEDOT:PSS film electrode.

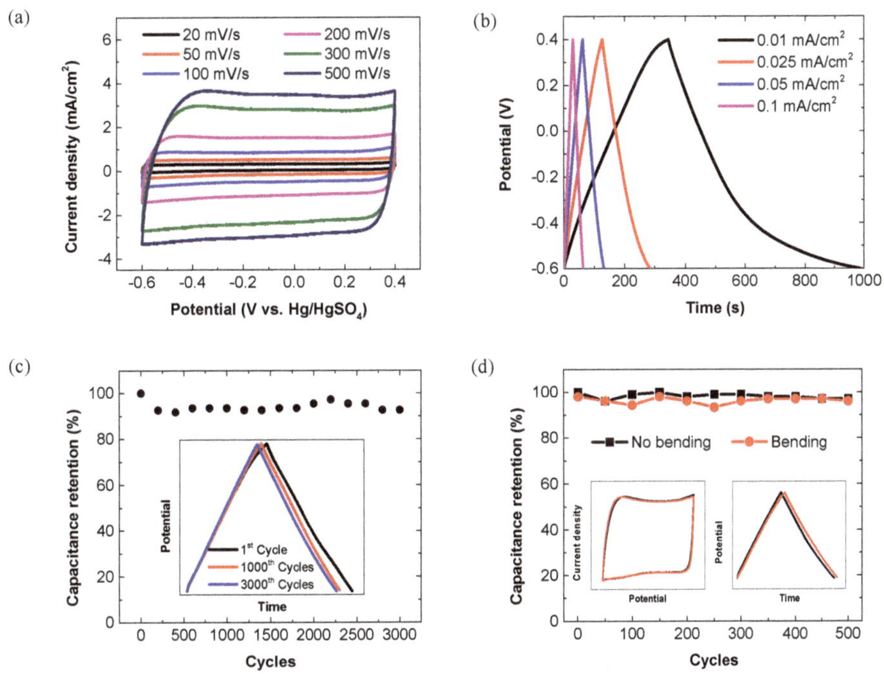

Figure 5. Electrochemical properties of the 4-cycle spin-coated PEDOT:PSS electrodes. (**a**) CV curves recorded at various scan rates. (**b**) GCD curves recorded at various current densities. (**c**) Cyclic stability at a current density of 0.2 mA/cm^2. Insets show the GCD curves of the 1st, 1000th and 3000th cycles. (**d**) Cyclic stability of the electrode without bending and under bending at 90°. Insets show the comparison of the corresponding CV curves (left) and GCD curves (right) tested at normal and bending states after 500 cycles.

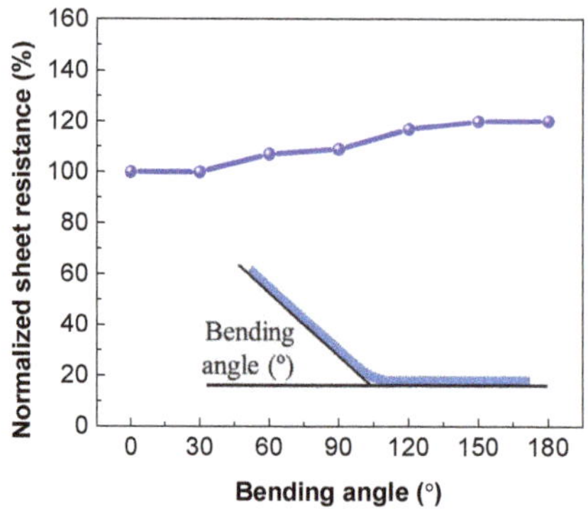

Figure 6. Sheet resistance of the 4-layer PEDOT:PSS film electrode as a function of the bending angle.

An all-solid-state symmetric supercapacitor was further constructed by sandwiching a H_3PO_4/PVA gel electrolyte between two PEDOT:PSS electrodes (Figure 7a). As shown in Figure 7b, the as-assembled supercapacitor exhibits an approximately rectangular shape in the CV curves at scan rates from 20 to 500 mV/s. The rectangular characteristic remains almost unchanged with the increasing scan rate, thereby indicating a rapid charge transport and a high rate performance of the device [42]. At current densities of 0.01–1 mA/cm^2, all the GCD curves exhibit a symmetric triangular shape (Figure 7c). The areal specific capacitance of the supercapacitor as a function of the current density is plotted in Figure 7d. As calculated from the discharge curves, the corresponding areal specific capacitances are 1.32 and 1.14 mF/cm^2 at the current densities of 0.01 and 1 mA/cm^2, respectively. Such rate capability (>86% retention) is greater over those of most other polymer-based all-solid-state supercapacitors in the literature [15,23,33]. Figure 7e presents the Nyquist plot of the device. At the low-frequency regions, the curve appears to be nearly vertical, which implies an ideal capacitive behavior of the device. The absence of the semicircle at high-frequency regions indicates the good electrical contact and the rapid electron transport in the all-solid-state device [51]. The long-term cycling stability is characterized by performing 3000 GCD cycles at the applied current density of 0.2 mA/cm^2. As shown in Figure 7f, there is no significant capacitance drop observed during the measurements. After 3000 charge/discharge cycles, the device achieves high capacitance retention of up to 96.8%. The 1st, 1000th and 3000th cycles are specifically highlighted in the insets of Figure 7f. The GCD curves after 1000 and 3000 cycles are nearly overlapped with the initial one, further demonstrating the outstanding cycling stability. To assess the flexibility of the all-solid-state supercapacitor, the CV and GCD measurements were further performed at normal and bending states. As shown in Figure 8, both the CV and GCD curves well remain in the original shapes, even at a highly bending state, revealing the excellent flexibility and mechanical robustness of the device.

The Ragone plot, which displays the relationship between energy density versus the power density, is commonly employed to evaluate the practical energy storage performance of a supercapacitor. As shown in Figure 9, the all-solid-state flexible supercapacitor based on the PEDOT:PSS electrodes delivers a high areal energy density of 0.183 μWh/cm^2 at an areal power density of 4.98 μW/cm^2, and maintains 0.131 μWh/cm^2 even at 453.5 μW/cm^2. Such power/energy densities are sufficient for powering burst communication for an integrated sensor [27]. The energy storage performance of our device is superior in comparison with those of many transparent and flexible PEDOT-based devices reported recently [23,27,33,52]. The excellent performance could be mainly attributed to the high electrical conductivity and good electrochemical property of the PEDOT:PSS films treated by the salt solution.

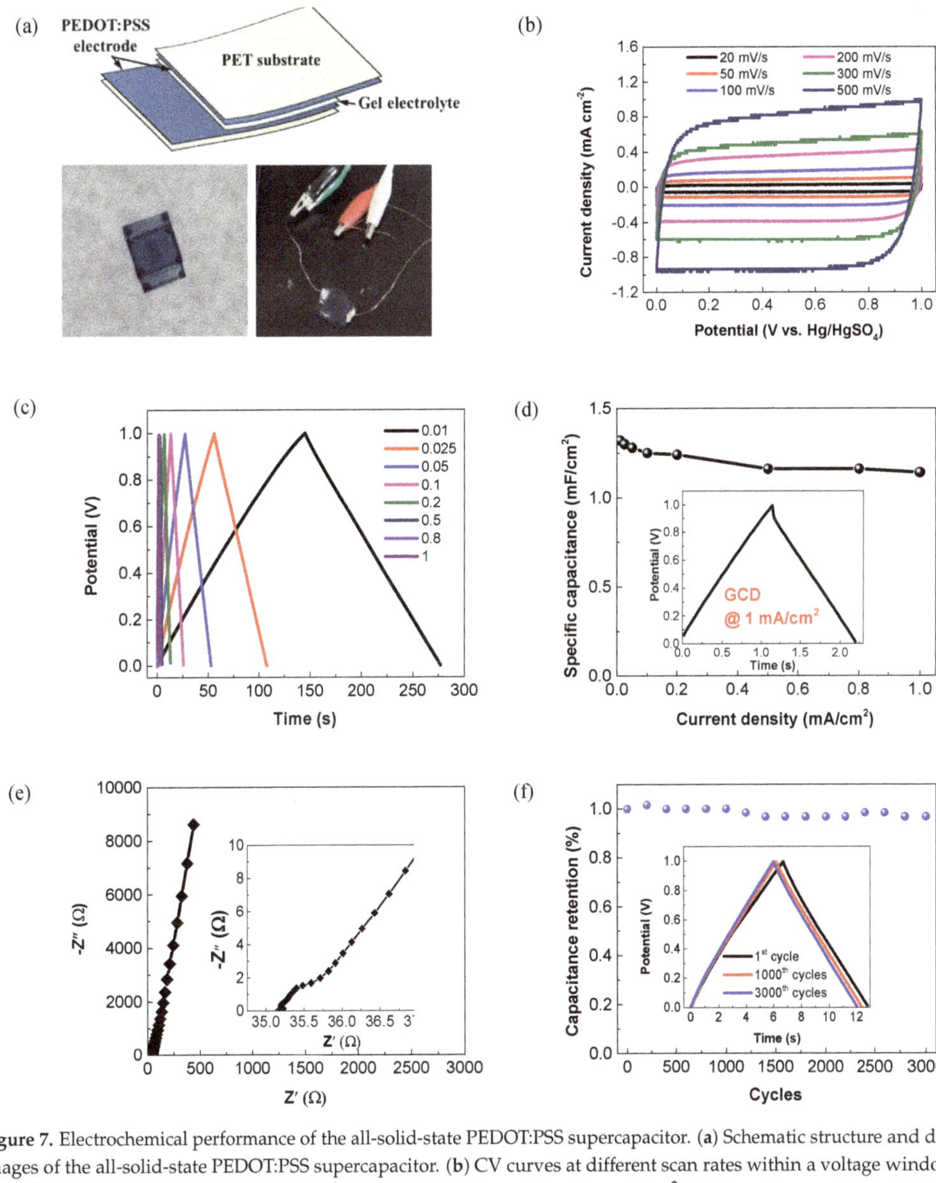

Figure 7. Electrochemical performance of the all-solid-state PEDOT:PSS supercapacitor. (**a**) Schematic structure and digital images of the all-solid-state PEDOT:PSS supercapacitor. (**b**) CV curves at different scan rates within a voltage window of 0–1.0 V. (**c**) GCD curves at different current densities (Unit of current density: mA/cm^2). (**d**) Specific capacitance at different current densities. (**e**) Nyquist plot with an enlarged view in the insert. (**f**) Cycling performance at a current density of 0.2 mA/cm^2.

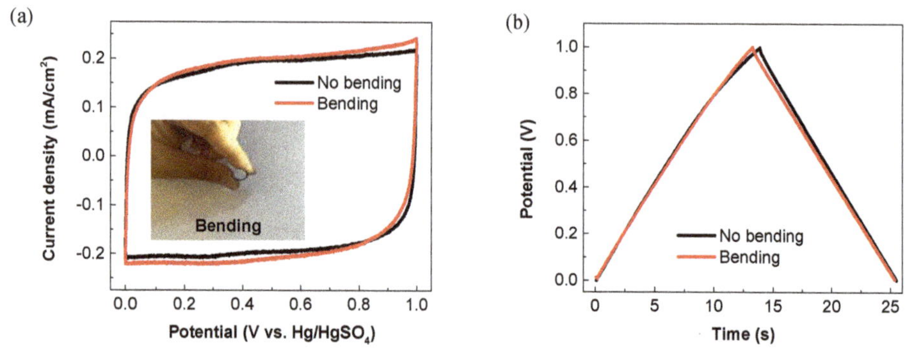

Figure 8. Flexibility of the all-solid-state PEDOT:PSS supercapacitor at normal and bending states. (**a**) CV curves (100 mV/s) and (**b**) GCD curves (0.1 mA/cm^2). Inset shows the digital image of the device at bending state.

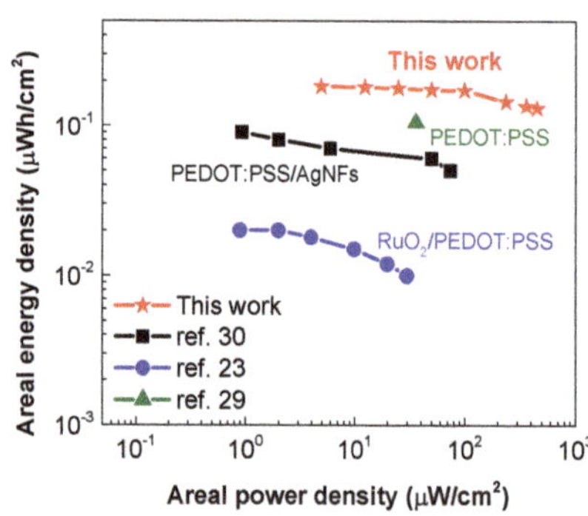

Figure 9. Power density versus energy density (Ragone plot) of transparent and flexible PEDOT-based symmetric supercapacitors in the literature.

4. Conclusions

In summary, transparent, flexible and highly conductive polymer film electrodes were successfully fabricated by spin-coating the commercial PEDOT:PSS dispersion on PET substrates followed by a mild post-treatment with $ZnCl_2$-DMF solution. The optical, electrical and electrochemical properties of the PEDOT:PSS film electrodes were optimized through a modulation of the cycle number of the spin-coating process. The PEDOT:PSS film electrodes obtained via 4-cycle spin-coating achieved an ideal balance between the electrochemical performance and the optical transparency, delivering a high areal specific capacitance of 3.92 mF/cm^2 at 0.1 mA/cm^2, a high capacitance retention of > 90% after 3000 cycles and a low ESR of only ~3 Ω with a good optical transmittance of exceeding 60%. Upon the construction of an all-solid-state transparent supercapacitor, the PEDOT:PSS film electrodes could be used as both the active electrodes and current collectors. Owing to their high electrical conductivity and good electrochemical performance, the device exhibited an areal capacitance of 1.32 mF/cm^2, a maximum power density of 453.5 μW/cm^2 and an energy density of 0.131 μWh/cm^2. Such power/energy densities were remarkably larger than most of the reported flexible and transparent PEDOT-based devices so far. The approach demonstrated in this work for fabricating the PEDOT:PSS transparent electrodes is mild and easy. The as-obtained PEDOT:PSS film electrodes and devices showed excellent electrochemical performance at good optical transparency, and thus, hold great promise for use as transparent electrodes for various flexible or wearable energy storage devices.

Author Contributions: The manuscript was written through contributions of all authors. All authors have given approval to the final version of the manuscript. Conceptualization, Z.F.; Methodology, X.G. and Z.F.; Formal analysis, X.G. and Z.F.; Investigation, X.G. and Z.F.; Project administration, Z.F.; Supervision, Z.F. and L.P.; Writing—original draft, X.G.; Writing—review and editing, Z.F. and L.P. All authors have read and agreed to the published version of the manuscript.

Funding: National Natural Science Foundation of China: 51803018, 51972039 and 51661145025; Fundamental Research Funds for the Central Universities: DUT21JC06; LiaoNing Revitalization Talents Program: XLYC 1902122.

Institutional Review Board Statement: Not applicable.

Conflicts of Interest: The authors declare no conflict of interest.

References

1. Cong, H.-P.; Ren, X.-C.; Wang, P.; Yu, S.-H. Flexible graphene–polyaniline composite paper for high-performance supercapacitor. *Energ. Environ. Sci.* **2013**, *6*, 1185–1191. [CrossRef]
2. Dong, X.-C.; Xu, H.; Wang, X.-W.; Huang, Y.-X.; Chan-Park, M.B.; Zhang, H.; Wang, L.-H.; Huang, W.; Chen, P. 3D Graphene–Cobalt Oxide Electrode for High-Performance Supercapacitor and Enzymeless Glucose Detection. *ACS Nano* **2012**, *6*, 3206–3213.
3. Liu, T.; Zhang, L.; You, W.; Yu, J. Core–Shell Nitrogen-Doped Carbon Hollow Spheres/Co3O4 Nanosheets as Advanced Electrode for High-Performance Supercapacitor. *Small* **2018**, *14*, 1702407. [CrossRef] [PubMed]
4. Castaings, A.; Lhomme, W.; Trigui, R.; Bouscayrol, A. Comparison of energy management strategies of a battery/supercapacitors system for electric vehicle under real-time constraints. *Appl. Energ.* **2016**, *163*, 190–200.
5. Zhang, Z.; Zhang, X.; Chen, W.; Rasim, Y.; Salman, W.; Pan, H.; Yuan, Y.; Wang, C. A high-efficiency energy regenerative shock absorber using supercapacitors for renewable energy applications in range extended electric vehicle. *Appl. Energ.* **2016**, *178*, 177–188. [CrossRef]
6. Soavi, F.; Bettini, L.G.; Piseri, P.; Milani, P.; Santoro, C.; Atanassov, P.; Arbizzani, C. Miniaturized supercapacitors: Key materials and structures towards autonomous and sustainable devices and systems. *J. Power Sources* **2016**, *326*, 717–725. [CrossRef]
7. Zhong, Y.; Zhang, X.; He, Y.; Peng, H.; Wang, G.; Xin, G. Simultaneously Armored and Active Graphene for Transparent and Flexible Supercapacitors. *Adv. Funct. Mater.* **2018**, *28*, 1801998. [CrossRef]
8. Chen, P.-C.; Shen, G.; Sukcharoenchoke, S.; Zhou, C. Flexible and transparent supercapacitor based on In2O3 nanowire/carbon nanotube heterogeneous films. *Appl. Phys. Lett.* **2009**, *94*, 043113. [CrossRef]
9. Yu, A.; Roes, I.; Davies, A.; Chen, Z. Ultrathin, transparent, and flexible graphene films for supercapacitor application. *Appl. Phys. Lett.* **2010**, *96*, 253105. [CrossRef]
10. Cai, G.; Darmawan, P.; Cui, M.; Wang, J.; Chen, J.; Magdassi, S.; Lee, P.S. Highly Stable Transparent Conductive Silver Grid/PEDOT:PSS Electrodes for Integrated Bifunctional Flexible Electrochromic Supercapacitors. *Adv. Energy Mater.* **2016**, *6*, 1501882. [CrossRef]
11. Sheng, H.; Zhang, X.; Ma, Y.; Wang, P.; Zhou, J.; Su, Q.; Lan, W.; Xie, E.; Zhang, C.J. Ultrathin, Wrinkled, Vertically Aligned Co(OH)2 Nanosheets/Ag Nanowires Hybrid Network for Flexible Transparent Supercapacitor with High Performance. *ACS Appl. Mater. Interfaces* **2019**, *11*, 8992–9001. [CrossRef]
12. Xu, J.-L.; Liu, Y.-H.; Gao, X.; Shen, S.; Wang, S.-D. Toward wearable electronics: A lightweight all-solid-state supercapacitor with outstanding transparency, foldability and breathability. *Energy Storage Mater.* **2019**, *22*, 402–409. [CrossRef]
13. Wang, Z.; Tammela, P.; Huo, J.; Zhang, P.; Strømme, M.; Nyholm, L. Solution-processed poly(3,4-ethylenedioxythiophene) nanocomposite paper electrodes for high-capacitance flexible supercapacitors. *J. Mater. Chem. A* **2016**, *4*, 1714–1722. [CrossRef]
14. Wu, C.; Kim, T.W.; Li, F.; Guo, T. Wearable Electricity Generators Fabricated Utilizing Transparent Electronic Textiles Based on Polyester/Ag Nanowires/Graphene Core–Shell Nanocomposites. *ACS Nano* **2016**, *10*, 6449–6457. [CrossRef] [PubMed]
15. Devarayan, K.; Lei, D.; Kim, H.-Y.; Kim, B.-S. Flexible transparent electrode based on PANi nanowire/nylon nanofiber reinforced cellulose acetate thin film as supercapacitor. *Chem. Eng. J.* **2015**, *273*, 603–609. [CrossRef]
16. Lee, K.; Lee, H.; Shin, Y.; Yoon, Y.; Kim, D.; Lee, H. Highly transparent and flexible supercapacitors using graphene-graphene quantum dots chelate. *Nano Energy* **2016**, *26*, 746–754. [CrossRef]
17. Chen, T.; Xue, Y.; Roy, A.K.; Dai, L. Transparent and Stretchable High-Performance Supercapacitors Based on Wrinkled Graphene Electrodes. *ACS Nano* **2014**, *8*, 1039–1046. [CrossRef] [PubMed]
18. Liu, X.Y.; Gao, Y.Q.; Yang, G.W. A flexible, transparent and super-long-life supercapacitor based on ultrafine Co3O4 nanocrystal electrodes. *Nanoscale* **2016**, *8*, 4227–4235. [CrossRef]
19. Liu, X.; Wang, J.; Yang, G. Transparent, flexible, and high-performance supercapacitor based on ultrafine nickel cobaltite nanospheres. *Appl. Phys. A* **2017**, *123*, 469. [CrossRef]
20. Chen, Y.; Fu, X.-Y.; Yue, Y.-Y.; Zhang, N.; Feng, J.; Sun, H.-B. Flexible and transparent supercapacitor based on ultrathin Au/graphene composite electrodes. *Appl. Surf. Sci.* **2019**, *467-468*, 104–111. [CrossRef]
21. Liu, X.; Wang, J.; Yang, G. In Situ Growth of the Ni3V2O8@PANI Composite Electrode for Flexible and Transparent Symmetric Supercapacitors. *ACS Appl. Mater. Interfaces* **2018**, *10*, 20688–20695. [CrossRef] [PubMed]

22. Zhu, W.-C.; He, P.-Q.; Tien, H.-C.; Liu, H.-L.; Chen, W.-C.; Lv, W.; Lee, W.-Y. Solvent-Enhanced Transparent Stretchable Polymer Nanocomposite Electrode for Supercapacitors. *ACS Appl. Energy Mater.* **2021**, *4*, 2266–2274. [CrossRef]
23. Singh, S.B.; Kshetri, T.; Singh, T.I.; Kim, N.H.; Lee, J.H. Embedded PEDOT:PSS/AgNFs network flexible transparent electrode for solid-state supercapacitor. *Chem. Eng. J.* **2019**, *359*, 197–207. [CrossRef]
24. Wu, B.; He, W.; Lu, M.; Li, Z.; Qiang, H. Fabrication and electrochemical properties of flexible transparent supercapacitor electrode materials based on cellulose nanofibrils and reduced graphene oxide. *Polymer Composites* **2020**, *41*, 1135–1144. [CrossRef]
25. Kumar, N.; Ginting, R.T.; Kang, J.-W. Flexible, large-area, all-solid-state supercapacitors using spray deposited PEDOT:PSS/reduced-graphene oxide. *Electrochim. Acta* **2018**, *270*, 37–47. [CrossRef]
26. Zhao, D.; Zhang, Q.; Chen, W.; Yi, X.; Liu, S.; Wang, Q.; Liu, Y.; Li, J.; Li, X.; Yu, H. Highly Flexible and Conductive Cellulose-Mediated PEDOT:PSS/MWCNT Composite Films for Supercapacitor Electrodes. *ACS Appl. Mater. Interfaces* **2017**, *9*, 13213–13222. [CrossRef] [PubMed]
27. Zhang, C.; Higgins, T.M.; Park, S.-H.; O'Brien, S.E.; Long, D.; Coleman, J.N.; Nicolosi, V. Highly flexible and transparent solid-state supercapacitors based on RuO2/PEDOT:PSS conductive ultrathin films. *Nano Energy* **2016**, *28*, 495–505. [CrossRef]
28. Yoonessi, M.; Borenstein, A.; El-Kady, M.F.; Turner, C.L.; Wang, H.; Stieg, A.Z.; Pilon, L. Hybrid Transparent PEDOT:PSS Molybdenum Oxide Battery-like Supercapacitors. *ACS Appl. Energy Mater.* **2019**, *2*, 4629–4639. [CrossRef]
29. Xia, Y.; Sun, K.; Ouyang, J. Solution-Processed Metallic Conducting Polymer Films as Transparent Electrode of Optoelectronic Devices. *Adv. Mater.* **2012**, *24*, 2436–2440. [CrossRef]
30. Chen, Y.; Zhu, X.; Yang, D.; Wangyang, P.; Zeng, B.; Sun, H. A novel design of poly (3,4-ethylenedioxythiophene):poly (styrene-sulfonate)/molybdenum disulfide/poly (3,4-ethylenedioxythiophene) nanocomposites for fabric micro-supercapacitors with favourable performances. *Electrochim. Acta* **2019**, *298*, 297–304. [CrossRef]
31. Zhang, H.; Cao, Y.; Chee, M.O.L.; Dong, P.; Ye, M.; Shen, J. Recent advances in micro-supercapacitors. *Nanoscale* **2019**, *11*, 5807–5821. [CrossRef]
32. Dubal, D.P.; Chodankar, N.R.; Kim, D.-H.; Gomez-Romero, P. Towards flexible solid-state supercapacitors for smart and wearable electronics. *Chem. Soc. Rev.* **2018**, *47*, 2065–2129. [CrossRef] [PubMed]
33. Cheng, T.; Zhang, Y.-Z.; Zhang, J.-D.; Lai, W.-Y.; Huang, W. High-performance free-standing PEDOT:PSS electrodes for flexible and transparent all-solid-state supercapacitors. *J. Mater. Chem. A* **2016**, *4*, 10493–10499. [CrossRef]
34. Ouyang, J.; Xu, Q.; Chu, C.-W.; Yang, Y.; Li, G.; Shinar, J. On the mechanism of conductivity enhancement in poly(3,4-ethylenedioxythiophene):poly(styrene sulfonate) film through solvent treatment. *Polymer* **2004**, *45*, 8443–8450. [CrossRef]
35. Xia, Y.; Sun, K.; Chang, J.; Ouyang, J. Effects of organic inorganic hybrid perovskite materials on the electronic properties and morphology of poly(3,4-ethylenedioxythiophene):poly(styrenesulfonate) and the photovoltaic performance of planar perovskite solar cells. *J. Mater. Chem. A* **2015**, *3*, 15897–15904. [CrossRef]
36. de Izarra, A.; Park, S.; Lee, J.; Lansac, Y.; Jang, Y.H. Ionic Liquid Designed for PEDOT:PSS Conductivity Enhancement. *J. Am. Chem. Soc.* **2018**, *140*, 5375–5384. [CrossRef]
37. Sun, K.; Zhao, B.; Murugesan, V.; Kumar, A.; Zeng, K.; Subbiah, J.; Wong, W.W.H.; Jones, D.J.; Ouyang, J. High-performance polymer solar cells with a conjugated zwitterion by solution processing or thermal deposition as the electron-collection interlayer. *J. Mater. Chem.* **2012**, *22*, 24155–24165. [CrossRef]
38. Fan, Z.; Ouyang, J. Thermoelectric Properties of PEDOT:PSS. *Adv. Electron. Mater.* **2019**, *5*, 1800769. [CrossRef]
39. Fan, Z.; Du, D.; Yu, Z.; Li, P.; Xia, Y.; Ouyang, J. Significant Enhancement in the Thermoelectric Properties of PEDOT:PSS Films through a Treatment with Organic Solutions of Inorganic Salts. *ACS Appl. Mater. Interfaces* **2016**, *8*, 23204–23211. [CrossRef]
40. Zhang, M.; Yu, X.; Ma, H.; Du, W.; Qu, L.; Li, C.; Shi, G. Robust graphene composite films for multifunctional electrochemical capacitors with an ultrawide range of areal mass loading toward high-rate frequency response and ultrahigh specific capacitance. *Energ. Environ. Sci.* **2018**, *11*, 559–565. [CrossRef]
41. Wang, D.; Fang, G.; Xue, T.; Ma, J.; Geng, G. A melt route for the synthesis of activated carbon derived from carton box for high performance symmetric supercapacitor applications. *J. Power Sources* **2016**, *307*, 401–409. [CrossRef]
42. Zou, Y.; Liu, R.; Zhong, W.; Yang, W. Mechanically robust double-crosslinked network functionalized graphene/polyaniline stiff hydrogels for superior performance supercapacitors. *J. Mater. Chem. A* **2018**, *6*, 8568–8578. [CrossRef]
43. Li, Y.; Ren, G.; Zhang, Z.; Teng, C.; Wu, Y.; Lu, X.; Zhu, Y.; Jiang, L. A strong and highly flexible aramid nanofibers/PEDOT:PSS film for all-solid-state supercapacitors with superior cycling stability. *J. Mater. Chem. A* **2016**, *4*, 17324–17332. [CrossRef]
44. Gao, W.; Singh, N.; Song, L.; Liu, Z.; Reddy, A.L.M.; Ci, L.; Vajtai, R.; Zhang, Q.; Wei, B.; Ajayan, P.M. Direct laser writing of micro-supercapacitors on hydrated graphite oxide films. *Nat. Nanotechnol.* **2011**, *6*, 496. [CrossRef] [PubMed]
45. Gilshteyn, E.P.; Amanbayev, D.; Anisimov, A.S.; Kallio, T.; Nasibulin, A.G. All-nanotube stretchable supercapacitor with low equivalent series resistance. *Sci. Rep.* **2017**, *7*, 17449. [CrossRef] [PubMed]
46. Li, N.; Huang, X.; Li, R.; Chen, Y.; Li, Y.; Shi, Z.; Zhang, H. Pseudocapacitive Transparent/Flexible Supercapacitor based on Graphene wrapped Ni(OH)2 Nanosheet Transparent Film Produced using Scalable Bio-inspired Methods. *Electrochim. Acta* **2016**, *219*, 61–69. [CrossRef]
47. Mohd Abdah, M.A.A.; Zubair, N.A.; Azman, N.H.N.; Sulaiman, Y. Fabrication of PEDOT coated PVA-GO nanofiber for supercapacitor. *Mater. Chem. Phys.* **2017**, *192*, 161–169. [CrossRef]
48. Kashani, H.; Chen, L.; Ito, Y.; Han, J.; Hirata, A.; Chen, M. Bicontinuous nanotubular graphene–polypyrrole hybrid for high performance flexible supercapacitors. *Nano Energy* **2016**, *19*, 391–400. [CrossRef]

49. Hong, K.; Yuk, J.; Kim, H.J.; Lee, J.Y.; Kim, S.; Lee, J.-L.; Lee, K.H. Electrospun polymer electrolyte nanocomposites for solid-state energy storage. *Compos. Part B Eng.* **2018**, *152*, 275–281. [CrossRef]
50. Luo, S.; Zhao, J.; Zou, J.; He, Z.; Xu, C.; Liu, F.; Huang, Y.; Dong, L.; Wang, L.; Zhang, H. Self-Standing Polypyrrole/Black Phosphorus Laminated Film: Promising Electrode for Flexible Supercapacitor with Enhanced Capacitance and Cycling Stability. *ACS Appl. Mater. Interfaces* **2018**, *10*, 3538–3548. [CrossRef]
51. Li, Z.; Cao, L.; Qin, P.; Liu, X.; Chen, Z.; Wang, L.; Pan, D.; Wu, M. Nitrogen and oxygen co-doped graphene quantum dots with high capacitance performance for micro-supercapacitors. *Carbon* **2018**, *139*, 67–75. [CrossRef]
52. Li, N.; Huang, X.; Zhang, H.; Li, Y.; Wang, C. Transparent and Self-Supporting Graphene Films with Wrinkled- Graphene-Wall-Assembled Opening Polyhedron Building Blocks for High Performance Flexible/Transparent Supercapacitors. *ACS Appl. Mater. Interfaces* **2017**, *9*, 9763–9771. [CrossRef]

membranes

MDPI

Article

Sludge Derived Carbon Modified Anode in Microbial Fuel Cell for Performance Improvement and Microbial Community Dynamics

Kaili Zhu, Yihu Xu, Xiao Yang, Wencai Fu, Wenhao Dang, Jinxia Yuan and Zhiwei Wang *

Key Laboratory of Clean Pulp & Papermaking and Pollution Control of Guangxi, College of Light Industry and Food Engineering, Guangxi University, Nanning 530004, China; zhukaili26@163.com (K.Z.); xuyihuzwj@163.com (Y.X.); yxiao0112@163.com (X.Y.); f1392597@163.com (W.F.); wenhaodang14@126.com (W.D.); yuan09@syr.edu (J.Y.)

* Correspondence: author: wangzhiwei@gxu.edu.cn

Abstract: The conversion of activated sludge into high value-added materials, such as sludge carbon (SC), has attracted increasing attention because of its potential for various applications. In this study, the effect of SC carbonized at temperatures of 600, 800, 1000, and 1200 °C on the anode performance of microbial fuel cells and its mechanism are discussed. A pyrolysis temperature of 1000 °C for the loaded electrode (SC1000/CC) generated a maximum areal power density of 2.165 ± 0.021 W·m^{-2} and a current density of 5.985 ± 0.015 A·m^{-2}, which is 3.017- and 2.992-fold that of the CC anode. The addition of SC improves microbial activity, optimizes microbial community structure, promotes the expression of c-type cytochromes, and is conducive to the formation of electroactive biofilms. This study not only describes a technique for the preparation of high-performance and low-cost anodes, but also sheds some light on the rational utilization of waste resources such as aerobic activated sludge.

Keywords: sludge carbon; extracellular polymeric substance; microbial fuel cell; electroactive biofilm; microbial community dynamics

Citation: Zhu, K.; Xu, Y.; Yang, X.; Fu, W.; Dang, W.; Yuan, J.; Wang, Z. Sludge Derived Carbon Modified Anode in Microbial Fuel Cell for Performance Improvement and Microbial Community Dynamics. *Membranes* **2022**, *12*, 120. https://doi.org/10.3390/membranes12020120

Academic Editor: Sophie Tingry

Received: 15 December 2021
Accepted: 17 January 2022
Published: 20 January 2022

1. Introduction

Microbial fuel cells (MFCs) can convert chemical energy in organic matter into electric energy by using the oxidation and metabolism mechanism of anaerobic bacteria on anodic electroactive biofilms (EABFs) [1,2]. Compared with other electrochemical cells, such as liquid flow cells and ordinary fuel cells, MFCs do not need external energy input and have high energy conversion efficiency [3,4]. MFCs can use all biodegradable organics and wastewater as fuels [2]. In wastewater treatment, MFCs can generate electrical energy while efficiently treating wastewater, which not only significantly reduces the operation cost of sewage treatment plants, but also makes efficient use of waste resources [3]. Traditional wastewater biological treatment approaches mainly involve anaerobic digestion and aerobic treatment technologies [5]. Anaerobic digestion technology is mainly applicable to high concentration wastewater [6,7]. Although it can produce fuel gases such as methane or hydrogen from organic wastewater, it also produces gases with no practical value such as carbon dioxide, hydrogen sulfide and nitrogen [5,8]. The energy utilization mode is complex and has significant requirements for production conditions. Aerobic treatment technology is mainly applicable to medium and low concentration wastewater. With large sludge production, aeration is required to maintain oxygen concentration, and the cost of sludge disposal and aeration is high [8]. MFCs have a wide application range, low sludge output and directly utilize electric energy. Thus, MFCs have incomparable technical advantages compared with traditional wastewater biological treatment technologies [3,7]. Therefore, the application of MFCs in wastewater treatment has broad prospects. However, recently, it was reported that the maximum current and power densities of MFCs are 3.85 mA·cm^2 and 0.66 mW·cm^2, respectively, and the power production is still relatively low, i.e., 2–4 orders of magnitude below that of chemical fuel cell [9]. In addition, the cost of MFCs is also relatively high, notably due to the commonly used proton exchange membrane and cathode platinum catalyst, which greatly increase the cost [10]. Thus, practical applications for microbial fuel cells are limited.

The anode, as the core of MFCs, plays a key role in the extracellular electron transfer (EET) between electroactive microorganisms (EAMs) and electrodes [8]. The properties of the anode materials significantly affect the growth of EABFs. An ideal anode material should have high biocompatibility and conductivity, cost efficiency, and easy commercial production [11]. Traditional carbon-based materials, such as carbon cloth (CC), carbon felt, carbon paper and graphite rod, have been widely used in MFCs. However, due to the low electrochemical activity of carbon-based materials, their application in MFC is limited [12]. Therefore, it is necessary to modify carbon-based materials to improve the power generation performance of MFCs. Given the poor durability of carbon paper [4], the high resistance of carbon felt [11] and the smooth surface of graphite rods [13], CC is often used to modify carbon-based materials because of its good durability, low resistance and easy modification [14]. Carbon-based nanomaterials, such as graphene, carbon nanotubes, conductive polymers and carbon nanoparticles have been widely used to modify carbon-based electrodes because of their good electrocatalytic activity and conductivity [11,12]. Liu et al., modified CC with tungsten carbide nanoparticles, which significantly improved the electrocatalytic performance of the resulting electrode. The power density reached 3.26 W·m^{-2}, which was 2.14 times that of the bare CC anode (1.52 W·m^{-2}) [15]. Li et al., modified a CC with polydopamine (PDA) and graphene oxide (rGO), which improved the hydrophilicity and significantly reduced charge transfer resistance. Its power density reached 2.047 W·m^{-2}, which was 6.1 times that of the bare CC electrode [10]. Liu et al., modified CC with graphene to enhance its biocompatibility. The power density and energy conversion efficiency were 2.7 times and 3 times higher than those of the bare CC electrode, respectively [16]. The above studies show that the modification of CC anodes can significantly improve the performance of MFCs. However, the preparation process of these nano materials is complex and yields are low [11]. It is estimated that the preparation cost of these nanomaterials is high, accounting for more than half of the total cost of MFCs [17]. Therefore, reducing the cost of MFC anode materials while maintaining excellent performance is very important for further practical applications of MFCs.

Recently, carbon materials derived from natural biomass materials have attracted extensive interest because of their low cost and sustainable resource utilization. Porous carbon derived from biomass wastes such as almond shells, pomelo peels, towel gourds, kenaf stems, silkworm cocoons and chestnut shells have been used to produce the anodes of MFCs, showing excellent power generation performance; such approaches also open up a new path for the utilization of natural waste [18–20]. Sewage sludge, a byproduct of wastewater treatment, is abundant but remains expensive to dispose of. According to statistical reports, China produces approximately 11.2 million tons of dry sludge every year, while the countries of the European Union produce 10.0 million tons [21]. The cost of sludge treatment and disposal is approximately 60% of the total operating cost of sewage treatment plants [22]. In addition, the sludge contains pathogenic bacteria, organic pollutants, heavy metals and other harmful substances, and its improper treatment can be harmful to human health and the environment. Compared to incineration, landfill, anaerobic digestion, and composting, pyrolysis treatment can effectively reduce the toxicity of dry sludge, yielding sludge carbon (SC) and high value-added fuel [23]. It is considered a safe, stable, and low-cost sludge treatment method. Previously, researchers have applied SCs prepared by one-step pyrolysis to lithium-ion batteries, supercapacitors, and environmental catalysts [23]. Sludge-derived carbon has been shown to be feasible as a low-cost conductive material. However, the effects of the physical and chemical properties of SC (such as specific surface area, porosity, functional groups and metal phase structure) on MFC anodes, especially on extracellular polymers (EPS) in anode biofilms, and the mechanisms of electron transfer, are still unclear. As an electrode material, SC still poses scientific and technical challenges.

The purpose of this study was to investigate the effects of SC characteristics on the EABFs of MFC anodes. The parameters assessed included power generation, wastewater treatment ability, electrochemical performance, EPS secretion, microbial activity, diversity, and metabolic pathways. It aimed to provide a theoretical basis for the practical application of sludge carbon in MFCs, and to lay out a safe and effective method for the utilization of sludge resources.

2. Materials and Methods

2.1. Preparation of Electrodes

The anode and cathode were made of carbon cloth, purchased from WENTE Co., Ltd., Nanjing, China. The CC for the anode substrate (1.0 × 2.0 cm^2) was relatively hydrophilic, with a thickness of 0.36 mm and a resistivity of less than 5 $m\Omega \cdot cm^2$. In order to increase the hydrophilicity of the anode carbon cloth, the treatment method was as follows: 10% nitric acid solution and 10% sulfuric acid solution were mixed in a 3:1 ratio; then, the carbon cloth was soaked in this solution for 12 h [15]. The CC for the cathode substrate was hydrophobic, with a thickness of 0.41 mm and a resistivity of less than 13 $m\Omega \cdot cm^2$.The preparation process of the anode was as follows: Aerobic activated sludge was collected from a sewage treatment plant in Nanning, China. The sludge was dried at 60 °C for 24 h and broken by grinding. It was then pyrolyzed at 600, 800, 1000, and 1200 °C for 2 h to obtain SC powder, filtered, and dried to obtain SC600, SC800, SC1000 and SC1200 samples. The obtained sludge carbon samples (1.5 mg cm^{-2}) were sprayed onto a carbon cloth to obtain SC600/CC, SC800/CC, SC1000/CC and SC1200/CC electrodes. Bare carbon cloth (CC) (1.0 × 2.0 cm^2) was used as the control electrode. The cathode was prepared to use activated carbon as the catalyst, 5 wt% Nafion solution as the binder, and carbon cloth as the collector [24]. The effective working area of the cathode CC was 4.91 cm^2.

2.2. MFC Construction

For air cathode microbial fuel cells, the elimination of the proton exchange membrane can significantly reduce the power output and cost of the cell. The reactors used in this study were 100 mL single-chamber membrane-free air cathode bottle MFCs. The MFC was inoculated on a super clean workbench. Firstly, 10 mL of pre-acclimated anaerobic granular sludge was sucked into the anode chamber with a disinfection syringe. Then, 80 mL of prepared artificial wastewater was sucked. The artificial wastewater was mainly composed of sodium acetate (1.5 $g \cdot L^{-1}$), phosphoric acid buffer solution (50 mM), trace elements (12.5 $mL \cdot L^{-1}$), and vitamin solution (5 $mL \cdot L^{-1}$) [24]. The formula of the nutrient solution is shown in the Tables S1–S3. Then, an external resistance of 1000 Ω was employed to connect the electrodes [24]. Intermittent water inlet mode was applied to the MFCs. The anode medium was replaced when the cell voltage passed below 50 mV. All experiments were conducted in triplicate, and average values were calculated. The external voltage generated by the MFCs was measured using a data acquisition instrument (Keithley6510, Cleveland, OH, USA). The polarization and power density curves were measured by gradually changing the external resistance (2000 Ω to 80 Ω) [7]. According to the formulas I = U/R and P = UI, the current and power under the corresponding resistances were calculated, respectively [25]. Cyclic voltammetry (CV) tests were performed in a three-electrode system using an electrochemical workstation (Chenhua, Shanghai). The prepared anode, a saturated calomel electrode (SCE), and a platinum wire were used as the working, reference and counter electrodes. The measurement range was −0.6~0.6 V, the scanning speeds were (1, 5, 10, 15 and 20 mV/s), and the static time was 2 s [9].

2.3. Materials Characterizations

The surface morphology and properties of the samples were characterized by scanning electron microscopy (SEM-EDX, SU8020, Hitachi hi tech, Tokyo, Japan), high-resolution electron microscopy (HR-TEM, FEI TECNAI G2 F30, FEI NanoPorts, Hillsboro, OR, USA), X-ray powder diffraction (XRD, RIGAKU D/MAX 2500V, Japan Science Corporation, Tokyo, Japan), X-ray photoelectron spectroscopy (XPS, ESCALAB 250XI+, Thermo Fisher Scientific, Waltham, MA, USA) and Raman spectroscopy (Raman, InVia Reflex, Renishaw, London, UK). The pore size distribution and specific surface area of the samples were measured by Barrett-Joyner-Halenda (BJH) method and Brunauer-Emmett-Teller (BET) theory (NOVA4200E, Cantar Instruments, Boynton Beach, FL, USA). The surface functional groups of the samples were characterized in the range of 4000–300 cm^{-1} by Fourier transform infrared spectroscopy (FTIR, TENSOR II, Bruker, Karlsruhe, Germany). A Diamond TG/DTA instrument (DTG-60(H), Hitachi hi tech, Tokyo, Japan) was used for a thermogravimetric (TG) analysis of the samples. Confocal scanning electron microscopy (CLSM, LSM800, zeroK Nanotech, Gaithersburg, MA, USA) was used to characterize the cell activity of the EABFs. Chemical oxygen demand (COD) was detected using spectrophotometry and a multiparameter water quality analyzer (Lianhua Technology Co., Ltd, Shanghai, China). The total suspended solids (TSS) in the anaerobic granular sludge was calculated by drying and weighing at 105 °C, while the volatile suspended solids (VSS) were calculated by calcination at 600 °C. The zeta potential of the samples was characterized using a zeta potential particle sizer (Nano-ZS90X, Marvin Instrument Equipment Co., Ltd, London, UK).

2.4. EPS Extraction and Analysis

Extracellular polymers (EPS) on anode biofilms were extracted by the water bath heating method [15]. After power generation, the CC was cut with sterile scissors and placed into centrifuge tubes containing 0.9% sodium chloride solution, rotated for 10 min, and centrifuged at 4000× *g* rpm for 15 min. The supernatant was discarded, replenished with sodium chloride solution, the above operations were repeated, and vibrated in an 80 °C water bath shaker for 30 min. Finally, the supernatant was centrifuged at 9000× *g* rpm for 10 min and filtered through a 0.22 μm membrane to obtain tightly bound EPS. The contents of polysaccharides, humic substances, protein, and outer membrane c-type cytochromes (OM c-Cyts) in EPS were detected using UV/visible spectrophotometer (Agilent 8453, Agilent Technology Co., Ltd., Santa Clara, CA, USA). EPS was characterized using a three-dimensional excitation-emission matrix (3D-EEM) fluorescence spectrometer (Hitachi F-7000, Hitachi hi tech, Tokyo, Japan). The main test parameters were as follows: the scanning ranges of the excitation spectrum (Ex) and emission spectrum (EM) were 220–500 nm and 220–550 nm, respectively, while the scanning step was 5 nm.

2.5. Microbial Community Analysis

The anodes were collected after power generation, and the genomic DNA was extracted from the biofilm samples using the E.Z.N.A.® Mag-Bind Soil m DNA Kit (OMEGA) according to the manufacturer's instructions [25]. The V3-V4 region of the 16S rRNA gene was amplified using polymerase chain reaction (PCR) with primers 338F (5′-ACTCCTACGG GAGGCAGCA-3′) and 806 R (5′-GGACTACHVGGGTWTCTAAT-3′) [26]. The same loading buffer volume was mixed with the PCR product, and electrophoresis was performed on a 2% agarose gel. The products were purified using a pre-QUS kit™. The PCR products were detected and quantified using a fluorometer. The NEXTFLEX Rapid DNA-Seq kit was used to build the libraries. After purification and quantification, the samples were tested for 16S rRNA gene sequencing based on the Illumina MiSeq platform. Based on the Kyoto Encyclopedia of Genes and Genomes (KEGG) database, the macrogenomic information of microbial metabolic function was predicted using the PICRUSt pipelines program [9].

3. Results

3.1. Characterization of Material

Figure 1 shows that the morphological changes in SC were significantly related to the carbonization temperature. The SC600 and SC800 samples showed disordered sheet structures. For the SC1000 sample, the surfaces were relatively rough, and carbon microspheres and small pores appeared due to the release of volatile substances and the sintering effect, which converted a large number of inorganic parts into mineral-like compounds and induced the carbon phase to warp some metals [27]. For SC1200 samples, high temperature enhanced the shrinkage of carbon, hindered the development of pores, led to the collapse or deformation of coke, and reduced the pore volume. Furthermore, it can be observed from the TEM images (Figure S1) that the morphology of SC changed significantly with the increase of pyrolysis temperature, with carbon particles appearing in SC1000 and SC1200 samples, which is consistent with the SEM results.

Figure 1. SEM images of SC samples under different carbonization temperature: (**a**) SC600, (**b**) SC800, (**c**) SC1000, and (**d**) SC1200.

As shown in Table 1, the pH, zeta potential, and ash content of sludge carbon was linearly related to the carbonization temperature. Increasing the pyrolysis temperature can increase the zeta potential of the sludge carbon surface. When the carbonization temperature was low, the measured zeta potential was negative because the original sludge was rich in negative groups such as carboxyl and hydroxyl groups, and the sludge carbon surface was negatively charged [21]. When the carbonization temperature was 1000 °C or above, the zeta potential was positive and the mud carbon surface was positively charged. This was because the oxygen-containing functional groups with negative charge were basically decomposed, and a large number of amino and metal groups were exposed on the carbon surface of the sludge [28]. The electrical properties of the material surface changed from negative to positive, which is useful for the electrical absorption of microorganizations with a negative charge. Another important feature of an ideal anode is high conductivity. The results showed that the SC1000 sample had the highest conductivity (57.43 $s \cdot m^{-1}$). In summary, the SC1000 sample may be the most suitable anode material. Its high conductivity and low absolute zeta potential value enable the adhesion of EAMs and the formation of EABFs. As shown in Figure S2, TGA showed a weight-loss temperature range of 200–1300 °C, indicating that sludge carbonization began at 200 °C and mainly ended at 1300 °C, and the mass was reduced by 51%. As shown in Figure 2, The FTIR spectra of all samples showed the following stretching vibrations: -OH of ethanol with a peak at 3450 cm^{-1}, C-H at 2849 cm^{-1}, N-H at 1650 cm^{-1}, unsaturated aldehyde with

C=O at 1406 cm^{-1}, C-O of amine at 1100 cm^{-1}, and C-H at 789 cm^{-1} and 2928 cm^{-1} [21]. The raw sludge was rich in carboxyl and hydroxyl groups [21]. As the carbonization temperature increased, the intensity of these peaks gradually decreased. This was due to the decomposition of oxygen-containing functional groups. Because of their negatively charged surface, oxygen-containing functional groups are not favorable for bacterial adhesion, thus affecting the electron transfer path [25].

Table 1. Physical parameters of SC samples.

Samples	Yield (%)	pH	Zeta Potential (mV)	Ash (%)	Volatiles (%)	Conductivity (s/m)
SC600	66.33	8.09	−19.20	78.60	20.30	13.87
SC800	63.63	8.56	−10.59	81.90	15.50	36.40
SC1000	55.93	8.98	1.39	83.20	12.70	57.43
SC1200	57.27	9.65	2.14	86.70	9.80	54.77

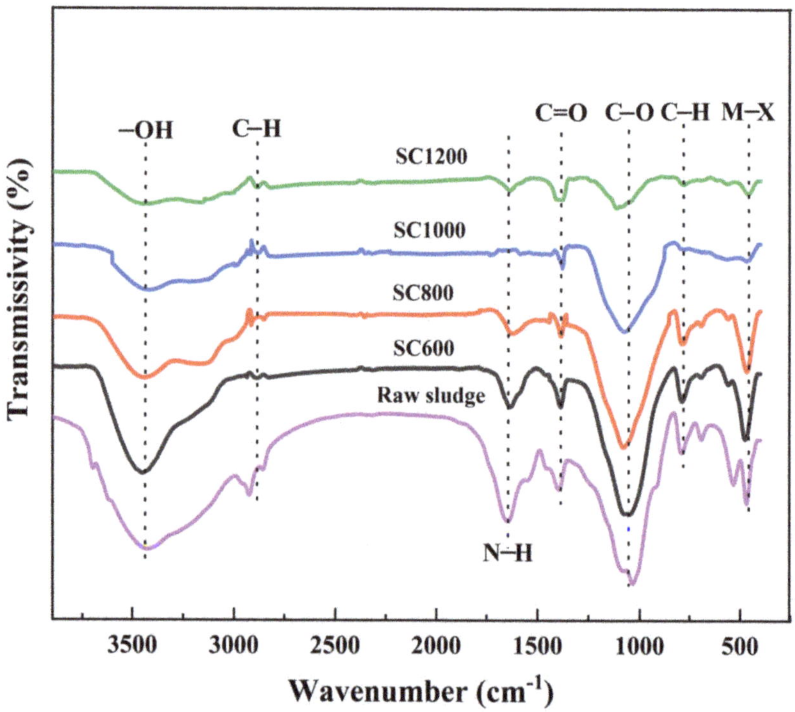

Figure 2. FTIR spectra obtained from raw sludge and SC samples.

As shown in Figure 3 and Table S4, the specific surface area and pore structure of SC were significantly affected by the carbonization temperature, which greatly influenced the catalytic activity of EABFs. The N_2 adsorption/desorption curves of the SC samples belonged to type IV isothermal curves, and there were notable H_3 hysteresis curves (Figure 3a). This adsorption hysteresis phenomenon is related to the mesoporous and microporous characteristics of the slit [29]. Table S4 summarizes the BET and pore-size distributions of the SC samples. The results showed that the specific surface areas of the SC samples gradually increased as the carbonization temperature increased. The specific surface area of the SC1000 sample was the largest (216.00 ± 0.005 $m^2 \cdot g^{-1}$). However, when the carbonization temperature was further increased to 1200 °C, the specific surface area of the sample decreased (176.55 ± 0.010 $m^2 \cdot g^{-1}$). The pore size distribution of the SC was further studied (Figure 3b and Table S1). There were two types of pores in all SC samples: micropores and mesopores, with an average size of approximately 8 nm. The total pore volume and mesopore number of the SC800 sample were the largest, while the total pore volume and mesopore volume of the SC1000 sample reduced, but the micropore volume increased, which may have been due to the collapse of some mesopores under high-temperature conditions, resulting in pore blockage [30]. It could be predicted that the special mesoporous and microporous structures produced on the surface of SC may provide active sites for the catalytic reaction and enhance the electrocatalytic ability, thus enhancing proton transfer and charge transfer [15,25,31]. The increase in the specific surface area also has a considerable effect on the decrease in the internal resistance [31]. In short, the SC1000 sample had a good specific surface area and rich pore structure for proton and electron transfer, substrate transport, and biofilm formation in MFCs [11,25,29].

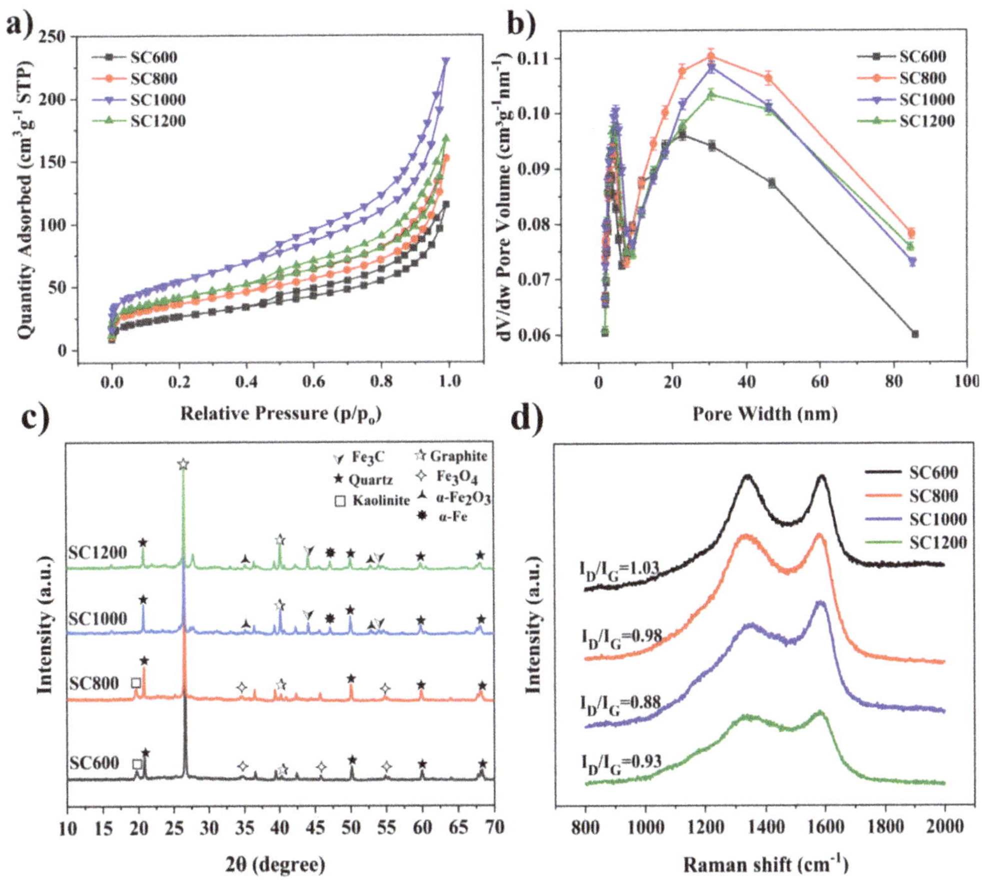

Figure 3. (**a**) N_2 adsorption-desorption isotherms, (**b**) pore size distribution, (**c**) powder XRD patterns and (**d**) Raman spectra of the SC samples.

As shown in Figure 3c, an XRD analysis of the structure and crystallinity of the SC samples at varying carbonization temperatures indicated pyrolysis differences in the samples at different temperatures. The characteristic diffraction peaks at 26.3° and 41.0° correspond to the (002) and (101) lattice planes of graphite carbon, respectively, confirming the existence of graphite carbon. The characteristic diffraction peaks at 40.9° and 54.1° corresponded to the (110) and (116) lattice planes of α-Fe_2O_3 (JCPDS. Card 33-0664). The characteristic diffraction peak at 35.4° corresponded to the (111) lattice plane of Fe_3O_4 (JCPDS. Card 72-2303). The characteristic diffraction peaks of Fe_3C (PDF No. 01-089-3689) are 43.7° and 54.7° [29,32]. As the carbonization temperature increased, the intensities of α-Fe_2O_3 and Fe_3O_4 decreased, while the intensities of graphite carbon and Fe_3C increased. The existence of graphite carbon and Fe_3C supported the conductivity of the SC [29]. The graphitization degree of SC was studied by testing the carbonized samples at different temperatures using Raman spectroscopy (Figure 3d). The SC samples showed two characteristic strong peaks at 1350 cm^{-1} and 1599 cm^{-1}, corresponding to the defect D and G peaks of graphite, respectively [25]. The strength ratios (I_D/I_G) of the D and G bands decreased as the carbonization temperature increased, indicating that the graphitization degree of SC improved. In addition, compared to the SC600 and SC800 samples, the SC1000 and SC1200 samples had wider peaks at 1350 cm^{-1}, indicating that the nanoparticles in the sludge carbon samples were smaller.

EDX was performed to detect the types and distribution of elements in SC1000 (Figure S3). According to the EDX results, SC1000 samples were rich in elements. In addition to the main elements, i.e., C, N, O, and Si, these samples also had low contents of metals such as K, Ca, Mg, Al, Fe, Cu, Zn, Ti, as well as trace amounts of P and S; these heteroatoms were evenly distributed. Among them, C, N, P, and S are conducive to enhancing the graphite properties and hydrophilicity of SC [25]. Al, Ca, and Mg are useful for the formation of biochar metal skeletons [23]. It is speculated that transition metal elements such as Fe, Ti, Cu and Zn can provide rich active sites for redox reactions and improve the electrocatalytic activity of EAMs [28,29]. The specific content of each element in the SC samples was determined using XPS (Table S5). Interestingly, with the increase in carbonization temperature, the contents of the other elements continued to decrease, except for C and S. This was mainly due to the loss of volatile substances during pyrolysis. In addition, with the increase in carbonization temperature, the contents of oxygen- and nitrogen-containing functional groups decreased, which was consistent with the FTIR results. It is well known that the reduction of oxygen content and pyridine groups enhance electron transfer [25,33]. The types and contents of the elements in the sludge carbon were further confirmed by XPS. Figure S4 shows the XPS full spectrum results of the SC samples carbonized at 600, 800, 1000, and 1200 °C. According to the results, the elements in the SC samples mainly comprised C, N, O, P, Si, Ca, Mg, Al, Fe, and P; this was consistent with the EDX test results.

Furthermore, changes in the chemical forms of the main elements, i.e., C, N, and Fe, in the SC samples prepared at different carbonization temperatures were analyzed by XPS. As shown in Figure 4, the XPS high-resolution spectrum of C 1s mainly included C-C and C=C graphite carbon with a binding energy of 284.80 eV, C-N with a binding energy of 285.78 eV, C-O with a binding energy of 286.88 eV and C=O with a binding energy of 289.12 eV [23]. It was further indicated that with the increase in carbonization temperature, the oxygen-containing functional groups gradually decreased, and most amorphous carbon was transformed into SP^2 graphite carbon and C-N. Graphite carbon improves the conductivity of electrodes, and the C-N bond could increase its hydrophilicity, reduce charge transfer resistance, and facilitate bacterial adhesion [10,12,31,32]. The XPS high-resolution spectra of N 1s mainly indicated pyridinic-N with a binding energy of 398.80 eV, graphitic-N with a binding energy of 401.50 eV, Fe-N with a binding energy of 399.60 eV, oxidized-N with a binding energy of 402.90 eV and pyrrolic-N with a binding energy of 400.69 eV [29,32]. When the carbonization temperature was lower than 1000 °C, the proportion of pyridinic-N and oxidized-N decreased, while the contents of graphitic-N, pyrrolic-N, and Fe-N increased as the temperature progressed. It is well known that graphitic-N can improve the conductivity of materials and accelerate electron transfer, and pyrrolic-N can improve the electrochemical reaction rate. Fe-N improves the electrocatalytic activity of EABFs and accelerates electron transfer [34]. The XPS high-resolution spectrum of Fe 2p showed that the oxidation state of Fe on the surface of the SC sample was complex, and that the difference was significant with the change of carbonization temperature. Due to the presence of various iron species, sludge carbon exhibits ferromagnetism, which is conducive to microbial adsorption [23]. Particularly, for the SC1000 sample, the presence of a Fe-C binding site with a binding energy of 720.7 eV confirmed the existence of Fe_3C. The surface of the CC anode was smooth, which was unfavorable for microbial adsorption (Figure S5). The surfaces of CC loaded with sludge carbon were relatively rough, which increased the microbial contact area. For SC1000/CC, the adhesion between the sludge carbon and CC was closer, which was better for microbial adsorption (Figure S6).

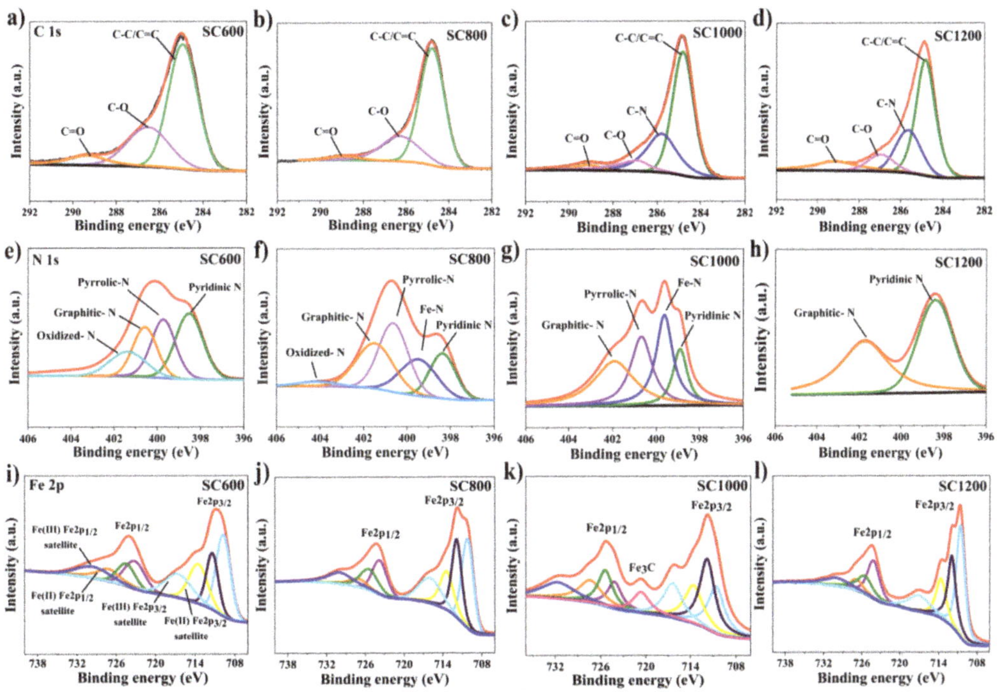

Figure 4. C1s spectra for (**a**) SC600, (**b**) SC800, (**c**) SC1000 and (**d**) SC1200; N1s spectra for (**e**) SC600, (**f**) SC800, (**g**) SC1000 and (**h**) SC1200; Fe2p spectra for (**i**) SC600, (**j**) SC800, (**k**) SC1000 and (**l**) SC1200.

3.2. Electrode Electrocatalytic Activity

CV analyzed the redox medium composition and redox potential of biofilms. All CV curves showed a typical S-type anode catalyst curve with sodium acetate as the substrate (Figure 5). This indicated the formation of electroactive biofilms and showed that all MFC systems may adopt similar electron transfer paths. However, except for the CC anode, the CV curves of other SC anodes showed a pair of redox main peaks, i.e., mainly redox pairs centered on −0.38 V (cathode) and−0.06 V (anode). The midpoint potential was−0.2 V, which is within the electron transfer activity range of outer membrane c-type cytochromes (OM c-Cyts), indicating that the electron transfer path between the biofilm and the electrode is mainly short-range direct electron transfer (DET) mediated by OM c-Cyts [35–37]. The capacitance area of the SC anodes was much larger than that of the CC anode. The above results show that the EABFs on SC electrodes have high electrocatalytic activity. In addition, the peak current density and capacitance area of the anodes revealed a significant linear relationship with the sludge carbonization temperature, and increased with increasing scanning speed.

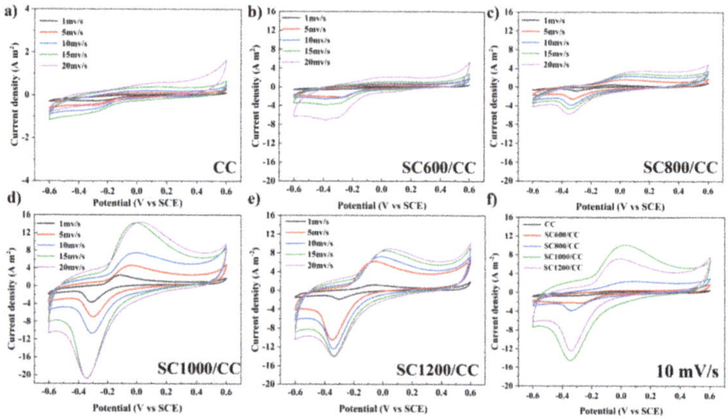

Figure 5. CV plots of (**a**) CC, (**b**) SC600/CC, (**c**) SC800/CC, (**d**) SC1000/CC, (**e**) SC1200/CC with 1, 5, 10 and 15 mV/s scanning rates and (**f**) CV plots of the electrodes with 10 mV/s scanning rates at three-electrode electrochemical systems after 54 d of batch mode operations.

3.3. MFC Performance

As shown in Figure 6a, in the first six cycles (approximately 25 days), the COD removal rates of all MFCs showed an increasing trend, and after the sixth cycle of power generation, the COD degradation rate stabilized. There was a correlation between COD removal efficiency and carbonization temperature. MFCs equipped with SC1000/CC had the largest COD removal efficiency (97.63 $\pm$ 0.039%), i.e., higher than that of CC (90.97 $\pm$ 0.035%). In addition, during stable power generation, the VSS/TSS of the granular sludge in MFCs equipped with SC electrodes was significantly higher than that of MFCs equipped with CC (Figure 6b), and the value of VSS/TSS increased with the increase of sludge carbonization temperature. This indicated that sludge carbon can increase the organic components in granular sludge, thus improving the microbial biomass and microbial activity. Interestingly, the trend regarding the COD removal efficiency of the MFC reactor was consistent with that of VSS/TSS. This may have been because the addition of SC promoted the enrichment of microorganisms, stimulated microorganisms to secrete EPS, and accelerated the carbohydrate metabolism of exoelectrogens and methanogens, thereby accelerating the degradation rate of organic matter [5]. Additionally, carbonization temperature also strongly affected the internal resistance, power density and current density of SC anodes (Figure 6c,d). According to the polarization curve, there was a significant difference in anode open circuit voltage. The open circuit voltage of the SC anode was 0.640 V (SC600/CC), 0.686 V (SC800/CC), 0.725 V (SC1000/CC) and 0.698 V (SC1200/CC), i.e., higher than that of the CC anode (0.626 V). Under the same voltage conditions, the current density of SC1000/CC was the largest, indicating that SC1000/CC has the highest electrochemical oxidation activity. The SC1000/CC anode generated a maximum areal power density of 2.165 $\pm$ 0.021 mW$\cdot$m^{-2} and current density of 5.985 $\pm$ 0.015 A$\cdot$m^{-2}, which was 3.017- and 2.992-fold that of the CC anode. The SC1000/CC anode had lower internal resistance and produced higher current density and power density, which meant the electroactive biofilm had higher electrocatalytic activity. As shown in Figure 7, the sludge carbonization temperature significantly affected the voltage output of the MFC system. Compared to the MFCs equipped with CC electrodes, the output voltage of the MFCs equipped with SC electrodes significantly improved. Among them, the MFCs equipped with SC1000/CC anodes showed the best power generation performance, with a maximum closed-circuit voltage of 0.501 V and an average power generation cycle of 146 h. The maximum closed-circuit voltage of the MFC equipped with CC anodes was only 0.323 V, and the average power generation cycle was 100 h.

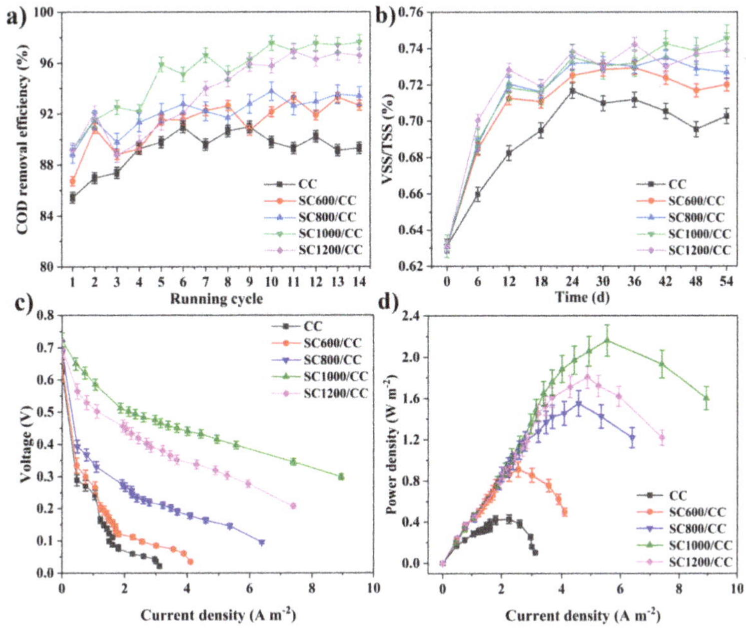

Figure 6. (**a**) COD removal efficiency of MFCs, (**b**) VSS/TSS of inoculated anaerobic granular sludge at MFC systems, (**c**) polarization curves, and (**d**) power density of MFCs.

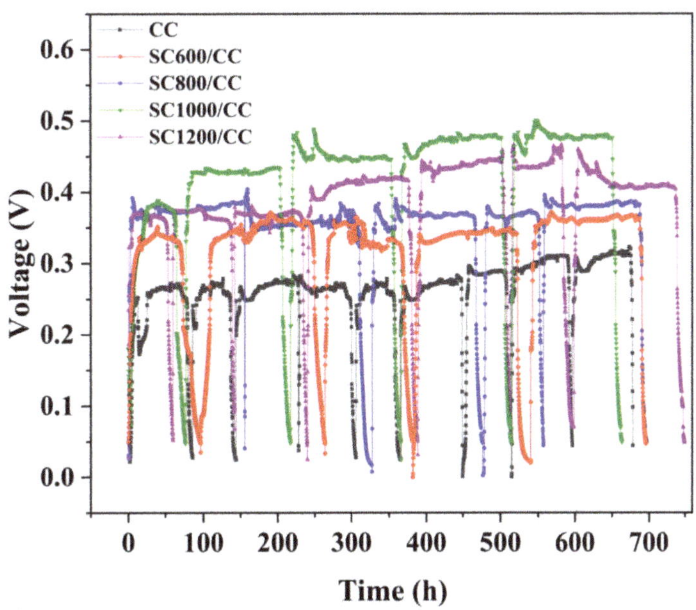

Figure 7. Reproducible cycles of output voltages produced in MFCs equipped with different anodes.

3.4. Anode Biofilm Characterizations

As shown in Figure S7, in the process of power generation, the modification of SC significantly promoted the secretion of EPS, further confirming that SC can improve microbial biomass and microbial activity. It is well known that in the biofilm-producing current, c-Cyts play a key role in interspecific electron transfer and extracellular electron transfer [36]. It is believed that the higher the concentration of c-Cyts, the higher the electron transfer efficiency between EAMs and electrodes [35]. Some studies have shown that the c-Cyts in EPS can closely bind to the active sites of transition metals Fe (III), Cu (II), and Zn (II) to promote electron transfer [23,38]. It can be seen in Figure 8a–f that in all MFCs, the absorbance of c-Cyts at a wavelength of 419 nm in the anode biofilm gradually increased with power generation, and that the addition of sludge carbon at different carbonization temperatures changed the absorbance of c-Cyts, indicating that sludge carbon affected the secretion of c-Cyts and further influenced the power generation. SC1000/CC had the highest absorbance at 419 nm, indicating that its c-Cyts concentration was the highest, so it can continuously absorb electrons from the embedded bacteria and transfer them to the electrode. As shown in Figure 8g–l, EPS extracted from the anode biofilm was analyzed using EEM fluorescence spectroscopy. Among the observed peaks, peak A (Ex/Em = 280–295/320–335 nm) was tryptophan-like acid and soluble microbial by-products, Peak B (Ex/Em = 380/435–450 nm) was humic-like acid, and peak C (Ex/Em = 420–425/450–470 nm) was coenzyme F_{420}, which plays an important role in the hydrogen nutrition pathway and is related to the methanogenic metabolic activity [5,39]. Peak D (Ex/Em = 230/325–330 nm) was protein-like. Compared to the inoculated sludge, the intensity of peaks A and B in the MFC anode biofilm gradually increased, and peak C appeared. Compared to CC, the intensities of peaks A, B, and C in the SC anode biofilms gradually increased, and peak D appeared. This was consistent with the conclusion in Figure S7, which further proved that the SC can change the composition and content of EPS. As shown in Figure S8, SC significantly promoted the increase in microbial biomass in the biofilm and the secretion of EPS. The EPS matrix covered carbon defects on the electrode surface and strengthened cell adhesion. As shown in Figures S8g–l and S9, SC-modified CC can improve microbial activity and promote the growth of EABFs. The ratio of live cells on the SC1000/CC anode biofilm was the largest, indicating that the anode had high biocompatibility.

3.5. Microbial Community Analysis

The α diversity represents the microbial community diversity on the anode EABFs. As shown in Table S6 and Figure S10, compared to the CC anode biofilm, the biofilm community diversity, richness, and the total number of species in the SC anode biofilms increased. These factors were linearly related to carbonization temperature, indicating that the SC has a significant impact on the community diversity of the anode biofilm. The effect of the SC characteristics on the microbial community structure of EABFs was further explored by high-throughput sequencing (Figure 9). SC significantly affected the community structure of the anode biofilm, thus directly affecting the power generation of MFCs. As shown in Figure 9a, at the phylum level of archaea, Euryarchaeota, and halobacteria were dominant. Among them, Euryarchaeota represent an irreplaceable functional microorganism in the anaerobic digestion process [40], and their numbers increased with an increase in the sludge carbonization temperature. In contrast, the content of halobacteria decreased with an increase in the sludge carbonization temperature, indicating that SC changes the structure of archaea, and that the characteristics of SC will selectively enrich functional microorganisms. At the genus level of archaea (Figure 9b), Methanobacterium and Methanosaeta were dominant. They can participate in EET and interspecific direct electron transfer (DIET) [5,39,40]. Methanosacrina is an obligate anaerobic bacterium that was significantly enriched in the SC1000/CC anode. It can possibly perform an EET. Hence, SC optimizes the archaeal community structure, selectively enriches functional microorganisms, and enhances the hydrogen methane production pathway.

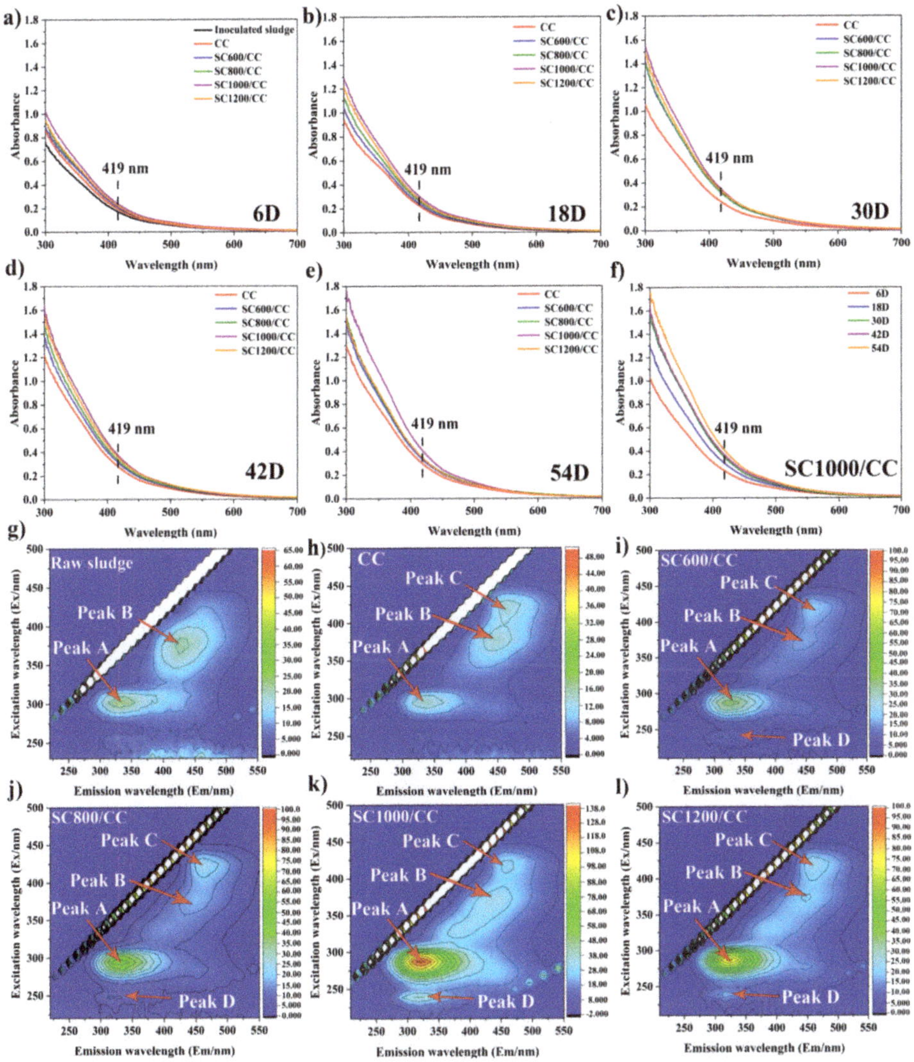

Figure 8. Absorbance of c-Cyts for inoculated sludge and different anodic biofilms at MFCs on (**a**) 6D, (**b**) 18D, (**c**) 30D, (**d**) 42D, and (**e**) 54D. Absorbance of c-Cyts for (**f**) SC1000/CC. EEM fluorescence spectra of EPS extracted from (**g**) inoculated sludge, (**h**) CC, (**i**) SC600/CC, (**j**) SC800/CC, (**k**) SC1000/CC, and (**l**) SC1200/CC anode biofilms at MFCs after 54 d of batch mode operations.

As shown in Figure 9c, at the phylum level, the microbial communities on all anodes were similar, but the relative abundance of dominant bacteria was significantly different. Compared to the CC anode, the relative abundance of Proteobacteria, Firmicutes, and Bacteroidetes in the SC anodes decreased, and there was a correlation with the carbonization temperature. In contrast, the relative abundance of Desulfobacterota, Synergistota, and Actinobacteria increased. These results showed that SC could promote a synergistic effect between exoelectrogens and Synergistota. Desulfobacteria and Actinobacteria are related to organic matter degradation and usually control the acetate oxidation community [41], which contains a large number of electrochemically-active species [2,42]. The relative abundance of Desulfobacterota on the SC1000/CC biofilm was the highest (35%), indicating that the biofilm was rich in electroactive bacteria.

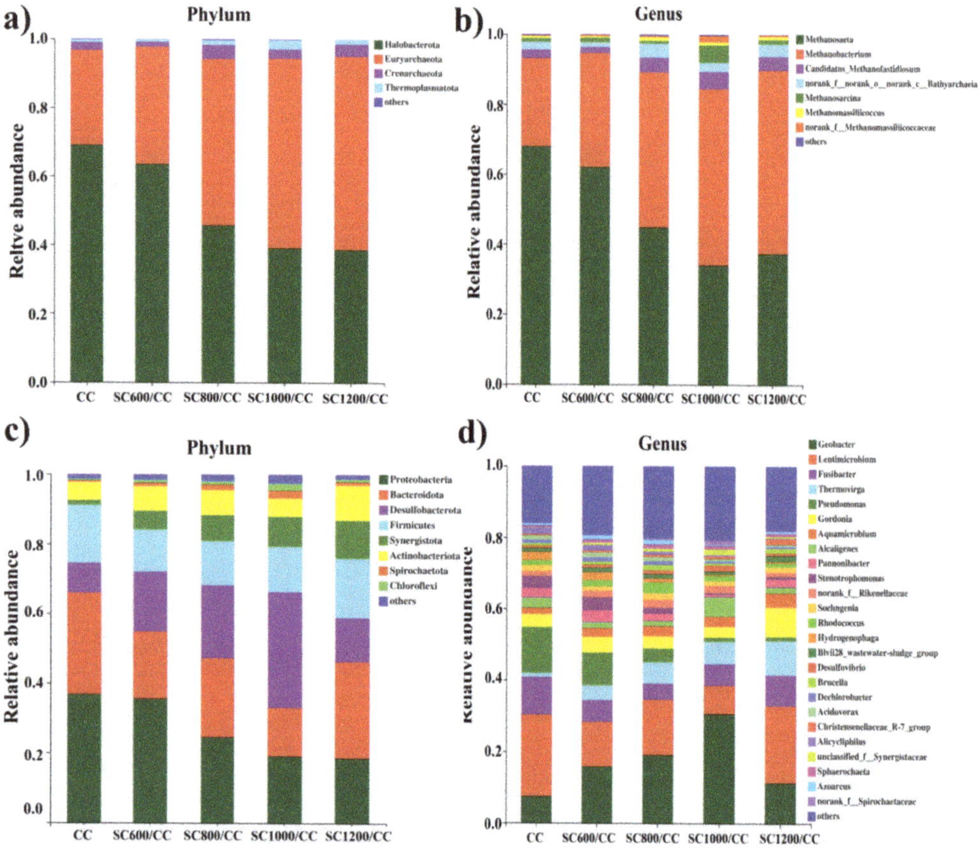

Figure 9. Microbial community structures of anodic biofilms of Archaea and bacteria attached on CC and SC anodes at the (**a**,**c**) phylum and (**b**,**d**) genus levels.

As shown in Figure 9d, at the genus level, SC increased the diversity of microbial community. The dominant bacteria in each MFC were Geobacter and Lentimicrobium. Geobacter is a common exoelectrogen that can secrete cytochromes and participate in direct electron transfer [35,37,42]. This may be reflected in the power generation performance of MFC systems. Direct electron transfer may occur between Geobacter and methanogens, which can cooperate in power generation [43]. Lentimicrobium is a strictly anaerobic gram-negative bacterium. Studies have shown that it may form a consortium with other EAMs to convert acetate into electric energy [42]. Interestingly, SC significantly increased the relative abundance of Thermovirga, which was positively correlated with carbonization temperature, and may produce cytochrome to participate in EET [37]. In contrast, the relative abundance of Pseudomonas was significantly reduced due to the presence of SC, which promoted EET by secreting phenazines [42]. The results showed that the electron transfer path of SC anode biofilms was mainly short-range DET of c-type cytochromes, which was consistent with the data presented in the CV curves in Figure 5. The above results show that the microbial community structure of EABFs is directly related to the characteristics of the anode materials, and that there may be electron transfer between Archaea and EAMs. The metabolic pathways of microorganisms determine the flow of electrons and protons, which affect the performance of electricity production. In metabolism clusters, all reactors used amino acid, carbohydrate, and energy metabolism as the main metabolic pathways (Figure S11). The SC1000/CC anode biofilm had the highest amino acid, carbohydrate, energy, and lipid metabolism. High carbohydrate metabolism indicates that microorganisms decompose organic matter quickly, which is related to increased output voltage [43].

4. Conclusions

The sludge carbon prepared by one-step pyrolysis had the advantages of high conductivity, good pore structure, high biocompatibility, high carbon content, rich heteroatom composition, and low cost. It significantly improved the activity and diversity of microorganisms on the anode biofilm, optimized the composition of the microbial community, regulated the metabolic pathway of microorganisms, promoted the secretion of EPS and the expression of cytochrome, and strengthened the electron transfer ability of EABFs. Therefore, an anode processed at 1000 °C generated a maximum areal power density of 2.165 ± 0.021 W·m^{-2} and current density of 5.985 ± 0.015 A·m^{-2}, which was 3.017- and 2.992-fold that of the CC anode. This study provides a theoretical basis for the practical application of sludge carbon in MFCs and provides a new direction for the rational utilization of biomass waste resources.

Supplementary Materials: The following supporting information can be downloaded at: https://www.mdpi.com/article/10.3390/membranes12020120/s1. Figure S1: TEM images of SC samples under different carbonization temperature: a) SC600, b) SC800, c) SC1000, and d) SC1200, Figure S2: TG-DSC plots of raw sludge under N2 atmosphere, Figure S3: EDX element mappings of the SC1000 sample, Figure S4: XPS survey spectrum of the SC samples under different carbonization temperature. Figure S5: SEM image of bare carbon cloth electrode, Figure S6: SEM images of loaded electrodes with different SC samples, Figure S7: The composition and content of extracellular polymer substances for inoculated sludge and different anode biofilms at MFCs, Figure S8: SEM and CLSM images of raw sludge and anodes biofilms from MFCs after 54 d of batch mode operations, Figure S9: Ratio of dead and alive cells for raw inoculated sludge and anode biofilms at MFCs, Figure S10: Venn diagram of the biofilms attached on CC and SC anodes, Figure S11: Predictive abundances of metagenomic functional genes on the anode biofilms based on PICRUSt of KEGG analysis, Table S1: PBS solution composition, Table S2: Trace element solution compositio, Table S3: Vitamin solution, Table S4: Specific area and total surface area of SC samples, Table S5: Atomic contents of SC samples, Table S6: Species diversity and abundance indexes.

Author Contributions: Conceptualization, K.Z. and Y.X.; methodology, X.Y.; validation, W.F. and J.Y.; formal analysis, K.Z.; investigation, W.F.; resources, Z.W.; data curation, K.Z.; writing—original draft preparation, K.Z.; writing—review and editing, K.Z. and Z.W.; supervision, W.D. and W.F.; funding acquisition, Z.W. All authors have read and agreed to the published version of the manuscript.

Funding: This research were funded by the National Natural Science Foundation of China, (grant number: 21868004), the Key Research and Development Plan of Guangxi Province (grant number: GuikeAB19259013), and Guangxi Natural Science Foundation (grant number: 2021GXNSFBA196094).

Institutional Review Board Statement: Not applicable.

Informed Consent Statement: Not applicable.

Data Availability Statement: Not applicable.

Acknowledgments: The authors thank Meiji biological platform for high-throughput sequencing of microorganisms.

Conflicts of Interest: The authors declare no conflict of interest.

References

1. Rabaey, K.; Read, S.T.; Clauwaert, P.; Freguia, S.; Bond, P.L.; Blackall, L.L.; Keller, J. Cathodic oxygen reduction catalyzed by bacteria in microbial fuel cells. *ISME J.* **2008**, *2*, 519–527. [CrossRef] [PubMed]
2. Logan, B.E.; Rabaey, K. Conversion of wastes into bioelectricity and chemicals by using microbial electrochemical technologies. *Science* **2012**, *337*, 686–690. [CrossRef] [PubMed]
3. Slate, A.J.; Whitehead, K.A.; Brownson, D.A.C.; Banks, C.E. Microbial fuel cells: An overview of current technology. *Renew. Sustain. Energy Rev.* **2019**, *101*, 60–81. [CrossRef]
4. Butti, S.K.; Velvizhi, G.; Sulonen, M.L.K.; Haavisto, J.M.; Koroglu, E.O.; Cetinkaya, A.Y.; Singh, S.; Arya, D.; Modestra, J.A.; Krishna, K.V.; et al. Microbial electrochemical technologies with the perspective of harnessing bioenergy: Maneuvering towards upscaling. *Renew. Sustain. Energy Rev.* **2016**, *53*, 462–476. [CrossRef]
5. Ma, H.T.; Wu, M.; Liu, H.; Wang, Z.W.; Guo, C.Y.; Wang, S.F. Study on enhancing sludge methanogenesis by adding acetylene black and effect on the characteristics & microbial community of anaerobic granular sludge. *RSC Adv.* **2019**, *9*, 23086–23095.
6. Selman, J.R. Materials science. Poison-tolerant fuel cells. *Science* **2009**, *326*, 52–53. [CrossRef]
7. Logan, B.E.; Hamelers, B.; Rozendal, R.A.; Schrorder, U.; Keller, J.; Freguia, S.; Aelterman, P.; Verstraete, W.; Rabaey, K. Microbial fuel cells: Methodology and technology. *Environ. Sci. Technol.* **2006**, *40*, 5181–5192. [CrossRef]
8. Santoro, C.; Arbizzani, C.; Erable, B.; Ieropoulos, I. Microbial fuel cells: From fundamentals to applications. *J. Power Sources* **2017**, *356*, 225–244. [CrossRef]
9. Cao, Z.Z.B.; Peng, L.; Hu, S.; Ding, M.; Song, F.; Guan, X.; Lee, C.K.; Huang, J.; Zhu, D.; Fu, X.; et al. Silver nanoparticles boost charge-extraction efficiency in shewanella microbial fuel cells. *Science* **2021**, *373*, 1336–1340. [CrossRef]
10. Li, Y.; Liu, J.; Chen, X.; Yuan, X.; Li, N.; He, W.; Feng, Y. Enhanced electricity generation and extracellular electron transfer by polydopamine–reduced graphene oxide (PDA–rGO) modification for high-performance anode in microbial fuel cell. *Chem. Eng. J.* **2020**, *387*, 123408. [CrossRef]
11. Zhao, C.E.; Gai, P.; Song, R.; Chen, Y.; Zhang, J.; Zhu, J.J. Nanostructured material-based biofuel cells: Recent advances and future prospects. *Chem. Soc. Rev.* **2017**, *46*, 1545–1564. [CrossRef]
12. Zhang, P.; Liu, J.; Qu, Y.; Li, D.; He, W.; Feng, Y. Nanomaterials for facilitating microbial extracellular electron transfer: Recent progress and challenges. *Bioelectrochemistry* **2018**, *123*, 190–200. [CrossRef]
13. Kamali, M.; Persson, K.M.; Costa, M.E.; Capela, I. Sustainability criteria for assessing nanotechnology applicability in industrial wastewater treatment: Current status and future outlook. *Environ. Int.* **2019**, *125*, 261–276. [CrossRef]
14. Liu, D.; Chang, Q.; Gao, Y.; Huang, W.; Sun, Z.; Yan, M.; Guo, C. High performance of microbial fuel cell afforded by metallic tungsten carbide decorated carbon cloth anode. *Electrochim. Acta* **2020**, *330*, 135243. [CrossRef]
15. Liu, J.; Qiao, Y.; Guo, C.X.; Lim, S.; Song, H.; Li, C.M. Graphene/carbon cloth anode for high-performance mediatorless microbial fuel cells. *Bioresour. Technol.* **2012**, *114*, 275–280. [CrossRef]
16. Ma, X.X.; Feng, C.H.; Zhou, W.J.; Yu, H. Municipal sludge-derived carbon anode with nitrogen- and oxygen-containing functional groups for high-performance microbial fuel cells. *J. Power Sources* **2016**, *307*, 105–111. [CrossRef]
17. Zou, L.; Lu, Z.; Huang, Y.; Long, Z.-E.; Qiao, Y. Nanoporous Mo_2C functionalized 3D carbon architecture anode for boosting flavins mediated interfacial bioelectrocatalysis in microbial fuel cells. *J. Power Sources* **2017**, *359*, 549–555. [CrossRef]
18. Wu, X.; Shi, Z.; Zou, L.; Li, C.M.; Qiao, Y. Pectin assisted one-pot synthesis of three dimensional porous NiO/graphene composite for enhanced bioelectrocatalysis in microbial fuel cells. *J. Power Sources* **2018**, *378*, 119–124. [CrossRef]
19. Lu, M.; Qian, Y.J.; Yang, C.C.; Huang, X.; Li, H.; Xie, X.J.; Huang, L.; Huang, W. Nitrogen-enriched pseudographitic anode derived from silk cocoon with tunable flexibility for microbial fuel cells. *Nano Energy* **2017**, *32*, 382–388. [CrossRef]
20. Li, M.; Li, Y.W.; Yu, X.L.; Guo, J.J.; Xiang, L.; Liu, B.L.; Zhao, H.M.; Xu, M.Y.; Feng, N.X.; Yu, P.F.; et al. Improved bio-electricity production in bio-electrochemical reactor for wastewater treatment using biomass carbon derived from sludge supported carbon felt anode. *Sci. Total Environ.* **2020**, *726*, 138573. [CrossRef]
21. Fu, Q.; Wang, D.; Li, X.; Yang, Q.; Xu, Q.; Ni, B.J.; Wang, Q.; Liu, X. Towards hydrogen production from waste activated sludge: Principles, challenges and perspectives. *Renew. Sustain. Energy Rev.* **2021**, *135*, 110283. [CrossRef]
22. Mian, M.M.; Liu, G.; Fu, B. Conversion of sewage sludge into environmental catalyst and microbial fuel cell electrode material. *Sci. Total Environ.* **2019**, *666*, 525–539. [CrossRef]
23. Logan, B.; Cheng, S.; Watson, V.; Estadt, G. Graphite fiber brush anodes for increased power production in air-cathode microbial fuel cells. *Environ. Sci. Technol.* **2017**, *41*, 3341–3346. [CrossRef]
24. Zhang, L.; He, W.; Yang, J.; Sun, J.; Li, H.; Han, B.; Zhao, S.; Shi, Y.; Feng, Y.; Tang, Z.; et al. Bread-derived 3D macroporous carbon foams as high performance free-standing anode in microbial fuel cells. *Biosens. Bioelectron.* **2018**, *122*, 217–223. [CrossRef]
25. Zou, L.; Qiao, Y.; Zhong, C.; Li, C.M. Enabling fast electron transfer through both bacterial outer-membrane redox centers and endogenous electron mediators by polyaniline hybridized large-mesoporous carbon anode for high-performance microbial fuel cells. *Electrochim. Acta* **2017**, *229*, 31–38. [CrossRef]
26. Gu, L.; Zhu, N.; Zhou, P. Preparation of sludge derived magnetic porous carbon and their application in Fenton-like degradation of 1-diazo-2-naphthol-4-sulfonic acid. *Bioresour. Technol.* **2014**, *118*, 638–642. [CrossRef]
27. Yuan, Y.; Yuan, T.; Wang, D.; Tang, J.; Zhou, S. Sewage sludge biochar as an efficient catalyst for oxygen reduction reaction in an microbial fuel cell. *Bioresour. Technol.* **2013**, *144*, 115–120. [CrossRef]

28. Ji, J.; Li, X.-Y.; Xu, J.; Yang, X.-Y.; Meng, H.-s.; Yan, Z.-R. Zn-Fe-rich granular sludge carbon (GSC) for enhanced electrocatalytic removal of bisphenol A (BPA) and Rhodamine B (RhB) in a continuous flow three-dimensional electrode reactor (3DER). *Electrochim. Acta* **2018**, *284*, 587–596. [CrossRef]
29. Feng, H.; Jia, Y.; Shen, D.; Zhou, Y.; Chen, T.; Chen, W.; Ge, Z.; Zheng, S.; Wang, M. The effect of chemical vapor deposition temperature on the performance of binder-free sewage sludge-derived anodes in microbial fuel cells. *Sci. Total Environ.* **2018**, *635*, 45–52. [CrossRef]
30. Li, X.; Hu, M.; Zeng, L.; Xiong, J.; Tang, B.; Hu, Z.; Xing, L.; Huang, Q.; Li, W. Co-modified MoO_2 nanoparticles highly dispersed on N-doped carbon nanorods as anode electrocatalyst of microbial fuel cells. *Biosens. Bioelectron.* **2019**, *145*, 111727. [CrossRef]
31. Jiang, P.Y.; Xiao, Z.H.; Li, S.H.; Luo, Z.N.; Qiu, R.; Wu, H.; Li, N.; Liu, Z.Q. Sulfuration of Fe-N/C porous nanosheets as bifunctional catalyst with remarkable biocompatibility for high-efficient microbial fuel cells. *J. Power Sources* **2021**, *512*, 331–335. [CrossRef]
32. Chen, Q.; Pu, W.; Hou, H.; Hu, J.; Liu, B.; Li, J.; Cheng, K.; Huang, L.; Yuan, X.; Yang, C.; et al. Activated microporous-mesoporous carbon derived from chestnut shell as a sustainable anode material for high performance microbial fuel cells. *Bioresour. Technol.* **2018**, *249*, 567–573. [CrossRef] [PubMed]
33. Xia, J.; Geng, Y.; Huang, S.; Chen, D.; Li, N.; Xu, Q.; Li, H.; He, J.; Lu, J. High-performance anode material based on S and N co-doped graphene/iron carbide nanocomposite for microbial fuel cells. *J. Power Sources* **2021**, *512*, 115–120. [CrossRef]
34. Gu, Y.; Srikanth, V.; Salazar-Morales, A.I.; Jain, R.; O'Brien, J.P.; Yi, S.M.; Soni, R.K.; Samatey, F.A.; Yalcin, S.E.; Malvankar, N.S. Structure of Geobacter pili reveals secretory rather than nanowire behavior. *Nature* **2021**, *597*, 430–434. [CrossRef] [PubMed]
35. Lovley, D.R. Bug juice: Harvesting electricity with microorganisms. *Nat. Rev. Microbiol.* **2006**, *4*, 497–508. [CrossRef]
36. Lovley, D.R. Electromicrobiology. *Annu. Rev. Microbiol.* **2012**, *66*, 391–409. [CrossRef]
37. Lim, J.W.; Park, T.; Tong, Y.W.; Yu, Z. The microbiome driving anaerobic digestion and microbial analysis. *Adv. Bioenergy* **2020**, *5*, 1–61.
38. Morita, M.; Malvankar, N.S.; Franks, A.E.; Summers, Z.M.; Giloteaux, L.; Rotaru, A.E.; Rotaru, C.; Lovley, D.R. Potential for Direct Interspecies Electron Transfer in Methanogenic Wastewater Digester Aggregates. *mBio* **2011**, *2*, e00159-11. [CrossRef]
39. Daghio, M.; Gandolfi, I.; Bestetti, G.; Franzetti, A.; Guerrini, E.; Cristiani, P. Anodic and cathodic microbial communities in single chamber microbial fuel cells. *New Biotechnol.* **2015**, *32*, 79–84. [CrossRef]
40. Logan, B.E.; Rossi, R.; Ragab, A.; Saikaly, P.E. Electroactive microorganisms in bioelectrochemical systems. *Nat. Rev. Microbiol.* **2019**, *17*, 307–319. [CrossRef]
41. Ding, J.; Lu, Y.Z.; Fu, L.; Ding, Z.W.; Mu, Y.; Cheng, S.H.; Zeng, R.J. Decoupling of DAMO archaea from DAMO bacteria in a methane driven microbial fuel cell. *Water Res.* **2017**, *110*, 112–119. [CrossRef]
42. Xiao, Y.; Zheng, Y.; Wu, S.; Zhang, E.H.; Chen, Z.; Liang, P.; Huang, X.; Yang, Z.H.; Ng, I.S.; Chen, B.Y.; et al. Pyrosequencing Reveals a Core Community of Anodic Bacterial Biofilms in Bioelectrochemical Systems from China. *Front. Microbiol.* **2015**, *6*, 1410. [CrossRef]
43. Deng, Q.J.; Su, C.Y.; Lu, X.Y.; Chen, W.Y.; Guan, X.; Chen, S.L.; Chen, M.L. Performance and functional microbial communities of denitrification process of a novel MFC-granular sludge coupling system. *Bioresour. Technol.* **2020**, *306*, 123173. [CrossRef]

membranes

Article

Influence of Degassing Treatment on the Ink Properties and Performance of Proton Exchange Membrane Fuel Cells

Pengcheng Liu, Daijun Yang *, Bing Li, Cunman Zhang and Pingwen Ming

Clean Energy Automotive Engineering Center, School of Automotive Studies, Tongji University, Shanghai 201804, China; 2011683@tongji.edu.cn (P.L.); libing210@tongji.edu.cn (B.L.); zhangcunman@tongji.edu.cn (C.Z.); pwming@tongji.edu.cn (P.M.)
* Correspondence: yangdaijun@tongji.edu.cn

Abstract: Degradation occurs in catalyst inks because of the catalytic oxidation of the solvent. Identification of the generation process of impurities and their effects on the properties of HSC ink and LSC ink is crucial in mitigating them. In this study, gas chromatography-mass spectrometry (GC-MS) and cyclic voltammetry (CV) showed that oxidation of NPA and EA was the primary cause of impurities such as acetic acid, aldehyde, propionic acid, propanal, 1,1-dipropoxypropane, and propyl propionate. After the degassing treatment, the degradation of the HSC ink was suppressed, and the concentrations of acetic acid, propionic acid, and propyl propionate plummeted from 0.0898 wt.%, 0.00224 wt.%, and 0.00046 wt.% to 0.0025 wt.%, 0.0126 wt.%, and 0.0003 wt.%, respectively. The smaller particle size and higher zeta potential in the degassed HSC ink indicated the higher utilization of Pt, thus leading to optimized mass transfer in the catalyst layer (CL) during working conditions. The electrochemical performance test result shows that the MEA fabricated from the degassed HSC ink had a peak power density of 0.84 W cm^{-2}, which was 0.21 W cm^{-2} higher than that fabricated from the normal HSC ink. However, the introduction of propionic acid in the LSC ink caused the Marangoni flux to inhibit the coffee ring effect and promote the uniform deposition of the catalyst. The RDE tests indicated that the electrode deposited from the LSC ink with propionic acid possessed a mass activity of 84.4 mA·mg_{Pt}^{-1}, which was higher than the 60.5 mA·mg_{Pt}^{-1} of the electrode deposited from the normal LSC ink.

Keywords: catalyst ink; PEMFC; rheology; catalyst layer; impurity

Citation: Liu, P.; Yang, D.; Li, B.; Zhang, C.; Ming, P. Influence of Degassing Treatment on the Ink Properties and Performance of Proton Exchange Membrane Fuel Cells. *Membranes* **2022**, *12*, 541. https://doi.org/10.3390/membranes12050541

Academic Editors: Marc Cretin, Sophie Tingry and Zhenghua Tang

Received: 24 March 2022
Accepted: 9 May 2022
Published: 22 May 2022

1. Introduction

Proton exchange membrane fuel cells (PEMFCs) have received significant research attention in recent decades, due to their high efficiencies, low operation temperature, and zero emissions [1–4]. Membrane electrode assemblies (MEAs), which comprise a proton exchange membrane (PEM), cathode and anode catalyst layers (CLs), microporous layers (MPLs), and gas diffusion layers (GDLs), are considered to be the heart of PEMFCs [5]. The complete working principle of an MEA consists of the following process: the oxidation reaction of H_2 at the anode catalyst layer (ACL) provides electrons to an external circuit and releases protons to the internal electrolyte, while the reduction reaction of O_2 at the cathode catalyst layer (CCL) receives electrons (from the external load) and protons (from the internal electrolyte). Both the CCL and ACL of an MEA are critical components of the system, because they represent energy conversion sites, where charge and mass transfer and the electrochemical reaction occur cointantaneously [6]. The cost, performance, and durability of PEMFCs are closely dependent on the structure and morphology of CLs, which face several challenges, such as the coupling effects of corrosion in a strong acid environment, humidity stress, thermal shock stress, and mechanical stress during the service period [7]. Therefore, the optimization of the CL microstructure is a considerably critical issue to ensure a high performance of PEMFCs.

Understanding and optimizing the preparation process of MEAs are imperative to obtaining CLs with the perfect microstructure for the most effective PEMFCs [8,9]. Generally, the process of preparing CLs involves the following procedures: (i) dispersion of the catalyst (such as Pt-loaded carbon) and the proton-conductive ionomer, which also works as a binder in a dispersion medium (such as water/NPA/isopropanol), in the CLs; (ii) coating of the catalyst ink on the PEM or GDLs; and (iii) drying to evaporate the dispersion medium [6,10]. The effects of the catalyst ink quality and process control exist throughout the above processes, which determine the CL microstructure and therefore the characteristics of the fuel cells [11]. Previous works have focused on the construction and optimization of the CL microstructure based on the coating process and ink formulation [5,12]. Coating and drying parameters influence the distribution of materials and pores in CLs and also have significant impacts on performance. Commonly, the ink formulation, including the alcohol content and type, ionomer content, and Pt dispersion, also affects the ink initial properties such as rheology, stability, and coatability, thereby exerting an influence on the fabrication of CLs [13–16]. However, the degradation of the catalyst ink quality after preparation also affects the catalyst ink viscosity, the size of agglomerates, which are a mixture of the catalyst and ionomer, the quality of the coated catalyst layer, and thus the performance of the fuel cells. Therefore, it is crucially necessary to understand the degradation process of catalyst inks and its impacts on the storage and rheologic properties, as well as well-constructed CLs.

Based on extensive studies on CLs, many researches have demonstrated that the microstructure of CLs was closely realated to the catalyst properties, which was dependent on the size of agglomerates [11]. Catalyst particles are generally found to be agglomerated, forming primary agglomerates with a particle size of 200–300 nm under the effect of van der Walls attractive force. Further agglomeration of primary aggregates happens to generate secondary agglomerates on the microscale. Additionally, the addition of an ionomer can reduce the size of agglomerates due to the electrostatic repulsion and steric hindrance interactions. The larger agglomerates become, the greater negative effect they will have on the construction of CLs, leading to a reduced output performance. Therefore, the state of the catalyst ink should be controlled at a uniform and stable condition to fabricate high-quality MEAs.

The change in composition of the catalyst ink will lead to a change in its properties, especially the generation of impurities. To date, only a handful of studies have described the generation of impurities in catalyst inks and their effects on the processability of catalyst inks [14,17,18]. For instance, some previous works have demonstrated the effects of various impurities, such as acids and aldehydes, on the agglomerate behavior of inks and ultimately on the final structure of CLs [14,17–20]. Uemura [14,18,19] used X-ray computed tomography to detect the presence of air bubbles and the third phase in a catalyst ink and proved the catalyst caused alcohol to decompose [19]. Kameya combined nuclear magnetic resonance (NMR) with magnetic resonance imaging (MRI) to analyze the internal state of an ink during the preparation process and detected the presence of air bubbles in the ink during the main mixing process [21]. In addition, ^{19}F NMR spectra revealed dramatic changes in the dispersion states of Nafion during the mixing period. Other previous studies targeting NPA oxidation on platinum electrodes in acid solutions have demonstrated that NPA is converted to propionic acid, whereas isopropanol is highly selectively converted to acetone, due to the difficulty in breaking the C−C bond [22,23]. Catalyzed oxidation of the dispersion medium and the deuterogenic reaction affect the state of the catalyst ink. These generated impurities induce the generation of larger agglomerates in the catalyst ink, and thus cracking of the CLs because of the capillary stress. Kumano [13] identified the structural parameters that control the dispersion state and stability of Pt/C agglomerates. In inks containing 48–75% of water, the amount of adsorbed ionomers decreased with decreasing water content, resulting in increases in the viscosity, storage modulus, and electrical conductivity. The adsorption rate of the ionomer into the Pt/C decreased, and the average size of agglomerates, viscosity, and storage modulus increased with the increase in the hydrophobicity of the solvent. The impurity produced in inks undoubtedly changes the hydrophobicity of the solvent and thus affects the properties of the ink. Hence, it is important to obtain an understanding of the generation process of impurities and their effects on processability.

HSC ink is often used in the coating procedure across industrial applications, whereas LSC ink is applied in the spraying procedure and rotating disk electrode (RDE) tests in laboratories [13,24]. High-quality RDE measurements need a thin, uniform film over the entire surface area of the glassy carbon to accurately evaluate the electroactivity of the catalyst [24]. The quality of the working electrode is delicately determined by the drying conditions, alcohol content and type, Pt dispersion, and surface state of the glassy carbon. Therefore, the effects of impurities on the properties of LSC ink require a detailed investigation, owing to their effect on the electrode structure, and the lack of clarity regarding their underlying mechanism of action. At present, there is limited cognition of the formation mechanism of impurities and their effects on the catalyst ink rheology and drying behavior. To fully explore the generation process of impurities, we first investigated the oxidation of the solvent and the effects of the temperature and atmosphere on catalyst ink degradation. In order to understand how impurities affect the rheology and drying process of catalyst inks, we present a comparative study of HSC ink before and after degassing treatment. The range of comparison includes rheological behavior and the property (ink)–structure (catalyst layer)–performance (MEA) relationship.The influence of impurities on the properties of LSC ink cannot be ignored because the LSC ink is widely used in the RDE experiments to evaluate the characteristics of catalysts. The effects of the introduction of propionic acid in the LSC ink on the microscale structure of the RDE were investigated using optimal microscopy, cyclic voltammetry (CV), and line sweep voltammetry (LSV). Based on the understanding of the impurity evolution process and the relationship between the impurity properties and ink quality for the two ink types, we provide insight into optimizing the preparation of catalyst inks to obtain excellent processability and coatability. This also helps in the construction of the desired catalyst layers.

2. Experimental Procedures

2.1. Preparation of Catalyst Ink

This section describes the procedures and instrumentation for the preparation of the HSC ink. Briefly, 5.7 g of Pt/C powder (Johnson Matthey, Alfa Aesar, Shanghai, China, Vulcan XC-72 with 60 wt% Pt and 20.5 g of Nafion® solution (Dupont™, New Castle, DE, USA, Nafion® PFSA polymer dispersions D-520) were mixed with 27 g of the NPA and ultrapure water (the ratio of NPA to ultrapure water was 1) dispersion medium. The process of fabricating the HSC catalyst ink followed the steps shown in Figure 1. Firstly, 5.7 g of catalyst powder was added to 13.5 g of water and stirred with a glass rod, and then 13.5 g of NPA and 20.5 g of Nafion solution were added successively (premixed process), followed by ultrasonic dispersion (35 kHz, 5 min, 15 °C). Secondly, the catalyst ink was homogenized by high-speed shear (1600 rpm, 30 min) and finally degassed using a magnetic stirrer at 30 rpm for 30 min at −0.1 MPa. The ink and raw material for the above process were contained in glass containers. The catalyst ink that was not degassed was measured after being kept for 24 h and marked as "I-ink", whereas the degassed ink was denoted "D-ink" and left to stand for at least 24 h before subsequent measurement. After preparation of the two catalyst inks, the container was filled with nitrogen as the protective gas.

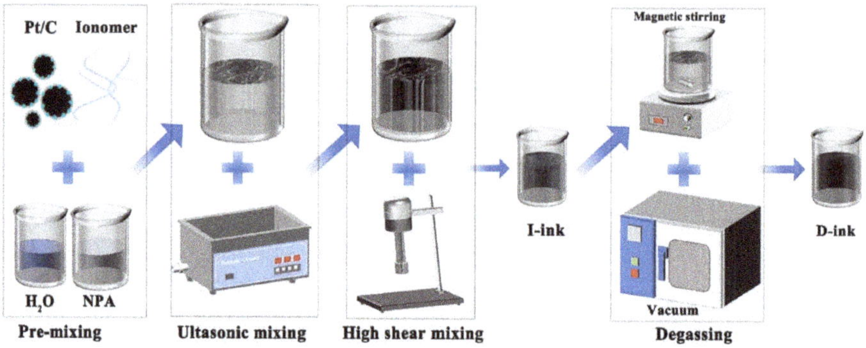

Figure 1. Schematic representation of the process of HSC ink preparation.

The LSC ink for oxygen reduction reaction (ORR) catalyzed activity was evaluated using a rotating disk electrode (RDE). Briefly, 2 mg of catalyst powder was added to a 1 mL mixture of 5 wt% Nafion® solution and NPA (volume ratio of 1:30), and the mixture was fully mixed by ultrasonication for 40 min (35 kHz, 15 °C). This LSC ink was denoted "N-ink". To characterize the effect of impurities on the LSC ink, we added 10 μL of propionic acid (playing the role of impurities) to the N-ink to produce a comparison ink known as "P-ink". A summary of the compositions of the HSC ink and LSC ink is shown in Table 1.

Table 1. Sample composition of catalyst inks.

Ink Type	Pt/C (g)	Deionized Water (g)	1-Propanol (g)	Nafion® Solution (g)	Solid Content (%)
HSC ink	5.7	14.082	13.705	20.517	12.45
LSC ink	0.002	0.000	0.967	0.0323	0.20

2.2. Electrochemical Evaluation

The HSC catalyst ink was directly coated on the proton exchange membrane (Gore, Newark, DE, USA, thickness of 18 μm) using a slot-die coating system. The slot die moved at a horizontal velocity of 10 mm/s above the proton exchange membrane with a coating gap of 100 μm. The baseplate temperature was maintained at 60 °C to remove the solvent from the wet film. The MEA was assembled by sandwiching a catalyst coating membrane between two pieces of gas diffusion layer (Freundenberg, Shanghai, China, H24CX483). The Pt loadings were controlled to be 0.4 $mg \cdot cm^{-2}$ and 0.2 $mg \cdot cm^{-2}$ in the cathode and anode, respectively. To evaluate the fuel cell performance, the polarization curve was measured with a 25 cm^2 three-serpentine cell fixture and tested with a fuel cell test system (Dalian New Sunrise Testing Technology Co., Ltd., Dalian, China, NSR-FTCS100B-1802-3). The temperature of cell was controlled at 75 °C. The stoichiometry of H_2/air was 1.5/2.5. The inlet gauge pressures of anode and cathode were maintained at 100 and 80 kPa. The relative humidity at the anode and cathode sides were both 55%. The electrochemical impedance spectra of single cells were recorded at 1.6 $A \cdot cm^{-2}$ using a scanning frequency from 10^3 to 0.1 Hz. A detailed electrochemical analysis of the oxidation behavior of NPA and EA under an acid environment was performed using a three-electrode cell system (PINE, AFCPRBE). This system comprises a thin film of catalyst ink deposited on a glassy carbon substrate as the working electrode (WE), a reversible hydrogen electrode (RHE) as the reference electrode, and a platinum sheet as the counter electrode. The electrolyte solution was de-aerated before each measurement with N_2 and O_2 for 30 min, and all electrochemical measurements were performed using an electrochemical workstation (CHI. Instrument company, Shanghai, China, 760E). The working electrode was prepared by transferring 10 μL of ink into the RDE (S = 0.196 cm^2), followed by natural drying. CV and LSV were carried out in 0.1 M $HClO_4$ at 25 °C. CV data were recorded at a potential range from 0.05 to 1.10 V, at a scanning rate of 0.05 $V \cdot s^{-1}$, in a N_2-saturated electrolyte solution. ORR polarization curves were obtained in an O_2-saturated electrolyte solution at a scanning rate of 0.005 $V \cdot s^{-1}$ and an RDE rotation rate of 1600 rpm. The electrochemical surface area (ECSA) of the WE was calculated based on the CV curves, using the following equation [25]:

$$ECSA = \frac{Q_H}{210 \times m_{Pt}} \tag{1}$$

where Q_H (mC) is the charge of hydrogen species' electro-adsorption peak; the value of 210 $\mu C \cdot cm^{-2}$ corresponds to monolayer adsorption of hydrogen atoms on a polycrystalline Pt; and m_{Pt} represents the mass load of Pt on the working electrode. ORR's catalytic activity, which is the kinetic current density at 0.9 V (vs. the RHE) from the LSV curve, was calculated based on the Koutecky–Levich (K–L) equation as follows [26]:

$$\frac{1}{i} = \frac{1}{i_k} + \frac{1}{i_d} \tag{2}$$

where i is the current density value measured at E = 0.9 V; i_d is the diffusion-limited current density at E = 0.4 V (vs. the RHE); and i_k represents the kinetic current. The specific mass activity (MA) of the catalyst is the kinetic current per unit mass loading of Pt [24,27].

2.3. GC-MS Instrumentation

Impurities in the catalyst ink were analyzed by GC-MS (Agilent, Shanghai, China, 7890B-5977B), which can detect various volatile components in a solution. Patterns of the mass spectra were analyzed using NIST-2008.

2.4. Rheological Measurements

The rheological property of the inks was measured using a stress-controlled rheometer (Anton Paar, Shanghai, China, MCR302), with a coaxial cylinder mold. Prior to measurement, the ink was kept quiescent at 25 °C for 5 min to remove any previous disequilibrium status and ensure that the constituent material established new equilibrium-status structures. A pre-shear treatment was first used to eliminate the shear history and ensure the repeatability of the test data. During this operation, the shear rate was controlled at 0.01 s^{-1} for 100 s. Thereafter, steady status flow measurements were carried out by step-wisely increasing the shear rate from 0.01 to 1000 s^{-1}, to test the viscosity function of the formulated inks. Three interval thixotropy tests were used to determine the structural regeneration of the HSC ink, and a typical step test with three intervals depicted as a time-dependent viscosity function was as follows: (1) the shear rate was kept at 0.1 s^{-1} for 60 s, at the beginning, to simulate the ink at rest; (2) the shear rate was maintained at 100 s^{-1} for 10 s to simulate the structural breakdown of the ink; (3) the rate was kept at 0.1 s^{-1} for 60 s to simulate the structural recovery of the ink at rest. Furthermore, the strain dependency of the storage modulus (G′) and loss modulus (G″) was applied to change the strain from 0.01 to 100% at 1 Hz after tests of the steady flow viscosity. All rheology experiments were performed at 25 ± 0.1 °C.

2.5. Measurement of Ink Cluster Size and Zeta Potential

The cluster size and zeta potential measurements of HSC were performed with dynamic light scattering (DLS) (Colloid Metrix, Shanghai, China, Nano-fiex) and a particle potential titrator (Colloid Metrix, Shanghai, China, Stabino), respectively. For testing purposes, 0.1 mL of the inks was diluted using 100 mL of a solution with the original solvent composition. The diluted inks were dispersed in an ultrasonic bath for 2 min prior to the DLS and zeta potential measurements.

2.6. Determination of Contact Angle and Deposition of LSC Ink

The morphology of the LSC ink (5 μL) drying on the RDE was measured using a digital microsystem (KEYENCE, Osaka, Japan, VH-S30B). The contact angle between the catalyst ink and the glass was determined, using the side view of the microscope, and the deposition process of the catalyst ink droplets was observed from the top view.

3. Results and Discussion

The total ion current (TIC) spectrum of I-ink, based on GC-MS detection, and the spectra of D-ink are shown in Figures 2 and S1; the signals of impurities classified by their corresponding mass spectra are illustrated in Figure 3; and the results of the quantitative analysis are listed in Table 2.

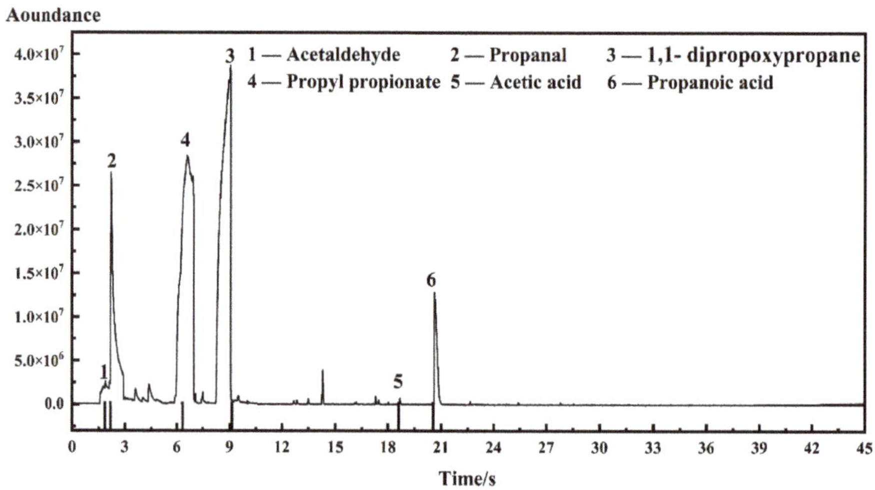

Figure 2. Total ion current (TIC) spectrum of I-ink.

The impurities in the ink were indexed as acetaldehyde, propanal, 1,1-dipropoxypropane, propyl propionate, acetic acid, and propanoic acid. The corresponding regions of the TIC spectrum were detected after 1.865, 2.194, 9.136, 6.338, 18.620, and 20.562 s. The presence of acetaldehyde was confirmed by the fragment ions (m/z) at 29, 43, 44, and 15 (Figure 3), whereas the propanal segments were ionized into m/z of 26, 27, 28, 29, 57, 58, and 59. Other impurities were identified through additional analysis of the mass spectra, such as propyl propionate, 1,1-dipropoxypropane, acetic acid, and propanoic acid. Interestingly, these impurities were found to simultaneously occur in I-ink and D-ink, albeit with significant differences in concentrations. Therefore, the degassing treatment efficiently suppressed the oxidation process of the solvent. Previous studies have demonstrated that Pt metal can catalyze solvents to produce complex oxidizing compounds, including acetaldehyde, propanal, acetic acid, and propanoic acid [22,28,29]. These catalytic products are then condensed to form esters. We observed significant differences in impurity concentrations between I-ink and D-ink. It is worth noting that the oxidation products exhibited more hydrophobic characteristics than the original solvent composition. Nafion®, a binder and stabilizer in catalyst inks, is essential for optimization of the properties of catalyst inks [30]. In fact, its hydrophobic backbone is attracted to the hydrophobic surface of the carbon support, whereas its hydrophilic sidechains are ionized to generate numerous ionic charges on the surfaces of the carbon support [31]. However, the presence of hydrophobic impurities improved the compatibility between the ionomer and solvent, thereby increasing the amount of free ionomer on the solvent. Consequently, this affected the interaction between the internal components and rheology of the catalyst ink [16,32]. Next, we investigated the effects of the atmosphere and temperature on the impurity generation process.

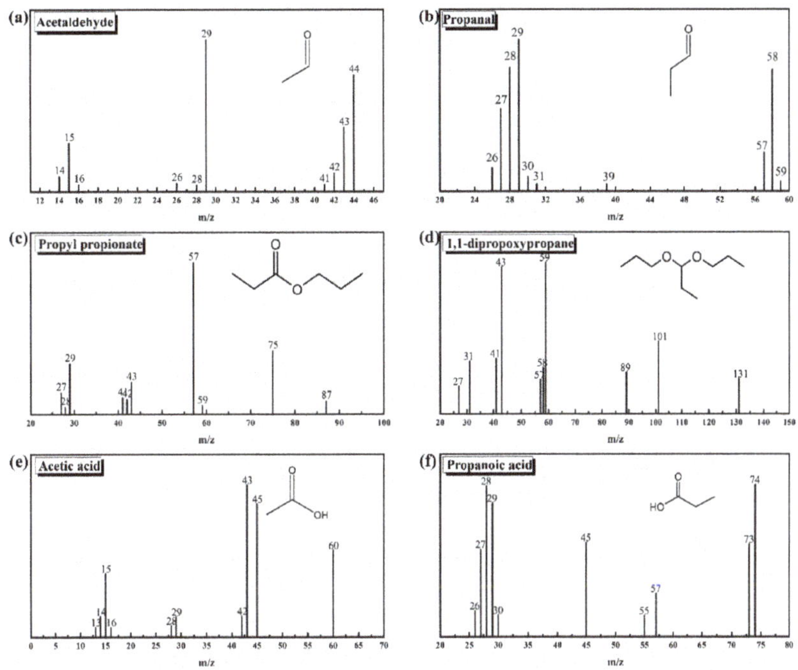

Figure 3. Mass spectra of impurities derived from electron impact ionization of I-ink: (**a**) acetaldehyde; (**b**) propanal; (**c**) propyl propionate; (**d**) 1,1-dipropoxypropane; (**e**) acetic acid; (**f**) propanoic acid.

Table 2. Summary of impurity concentrations of the catalyst inks derived from GC-MS data.

Sample	Acetic Acid	Propionic Acid	Propyl Propionate
I-ink	0.0898 wt.%	0.0224 wt.%	0.00046 wt.%
D-ink	0.0025 wt.%	0.0126 wt.%	0.0003 wt.%

The CV of Pt/C electrodes in 0.05 M H_2SO_4 + 0.1 M NPA solution at 0.002 V·s^{-1} shows the electro-oxidation process of NPA (Figure 4a). The first oxidation peak was detected at 0.90V vs. the RHE in the O_2-purged solution, in the positive scanning process. In contrast, this peak was found at 0.95 V vs. the RHE in the N_2-purged solution, due to the overpotential required to overcome the concentration polarization caused by the lack of O_2. Both oxidation peaks corresponded to the poisonous intermediate formation in the NPA oxidation reaction [28]. However, an increase in the potential generated the second oxidation peak at 1.29 V in the O_2-purged solution, whereas a similar peak was observed at 1.35 V in the N_2-purged solution. This oxidation peak indicated the formation of reaction intermediates during NPA oxidation. Furthermore, several higher peaks' current densities were recorded in the O_2-purged solution relative to the N_2-purged solution, because sufficient electro-oxidation of NPA produces a larger reaction current. A similar phenomenon was observed during the oxidation of alcohol (Figure 4b). The oxidation peaks' current densities of NPA and EA are summarized in Table 3. The high impurity concentration of I-ink derived from quantitative analysis of GC-MS also supported this phenomenon (Table 2). As previously mentioned, the CVs of NPA and EA in the O_2 atmosphere exhibited a significantly higher oxidation overpotential and smaller oxidation current density, suggesting that the anoxic environment can reduce the intensity of solvent oxidation reactions.

The effects of the temperature on the electro-oxidation of NPA and EA are discussed in the Supporting Information, and the CVs of NPA and EA under various temperatures are shown in Figure S2. Summarily, high temperatures promoted the solvent's oxidation behavior, suggesting the need to regulate the N_2 atmosphere and control the temperature for an effective reduction in solvent electro-oxidation during the ink preparation process. A summary of the mechanism underlying the formation of impurities in the catalyst inks is shown in Figure 5. Briefly, EA and NPA can be oxidized to their respective aldehydes and acids, in the presence of platinum catalysis; meanwhile, propyl propionate is generated by esterification of propionic acid and propanol. Specifically, the aldolization reaction of NPA and propanal, via Pt catalysis, is the cause of 1,1-dipropoxypropane [14].

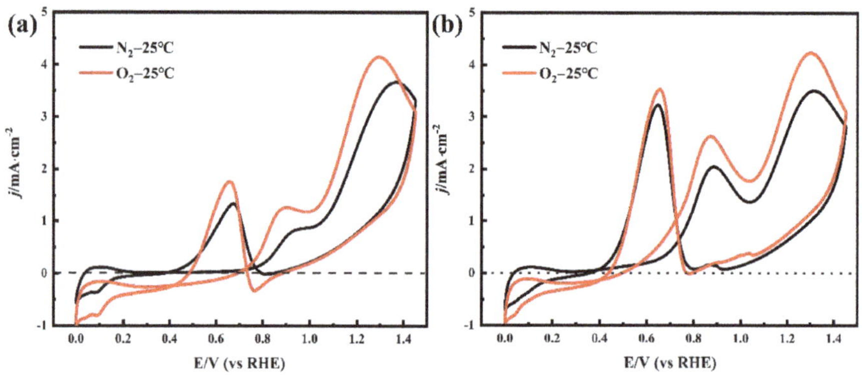

Figure 4. (**a**) Cyclic voltammograms recorded with a Pt/C electrode in 0.05 M H_2SO_4 + 0.1 M NPA solution and (**b**) 0.1 M EA solution, with a scan rate of 20 mV·s^{-1} and a temperature of 25 °C.

Table 3. Current densities of oxidation peaks recorded from the CV of NPA and EA.

Condition	NPA's Oxidation Current Density/mA cm^{-2}			EA's Oxidation Current Density/mA cm^{-2}		
	Peak I	Peak II	Peak III	Peak I	Peak II	Peak III
N_2-25 °C	0.855	3.66	1.338	2.046	3.506	3.231
O_2-25 °C	1.266	4.148	1.758	2.634	4.241	3.532

Rheological characterization of the catalyst ink is an essential index for each step during MEA fabrication. The catalyst ink is taken as the working fluid in a slot die, and its viscosity is perceived as the most crucial rheological property during the coating procedure [33,34]. It directly influences the behavior of the ink formulation during mixing and production of the wet catalyst layer [35]. The relationship between viscosity and shear stress is shown in Figure 6a. Obviously, shear thinning behavior occurred in D-ink and I-ink, which means the viscosity was negatively correlated with the shear rate. Catalyst inks are multi-component, complex solid–liquid mixtures that consist of a catalyst, ionomer, and solvent medium. The dynamic viscosity at low shear rate stages is an index of the settling degree of the solid content, while that at high shear rate stages is an index of the coating processability [36]. Both inks showed a high viscosity at a low shear rate stage, which is preferred owing to the lack of significance in the settlement of the solid content. Notably, at a higher shear rate stage, the strong shear rate force tended to destroy the microstructure of the catalyst inks, which subsequently realigned the internal structure and significantly reduced the viscosity. This behavior means that D-ink and I-ink are non-Newtonian fluids, a property that is quite suitable for the actual production process. At the ink storage stage, the particles in the ink were subjected to external forces, including gravity and shear forces. The shear rate ranged from 10^{-6} to 10^{-2} s^{-1}. The high viscosity indicates the excellent anti-sedimentation properties of these inks. During coating, the fluid with a high shear rate requires a low viscosity. After coating, the advection of ink occurs on the proton exchange membrane under surface tension and the action of gravity. The viscosity of I-ink was slightly higher than that of D-ink in most of the shear rate scope, but with the shear rate increasing, the gap between the viscosities of the inks gradually narrowed. This implies that the network structure of the catalyst and binder grew after degassing, and dispersion states in the ink were changed. The rising hydrophobicity of the solvent caused an increase in free ionomers, which subsequently increased the viscosity [37,38]. The difference in the network structure's strength between these two inks was further evidenced by the hysteresis flow curves, as shown in Figure 6b,c. Notably, hysteresis phenomena, where different shear stress values appeared in the positive scan and negative scan, were observed in the inks. Moreover, any destabilization of this steady state would destroy the ionomer structure, owing to entanglements of ionomer chains and fluctuations in the arrangement of the catalyst particles brought about by changes in shear and relaxation processes [39,40]. The degassing treatment improved the adsorption of the ionomer into the catalyst, due to the removal of the microbubbles in the aggregates.

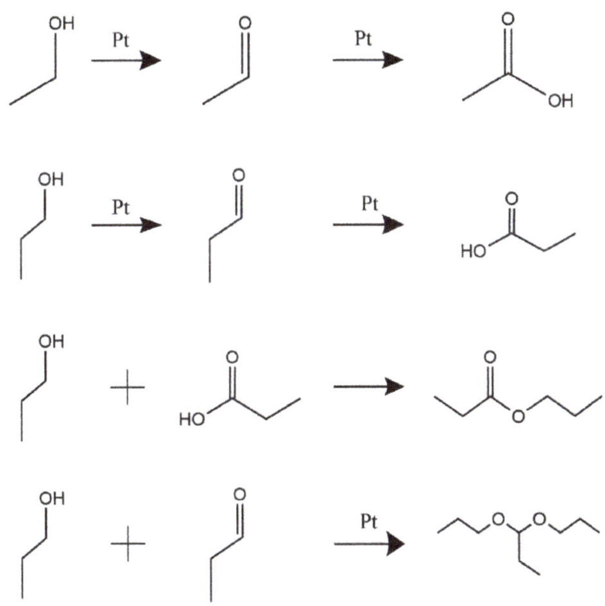

Figure 5. A schematic illustration of the proposed NPA and EA conversion process.

Furthermore, the continuous increase in bridging within the catalyst, yield stress, level of shear thinning, and equilibrium G′ within the inks' linear viscoelastic regime were all strengthened, and both types of inks exhibited a shear stress plateau at shear rates from 1 s^{-1} to 10 s^{-1}, indicating the existence of yield stress in both inks (Figure 6b,c). We calculated the numerical value of the yield stress by averaging the initial five points in this stress plateau and found a higher value in I-ink (2.5 Pa) than in D-ink (1.7 Pa). The decrease in the yield stress of the inks contributed to their self-leveling, which suppressed the uneven thickness distribution in the catalyst layer [5,41].

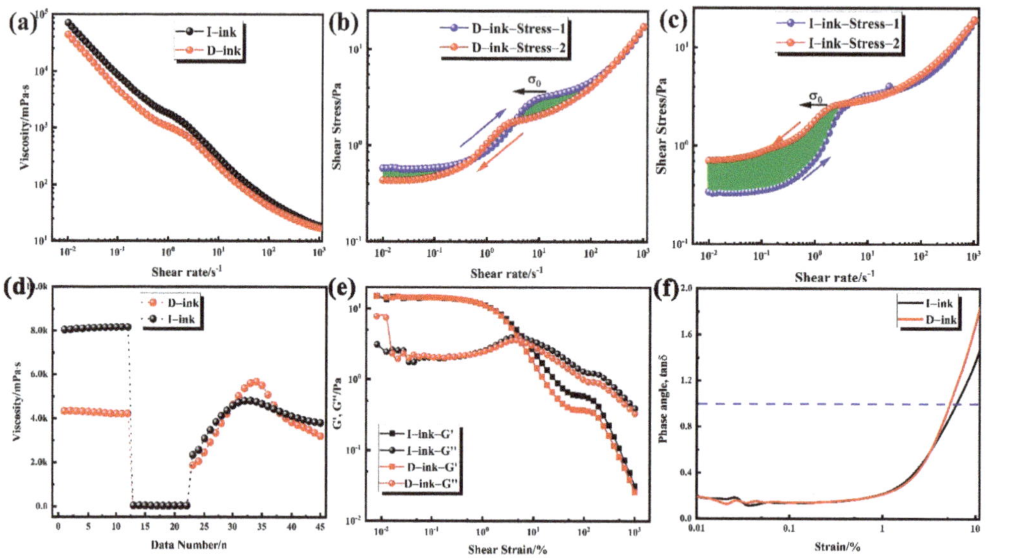

Figure 6. (**a**) Dynamic viscosity data of tested catalyst inks; (**b**,**c**) shear stress as a function of the shear rate for I-ink and D-ink; (**d**) three interval thixotropy test of catalyst inks, and amplitude oscillation test of the inks; (**e**,**f**) strain-dependent storage modulus (G′) and loss modulus (G″) at 1 Hz and phase angle.

According to the determined yield stress, the level of shear thinning of the inks can be quantified by implementing the Herschel–Bulkley model [42,43]:

$$\sigma = \sigma_0 + K\gamma^n \quad (3)$$

where σ and σ_0 represent the measured shear stress and yield stress measured at a specific shear rate (Pa), respectively; K denotes the consistency index (Pa·s^n); γ is the shear rate (s^{-1}); and n is the dimensionless flow index. Only information obtained from a shear rate above 2 s^{-1} was considered in the modeling procedure, because this range of the shear rate matches the actual coating process. The consistency index indicates the degree of viscous contribution during the increase in the shear rate, whereas the function of shear stress and the shear rate of the catalyst inks were presented by the flow index [44]. As shown in Table 4, the results show that I-ink had a significantly higher consistency index than D-ink because I-ink exhibited a significantly higher level of shear thinning and was more viscous than D-ink. All of the inks had $n < 1$, a phenomenon that corresponds to the shear thinning behavior as illustrated in Figure 6a. Notably, a small dimensionless flow index resulted in stronger shear thinning behavior [42], and the gap in viscosity between D-ink and I-ink almost disappeared when the shear rates were increased to about 100 s^{-1}. On the other hand, an increase in the low-shear viscosity (LSV) resulted in a coating layer with a sharper edge, implying less cut-off waste during subsequent processing [45].

Table 4. Calculated Herschel–Bulkley parameters for the catalyst inks.

Ink	Yield Stress, σ_0 (Pa)	Consistency Index, κ(Pa·s^n)	Flow Index, n	R^2
I-ink	2.49505	0.08131	0.76772	0.99982
D-ink	1.6832	0.05865	0.80801	0.99971

Catalyst ink coating entails a high-shear-rate process, while self-leveling of the ink onto the PEM is a low-shear-rate process. The essential requirements for this structural regeneration process include: (1) applying a slow reconstruction rate for good leveling; and (2) ensuring the rate is not too slow to prevent sagging and to allow a sufficient wet layer thickness and flatness. To investigate this time-dependent behavior, we performed a rotational test with three intervals and present the result as a time-dependent viscosity function (Figure 6d). At the first stage, a very low shear rate ($0.1\ s^{-1}$) was used to simulate behavior at rest, and as the hydrophobicity of the dispersion solvent increased, the viscosity of D-ink became lower than that of I-ink. This difference in viscosity resulted from the increase in the free ionomer and generated aggregation, which was related to the change in the rate of adsorption of the ionomer into the catalyst [13,38,46]. An increase in the shear rate to $100\ s^{-1}$ (stage 2) caused the strong shear to simulate the structural breakdown of the catalyst inks during the coating process [47]. Moreover, both ink types exhibited very low viscosity due to the shear thinning behavior. At the final stage, the low shear rate simulated structural regeneration for the ink self-leveling process, although the structural strength and viscosity of the inks gradually recovered with time. The thixotropic recovery rates of D-ink and I-ink were 75.8 and 46.6%, respectively. We hypothesized that the high ionomer adsorption on the catalyst's surface strengthened the interaction between the ionomer and the catalyst, and this behavior was also observed in the lithium-ion battery field [47,48].

Next, we used oscillatory shear to investigate the inks' microstructure, and amplitude sweep to characterize the inks' linear viscoelastic regime (LVR). Subsequently, we applied the LVR to accurately measure the breakdown of the network structure and acquire the structural strength of the initial state [42]. The results reveal lower G'' values for both inks at the low-strain region compared to G', indicating an elastic-dominant response property of the inks (Figure 6e) [42]. At a strain range of 5% to 7%, the G' value fell below that of G'', indicating that only a slight increase in the shear strain could promote the inks' shift from elastic-dominant to viscous-dominant [49]. The phase angle δ of the ink, which is calculated using Equation (4) below, is shown in Figure 6f.

$$\delta = \tan^{-1}\left(\frac{G''}{G'}\right) \tag{4}$$

Briefly, the value of δ for D-ink reached 1 at a faster rate, indicating that the gel–sol transformation occurred more easily. For D-ink, the better self-leveling effect, during the drying procedure, resulted in the gradual development of pores in the wet catalyst layer. Therefore, a homogenized cavity structure is beneficial to the reduction in capillary stress during the drying process. In contrast, a wet film with low self-leveling after drying generates a hierarchic pore structure with larger fluctuations, thereby enhancing capillary stress and increasing the risk of CL cracks [46,50–53].

Figure 7a shows the cell polarization curves of the MEAs fabricated from different HSC inks. According to Figure 7a, the MEA prepared from D-ink exhibited excellent improvements compared with that made from I-ink. At the electrochemical polarization control region, the voltage of the two MEAs showed no distinct differences, which can be attributed to the catalyst having the same catalytic intrinsic activity in both MEAs. With the increase in the current density, the gap in output voltages of the different MEAs expanded. Especially under high current densities, mass transfer loss led to a significant performance reduction for the MEA fabricated with I-ink. The electrochemical impedance spectra were recorded to analyze the H_2/air performance and fitting using an equivalent circuit (Figure 7b) [54]. R_Ω denotes the ohmic resistance of the cell. R_{anode} and $R_{cathode}$ are faradaic resistances, which represent the kinetics of the electrochemical reactions occurring on the anode and cathode, respectively. The finite Warburg circuit element (W_{mt}) is used to reflect the mass transport loss on the cathode side. As observed, the MEA performance improved with the decrease in the impedance arc. The fitted values of R_Ω were $0.0043\ \Omega$ and $0.0045\ \Omega$ for D-ink and I-ink, respectively. The impedance spectra consist of semicircles in the high-, medium-, and low-frequency regions, and each of these semicircles corresponds to the resistances of anode activation, cathode activation, and mass transport. The increments in the semicircles in the medium- and low-frequency regions reflect the greater resistance of the activation kinetics and mass transport. The R_{anode} values for D-ink and I-ink were similar, with 0.0211 Ω and 0.0223 Ω, respectively, due to them having the same anode catalyst layer. The MEA fabricated with I-ink exhibited a higher $R_{cathode}$ and R_{mt} than that fabricated with D-ink. The $R_{cathode}$ values decreased from 0.0604 Ω to 0.0353 Ω, and the R_{mt} values varied from 0.0255 Ω to 0.0183 Ω, with the degassing treatment of the ink. Therefore, the cathodic mass transport process and ORR kinetics dominated the H_2/air performance. Jian Xie [12] reported a similar phenomenon where the increase in the NPA ratio in the solvent could intensify the resistance of the ORR kinetics and mass transport limitations.

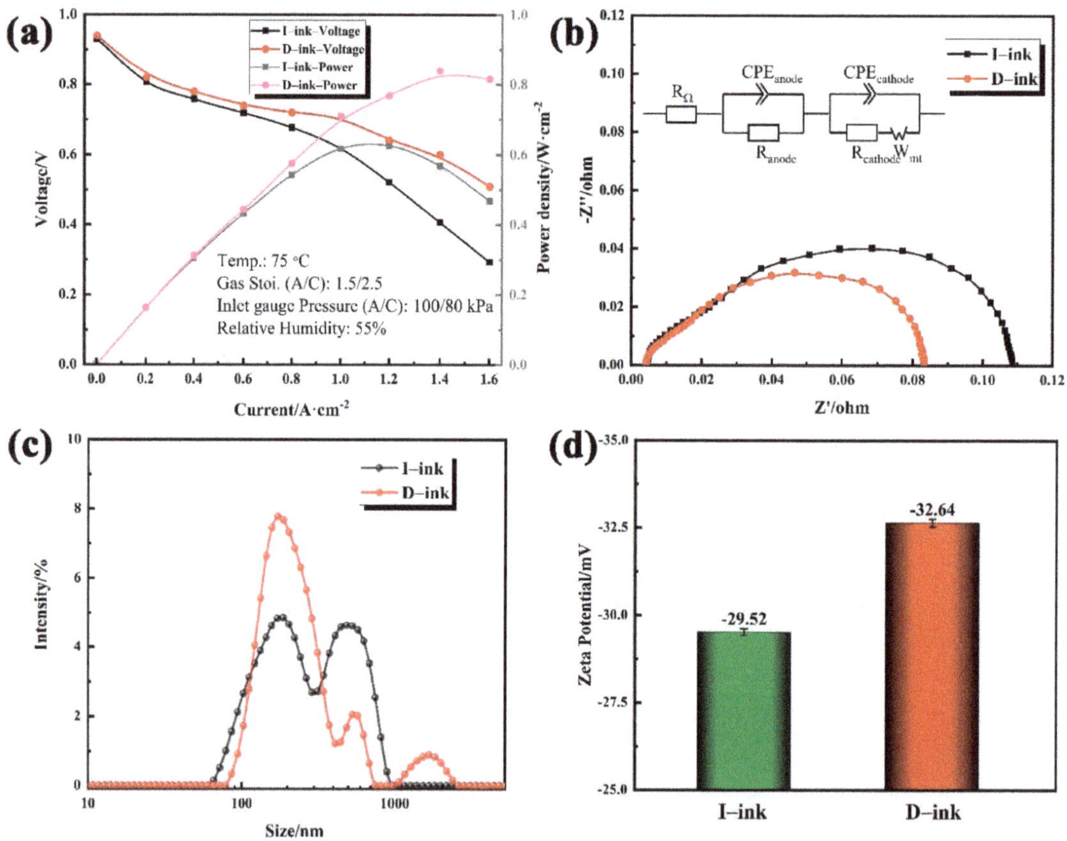

Figure 7. (**a**) H_2/air polarization curves of membrane electrode assemblies fabricated with HSC inks, and (**b**) corresponding Nyquist plots obtained at 1.6 A·cm^{-2} from 0.1 Hz to 1 kHz; (**c**) size distribution and (**d**) zeta potential of I-ink and D-ink.

To better understand how the impurities affect the MEA performance, the catalyst cluster size distributions and zeta potential in different inks, which determine the catalyst/ionomer interface and CL structure, need to be studied. The microscale mass transport in the CLs depends on the aggregate structure and ionomer distribution [55]. As evidently shown in Figure 7c, the intensity signals in I-ink showed a double-peak structure, indicating quite a few of the larger clusters. On the contrary, the intensity signals of D-ink were concentrated in small-size regions. The average diameter of I-ink clusters reached 244.7 nm, while the size of the D-ink clusters was 199.7 nm. The zeta potential results show the stability differences for D-ink (−32.64 mV) and I-ink (−29.52 mV). A hydrophobic impurity causes the ionomer to adsorb into the solvent and desorb from the catalyst. This reduces the adsorption capacity of the ionomer, leads to the inhibition of steric hindrance between the clusters, and increases the risk of cluster agglomeration.

Apart from electrochemical contamination, impurities also affect the drying behavior of LSC ink. High-quality ORR tests require a thin, uniform film over the entire surface area of the GC electrode [56–58]. However, the catalyst dispersed in a drying ink drop migrates towards the edge of the ink drop to form a "coffee ring" [59,60]. The effects of impurities on catalyst activity were tested in a half-cell using the as-prepared ink coating on a GC electrode. Deegan et al. [59,61,62] postulated that the "coffee ring" effect occurs because the evaporation rate at the edge of droplet is higher than that at the center, resulting in an outward capillary flow within the droplet. This, in turn, transfers the suspended particles to the edge of the droplet and deposits them into a ring at the edge. As shown in Figure 8a, the "coffee ring" phenomenon appeared in the N-ink electrode, leading to an ununiform distribution of the catalyst, but this phenomenon was alleviated in the P-ink electrode. Notably, the introduced propionic acid has a higher boiling point and lower surface tension than the original solvent. As evaporation proceeds, the water evaporation rate at the edge of the droplet exceeds that at the center, whereas the evaporation rate of the propionic acid at the edge of the droplet becomes slower. Therefore, the propionic acid gradually becomes enriched at the edge. The difference in surface tension between the edge and center of the droplet creates the Marangoni effect [63]. On this basis, the enhanced Marangoni flow moves the catalyst particles radially from the edge to the center of the droplet surface, thereby inhibiting the "coffee ring" effect. Therefore, an RDE with a uniform catalyst deposition layer shows a better ORR performance.

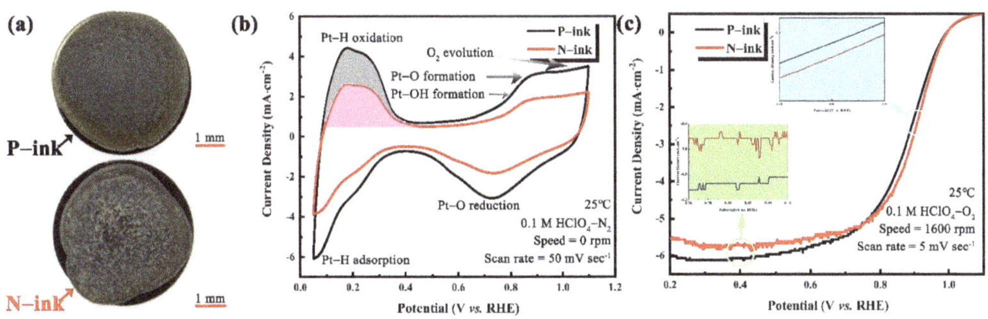

Figure 8. (a) Optical photograph and ORR performance of the electrodes in an RDE: (b) CV and (c) the corresponding ORR polarization curves.

The CV curves of electrodes from N-ink and P-ink are shown in Figure 8b. The anodic H waves and the cathodic H waves in the CV represent the H from the electrochemical desorption and adsorption process, respectively [24,27,64]. The H_{upd} charge is estimated after the conventional correction for the pseudocapacity seen in the double-layer region by a straight line. As an electron is transferred during the oxidation of the adsorbed H_{upd}, the charge of Hupd is therefore given by $Q_H = \frac{\int idE}{v}$ with the potential E, the sweep rate v, and the current i. The amounts of H_{des} in the curves of P-ink and N-ink, with double-layer charges subtracted, are 1.58 mC and 2.83 mC, respectively. In agreement with the report of Garsany et al. [24], the film quality affects the electrochemical surface area measurement, with an ECSA of 62.7 $m^2 \cdot g_{Pt}{}^{-1}$ for the electrode fabricated from N-ink, compared to 112.2 $m^2 \cdot g_{Pt}{}^{-1}$ for the electrode fabricated from P-ink (cf. Table 5—data from tables). Notably, the data shown in Figure 8c indicate that the electrode deposited by P-ink had a higher mass activity (84.4 $mA \cdot mg_{Pt}{}^{-1}$) than that deposited by N-ink. Furthermore, the normal LSC ink exhibited lower electrochemical properties of electrode deposition than the LSC ink with propionic acid, under similar conditions of the catalyst ink formation and drying process. The superior catalyst performance of the N-ink electrode was attributed to the morphology of the deposited catalyst, due to the effect of impurities on the ink [65]. The records of the drying process for N-ink and P-ink are presented in the Supporting Information to reveal the drying behavior of the ink droplets on the glass substrate.

Table 5. Properties of electrodes deposited by N-ink and P-ink.

Sample	ECSA ($m^2 \cdot g_{Pt}{}^{-1}$)	MA ($mA \cdot mg_{Pt}{}^{-1}$)
P-ink	112.2	84.4
N-ink	62.7	60.5

4. Conclusions

In the traditional ink preparation procedure, the catalytic oxidation of alcohols and its effects on the quality of the catalyst ink are generally ignored. In this study, we explored the mechanism by which impurities are generated, and the effects of such impurities on HSC ink and LSC ink. The GC-MS results indicate that the impurities in the inks included propionic acid, acetic acid, propanal, acetaldehyde, propyl propionate, and 1,1-dipropoxypropane. Together with the electrochemical behavior of NPA and EA, the impurity evolution process is as follows: NPA and EA are catalytically oxidized to propionic acid and acetic acid. Then, they are under esterification and aldolization reactions. The strength of the catalytic reaction is strongly correlated with the oxygen atmosphere and temperature. This result demonstrates the importance of temperature control and degassing treatment during the ink preparation process.

The effects of impurities on the HSC ink and LSC ink were identified, and the production of impurities was suppressed by the degassing treatment, causing the concentrations of acetic acid, propionic acid, and propyl propionate to decrease from 0.0898 wt.%, 0.00224 wt.%, and 0.00046 wt.% to 0.0025 wt.%, 0.0126 wt.%, and 0.0003 wt.%, respectively. For the HSC ink, the viscosity and yield stress of D-ink were lower than those of I-ink, and its structural resilience, as a result of the stronger interactions between the ionomer and catalyst, was 29.2 % higher than that of I-ink. For the LSC ink, the addition of propionic acid reduced the surface tension of the original solvent, thereby suppressing the "coffee ring" effect. This creates a thin, uniform catalyst deposition layer. Based on this, the ECSA of electrodes derived from N-ink was 62.7 $m^2 \cdot g_{Pt}{}^{-1}$, but the parameter for P-ink reached 112.2 $m^2 \cdot g_{Pt}{}^{-1}$. Therefore, the inhibition of impurity generation in HSC ink will provide better coatability, but the introduction of impurities will promote the uniform deposition of the catalyst in LSC ink. In the end, extending this study to the industrial scale will be particularly valuable for the improvement of the electric generation performance of PEMFCs, which is accessible by controlling of the HSC ink quality to establish the desired catalyst layer microstructure.

Supplementary Materials: The following supporting information can be downloaded at: https://www.mdpi.com/article/10.3390/membranes12050541/s1, Figure S1: Total ion current (TIC) spectrum of D-ink; Figure S2: Cyclic voltammograms recorded with Pt/C electrode in 0.05 M H_2SO_4 + 0.1 M NPA solution (a) and 0.1 M EA solution (b) with a scan rate of 20 $mV \cdot s^{-1}$ under N_2 atmosphere. Figure S3: A side view of catalyst ink droplets at the initial state and top view of ink droplets with the evolution of time for N-ink (a) and P-ink (b).

Author Contributions: Conceptualization, P.L. and D.Y.; methodology, P.L. and D.Y.; software, P.L.; validation, C.Z. and P.M.; formal analysis, D.Y.; investigation, D.Y.; resources, D.Y.; data curation, P.L.; writing—original draft preparation, P.L.; writing—review and editing, P.L. and D.Y.; visualization, P.L.; supervision, D.Y.; project administration, C.Z. and P.M.; funding acquisition, B.L. All authors have read and agreed to the published version of the manuscript.

Funding: This work was supported by National Nature Science Foundation of China (No. 52176198).

Institutional Review Board Statement: Not applicable.

Informed Consent Statement: Not applicable.

Data Availability Statement: Not applicable.

Conflicts of Interest: The authors declare no conflict of interest.

References

1. Ioroi, T.; Siroma, Z.; Yamazaki, S.I.; Yasuda, K. Electrocatalysts for pem fuel cells. *Adv. Energy Mater.* **2018**, *9*, 1801284. [CrossRef]
2. Wang, Y.; Ruiz Diaz, D.F.; Chen, K.S.; Wang, Z.; Adroher, X.C. Materials, technological status, and fundamentals of pem fuel cells—A review. *Mater. Today* **2020**, *32*, 178–203. [CrossRef]
3. Son, T.Y.; Kim, T.H.; Nam, S.Y. Crosslinked pore-filling anion exchange membrane using the cylindrical centrifugal force for anion exchange membrane fuel cell system. *Polymers* **2020**, *12*, 2758. [CrossRef] [PubMed]
4. Qaisrani, N.A.; Ma, Y.; Ma, L.; Liu, J.; Gao, L.; Li, L.; Gong, S.; Yan, X.; Zhang, F.; He, G. Facile and green fabrication of polybenzoxazine-based composite anion-exchange membranes with a self-cross-linked structure. *Ionics* **2018**, *24*, 3053–3063. [CrossRef]
5. Guo, Y.; Yang, D.; Li, B.; Yang, D.; Ming, P.; Zhang, C. Effect of dispersion solvents and ionomers on the rheology of catalyst inks and catalyst layer structure for proton exchange membrane fuel cells. *ACS Appl. Mater. Interfaces* **2021**, *13*, 27119–27128. [CrossRef]
6. Fouzaï, I.; Gentil, S.; Bassetto, V.C.; Silva, W.O.; Maher, R.; Girault, H.H. Catalytic layer-membrane electrode assembly methods for optimum triple phase boundaries and fuel cell performances. *J. Mater. Chem. A* **2021**, *9*, 11096–11123. [CrossRef]
7. Chu, T.; Zhang, R.; Wang, Y.; Ou, M.; Xie, M.; Shao, H.; Yang, D.; Li, B.; Ming, P.; Zhang, C. Performance degradation and process engineering of the 10 kw proton exchange membrane fuel cell stack. *Energy* **2021**, *219*, 119623. [CrossRef]
8. Yoshino, S.; Shinohara, A.; Kodama, K.; Morimoto, Y. Fabrication of catalyst layer with ionomer nanofiber scaffolding for polymer electrolyte fuel cells. *J. Power Sources* **2020**, *476*, 228584. [CrossRef]
9. Chen, L.; Zhang, R.; Kang, Q.; Tao, W.-Q. Pore-scale study of pore-ionomer interfacial reactive transport processes in proton exchange membrane fuel cell catalyst layer. *Chem. Eng. J.* **2020**, *391*, 123590. [CrossRef]
10. Stähler, M.; Stähler, A.; Scheepers, F.; Carmo, M.; Stolten, D. A completely slot die coated membrane electrode assembly. *Int. J. Hydrogen Energy* **2019**, *44*, 7053–7058. [CrossRef]
11. Li, B.; Liu, Y.; Guo, Y.; Yang, D.; Yang, D.; Ming, P.; Zhang, C. Controlling the microscopic morphology and permeability of catalyst layers in proton exchange membrane fuel cells by adjusting catalyst ink agglomerates. *Int. J. Hydrogen Energy* **2021**, *46*, 32215–32225. [CrossRef]
12. Gong, Q.; Li, C.; Liu, Y.; Ilavsky, J.; Guo, F.; Cheng, X.; Xie, J. Effects of ink formulation on construction of catalyst layers for high-performance polymer electrolyte membrane fuel cells. *ACS Appl. Mater. Interfaces* **2021**, *13*, 37004–37013. [CrossRef] [PubMed]
13. Kumano, N.; Kudo, K.; Akimoto, Y.; Ishii, M.; Nakamura, H. Influence of ionomer adsorption on agglomerate structures in high-solid catalyst inks. *Carbon* **2020**, *169*, 429–439. [CrossRef]
14. Uemura, S.; Yoshida, T.; Koga, M.; Matsumoto, H.; Yang, X.; Shinohara, K.; Sasabe, T.; Hirai, S. Ink degradation and its effects on the crack formation of fuel cell catalyst layers. *J. Electrochem. Soc.* **2019**, *166*, F89–F92. [CrossRef]
15. Balu, R.; Choudhury, N.R.; Mata, J.P.; de Campo, L.; Rehm, C.; Hill, A.J.; Dutta, N.K. Evolution of the interfacial structure of a catalyst ink with the quality of the dispersing solvent: A contrast variation small-angle and ultrasmall-angle neutron scattering investigation. *ACS Appl. Mater. Interfaces* **2019**, *11*, 9934–9946. [CrossRef] [PubMed]
16. Pan, W.; Chen, Z.; Yao, D.; Chen, X.; Wang, F.; Dai, G. Microstructure and macroscopic rheology of microporous layer nanoinks for pem fuel cells. *Chem. Eng. Sci.* **2021**, *246*, 117001. [CrossRef]
17. Uemura, S.; Yoshida, T.; Koga, M.; Matsumoto, H.; Shinohara, K.; Hirai, S. Ink degradation phenomena and its impact on crack formation of fuel cell catalyst. *ECS Trans.* **2018**, *86*, 151–156. [CrossRef]
18. Uemura, S.; Sakai, K.; Sasabe, T.; Matsumoto, H.; Sugimori, H.; Shinohara, K.; Hirai, S. Effect of reaction products on the pefc catalyst ink property and catalyst layer quality. *ECS Trans.* **2020**, *98*, 61–65. [CrossRef]
19. Uemura, S.; Kameya, Y.; Iriguchi, N.; Yoshida, T.; Shinohara, K.; Hirai, S. Communication—investigation of catalyst ink degradation by x-ray ct. *J. Electrochem. Soc.* **2018**, *165*, F142–F144. [CrossRef]
20. Uemura, S.; Sasabe, T.; Sakai, K.; Shinohara, K.; Hirai, S. Relation between degradation reaction and mixing at fuel cell catalyst ink fabrication process. *ECS Trans.* **2019**, *92*, 183–187. [CrossRef]
21. Kameya, Y.; Iriguchi, N.; Ohki, M.; Yokoyama, K.; Sugawara, S.; Sugimori, H.; Uemura, S.; Sasabe, T.; Yoshida, T.; Hirai1, T. Mri and $^{1}h/^{19}f$ nmr investigation of dispersion state of pefc catalyst ink. *ECS Trans.* **2017**, *80*, 819. [CrossRef]
22. Pastor, E.; Wasmus, S.; Iwasita, T.; Arévalo, M.C.; González, S.; Arvia, A.J. Spectroscopic investigations of c_3 primary alcohols on platinum electrodes in acid solutions. *J. Electroanal. Chem.* **1993**, *350*, 97–116. [CrossRef]
23. Pastor, E.; González, S.; Arvia, A.J. Electroreactivity of isopropanol on platinum in acids studied by dems and ftirs. *J. Electroanal. Chem.* **1995**, *395*, 233–242. [CrossRef]
24. Garsany, Y.; Baturina, O.A.; Swider-Lyons, K.E.; Kocha, S.S. Experimental methods for quantifying the activity of platinum electrocatalysts for the oxygen reduction reaction. *Anal. Chem.* **2010**, *82*, 6321–6328. [CrossRef]

25. Tang, X.; Zeng, Y.; Cao, L.; Yang, L.; Wang, Z.; Fang, D.; Gao, Y.; Shao, Z.; Yi, B. Anchoring ultrafine pt nanoparticles on the 3d hierarchical self-assembly of graphene/functionalized carbon black as a highly efficient oxygen reduction catalyst for pemfcs. *J. Mater. Chem. A* **2018**, *6*, 15074–15082. [CrossRef]
26. Luo, L.; Zhu, F.; Tian, R.; Li, L.; Shen, S.; Yan, X.; Zhang, J. Composition-graded pdxni1−x nanospheres with pt monolayer shells as high-performance electrocatalysts for oxygen reduction reaction. *ACS Catal.* **2017**, *7*, 5420–5430. [CrossRef]
27. Mayrhofer, K.J.J.; Strmcnik, D.; Blizanac, B.B.; Stamenkovic, V.; Arenz, M.; Markovic, N.M. Measurement of oxygen reduction activities via the rotating disc electrode method: From pt model surfaces to carbon-supported high surface area catalysts. *Electrochim. Acta* **2008**, *53*, 3181–3188. [CrossRef]
28. Umeda, M.; Sugii, H.; Uchida, I. Alcohol electrooxidation at pt and pt–ru sputtered electrodes under elevated temperature and pressurized conditions. *J. Power Sources* **2008**, *179*, 489–496. [CrossRef]
29. Tu, K.; Li, G.; Jiang, Y. Effect of temperature on the electrocatalytic oxidation of ethanol. *Acta Phys.-Chim. Sin* **2020**, *36*, 1906026. [CrossRef]
30. Kusoglu, A.; Weber, A.Z. New insights into perfluorinated sulfonic-acid ionomers. *Chem. Rev.* **2017**, *117*, 987–1104. [CrossRef]
31. Woo, S.; Lee, S.; Taning, A.Z.; Yang, T.-H.; Park, S.-H.; Yim, S.-D. Current understanding of catalyst/ionomer interfacial structure and phenomena affecting the oxygen reduction reaction in cathode catalyst layers of proton exchange membrane fuel cells. *Curr. Opin. Electrochem.* **2020**, *21*, 289–296. [CrossRef]
32. Khandavalli, S.; Iyer, R.; Park, J.H.; Myers, D.J.; Neyerlin, K.C.; Ulsh, M.; Mauger, S.A. Effect of dispersion medium composition and ionomer concentration on the microstructure and rheology of fe-n-c platinum group metal-free catalyst inks for polymer electrolyte membrane fuel cells. *Langmuir* **2020**, *36*, 12247–12260. [CrossRef] [PubMed]
33. Berlinger, S.A.; Garg, S.; Weber, A.Z. Multicomponent, multiphase interactions in fuel-cell inks. *Curr. Opin. Electrochem.* **2021**, *29*, 100744. [CrossRef]
34. Hatzell, K.B.; Dixit, M.B.; Berlinger, S.A.; Weber, A.Z. Understanding inks for porous-electrode formation. *J. Mater. Chem. A* **2017**, *5*, 20527–20533. [CrossRef]
35. Dixit, M.B.; Harkey, B.A.; Shen, F.; Hatzell, K.B. Catalyst layer ink interactions that affect coatability. *J. Electrochem. Soc.* **2018**, *165*, F264–F271. [CrossRef]
36. Lee, G.-W.; Ryu, J.H.; Han, W.; Ahn, K.H.; Oh, S.M. Effect of slurry preparation process on electrochemical performances of $licoo_2$ composite electrode. *J. Power Sources* **2010**, *195*, 6049–6054. [CrossRef]
37. Mizukawa, H.; Kawaguchi, M. Effects of perfluorosulfonic acid adsorption on the stability of carbon black suspensions. *Langmuir* **2009**, *25*, 11984–11987. [CrossRef]
38. Lee, J.-H.; Paik, U.; Choi, J.-Y.; Kim, K.K.; Yoon, S.-M.; Lee, J.; Kim, B.-K.; Kim, J.M.; Park, M.H.; Yang, C.W.; et al. Dispersion stability of single-walled carbon nanotubes using nafion in bisolvent. *J. Phys. Chem. C* **2007**, *111*, 2477–2483. [CrossRef]
39. Cho, K.Y.; Kwon, Y.I.; Youn, J.R.; Song, Y.S. Evaluation of slurry characteristics for rechargeable lithium-ion batteries. *Mater. Res. Bull.* **2013**, *48*, 2922–2926. [CrossRef]
40. Kim, T.H.; Jang, L.W.; Lee, D.C.; Choi, H.J.; Jhon, M.S. Synthesis and rheology of intercalated polystyrene/na+-montmorillonite nanocomposites. *Macromol. Rapid Commun.* **2002**, *23*, 191–195. [CrossRef]
41. Du, S.; Li, W.; Wu, H.; Abel Chuang, P.-Y.; Pan, M.; Sui, P.-C. Effects of ionomer and dispersion methods on rheological behavior of proton exchange membrane fuel cell catalyst layer ink. *Int. J. Hydrogen Energy* **2020**, *45*, 29430–29441. [CrossRef]
42. Hawley, W.B.; Li, J. Beneficial rheological properties of lithium-ion battery cathode slurries from elevated mixing and coating temperatures. *J. Energy Storage* **2019**, *26*, 100994. [CrossRef]
43. Zhao, J.; He, N. A mini-review of embedded 3d printing: Supporting media and strategies. *J. Mater. Chem. B* **2020**, *8*, 10474–10486. [CrossRef] [PubMed]
44. Björn, A.; de La Monja, P.S.; Karlsson, A.; Ejlertsson, J.; Svensson, B.H. *Rheological Characterization*; InTech: London, UK, 2012.
45. Bitsch, B.; Dittmann, J.; Schmitt, M.; Scharfer, P.; Schabel, W.; Willenbacher, N. A novel slurry concept for the fabrication of lithium-ion battery electrodes with beneficial properties. *J. Power Sources* **2014**, *265*, 81–90. [CrossRef]
46. Kumano, N.; Kudo, K.; Suda, A.; Akimoto, Y.; Ishii, M.; Nakamura, H. Controlling cracking formation in fuel cell catalyst layers. *J. Power Sources* **2019**, *419*, 219–228. [CrossRef]
47. Su, F.-Y.; Dai, L.-Q.; Guo, X.-Q.; Xie, L.-J.; Sun, G.-H.; Chen, C.-M. Micro-structure evolution and control of lithium-ion battery electrode laminate. *J. Energy Storage* **2017**, *14*, 82–93. [CrossRef]
48. Bauer, W.; Nötzel, D. Rheological properties and stability of nmp based cathode slurries for lithium ion batteries. *Ceram. Int.* **2014**, *40*, 4591–4598. [CrossRef]
49. Lanceros-Méndez, S.; Costa, C.M. *Printed Batteries: Materials, Technologies and Applications*; John Wiley & Sons: Hoboken, NJ, USA, 2018.
50. Kiennemann, J.; Chartier, T.; Pagnoux, C.; Baumard, J.F.; Huger, M.; Lamérant, J.M. Drying mechanisms and stress development in aqueous alumina tape casting. *J. Eur. Ceram. Soc.* **2005**, *25*, 1551–1564. [CrossRef]
51. Koos, E. Capillary suspensions: Particle networks formed through the capillary force. *Curr. Opin. Colloid Interface Sci.* **2014**, *19*, 575–584. [CrossRef]
52. Velankar, S.S. A non-equilibrium state diagram for liquid/fluid/particle mixtures. *Soft Matter* **2015**, *11*, 8393–8403. [CrossRef]
53. Kralchevsky, P.A.; Denkov, N.D. Capillary forces and structuring in layers of colloid particles. *Curr. Opin. Colloid* **2001**, *6*, 383–401. [CrossRef]

54. Garsany, Y.; Atkinson, R.W.; Sassin, M.B.; Hjelm, R.M.E.; Gould, B.D.; Swider-Lyons, K.E. Improving pemfc performance using short-side-chain low-equivalent-weight pfsa ionomer in the cathode catalyst layer. *J. Electrochem. Soc.* **2018**, *165*, F381–F391. [CrossRef]
55. Shukla, S.; Bhattacharjee, S.; Weber, A.Z.; Secanell, M. Experimental and theoretical analysis of ink dispersion stability for polymer electrolyte fuel cell applications. *J. Electrochem. Soc.* **2017**, *164*, F600–F609. [CrossRef]
56. Kocha, S.S.; Shinozaki, K.; Zack, J.W.; Myers, D.J.; Kariuki, N.N.; Nowicki, T.; Stamenkovic, V.; Kang, Y.; Li, D.; Papageorgopoulos, D. Best practices and testing protocols for benchmarking orr activities of fuel cell electrocatalysts using rotating disk electrode. *Electrocatalysis* **2017**, *8*, 366–374. [CrossRef]
57. Garsany, Y.; Ge, J.J.; St-Pierre, J.; Rocheleau, R.; Swider-Lyons, K.E. Standardizing thin-film rotating disk electrode measurements of the oxygen reduction activity of pt/c. *ECS Trans.* **2013**, *58*, 3–14. [CrossRef]
58. Garsany, Y.; Singer, I.L.; Swider-Lyons, K.E. Impact of film drying procedures on rde characterization of pt/vc electrocatalysts. *J. Electroanal. Chem.* **2011**, *662*, 396–406. [CrossRef]
59. Deegan, R.D.; Bakajin, O.; Dupont, T.F.; Huber, G.; Nagel, S.R.; Witten, T.A. Capillary flow as the cause of ring stains from dried liquid drops. *Nature* **1997**, *389*, 827–829. [CrossRef]
60. Zhao, J.; Li, X.; Liu, Z.S. The effect of ink dilution and evaporation on the microstructures of catalyst layers in polymer electrolyte membrane fuel cells. *Int. J. Energy Res.* **2019**, *43*, 6799–6811. [CrossRef]
61. Deegan, R.D. Pattern formation in drying drops. *Phys. Rev. E* **2000**, *61*, 475–485. [CrossRef]
62. Deegan, R.D.; Bakajin, O.; Dupont, T.F.; Huber, G.; Nagel, S.R.; Witten, T.A. Contact line deposits in an evaporating drop. *Phys. Rev. E* **2000**, *62*, 756–765. [CrossRef]
63. Hu, H.; Larson, R.G. Marangoni effect reverses coffee-ring depositions. *J. Phys. Chem. B* **2006**, *110*, 7090–7094. [CrossRef] [PubMed]
64. Li, W.; Lane, A.M. Resolving the hupd and hopd by dems to determine the ecsa of pt electrodes in pem fuel cells. *Electrochem. Commun.* **2011**, *13*, 913–916. [CrossRef]
65. Garsany, Y.; Ge, J.; St-Pierre, J.; Rocheleau, R.; Swider-Lyons, K.E. Analytical procedure for accurate comparison of rotating disk electrode results for the oxygen reduction activity of pt/c. *J. Electrochem. Soc.* **2014**, *161*, F628–F640. [CrossRef]

membranes

Article

Conversion Kinetics and Ionic Conductivity in Na-β″-Alumina + YSZ (Naβ″AY) Sodium Solid Electrolyte via Vapor Phase Conversion Process

Liangzhu Zhu [1,2,*] **and Anil V. Virkar** [1,*]

1 Materials Science & Engineering Department, University of Utah, Salt Lake City, UT 84112, USA
2 Key Laboratory of Advanced Fuel Cells and Electrolyzers Technology of Zhejiang Province, Ningbo Institute of Materials Technology and Engineering, Chinese Academy of Sciences, Ningbo 315201, China
* Correspondence: zhuliangzhu@nimte.ac.cn (L.Z.); anil.virkar@utah.edu (A.V.V.)

Abstract: Sodium ion batteries have been receiving increasing attention and may see potential revival in the near future, particularly in large-scale grid energy storage coupling with wind and solar power generation, due to the abundant sodium resources, low cost, and sufficiently high energy density. Among the known sodium ion conductors, the Na-β″-alumina electrolyte remains highly attractive because of its high ionic conductivity. This study focuses on the vapor phase synthesis of a Na-β″-Alumina + YSZ (Naβ″AY) composite sodium electrolyte, which has higher mechanical strength and stability than conventional single phase β″-Alumina. The objectives are the measurement of conversion kinetics through a newly developed weight-gain based model and the determination of sodium ionic conductivity in the composite electrolyte. Starting samples contained ~70 vol% α-Alumina and ~30 vol% YSZ (3 mol% Y_2O_3 stabilized Zirconia) with and without a thin alumina surface layer made by sintering in air at 1600 °C. The sintered samples were placed in a powder of Na-β″-alumina and heat-treated at 1250 °C for various periods. Sample dimensions and weight were measured as a function of heat treatment time. The conversion of α-Alumina in the α-Alumina + YSZ composite into Naβ″AY occurred by coupled diffusion of sodium ions through Na-β″-alumina and of oxygen ions through YSZ, effectively diffusing Na_2O. From the analysis of the time dependence of sample mass and dimensions, the effective diffusion coefficient of Na_2O through the sample, D_{eff}, was estimated to be 1.74×10^{-7} cm^2 s^{-1}, and the effective interface transfer parameter, k_{eff}, was estimated as 2.33×10^{-6} cm s^{-1}. By depositing a thin alumina coating layer on top of the bulk composite, the chemical diffusion coefficient of oxygen through single phase Na-β″-alumina was estimated as 4.35×10^{-10} cm^2 s^{-1}. An AC impedance measurement was performed on a fully converted Naβ″AY composite, and the conductivity of the composite electrolyte was 1.3×10^{-1} S cm^{-1} at 300 °C and 1.6×10^{-3} S cm^{-1} at 25 °C, indicating promising applications in solid state or molten salt batteries at low to intermediate temperatures.

Keywords: sodium β″-alumina; Naβ″AY; sodium electrolyte; sodium solid-state battery; vapor phase process; sodium batteries

Citation: Zhu, L.; Virkar, A.V. Conversion Kinetics and Ionic Conductivity in Na-β″-Alumina + YSZ (Naβ″AY) Sodium Solid Electrolyte via Vapor Phase Conversion Process. *Membranes* **2022**, *12*, 567. https://doi.org/10.3390/membranes12060567

Academic Editor: Jijeesh Ravi Nair

Received: 3 May 2022
Accepted: 26 May 2022
Published: 30 May 2022

1. Introduction

Na-β″-alumina (more commonly referred to as β″-alumina) is a sodium ion conductor, which is used as a solid electrolyte in sodium-sulfur batteries, sodium-nickel chloride batteries, and in alkali metal-based thermoelectric converters [1–3]. While significant attention to Na-β″-alumina dates back as early as in the 1960s when sodium-sulfur batteries were originally developed by Weber and Kummer at the Ford Motor Company [1], the relatively high working temperature in a molten sodium-sulfur battery has significantly limited their applications and development, and soon, they were almost completely replaced by more popular lithium ion batteries in mobile applications. However, with increasing demands on large-scale energy storage due to climate change and the bloom of a solid-state battery, applications of Na-β″-alumina are seemingly rising again because of the abundance of sodium as compared to lithium sources and the high ionic conductivity of about 1 S cm^{-1} in single crystal Na-β″-alumina and 0.2–0.4 S cm^{-1} in polycrystalline β″-alumina at 300 °C [4–10]. For example, recently, Fertig et al. have provided a detailed review on the potential revival of Na-β″-alumina for sodium solid-state batteries [4]. Ligon et al. have reported large planar Na-β″-alumina solid electrolytes (150 mm in diameter) for next generation Na-batteries [5]. Lee et al. have demonstrated a full-scale pilot model of a cost-effective sodium-nickel-iron chloride battery over 40 Ah using a Na-β″-alumina electrolyte [9]. Zhu et al. have reported that by ion exchange, a Na-β″-alumina-containing composite electrolyte may be ion exchanged with molten salts to lithium-ion or silver-ion conducting electrolytes [11], which may further expand the application of Na-β″-alumina in other types of batteries.

The conventional process for the fabrication of dense samples of Na-β″-alumina consists of first calcining a mixture of Na_2CO_3, α-alumina, and $LiNO_3$ (or MgO) at ~1250 °C in air, which leads to a powder mixture containing Na-β″-alumina, Na-β-alumina (β-alumina), and some $NaAlO_2$. Powder compacts are then sintered in sealed platinum or MgO crucibles at ~1600 °C [12]. Sintering in sealed containers suppresses the loss of Na_2O through the vapor phase. Densification occurs by a transient liquid phase mechanism. Sintered samples are subsequently heat-treated at a lower temperature (~1400 °C) to convert β-alumina into β″-alumina by reacting with $NaAlO_2$. Usually, a small amount of $NaAlO_2$ remains unreacted in β″-alumina as a thin film along grain boundaries, which makes it susceptible to degradation due to moisture in the atmosphere. Thus, Na-β″-alumina made by the conventional process is stored in desiccators.

In concept, if an already-sintered α-alumina can be converted into Na-β″-alumina by reacting it with Na_2O, it may be possible to avoid the formation of $NaAlO_2$ along the grain boundaries, thus making it moisture-resistant. A possible vapor process for the conversion of sintered α-alumina into Na-β″-alumina involves exposing it to a vapor containing Na_2O. The approximate composition of Na-β″-alumina is Na_2O ~$6Al_2O_3$. For the conversion of α-alumina into Na-β″-alumina, the reaction is

$$Na_2O + {\sim}6Al_2O_3 \rightarrow Na_2O\ {\sim}6Al_2O_3 \quad (1)$$

If a fully dense sintered Al_2O_3 is exposed to Na_2O vapor, a thin layer of Na-β″-alumina forms on the surface. Further conversion of the interior α-alumina requires the transport of Na_2O (as coupled (ambipolar) transport of $2Na^+$ and O^{2-}) through the formed Na-β″-alumina (Figure A1a in Appendix A). While the Na^+ diffusivity through Na-β″-alumina is high, O^{2-} diffusivity is very low. Thus, the conversion kinetics are dictated by the diffusion coefficient of oxygen ions in Na-β″-alumina and are very sluggish. Furthermore, the conversion kinetics are diffusion-limited (parabolic). At 1300 °C, for example, the thickness of α-alumina converted into Na-β″-alumina in 16 h was only about 25 μm [13].

In the novel vapor phase process, a two-phase composite of α-alumina and YSZ, with both phases being contiguous, is exposed to Na_2O vapor [14]. Once α-alumina on the surface is converted to Na-β″-alumina, subsequent conversion of the interior α-alumina to Na-β″-alumina involves the transport of Na_2O, such that $2Na^+$ transports through the formed Na-β″-alumina, while the O^{2-} transports through the YSZ phase (Figure 1b). More specifically, the effective diffusion of Na_2O through the two phase mixture occurs by a coupled (ambipolar) transport of $2Na^+$ through Na-β″-alumina and O^{2-} through YSZ. Since the diffusion coefficient of O^{2-} through YSZ is much higher than that of O^{2-} through Na-β″-alumina, the kinetics of conversion of α-alumina + YSZ into Na-β″-alumina + YSZ is much faster than the conversion of single phase α-alumina into Na-β″-alumina. The resulting two-phase composite of Naβ″AY is also much stronger than conventional Na-β″-alumina and is moisture-resistant.

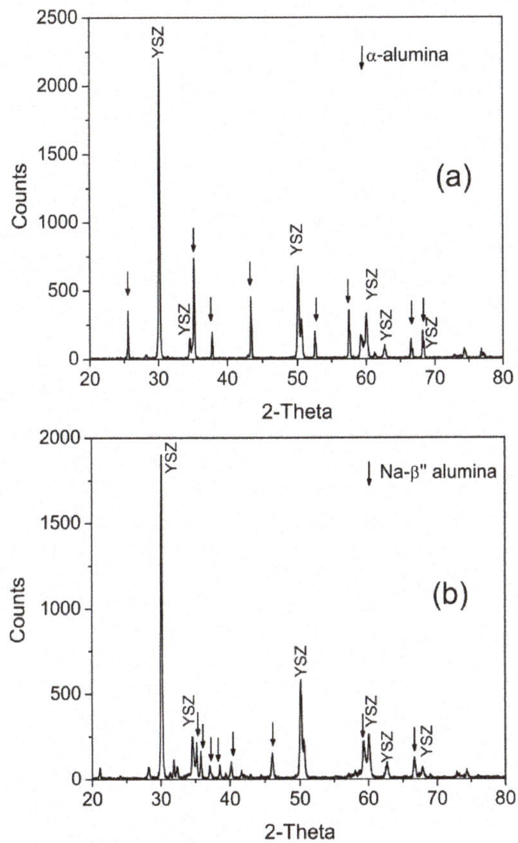

Figure 1. XRD patterns of α-Alumina + YSZ composites (**a**) before and (**b**) after subjecting to vapor phase conversion at 1250 °C.

Since the invention of the novel vapor phase process patented by Virkar et al. [14], it has been gaining attention in Na batteries [15–19]. A recent study has shown that a planar sodium nickel chloride battery demonstrated a specific energy density of as high as 350 Wh kg^{-1}, operated at 190 °C over 1000 cycles. The composite electrolyte adopted in that battery consisted of α-alumina and 8YSZ, also with a volume ratio of 7:3 as starting materials, and then was converted via the vapor phase conversion process. Such a composite electrolyte also offered high mechanical strength. A flexural strength of higher than 300 MPa has been reported in literature [20,21], making them an excellent Na-conducting membrane in Na batteries.

The only available model reported so far on the kinetics of this novel vapor phase conversion process is the one reported by Parathasarathy and Virkar [22], where the kinetics of conversion of α-alumina + YSZ composites of various grain sizes over a range of temperatures between 1250 °C and 1400 °C were investigated. The experimental procedure involved packing sintered α-alumina + YSZ samples into Na-β″-alumina powder, heat treating at a given temperature for a period of time, cooling down to room temperature, and grinding/polishing the sample to measure (using a microscope) the conversion thickness $x(t)$ as a function of cumulative heat treatment time t. The process thus is tedious, as it requires repeated grinding and polishing. Furthermore, this involves the destruction of the sample. Under some situations, where both conversion fraction and conductivity as a function of conversion time are of interest, the above thickness-based model may be insufficient, since it requires cutting of the sample. Therefore, an alternate model needs to be developed to fulfill the routine examination of this vapor phase process.

The objective of the present work was to measure the sample weight and the external dimensions after various thermal treatments so that the kinetics of conversion could be measured more effectively without destroying the sample and, at the same time, minimizing the effort required in the actual measurements.

An equally important model was also developed and experimentally verified to measure the chemical diffusion coefficient of oxygen through single phase -β″-alumina by depositing a thin layer of α-alumina on the α-alumina + YSZ composite prior to conversion. The α-alumina-coated samples were subjected to the same conversion treatment as the uncoated samples. Both of the newly developed models may be used as effective tools for future studies and applications of the vapor phase process. Note that the concepts and models developed here may well fit to other binary or possibly ternary diffusion system where diffusion kinetics of certain mobile ions, such as oxygen-ion, proton, metal ions, etc., are of interest.

2. Materials and Methods

2.1. *Preparation of α-Alumina + YSZ Samples*

High purity α-alumina (AKP-53, Sumitomo Chemical) and 3YSZ (TZ-3Y, Tosoh Corporation, Tokyo, Japan) were mixed in a volume ratio of alumina:3YSZ of 7:3. A slurry of the powder mixture was made in distilled water with a small amount of ammonium poly methyl methacrylate (DARVAN C-N, R. T. Vanderbilt, Inc., New York, NY, USA), added as a dispersant and as a stabilizer. The slurry was planetary milled for 12 h, dried, and then heated to 500 °C for 5 h. The dried powder was sieved through a 70-mesh screen to remove any large agglomerates. A few grams of powder was placed in a circular die and uniaxially pressed under a force of 5 tons, followed by cold isostatic pressing at 30,000 psi. The disc was then pre-sintered at 1200 °C for 2 h and then polished on both sides to a finish of ~50 nm using an alumina suspension in water. After polishing, selected samples were either spin-coated or dip-coated with an alumina suspension depending on the desired thickness. After slowly heating the samples to 1000 °C to burnout the organic components, all of the samples were sintered in air at 1600 °C for 5 h. The weight and the dimensions (diameter and thickness) were measured as the baseline information.

2.2. *Formation of Na-β″-Alumina + YSZ Samples via Vapor Phase Conversion*

Sintered discs were then buried in Na-β″-alumina powder (Materials and Systems Research, Inc., Salt Lake City, UT, USA) in an alumina crucible and covered with a lid. The crucible was placed in a furnace in air and heated to 1250 °C (5 °C min^{-1} ramp up rate), maintained for a period of time at temperature, and rapidly cooled to room temperature (>15 °C min^{-1} ramp down rate). The objective was to ensure that for most of the reaction time, the samples were at the heat treatment temperature. Since the process of vapor phase conversion of α-alumina + YSZ into Na-β″-alumina + YSZ is thermally activated, most of the conversion occurred isothermally at 1250 °C, and any conversion during heating to 1250 °C and cooling down from the heat treatment temperature could be neglected. This process was repeated several times until a maximum cumulative time of ~84 h. The dimensions of the samples and their weights were carefully recorded after each thermal treatment. For weight measurement, a balance with a resolution of 0.0001 g and repeatability <0.0002 g was used. The digital caliper used for dimension measurements had a resolution of 0.01 mm with negligible deviation in repeat measurements. In order to examine the conversion front at different conversion times, a small part of one of the samples was cut from the edge after each conversion period and fine polished for microstructure characterization. Samples used for weight measurements were not cut.

2.3. Characterization

Similar samples were prepared for X-ray diffraction (XRD) and scanning electron microscopy (SEM). An Energy Dispersive Spectroscopy (EDS) elemental line scan was performed on a partially converted sample to examine the α-alumina + YSZ/Na-β″-alumina + YSZ interface. Density measurements were performed by the standard fluid immersion method on an as-sintered α-alumina + YSZ sample and a fully converted Naβ″AY sample. Conductivity was measured in air as a function of temperature using an AC Electrochemical Impedance Spectroscopy (EIS) with gold paste as electrodes. Measurements were conducted over a frequency range from 10 Hz to 1 MHz. The high frequency intercept was taken as the measure of conductivity.

3. Results and Discussion

3.1. X-ray-Diffraction

Figure 1 shows XRD patterns of an as-sintered α-alumina + YSZ sample and a fully converted Naβ″AY sample. In Figure 1a, all peaks are identified as belonging to either α-alumina or YSZ. Peaks belonging to α-alumina have been identified by arrows. In Figure 1b, which is for the converted sample, peaks belonging to Na-β″-alumina have been identified by arrows. No peaks belonging to α-alumina were observed in the converted sample. Some of the peaks belonging to Na-β-alumina overlapped with those of Na-β″-alumina. Thus, the presence of some Na-β-alumina cannot be ruled out.

3.2. SEM and EDS Analysis

Figure 2a shows an SEM image of a polished section of a sample that had been heat-treated at 1250 °C for 4 h by packing in Na-β″-alumina powder. The converted region is darker in color. The average conversion thickness measured was ~110 µm. Figure 2b shows an EDS scan from the surface of the sample, across Na-β″-alumina + YSZ/α-alumina + YSZ interface and into the α-alumina + YSZ unconverted region. While the EDS scan was not smooth and showed considerable scatter, the demarcation between the converted region and the pristine region is clear.

Figure 3 shows SEM micrographs of the cross-sections of an uncoated sample, a sample with a ~2.5 µm-thick layer Na-β″-alumina (initially, an alumina layer of 2 µm-thickness) and a sample with a ~15 µm-thick layer of Na-β″-alumina (initially, an alumina layer of 12 µm-thickness) after conversion treatment for 4 h, 4 h, and 84 h, respectively. The micrographs in (b) and (c) thus correspond to samples that were initially coated with α-alumina. These layers fully converted to Na-β″-alumina. The interiors of all three samples contained Na-β″-alumina + YSZ. The dark phase is Na-β″-alumina; the light phase is YSZ. The Na-β″-alumina and YSZ phases were both contiguous, as would be required for coupled transport of Na_2O to occur through the two phase mixture: Na^+ through Na-β″-alumina and O^{2-} through YSZ [22].

Figure 4 compares the conversion thicknesses of the alumina-coated (~2.5 µm Na-β″-alumina after conversion) portion of the sample, with part of the region that was uncoated, after conversion at 1250 °C for 6 h. The conversion thickness corresponding to the non-coated region was about twice that of the coated region. This shows that the presence of a 2.5-µm Na-β″-alumina surface coating substantially lowered the conversion kinetics.

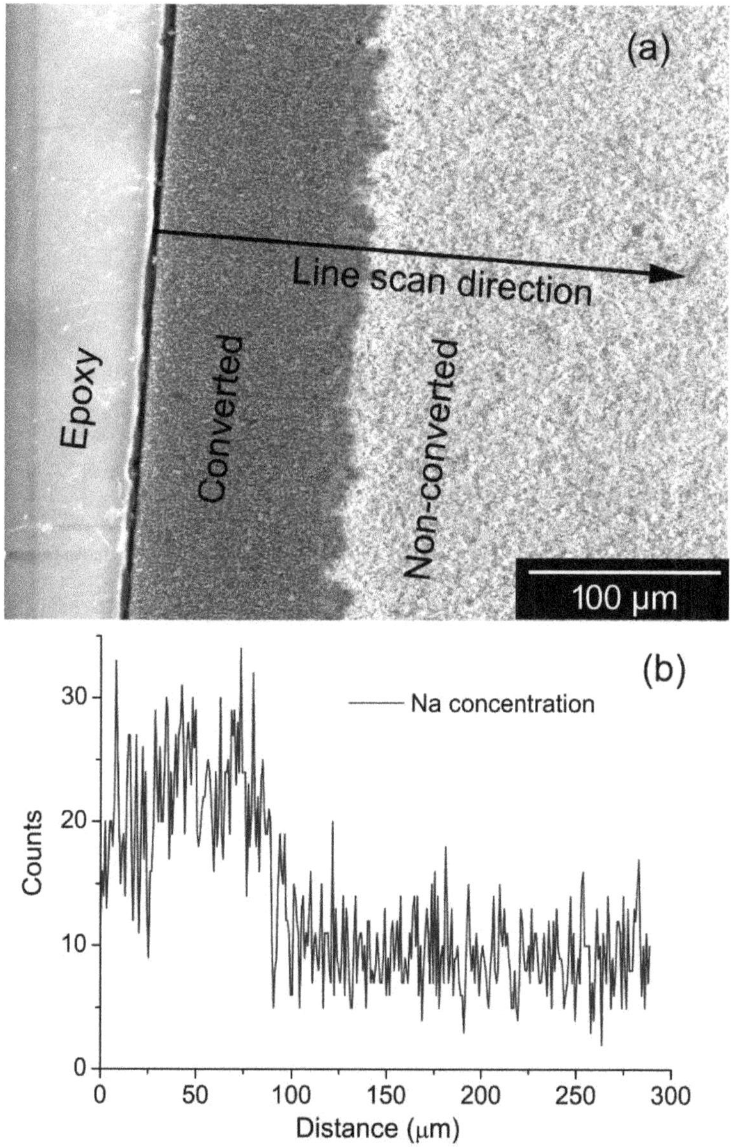

Figure 2. (**a**) An SEM image showing the converted and the pristine regions of the initially α-alumina + YSZ sample after 4 h of conversion at 1250 °C. (**b**) The corresponding EDS line scan is indicated by the arrow.

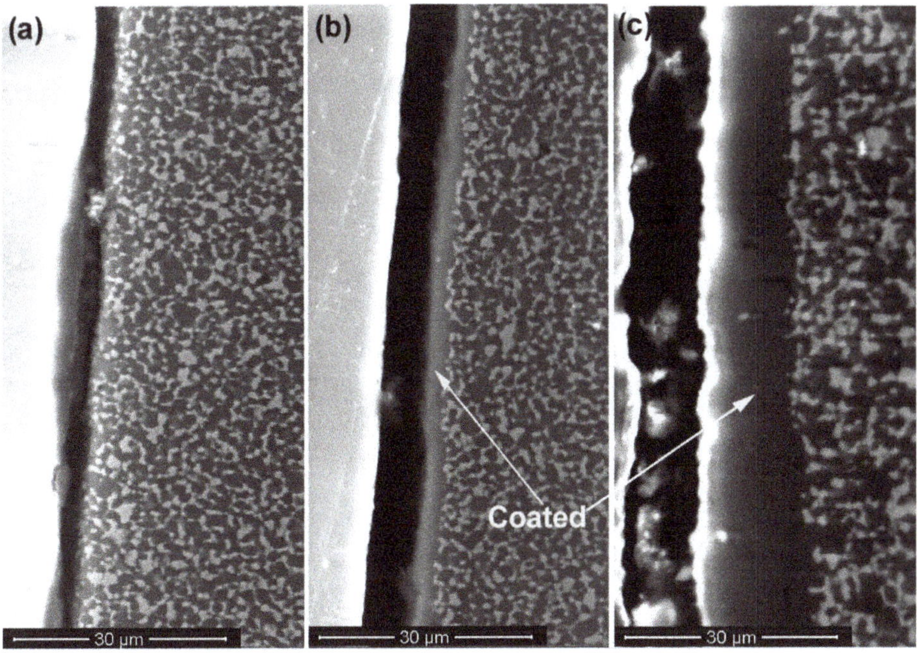

Figure 3. SEM images showing the enlarged view at the interface of (**a**) The uncoated region, (**b**) A ~2.5 μm Na-β″-alumina-coated sample, and (**c**) A ~15 μm Na-β″-alumina-coated sample.

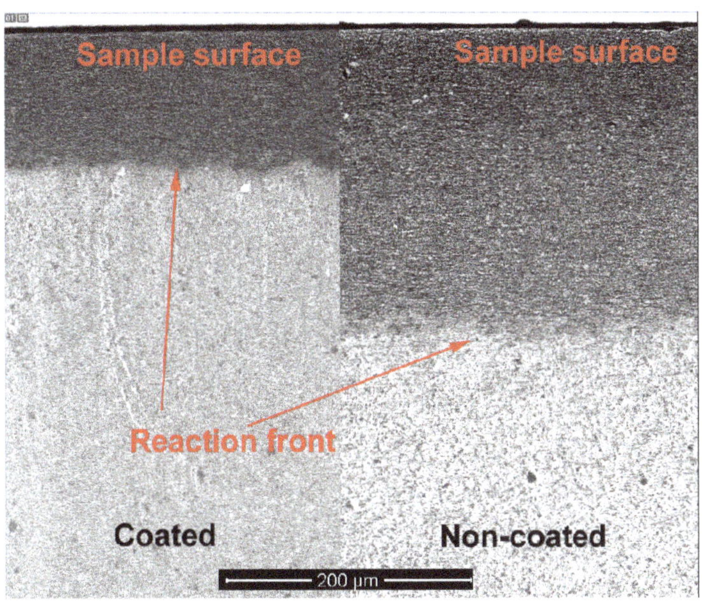

Figure 4. SEM images showing the conversion fronts for the uncoated and 2.5 μm Na-β″-alumina-coated regions after 6 h of conversion.

3.3. Estimation of the Kinetic Parameters on the Uncoated Samples

Appendix A gives the relevant equations describing the kinetics of conversion and the chemical diffusion coefficients. Equation (A1) gives the conversion thickness x corresponding to isothermal treatment time of t in terms of the effective diffusion coefficient, D_{eff}, and the effective interface transfer parameter, k_{eff}. Appendix B gives the kinetic equation in terms of the sample dimensions and the increase in weight as a function of time. The following describes an example of the estimation of the kinetic parameters on an uncoated sample.

The density of the as-sintered α-alumina + YSZ samples, ρ_{AY}, was measured as 4.59 g cm^{-3}. The initial thickness of the α-Al_2O_3 + YSZ disc sample was 3.64 mm, and the initial diameter was 29.54 mm. Figure 5 shows the measured weight, the thickness, and the diameter changes in percent with respect to the initial measurements. The initial weight of the sample, m_o, was 11.2452 g. After conversion for a cumulative time of 84 h at 1250 °C, the sample weight increased to 11.9302 g. That is, the sample weight increased by 685 mg after a cumulative conversion time of 84 h at 1250 °C. After 72 h, the weight increased only slightly. This indicated that the sample had fully or very close to fully converted after 84 h.

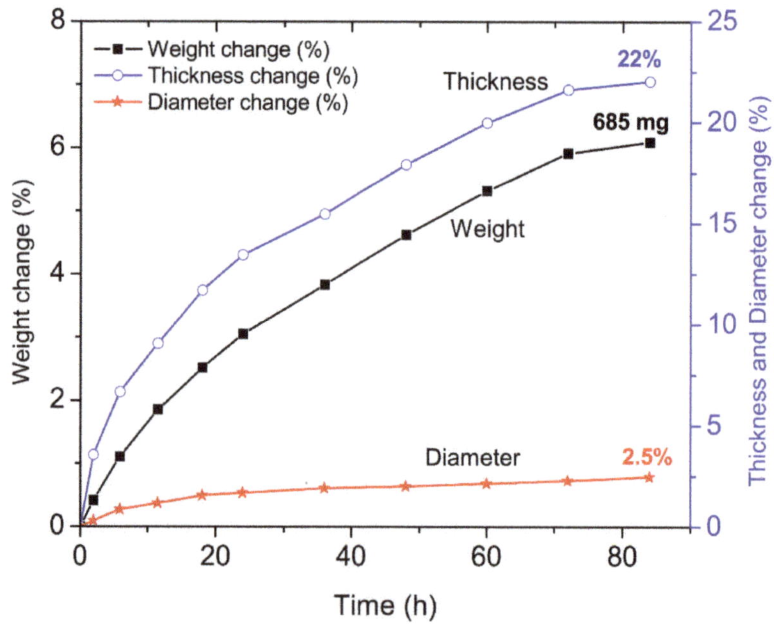

Figure 5. Measured weight and geometry change of an uncoated sample as a function of conversion time.

At 1250 °C. The sample cross-section confirmed that it had fully converted to Na-β″-alumina + YSZ. Thus, the final weight of the sample of 11.9302 g was taken as the weight after full conversion, that is, $m(\infty) = 11.9302$ g. The density of the fully converted thinner sample prepared using the same procedure, $\rho_{\beta''Y}$, was measured as 3.94 g cm^{-3}. From the measured values of m_o, $m(\infty)$, ρ_{AY}, and $\rho_{\beta''Y}$, the ratio $\frac{x'(t)}{x(t)}$ corresponding to Equation (A16) was measured as

$$\frac{x'(t)}{x(t)} = \frac{l_0}{l(\infty)} = \frac{\rho_{AY} m_o}{\rho_{\beta''Y} m(\infty)} = 0.81 \tag{2}$$

The ratio

$$\frac{m_o}{m(\infty)} = 0.94 \tag{3}$$

is used in Equation (A18).

The thickness measured after conversion (84 h) was 4.45 mm. Thus, $\frac{l_o}{l(\infty)}$ based on the thickness measurement is

$$\frac{l_o}{l(\infty)} = 0.82 \tag{4}$$

The increase in thickness was ~22%. The corresponding increase in diameter was only ~2.5%. For this reason, the assumption that most of the dimensional change occurs along the thickness direction is reasonable. Based on the weight gain, it is concluded that the stoichiometry of Na-β″-alumina is almost exactly $Na_2O{\cdot}6Al_2O_3$.

The kinetic equation in terms of weight change given in Equation (A19) is reproduced here

$$\frac{(\Delta m(t))^2}{4D_{eff}\left(A\rho_{\beta''}Y\left(1-\frac{m_o}{m(\infty)}\right)\right)^2} + \frac{\Delta m(t)}{2k_{eff}\left(A\rho_{\beta''}Y\left(1-\frac{m_o}{m(\infty)}\right)\right)} = t \tag{5}$$

The same equation can be written as

$$\Delta m(t) = 4D_{eff}\left(A\rho_{\beta''}Y\left(1-\frac{m_o}{m(\infty)}\right)\right)^2\frac{t}{\Delta m(t)} - \frac{2D_{eff}}{k_{eff}}\left(A\rho_{\beta''}Y\left(1-\frac{m_o}{m(\infty)}\right)\right) \tag{6}$$

In Equations (5) and (6), the $\Delta m(t)$ is in g. Equation (6) shows that a plot of $\Delta m(t)$ vs. $\frac{t}{\Delta m(t)}$ should be linear, with the slope given by $4D_{eff}\left(A\rho_{\beta''}Y\left(1-\frac{m_o}{m(\infty)}\right)\right)^2$ and the intercept given by $-\frac{2D_{eff}}{k_{eff}}\left(A\rho_{\beta''}Y\left(1-\frac{m_o}{m(\infty)}\right)\right)$. Thus, from the slope and the intercept, one should be able to estimate D_{eff} and k_{eff}. Figure 6a gives a plot of $\frac{\Delta m(t)}{A}$ vs. $\frac{t}{\left(\frac{\Delta m(t)}{A}\right)}$. Thus, the slope is $4D_{eff}\left(\rho_{\beta''}Y\left(1-\frac{m_o}{m(\infty)}\right)\right)^2$, and the intercept is $-\frac{2D_{eff}}{k_{eff}}\left(\rho_{\beta''}Y\left(1-\frac{m_o}{m(\infty)}\right)\right)$. As seen in Figure 6a, the plot is linear. The corresponding slope was 4.99×10^{-8} g^2 cm^{-4} s^{-1}, and the intercept was -0.04 g cm^{-2}. It is significant that the intercept was negative, as required. The corresponding estimated values of the kinetic parameters are $D_{eff} \cong 1.74 \times 10^{-7}$ cm^2 s^{-1} and $k_{eff} \cong 2.33 \times 10^{-6}$ cm s^{-1}. The experimental data were directly fitted to Equation (5) by a polynomial fitting, as shown in Figure 6b. The estimated values were $D_{eff} \cong 1.76 \times 10^{-7}$ cm^2 s^{-1} and $k_{eff} \cong 2.29 \times 10^{-6}$ cm s^{-1}, which were in good agreement with the results of the linear plot in Figure 6a. The estimated value of D_{eff} was in good agreement with that measured by Parthasarrathy and Virkar [22] based on conversion thickness, which was $D_{eff} \cong 1.5 \times 10^{-7}$ cm^2 s^{-1}. The k_{eff} measured by Parthasarathy and Virkar [22] ranged between ~3.6×10^{-7} cm s^{-1} for large-grained samples to ~1.0×10^{-6} cm s^{-1} for fine-grained samples. The k_{eff} estimated in the present work was thus between 3- and 10-times larger than in the study by Parthasarathy and Virkar. Given the possible differences in microstructures, the agreement is deemed reasonable. The present work thus shows that weight measurements can be used to estimate both kinetic parameters without having to section and polish after each thermal treatment.

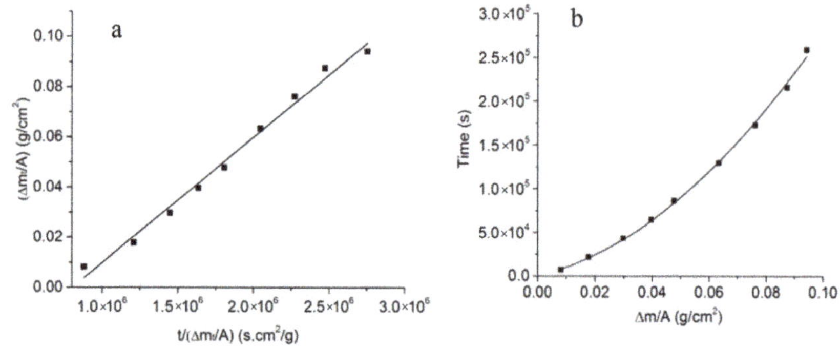

Figure 6. (**a**) A plot of $\frac{\Delta m(t)}{A}$ vs. $\frac{t}{\left(\frac{\Delta m(t)}{A}\right)}$ for the uncoated sample. From the slope and the intercept, both D_{eff} and k_{eff} can be determined. (**b**) A plot of t vs. $\frac{\Delta m(t)}{A}$ for the uncoated sample by a second order polynomial fitting with the intercept set as 0. From the 1st order and 2nd order coefficients, both k_{eff} and D_{eff} can be determined.

3.4. Effects of Na-β″-Alumina Coating on Conversion Kinetics

Figure 7 compares the measured weight changes for the uncoated, 2.5 μm coated, and 15 μm coated (thicknesses correspond to the formed Na-β″-alumina after conversion) disc samples. For the uncoated and 15 μm coated samples, the figure shows the actual measured weight changes (in percent) as a function of time. For the sample with a 2.5 μm coating, the actual conversion thickness was measured as a function of time. In order to graph the data on the same plot, the expected weight changes in the 2.5 μm coated sample were calculated from the measured conversion thickness using Equation (A15). In Figure 7, these data are shown by a dashed line. The uncoated sample showed a much higher weight percentage change than the coated samples. As seen in the figure, the uncoated sample exhibited the largest weight gain (faster kinetics), and the 15 μm coated sample showed the lowest (slowest kinetics of the three samples tested). Also significant is the observation that over the duration of the tests, the weight gain vs. time plots were linear for both of the coated samples, indicating that the kinetics can be described as interface-controlled.

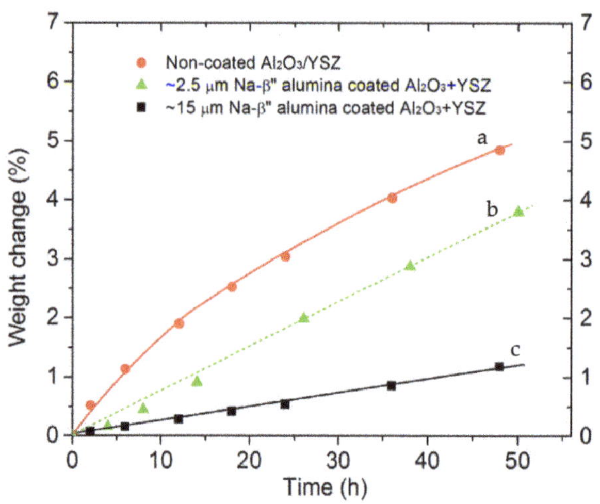

Figure 7. Comparison in weight change for (**a**) uncoated, (**b**) ~2.5 μm Na-β″-alumina-coated, and (**c**) ~15 μm Na-β″-alumina-coated samples as a function of conversion time at 1250 °C.

3.5. Estimation of the Kinetic Parameter for the Coated Samples

Figure 8a,b show plots of $\frac{\Delta m_t}{A}$ vs. t for the two coated samples. Note that these plots are linear, indicating that the kinetics is interface-controlled. Equation (A19) then becomes

$$\frac{\Delta m(t)}{A} \cong 2k_{eff}\left(\rho_{\beta''Y}\left(1-\frac{m_o}{m(\infty)}\right)\right)t \tag{7}$$

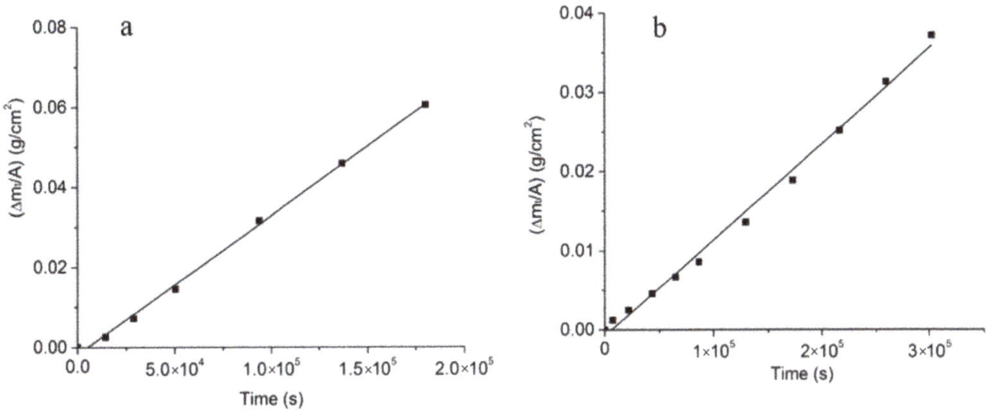

Figure 8. (**a**) A plot of $\frac{\Delta m(t)}{A}$ vs. t for a sample with a 2.5 μm Na-β″-alumina surface coating. From the slope, the k_{eff} is obtained. (**b**) A plot of $\frac{\Delta m(t)}{A}$ vs. t for a sample with a 15 μm Na-β″-alumina surface coating. From the slope, the k_{eff} is obtained.

Thus, the slope is given by

$$2k_{eff}\left(\rho_{\beta''Y}\left(1-\frac{m_o}{m(\infty)}\right)\right) \tag{8}$$

from which the k_{eff} can be estimated. The estimated values of k_{eff} were 6.45×10^{-7} cm s^{-1} for the sample with a 2.5 μm Na-β″-alumina coating and 2.26×10^{-7} cm s^{-1} for the sample with a 15 μm Na-β″-alumina coating. This k_{eff} has a number of series contributions, as discussed below.

Figure 9 shows a schematic variation of the chemical potential of Na_2O from the gas phase, $\mu^I_{Na_2O}$; just inside the single phase Na-β″-alumina close to the gas phase, $\mu^1_{Na_2O}$; in the single phase Na-β″-alumina close to the single phase Na-β″-alumina/Na-β″-alumina + YSZ interface, $\mu^2_{Na_2O}$; just inside the Na-β″-alumina + YSZ close to the single phase Na-β″-alumina/Na-β″-alumina + YSZ interface, $\mu^3_{Na_2O}$; inside the Na-β″-alumina + YSZ close to the Na-β″-alumina + YSZ/α-alumina + YSZ interface, $\mu^4_{Na_2O}$; and inside the α-alumina + YSZ close to the Na-β″-alumina + YSZ/α-alumina + YSZ interface, $\mu^{II}_{Na_2O}$. The transfer of Na_2O across the three interfaces can be described by three interface kinetic parameters, namely, k_{I1}, k_{23}, and k_{4II}. The transport of Na_2O through single-phase Na-β″-alumina is dictated by the chemical diffusion coefficient of Na_2O, $\widetilde{D}^{\beta''}_{Na_2O}$. However, this transport occurs through a layer of fixed thickness, δ. Thus, it reflects as an interface step, with the corresponding k_{12} given by $\frac{\widetilde{D}^{\beta''}_{Na_2O}}{\delta}$. The measured interface parameter, k_{eff}, is thus related to these parameters by the following equation.

$$\frac{1}{k_{eff}} = \frac{1}{k_{I1}} + \frac{1}{k_{23}} + \frac{1}{k_{4II}} + \frac{\delta}{\widetilde{D}^{\beta''}_{Na_2O}} \tag{9}$$

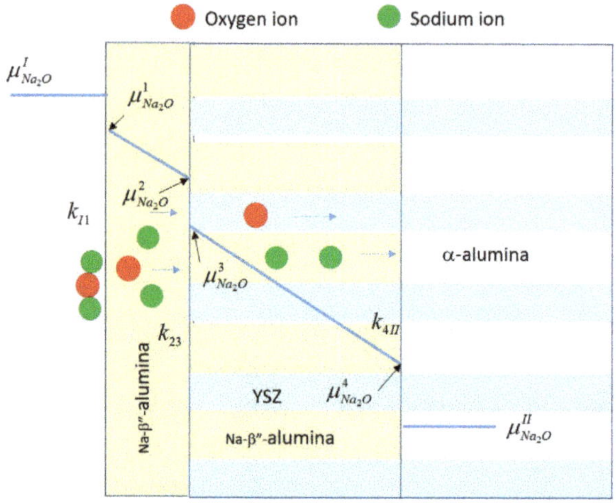

Figure 9. A schematic showing the variation of the chemical potential of Na_2O, μ_{Na_2O}, through a sample with surface layer of Na-β"-alumina of thickness, δ.

The preceding suggests that a plot of $\frac{1}{k_{eff}}$ vs. layer thickness, δ, should be a straight line with the slope given by $\frac{1}{\tilde{D}^{\beta''}_{Na_2O}}$ and the intercept given by $\frac{1}{k_{I1}} + \frac{1}{k_{23}} + \frac{1}{k_{4II}}$. Figure 10 is a plot of $\frac{1}{k_{eff}}$ vs. δ. In this study, measurements were conducted on samples with only two different thicknesses. Thus, there are only two data points. The straight line shown is thus through these two points. Furthermore, plotted on the same plot, however, is $\frac{1}{k_{eff}}$ for the uncoated sample, for which $\delta = 0$. For the uncoated sample, the relevant equation is

$$\frac{1}{k_{eff}} = \frac{1}{k_{I3}} + \frac{1}{k_{4II}} \tag{10}$$

where k_{I3} corresponds to the interface transfer parameter at the gas phase/Na-β"-alumina + YSZ interface.

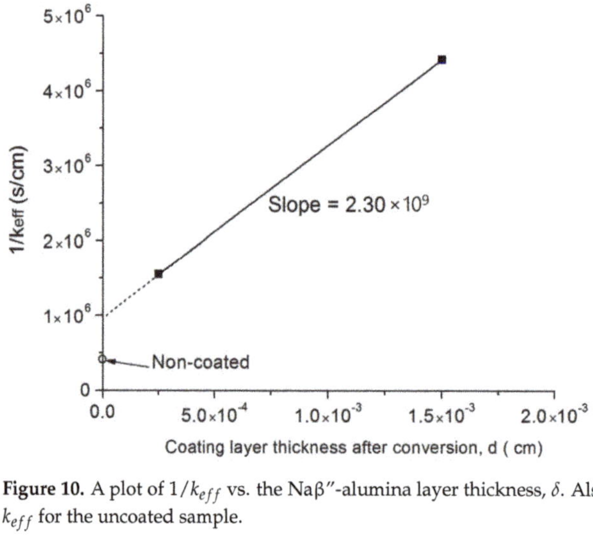

Figure 10. A plot of $1/k_{eff}$ vs. the Naβ"-alumina layer thickness, δ. Also shown in the figure is the k_{eff} for the uncoated sample.

Note that

$$\frac{1}{k_{I3}} + \frac{1}{k_{4II}} \neq \frac{1}{k_{I1}} + \frac{1}{k_{23}} + \frac{1}{k_{4II}} \tag{11}$$

From Figure 10, it is observed that the

$$\frac{1}{k_{I3}} + \frac{1}{k_{4II}} < \frac{1}{k_{I1}} + \frac{1}{k_{23}} + \frac{1}{k_{4II}} \tag{12}$$

or

$$\frac{1}{k_{I3}} < \frac{1}{k_{I1}} + \frac{1}{k_{23}} \tag{13}$$

or

$$k_{I3} > \frac{k_{I1}k_{23}}{k_{I1} + k_{23}} \tag{14}$$

At this stage, it is not possible to separately determine the different interface transfer parameters.

From the slope, the chemical diffusion coefficient of Na_2O through single phase Na-β″-alumina is given by $\widetilde{D}^{\beta''}_{Na_2O} \cong 4.35 \times 10^{-10}$ cm^2 s^{-1}. By contrast, the D_{eff} determined for coupled transport of Na_2O as $2Na^+$ through Na-β″-alumina and O^{2-} through YSZ was ~1.74×10^{-7} cm^2 s^{-1}. That is, $\widetilde{D}^{\beta''/YSZ}_{Na_2O}$ was over 400 times larger than $\widetilde{D}^{\beta''}_{Na_2O}$. As given in Equation (A6), D_{eff} for the two-phase sample is given by

$$D_{eff} = \widetilde{D}^{\beta''/YSZ}_{Na_2O} \frac{2V_m}{f} \Delta C^{\beta''}_{Na_2O} \tag{15}$$

wherein

$$\widetilde{D}^{\beta''/YSZ}_{Na_2O} = \frac{C^{YSZ}_{O^{2-}} D^{YSZ}_{O^{2-}} V_{YSZ}}{C^{\beta''}_{Na_2O}} \frac{1}{k_B T} \left(\frac{\partial \mu_{Na_2O}}{\partial \ln C^{\beta''}_{Na_2O}} \right) \tag{16}$$

That is, the $\widetilde{D}^{\beta''/YSZ}_{Na_2O}$ is proportional to the diffusion coefficient of O^{2-} in YSZ, namely, $D^{YSZ}_{O^{2-}}$. However, in single-phase Na-β″-alumina, the Na_2O chemical diffusion coefficient is given by

$$\widetilde{D}^{\beta''}_{Na_2O} = C^{\beta''}_{O^{2-}} D^{\beta''}_{O^{2-}} \frac{1}{k_B T} \left(\frac{\partial \mu^{\beta''}_{Na_2O}}{\partial C^{\beta''}_{Na_2O}} \right) \tag{17}$$

That is, the $\widetilde{D}^{\beta''}_{Na_2O}$ is proportional to the diffusion coefficient of O^{2-} in Na-β″-alumina. It is well known that oxygen diffusion is much faster in YSZ, an oxygen ion conductor, than in Na-β″-alumina. The present results are consistent with this expectation. Table 1 lists the measured k_{eff} and D_{eff} for alumina-coated and non-coated samples.

Table 1. Summary of measured k_{eff} and D_{eff} for alumina-coated and non-coated samples.

Samples	k_{eff} (cm s^{-1})	D_{eff} (cm^2 s^{-1})
Non-coated α-alumina/YSZ composite	2.33×10^{-6}	1.74×10^{-7}
~2.5 μm Na-β″-alumina-coated α-alumina + YSZ composite	6.45×10^{-7}	1.74×10^{-7}
~15 μm Na-β″-alumina-coated α-alumina + YSZ composite	2.26×10^{-7}	1.74×10^{-7}

3.6. Measurement of Ionic Conductivity

Conductivity was measured on a disc sample of 0.9-mm-thickness that had been converted at 1250 °C for 36 h and a disc sample of 3.7-mm-thickness that had been converted at 1250 °C for 108 h. The heat treatment was long enough to ensure that the entire samples converted into Naβ″AY. Figure 11 shows an Arrhenius plots for the Naβ″AY composite and along with some common Na and Li electrolytes for better comparison [23–29]. Over the range of temperatures measured, the oxygen ion conductivity of YSZ was much lower than the sodium ion conductivity of Na-β″-alumina. Thus, the measured conductivity is attributed exclusively to sodium ion conduction. The measured activation energies of ~23.2 kJ mol^{-1} (after 36 h conversion) and ~21.2 k J mol^{-1} (after 108 h conversion) were also in accordance with the previously reported activation energy of Na^+-ion conduction, especially in fine-grained Na-β″-alumina [10]. At 300 °C, the resistivity was measured as ~16 Ω cm (conductivity of 6.3×10^{-2} S cm^{-1}) and ~8 Ω cm (conductivity of 1.3×10^{-1} S cm^{-1}) for the samples converted for 36 h and 108 h, respectively. The calculated conductivity based on the best linear fitting of the Arrhenius plot at 25 °C was 1.4×10^{-3} and 1.6×10^{-3} S cm^{-1} for the samples converted for 36 h and 108 h, respectively.

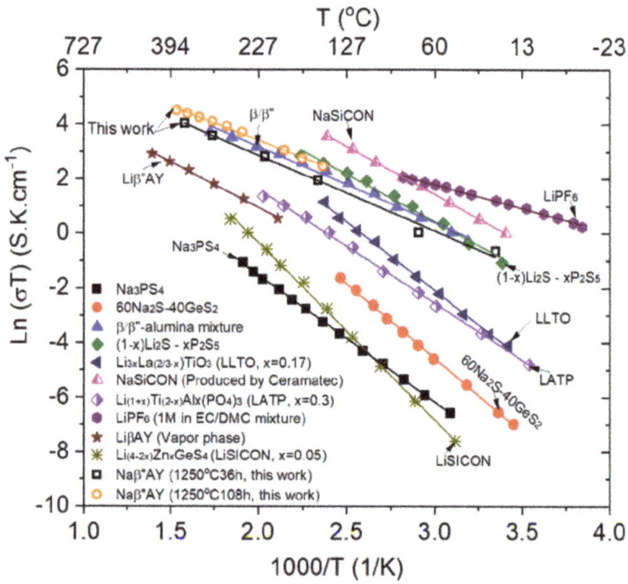

Figure 11. An Arrhenius plot of the measured conductivity of NaβAY and some common Na and Li electrolytes. Conductivity sources: Na_3PS_4 [23], $60Na_2S$-$40GeS_2$ [23], β/β″-alumina mixture [15], $(1-x)Li_2S - xP_2S_5$ [23], $Li_{3x}La_{(2/3-x)}TiO_3$ (LLTO, x = 0.17) [27], NaSiCON (Produced by Ceramatec) [23], $Li_{(1+x)}Ti_{(2-x)}Al_x(PO_4)_3$ (LATP, x = 0.3 [29], $LiPF_6$ (1M in EC/DMC mixture) [26], Liβ″AY (Vapor phase) [25], $Li_{(4-2x)}Zn_xGeS_4$ (LiSICON, x = 0.05) [28].

An interesting feature revealed by this combined plot is that the activation energies of the listed electrolytes do not change vastly regardless of the crystal structure or conduction mechanisms. In comparison with a liquid electrolyte, most of the solid electrolytes show better conductivity at a higher temperature regime. The comparison in the solid electrolytes indicates that the Naβ″AY composite electrolyte showed higher conductivity than most of the listed electrolytes, other than the mixed β/β″-alumina electrolyte and the NaSICON-type Na electrolyte. However, when mechanical strength is of a concern, Naβ″AY (~300 MPa flexural strength [20,21]) may be a better choice, due to a higher mechanical strength than that of NaSICON electrolyte, with ~100 MPa flexural strength [30]. Therefore, the Na″βAY composite produced by the vapor phase conversion process seems a promising solid electrolyte for Na battery applications. It is to be noted that although the Li-β″-alumina + YSZ electrolyte (Liβ″AY) synthesized by the vapor phase conversion process showed lower conductivity as compared with the Naβ″AY electrolyte, its conductivity was still higher than a few of the listed Li solid electrolytes.

4. Conclusions

The vapor phase conversion of α-alumina + YSZ into Na-β″ alumina + YSZ (Naβ″AY) was investigated using the measurement of weight gain and external sample dimensions as a function of time. Vapor phase conversion studies were also conducted on samples coated with thin layers of α-alumina on the one surface, which during the process, first converted into a layer of Na-β″-alumina. In the coated samples, the transport of Na_2O occurred through single phase Na-β″-alumina prior to the conversion of the interior α-alumina in the α-alumina + YSZ composite. From the measurement of conversion kinetics of the uncoated sample, the effective diffusion coefficient, D_{eff}, and effective interface transfer parameter, k_{eff}, were determined. At 1250 °C, the measured D_{eff} was 1.74×10^{-7} cm^2 s^{-1}, in agreement with the previously reported results of Parthasarathy and Virkar [22]. The measured k_{eff} was 2.33×10^{-6} cm s^{-1}, which is within an order of magnitude of the values reported by Parthasarathy and Virkar [22]. Some difference may be related to the possible differences in microstructural details. The D_{eff} is a measure of the chemical diffusion coefficient of Na_2O in the two-phase Naβ″AY composite, $\widetilde{D}_{Na_2O}^{\beta''/YSZ}$, which is proportional to the oxygen ion diffusion in YSZ.

The measured kinetics of conversion of the coated samples were found to be linear in time, indicating interface-controlled kinetics. Since the chemical diffusion of Na_2O must first occur through the surface Na-β″-alumina layer, it reflects as an interface step which is proportional to the oxygen diffusion coefficient through single-phase Na-β″-alumina. From the dependence of conversion kinetics on the thickness of the surface Na-β″-alumina layer, the chemical diffusion coefficient of Na_2O through Na-β″-alumina, $\widetilde{D}_{Na_2O}^{\beta''}$, was estimated. The estimated value was 4.35×10^{-10} cm^2 s^{-1}, which is over 400-times smaller than $\widetilde{D}_{Na_2O}^{\beta''/YSZ}$. The measured $\widetilde{D}_{Na_2O}^{\beta''}$ was proportional to the oxygen diffusion coefficient through Na-β″-alumina. The present work thus provides a method to determine the chemical diffusion coefficient of Na_2O through a Naβ″AY composite, as well as through single-phase Na-β″-alumina. This method may have applicability to other systems.

The measured highest conductivity of the converted Naβ″AY was 1.6×10^{-3} S cm^{-1} at 25 °C and 1.3×10^{-1} S cm^{-1} at 300 °C. These values are higher than many single-phase Li or Na solid electrolytes, indicating the Naβ″AY composite synthesized by the vapor phase conversion process may be used as a promising electrolyte membrane for Na batteries.

Author Contributions: Methodology and original draft preparation, L.Z.; Conceptualization, formal analysis, writing-review and editing, L.Z. and A.V.V. All authors have read and agreed to the published version of the manuscript.

Funding: This work was supported in part by the National Science Foundation, under Grant No. DMR-1407048, and in part by the US Department of Energy, under Grant Number DE-FG02-06ER46086.

Data Availability Statement: The data that support the plots within this paper are available from the corresponding authors upon reasonable request.

Acknowledgments: This work was supported in part by the National Science Foundation, under Grant No. DMR-1407048, and in part by the US Department of Energy, under Grant Number DE-FG02-06ER46086. The authors would like to thank the support from Fuel Cell Research Team and Key Laboratory of Advanced Fuel Cells and Electrolyzers Technology of Zhejiang Province, Ningbo Institute of Materials Technology and Engineering, Chinese Academy of Sciences; Ningbo Municipal Government and the Chinese Academy of Science (Beijing).

Conflicts of Interest: The authors declare no conflict of interest.

Appendix A. Vapor Phase Conversion Kinetics and Transport Parameters

The vapor phase conversion of α-Alumina + YSZ (or with a thin α-alumina coating layer) to Na-β″-alumina/YSZ consists of three steps in series. (a) Step 1: Na_2O from the vapor phase is adsorbed on the Na-β″ alumina + YSZ (or pure Na-β″-alumina for an alumina-coated sample) surface. (b) Step 2: coupled diffusion of $2Na^+$ and O^{2-} through converted Na-β″-alumina + YSZ in the composite or through, first, a thin layer of Na-β″ alumina single phase, followed by coupled diffusion through a Na-β″ alumina + YSZ composite for the coated sample. (c) Step 3: conversion of α-Alumina + YSZ to Na-β″ alumina + YSZ composite at the reaction front; this requires lateral transport of O^{2-} ions arriving from the YSZ phase through Na-β″-alumina, α-Alumina, or along the α-Alumina/Na-β″-alumina interface, as discussed by Parthasarathy and Virkar [22]. Since in both coated and non-coated samples, the reaction front is the same, the interfacial kinetic parameter at this interface should be the same. This transport at the reaction front is similar to the conversion of γ-Fe into pearlite (α-Fe + Fe_3C) in low carbon steels [31].

The kinetics of the conversion process of α-alumina + YSZ into Na-β″-alumina + YSZ was modeled by Parathasarthy and Virkar [22]. Figure A1b shows a schematic. Isothermal conversion kinetics could be described by an equation of the form

$$\frac{x^2}{D_{eff}} + \frac{x}{k_{eff}} = t \tag{A1}$$

where x is the thickness converted in time t, D_{eff} is the effective diffusion coefficient of Na_2O for ambipolar transport of Na_2O through the composite with $2Na^+$ transporting through Na-β″-alumina and O^{2-} transporting through the YSZ, and k_{eff} is the interface transfer parameter that includes processes at the vapor phase/Na-β″-alumina + YSZ interface and also at the Na-β″-alumina + YSZ/α-alumina + YSZ reaction front. The D_{eff} is given by

$$D_{eff} = \frac{2V_{YSZ}C^{YSZ}_{O^{2-}}D^{YSZ}_{O^{2-}}\left(\mu^{I}_{Na_2O} - \mu^{II}_{Na_2O}\right)V_m}{k_B T f} \tag{A2}$$

where V_m is the molar volume of Na-β″-alumina, $\mu^{I}_{Na_2O}$ is the chemical potential of Na_2O in gas phase at the Na_2O-gas phase/Na-β″-alumina interface, $\mu^{II}_{Na_2O}$ is the chemical potential of Na_2O in α-alumina at the Na-β″-alumina + YSZ/α-alumina + YSZ interface, and f is a factor which includes the dimensional change that occurs when α-alumina in the composite converts into Na-β″-alumina. We will write Equation (A2) as

$$D_{eff} = \frac{2V_{YSZ}C^{YSZ}_{O^{2-}}D^{YSZ}_{O^{2-}}\left(\mu^{I}_{Na_2O} - \mu^{II}_{Na_2O}\right)\left(C^{\beta'' I}_{Na_2O} - C^{\beta'' II}_{Na_2O}\right)V_m}{k_B T\left(C^{\beta'' I}_{Na_2O} - C^{\beta'' II}_{Na_2O}\right)f} \tag{A3}$$

or

$$D_{eff} = \frac{2V_{YSZ}C^{YSZ}_{O^{2-}}D^{YSZ}_{O^{2-}}\left(C^{\beta'' I}_{Na_2O} - C^{\beta'' II}_{Na_2O}\right)V_m}{k_B T f}\left(\frac{\partial\mu^{\beta''}_{Na_2O}}{\partial C^{\beta''}_{Na_2O}}\right) \tag{A4}$$

or

$$D_{eff} = \frac{2V_{YSZ}C^{YSZ}_{O^{2-}}D^{YSZ}_{O^{2-}}}{k_B T}\left(\frac{\partial\mu^{\beta''}_{Na_2O}}{\partial C^{\beta''}_{Na_2O}}\right)\frac{V_m}{f}\Delta C^{\beta''}_{Na_2O} \tag{A5}$$

where $\Delta C^{\beta''}_{Na_2O}$ denotes the difference in Na_2O concentration in Na-β″-alumina from the Na_2O-gas phase/Na-β″-alumina boundary to the Na-β″-alumina/α-alumina boundary. A comparison of Equations (A2) and (A5) shows that

$$D_{eff} = \tilde{D}^{\beta''/YSZ}_{Na_2O}\frac{2V_m}{f}\Delta C^{\beta''}_{Na_2O} \tag{A6}$$

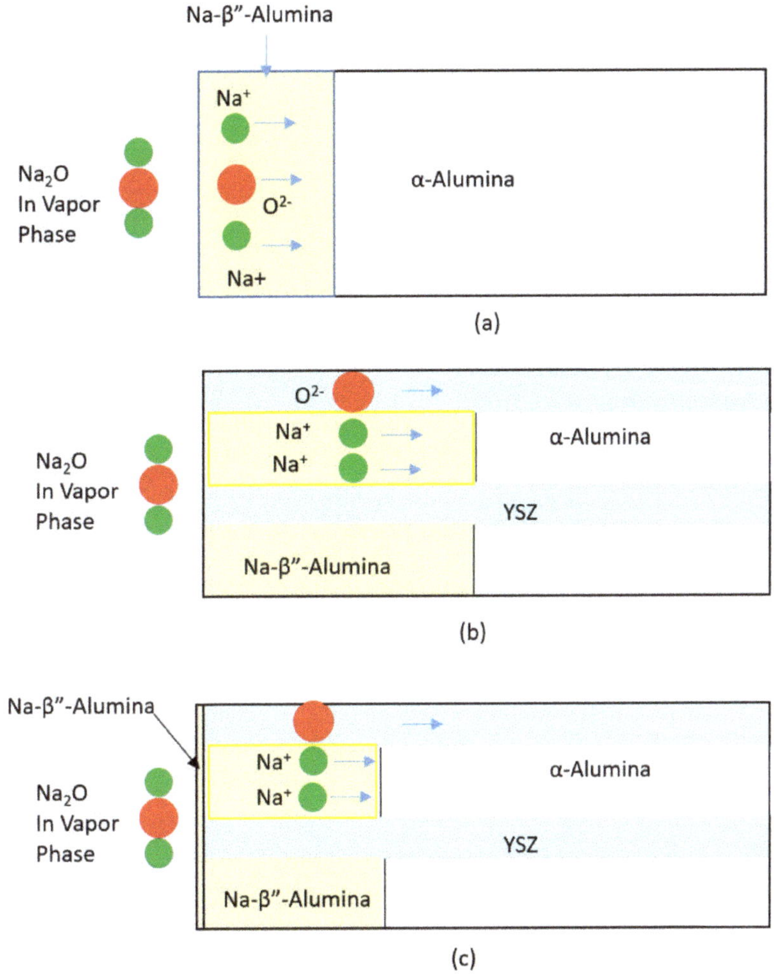

Figure A1. Schematics showing vapor phase conversion in (**a**) single phase α-alumina, (**b**) α-alumina + YSZ composite, and (**c**) α-alumina-coated α-alumina + YSZ composite. The schematics do not show grain structure in this one-dimensional model. The widths of YSZ, α-alumina, and Na-β″-alumina are essentially equal to the respective volume fractions.

Thus, the estimated D_{eff} measured from conversion kinetics is a measure of the chemical diffusion coefficient of Na_2O through the two-phase Na-β″-alumina + YSZ composite, namely, $\tilde{D}_{Na_2O}^{\beta''/YSZ}$, since $\frac{2V_m}{f}\Delta C_{Na_2O}^{\beta''}$ is a dimensionless constant. This chemical diffusion coefficient is proportional to the O^{2-} diffusion coefficient through YSZ.

The k_{eff} also depends on the chemical potentials of Na_2O. Equation (A1) assumes that the conversion thickness at the beginning of the isothermal process is negligible.

Figure A1a shows a schematic of the conversion of single phase α-alumina into Na-β″-alumina by the vapor phase process. In this case, the effective diffusion coefficient, as determined from conversion kinetics, would given by

$$D_{eff} = \widetilde{D}^{\beta''}_{Na_2O}\frac{2V_m}{f'}\Delta C^{\beta''}_{Na_2O} \tag{A7}$$

where f' is the factor that includes the dimensional change when α-alumina converts into Na-β″-alumina. As given in Equation (A7), the effective diffusion coefficient is proportional to the chemical diffusion coefficient of Na_2O through Na-β″-alumina, which is effectively proportional to the O^{2-} diffusion coefficient through single phase Na-β″-alumina. This is expected to be very low compared to the O^{2-} diffusion coefficient through YSZ. Thus, we expect that

$$\widetilde{D}^{\beta''/YSZ}_{Na_2O} >> \widetilde{D}^{\beta''}_{Na_2O} \tag{A8}$$

Figure A1c schematically shows the conversion process in thin α-alumina-coated α-alumina + YSZ composites. Initially, the thin α-alumina layer is converted into Na-β″-alumina, the kinetics of which is governed by the effective diffusion coefficient, corresponding to transport through single-phase Na-β″-alumina, which is governed by oxygen diffusion through single-phase Na-β″-alumina. Once this layer is fully converted into Na-β″-alumina, the conversion of α-alumina in the interior α-alumina + YSZ composite begins. This conversion of the interior α-alumina into Na-β″-alumina occurs by coupled transport of $2Na^+$ through Na-β″-alumina and of O^{2-} through YSZ. However, this conversion requires a continual supply of Na_2O, which must transport through the thin surface layer of the single-phase Na-β″-alumina of a fixed thickness. Because of the fixed thickness of this layer, the overall conversion kinetics contain an additional 'interface' transfer step, corresponding to Na_2O diffusion through the single-phase Na-β″-alumina of a fixed thickness, δ. That is, the transport of Na_2O through the single-phase Na-β″-alumina layer is equivalent to an additional interface step given by

$$k_{\beta''} \propto \frac{D^{\beta''}_{O^{2-}}}{\delta} \tag{A9}$$

The larger the layer thickness, the smaller is the $k_{\beta''}$, and the slower is the kinetics. This suggests that the kinetics of conversion should be slower in samples coated with a Na-β″-alumina layer and that by comparing the kinetics of conversion of samples initially coated with two different thicknesses of alumina layers, it should be possible to estimate the diffusion coefficient of O^{2-} through a single phase Na-β″-alumina.

Appendix B. Study of Kinetics of Vapor Phase Conversion by Weight Measurement

Figure A2 shows a schematic of the conversion process of an α-alumina + YSZ composite. The following analysis assumes that the samples are in the form of thin discs. When α-alumina is converted into Na-β″-alumina, there is both an increase in weight, since Na_2O reacts with Al_2O_3 to form Na-β″-alumina (Na_2O~$6Al_2O_3$), and an increase in volume, since the density of Na-β″-alumina is lower than that of Al_2O_3, and it also includes Na_2O, which is an additional reason for the increase in volume. Prior work has shown that if the disks are sufficiently thin, most of the dimensional change occurs in the thin direction, and the disk diameter remains essentially unchanged. If the initial sample thickness is l_o, the cross-sectional area is A; then, the initial sample volume is $V_o = Al_o$. The density of the α-alumina + YSZ composite is ρ_{AY}, and the density of the Na-β″-alumina is $\rho_{\beta'' Y}$. The initial mass of the sample is

$$m_o = \rho_{AY}V_o = \rho_{AY}Al_o \tag{A10}$$

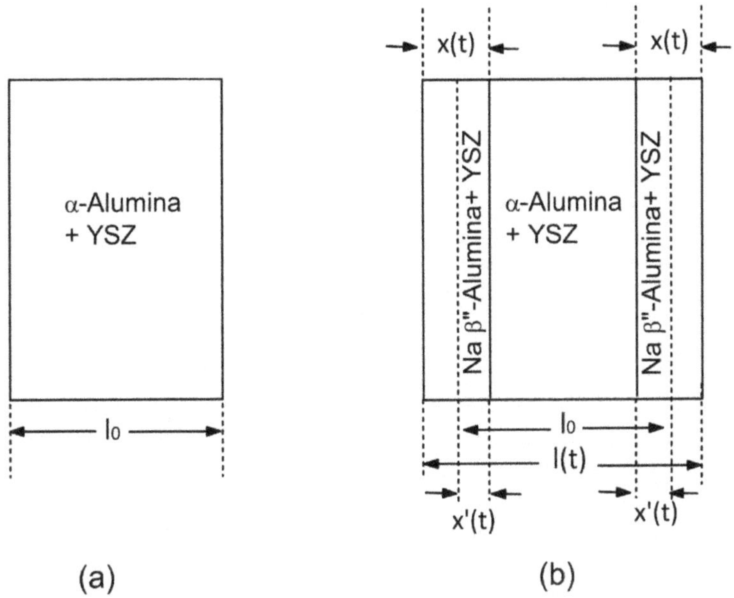

Figure A2. Schematics showing (**a**) a sintered α-alumina + YSZ sample before conversion and (**b**) after partial conversion. $x'(t)$ is the initial thickness of α-alumina + YSZ which converted into Na-β″-alumina + YSZ of thickness $x(t)$, in a conversion time of t.

After time t, the thickness $x'(t)$ of the initial sample on both surfaces converts to Na-β″-alumina + YSZ of thickness $x(t)$. Thus, the sample thickness at time t becomes

$$l(t) = l_o + 2x(t) - 2x'(t) \tag{A11}$$

and the volume at time t is

$$V(t) = Al(t) = A(l_o + 2x(t) - 2x'(t)) \tag{A12}$$

The corresponding mass at time t is

$$m(t) = A\left\{\rho_{AY}(l_o - 2x'(t)) + \rho_{\beta''Y}2x(t)\right\} = \rho_{AY}Al_o + 2A\left(x(t)\rho_{\beta''Y} - x'(t)\rho_{AY}\right) \tag{A13}$$

or

$$m(t) = m_o + \Delta m(t) \tag{A14}$$

where

$$\Delta m(t) = 2A\left(x(t)\rho_{\beta''Y} - x'(t)\rho_{AY}\right) \tag{A15}$$

is the increase in sample weight after time t. Note that

$$\frac{x'(t)}{x(t)} = \frac{l_o}{l(\infty)} = \frac{m_o\rho_{\beta''Y}}{m(\infty)\rho_{AY}} \tag{A16}$$

where the final thickness after full conversion is $l(\infty)$, and the final mass is $m(\infty)$.

Thus,

$$\Delta m(t) = 2Ax(t)\rho_{\beta''Y}\left(1 - \frac{m_o}{m(\infty)}\right) = 2Ax(t)\rho_{\beta''Y}\left(1 - \frac{\rho_{AY}l_o}{\rho_{\beta''Y}l(\infty)}\right) \tag{A17}$$

The ρ_{AY}, l_o, and m_o are measured on the initial sample. The $\rho_{\beta''Y}$ is measured on a fully converted sample. Thus, both $l(\infty)$ and $m(\infty)$ can be estimated. The preceding implies that it is not necessary to carry out full conversion to achieve the final weight, $m(\infty)$, and the final thickness, $l(\infty)$, in order to use the kinetic equation given in (A1). In the present work, however, the sample was heat-treated for a long enough time to achieve the final weight and thickness. Thus, Equation (A17) is used in the analysis.

From Equation (A17), the conversion thickness in terms of weight gain is given by

$$x(t) = \frac{\Delta m(t)}{2A\rho_{\beta''Y}\left(1 - \frac{m_o}{m(\infty)}\right)} = \frac{\Delta m(t)}{2A\rho_{\beta''Y}\left(1 - \frac{\rho_{AY}l_o}{\rho_{\beta''Y}l(\infty)}\right)} \quad \text{(A18)}$$

Equations (A17) and (A18) show that the conversion thickness $x(t)$ at time t can be given in terms of change in weight, $\Delta m(t)$, and the densities of the initial sample, (ρ_{AY}), and after conversion to Na-β"-alumina, ($\rho_{\beta''Y}$), the sample cross sectional area, A; the initial and final sample thicknesses, l_o and $l(\infty)$, respectively; and the initial and final masses, m_o and $m(\infty)$, respectively. Therefore, Equation (A1) can be written as

$$\frac{(\Delta m(t))^2}{4D_{eff}\left(A\rho_{\beta''Y}\left(1 - \frac{m_o}{m(\infty)}\right)\right)^2} + \frac{\Delta m(t)}{2k_{eff}\left(A\rho_{\beta''Y}\left(1 - \frac{m_o}{m(\infty)}\right)\right)} = t \quad \text{(A19)}$$

Thus, by measuring $\Delta m(t)$ as a function of time t and fitting experimental data to Equation (A19), it should be possible to obtain both kinetic parameters, D_{eff} and k_{eff}. Alternatively, Equation (A19) may be linearized as follows

$$\Delta m(t) = 4A^2(\rho_{\beta''Y} - \rho_{AY}\lambda)^2 D_{eff}\frac{t}{\Delta m(t)} - 2A(\rho_{\beta''Y} - \rho_{AY}\lambda)\frac{D_{eff}}{k_{eff}} \quad \text{(A20)}$$

A plot of $\Delta m(t)$ versus $\frac{t}{\Delta m(t)}$ is thus expected to be linear. Then, from the slope and the intercept, it should be possible to obtain both D_{eff} and k_{eff}.

References

1. Weber, N.; Kummer, J.T. Sodium-sulfur Secondary Battery. *Annu. Power Sources Conf.* **1967**, *21*, 3–39.
2. Sudworth, J.L. Zebra batteries. *J. Power Sources* **1994**, *51*, 105–114. [CrossRef]
3. Hunt, T.K.; Weber, N.; Cole, T. High efficiency thermoelectric conversion with beta"-alumina electrolytes, the sodium heat engine. *Solid State Ion.* **1981**, *5*, 263–265. [CrossRef]
4. Fertig, M.P.; Skadell, K.; Schulz, M.; Dirksen, C.; Adelhelm, P.; Stelter, M. From High- to Low-Temperature: The Revival of Sodium-Beta Alumina for Sodium Solid-State Batteries. *Batter. Supercaps* **2022**, *5*, e202100131. [CrossRef]
5. Ligon, S.C.; Bay, M.C.; Heinz, M.V.; Battaglia, C.; Graule, T.; Blugan, G. Large Planar Na-β"-Al_2O_3 Solid Electrolytes for Next Generation Na-Batteries. *Materials* **2020**, *13*, 433. [CrossRef]
6. Yao, Y.; Wang, X.; Dong, C.; Chen, J.; Wang, D.; Mao, Z. Constructing effective interface for room-temperature Beta-Al_2O_3 based sodium metal batteries. *J. Power Sources* **2022**, *523*, 231034. [CrossRef]
7. Ligon, S.C.; Blugan, G.; Bay, M.C.; Battaglia, C.; Heinz, M.V.; Graule, T. Performance analysis of Na-β"-Al_2O_3/YSZ solid electrolytes produced by conventional sintering and by vapor conversion of α-Al_2O_3/YSZ. *Solid State Ion.* **2020**, *345*, 115169. [CrossRef]
8. Bay, M.C.; Wang, M.; Grissa, R.; Heinz, M.V.; Sakamoto, J.; Battaglia, C. Sodium Plating from Na-β"-Alumina Ceramics at Room Temperature, Paving the Way for Fast-Charging All-Solid-State Batteries. *Adv. Energy Mater.* **2020**, *10*, 1902899. [CrossRef]
9. Lee, D.G.; Ahn, B.M.; Ahn, C.W.; Choi, J.H.; Lee, D.H.; Lim, S.K. Fabrication of a Full-Scale Pilot Model of a Cost-Effective Sodium Nickel-Iron Chloride Battery Over 40 Ah. *J. Electrochem. Sci. Technol.* **2021**, *12*, 398–405. [CrossRef]
10. Sudworth, J.; Tiley, A.R. *Sodium Sulfur Battery*, 1st ed.; Springer: Berlin/Heidelberg, Germany, 1985.
11. Zhu, L.; Virkar, A. Sodium, Silver and Lithium-Ion Conducting β"-Alumina + YSZ Composites, Ionic Conductivity and Stability. *Crystals* **2021**, *11*, 293. [CrossRef]
12. Youngblood, G.E.; Virkar, A.V.; Cannon, W.R. Sintering Processes and Heat Treatment Schedules for Conductive, Lithia-stabilized beta"-Al_2O_3. *Am. Ceram. Soc. Bull.* **1977**, *56*, 206–210, 212.
13. Virkar, A.V.; Armstrong, T.J.; Weber, N.; Fung, K.-Z.; Jue, J.-F. Role of coupled transport in the fabrication of sodium beta-alumina containing ceramics by a vapor phase process. *Electrochem. Soc. Proc.* **2002**, *2002-5*, 200–211.

14. Virkar, A.V.; Jue, J.F.; Fung, K.Z. Alkali-Metal-β- and β″-Alumina and Gallate Polycrystalline Ceramics and Fabrication by a Vapor Phase Method. Google Patent US-6632763-B2, 14 October 2003.
15. Li, G.; Lu, X.; Kim, J.Y.; Meinhardt, K.D.; Chang, H.J.; Canfield, N.L.; Sprenkle, V.L. Advanced intermediate temperature sodium–nickel chloride batteries with ultra-high energy density. *Nat. Commun.* **2016**, *7*, 10683. [CrossRef] [PubMed]
16. Li, G.; Lu, X.; Kim, J.Y.; Lemmon, J.P.; Sprenkle, V.L. Cell degradation of a Na-NiCl2 (ZEBRA) battery. *J. Mater. Chem. A* **2013**, *1*, 14935–14942. [CrossRef]
17. Wang, L.; Lu, Y.; Liu, J.; Xu, M.; Cheng, J.; Zhang, D.; Goodenough, J.B. A superior low-cost cathode for a Na-Ion battery. *Angew. Chem. Int. Ed.* **2013**, *52*, 1964–1967. [CrossRef] [PubMed]
18. Lu, X.; Li, G.; Kim, J.Y.; Meinhardt, K.D.; Sprenkle, V.L. Enhanced sintering of β″-Al_2O_3/YSZ with the sintering aids of TiO2 and MnO2. *J. Power Sources* **2015**, *295*, 167–174. [CrossRef]
19. Lu, X.; Kirby, B.W.; Xu, W.; Li, G.; Kim, J.Y.; Lemmon, J.P.; Sprenkle, V.L.; Yang, Z. Advanced intermediate-temperature Na–S battery. *Energy Environ. Sci.* **2012**, *6*, 299–306. [CrossRef]
20. Lin, G.-Y.; Virkar, A.V. Development of Surface Compressive Stresses in Zirconia-Alumina Composites by an Ion-Exchange Process. *J. Am. Ceram. Soc.* **2004**, *84*, 1321–1326. [CrossRef]
21. Canfield, N.L.; Kim, J.Y.; Bonnett, J.F.; Pearson, R.; Sprenkle, V.L.; Jung, K. Effects of fabrication conditions on mechanical properties and microstructure of duplex β″-Al_2O_3 solid electrolyte. *Mater. Sci. Eng. B* **2015**, *197*, 43–50. [CrossRef]
22. Parthasarathy, P.; Virkar, A.V. Vapor Phase Conversion of α-Alumina + Zirconia Composites into Sodium Ion Conducting Na-β″-Alumina + Zirconia Solid Electrolytes. *J. Electrochem. Soc.* **2013**, *160*, A2268–A2280. [CrossRef]
23. Vignarooban, K.; Kushagra, R.; Elango, A.; Badami, P.; Mellander, B.-E.; Xu, X.; Tucker, T.; Nam, C.; Kannan, A. Current trends and future challenges of electrolytes for sodium-ion batteries. *Int. J. Hydrogen Energy* **2016**, *41*, 2829–2846. [CrossRef]
24. Fergus, J.W. Ceramic and polymeric solid electrolytes for lithium-ion batteries. *J. Power Sources* **2010**, *195*, 4554–4569. [CrossRef]
25. Koh, J.-H.; Weber, N.; Virkar, A.V. Synthesis of lithium-beta-alumina by various ion-exchange and conversion processes. *Solid State Ion.* **2012**, *220*, 32–38. [CrossRef]
26. Dahbi, M.; Ghamouss, F.; Tran-Van, F.; Lemordant, D.; Anouti, M. Comparative study of EC/DMC LiTFSI and LiPF6 electrolytes for electrochemical storage. *J. Power Sources* **2011**, *196*, 9743–9750. [CrossRef]
27. Jimenez, R.; Rivera, A.; Varez, A.; Sanz, J. Li mobility in Li0.5−xNaxLa0.5TiO3 perovskites ($0 \leq x \leq 0.5$) Influence of structural and compositional parameters. *Solid State Ion.* **2009**, *180*, 1362–1371.
28. Knauth, P. Inorganic solid Li ion conductors: An overview. *Solid State Ion.* **2009**, *180*, 911–916. [CrossRef]
29. Kosova, N.V.; Devyatkina, E.T.; Stepanov, A.P.; Buzlukov, A.L. Lithium conductivity and lithium diffusion in NASICON-type Li1+xTi2−xAlx(PO4)3 (x = 0; 0.3) prepared by mechanical activation. *Ionics* **2008**, *14*, 303–311. [CrossRef]
30. Kim, J.; Jo, S.H.; Bhavaraju, S.; Eccleston, A.; Kang, S.O. Low temperature performance of sodium–nickel chloride batteries with NaSICON solid electrolyte. *J. Electroanal. Chem.* **2015**, *759 Pt 2*, 201–206. [CrossRef]
31. Fine, M.E. *Introduction to Phase Transformations in Condensed Systems*; Macmillan: New York, NY, USA, 1964.

electrochem

Article

Fault Structural Analysis Applied to Proton Exchange Membrane Fuel Cell Water Management Issues

Etienne Dijoux [1,2,*], Nadia Yousfi Steiner [2], Michel Benne [1], Marie-Cécile Péra [2] and Brigitte Grondin-Perez [1]

[1] ENERGY Lab—LE2P, University La Reunion, 97415 Saint-Denis, France; michel.benne@univ-reunion.fr (M.B.); Brigitte.grondin@univ-reunion.fr (B.G.-P.)
[2] FEMTO-ST Institute, FCLAB, University Bourgogne Franche-Comté, 90000 Belfort, France; nadia.steiner@univ-fcomte.fr (N.Y.S.); marie-cecile.pera@univ-fcomte.fr (M.-C.P.)
* Correspondence: etienne.dijoux@univ-reunion.fr

Abstract: Proton exchange membrane fuel cells are relevant systems for power generation. However, they suffer from a lack of reliability, mainly due to their structural complexity. Indeed, their operation involves electrochemical, thermal, and electrical phenomena that imply a strong coupling, making it harder to maintain nominal operation. This complexity causes several issues for the design of appropriate control, diagnosis, or fault-tolerant control strategies. It is therefore mandatory to understand the fuel cell structure for a relevant design of these kinds of strategies. This paper proposes a fuel cell fault structural analysis approach that leads to the proposition of a structural graph. This graph will then be used to highlight the interactions between the control variables and the functionalities of a fuel cell, and therefore to emphasize how changing a parameter to mitigate a fault can influence the fuel cell state and eventually cause another fault. The final aim of this work is to allow an easier implementation of an efficient and fault-tolerant control strategy on the basis of the proposed graphical representation.

Keywords: fuel cell fault structural analysis; diagnosis; fault tolerant control; fuel cell cathode water management

Citation: Dijoux, E.; Steiner, N.Y.; Benne, M.; Péra, M.-C.; Grondin-Perez, B. Fault Structural Analysis Applied to Proton Exchange Membrane Fuel Cell Water Management Issues. *Electrochem* **2021**, 2, 604–630. https://doi.org/10.3390/electrochem2040038

Academic Editors: Marc Cretin, Sophie Tingry and Zhenghua Tang

Received: 30 September 2021
Accepted: 19 October 2021
Published: 1 November 2021

1. Introduction

Proton exchange membrane fuel cells (PEMFCs) are efficient and clean energy supply systems. However, they are subject to the occurrence of various faults, which decreases their reliability. Faults are inordinate phenomena that degrade a system's performance more or less rapidly and substantially [1]. Their occurrence can be attributed to several factors (exogenous and endogenous). Both exogenous factors, such as gas purity or demanding load profile, and endogenous factors, such as poor internal design or natural aging, can lead to fault occurrence and, therefore, to fuel cell damage. The operating conditions need to be adjusted to mitigate the faults. Moreover, PEMFCs are nonlinear, multivariate, and strongly coupled systems, which complicates their ability to be maintained under normal operation. Therefore, it is necessary to highlight the coupling of the PEMFCs' parameters to facilitate the understanding of the fault occurrence process. Indeed, an exhaustive analysis of the variables' effects and interactions inside the system is a major issue to be considered to set an efficient fault-tolerant control (FTC) strategy.

The literature provides different approaches to modeling and analyzing a system in order to understand the influences of its variables and their interactions. Noyes [1] highlighted two types of methodologies that allow this analysis: statistical and functional. On the one hand, a statistical analysis consists in the observation of events [2,3] with the aim of making assumptions to predict events in similar situations. For instance, Bayesian statistical analysis [2] aims to process small datasets. Indeed, this approach allows one to obtain relevant information by limiting costly observations. The obtained information is then iteratively refined according to a Bayesian law. On the other hand, functional analysis, focuses on all functions ensured by the system and their influence on the occurrence of a fault. These approaches are usually based on a graphical formalism that leads to the specification of good operating conditions.

Several methods for functional analysis can be found in the literature, such as FAST (Function Analysis System Technique), SADT (Structured Analysis and Design Technique), or FTA (Fault Tree Analysis).

FAST is a graphical representation of a system's functions that answers the following questions: why does a function have to be ensured? How is this function ensured? When must the function be ensured? Other functional analysis methods also exist in the literature [4–6], and they consist in analyzing a system with measurements to observe its structural and functional modifications.

The SADT [7,8] method is a graphical tool associated with a top-down analysis method. This method is used to decompose a function with a functionally oriented methodology. It consists of modeling the process by breaking it down into subsets. A data-based or function-based diagram is then created to model the process of each subset.

FTA is also a functionally based analysis that allows one to perform a failure mode analysis. It is a top-down approach: a top event is considered, and all combinations of sub-events that lead to it are determined [9].

The top event is reached with a combination of several sub-events. Sub-events have their origin in the combination of basic events. Fault trees, therefore, bring information about the variables (basic events) that are involved in a specific fault occurrence.

To summarize, on the one hand, FAST uses a diagram to organize the ways of thinking, acting, or talking. It enables the development of technical solutions according to a functional logic. However, this method does not consider the system complexity or the coupling phenomena of internal system variables. On the other hand, the SADT method takes into consideration the complexity and allows the analytical decomposition of the system according to a hierarchical structure. However, this approach does not allow links between transitions of operating conditions, such as from normal condition to a faulty one. On the contrary, FTA is relevant in understanding how a fault can occur with a combination of basic events. Indeed, FTA allows the link between each transition of the operating conditions. With logic gates and sub-events, it describes all paths that lead to the appearance of the fault. FTA highlights the information about fault occurrence and, therefore, the relevant variables for fault mitigation. Fuel cells are strongly coupled systems, and several variables have mutual interactions that are not highlighted by FTA. For this reason, a new analysis methodology, which is called Fault Structural Analysis (FSA), is applied to these systems. Indeed, the FSA allows describing the system's structure and highlighting all variables that influence the fuel cell operating conditions. This approach is also relevant for designing fault tolerant control strategies because it is helpful for fault mitigation and system monitoring process. For instance, as presented in [10], authors present their strategy for fuel cell fault mitigation. Their work consists of gathering information about the PEMFC state of health through the remaining useful lifetime. The approach is based on the analysis of the system nominal and faulty conditions which are provided by a key variable behavior. This strategy is thus highly dependent on the relevant choice of the key variable that should be subject to a study of its field of action in the fuel cell for more efficiency. In [11], Yang et al. try to improve the PEMFC reliability with the implementation of a robust fault observer for air management system fault diagnosis. Once again, the choice of the estimated variable is a key factor for their diagnosis tool. Indeed, the implementation of their strategy depends on sensitivity of the diagnosis variables to the fuel cell functionalities which are subject to faulty conditions. In [12], authors proposed a fuel cell health management system. They used the electrochemical impedance spectroscopy (EIS) in a fault tolerant control strategy in order to diagnose the water management faults. The drawback of EIS diagnosis tools lies in their low computational time, their offline operating mode and the cost of the used equipment. To avoid this problem, a solution for the implementation of a diagnostic tool can be based, for each relevant variable, on identifying the one most influenced by each faulty condition. Another study proposed by Rubio et al. [13], consist of the implementation of a fuzzy model to determine the water dehydration in a PEMFC. The real-time aspect of the strategy involves the use of fast response time of the control variable. The current, the flow rate and the voltage are thus used in the strategy for the fuel cell hydration characterization. This study only considers fast response time variables for the diagnosis tool, but the studied phenomena have low, medium and high frequency behavior. In the case of the introduction of variables which are influenced on the overall spectrum, authors would improve their strategy efficiency.

The FSA design leads to a graphical representation which is achieved in three steps: (i) identification of the system's functionalities and definition of each control variable, (ii) finding the constraints (restriction of the system functionalities) that are influenced by system's variables and (iii) designing the structural graph.

Water management faults is a recurrent faulty condition for PEMFC systems. Indeed, they can lead to severe performance losses and in some cases to irreversible degradation. Therefore, the FSA approach that leads to a structural graph considers two faulty conditions related to the water management: flooding and membrane drying out. The introduction of these two faults in the structural graph will underline the fuel cell functionalities impacted by their occurrence, and also highlight the available control variables which can be used for their mitigation in the case of an FTC strategy. FSA leads thus to the following contributions:

- Describe the fuel cell structure only with a graph,
- Highlight all variables which influence the fuel cell functionalities and therefore its operating conditions,
- Underline the links between the fuel cell functionalities and faults,
- Highlight the relevant control variables which can be used for fault mitigation

This paper is organized as follows: The first section is dedicated to the PEMFC water management issues. The second section proposes a structural analysis which allows highlighting all couplings between the system variables. Then, in the third section the structural analysis approach is defined and applied on the PEMFC system. The two last sections discuss and conclude about the structural analysis approach applied to a PEMFC system.

2. Water Management Faults

PEMFC systems may be subjected to different faulty operating modes. In [14], authors define fault as a decrease in system performance caused by improper major or minor fuel cell operation. These kinds of operations could lead to a permanent loss of the fuel cell performance due to the occurrence of faults or fuel cell ageing. This paper only focuses on faults which lead to fuel cells performance losses. These faults can be classified according to some criteria like: effects, response time, recovery property (the loss of performance can be totally recovered or not) or location.

Fuel cell faults can be detected and isolated, with the use of appropriate fault diagnosis tools. The literature relates several diagnosis techniques and many of them are used for PEMFCs water management issues. For instance, Lu et al. [15] proposed a fault diagnosis based on a fast electrochemical impedance spectroscopy (EIS) measurement. The developed tool allows an on-line flooding and drying out diagnosis but the authors underline that it cannot rules on the fuel cell state of health (SoH) in the case of multiple fault occurrence. Regarding their experiment, the multiple fault occurrence happens when the recovering time between each faulty condition is not enough. Therefore, in order to have a complete fault deletion and to improve the diagnosis tool performance, it is therefore relevant to know what the involved variables during a fault mitigation process are. Another example of a diagnosis tool consists of a model-based observer for fuel cell internal states estimation [16]. Authors aim to resolve the unmeasured internal variables issue thanks to a virtual sensor based on observer. This work highlights the need to understand the fuel cells operation mechanism through the internal state estimation and by the coupled variables involved in the change of the fuel cell SoH. In the follow-up, Alves-Lima et al. [17]

have worked on a quantitative video-rate hydration imaging of Nafion. Their results have shown that membrane water content is correlated to several factors: the membrane thickness, the fuel cell temperature (the room temperature in their study) and the water desorption process are major factors that influence the membrane hydration. Once again, the knowledge of the fuel cell internal variables are major issues to maintain the membrane under nominal hydration. In [18] authors use an EIS measurement to characterize the impact of the membrane water management on the performance of PEMFC commercial stacks. Their goal is to understand what the effects of the inlet gas water content on the fuel cell operations via the analysis of the Nyquist plot for fuel cell stack and single cells are.

These papers underline the need of knowledge about the internal fuel cell variables and how they have mutual influences. Indeed, as shown in [17], a modification of only one internal variable value could lead to the system destabilization. For this reason, the FSA analysis is focused on all the variables involved in the fuel cell to highlight extensively and systematically the multilateral effects of water management faults inside the fuel cell.

Cell flooding and membrane drying out are two possible consequences of improper water management. Flooding is caused by an accumulation of liquid water either in the diffusion layer of the electrodes, the bipolar plate channels or the feeding lines that limits the access of the reactants to the catalyst sites, and then decreasing the electrochemical reaction rates. Membrane drying out is the result of an insufficient hydration of the PEMFC membrane, thus increasing its proton resistivity.

Li et al. [19] identified flooding as the most recurrent PEMFCs' fault and point out that the cathode—being the place of water production—is particularly affected. The antagonist phenomenon is the drying out. It can occur when the gases relative humidity is too low [20,21], the input gas flow rate is too high, the operating temperature is too high or when these improper operating parameters are combined.

To better understand these faults, a functional analysis of PEMFC water management has been investigated through the Fault Tree Analysis in [11], among the possible approaches presented previously. The FTA is reported in Figure 1.

The FTA approach provides information about some coupling phenomena between variables inside the system. For instance, the temperature (T) influences the gas relative humidity. However, it is also linked to other variables which can be used to settle a fault tolerant control law. It is therefore important to understand how the different control variables are coupled to reduce the risk of unexpected phenomena. The proposed FSA approach aims to bring out this information in an explicit and systematic way via the design of diagrams called structural graphs.

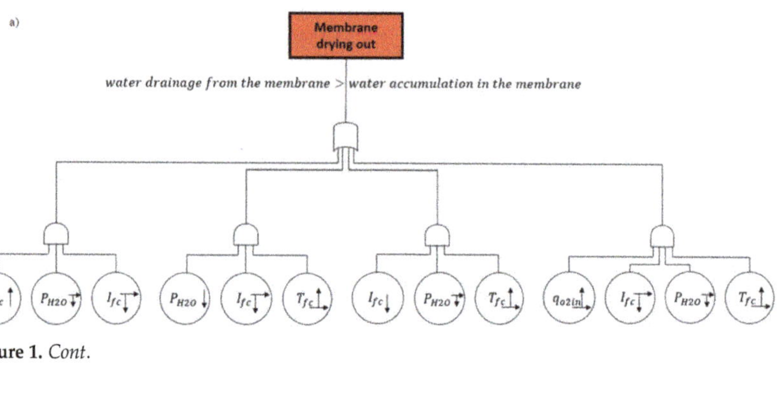

Figure 1. *Cont.*

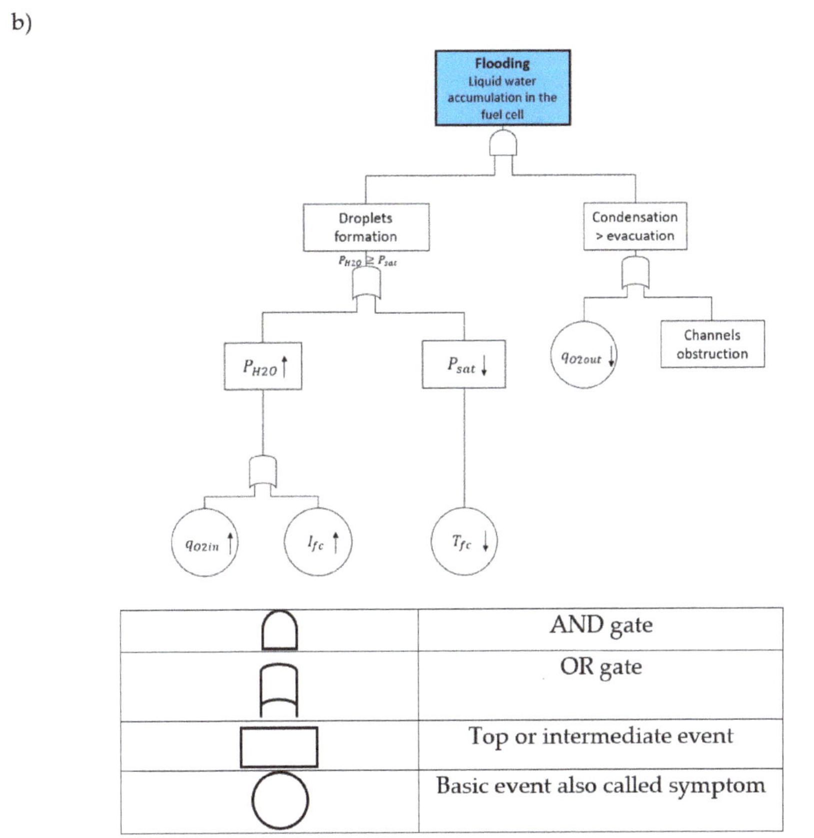

Figure 1. FTA applied on membrane drying out (**a**) and on a flooding (**b**) reproduced with the permission from Yousfi Steiner et al., 2021 [9].

3. Fault Structural Analysis: Definition and Objectives

FSA is a low-level representation of a system behavior that allows highlighting the operating conditions that potentially lead to a fault occurrence and the variables that could be used to mitigate this fault. The representation is based on connections between the system variables through a bipartite graph. All system features are described by a set of constraints, viewed as restrictions on the system functionalities, and the violation of one of them that indicates a fault occurrence.

3.1. Dynamical Systems

Structural analysis considers only the structural information, highlighting the variable involved in the studied phenomenon. A representation is given in Figure 2.

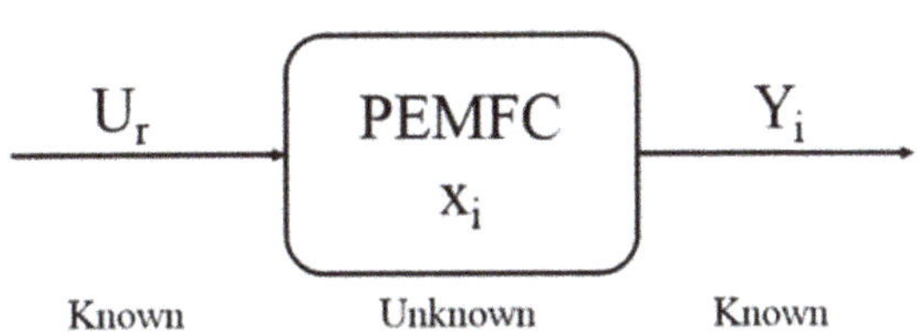

Figure 2. Known and unknown variables representation.

Inputs U_r are defined as: U_r with r ϵ {1,2, . . . ,n}

Outputs Y_i are defined as: Y_i with i ϵ {1,2, . . . ,k}

Unknown states, which are not directly measured but could be estimated from the known ones, they are defined by: x_j with j ϵ {1,2, . . . ,l}.

3.2. *Bipartite Graph*

Beauguitte [22] defined the bipartite graph as a structure that displays relationships between two separate sets of vertices, describing system characteristics, variables, and constraints. For this reason, in the next section the bipartite graph will be called structural graph. The author underlines that vertices can be separated according to their contribution to an event such as the occurrence of a particular fault.

The links between vertices are usually not oriented and represent the system's structure which is noted as: G = (C U Z, Γ), where G is the structure of the bipartite graph and U the union operator. Z is the set of characteristic variables: Z {z1, z2, . . . , zn}. C is the set of constraints: C = {c1, c2, . . . , cm}. Arcs that connect each vertex (constraints/variables) are noted:

$$\Gamma = \{(c\_i,z\_j) \dashv \mid z\_j \text{ exist in } c\_i, z\_j \in Z, i \in [1,m], j \in [1,n]\}. \quad (1)$$

An illustration of a bipartite graph is given in Figure 3.

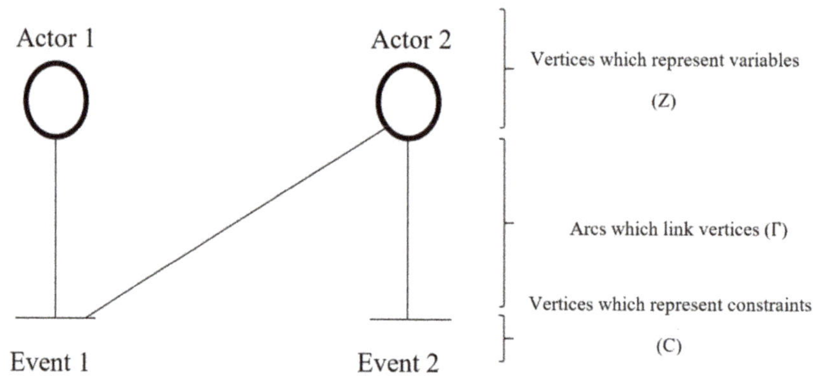

Figure 3. Illustration of a bipartite graph.

It is possible to create an incidence matrix M that represents the bipartite graph and consists of a Boolean matrix where the rows are the constraints and the columns the characteristic variables:

$$M = \{m_{i,j} \mid m_{i,j} = 1 \text{ if } (c_i,v_j) \in \Gamma, 0 \text{ else}\} \quad (2)$$

3.3. *Differentiation*

The studied phenomena (faults) are time dependent. Therefore, a dynamic model is mandatory in order to take into account time dependent variables. Three options can be considered:

Option 1: Considering x (variable of the studied system) and $\dot{x}$ as the same variable and treating the dynamical equations in the same way as the static ones.

Option 2: Considering x and x = dx/dt as structurally distinct and using the model with the explicit differential equation: dx/dt ([23]).

Option 3: Considering x and x = dx/dt as structurally distinct and proceeding to the structural differentiation of the initial model ([24]).

The choice to use the explicit differential equation $\dot{x}$ is preferred in our study to take into consideration the system dynamic behavior with the present and past values (option 2).

3.4. Representation of Faults

As said before, faults are abnormal phenomena which decrease the system performance and can lead to its degradation. It is thus mandatory to detect their occurrence to proceed to a fault mitigation strategy. In case a diagnosis tool is used, several levels of a priori knowledge must be considered.

- A low level of knowledge, that only allows specifying the system functionalities (i.e., the function which describes the operations) which are influenced by the faults' occurrences. The description of the cause-and-effect relationship is not mandatory.
- A medium level of knowledge for which an analytical description of the cause-and-effect relationships between system functionalities and faults is available. In this case, variables that describe the faults must be integrated into the model.
- A high level of knowledge for which a fault model is specified.

For this work, a low level of knowledge is considered. Indeed, accurate fault model is not mandatory. Only the functionalities that are influenced by their occurrence are needed whereas the cause-and-effect relationship is not.

The FSA approach and objectives are now explained. The structural graph design process is therefore described by the following steps: (i) find a model of the system which describes its functionalities, (ii) identify the system constraints in order to get all the variables which have an influence on the system operation, (iii) create an incidence matrix for the design of the structural graph. The next section consists in choosing a PEMFC model to guide the structural graph design.

4. PEMFC Functionalities and Control Variables

4.1. PEMFC System

A PEM fuel cell is composed of an anode, supplied with hydrogen, and a cathode, supplied with oxygen (pure or from the air). At the anode side, one molecule of hydrogen is oxidized thanks to a catalyst made of platinum (Pt). It allows the formation of two protons and two electrons:

$$H_2 \rightarrow 2H^+ + 2e^- \quad (3)$$

At the cathode side, oxygen is reduced by the protons thanks to a catalyst made of platinum (Pt). Its reduction allows the formation of water through the electrochemical reaction:

$$1/2\, O_2 + 2H^+ + 2e^- \rightarrow H_2O \quad (4)$$

The electrons move from the anode to the cathode through an external electric circuit. These two electrodes are separated by a proton exchange membrane that is impermeable to reactant gases. The supply of reactants to the PEMFC and the exhausts are carried in and out via channels in bipolar or mono-polar plates as represented in [25].

For the next sections, the studied system is a PEMFC stack. All cells are supposed to have the same mean behavior, and the reactant supply system is represented on the scheme on Figure 4:

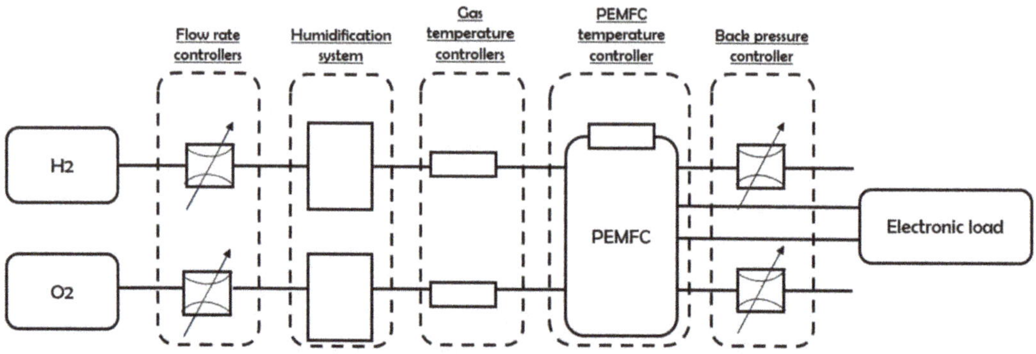

Figure 4. Example of humidified system synoptic of a PEMFC.

The Figure 4 represents the main elements of the PEMFC system that will be involved in fault generations and then monitored for their mitigation. Indeed, the system is supplied with hydrogen and oxygen and the gas flows are controlled with two controllers which manage to supply the PEMFC with the good ratio of reactants. The reactants are humidified with the humidification system. The PEMFC input gas relative humidity is adjusted with a heated line. The PEMFC pressure is controlled with two back pressure controllers. To simplify the representation, in the following, the considered oxidant is pure oxygen, but the approach would be the same with air.

4.2. PEMFC Modeling with an Energetic Macroscopic Representation (EMR)

To perform an FSA for PEMFCs water management faults, a zero-D fuel cell model is considered. The energetic macroscopic representation (EMR) proposed in [26] is used to identify the system functionalities. This macroscopic model is chosen because it highlights the system's control variables. Indeed, the structural analysis presented in this paper aims to bring the relevant control variables which could mitigate a fault occurrence. Other multidimensional models would lead the same analysis, but they involve more variables which are not measurable or controllable. Therefore, this kind of model are useless for the aim of fault mitigation.

The EMR representation is a quasi-static model involving differential equations that make possible the study of time-dependent systems. It also specifies the equations that describe the system's functionalities that could be submitted to a faulty condition. The principle of the EMR consists in considering the different power conversions and the cause-consequence relationship. Then, each element is linked to the others through a couple of action/reaction variables which product is a power. The instantaneous conversion and the accumulation of power are distinguished.

The PEMFC is modelled as several sub systems linked together in order to create a complete fuel cell model. This simple model and few additional equations are relevant to perform the ASD because it allows taking into account the variables of interest.

This representation is composed of three parts.

First, the fluidic part is dedicated to the gas channels located between the gas tank and the reaction sites. An electric analogy is used with pressure is assumed to be a potential and volume flow rates are assumed to be currents. Using electrical analogy:

- the system is also composed of a distribution resistance RdO21, an exhaust resistance RdO22 and a hydraulic capacity Ch (for gas accumulation in the circuit),
- and the current generator represents the consumption of the reactant or the generation of the products.

The Figure 5 shows the analogy between the pipe and the RC electrical circuit [27].

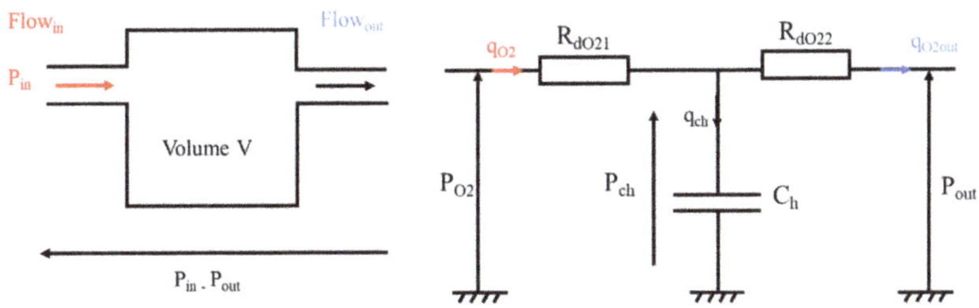

Figure 5. Analogy between pipe of the hydraulic circuit and an electric RC circuit.

The oxygen circulation qO2 is enabled by fuel cell input/output pressure difference. The consumed (resp. produced) gases' flows qcO2 are calculated from the Faraday law and the number of electrons exchanged nO2 counted negatively (resp. positively). Partial pressures at each electrode are imposed by the gas accumulation represented by qch. Ifc is the electric current of the fuel cell, N is the number of cells.

$$PO2 = PscO2 + RdO21\ qO2 \tag{5}$$

$$qO2out = (PscO2 - PsO2)/RdO22 \tag{6}$$

$$(dPscO2)/dt = 1/Ch\ (qO2 - qcO2 - qO2out) \tag{7}$$

$$qcO2 = \pm N * Ifc/(nO2 * F) * (R * Tfc)/PO2 \tag{8}$$

$$qO2out = qO2 + qcO2 \tag{9}$$

In Equation (7) the derivative term is assumed to be structurally different from PscO2 (Cf. Section c).

The second part of the EMR is the electrochemical part. The Nernst potential En is based on the computation of a thermodynamic potential E0 and the influence of the partial pressures and the temperature:

$$En = E0 + \Delta E \tag{10}$$

with,

$$E0 = \alpha + \beta * Tfc + \gamma * Tfc\hat{}2 + \delta * Tfc\hat{}3 + \nu * Tfc * \ln(Tfc) \tag{11}$$

and,

$$\Delta E = Acd * \ln\ (PscH2/P0) + Bcd * \ln(PscO2/P0) \tag{12}$$

where α, β, γ, ν, Acd(Tfc) and Bcd(Tfc) are model adjustment variables. Tfc is the fuel cell temperature and the chosen model does consider a cooling system. E0 is the thermodynamic potential of the PEMFC [26]. The voltage drop ΔV calculation (activation, concentration and ohmic losses) is then carried out:

$$\Delta V = 〖\Delta V〗\_act + 〖\Delta V〗\_conc + 〖\Delta V〗\_ohm \tag{13}$$

The PEMFC voltage is thus:

$$V\_M = E\_n - \Delta V \tag{14}$$

Each loss is expressed as follows, using In the cross over current, I0 the exchange current, and Il the limit current:

$$\Delta Vact = A * Tfc * \ln((Ifc + In)/I0) \tag{15}$$

$$\Delta Vconc = B * Tfc * \ln(1 - Ifc/Il) \tag{16}$$

$$\Delta Vohm = Rm * Ifc \tag{17}$$

The third part is the electric impedance of the cell, represented by the block "Charge double layer" but as the fast-electric dynamic is not considered in the analysis, this block is not considered. Considering N the number of stack cells, the stack voltage is expressed as follows:

$$Vfc = N * (VM + Vc) \tag{18}$$

Thanks to this fuel cell model, it is possible to define the control variables for the system functionalities. All variables that influence the PEMFC functionalities are supposed to be known. The next step consists of defining the constraints that can be influenced by a fault occurrence.

5. Constraints of PEMFC

5.1. Structural Analysis of a PEMFC

The structural analysis design is focused on the cathode fluidic part because it is the location of the water production. The cathode area therefore represents the major issue regarding fuel cells water management. It is located between the gas tank and the reaction sites. This part is modeled with the electric analogy represented on Figure 5. The oxygen circuit is composed of a fluidic resistance Rd_{O21}, an exhaust resistance Rd_{O22}, a hydraulic capacity Ch_{O2} and the consumed oxygen flow generator. The water circuit is composed of the same type of elements, with the produced oxygen flow generator. The input gas flow q_{O2} is imposed by a flow controller, the values of the input water flow q_{H2Oin} is imposed by the humidification system depending on the controlled humidity rate and the temperature of the fuel cell T_{fc}. The pressure at the exhaust is the atmospheric pressure.

The gas flow inside the PEMFC supply channels is considered as the first event for the normal fuel cell operation. This first event, called constraint for the structural analysis, is the first constraint noted C1. However, this gas circulation can be influenced by the system variables given in the following equations:

$$P_{O2in} - P_{O2out} = R_{dO21} * q_{O2in} + Rd_{O22} * q_{O2in} - q_{O2ca} \tag{19}$$

where P_{O2in} and P_{O2out} are the pressure of the input and output gas respectively. Regarding the water circuit the equation becomes:

$$P_{H2Oin} - P_{H2Oout} = R_{dH2O1} * q_{H2Oin} + R_{dH2O2} * q_{H2Oin} + q_{H2Oca}$$

The cathode pressure drop is expressed as follows:

$$\Delta P = P_{tot\_in} - P_{tot\_out}$$

Using the humidified input gas flow, which is actually the total input flow (q_{O2hum_in}), the pressure drop expression becomes:

$$\Delta P = R_{dO2hum_1} q_{O2hum_{in}} + R_{dO2hum_2}\left(q_{O2hum_{in}} - q_{O2_{ca}} + q_{H2O_{ca}}\right)$$

Thus:

$$C1 : q_{O2hum_{in}} = \frac{1}{R_{dO2hum_1} + R_{dO2hum_2}}\left(\Delta P + R_{dO2hum_2} q_{O2_{ca}} - R_{dO2hum_2} q_{H2O_{ca}}\right) \tag{20}$$

The humidified gas inside the channels crosses the gas diffusion layer (GdL) for the purpose of air diffusion to the catalytic site. However, the GdL gas concentration has to be kept at a high value for a good and safe catalytic feeding. The gas amount accumulated in the GdL is thus the third constraint C2. The variables which have an influence on this gas amount inside the GdL appear in the expression of the hydraulic capacity of the GdL Ch:

$$C_h = \frac{1}{\dot{P}_{O2_{ca}} + \dot{P}_{H2O_{ca}}}\left(q_{O2} + q_{H2O_{in}} - q_{cO2} + q_{H2O_{ca}} - q_{O2_{out}} - q_{H2O_{out}}\right) \tag{21}$$

where:

$$q_{O2hum_{out}} = \frac{P_{O2_{ca}} + P_{H2Oca} - P_{O2_{out}} - P_{H2O_{out}}}{R_{dO2hum_2}} \tag{22}$$

hence:

$$C2: \ C_h = \frac{1}{\dot{P}_{O2_{ca}} + \dot{P}_{H2O_{ca}}}\left(q_{O2_{in}} + q_{H2O_{in}} - q_{O2_{ca}} + q_{H2O_{ca}} - \frac{P_{O2_{ca}} + P_{H2Oca} - P_{O2_{out}} - P_{H2O_{out}}}{R_{dO2hum_2}} - q_{H2O_{out}}\right) \tag{23}$$

It should be noted that the constraint C2 has also the fuel cell temperature as input variable. Indeed, the flowrates are expressed as volume so they depend on Tfc.

At the catalytic sites, and for high current densities, the kinetic of reactions increases and the gas consumption intensifies. During this operating condition, the quantity of O2 species decreases and the steam production increases. Therefore, the fuel cell voltage decreases. The steam partial pressure at the catalytic site is thus the fifth constraint C3 because it has to be under the saturation pressure value to avoid water condensation. The variables which influence this constraint are expressed by the following inequality relationship which represent the steam partial pressure at the catalyst (PH2Oca) and the saturation pressure (Psat):

$$C3: \ P_{H2Oca} \leq P_{sat} \tag{24}$$

Then as depicted in [28], there is a thermodynamic equilibrium between the GdL [29] water content (λ_{GdL}) and the membrane water content λ_m ([30]). The constraint C4 represents this equilibrium. Variables which have an influence on C4 are expressed below ([30]):

$$C4: \ \lambda_{GdL} = a_1 + a_2\left(\frac{P_{H2O_{ca}}}{P_{sat}}\right) - a_3\left(\frac{P_{H2O_{ca}}}{P_{sat}}\right)^2 + a_4\left(\frac{P_{H2O_{ca}}}{P_{sat}}\right)^3 \tag{25}$$

Regarding the membrane, it has to be hydrated for an appropriate fuel cell operation. An electro osmotic flow (qosm) allows the membrane water supply via a protonic load flow [10]. This flow must get a sufficient value for a good membrane hydration. This is the fifth constraint C5 which involves the electro osmotic flow. Then, a diffusive flow (qdiff) through the membrane also exists and modifies the water content of the membrane. Like the electro osmotic flow, it has to get a relevant value for a good membrane hydration. This value constitutes the sixth constraint C6. The constraint C5 and C6 are expressed as below:

$$q_{H2Oca} = q_{osm} + q_{diff} \tag{26}$$

with:

$$C5: \ q_{osm} = \lambda_m \tau_0 \frac{I_{fc}}{F} \tag{27}$$

and:

$$C6: \ q_{diff} = -D_m \frac{\rho_{dry}}{M_m}\frac{d\lambda_m}{dx} \tag{28}$$

An optimal membrane hydration is mandatory for the nominal operation of the PEMFC. Its water content is related to the water concentration inside the membrane. A constraint C7 is thus considered for the membrane water content. This constraint is influenced by the water concentration variation and is expressed as below ([31]):

$$C7: \ \lambda_m = \frac{M_m}{\rho_{dry}} c_{H2O} \tag{29}$$

where c_{H2O} is the membrane water concentration which depends on the fuel cell temperature.

The membrane hydration also depends on the thermodynamic equilibrium; a relationship between the GdL water content (λGdL) and the membrane water content λ_m exists. The constraints C5 and C6 are therefore linked to the fuel cell temperature.

The last constraint C8 is also about the membrane water content. Indeed, the lower the membrane water content is, the higher the ohmic resistance is. This resistance can be expressed as below ([31]):

$$C8: R_m = \frac{t_m}{\left(b_1 \exp\left(b_2\left(\frac{1}{303} - \frac{1}{T_{fc}}\right)\right)\right)} \tag{30}$$

where, t_m represents the membrane thickness, b_1 and b_2 are coefficients that depend on fuel cell being tested and b_1 depends on the membrane water content ([31]):

$$b_1 = b_{11}\lambda_m - b_{12} \tag{31}$$

Then, the higher the membrane resistance is, the higher the ohmic losses are. The ohmic losses is written as:

$$C8: \Delta V_{ohm} = R_m I_{fc} \tag{32}$$

The incidence matrix can now be set in order to create the PEMFC structural graph.

5.2. Incidence Matrix of the Structural Analysis

Based on the extraction of the PEMFC constraints, it is possible to create an incidence matrix A ($a_{i,j}$) that allows to link vertices (variables/constraints) and arcs. It contains n rows and m columns:

- $a_{i,j}$ is +1, if the numbered arc j admits the vertex i as origin,
- $a_{i,j}$ is −1, if the numbered arc j admits the vertex i as the arrival,
- $a_{i,j}$ is 0 in other cases.

The incidence matrix is represented in the Table 1.

Table 1. Incidence matrix of the structural analysis.

	C1	C2	C3	C4	C5	C6	C7	C8
T_{fc}			1	1	1	1	1	1
q_{O2in}		1						
q_{H2O}		1						
q_{hum_in}	1							
P_{tot_in}	1		1	1				
P_{O2_out}		1						
P_{O2_ca}		1						
q_{H2O_ca}	1	1			1	1		
P_{H2O_out}		1						
P_{tot_out}	1		1	1				
q_{H2O_out}		1						
q_{O2_ca}	1	1						
P_{H2O_ca}		1	1	1				
$\dot{P}_{H2O\_ca}$		1						
$\dot{P}_{O2\_ca}$		1						
V_{fc}								1
I_{fc}					1	1		1

Table 1. *Cont.*

	C1	C2	C3	C4	C5	C6	C7	C8
q_{osm}					1	1		
q_{diff}					1	1		
λ_m					1	1	1	1
λ_{GDL}				1		1		
C_{H2O}							1	
R_m								1

The control variables are separated from the others with a double vertical line. The structural graph is designed on the basis of the incidence matrix of the Table 1. It is represented on the Figure 6 below.

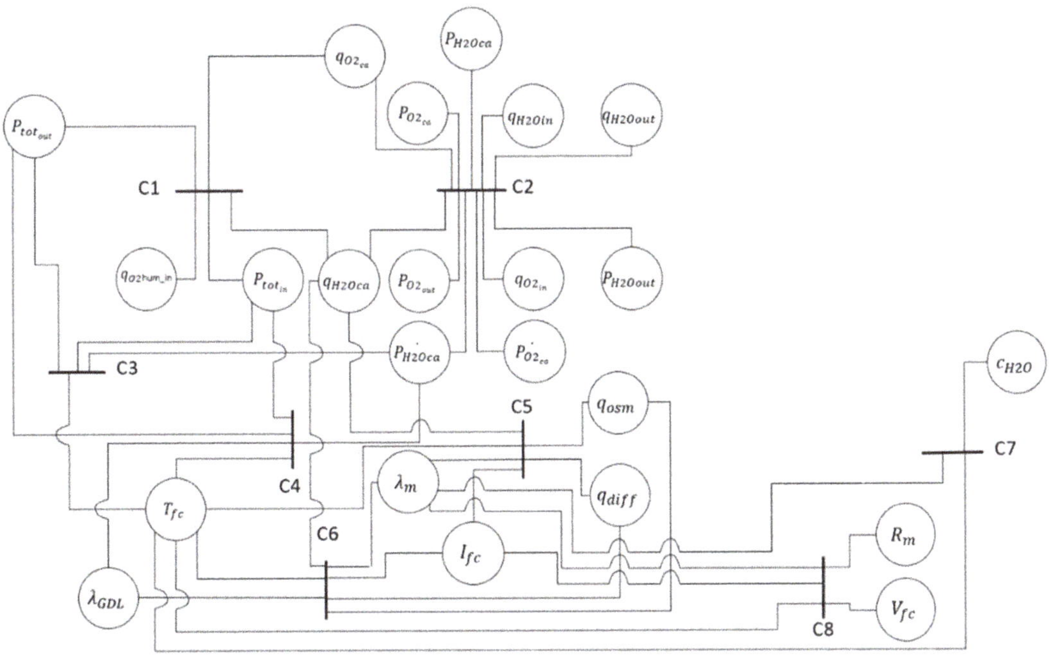

Figure 6. Structural graph of the PEMFC.

The next step consists of adding the PEMFC faults on the structural graph in order to represent their interaction inside the PEMFC.

5.3. Fuel Cell Flooding Structural Analysis

A PEMFC flooding can occur in two areas. Both inside the GdL with a water droplet accumulation which reduces the catalytic site reactant feeding, and inside the channels by propagation of water droplet accumulation. This water accumulation can also appear directly inside the supply channels and reduce the fuel cell reactant feeding. To integrate the flooding in the structural analysis, the fault is assumed to be a variable which has influence on the PEMFC constraints defined above. For this purpose, a new variable (F_{flood}) which represents a flooding occurrence is added. F_{flood} can therefore be expressed as below:

$$F_{flood} = \frac{V_{electrode\_available}}{V_{geo\_electrode}} \tag{33}$$

with, $V_{electrode\_available}$ is the global volume available in the compartment (channels + GdL) and $V_{geo\_electrode}$ the geometrical volume of the electrode. This variable is set to 1 in case of optimal hydration and to 0 when completely clogged.

The flooding variable has an influence on the supply channels. The new variable is thus added to the constraint C1 which is linked to PEMFC supply channels:

$$C1:\ q_{O2hum_{in}} = \frac{F_{flood}}{R_{dO2hum_1} + R_{dO2hum_2}}\left(\Delta P + R_{dO2hum_2} q_{O2_{ca}} - R_{dO2hum_2} q_{H2O_{ca}}\right) \tag{34}$$

The flooding variable has also an influence on the GdL and thus on the constraints C2 and C4. These two constraints become:

$$C2:\ C_h = \frac{F_{flood}}{\dot{P}_{O2_{ca}} + \dot{P}_{H2O_{ca}}}\left(q_{O2_{in}} + q_{H2O_{in}} - q_{O2_{ca}} + q_{H2O_{ca}} - \frac{P_{O2_{ca}} + P_{H2Oca} - P_{O2_{out}} - P_{H2O_{out}}}{R_{dO2hum_2}} + q_{H2O_{out}}\right) \tag{35}$$

$$C4:\ \lambda_{GdL} = F_{flood}\left(a_1 + a_2\left(\frac{P_{H2O_{ca}}}{P_{sat}}\right) - a_3\left(\frac{P_{H2O_{ca}}}{P_{sat}}\right)^2 + a_4\left(\frac{P_{H2O_{ca}}}{P_{sat}}\right)^3\right) \tag{36}$$

The incidence matrix is updated with the flooding variable as represented in the Table 2.

Table 2. Incidence matrix updated with the flooding variable.

	C1	C2	C3	C4	C5	C6	C7	C8
T_{fc}			1	1	1	1	1	1
q_{O2in}		1						
q_{H2O}		1						
q_{hum_in}	1							
P_{tot_in}	1		1	1				
P_{O2_out}		1						
P_{O2_ca}		1						
q_{H2O_ca}	1	1			1	1		
P_{H2O_out}		1						
P_{tot_out}	1		1	1				
q_{H2O_out}		1						
q_{O2_ca}	1	1						
P_{H2O_ca}		1	1	1				
$\dot{P}_{H2O\_ca}$		1						
$\dot{P}_{O2\_ca}$		1						
V_{fc}								1
I_{fc}					1	1		1

Table 2. *Cont.*

	C1	C2	C3	C4	C5	C6	C7	C8
q_{osm}					1	1		
q_{diff}					1	1		
λ_m					1	1	1	1
λ_{GDL}				1		1		
C_{H2O}							1	
R_m								1
F_{flood}	1	1						

The structural graph with the flooding variable is represented on the Figure 7:

The next step consists in introducing the membrane drying out variable inside the structural analysis. The goal is to identify the constraints which are influenced with its occurrence.

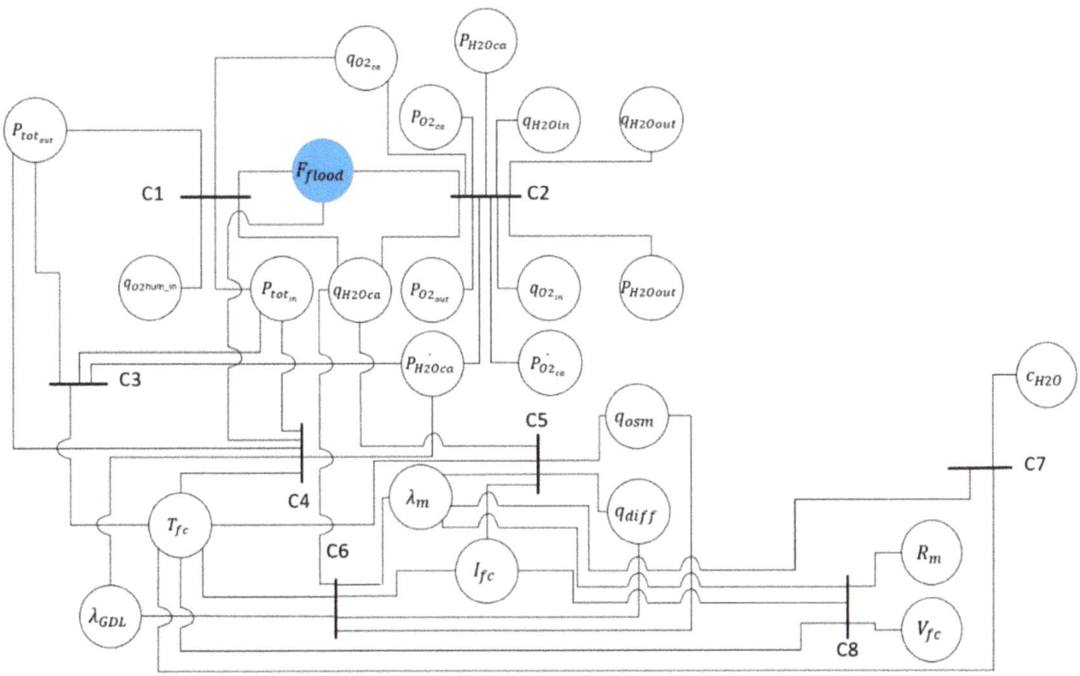

Figure 7. Structural graph with the flooding variable.

5.4. Membrane Drying out Structural Analysis

The membrane drying out fault results from a decrease of the membrane water content. Therefore, it has an influence on constraints C7 and C8 that involve the membrane water content variable (λ). The fault can thus be represented by a variable which is expressed as below:

$$F_{dry} = \frac{V_{abs\_membrane}}{V_{abs\_tot\_membrane}} \quad (37)$$

where, $V_{abs\_membrane}$ is the volume absorbed by the membrane and $V_{abs\_tot\_membrane}$ is the total volume absorbable by the membrane. F_{dry} has a value between 0 and 1.

This variable can now be introduced in the constraint C9:

$$C9: \lambda_m = \frac{M_m \cdot F_{dry}}{\rho_{dry}} c_{H2O} \quad (38)$$

The incidence matrix is updated with the membrane drying out variable as represented on the Table 3.

Table 3. Incidence matrix updated with the membrane drying out variable.

	C1	C2	C3	C4	C5	C6	C7	C8
T_{fc}			1	1	1	1	1	1
q_{O2in}		1						
q_{H2O}		1						
q_{hum_in}	1							
P_{tot_in}	1		1	1				
P_{O2_out}		1						
P_{O2_ca}		1						
q_{H2O_ca}	1	1			1	1		
P_{H2O_out}		1						
P_{tot_out}	1		1	1				
q_{H2O_out}		1						
q_{O2_ca}	1	1						
P_{H2O_ca}		1	1	1				
$\dot{P}_{H2O\_ca}$		1						
$\dot{P}_{O2\_ca}$		1						
V_{fc}								1
I_{fc}					1	1		1
q_{osm}					1	1		
q_{diff}					1	1		
λ_m					1	1	1	1
λ_{GDL}				1		1		
C_{H2O}							1	
R_m								1
F_{flood}	1	1						
F_{dry}							1	

The structural graph with the membrane drying out variable is represented on the Figure 8:

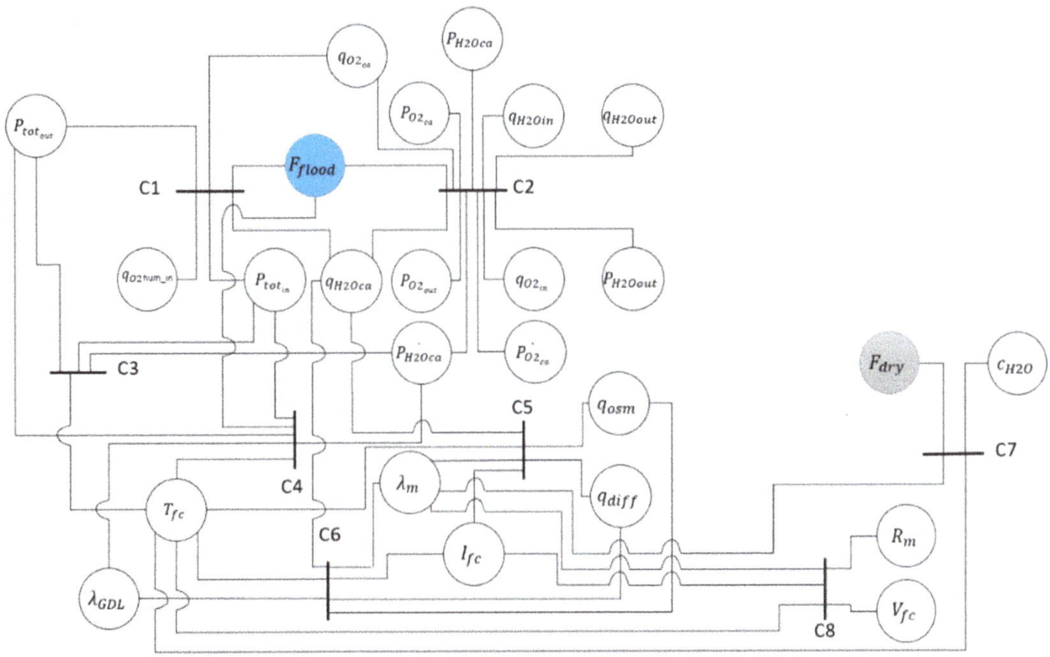

Figure 8. Structural graph with the membrane drying out variable.

6. FSA in Experimental Context

The experimental FSA illustration is made through the mitigation of fuel cell flooding fault. Indeed, an active fault tolerant control (AFTC) strategy which is detailed in another work [32], is applied on a single cell fuel during a flooding.

AFTC strategies, compared to other kinds of FTC strategies, differs mainly by their structure. Indeed, in a previous literature review [1] it has been highlighted that there are two kinds of FTC strategies: Active or Passive ones. Passive strategies (PFTC) consist of a robust controller design for fault mitigation. In this case, the controller is designed by considering the fault as a disturbance. This structure allows to not use diagnosis tools. But as more is important the number of faults that should be mitigated as more is the PFTC design complexity. To reduce this complexity and instead of use a unique robust controller, the AFTC structure decomposed the mitigation process to several steps. For instance, in [33] authors proposed a three-modules fault mitigation process on a powertrain city bus. Authors used as first module a diagnosis tool for fault detection whereas the second module is composed of decision-making part. Finally, the third module is composed by a set of controllers which implement on the powertrain city bus the mitigation strategy computed by the second module. Another work in [23] where authors proposed another application of the AFTC strategy. In this case the purpose is to address the water management issues. They manage to couple a fault detection and isolation (FDI) algorithm with a reconfiguration mechanism and an adjusting controller. The FDI process is a machine learning based whereas a self-tuning PID is implemented as the control part. Authors also highlight the advantages of the self-tuning PID which shows robustness against noise and model uncertainties. Wu et al. proposed in another fault mitigation process which also consists of an implementation of a three-module AFTC to diagnose a PEMFC fault occurrence, to decide on a relevant mitigation action, and to apply the strategy on the fuel cell through a control set. The main purpose of the method is the distribution of the complexity of the AFTC design into the three modules which significantly simplifies its implementation. In [13] authors proposed a model-based AFTC strategy for PEMFC temperature sensor fault. They used a FDI process with the aim of real time fault diagnosis with a sliding mode controller for the fuel cell thermal management.

The advantage of the use of AFTC structure lies in its modular aspect. Indeed, it allows the decomposition of the mitigation process into several steps. Each step being dedicating to a different task, it reduces the complexity of the strategy.

6.1. Experimental Set Up

The fault tolerant control strategy is applied on a test bench of the FCT brand [34]. It manages to supply a fuel cell with hydrogen at the anode and oxygen or air at the cathode. The experimentations are carried out only with oxygen. The tested fuel cell is a 50 W unit which is composed of N117 ION POWER single-cell of 50 cm^2 [35]. The synopsis of test bench is given on the Figure 4.

The Figure 4 shows that the test bench is mounted with two flow rate controllers in order to regulate the input reactant flows. Then, two humification systems are placed on O_2 and H_2 feeding lines in order to control their relative humidity thanks to two gas temperature controllers. At the event of the fuel cell, two back pressures controllers are used at the anode and cathode lines to manage the fuel pressure. An electronic load is finally connected to the FC electric terminals which can absorb a power of 1.8 kW. The whole test bench is monitored with a LabVIEW virtual instrument.

6.2. Flooding Generation Test

The experiment consists of mitigating a flooding occurrence with an AFTC strategy. It will modify the fuel cell operating conditions by changing some control variables iteratively. The selection of the control variables is based on the previous structural graph and on the available sensors and actuators on the test bench. The selected control variables are the: input cathode gas flow rate (q_{O2ca}); fuel cell temperature (T_{fc}); inlet gas flow relative humidity (q_{H2Oca}).

The structural graph shows that the flooding affects the fuel gas channels and the GDL on the fuel gas channels and on the GDL. The control variables q_{O2ca} and q_{H2Oca} have an influence on the gas channels whereas Tfc influences the GDL.

In the experiment, the flooding is generated by introducing liquid water in the cell from the canalization which is between the humidifier and the fuel cell. Indeed, on Figure 4 the gas at the outlet of the humidification system is temperature-regulated in order to reach the relative humidity setpoint. With the temperature controller, it is possible to condense the steam. Then the condensed water goes into the fuel cell to cause a flooding inside the electrode.

The used algorithm for the AFTC strategy is based on an iterative modification of the operating condition with the modification of the selected control variables. The Figure 9 is an illustration of the AFTC mechanism.

6.3. AFTC Application to Flooding

The used AFTC strategy has been selected from a previous literature review [1] which is constituted of three modules: diagnosis; decision; control. The diagnosis module is used for the flooding identification based on some outputs fuel cell measurements. Then, if a fault occurred, a decision process is launched through the decision module. At this step, a mitigation strategy is computed for a fast and sustainable fault mitigation. The decision strategy is finally implemented on the fuel cell through the control module which is composed of a set of controllers.

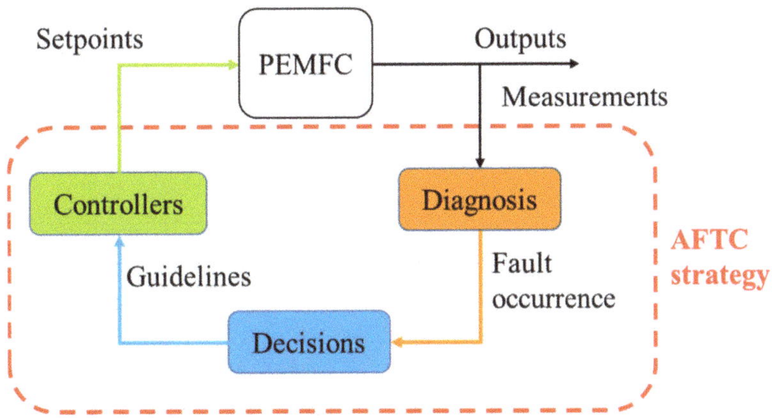

Figure 9. Iterative AFTC strategy.

In this work, the diagnosis consists of identifying a flooding occurrence by monitoring the fuel cell pressure drop and the voltage. The decision process provides a decision regarding the fault occurrence to proceed fast mitigation with minimal change in the operating point. The guidelines which are the output of the decision process are transmitted to all controllers to apply the mitigation strategy on the fuel cell.

The AFTC iterative process leads to the setup of two testing cases for the flooding mitigation. Figure 10 represents these testing cases.

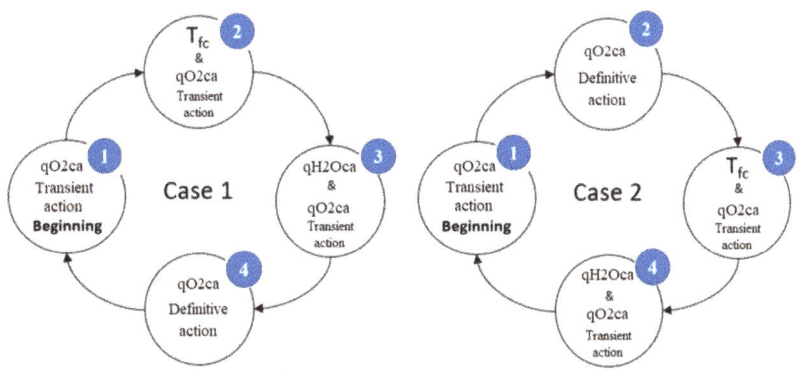

Figure 10. Two possible test cases for flooding mitigation.

In both cases, the three selected control variables are used by the AFTC strategy for fault mitigation. If a fault is identified by the diagnosis block, one of the variable values is modified. For instance, in the case 1, the strategy starts with a correction of the gas flow rate qO2ca (1). This is referred to as a Transient action because qO2ca is temporally modified until the flooding is mitigated. The second step of the decision process (2) is an action on Tfc and on qO2ca. Here, Tfc is permanently modified to a new value. In parallel with the action on Tfc, another Transient action on qO2ca is triggered. The third step (3) of the case 1 consists of a permanent modification of the input gas water content (qH2Oca). This action is also supported by a Transient action on qO2ca. The fourth step (4) consists of a permanent change of qO2ca. Here, there is no Transient action.

The test case 2 consists of a different order of the corrective actions. It is composed of the same actions as for the case 1, in a different sequence; except for the first action which is a Transient action on qO2ca.

6.4. Experimental Validation of the AFTC Strategy Based on the Variables Extracted from the Structural Graph

A flooding mitigation process based on the test case 1 has been tested on a single fuel cell. Results are depicted in Figure 11a–c.

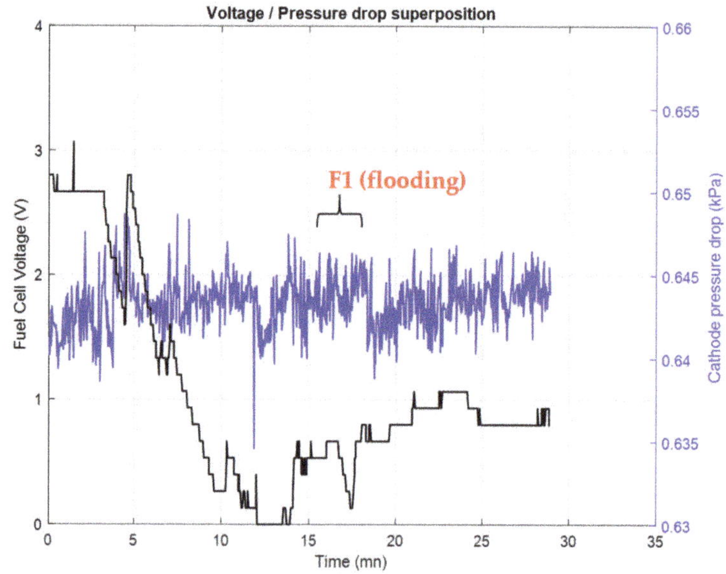

(**a**) Fuel cell voltage and cathode pressure drop superposition

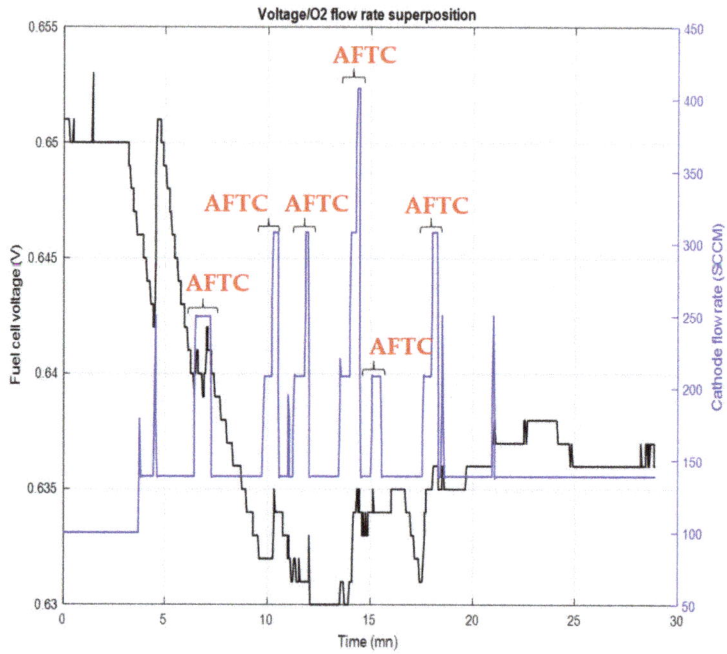

(**b**) Fuel cell voltage and input gas flow superposition

Figure 11. *Cont.*

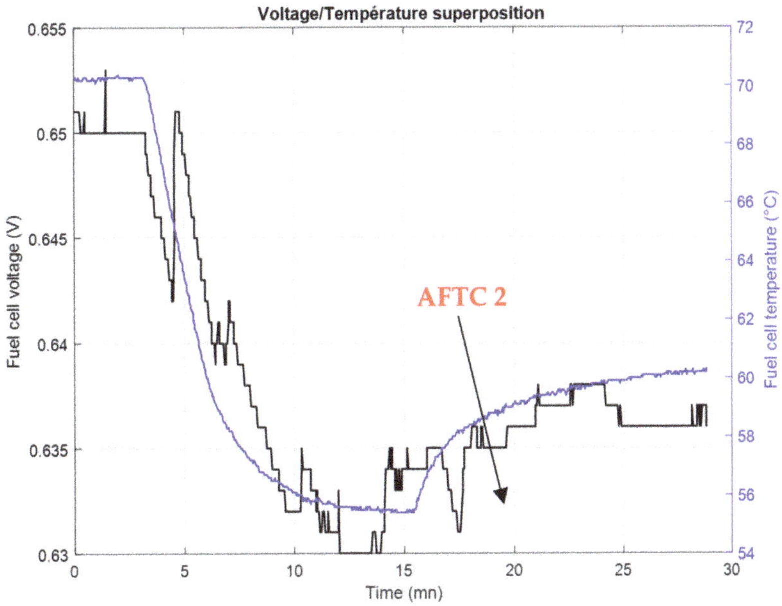

(**c**) Fuel cell voltage and temperature superposition

Figure 11. Pressure drop (**a**), oxygen flow rate (**b**) and temperature (**c**) superposition with the fuel cell voltage for the test case 1.

The Figure 11a shows the superposition of the fuel cell voltage with the cathode pressure drop (PD). The PD evolution is used for flooding diagnosis. For instance, F1 corresponds to an increase in PD, which means flooding is in progress.

On Figure 11b all the applied Transient actions on O2 flow rate qO2ca are marked with the AFTC. Each increase in qO2ca aims to mitigate the flooding. When the flooding is mitigated (not diagnosed anymore), the qO2ca value is reset to its value before the Transient action triggering.

On Figure 11c, the second mitigation action of the test case 1, that is the fuel cell temperature is triggered (AFTC 2) at the same time as a Transient action

Figure 12 represents a flooding mitigation process based on the test case 2.

The Figure 12 represents the experimentation applied to the test case 2. Here the first mitigation action is always based on a Transient action, on qO2ca. The second action is also based on the same control variable qO2ca. The increase in qO2ca (AFTC 3) with the Transient action manages to mitigate the flooding by causing the liquid water to drain from the fuel cell.

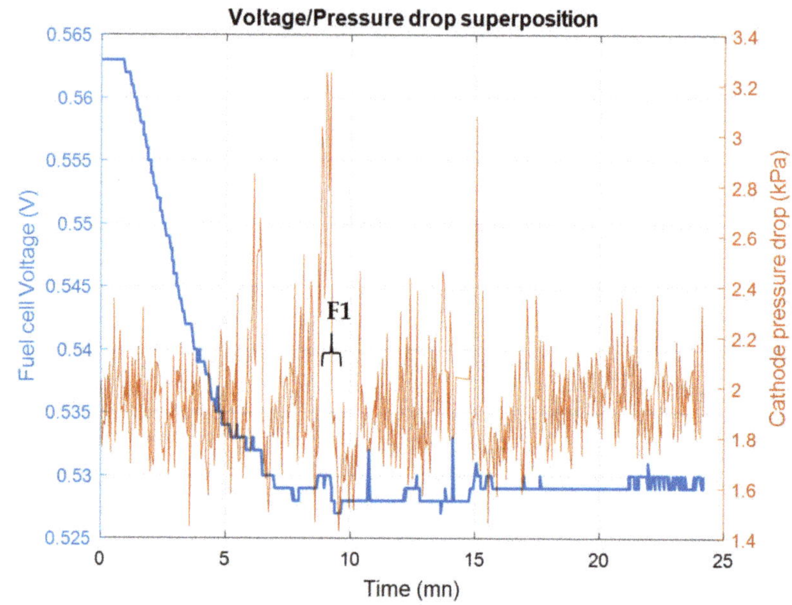

(**a**) Fuel cell voltage and cathode pressure drop superposition

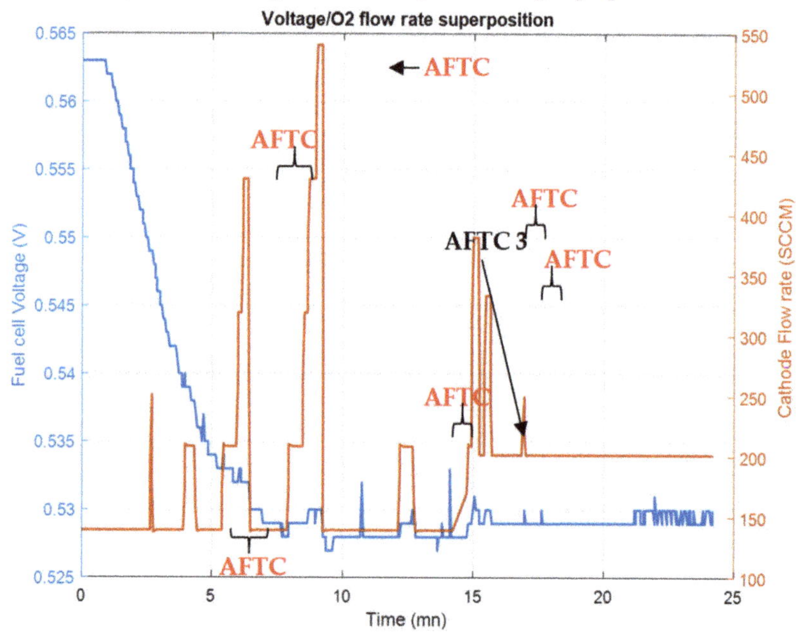

(**b**) Fuel cell voltage and input gas flow superposition

Figure 12. *Cont.*

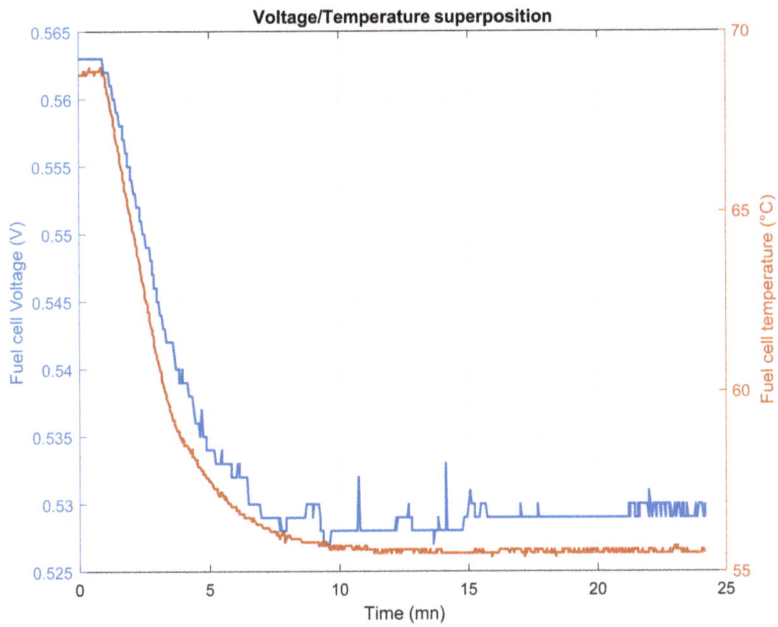

(**c**) Fuel cell voltage and temperature superposition

Figure 12. Pressure drop (**a**), oxygen flow rate (**b**) and temperature (**c**) superposition with the fuel cell voltage for the test case 2.

The test results are summarized in the Table 4.

Table 4. Test results for flooding mitigation for the tests cases 1 and 2.

Flooding	Case 1	Case 2
Number of Transient decisions	6	6
Number of permanent decisions	1	1
Permanent decision triggering time	11 mn	15 mn
Flooding mitigation time	18 mn	15 mn
Voltage loss	−3.1%	−6.21%
Voltage recovery	+1%	+0.35%

6.5. Experimental Analysis

The test of the case 1 and 2 allows the flooding mitigation in about 15 to 18 min and for the same number of Transient actions and permanent actions. The main difference concerns the voltage recovery which is three times fastest in the case of an action on temperature rather than on the cathode flow rate.

Indeed, regarding the structural graph it appears that a flooding acts on the gas feeding channels and on the GDL. Actions on qO2ca allow the water to drain from the fuel cell channels but do not allow the water to drain from the GDL. In the same way, Tfc manages to remove water from the GDL and has no effects on the fuel cell channels. For these reasons, the case 1 which is a mix of actions on the channels and GDL through the qO2ca and Tfc, is a more efficient way for flooding mitigation than case 2 which only acts on the channels.

7. Discussion

The structural analysis gives a graphical representation of all PEMFC variables that have an influence on the fuel cell operating conditions. It highlights the coupling of variables inside the PEMFC and leads to the design of a structural graph. The structural graph of the PEMFC on the Figure 8 shows that the fuel cell temperature appears as the most coupled variable. It has an influence on the steam partial pressure at the catalyst site (C3), on the thermodynamic equilibrium of the GdL and the membrane water content (C4), on the electro osmotic and diffusive flows (C5 and C6) and on the membrane water content (C7 and C8). Other variables such as the membrane water flow rate and the input steam flow rate appear to be also strongly coupled. This information, brought by the structural graph, is very relevant to understand the PEMFC complexity and why it is so challenging to maintain fuel cell systems under nominal operating conditions.

Then, the flood variable (F_{flood}), which is considered as a fuel cell variable, is integrated in the structural graph. This variable has an influence on the input gas circulation inside the supply channels (C1) and on the GdL gas volume (C2). In case of a design of a flooding mitigation strategy, the structural graph gives the influenced constraints in order to get the appropriate control variables. The structural graph shows that the input gas pressure (Ptot) and flow rate have an influence on the gas feed channel and can lead to the mitigation of the flooding. Regarding a flooding diagnosis tool inside the gas feed channels, the measurement of the inlet gas pressure at the inlet and at the catalytic sites are relevant variables for the design. Indeed, when there is water accumulation inside the gas feed channels, the gas pressure difference between the inlet and catalytic sites increases. However, the pressure at the catalytic site is not measurable and for this reason the pressure difference is determined with the inlet and outlet pressures.

Regarding the flooding in the gas diffusion layer, the analysis conducted gives no relevant variable to diagnose the GdL flooding. This is due to the scale of the model which does not provides the fuel cell behavior inside the GdL and does not discriminate a clogging in the channels or in the GdL. However, the humidified inlet gas and the fuel cell temperature are relevant control variables for the GdL flooding mitigation with a fault mitigation strategy. Indeed, by decreasing the steam injected inside the fuel cell the water accumulation inside the GdL is restrained. An increase of the fuel cell temperature modifies the relative humidity inside the fuel cell and makes the water droplets evaporate.

Concerning the membrane drying out variable on the Figure 8, the structural graph shows that it is related to the membrane water content. The fuel cell current and temperature appear as relevant control variables on the structural graph for a mitigation strategy. These variables have an influence on the membrane water content and can be used in a fault mitigation strategy to alleviate the drying out occurrence. Indeed, the decrease of the fuel cell temperature leads to rehydrate the membrane by an increase of the relative humidity inside the fuel cell. The change of the current value has also an influence on the electroosmotic flow and thus on the water dragged through the membrane which participates to its rehydration. For this faulty condition, the fuel cell voltage, water content and current are relevant variables to set a membrane drying out diagnosis tool.

Therefore, the PEMFC FSA leads to the highlighting of the internal variable coupling. The integration of the fault variables in the structural graph allows understanding the fault process and their location. The analysis underlines the PEMFC constraints that are directly influenced by their occurrence and shows the relevant variables that can be used in FTC strategies.

8. Conclusions

The structural analysis approach aims to synthesize the known information about the PEMFC structure and water management issues, leading to flooding and drying out faults. Therefore, the analysis allows describing the system functionalities and their variables. It allows the graphical representation of the fuel cell strong coupling and provides information about relevant variables for the design of a diagnosis tool for flooding and membrane drying out. This analysis also highlights the relevant variables for the design of FTC strategies. Indeed, in case of the development of PEMFC monitoring or mitigation tools, the knowledge of the relevant variables that have to be used and given by the FSA is very helpful. The relevance of the FSA and particularly the graph is shown by the experimental mitigation of a flooding occurrence on a single fuel cell. Indeed, the experiment shows that it is very important to consider which control variable has an influence on which fuel cell functionalities in order to introduce in the FTC strategy enough control variables for fault mitigation.

As discussed in the previous section, the information contained in the FSA depends on the accuracy of the chosen model. But if a higher level was used for the structural graph design, maybe other control variables could appear in the structural graph. For this work and for the used AFTC strategy, only a low level of knowledge was needed because for the FTC design, the fact that not all variables can be measured nor estimated has to be kept in mind to lead to an implementable solution on the used test bench. Hence, there is a trade-off to be found between the different levels of knowledge regarding the selected models.

The literature also reports some other fuel cell faults like CO poisoning, hydrogen and oxygen starvation and low cathode and anode stoichiometry. The next step of the FSA design is therefore to complete the structural graph by adding these faulty conditions.

Author Contributions: Conceptualization, E.D., N.Y.S., M.B. and M.-C.P.; methodology, E.D., N.Y.S., M.B. and M.-C.P.; validation, E.D., N.Y.S., M.B. and M.-C.P., B.G.-P.; formal analysis, E.D., N.Y.S. and M.-C.P.; investigation, E.D., N.Y.S., M.B. and M.-C.P.; writing—original draft preparation, E.D.; writing—review and editing, E.D., N.Y.S., M.B. and M.-C.P.; supervision, E.D., N.Y.S., M.B., B.G.-P. and M.-C.P. All authors have read and agreed to the published version of the manuscript.

Funding: This research received no external funding.

Institutional Review Board Statement: Not applicable.

Informed Consent Statement: Not applicable.

Data Availability Statement: Not applicable.

Acknowledgments: This work has been supported by the Région Réunion and the Région Bourgogne Franche-Comté and by the EIPHI Graduate School (contract ANR-17-EURE-0002).

Conflicts of Interest: The authors declare no conflict of interest.

References

1. Dijoux, E.; Steiner, N.Y.; Benne, M.; Péra, M.C.; Pérez, B.G. A review of fault tolerant control strategies applied to pro-ton exchange membrane fuel cell systems. *J. Power Sources* **2017**, *359*, 119–133. [CrossRef]
2. Berger, J.O. Statistical Decision Theory and Bayesian Analysis-Google Books. 21 August 1985. Available online: https://books.google.fr/books (accessed on 21 October 2021).
3. Kwon, O.-J.; Shin, H.-S.; Cheon, S.-H.; Oh, B.S. A study of numerical analysis for PEMFC using a multiphysics program and statistical method. *Int. J. Hydrogen Energy* **2015**, *40*, 11577–11586. [CrossRef]
4. Prifling, B.; Ridder, A.; Hilger, A.; Osenberg, M.; Manke, I.; Birke, K.P.; Schmidt, V. Analysis of structural and functional aging of electrodes in lithium-ion batteries during rapid charge and discharge rates using synchrotron tomography. *J. Power Sources* **2019**, *443*, 227259. [CrossRef]
5. Göbel, M.; Godehardt, M.; Schladitz, K. Multi-scale structural analysis of gas diffusion layers. *J. Power Sources* **2017**, *355*, 8–17. [CrossRef]
6. Tsukasaki, H.; Otoyama, M.; Mori, Y.; Mori, S.; Morimoto, H.; Hayashi, A.; Tatsumisago, M. Analysis of structural and thermal stability in the positive electrode for sulfide-based all-solid-state lith-ium batteries. *J. Power Sources* **2017**, *367*, 42–48. [CrossRef]

7. Delphine Mathilde Cosme. L'analyse Fonctionnelle: L'arborescence des Fonctions et Solutions Techniques. Available on line: https://www.techniques-ingenieur.fr/fiche-pratique/genie-industriel-th6/pratique-de-la-conception-industrielle-dt52/l-analyse-fonctionnelle-representer-l-arborescence-des-fonctions-pour-trouver-les-solutions-techniques-adaptees-0617/ (accessed on 21 October 2021).
8. Nurdiansyah, Y.; Wijayanto, F. The Design of E-Commerce System in the Shrimp Paste Industry using the Method of Structured Analysis and Design Technique (SADT) to Increase Marketing. In Proceedings of the 3rd International Conference on Electrical Systems, Technology and Information (ICESTI 2017), Bali, Indonesia, 26–29 September 2017; Volume 164, p. 01049. [CrossRef]
9. Steiner, N.Y.; Hissel, D.; Moçotéguy, P.; Candusso, D.; Marra, D.; Pianese, C.; Sorrentino, M. Application of Fault Tree Analysis to Fuel Cell diagnosis. *Fuel Cells* **2012**, *12*, 302–309. [CrossRef]
10. Gallo, M.; Costabile, C.; Sorrentino, M.; Polverino, P.; Pianese, C. Development and application of a comprehensive mod-el-based methodology for fault mitigation of fuel cell powered systems. *Appl. Energy* **2020**, *279*, 115698. [CrossRef]
11. Yang, D.; Wang, Y.; Chen, Z. Robust fault diagnosis and fault tolerant control for PEMFC system based on an augment-ed LPV observer. *Int. J. Hydrog. Energy* **2020**, *45*, 13508–13522. [CrossRef]
12. Kamal, E.; Aitouche, A. Fuzzy observer-based fault tolerant control against sensor faults for proton exchange membrane fuel cells. *Int. J. Hydrog. Energy* **2020**, *45*, 11220–11232. [CrossRef]
13. Yan, C.; Chen, J.; Liu, H.; Lu, H. Model-Based Fault Tolerant Control for the Thermal Management of PEMFC Systems. *IEEE Trans. Ind. Electron.* **2020**, *67*, 2875–2884. [CrossRef]
14. Piechowiak, S. Intelligence Artificielle et Diagnostic. *Tech. Ing.* **2003**. [CrossRef]
15. Lu, H.; Chen, J.; Yan, C.; Liu, H. On-line fault diagnosis for proton exchange membrane fuel cells based on a fast electro-chemical impedance spectroscopy measurement. *J. Power Sources* **2019**, *430*, 233–243. [CrossRef]
16. Yuan, H.; Dai, H.; Wei, X.; Ming, P. Model-based observers for internal states estimation and control of proton exchange membrane fuel cell system: A review. *J. Power Sources* **2020**, *468*, 228376. [CrossRef]
17. Alves-Lima, D.F.; Letizia, R.; Degl'Innocenti, R.; Dawson, R.; Lin, H. Quantitative video-rate hydration imaging of Nafi-on proton exchange membranes with terahertz radiation. *J. Power Sources* **2020**, *450*, 227665. [CrossRef]
18. Moçotéguy, P.; Ludwig, B.; Beretta, D.; Pedersen, T. Study of the impact of water management on the performance of PEMFC commercial stacks by impedance spectroscopy. *Int. J. Hydrog. Energy* **2020**, *45*, 16724–16737. [CrossRef]
19. Li, H.; Tang, Y.; Wang, Z.; Shi, Z.; Wu, S.; Song, D.; Zhang, J.; Fatih, K.; Zhang, J.; Wang, H.; et al. A review of water flooding issues in the proton exchange membrane fuel cell. *J. Power Sources* **2008**, *178*, 103–117. [CrossRef]
20. Barbir, F.; Gorgun, H.; Wang, X. Relationship between pressure drop and cell resistance as a diagnostic tool for PEM fuel cells. *J. Power Sources* **2005**, *141*, 96–101. [CrossRef]
21. Hasheminasab, M.; Kermani, M.; Nourazar, S.; Khodsiani, M. A novel experimental based statistical study for water management in proton exchange membrane fuel cells. *Appl. Energy* **2020**, *264*, 114713. [CrossRef]
22. Beauguitte, L. *L'analyse des Graphes Bipartis*; L'archive ouverte HAL-SHS: Paris, France, 2013.
23. Blanke, M.; Kinnaert, M.; Lunze, J.; Staroswiecki, M.; Schröder, J. *Diagnosis and Fault-Tolerant Control with Contributions by Jochen Schröder*; Springer: Heidelberg, Germany, 2006.
24. Düştegör, D.; Frisk, E.; Cocquempot, V.; Krysander, M.; Staroswiecki, M. Structural analysis of fault isolability in the DAMADICS benchmark. *Control Eng. Pract.* **2006**, *14*, 597–608. [CrossRef]
25. Stevens, P.; Lepmi, P.; Grenoble, I.; Lamy, C.; Cassir, M. Piles à Combustible Frédéric Novel-Cattin Abdel Hammou. Available online: https://www.techniques-ingenieur.fr/base-documentaire/energies-th4/accumulateurs-d-energie-42243210/piles-a-combustible-d3340/ (accessed on 21 October 2021).
26. Boulon, L.; Hissel, D.; Bouscayrol, A.; Péra, M.C. From modeling to control of a PEM fuel cell using energetic macroscop-ic representation. *IEEE Trans. Ind. Electron.* **2010**, *57*, 1882–1891. [CrossRef]
27. Chnani, M.; Maker, H.; Candusso, D.; Péra, M.C.; Daniel, H. Electrical analogy modelling of PEFC system fed by a compressor. In Proceedings of the European Fuel Cell Forum, Luzern, Switzerland, 4 July 2005.
28. Mularczyk, A.; Michalski, A.; Striednig, M.; Herrendörfer, R.; Schmidt, T.J.; Büchi, F.N.; Eller, J. Mass transport limitations ofwater evaporation in polymer electrolyte fuel cell gas diffusion layers. *Energies* **2021**, *14*, 2967. [CrossRef]
29. Zhou, C.; Guo, L.; Chen, L.; Tian, X.; He, T.; Yang, Q. Pore-Scale Modeling of Air–Water Two Phase Flow and Oxygen Transport in Gas Diffusion Layer of Proton Exchange Membrane Fuel Cell. *Energies* **2021**, *14*, 3812. [CrossRef]
30. Hinatsu, J.T.; Mizuhata, M.; Takenaka, H. Water Uptake of Perfluorosulfonic Acid Membranes from Liquid Water and Water Vapor. *J. Electrochem. Soc.* **1994**, *141*, 1493–1498. [CrossRef]
31. Pukrushpan, J.T. *Modeling and Control of Fuel Cell Systems and Fuel Processors*; University of Michigan: Ann Arbor, MI, USA, 2003.
32. Dijoux, E.; Steiner, N.Y.; Benne, M.; Pérez, B.G.; Marie-Cécile Péra Université de La Réunion. Contrôle Tolérant Aux Défauts Appliqué Aux Systèmes Pile à Combustible à Membrane Échangeuse de Protons (Pemfc) Ecole Doctorale N°542. 2019. Available online: https://tel.archives-ouvertes.fr/tel-02307308 (accessed on 25 February 2020).
33. Xu, L.; Li, J.; Ouyang, M.; Hua, J.; Li, X. Active fault tolerance control system of fuel cell hybrid city bus. *Int. J. Hydrog. Energy* **2010**, *35*, 12510–12520. [CrossRef]

34. Single Cell Hardware. Available online: http://fuelcelltechnologies.com/single-cell-hardware (accessed on 1 May 2020).
35. MEA for Expandable PEM Research Test Cell-50 cm^2. Available online: https://www.fuelcellstore.com/fuel-cell-components/membrane-electrode-assembly/mea-expandable-test-cell-50 (accessed on 3 May 2020).

MDPI
St. Alban-Anlage 66
4052 Basel
Switzerland
Tel. +41 61 683 77 34
Fax +41 61 302 89 18
www.mdpi.com

MDPI Books Editorial Office
E-mail: books@mdpi.com
www.mdpi.com/books

www.ingramcontent.com/pod-product-compliance
Lightning Source LLC
Chambersburg PA
CBHW061506220926
51850CB00004B/181
* 9 7 8 3 0 3 6 5 6 9 0 8 6 *